636.085 UND

THE MINERAL NUTRITION OF LIVESTOCK

3rd Edition

SHOLT COLLEGE HAMPSHIRE
and Information Centre
returned on or before
date below

WITHDRAWN

THE MINERAL NUTRITION OF LIVESTOCK
3rd Edition

E.J. UNDERWOOD (deceased)

and

N.F. SUTTLE

Moredun Research Institute
Pentland Science Park
Bush loan
Penicuik
Midlothian EH26 0PZ
UK

WITHDRAWN

CABI *Publishing*

THE RESOURCE CENTRE, SPARSHOLT COLLEGE, HAMPSHIRE

ACCESSION No: 034182 WITHDRAWN

CLASS NUMBER: 636.085 UND

CABI *Publishing* is a division of CAB *International*

CABI Publishing
CAB International
Wallingford
Oxon OX10 8DE
UK

Tel: +44 (0)1491 832111
Fax: +44 (0)1491 833508
Email: cabi@cabi.org
Web site: http://www.cabi.org

CABI Publishing
10 E. 40th Street
Suite 3203
New York, NY 10016
USA

Tel: +1 212 481 7018
Fax: +1 212 686 7993
Email: cabi-nao@cabi.org

© CAB *International* 1999. All rights reserved. No part of this publication may be reproduced in any form or by any means, electronically, mechanically, by photocopying, recording or otherwise, without the prior permission of the copyright owners.

A catalogue record for this book is available from the British Library, London, UK.

Library of Congress Cataloging-in-Publication Data
Underwood, Eric J. (Eric John), 1905–
The mineral nutrition of livestock / E.J. Underwood and N.F. Suttle. -- 3rd ed.
p. cm.
Includes bibliographical references (p.) and index.
ISBN 0-85199-128-9
1. Minerals in animal nutrition. I. Suttle, N. F. II. Title.
SF98.M5U5 1999
636.08′527--dc21 99–11802
CIP

HB ISBN 0 85199 128 9
PB ISBN 0 85199 557 8

First printed 1999
Reprinted with corrections 2001

Typeset in Garamond by Columns Design Ltd, Reading.
Printed and bound in the UK by Biddles Ltd, Guildford and King's Lynn.

Contents

WITHDRAWN

Preface

It was with some trepidation that, some 7 years ago, I accepted an invitation from CAB International to revise the late Eric Underwood's 1980 text for *The Mineral Nutrition of Livestock*. As a 'raw' post doctorate I had reviewed the third (1971) edition of his other major publication *Trace Elements in Human and Animal Nutrition* for CAB International and remember likening him to Goldsmiths 'Village Teacher' – 'that one small head could carry all he knew'! I also had the privilege of hearing him sum up the proceedings of the first two international symposia on Trace Element Metabolism in Animals (TEMA 1 and 2) in Aberdeen (1968) and Wisconsin (1971): his ability to quickly distil the packed and complex proceedings of such gatherings and present the highlights with a touch of humour were truly remarkable. However, trace elements were *only half* of his repertoire and this book gathered together the essentials of macro- as well as trace element nutrition while focusing entirely on the practical needs of animals. My task has been simplified by the excellent foundation that Eric laid, the advances in information technology which now expedite any literature search and – sadly – a dramatic reduction in funding for applied research relating to animal nutrition in most developed countries, which took place towards the end of the last decade. Thus, the series of TEMA symposia, now approaching their 10th anniversary, were soon renamed *Trace Elements in Man and Animals* and the 'Animal' contribution has shrunk out of all recognition: this despite the fact that natural mineral imbalances frequently afflict animals and animal products are still important sources of minerals for man. The aim of the book must therefore, remain unchanged from that of the second edition 'meeting the needs of under-graduate and graduate students of nutrition in colleges of agricultural science, animal husbandry and veterinary science, of teachers and research workers in animal nutrition, of agricultural and veterinary extension officers in developed as well as developing regions of the world and of progressive live-stock producers, wherever situated, who wish to apply modern scientific knowledge of mineral nutrition to their own enterprises'.

It was also with trepidation that I made increasingly bold changes to the structure of Eric's second edition, beginning by breaking the almost legendary links between calcium and phosphorus, and between sodium and potassium. There has been an explosion of interest in phosphorus, and the distinctive features of its metabolism which emerged merited a separate chapter: likewise, the contrasts between sodium and potassium become clearer when they are physically (though not physiologically) parted. Sulphur is now also given a chapter of its own, without demeaning the significance of its interactions with copper and molybdenum that were a feature of the second edition. The newer essential elements, boron, chromium, lithium, nickel, molybdenum, tin and vanadium have little practical nutritional significance, but this alone is worth emphasis and it is given under a new corporate heading of 'Occasionally Beneficial Elements'. With fluorosis being an increasingly rare occurrence and other mineral toxicities causing increasing concern, coverage of fluorine is now given along with five other 'Essentially Toxic Elements', aluminium, arsenic, cadmium, mercury and lead in a single chapter.

In the process of getting experts to critically review each draft chapter, I received the kind of comment that every author dreads: 'of course I would not have structured the chapter that way'! This prompted a major rethink and a decision to depart from the original chapter structure which began with 'Functions', to one which begins with 'Dietary sources' of a given mineral and proceeds through 'Metabolism' to 'Function' and thence to the various aspects of dysfunction and its avoidance. Putting feed sources 'up front' showed just how sketchy the information was for most minerals, but it is better to reveal this than ignore it. Mineral imbalances are understood and controlled by exploiting sources and not prejudiced by skirting round numerous functions which never become rate-limited or health-limiting.

Two recurring themes have been introduced to cope with the difficulties of assessing the likelihood of mineral-responsive disorders occurring on farms. One concerns the smooth, sequential patterns of biochemical change which usually accompany transition from a state of normality to one of dysfunction and underlie the development of most disorders. Arising from this is the recognition that all biochemical measures of mineral status, from diet to urine, from erythrocyte to mitochondrion, from enzyme activity to spin-trapped electron, must be interpreted using a marginal band or 'grey area' which recognizes the imprecise link between what is measured and what is 'felt' functionally by the animal and financially by the farmer. The final revision was the addition of a concluding chapter on the role of 'Supplementation Trials' in resolving these continuing problems of defining whether or not mineral deficiencies have become sufficiently severe to impair performance and merit routine intervention.

Since the first two editions were published, increasing pressure has been brought upon scientists to use international S.I. units of measurement when publishing results, i.e. a molar rather than mass basis. Compliance has varied from country to country due in part to opposing pressure from regulators of feed composition for animals and man who generally set limits in terms of mass. A compromise has been struck in this revision by using mass as the basis for reporting the mineral composition of feeds and for those minerals in animal tissues which are noted for the risks they might pose to man (e.g. arsenic, lead and cadmium): S.I. units are generally used for the assessment of mineral status of animal tissues and fluids. Whichever system is used, a conversion factor is usually close at hand. Comments and suggestions for inclusion or omission in any subsequent revision can be sent by e-mail to SUTTN@MRI.SARI.ac.UK or suttle_hints@hotmail.com.

I am indebted to the Moredun Foundation for Animal Health and Welfare for financial support and library facilities without which the revision would not have been possible. I am also grateful to the following colleagues for scrutinizing particular chapters: Dennis Scott (Rowett Research Institute (RRI), Aberdeen), Colin Whitehead (Roslin Institute, Roslin) and Ronald Horst (Iowa State University, Iowa) on calcium and phosphorus; Andrew Sykes (Lincoln University, New Zealand) on magnesium and cobalt; Clive Phillips (School of Veterinary Medicine, Cambridge) and Paul Chiy (University of Wales, Bangor) on sodium, potassium and chlorine; Glen Kennedy (Veterinary Research Laboratories, Belfast) on cobalt; Bob Orskov (RRI) on sulphur; Alan MacPherson (Scottish Agricultural Colleges, Auchincruive) on copper; Ian Bremner (RRI) on iron and cadmium; Neville Grace (New Zealand AgResearch, Palmerston North) on manganese; John Arthur (RRI) on iodine and selenium; Roger Sunde (University of Columbia, Missouri) and Doug Jones (Moredun Research Institute, Penicuik) on selenium; John Chesters (RRI) on zinc; Jerry Spears (North Carolina University, USA) on the 'Occasionally Beneficial Elements'; and Chris Livesey (Veterinary Laboratories Agency, Weybridge) on the 'Essentially Toxic Elements'. Without their vigilance and advice, this revision would have been sadly lacking. I am even more indebted to my typist-cum-research assistant, Jocelyn Brebner, for her unbelievable patience and cheerfulness through seemingly endless redrafts. Finally, I must thank my equally patient wife, Tilly, who thought I had retired 3 years ago!

N.F. Suttle

1 General Introduction

All animal tissues and all feeds contain inorganic or mineral elements in widely varying amounts and proportions. These inorganic elements constitute the ash that remains after ignition. They exist in the ash mostly as oxides, carbonates and sulphates, so that the percentage of total ash is higher than the sum of the inorganic elements individually determined, although losses of some volatile forms may occur during ashing. Before the middle of the 19th century, only the most nebulous ideas existed as to the nature, origin and functions of the mineral constituents of plant and animal tissues. In 1875, Sir Humphrey Davy identified the element potassium in the residues of incinerated wood and gave it the name 'pot ash'. The earliest demonstration of the nutritional significance of minerals came from Fordyce (1791), who showed that canaries on a seed diet required a supplement of 'calcareous earth' to remain healthy and produce eggs. Later, Boussingault (1847) obtained the first experimental evidence that cattle have a dietary need for common salt and Chatin (1850–1854) revealed the relation of environmental iodine deficiency to the incidence of endemic goitre in humans and animals.

Throughout the second quarter of the 20th century, outstanding advances were made in our understanding of the nutritional significance of minerals. These advances came from applied investigations in areas where animals suffered from deficiencies, toxicities or imbalances of certain minerals and from more basic studies with laboratory animals fed on specially purified diets. The latter continued into the third quarter, with animals maintained in plastic isolators so that atmospheric contamination was virtually excluded (Smith and Schwarz, 1967), a procedure which eventually extended the list of essential minerals. The final quarter of this century has seen the application of molecular biology to studies of mineral metabolism and function (O'Dell and Sunde, 1997). As a result, the complex mechanisms by which minerals are safely transported across membranes are being unravelled. For potassium alone, there are at least ten different membrane transport mechanisms. Genes controlling the synthesis of key metalloproteins, such as metallothionein and superoxide dismutases, have been isolated. Deficiencies of zinc have been

© CAB *International* 1999. *Mineral Nutrition of Livestock*
(E.J. Underwood and N.F. Suttle)

found to influence the expression of genes controlling the synthesis of molecules which do not contain zinc (Chesters, 1992). The induction of the mRNA for transport and storage proteins is proving to be a sensitive indicator of copper (Wang *et al.*, 1996) and selenium status (Weiss *et al.*, 1996). New functions have been revealed, such as the role of selenium in regulating iodine metabolism.

Essentiality of Minerals

In 1981, 22 mineral elements were believed to be 'essential' for the higher forms of animal life (Underwood, 1981). These comprised seven major or macronutrient minerals – calcium, phosphorus, potassium, sodium, chlorine, magnesium and sulphur – and 15 trace or micronutrient mineral elements – iron, iodine, zinc, copper, manganese, cobalt, molybdenum, selenium, chromium, tin, vanadium, fluorine, silicon, nickel and arsenic. Since then, aluminium, lead and rubidium have been shown to be beneficial in some circumstances. The essentiality of the last nine, often referred to as the 'newer' trace elements, is based largely on growth effects with laboratory animals reared in highly specialized conditions. So far, the 'newer' elements have not been shown to have any practical significance in the nutrition of domestic livestock (see Chapters 17 and 18), but experience with selenium suggests that such possibilities should not be dismissed. This element was initially considered to be only of scientific interest but, within a few years of the first demonstration of its value in the prevention of liver necrosis in rats, a range of selenium-responsive diseases in farm animals were observed in several countries. Little is still known of the amounts, forms and movements of the newer trace elements in soils, plants and animals or of the physiological availability of the forms in which they occur. In addition to the 25 'essential' minerals, all plant and animal tissues contain a further 20–30 mineral elements, mostly in small and variable concentrations. No vital functions have been found for them and they are believed to be merely accidental or adventitious constituents, reflecting the contact of the animal with its environment. As nutritional and biochemical techniques continue to become more refined, some of these 'accidental' mineral elements will probably become additions to the 'essential' list.

Mineral Nutrition of Livestock

Mineral elements exist in the cells and tissues of the animal body in a variety of functional, chemical combinations and in characteristic concentrations, which vary with the element and the tissue. The concentrations of essential elements must usually be maintained within quite narrow limits if the functional and structural integrity of the tissues is to be safeguarded and the growth, health and productivity of the animal are to remain unimpaired.

Continued ingestion of diets that are deficient, imbalanced or excessively high in a mineral induces changes in the form or concentration of that mineral in the body tissues and fluids, so that it falls below or rises above the tolerable limits. In such circumstances, biochemical lesions develop, physiological functions are affected adversely and structural disorders may arise, in ways which vary with the element, the degree and duration of the dietary deficiency (Chesters and Arthur, 1988) or toxicity and the age, sex and species of animal involved. Homeostatic mechanisms in the body can be brought into play which delay or minimize the onset of such diet-induced changes. Ultimate prevention of the changes requires that the animal be supplied with a diet that is palatable and non-toxic and which contains the required minerals, as well as other nutrients, in adequate amounts, proper proportions and available forms.

Large numbers of livestock in many parts of the world consume diets that do not meet these exacting requirements (McDowell *et al.*, 1993). In consequence, nutritional disorders arise, which range from acute or severe mineral deficiency or toxicity diseases, characterized by well-marked clinical signs, pathological changes and high mortality, to mild and transient conditions, difficult to diagnose with certainty and expressed merely as unthriftiness or unsatisfactory growth, production and fertility. Mild deficiencies or toxicities assume great importance in the nutrition of livestock because of their extent and the ease with which they can be confused with the effects of semistarvation due to underfeeding, protein deficiency and various types of parasitic infestation. In many parts of the world, animal productivity is limited primarily by shortages of available energy and protein, infectious and parasitic disease and genetic inadequacies in the animal. As those limitations are increasingly rectified, local mineral deficiencies and imbalances are likely to become more apparent and more critical (Suttle, 1991).

Inadequate or excessive intakes of a single mineral element are uncommon in most natural environments. They are often exacerbated or ameliorated, i.e. 'conditioned', by the extent to which other dietary components interact metabolically with the mineral. The significance of mineral and nutrient balance is exemplified in the chapters on individual elements that follow. With copper, the question of mineral balance or dietary ratios is of crucial importance because of the potent influence of molybdenum and sulphur on copper retention, but it should be recognized that metabolic interactions which significantly affect minimum requirements and maximum tolerances are widespread among the mineral elements.

The incidence and severity of mineral malnutrition in livestock can be further influenced, both directly and indirectly, by climatic factors, such as sunlight and rainfall. Sunlight promotes vitamin D formation in the animal, which in turn facilitates calcium and phosphorus absorption. The phosphorus concentrations in herbage plants fall with increasing maturity and with the shedding of seed. In any area, the relative lengths of the dry, mature period (low herbage phosphorus) and of the green, growing period (high herbage phosphorus) are determined largely by incidence of rainfall. Climatic or seasonal conditions thus influence the occurrence of phosphorus deficiency in grazing stock where appropriate remedial measures are not imposed. Heavy rainfall, resulting in waterlogging, also increases the availability of some soil minerals to plants, notably cobalt and molybdenum, so affecting the concentrations of those elements in the grazed herbage. These aspects are considered more appropriately in later chapters.

The Functions of Minerals

Four broad types of function for minerals exist – structural, physiological, catalytic and regulatory – although they are not exclusive to particular elements and many may be discharged by the same element in the same individual.

1. *Structural*: minerals can form structural components of body organs and tissues, exemplified by minerals such as calcium, phosphorus, magnesium, fluorine and silicon in bones and teeth and phosphorus and sulphur in muscle proteins. Minerals such as zinc and phosphorus can also contribute structural stability to the molecules and membranes of which they are part.
2. *Physiological*: minerals occur in body fluids and tissues as electrolytes, concerned with the maintenance of osmotic pressure, acid–base balance, membrane permeability and tissue irritability; sodium, potassium, chloride, calcium and magnesium in blood, cerebrospinal fluid and gastric juice provide examples of such functions.
3. *Catalytic*: minerals can act as catalysts in enzyme and hormone systems, as integral and specific components of the structure of metalloenzymes or as less specific activators within those systems. The number and variety of metalloenzymes that have been identified have increased greatly during the last two decades. An indication of the wide range and functional importance of the metalloenzymes is given in Table 1.1.
4. *Regulatory*: in recent years, minerals have been found to regulate cell replication and differentiation; calcium, for example, influences signal transduction and zinc influences transcription, adding to long-established regulatory roles, such as that of the element iodine as a constituent of thyroxine.

In metalloenzymes, the metal is firmly attached to the protein moiety, with a fixed number of metal atoms per mole of protein. The metal cannot be removed without loss of enzyme activity and usually cannot be replaced by any other metal. However, the native zinc atoms in several zinc enzymes can be substituted by cobalt and cadmium without complete loss of activity (Vallee, 1971), and individual metalloenzymes are not always the domain of a single metal. Superoxide dismutase, which catalyses the dismutation of the superoxide free radical, may contain copper plus zinc or manganese, depending on its source, as described in the chapters on those elements. The concentrations and activities of mineral–enzyme associations in particular

Table 1.1. Some important metalloenzymes in livestock.

Metal	Enzyme	Function
Iron	Succinate dehydrogenase	Aerobic oxidation of carbohydrates
	Cytochromes *a*, *b* and *c*	Electron transfer
	Catalase	Protection against H_2O_2
Copper	Cytochrome oxidase	Terminal oxidase
	Lysyl oxidase	Lysine oxidation
	Ceruloplasmin (ferroxidase)	Iron utilization: copper transport
	Superoxide dismutase	Dismutation of superoxide radical O_2^-
Zinc	Carbonic anhydrase	CO_2 formation
	Alcohol dehydrogenase	Alcohol metabolism
	Carboxypeptidase A	Protein digestion
	Alkaline phosphatase	Hydrolysis of phosphate esters
	Nuclear poly(A) polymerase	Cell replication
	Collagenase	Wound healing
Manganese	Pyruvate carboxylase	Pyruvate metabolism
	Superoxide dismutase	Antioxidant by removing O_2^-
	Glycosylaminotransferases	Proteoglycan synthesis
Molybdenum	Xanthine dehydrogenase	Purine metabolism
	Sulphite oxidase	Sulphite oxidation
	Aldehyde oxidase	Purine metabolism
Selenium	Glutathione peroxidases (four)	Removal of H_2O_2 and hydroperoxides
	Type I and III deiodinases	Conversion of thyroxine to active form

H_2O_2, hydrogen peroxide; CO_2, carbon dioxide.

cells and tissues have, in some instances, been related to the manifestations of deficiency and toxicity of those elements in the animal body. In some cases, serious clinical and pathological disorders arise as a consequence of dietary mineral abnormalities which cannot as yet be explained in such biochemical terms (Chesters and Arthur, 1988). Two of the mineral elements, iodine and cobalt, are remarkable because, on present evidence, their entire functional significance can be accounted for by their presence in single compounds, thyroxine and vitamin B_{12}, respectively; both compounds, and therefore iodine and cobalt, are nevertheless involved in a range of metabolic processes.

The functions of calcium and phosphorus are dominated so quantitatively by their requirements for the mineral base of the skeletal tissues that their manifold activities in the soft tissues and fluids of the body have been neglected by nutritionists concerned with dietary needs. These two minerals combine in the bones to provide strength, shape and rigidity, protecting the soft tissues and giving attachment to the muscles, while simultaneously forming a storage depot from which they can be mobilized for nutritional and metabolic emergencies, regulating the amounts in the blood and buffering against prolonged dietary inadequacies. Regulation is vital, because calcium plays an important intracellular role involving cell signalling and an extra-cellular role in transmitting nerve impulses. Phosphorus participates in a wide range of metabolic reactions involving energy transfer. Every physiological event involving gain or loss of energy and almost every form of energy exchange in the cell include the making or breaking of high-energy phosphate bonds. In addition, phosphorus is an integral part of protein molecules and of the nucleic acids and their derivatives that are vitally concerned in cell replication and the transmission of the genetic code.

The functions performed by minerals can only be fulfilled if the finite amounts ingested are sufficient to keep pace with the growth and development of the body and the reproduction of the species and to replace minerals that are 'lost', either as 'harvested' products or insidiously during the process of living.

Net Mineral Needs for Maintenance and Work

Maintenance

The basic maintenance requirement comprises the nutrients needed to keep intact the tissues of an animal which is not growing, working, reproducing or yielding any product. Body maintenance involves the performance of internal work in digestion, circulation, respiration and other vital processes, together with some external work in the ordinary movements of the animal. There is, in addition, a 'productivity increment', which contributes to the total maintenance requirement (M) because growth, work, reproduction or yield of a product increases the amount of food eaten and hence the amount of internal work done in utilizing the additional nutrients (Milligan and Summers, 1986). The difference between organic and inorganic nutrients in respect of maintenance needs stems from the different metabolic fates which the two types of nutrients experience. Organic nutrients enter metabolic pools, from which they are lost as heat, or converted to end-products of metabolism, simpler in form, which are excreted through the normal channels, i.e. no longer available as sources of energy or protein. In contrast, inorganic ions liberated in the course of metabolism are not changed and remain as available for the re-formation of their functional combinations as are the inorganic ions absorbed from the alimentary tract. Recycling is not complete, however, and there is an inescapable 'leakage' of minerals through the kidneys, intestinal mucosa, digestive glands and skin, which must be replaced. Maintenance requirements vary appreciably for different elements. Endogenous losses of calcium and phosphorus, for instance, are substantial in the adult ruminant (approximately 6 g of each mineral for a 500 kg cow). On the other hand, sodium and chlorine are efficiently conserved, so that the adult maintenance requirements for these elements are usually extremely small, except in conditions of excessive sweating. Though small in amount, the maintenance requirement for manganese is probably of a similar order to that for growth at certain stages of development.

Faecal endogenous losses

For most elements and most situations, the major component of M is the endogenous loss of minerals via the faeces (FE). The amounts that are unavoidably lost are hard to measure if the faeces either serves as a route of excretion for a mineral absorbed in excess of need, as it does for phosphorus and manganese for example, or as a means of conserving a mineral during deficiency, as it does for sodium and iron. An early approach was to determine 'minimum endogenous loss' from the intercept of the regression of faecal excretion against intake for a given mineral, i.e. the faecal loss at zero mineral intake. Estimates of the maintenance requirement of ruminants for phosphorus (P) were reduced from 40 (ARC, 1965) to 12 mg P kg^{-1} live weight (LW) by the Agricultural Research Council (ARC, 1980) by adopting the minimum FE approach. Since no animal can survive for long without an essential mineral, animals need to do more than replace the minimum FE. However, failure to acknowledge that it is unnecessary to replace the entire FE, regardless of intake, leads to an overestimation of M. The most useful value is that FE which occurs at the minimum mineral intake needed to sustain optimal (usually maximal) production or zero balance in the case of non-producing livestock. Given that replenishment of skeletal phosphorus reserves during the non-productive dry period is vital for sustaining the next lactation, it is important to define M accurately for this element. Furthermore, the maximal efficiency with which a mineral can be absorbed from the gut cannot be ascertained unless the flow of mineral in the opposite direction, i.e. FE, is known.

Influence of food intake

Maintenance requirements were expressed on a body-weight basis, but there is now evidence that, for calcium and phosphorus at least, M for ruminants is a function of food intake (AFRC, 1991). The relationship with food intake is to be expected, given that FE is derived from sloughed mucosal cells and unresorbed digestive secretions, and it would be surprising if food intake does not also affect M for most minerals. Relationship to food intake rather than body weight will mean that M is relatively high at times of high food intake, such as lactation, and when diets of low digestibility are fed: in contrast, M should fall at times of low food intake, except for periods of inappetence caused by heat stress, when more minerals are lost in sweat.

Work

With grazing stock, movement may raise maintenance needs for energy by 10–20% or more above those of the animal at rest. With working animals, such as horses or bullocks used for draught or transport purposes, the energy requirements can be increased several times above maintenance. The extra food required to meet these requirements will usually supply the animal with sufficient additional minerals, even the cow, which is increasingly used for draught purposes, except perhaps for sodium. Hard physical work, especially in hot conditions, greatly increases sweating and therefore losses of sodium and potassium; the dietary requirement for salt is thereby raised. No other specific mineral requirements for physical work have been reported. No significant change in calcium or phosphorus balance or in losses of these minerals from the body was observed in horses performing light, medium or hard farm work, relative to those of the same horses performing no such work. However, there is a growing belief that increased consumption of oxygen during exercise leads to increased requirements for elements, such as selenium, which are involved in antioxidant defence (see Chapter 15) (Avellini *et al.*, 1995).

Net Mineral Requirements for Reproduction

The mineral requirements for reproduction (R) in mammals are usually equated to the mineral content of the fetus and products of conception (placenta, uterus and fetal fluids) and therefore increase exponentially to reach a peak in late gestation (Fig. 1.1). For the twin-bearing ewe, the calcium requirement in late gestation actually exceeds that of lactation, leading to important contrasts in the period of vulnerability to calcium deprivation when compared with the dairy cow (see Chapter 4). There is also a small additional requirement for growth of mammary tissue and the

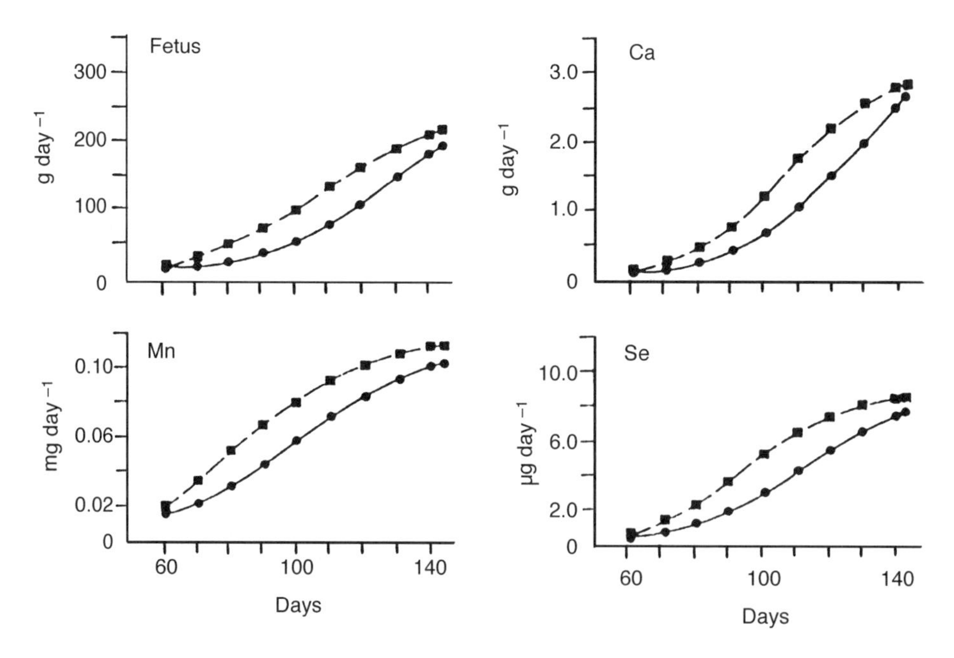

Fig. 1.1. Patterns of fetal accumulation of calcium, manganese and selenium in singleton (●) and twin-bearing ewes (■) during gestation reflect fetal weight gain (from Grace *et al.*, 1986).

accumulation of colostrum prior to parturition. For those minerals where the generosity of maternal nutrition determines the size of the fetal reserve (e.g. copper and selenium (Langlands *et al.*, 1984)), it may be prudent, but not essential, to allow for maximal fetal retention.

Net Mineral Requirements for Production

The net mineral requirements for production (P) are given by the mineral content of each unit of production, such as weight gain (WG), milk yield (L) or fleece growth (F), and are usually taken to remain constant. However, for elements such as calcium and phosphorus, which are far richer in bone than in soft tissue, requirements for growth diminish as animals mature, because bone makes a progressively smaller contribution to each unit of live-weight gain (AFRC, 1991). Mineral requirements for production are affected by the species or breed of animal, the intensity or rate of production permitted by other constituents of the diet – notably energy – and by the environment. Chicks and weaned pigs consume similar types of diet, but the faster-growing broiler chick requires nearly twice the dietary concentrations of calcium and manganese required by pigs. High-yielding dairy cows obviously require much more dietary calcium and phosphorus than low-yielding cows, because of the richness of milk in these elements. However, the levels necessary in the dry ration do not rise, because total dry matter (DM) intakes increase with rising productivity of the cow as rapidly as do mineral requirements (AFRC, 1991). The phosphorus requirements of hens tend to follow a similar pattern with onset of egg production, remaining a constant proportion of the diet, but calcium requirements greatly increase. A non-laying hen can normally meet its calcium (Ca) needs from a diet containing 5 g Ca kg^{-1} DM, whereas some eight to ten times this concentration is necessary for a hen laying one egg per day.

Mineral intakes must ideally be sufficient to ensure the maintenance of adequate mineral reserves of the body tissues and adequate amounts in edible products. The animal body can sometimes adjust to suboptimal intakes by reducing the amount of a mineral in its products. Thus quality, such as the shell strength of eggs and the tensile strength of wool fibres, may be reduced in order to maintain other more essential functions. This is clearly undesirable, so that assessment of mineral needs usually includes determination of the minerals in the tissues, fluids and products, as well as such gross criteria as weight gains, milk yields and so on. Milk is an exception, in that normal mineral concentrations are usually maintained during deficiency, priority being given to the mineral nutriture of the new generation at the expense of the mother. Where the provision of excess mineral increases concentrations in milk (e.g. iodine), it is unnecessary to allow for replacement of all the secreted mineral.

Gross Mineral Requirements

Net mineral requirements underestimate the dietary needs of livestock for minerals, because ingested minerals are incompletely utilized, due to limits upon their absorption from the gut. A basic or minimum requirement for any mineral can be conceived as one in which all the dietary conditions affecting that mineral are optimum. Since these exacting conditions rarely apply, there can be no single requirement but rather a series of requirements, depending on the extent to which 'conditioning' factors are present in a particular grazing or ration. By the same reasoning, there must be a series of maximum 'safe' dietary levels, depending on the extent to which other minerals or compounds are affecting the absorption, retention and excretion of a mineral consumed in excess of need (Fig. 1.2). Differences in the chemical and physical forms in which a mineral is present also affect requirements and tolerances and are considered in appropriate detail in subsequent chapters. The chemical form of the mineral acquires particular nutritional significance for non-ruminant livestock, such as pigs and poultry. For example, phytate phosphorus and zinc bound to phytate are poorly absorbed in non-ruminants but are well used by sheep and cattle. Physical form may be important for mineral supplements, such as magnesium oxide, which can vary in particle size and hence in nutritive value.

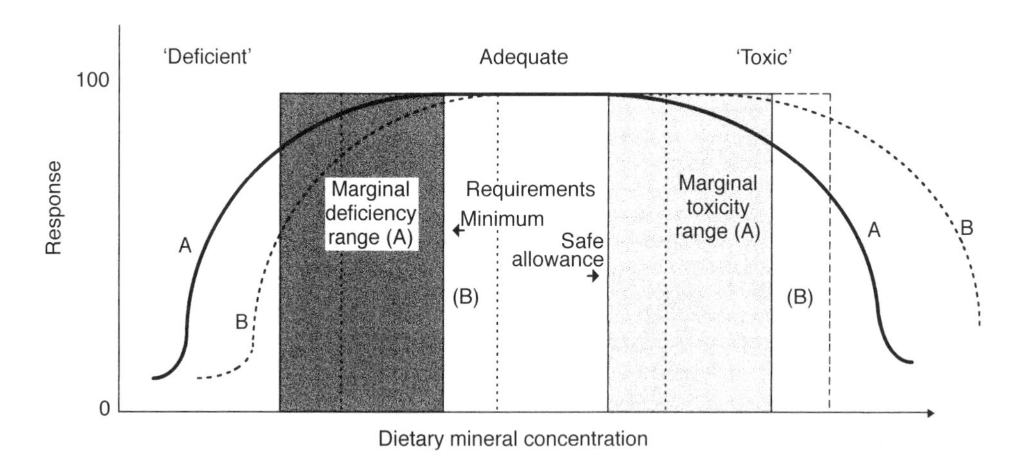

Fig. 1.2. Schematic dose–response relationship between mineral supply and animal production showing marginal bands between adequate and inadequate or toxic dietary concentrations. The graph and marginal ranges move to the right as absorbability of the mineral source declines; thus A represents the more and B (dashed line) the lesser absorbable of two mineral sources. 'Requirements' are variously set within the central adequate band from minimum requirements to safe allowances, depending on the extent to which absorbability and other variables are taken into consideration.

Estimates of gross mineral requirements

There are two different but complementary ways for estimating gross mineral requirements (GR).

Factorial estimates

The GR of livestock can be determined by factorial models, which summate the components of net requirement and divide the total by an absorption coefficient (A), thus allowing for inefficient use of the dietary mineral supply. Thus, for a lactating ewe:

$$GR = (M + L + F)/A$$

where M, L and F are the net requirements for maintenance, lactation and fleece growth, respectively. Application of the factorial method was pioneered by ARC (1965) in their study of the nutrient requirements of ruminants. The main advantage of the factorial approach is that requirements can be predicted for a wide range of production circumstances, provided that reliable data are available for each model component. The general strengths and weaknesses of factorial models of mineral requirements have been discussed by Suttle (1985) and White (1996). The output from factorial models will form the basis of many of the requirements tabulated in later chapters.

Dietary estimates

Mineral requirements can also be estimated by feeding livestock with diets providing a range of mineral inputs above and below the minimum requirement and measuring responses in a relevant variable, such as growth rate or blood composition. However, five or more different mineral input levels may be needed to define the optimum requirement precisely and the result can still depend on the statistical model used to describe the response (Remmenga *et al.*, 1997). A further difficulty is that it is often necessary to use purified ingredients to obtain sufficiently low mineral inputs; if the diet then lacks naturally occurring antagonists, the results will underestimate requirements on natural diets (Suttle, 1985; White, 1996). Another difficulty is that it is impractical to allow for requirements which vary with time. Where the requirement for production is high relative to that for maintenance, as it is for iron in growing livestock, requirements will fall with time when expressed as a proportion of a steadily increasing food intake if growth rate remains constant (Suttle, 1985). The dietary approach has demonstrated much higher zinc (Zn) requirements for turkey poults than chicks on the same semipurified diet (25 vs. 18 mg Zn kg^{-1} DM (Dewar and Downie, 1984)); the difference may be partly attributable to the higher food conversion efficiency of the turkey poult.

Criteria of Adequacy

The criterion of adequacy employed is an important determinant of the estimated requirement of a mineral. As the amount available to the animal becomes insufficient for all the metabolic processes in which the element participates and after depletion of body reserves, certain of these processes are adversely affected while others remain initially unimpaired. For instance, the processes of pigmentation and keratinization of wool appear to be the first to be affected by a low copper status in sheep. At certain rates of copper intake, no other function dependent on copper appears to be affected. If wool quality is taken as the criterion of adequacy, it follows that the copper requirement of the sheep is higher than if growth rate and blood haemoglobin content are used as criteria. The minimum zinc requirements for spermatogenesis and testicular development in young male sheep are significantly higher than they are for body growth (Underwood and Somers, 1969). If body growth is taken as the criterion of adequacy, as it would be with lambs destined for slaughter at an early age, the zinc requirements would clearly be lower than for similar animals kept for breeding. When production traits are used, optimum mineral intakes are easily defined, because the optimum input is that which sustains maximum performance, e.g. growth. When biochemical traits are used, optimum intakes are less easily defined. For example, Dewar and Downie (1984) concluded that the optimum zinc concentrations for chicks were 18, 24 and 27 mg kg^{-1} DM, respectively, for growth, plasma zinc and tibia zinc concentrations. Adequacy of phosphorus nutrition can be assessed on the basis of body growth, bone dimensions, bone composition, bone histology or bone ash, growth-plate hypertrophy giving the most sensitive estimates of requirement in turkey poults (Fig. 1.3; Qian *et al.*, 1996). The complementary nature of factorial and dietary approaches to the estimation of mineral requirements arises because the latter is needed to define the conditions under which components of the factorial model are measured. The endogenous losses of phosphorus (i.e. the value of M for the model) at optimal growth are almost certainly lower than those commensurate with optimal bone strength.

Requirements for Breeding Stock

Particular problems arise in defining requirements for reproduction and lactation when the mother gives priority to offspring and any deficit between daily need and dietary provision is met by drawing on maternal mineral reserves. A classic experiment was performed by Kornegay *et al.* (1973), in which sows were fed constant daily amounts of calcium and phosphorus (10.3 and 11.0 g day^{-1}, respectively) throughout five reproductive cycles. The allowance was grossly inadequate for lactation but sufficient to allow some recovery of reserves between lactations. No harm was done to their offspring and even the fifth generation remained biochemically normal. The only production trait

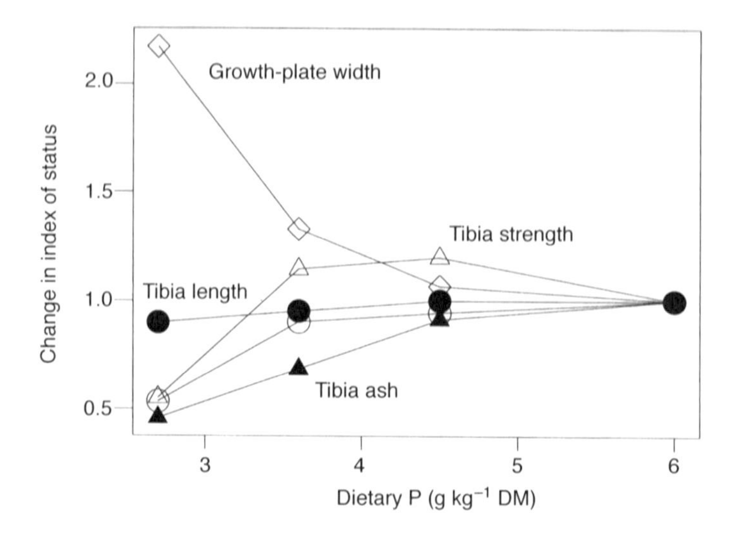

Fig. 1.3. Measures of bone quality in turkey poults vary in sensitivity to dietary phosphorus (P) supply and assessments of 'optimal' P supply vary accordingly: sensitivity of growth rate is shown by the open circles (values relative to those at maximal P intake; data from Qian *et al.*, 1996).

to suffer was sow longevity, which was reduced chiefly by the gradual development of lameness and other leg abnormalities. Longevity is an important trait in economic terms and the lengthy critical experiments required to define mineral requirements for lifetime production of the breeding female have yet to be carried out for any species, including the sow, for any mineral.

Variation Among Tables of Mineral Requirements

Considerable confusion has arisen from the use of different criteria and assumptions by national authorities when calculating mineral requirements. An attempt was made in 1983 to achieve an international consensus on the mineral and other nutrient requirements of poultry (Blair *et al.*, 1983) and revealed marked discrepancies; some subsequent revisions by respected authorities are unjustifiable (see Chapters 14 and 15). Similar variation has recently been noted in the mineral requirements of sheep (Table 1.2; White, 1996). Some authorities have estimated minimum requirements, others average requirements and some safe allowances, which cater for the most adverse individual circumstances and leave no animal at risk from deficiency. Each type of calculation has its uses and its difficulties. Safe allowances are essential for those who prepare compound feeds, but there is rarely sufficient knowledge of the relevant variables for safety margins to be calculated with precision. Furthermore, avoidance of deficiency may not be compatible with

Table 1.2. Variation in recommended mineral requirements for sheep between countries (from White, 1996).

	Ca	P	Cu	Se
	(g kg^{-1} DM)		(mg kg^{-1} DM)	
New Zealand (1983)	2.9	2.0	5.0	0.03
United States (1985)	2.0–8.2	1.6–3.8	7–11	0.1–0.2
Australia (1990)	1.5–2.6	1.3–2.5	5.0	0.05
United Kingdom (1980)	1.4–4.5	1.2–3.5	1[a]–8.6	0.03–0.05

[a] The low requirement is a unique recognition of the distinctive low mineral requirements of milk-fed animals.

avoidance of toxicity, e.g. in providing copper and phosphorus to breeds of sheep that are vulnerable to copper poisoning and urinary calculi, respectively. Average requirements are useful for those assessing forages as sources of minerals for grazing livestock but cannot be used for diagnosing deficiencies or toxicities, because of variations in absorption and tolerance (Suttle, 1991). In using tables of requirements, the methods and assumptions underlying the data should be checked. Where authors and working parties fail to indicate a 'marginal' band between the 'adequate' and 'inadequate', the user should interpolate one to allow for the inevitable uncertainty surrounding precise mineral needs in a particular context. Marginal bands will be adopted throughout this book.

Expression of Requirements

The requirements of animals for minerals can be expressed in several ways: in amounts per day or per unit of product, such as milk, eggs or weight gain; in proportions, e.g. percentage, parts per million (ppm), mass $mass^{-1}$ (e.g. mg kg^{-1}) or moles (sometimes micro- or millimoles) kg^{-1} DM of the whole diet. Required amounts are more precise but not invariably independent of variations in total food intake, as illustrated by the latest estimates of calcium and phosphorus requirements for ruminants (AFRC, 1991). Proportions have the merit of simplicity and have obvious practical advantages, so long as the total diet is palatable, but they are of limited value when the regular *daily* intake of large amounts of mineral is essential, as in the case of calcium for the laying hen. Whether expressed as amounts or concentrations, requirements can be greatly influenced by factors that limit the absorption and utilization of the mineral in question.

References

AFRC (1991) *Technical Committee on Responses to Nutrients, Report* No. 6: a reappraisal of the calcium and phosphorus requirements of sheep and cattle. *Nutrition Abstracts and Reviews* 61, 573–612.

ARC (1965) *The Nutrient Requirements of Farm Livestock,* No. 2. *Ruminants.* HMSO, London.

ARC (1980) *The Nutrient Requirements of Ruminants.* Commonwealth Agricultural Bureaux, Farnham Royal, Slough, UK.

Avellini, L., Silvestrelli, M. and Gaiti, A. (1995) Training-induced modifications in some biochemical defences against free radicals in equine erythrocytes. *Veterinary Research Communications* 19, 179–184.

Blair, R., Daghir, N.J., Morimoto, H., Peter, V. and Taylor, T.G. (1983) International nutrition standards for poultry. *Nutrition Abstracts and Reviews, Series B* 53, 673–703.

Boussingault, J.B. (1847) *Annals of Chemistry and Physics, 3rd Ser.* 19, 117; 22, 116; *Comptes Rendus des Séances de l'Académie des Sciences* 25, 72Y. Cited by McCollum, E.V. (1957) *A History of Nutrition.* Houghton Mifflin, Boston.

Chatin, A. (1850–1854) Recherche de l'iode dans l'air, les eaux, le sol et les produits alimentaires des Alpes de la France. *Comptes Rendus des Séances de l'Académie des Sciences* 30–39.

Chesters, J.K. (1992) Trace elements and gene expression. *Nutrition Reviews* 50, 217–223.

Chesters, J.K. and Arthur, J.R. (1988) Early biochemical defects caused by dietary trace element deficiencies. *Nutrition Research Reviews* 1, 39–56.

Dewar, W.A. and Downie, J.N. (1984) The zinc requirements of broiler chicks and turkey poults fed on purified diets. *British Journal of Nutrition* 51, 467–477.

Fordyce, G. (1791) *Treatise on the Digestion of Food,* 2nd edn. London. Cited by McCollum, E.V. (1957) *A History of Nutrition.* Houghton Mifflin, Boston.

Grace, N.D., Watkinson, J.H. and Martinson, P. (1986) Accumulation of minerals by the foetus(es) and conceptus of single- and twin-bearing ewes. *New Zealand Journal of Agricultural Research* 29, 207–222.

Kornegay, E.T., Thomas, H.R. and Meacham, T.N. (1973) Evaluation of dietary calcium and phosphorus for reproducing sows housed in total confinement on concrete or in dirt lots. *Journal of Animal Science* 37, 493–500.

Langlands, J.P., Bowles, J.E., Donald, G.E. and Smith, A.J. (1984) Deposition of copper, manganese, selenium and zinc in 'Merino' sheep. *Australian Journal of Agricultural Research* 35, 701–707.

McDowell, L.R., Conrad, J.H. and Hembry, F.G. (1993) *Minerals for Grazing Ruminants in Tropical Regions,* 2nd edn. Centre for Tropical Agriculture, University of Florida, Gainesville, pp. 53–55.

Milligan, L.P. and Summers, M. (1986) The biological basis of maintenance and its relevance to assessing responses to nutrients. *Proceedings of the Nutrition Society* 45, 185–193.

O'Dell, B.L. and Sunde, R.A. (1997) Introduction. In: O'Dell, B.L. and Sunde, R.A. (eds) *Handbook of Nutritionally Essential Minerals.* Marcel Dekker, New York, pp. 8–11.

Qian, H., Kornegay, E.T. and Veit, H.P. (1996) Effects of supplemental phytase and phosphorus on histological, mechanical and chemical traits of tibia and performance of turkeys fed on soybean meal-based, semi-purified diets high in phytate phosphorus. *British Journal of Nutrition* 76, 263–272.

Remmenga, M.D., Milliken, G.A., Kratzer, D., Schwenke, J.R. and Rolka, H.R. (1997) Estimating the maximum effective dose in a quantitative dose–response experiment. *Journal of Animal Science* 75, 2174–2183.

Smith, J.C. and Schwarz, K. (1967) A controlled environment system for new trace element deficiencies. *Journal of Nutrition* 93, 182–188.

Suttle, N.F. (1985) Estimation of requirements by factorial analysis: potential and limitations. In: Mills, C.F., Bremner, I. and Chesters, J.K. (eds) *Proceedings of the Fifth International Symposium on Trace Elements in Man and Animals.* Commonwealth Agricultural Bureaux, Farnham Royal, UK, pp. 881–883.

Suttle, N.F. (1991) Mineral supplementation of low quality roughages. In: *Proceedings of Symposium on Isotope and Related Techniques in Animal Production and Health.* International Atomic Energy Commission, Vienna, pp. 101–144.

Underwood, E.J. (1981) *The Mineral Nutrition of Livestock*, 2nd edn. Commonwealth Agricultural Bureaux, Farnham Royal, UK, p. 1.

Underwood, E.J. and Somers, M. (1969) Studies of zinc nutrition in sheep. I. The relation of zinc to growth, testicular development and spermatogenesis in young rams. *Australian Journal of Agricultural Research* 20, 889–897.

Vallee, B.L. (1971) Spectral characteristics of metals in metalloenzymes. In: Mertz, W.A. and Cornatzer, W.E. (eds) *Newer Trace Elements in Nutrition.* Marcel Dekker, New York, pp. 33–50.

Wang, Y.R., Wu, J.Y.J., Reaves, S.K. and Lei, K.Y. (1996) Enhanced expression of hepatic genes in copper-deficient rats detected by the messenger RNA differential display method. *Journal of Nutrition* 126, 1772–1781.

Weiss, S.L., Evenson, J.K., Thompson, K.M. and Sunde, R.A. (1996) The selenium requirement for glutathione peroxidase mRNA is half of the selenium requirement for glutathione peroxidase activity in female rats. *Journal of Nutrition* 126, 2260–2267.

White, C.L. (1996) Understanding the mineral requirements of sheep. In: Masters, D.G. and White, C.L. (eds) *Detection and Treatment of Mineral Nutrition Problems in Grazing Sheep.* ACIAR Monograph No. 37, Canberra, pp. 15–29.

2 Natural Sources of Minerals

In most circumstances, farm animals derive a high proportion of their mineral nutrients from the feeds and forages that they consume. For this reason, the factors that determine the mineral content of the vegetative parts of plants and their seeds are the factors that basically determine the mineral intakes of livestock. The concentration of all minerals in crop and forage plants depends on four basic interdependent factors: (i) the genus, species or strain (variety); (ii) the type of soil on which the plant grows; (iii) the climatic or seasonal conditions during growth; and (iv) the stage of maturity of the plants. The extent to which these factors affect the concentration of a mineral element in the plant tissues varies with different minerals and with the treatments imposed by humans in their efforts to increase crop or pasture yields. Such treatments involve the use of fertilizers, soil amendments and irrigation water and include the breeding and selection of higher-yielding cultivars, which may differ significantly in mineral composition from the varieties they supplant. Inorganic compounds of geological or industrial origin are used freely and increasingly to supplement the minerals supplied by feeds and forages, especially for modern farm animals of high genetic productivity. Feed supplements from animal and fish sources, usually by-products of abattoirs and processing factories, given primarily to supply extra protein, can be additional sources of minerals. However, the factors that govern the mineral composition of plants must be considered first.

Genetic Differences in Mineral Composition Among Plants

Adaptation to extreme environments

The most striking evidence of genetic influence on mineral composition is provided by certain genera and species growing in enriched soils which carry concentrations of particular elements often several orders of magnitude higher than those of other species growing in the same extreme conditions. Thus, the saltbushes (*Atriplex* spp.) and blue-bushes (*Kochia* spp.) carry

© CAB *International* 1999. *Mineral Nutrition of Livestock*
(E.J. Underwood and N.F. Suttle)

many times the sodium and chlorine contents of most other herbage plants. The former species typically contain 8–14% sodium chloride (NaCl) in their dry matter (DM), compared with only about one-hundredth of these concentrations in common pasture plants. Similarly, certain species of *Astragalus* growing on seleniferous soils contain 5000 mg selenium (Se) kg^{-1} DM or more, compared with < 20 mg kg^{-1} DM in common herbage species supported by the same soils. The significance of selenium 'accumulator' plants to the problem of selenosis in grazing livestock is considered in Chapter 15. In some plant species, strontium contents as high as 26 g kg^{-1} DM have been reported, compared with 0.1–0.2 g kg^{-1} DM in other species from similar strontium-rich soils (Bowen and Dymond, 1955), but such high contents have no known nutritional significance to animals.

Legumes and grasses

Less spectacular but more nutritionally important differences in mineral content exist between and among legumes and grasses. Leguminous species are generally substantially richer in macroelements than grasses growing in comparable conditions, and this difference applies to both temperate pasture species and tropical forages. For example, Minson (1990) reported mean calcium concentrations of 14.2 and 10.1 g kg^{-1} DM, respectively, in temperate and tropical legumes, as against 3.7 and 3.8 g kg^{-1} DM in the corresponding grasses. A mixed sward of *Lolium perenne* and *Trifolium repens* had consistently higher concentrations of calcium, magnesium and potassium than a pure sward of *L. perenne*, in trials lasting 4 years at 16 sites, but sodium, sulphur and phosphorus concentrations were close with no nitrogen applied (Hopkins *et al.*, 1994). The trace elements, notably iron, copper, zinc, cobalt and nickel, are also generally higher in leguminous than in graminaceous species grown in temperate climates, with copper and zinc higher in a mixed than in a grass sward (Burridge *et al.*, 1983; Hopkins *et al.*, 1994). However, these differences are narrowed when the soil is low in available mineral and they can be reversed (e.g. copper) in tropical climates (Minson, 1990). Grasses and cereals commonly carry higher concentrations of manganese and silicon than legumes grown in similar conditions (see Underwood, 1977). The position of sodium is of interest. Many tropical legumes are exceedingly low in sodium, 50% containing less than 4 g kg^{-1} DM (Minson, 1990). Considerable variability has been observed in the sodium contents of common temperate pasture species in New Zealand (Sherrell, 1978). Furthermore, two distinct types of species were found: natrophilic, with high sodium concentrations in the shoots and leaves and low in the roots, and natrophobic, with the reverse distribution (Smith *et al.*, 1978). Where natrophobic species, such as red clover and timothy, form a substantial part of the animal's diet away from the coast, they could pose a nutritional problem, because of the low sodium content in their aerial (edible) parts. Herbs generally contain higher mineral concentrations than cultivated plants (e.g. Wilman and Derrick, 1994).

Variation among grasses and forages

In a study of 58 East African grasses, grown together on the same soil type and sampled at the same growth stage, the following range of concentrations was found: total ash, 40–122, calcium, 0.9–5.5 and phosphorus, 0.5–3.7 g kg^{-1} DM (Dougall and Bogdan, 1958). In an investigation of 17 North American grass species, grown together on a sandy loam soil and sampled at similar growth stages, the concentrations (mg kg^{-1} DM) ranged from 0.05 to 0.14 for cobalt, from 4.5 to 21.1 for copper and from 96 to 815 for manganese (Beeson *et al.*, 1947). While some of the variation may be due to soil contamination, this in turn may be determined by genetic characteristics of the species, such as its growth habit (upright or prostrate). Jumba *et al.* (1996) concluded that botanical composition was far more important than soil origin and composition in determining the mineral content of tropical grasses in western Kenya. Figure 2.1 summarizes the data and shows that the introduced species, Kikuyu, was richer in most macro- and trace elements than the native species. Strain or varietal, i.e. intraspecific, differences in the mineral composition of plants are highlighted by New Zealand experience with ryegrass. Short-rotation hybrid (H1) ryegrass, bred and selected specifically for high DM yield, was found to contain only about one-tenth of the iodine concentration of perennial ryegrass (*L. perenne*) at both low and

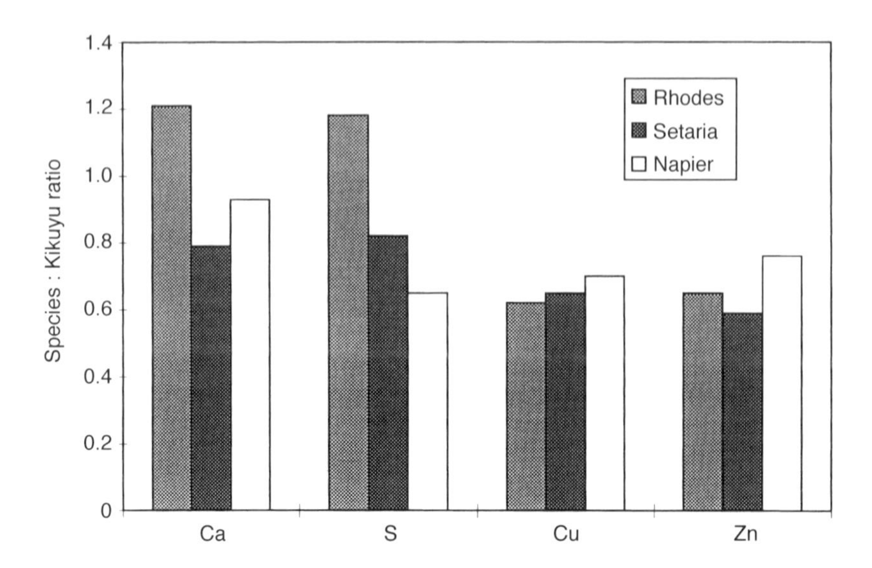

Fig. 2.1. Species differences in some macro- and trace element concentrations of three tropical grasses harvested in the dry season in Western Kenya compared with Kikuyu grass (given value of 1) (actual values for Kikuyu were 1.4 g Ca and 1.7 g S kg^{-1} DM: 6.0 mg Cu and 30.1 mg Zn kg^{-1} DM; differences for other common minerals were generally non-significant: data from Jumba *et al.*, 1996).

high soil iodine levels (Johnson and Butler, 1957). Botanical composition of the pasture was again a more important determinant of iodine intakes by grazing stock than the soil type. In a study of forages grown on a silt loam in Florida, selenium concentrations were twice as high in Bermuda grass as in Bahia grass and magnesium levels 60% higher in sorghum silage than in maize silage (Kappel *et al.*, 1983).

Variation among grains and seeds

Genetic differences in the mineral composition of the vegetative parts of plants are not necessarily paralleled by comparable differences in the seeds. The seeds of leguminous plants and the oil-seeds used as dietary protein supplements are nevertheless invariably richer in most minerals than are the seeds of grasses and cereals, which are inherently low in calcium and sodium. Fivefold differences exist in the manganese contents of cereal grains (Underwood *et al.*, 1947; Gartner and Twist, 1968), typical values being wheat and oats, 35–40, barley and sorghum, 14–16 and maize 5–8 mg kg^{-1} DM; maize is also low in cobalt. Species differences in the manganese (Mn) content of lupin seeds are even greater; the seeds of white lupins (*Lupinus albus*) have been reported to contain potentially toxic levels (817–3397 mg Mn kg^{-1} DM), 10–15 times as much as in other species of lupins growing on the same sites in Western Australia (Gladstones and Drover, 1962). *Lupinus albus* seed is also far richer in cobalt than *Lupinus augustifolius* (240 vs. 74 μg kg^{-1} DM); zinc (Zn) and selenium concentrations are similar in the two lupin species (29–30 mg Zn kg^{-1} and 57–89 μg Se kg^{-1} DM) but much higher than in wheat grain harvested from the same sites (22 mg Zn kg^{-1} and 23 μg Se kg^{-1} DM) (White *et al.*, 1981). Wide variability in the mineral concentrations of pasture grass and legume seeds has also been observed in New Zealand (Sherrell and Smith, 1978), although some of this variability arises from soil differences.

Effects of Soils and Fertilizers on Plant Minerals

Mineral concentrations in plants generally reflect the adequacy with which the soil can supply absorbable mineral to their roots. However, plants react to inadequate supplies of available minerals in the soil by limiting their growth, reducing the concentration of the deficient elements in their tissues or, more commonly, reducing growth and concentration simultaneously. The extent to which a particular response occurs varies with different minerals, with different plant species or varieties and with the soil and climatic conditions. Nevertheless, the primary reason for the existence of areas of mineral deficiencies in grazing animals, such as those of phosphorus, sodium, cobalt and selenium, is that the soils of the areas are inherently low in plant-available supplies of these minerals. Large differences exist between the mineral needs of plants for growth and those of animals dependent on these plants. For example, the growth requirements of plants for iodine, selenium and cobalt are negligible and the required treatments are therefore those that raise the herbage concentrations of these elements to an extent satisfactory for livestock. Applications of magnesium compounds are sometimes necessary to raise the magnesium concentration in pasture to meet the intense needs of cows for this element during early lactation, although they usually have no effect on pasture yield (see Chapter 6). The position with potassium and manganese is quite different; these two minerals are required by plants and animals, but, even on deficient soils which exhibit substantial crop or pasture yield responses to applications of fertilizer containing these elements, the amounts carried by the untreated herbage are generally adequate for the requirements of grazing stock.

Soil characteristics

Mineral uptake by plants and hence their mineral composition are greatly influenced by soil pH (Fig. 2.2), particularly in legumes. Molybdenum uptake by plants increases as soil pH rises, and soils on which copper deprivation occurs in animals, due to high concentrations of molybdenum in the herbage (see Chapter 11), are mostly derived from clays, shales and limestones and are alkaline or calcareous. In contrast, molybdenum deficiency in leguminous

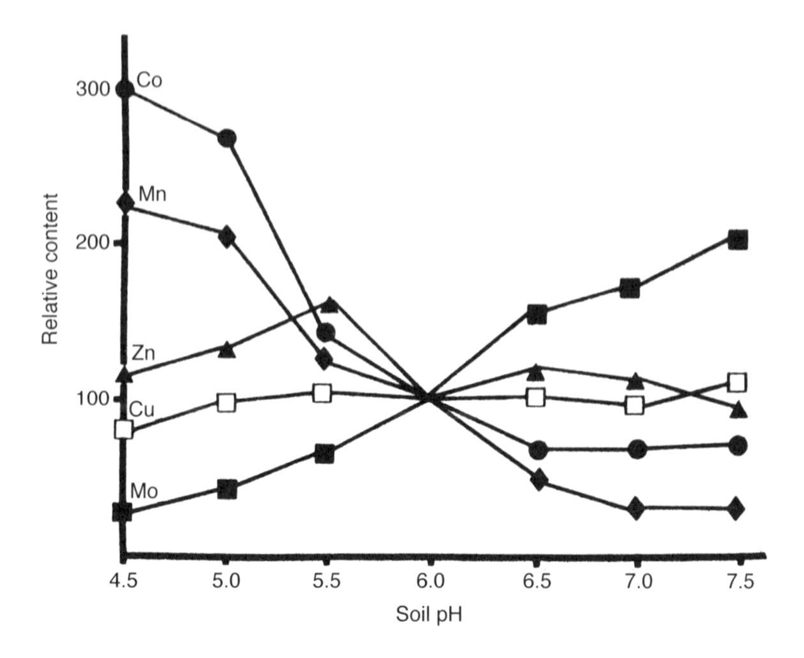

Fig. 2.2. Concentrations of cobalt and manganese in plants decrease markedly as soil pH increases while those of molybdenum increase and those of copper and zinc show little change (values at pH 6.0 were (mg kg^{-1} DM): Co 0.03; Cu 3.1; Mn 51.0; Mo 2.7; Zn 23 and given a relative value of 1.0: data for ryegrass in Aberdeenshire from COSAC, 1982).

plants, with low plant molybdenum concentrations, usually arises on soils with an acid reaction. The application of lime and sulphur can, respectively, raise or lower soil pH and so change the availability of particular minerals to plants. Liming is an important part of the process of improving the quality of hill pastures, but the associated increase in pasture molybdenum concentrations can induce copper deficiency in grazing sheep (Suttle and Jones, 1986). The absorption of nickel, cobalt and manganese by plants is favoured by acid conditions (Mitchell, 1957). The cobalt, molybdenum and manganese contents of pasture plants, in contrast, are all greatly increased by soil waterlogging (Adams and Honeysett, 1964; Burridge *et al.*, 1983). Thus, soil conditions greatly influence the value of forage plants as sources of minerals for grazing livestock. Furthermore, they influence the success of the farmer's attempts to improve mineral supplies by the application of fertilizers.

Phosphorus fertilizers

Most soils in the tropics supply insufficient phosphorus for maximum crop or pasture growth, and yields can be increased by applying phosphate fertilizers at 20–50 kg ha^{-1} (Jones, 1990). Herbage phosphorus concentrations are increased from 0.2–1.2 to 0.7–2.1 g kg^{-1} DM, depending on the site, but this does not necessarily result in herbage rich enough in phosphorus to meet the requirements of sheep and cattle at all times. Australian research has established that superphosphate applications to pastures, over and above those required for maximum plant growth responses, can result in herbage of improved palatability and digestibility, giving significantly greater weight gains in sheep and cattle, wool yields and lamb and calf crops, but expected responses do not always materialize (Winks, 1990). A further complication can arise because the botanical composition of leguminous tropical pasture can be changed by applying phosphatic fertilizer (Coates *et al.*, 1990). The position with grains and seeds is similar to that just described for whole plants. The phosphorus concentrations in wheat, barley and oat grains from soils low in available phosphorus (P) are generally only 50–60% of those from more fertile soils, but they can be raised from 2–3 to 4–5 g P kg^{-1} DM by superphosphate applications, which, at the same time, increase grain yields.

Trace element and sulphur fertilizers

In some areas, small applications of molybdenum to deficient soils markedly increase legume yields and herbage molybdenum and protein concentrations. The increase in protein is usually advantageous to the grazing animal but the increases in molybdenum are of no value, except where copper intakes are high. In the latter circumstances, copper retention is depressed and the chances of chronic copper poisoning are reduced but where copper concentrations are low, copper deficiency may be induced in sheep and cattle. The zinc and selenium content of grains and pastures reflect the soil status of these minerals and fertilizer usage. Wheat grain from zinc-deficient soils in Western Australia averaged only 16 mg Zn kg^{-1} DM, compared with 35 mg

kg^{-1} DM for wheats from the same soil fertilized with zinc oxide at 0.6 kg ha^{-1} (Underwood, 1977). The effect of soil selenium status on selenium in wheat grain is striking: median values of 0.80 and 0.05 mg Se kg^{-1} DM have been reported from soils high and low in selenium, respectively, in the USA (Scott and Thompson, 1971) and even lower values (down to 0.005) occur in parts of New Zealand, Finland and Sweden, where the soils are extremely deficient. A single application of selenium fertilizer has increased pasture selenium from around 0.02 to 0.06 mg kg^{-1} DM or more for over 2 years and improved sheep production in Western Australia (see Chapter 15). The selenium concentration in sugar-cane tops was 6–14 times higher in seleniferous than in non-seleniferous areas in India, but was reduced from 15.2 to 5.1 mg kg^{-1} DM by applying gypsum as a fertilizer at 1 ton ha^{-1} (Dhillon and Dhillon, 1991). Much smaller applications of gypsum are used to raise sulphur levels in pastures (Chapter 9). The nature of the soil and its treatment are thus important determinants of the value of seeds and forages as sources of minerals for animals.

Nitrogenous and potassic fertilizers

There is a widely held view that the 'improvement' of permanent pasture by reseeding and applying nitrogenous fertilizer increases risks of mineral deficiencies occurring in grazing livestock. A recent comprehensive study, involving 16 sites in England and Wales over 4 years, did not appear to support this generalization: the concentrations of three minerals (magnesium (Mg), sodium (Na) and Zn)) rose, four (calcium (Ca), Mn, molybdenum (Mo) and sulphur (S)) fell and three (potassium (K), cobalt (Co) and copper (Cu)) were not consistently changed. The authors (Hopkins *et al.*, 1994) concluded that only where availability fell (for Cu and Mg) was the risk of deficiency increased. However, fertilizers are applied to increase yields of forage and crop, and their use increases the rate at which minerals are 'exported' in products from the site of application and thus increase risk of mineral deficiency (see final section). Heavy applications of nitrogenous fertilizers can depress legume growth, so reducing the overall calcium content of the pasture, due to the higher calcium content of leguminous than of graminaceous species. Heavy applications of potassium fertilizers can raise herbage yields and potassium contents, while at the same time depressing herbage magnesium and sodium.

The Influences of Climate, Season and Stage of Maturity

Plants mature partly in response to internal factors inherent in their genetic make-up and partly in response to external factors, notably climate and season, which can be modified by irrigation and management practices. The tendency for forage concentrations of copper to increase and selenium to decrease with increasing altitude (Jumba *et al.*, 1996) is probably a reflection of the influence of rainfall. White *et al.* (1981) found a negative correlation between rainfall and the selenium concentration in wheat grain. The phosphorus and potassium contents of crop and forage plants decline markedly with advancing maturity but seasonal fluctuations in phosphorus concentrations are usually less in legumes than in grasses (Coates *et al.*, 1990). The concentrations of magnesium, zinc, copper, manganese, cobalt, nickel, molybdenum and iron also fall, but rarely to the same extent as phosphorus and potassium, whereas the concentration of silicon usually increases as the plant matures. Decreases in mineral concentrations with advancing maturity are usually reflections of increases in the proportion of stem to leaf and old to new leaves, stems and old leaves having lower mineral concentrations than young leaves (Minson, 1990), but there are exceptions (Fig. 2.3: Mn). Furthermore, seasonal fluctuations in mineral composition (Ca, P, Mg, K, Mn, Cu and Se) persisted when a Bahia grass pasture was kept in an immature state by rotational grazing and clipping (Kappel *et al.*, 1983). Contrasts between wet-season and dry-season forages have been reported in Florida, with wet-season forage the higher in potassium, phosphorus and magnesium by 110, 60 and 75%, respectively (Kiatoko *et al.*, 1982). Selenium concentra-

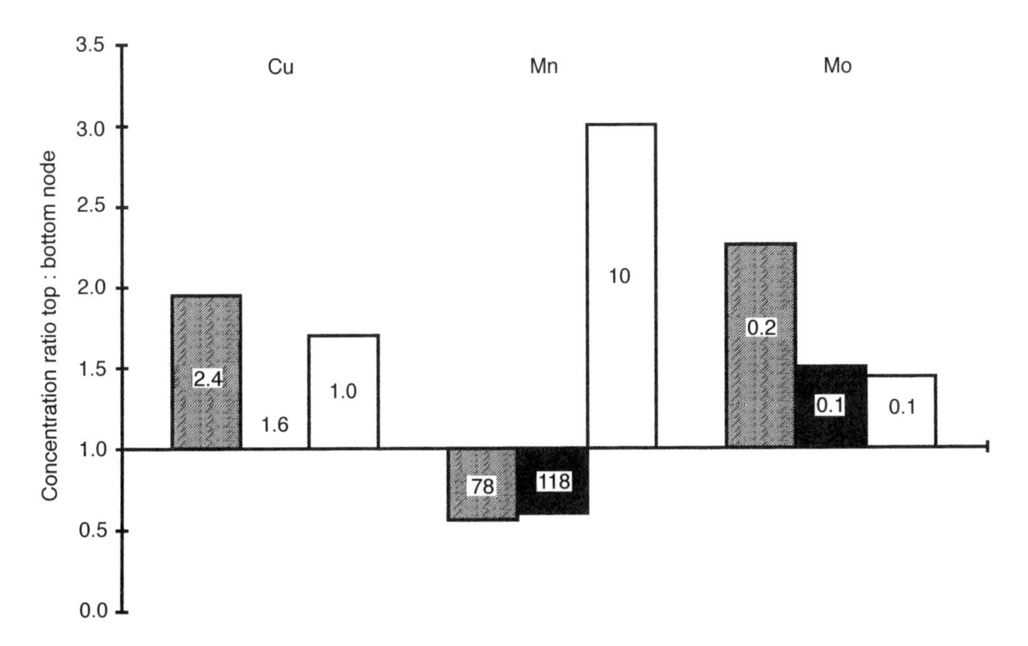

Fig. 2.3. Concentrations of minerals in plants vary from part to part and with state of maturity: values for leaf (grey), sheath (black) and stem (white) are given in different columns and height of column indicates proportionate difference between parts from the top (youngest) and bottom (oldest) nodes of wheat plants (actual values for bottom node given within columns in mg kg^{-1} DM; Burridge *et al.*, 1983).

tions in sugar cane are much higher in the tops than in the cane (5.7–9.5 vs. 1.8–2.1 mg kg^{-1} DM) on seleniferous soils, but values decline with maturity (Dhillon and Dhillon, 1991).

The shedding of the seed is normally responsible for losses of many minerals, so that the material remaining, e.g. the straw, is a poor source (Suttle, 1991). Furthermore, standing straw is subject to leaching of phosphorus and potassium (see Chapter 8). In a study of two Australian barley crops, the average phosphorus concentration of the straw fell from 0.7 to 0.4 g kg^{-1} DM and that of potassium from 9 to the remarkably low level of 0.9 g kg^{-1} DM, while standing in the field from harvesting time in December to the following May. No comparable or significant decline in copper, zinc and manganese concentrations was apparent in these circumstances. Calcium and silicon, in contrast, are higher in cereal straw than they are in the grain.

Soil Ingestion

Soil and dust contamination of herbage can at times provide a further significant source of minerals to grazing animals, especially when grazing intensity is high or when pasture availability is low. Soil intake can rise to 163 g day^{-1} in sheep in arid conditions (Vaithiyanathan and Singh, 1994) and can constitute 10–25% of total DM intake by out-wintered sheep and cattle (Healy *et al.*, 1974; Thornton, 1974). With elements such as cobalt, fluorine, iodine and selenium, which occur in soils in concentrations usually much higher than those of the plants growing on them, soil ingestion can be beneficial to the animal (Grace *et al.*, 1996). In contrast, the copper antagonists iron, molybdenum and zinc are biologically active in soils (Suttle *et al.*, 1975) and their ingestion on herbage may be a factor in the aetiology of hypocuprosis in cattle and ‘swayback’ (copper deficiency) in sheep (Suttle *et al.*, 1984). Soil ingestion can also provide an important route for toxic elements, such as lead (see Chapter 18) and cadmium (Bramley, 1990), to accumulate in the tissues of grazing livestock. Contamination of forage samples with soil commonly causes anomalously high cobalt, iron and manganese concentrations, and is confirmed by high titanium values.

Atmospheric Inputs

Significant amounts of minerals can enter farm systems from the atmosphere. A New Zealand study found that inputs of sulphur in rainfall varied from 0.5 to 15.0 kg ha^{-1}, decreasing exponentially with distance from the coast (Ledgard and Upsdell, 1991); similar gradients exist for iodine and sodium (see Chapters 12 and 7). Inputs of sulphur from industrial sources also occur but are declining in many countries under the influence of stricter controls on pollution. Contamination with fluorine-containing dusts and fumes from industrial sources and the dispersal of wastes from mining activities can also pose a further hazard to livestock in some areas, as discussed in Chapter 18.

Representative Values for Mineral Composition of Feeds

Crops and forages

Given the many factors which can influence the mineral composition of a particular crop or forage, there are considerable difficulties in deriving representative values and limitations to the uses which can be made of those values. For many years, it has been customary to use National Research Council (NRC) values for representative purposes (McDowell, 1992), but these are in need of revision (Berger, 1995). Workers in Alberta (Suleiman *et al.*, 1997) derived a local database which indicated a far greater risk of copper and lesser risk of zinc deficiencies in Albertan cattle than was evident from the NRC standards. They also noted large coefficients of variation attached to many of the mean values, those for lucerne hay ranging from 23.8 to 107.4% (for phosphorus and manganese, respectively, with $n > 220$), and suggested a need for frequent sampling. The problem is summarized in Table 2.1, which includes recent data from the UK, Canada and the USA. For barley grain, there is a general lowering of copper and increase in zinc concentration, compared with the old NRC standard. For lucerne hay and maize silage, there are no 'global' trends, but these may have been masked by some extreme values in the case of manganese and zinc. When the population distributions are clearly skewed, median values are more representative and may show closer agreement between states or countries and with historical data sets. The adequacy of the mineral supply from a particular compound ration will rarely be predictable with sufficient accuracy from tables of mineral composition for the principal feeds.

Crop by-products

Extensive use has been made of crop residues, such as oil-seed cakes as sources of energy, protein and fibre in livestock diets and their inclusion can lead to enrichment of the mineral content of the whole diet. Crop by-products vary widely in mineral composition but are generally most useful as sources of phosphorus (except for beet pulp and sorghum hulls). The problem of variable mineral composition within a feed type is even more important for crop by-products than with the main crop (Arosemena *et al.*, 1995). Data for three by-products is presented in Table 2.2 to illustrate the effect of the industrial process on mineral content of the by-product. Differences between nutrients from the brewery and distillery industries in the USA and UK derive partly from the different grains used. The high and variable copper concentration in distillery by-products from the UK arises from the widespread use of copper stills for the distillation process. Differences between countries for soybean meal are small, because the USA is the major supplier to the UK. Note that, for a given source of by-product,

Table 2.1. Recent data for the mean (SD) mineral concentrations in three major foodstuffs from three different countries compared with the widely used NRC (1982) database US–Canadian Feed Composition Tables.

			Macroelement (g kg^{-1} DM)				Trace element (mg kg^{-1} DM)			
Feed	IFN	Source	Ca	P	K	Mg	Cu	Mn	Se	Zn
Barley	4–07–939	CAN	0.6(0.2)	3.7 (0.6)	5.4 (1.8)	1.5 (0.2)	5.4 (2.4)	18.9 (6.7)	0.11 (0.09)	42.3 (10.5)
	4–00–549	USA	0.7 (0.02)	3.8 (0.07)	6.3 (0.11)	1.4 (0.02)	7.0 (2.0)	22.0 (6.0)	–	38.0 (7.0)
	4–00–549	UK	0.9 (0.6)	4.0 (0.5)	5.0 (0.7)	1.2 (0.2)	4.2 (1.5)	18.5 (3.6)	0.10 (–)	32.5 (8.5)
		NRC	0.5	3.7	4.7	1.5	9.0	18.0	–	17.1
Lucerne hay	1–00–059	CAN	17.9 (5.3)	2.1 (0.5)	17.5 (4.9)	3.1 (0.9)	6.1 (2.5)	43.7 (41.8)	0.27 (0.29)	25.4 (11.7)
	1–00–059	USA	13.1 (3.3)	3.0 (0.6)	26.3 (4.4)	2.7 (0.8)	4.4 (6.3)	22.7 (13.9)	–	13.4 (17.5)
	1–00–078	UK								
		NRC	14.2	2.2	25.4	3.3	11.1	31.0	–	24.8
Maize silage	3–00–216	CAN	2.6 (1.2)	2.3 (0.5)	11.2 (2.6)	2.5 (0.9)	4.4 (2.2)	37.8 (18.8)	0.04 (0.02)	34.0 (13.1)
	3–28–250	USA	2.5 (0.14)	2.3 (0.6)	10.8 (3.3)	1.8 (0.4)	3.0 (2.7)	17.1 (20.3)	–	12.2 (18.8)
	3–02–822	UK	4.3 (2.0)	2.6 (1.2)	12.3 (4.1)	2.2 (0.7)	5.2 (0.75)	14.6 (7.2)	–	45.3 (13.9)
		NRC	2.3	2.2	9.9	1.9	10.1	30.0	–	21.0

CAN, data from Alberta (Sulieman *et al.*, 1997); USA, pooled data from the states of New York, Indiana, Idaho and Arizona (Berger, 1995); UK, data from the Ministry of Agriculture, Fisheries and Food (MAFF, 1990).

Table 2.2. Mineral composition of crop by-products can vary with source and nature of the industrial process.

Crop by-product	IFN	Source	Ca (g kg^{-1} DM)	P	Cu (mg kg^{-1} DM)	Zn
Brewers'	5–02–141	USA (maize)	3.3 (1.2)	5.9 (0.8)	11.2 (5.0)	97.0 (16.0)
grains	5–00–517	USA (barley)	3.5 (1.4)	5.1 (1.0)	18.7 (9.2)	72.8 (12.5)
Distillers'	5–28–236	USA (maize)	2.9 (1.5)	8.3 (1.7)	9.7 (9.0)	73.5 (47.0)
dried grains	5–12–185	UK (barley)	1.7 (0.3)	9.6 (0.8)	40.5 (16.8)	55.4 (4.7)
Soybean	5–04–612	USA (dehulled)	4.1 (2.9)	7.2 (2.8)	17.7 (7.0)	69.5 (141)
meal (extracted)	5–04–604	UK	3.9 (1.6)	7.4 (0.4)	15.8 (2.6)	49.0 (9.9)

USA, data pooled from four states (Berger, 1995); UK, data collected in England (MAFF, 1990). IFN, International Food Number.

only phosphorus values have sufficiently low variability for useful predictions to be made of contribution of mineral from the by-product to the total ration. In tropical countries, by-products of the sugar-cane industry, such as bagasse and molasses, now form the basis of successful beef and milk production enterprises. The availability of crop by-products is likely to increase greatly in tropical countries, as dual-purpose crops are grown for fuel as well as feed production (Leng and Preston, 1984). Their feed value will only be realized if any shortcomings in mineral content are recognized and overcome.

Drinking-water

The drinking-water is not normally a major source of minerals to livestock, although there are exceptions. Sulphur concentrations in water from deep aquifers can reach 600 mg l^{-1}, adding 3 g S kg^{-1} DM to the diet as sulphates; this is far more than any nutritional requirement and may even create problems by inducing copper deficiency (Smart *et al.*, 1986). In most natural fluorosis areas, such as those in the semi-arid interior of Argentina, the water-supplies and not the feed are mainly responsible for supplying toxic quantities of fluoride (see Chapter 18), and, in some parts of the world, the water available to animals is so saline that sodium and chlorine are ingested in quantities well beyond the requirements for these elements (see Chapter 7). Some 'hard' waters also supply significant amounts of calcium, magnesium and sulphur and occasionally of other minerals. Individual daily water consumption is highly variable, as is the mineral composition of different drinking-water sources. A New Zealand study with grazing beef cattle reported a fourfold variation in individual water intakes (Wright *et al.*, 1978). Meaningful 'average' mineral intakes from the drinking-water are therefore impossible to calculate (Shirley, 1978).

Milk

The most important animal source of minerals for species which suckle young is the milk, and this is underlined by the fact that the first milk, or colostrum, is rich in minerals relative to other constituents and relative to later milk.

Stage of lactation and species effects

The decline in mineral concentrations in bovine colostrum over the first 3 days after parturition is shown in Table 2.3. Kume and Tanabe (1993) showed that, by the fourth calving, concentrations of calcium, phosporus, magnesium and zinc in the first colostrum had fallen to 72–83% of the level found at the first calving: there was no effect of parity after 4 days' lactation, however. The length of time that suckled offspring remain with their mothers varies widely, according to species and management practice, but would commonly be 42 days for piglets, 120 days for lambs, 150 days for kids, 180 days for foals and 300 days for beef calves. The ash content of the milk tends to decline with the lactation length of the species. The sow and the ewe secrete milk of the highest ash content (almost 10 g l^{-1}); they are followed by the goat and the cow (7–8 g l^{-1}). The mare secretes milk of exceptionally low ash content (about 4 g ash l^{-1}) but the volume secreted each day is high (20 l). Relative growth rate of the suckled offspring also declines with lactation length, so that mineral concentration tends to match need across species. The mineral composition of milk also varies with the stage of lactation and the incidence of various diseases. Representative values for the mineral constituents of the main milk of various species are given in Table 2.4 and show similar species differences to those in ash concentration. Milk is clearly a rich source of calcium,

Table 2.3. Mean colostrum yield and mineral concentrations in colostrum of 21 cows shortly after parturition (Kume and Tanabe, 1993).

	Time after parturition (h)				
	0	12	24	72	SD
Colostrum yield (kg day^{-1})	11.7	–	14.9	21.6	2.4
Ca (g l^{-1})	2.09	1.68	1.43	1.25	0.19
P (g l^{-1})	1.75	1.43	1.25	1.01	0.17
Mg (g l^{-1})	0.31	0.21	0.15	0.11	0.04
Na (g l^{-1})	0.69	0.64	0.58	0.53	0.008
K (g l^{-1})	1.48	1.49	1.58	1.50	0.12
Fe (mg l^{-1})	2.0	1.5	1.2	1.1	0.6
Zn (mg l^{-1})	17.2	10.3	6.4	5.2	2.5
Cu (mg l^{-1})	0.12	0.09	0.08	0.08	0.03
Mn (mg l^{-1})	0.06	0.04	0.03	0.02	0.02

SD, standard deviation.

Table 2.4. Mineral concentrations in the main milk of farm animals (representative values).

	g l^{-1}					mg l^{-1}				
	Ca	P	Mg	K	Na	Cl	Zn	Fe	Cu	Mn
Cow	1.2	1.0	0.1	1.5	0.5	1.1	4.0	0.5	0.15	0.03
Ewe	1.9	1.5	0.2	1.7	0.5	1.4	4.0	0.5	0.25	0.04
Goat	1.4	1.2	0.2	1.7	0.4	1.5	5.5	0.4	0.15	0.08
Sow	2.7	1.6	0.1	1.0	0.03	0.9	5.0	1.5	0.70	0.15
Mare	1.0	0.6	0.1	0.7	–	0.2	1.5	0.7	0.20	0.05
Buffalo	1.8	1.2	–	–	–	0.6	–	–	–	–
Llama[a]	1.7	1.2	0.15	1.2	0.27	0.7	–	–	–	–

[a] Morin *et al.* (1995).

phosphorus, potassium, chlorine and zinc, but is much less satisfactory as a source of magnesium, iron, copper and manganese.

Diet

The effects of diet on the composition of milk vary greatly for different minerals (Kirchgessner *et al.*, 1967). Dietary deficiencies of calcium, phosphorus, sodium and iron are reflected in a diminished yield of milk but not in the concentrations of these minerals in the milk that is secreted. In copper and iodine deficiencies, in contrast, there can be a marked fall in concentration of these minerals in the milk produced. In cobalt deficiency, concentrations of vitamin B_{12} in milk decrease and can be boosted by cobalt or vitamin B_{12} supplements. Iodine and molybdenum concentrations in milk can also be increased beyond the normal by increased dietary intakes of the element, but those of copper, manganese, molybdenum and zinc cannot. High fluoride intakes by sheep and cattle also have little effect on the fluorine content of the milk produced.

Animal By-products

Bovine milk and the by-products of butter and cheese manufacture – skimmed milk, buttermilk and whey – are among the most important animal by-products used in livestock feeding. Skimmed milk and buttermilk differ little in mineral composition from the milk from which they are made, since little of the minerals is separated with the fat in cream and butter manufacture, and they vary little from source to source. These materials, in either their liquid or dried forms, therefore constitute valuable mineral, as well as protein, supplements to cereal grain diets for pigs and poultry, particularly in view of their high calcium content (Table 2.5). Whey is not as rich in calcium and phosphorus as is separated milk or buttermilk, because much of the phosphorus and a proportion of the calcium and other minerals in the milk separate with the curd in cheese manufacture. The amounts left are neverthe-

Table 2.5. Macromineral concentrations (g kg^{-1} DM) in animal by-products (source mostly MAFF (1990) or NRC as cited by McDowell (1992)).

	Ca	P	Mg	Na	K	S
Battery waste	32	18	0.5	–	17	13
Broiler waste	93	25	0.6	–	23	0.2
Blood meal	0.6 (0.26)[a]	1.5 (0.19)	0.2 (0.01)	3.6 (0.32)	1.9 (0.12)	8.4 (0.25)
Buttermilk	14.4	10.1	5.2	9.0	9.0	0.9
Feather meal	5.6 (0.07)	3.1 (0.06)	0.4 (0.04)	1.4 (0.01)	1.5 (0.01)	18 (0.6)
Fish-meal	56 (6.0)	38 (14.5)	2.3 (0.3)	11.2 (1.5)	10.2 (1.3)	5.0
Meat- and bone-meal	90 (13)	43 (6.5)	2.2 (0.2)	8.0 (1.0)	5.2 (0.6)	9.2 (0.72)
Whey	9.2	8.2	2.3	7.0	12	11.5

[a] Standard deviations given in parentheses are for MAFF data.

less sufficient to make this carbohydrate concentrate a valuable source of many of the mineral nutrients.

Most of the animal products used primarily as protein supplements contribute significant quantities of the mineral nutrients, especially calcium, phosphorus, iron, zinc and selenium (Table 2.5). Mineral content varies greatly with the source materials from which they are made and with the processing methods employed. Thus blood-meal, liver-meal and whale-meal contain only 2–3% total ash and are low in calcium and phosphorus, but they can be valuable sources of iron, copper, selenium and zinc. Commercial meat-meals and fish-meals, on the other hand, are usually rich in calcium, phosphorus, magnesium and zinc, depending on the proportion of bone that they contain. The ash content of meat- and bone-meals varies from 4 to 25% and most of this ash consists of calcium and phosphorus. A typical European, North American or Australian meat-meal contains 6–8% Ca and 3–4% P; in other words, they contain about 100 times the concentration of calcium and ten times the concentration of phosphorus of the cereal grains which they commonly supplement. Fish-meal often contains appreciable amounts of sodium chloride, as well as calcium and phosphorus.

Mineral Supplements

Ideally, mineral supplements should be used only when requirements cannot be met with adequacy and safety by the judicious selection and combination of available feeds alone. However, this requires knowledge of the mineral composition of feeds, and sucess depends heavily on the appropriate database being available (Arosemena *et al.*, 1995). The addition of protein concentrates to a grain mixture raises its content of such minerals as calcium, phosphorus, zinc and iodine. Bran is freely available on many farms and an excellent source of phosphorus for ruminants (Suttle, 1991). While the substitution of a plant protein source for animal products, such as meat-meal or fish-meal, can result in lower availability of some minerals for pigs and poultry, notably of zinc and phosphorus, because of the presence of fibre and phytate, bran may be beneficial because it contains phytases (see Chapter 5). In practice, mineral supplements are added routinely to home-mixed and commercially compounded rations as an insurance against the inclusion of components which deviate from the norm of mineral composition. In some circumstances, mineral supplements are always necessary, because the pastures or feeds are abnormal in mineral composition as a consequence of local soil and climatic effects. A wide range of inorganic mineral supplements, covering all essential minerals are now available (see Appendix Table 2, p. 600) and are increasingly used to fortify rations because of increasing rates of animal production, decreasing availability and acceptability of animal by-products in feed formulation and increasing use of industrial products, such as urea, which replaces protein in feeds for ruminants without providing minerals. Mineral supplementation is an essential adjunct to urea supplementation and plays a vital role in increasing the nutritive value of low-quality roughages and crop by-products in developing countries (Suttle, 1991). The general principles governing the choice and use of mineral supplements are considered in the next chapter.

The Availability of Minerals to Animals

The evaluation of feeds and feed supplements as sources of minerals depends not only on the total mineral content or concentration but also on how much can be absorbed from the gut and used by the animal's cells and tissues. This, in turn, depends on the age and species of the animal; the intake of the mineral relative to need; the chemical form in which the mineral is ingested; the amounts and proportions of other dietary components with which it interacts metabolically; and environmental factors, such as the accessibility and intensity of sunlight (for reviews, see Hazell, 1985; Ammerman *et al.*, 1995). Assessment of the physiological availability of minerals in feeds or feed supplements presents difficulties that do not exist with the organic nutrients. Ordinary digestibility trials are of limited value, because the faeces constitute an important and variable pathway of excretion of many minerals and therefore contain minerals that have previously been absorbed, in addition to the proportion that has escaped absorption from the diet while in the gastrointestinal tract. Balance studies, which measure the difference between the amount of the mineral appearing in the urine and faeces over a specified period and the amount ingested, can therefore be misleading indicators of availability; furthermore, the influence of body stores may change as the study progresses and some losses (e.g. by sweating and exhalation) may go undetected. Extensive detail on the design and evaluation of experiments for assessing mineral availability are given by Littell *et al.* (1995, 1997).

Measures of relative availability

Comparisons of mineral retention ('comparative balance') from two or more sources at two or more rates of intake have been used to assess the relative nutritional value of minerals in those sources from the ratio of the linear responses in retention. Partial body retention is commonly used; for example, haemoglobin responses have been used to assess the availability of iron (Saylor and Finch, 1953), bone growth to assess calcium sources (Lengemann *et al.*, 1957) and increases in liver copper to compare copper sources for chicks and lambs (Pott *et al.*, 1994). The results obtained can, however, vary, depending on the experimental conditions chosen (Suttle, 1983, 1985). Comparisons of rates of urinary excretion and milk secretion between two mineral sources can also be used to assess relative availabilities, provided that the losses are linearly related to dose and there are no differences between sources in the partition of absorbed mineral prior to excretion or secretion. Thus magnesium availability to sheep has been assessed from rates of urinary magnesium excretion. This 'comparative loss' approach has been modified by the use of double-radioisotope techniques, in which one source of the element is labelled with a radioisotope and then compared with another source labelled with a different radioisotope of the same element and given simultaneously. Lengemann (1969) compared iodate and iodide as sources for milk iodine (I) in lactating goats given sodium iodide ($Na^{131}I$) and sodium iodate ($Na^{125}IO_3$) daily or as a single oral dose. It was concluded from the $^{125}I:^{131}I$ ratio of the milk that iodate had 0.86 of the value of iodide as a source of milk iodine. By a similar procedure, the iodine of 3,5-diiodosalicylic acid was claimed to be only 20% as available as that of sodium iodide (Aschbacher *et al.*, 1966). Inferences can also be drawn about the value of different sources of minerals from the relative amounts that are required to ameliorate, prevent or potentiate signs of deficiency in the animal. Cantor *et al.* (1975) assessed the value of plant and animal feed sources of selenium for chicks from their ability to prevent exudative diathesis and raise activities of the seleno-enzyme, glutathione peroxidase. Using a chick growth assay, Odell *et al.* (1972) reported that zinc in wheat, fish-meal and non-fat milk was 59, 75 and 82% as available, respectively, as that in zinc carbonate. Complications can arise when natural feeds are added to semipurified diets as the major source of a given mineral, because they provide variable amounts of other nutrients, which might also influence growth (Suttle, 1983).

Measures of absolute availability

The above techniques, whether radioactive or not, normally yield only qualitative assessments relative to a standard source, assumed to be highly, if not completely, available (e.g. ferrous sulphate ($FeSO_4$) as an iron (Fe) source). However, repletion techniques can be modified to give quantitative data (Suttle, 1985). To obtain quantitative assessments of absorption, radioisotopes have to be applied in special ways. Refinements were first introduced to differentiate between the unabsorbed dietary mineral fraction of the faeces and the fraction that has been absorbed and re-excreted by the gut, i.e. the endogenous faecal mineral (FE). The essential calculations (following a single parenteral injection of radioisotope, total collection of faeces and measurements of total excretion of both radioisotope and stable mineral) were first outlined by Comar (1956), using ^{45}Ca; they were:

$$\text{Endogenous faecal Ca} = \frac{\text{Specific activity of faeces}}{\text{Specific activity of endogenous Ca}} \times \text{Total faecal Ca}$$

where specific activity is the ratio of radioisotopic to stable mineral concentration (hence the term 'radioisotopic dilution technique'). A crucial assumption has to be made, in that the specific activity of endogenous mineral is predicted from samples taken from an accessible pool with which the endogenous mineral is supposedly in equilibrium (e.g. plasma or urine). Having calculated faecal endogenous Ca (FE_{Ca}), calcium absorption coefficients are than derived as follows:

$$\text{Ca absorption coefficient} = \frac{\text{Ca intake} - (\text{Faecal Ca} - FE_{Ca})}{\text{Ca intake}}$$

The technique is sometimes modified by giving the radioisotope orally and intravenously, either sequentially to the same animal or simultaneously to matched groups. Where two isotopes of the same element are suitable (e.g. ^{45}Ca and ^{47}Ca), they can be given simultaneously by alternative routes to the same animal. Estimates of intake and faecal excretion may then be confined to the radioisotope and 'intake' is the dose of radioisotope given; this technique is useful for studies with elements which readily contaminate the average experimental environment, polluting both feed and faeces (e.g. iron, zinc and copper). The term 'comparative radioisotope balance' technique more accurately describes a further modification, in which retention of the radioisotope is measured in a whole-body monitor and route of excretion ignored. Thus Heth and Hoekstra (1965) gave ^{65}Zn as the chloride or as a glycine complex to growing rats in the feed or by intramuscular injection. Between 100 and 250 h after administration of the marker, the decline in radioactivity was a simple exponential function, with identical slopes for the two types of dosing (Fig. 2.4). By extrapolating these lines to zero time, it was estimated that 43% of the oral ^{65}Zn was absorbed.

Compartmental modelling

Yet another refinement of the radioisotope approach requires sophisticated compartmental analysis of changes in radioisotope activity in the bloodstream after a single dose, using computer programmes such as SAAM[27]. From the rate of dilution of radioisotope in the plasma caused by 'influx' of stable mineral by absorption and mobilization of body stores and 'outflow' by tissue uptake and endogenous loss, estimates can be made of the rate at which the stable mineral is entering the blood plasma pool from the gut (e.g. from phosphorus in herbage eaten by sheep (Grace, 1981)) and flowing to and from compartments (e.g. selenium in hay (Krishnamurti *et al.*, 1997)).

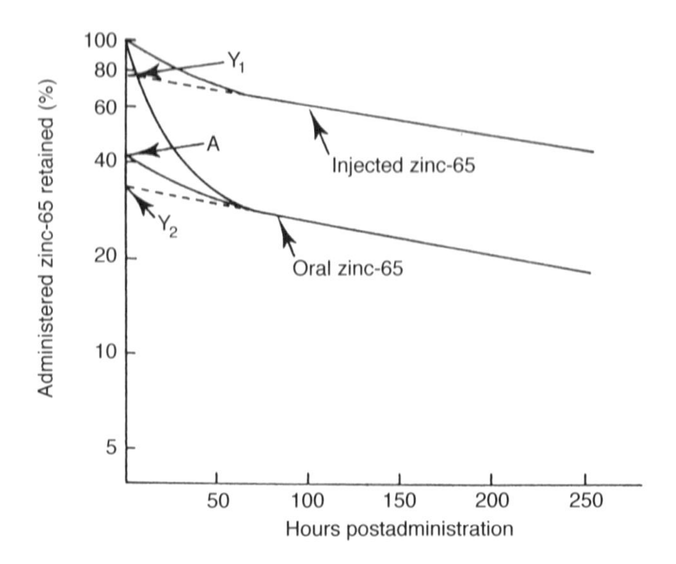

Fig. 2.4. A comparison of retention curves for ^{65}Zn given in the feed and by intramuscular injection. Percentage absorption (A) of oral ^{65}Zn is calculated by dividing Y_2 (the *y* intercept of the extrapolated retention curve for oral ^{65}Zn) by Y_1 (the *y* intercept of the extrapolated retention curve for injected ^{65}Zn) and multiplying by 100 (from Heth and Hoekstra, 1965).

However, complexity is no guarantor of accuracy and, in the case of selenium, failure to recognize key 'molecular' compartments leads to results which cannot yet be trusted (see Chapter 15).

Stable isotopes

The most recent development in the assessment of availability has been the use of stable or non-radioactive isotopes as tracers (Turnlund, 1989). Suitability depends on the existence of isotopes of the element of different natural abundance; the natural isotope ratio is then perturbed by dosing the animals with the least abundant isotope and the same dilution principles as those used for radioisotopes are applied (see Fairweather-Tait *et al.*, 1989). The stable isotope approach has the advantage of avoiding the hazards and safety constraints associated with radioisotopes; two disadvantages are the need for expensive capital equipment (e.g. inductively coupled 'plasma' generator plus mass spectrometer) and the slow (chromatographic) throughput of samples.

Problems with isotopic markers

All isotopic tracer techniques have two further potential pitfalls, which may lead to false estimates of the value of a mineral source. First, they may fail to be representative of the mineral in the feed which they are supposed to trace or mimic; the lower the 'true' availability of the feed source, the greater is the likelihood of underestimation. Secondly, the full potential of the mineral source may be masked by an animal capable of absorbing only what it needs; this applies to calcium, zinc and iron, for example (Suttle, 1985). Accuracy of the tracer might theoretically be improved by growing the feed in isotope- (radioactive or stable) enriched nutrient solutions, a technique known as 'intrinsic labelling'. However, intrinsic labelling is laborious and expensive and has yet to demonstrate marked differences when compared with extrinsic labelling for most trace elements, selenium, chromium and cadmium being the exceptions (Johnson *et al.*, 1991). *In vitro* measurements of the completeness of labelling under conditions of simulated digestion might usefully and economically predict the accuracy of a potential tracer (Schwarz *et al.*, 1982). When the tracer is given parenterally, the problem of poor representation is minimized, because absorption partitions the mineral into absorbed and non-absorbable forms and the former will generally be more readily labelled than the heterogeneous forms in the diet. Care must be taken not to administer so much tracer that the homeostatic control mechanisms are grossly disturbed; this problem is of particular concern in studies using stable isotopes, where sufficient tracer must be given to perturb the normal isotope distribution.

Problems introduced by the animal

The problem of absorption according to need is theoretically overcome by measuring absorption with mineral intakes at or below need (AFRC, 1991). However, need is determined by availability and ideal mineral intakes may therefore vary from source to source. There are a growing number of false estimates of availability in the literature, mostly involving chelated sources of trace elements (e.g. Mn, Chapter 14), in which the use of different ranges of intake for different mineral sources has led to animal influences being interpreted as hyperavailability of a particular source. The best approach is to assess each source at progressively lower intakes and note the maximal absorptive efficiency attained (AFRC, 1991). The term 'absorbability' should be used to describe values obtained by this type of approach and to distinguish these proper assessments of the full potential of a mineral source from the sundry measurements covered in the past by the term 'availability'.

Gaps in knowledge

The application of tracer and other techniques to the assessment of the value of mineral sources has led to a few major advances. In the experiments of Heth and Hoekstra (1965), a significant depression in zinc absorption was demonstrated when the calcium content of the diet was raised. In others, the calcium of milk was shown to be better absorbed by calves than the calcium of a hay and grain diet. Differences in the availability of various dietary forms of selenium are also well established. Cantor *et al.* (1975) found that, in most of the feeds of plant origin, the selenium was readily available (values ranging from 60% to 90%) to chicks, whereas selenium in animal products tested was less than 25% available. However, these isolated examples merely illustrate how much remains to be learned of absorbability and its relationship to the chemical forms in which minerals occur. Why is fresh herbage less effective in promoting body copper stores in cattle and sheep than hay or dried herbage of equivalent total copper content (Hartmans and Bosman, 1970; Suttle, 1983), for example? Only sophisticated, physiologically and biochemically sound approaches to the use of tracers (e.g. Buckley, 1988) will provide the necessary information on these and other intriguing questions concerning mineral absorbability, arguably the most important determinant of mineral deficiency diseases. For a detailed text on the practical assessment of mineral availability, the reader should consult Ammerman *et al.* (1995).

Interactions Between Minerals

Interactions between minerals are a major cause of variation in availability and thus influence the nutritive value and potential toxicity of a particular source. An interaction is demonstrated when responses to a given element are studied at two or more levels of another element and combined effects cannot be predicted by simply adding up the main effects of each element. The simplest example is provided by the 2 × 2 factorial experiment and an early indication of interaction is provided by arranging the results in a two-way table (Table 2.6). Striking main effects are given by the difference in row and column totals, but the interaction indicated by differences in diagonal totals is equally pronounced. Large groups are often needed to show significant interactions, because the interaction term has few degrees of freedom in the analysis of variance (only 1 degree of freedom (d.f.) in a 2 × 2 experiment). The frequency with which physiological interactions between minerals occur has, therefore, probably been underestimated.

Table 2.6. Outcome of a 2 × 2 factorial experiment in which the separate and combined antagonisms of iron and molybdenum towards copper were investigated, using liver copper concentrations to assess status (Bremner *et al.*, 1987). Difference between totals on the diagonals (in parentheses) indicates strength of interaction equal to main effects (i.e. differences between row or column totals).

		Fe treatment 0		Fe treatment +	Mo totals
Mo treatment	0	1.18	(0.52)	0.35	1.53
	+	0.17	(1.32)	0.14	0.31
	Fe totals	1.35		0.49	1.84 mmol kg^{-1} DM

Mechanisms

Interactions can affect each step in the metabolism of an element, be it absorption, transport, cellular uptake, intracellular function, storage or excretion. Sometimes several steps are affected simultaneously. The primary mechanisms whereby minerals interact are:

- the formation of unabsorbable complexes between dissimilar ions in the gut (e.g. metal phytates: Chapters 5 and 16);
- competition between similar ions for metabolic pathways (e.g. sulphate (SO_4^{2-}) and molybdate (MoO_4^{2-}): Chapter 17);
- induction of non-specific metal-binding proteins (e.g. metallothionein by Cu, Zn or cadmium (Cd): Chapter 18).

The net effect of an interaction is usually a reduced retention of one mineral under the influence of another, but there may be only localized accumulations of the affected element in organs such as the kidney when the interactions occur postabsorptively. Occasionally, interactions may be beneficial, as in the case of enhancement of iron utilization by small copper supplements, but large copper supplements can increase iron requirements (Chapter 11).

Outcome

The nutritional value of a particular mineral source is determined by the sum of all the interactions with other dietary constituents, mineral and non-mineral. Few attempts have been made to quantify the outcome of interactions between minerals, notable exceptions being the interactions between potassium and magnesium (Chapter 6) and between copper, molybdenum and sulphur (Chapter 11) in ruminant nutrition and the influence of phytates on zinc absorption in non-ruminants (Chapter 16). Prediction is particularly difficult in the case of three elements, such as copper, cadmium and zinc, which are mutually antagonistic. Thus Campbell and Mills (1979) anticipated that, since cadmium and zinc each separately impaired copper metabolism in sheep, addition of both elements would severely compromise copper status; however, zinc (+120 mg kg^{-1} DM) apparently protected lambs from the copper-depleting effects of cadmium (+3 mg kg^{-1} DM), and exceedingly high levels of zinc (750 mg kg^{-1} DM) were needed to exacerbate copper depletion (Bremner and Campbell, 1980). When each element can negatively affect the other, a supplement of any one element may have a positive net effect, either by countering an antagonism against itself or by neutralizing the antagonistic properties of the second element upon the third (Fig. 2.5). The clinical outcome of such three-way interactions will depend upon which element is the first to become rate-limiting; this, in turn, will be determined by the strength of each separate antagonism and will change with the level at which each antagonist is employed. Thus cadmium is capable of inducing either copper or zinc deficiencies. The complex interactions between cadmium, copper and zinc are considered further in Chapter 18. Prediction of three-way interactions will also be difficult if there is a shared potentiator. Thus, sulphur

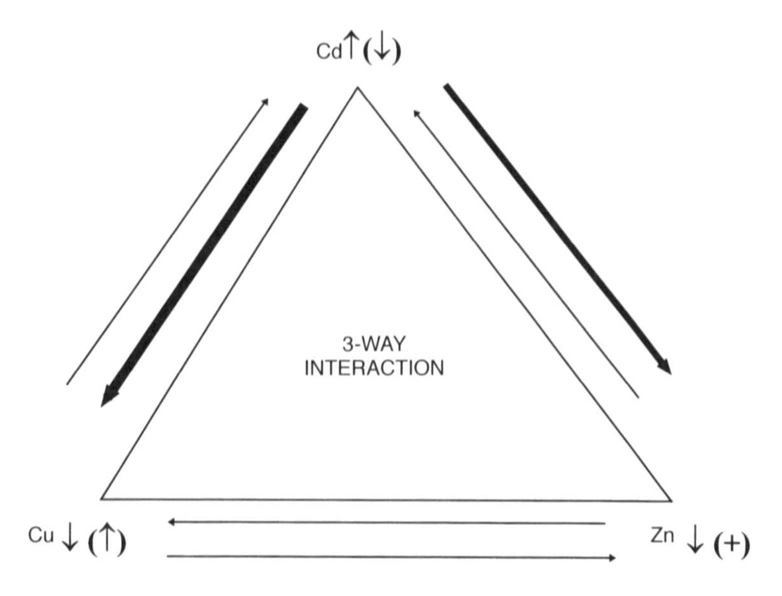

Fig. 2.5. Mutual antagonisms between Cu, Cd and Zn lead to complex three-way interactions: raising the level of one interactant can lower the status of the other two, as shown for Cd; however, inequalities in the strength of particular interactions, such as the strong antagonism of Cu by Cd, may mean that supplements of Zn (+) may raise (↑) Cu status by reducing (↓) the Cd × Cu interaction.

(as sulphide) is necessary for both the iron- and the molybdenum-mediated antagonism of copper in ruminants (Chapter 11). When iron and molybdenum are added together in a diet of marginal copper content, there is not an additive antagonism towards copper (Table 2.6; Bremner *et al.*, 1987), possibly because iron and molybdenum compete for sulphide.

Mineral Cycles

The natural mineral supply to livestock is clearly the outcome of a complex chain of events, as summarized in Fig. 2.6, and assessments of mineral status of the soil, the plant and the animal can give conflicting results as to adequacy of supply (e.g. Kiatoko *et al.*, 1982). The net flow of utilizable mineral to the grazing animal, in particular, is likely to vary widely from season to season and from year to year. Where mineral nutrients in herbage are marginal in respect of animal requirements, changes in concentrations, brought about by atmospheric, climatic or seasonal influences and by plant maturity and seed shedding, can obviously be significant factors in the incidence or severity of deficiency states in livestock wholly or largely dependent on those plants. For example, it has long been recognized that the risk of 'swayback' in newborn lambs deprived of copper is high after mild winters in the UK; this is probably due to decreased use of supplementary

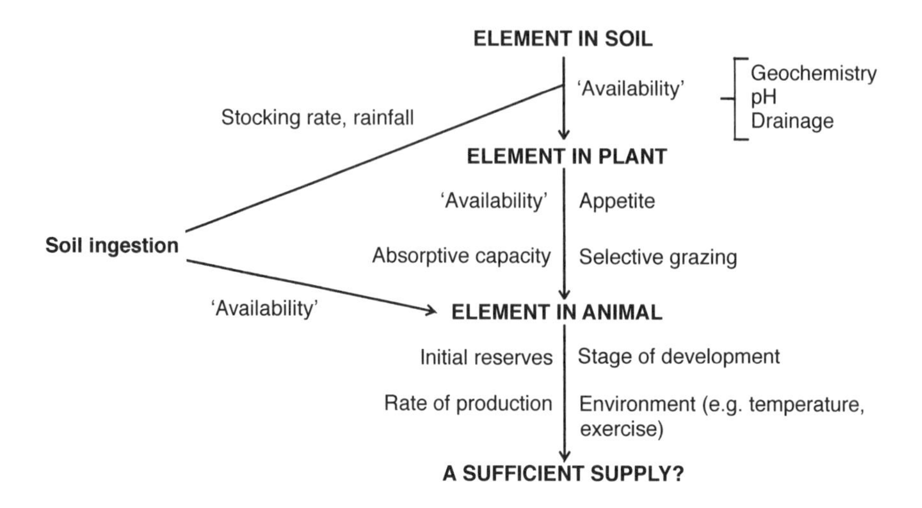

Fig. 2.6. Summary of the many and varied factors in addition to plants which can influence the flow of an element from the soil to the grazing animal and whether or not the supply will meet the animal requirement.

feeds of higher available copper content and the increased ingestion of soil (containing copper antagonists) during winters of low snow cover. Furthermore, there is or can be substantial recycling of mineral supplies via excreta, and there are also withdrawals of minerals from the farm ecosystem each year in harvested crops and livestock products, which are variably replaced. Changes in the husbandry system can, therefore, shift mineral balances substantially. Figure 2.7 illustrates the way in which variation in the amounts of nitrogen fertilizer used changes the phosphorus cycle in a dairy enterprise on tropical grass pasture. The influence of fertilizer inputs of a potentially toxic mineral, cadmium, on the cycling and accumulation of that element in soil are discussed in Chapter 18. It is important to appreciate the cyclical nature of mineral nutrition before breaking problems down into small compartments, and there is a need to frequently reassess the adequacy of mineral supplies experienced by the animal, a subject addressed in the next chapter.

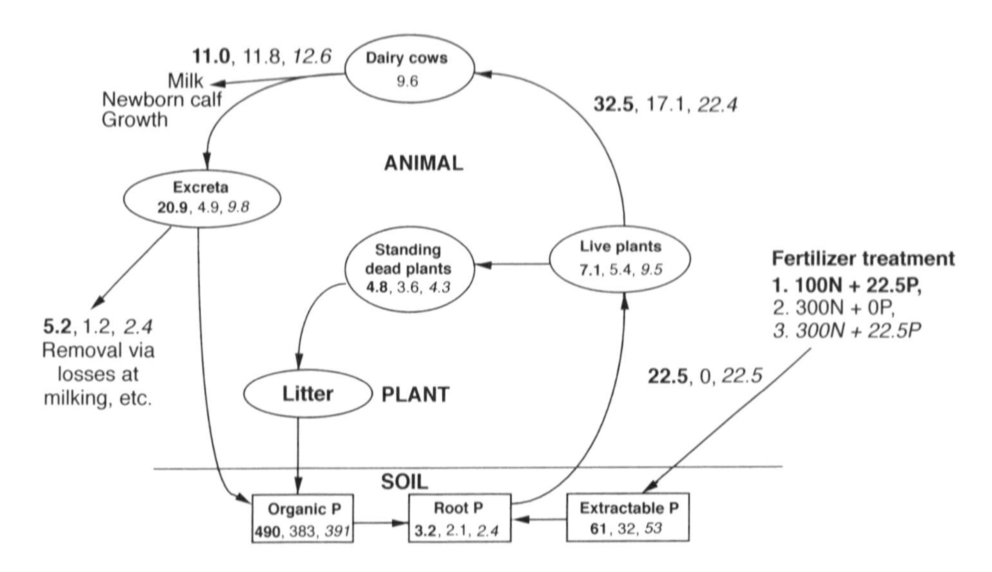

Fig. 2.7. Example of a mineral cycle showing pool sizes and flows of phosphorus (kg P ha^{-1}) for a dairy herd grazing tropical pasture under three different fertilizer regimens (after Davison *et al.*, 1997): raising the amount of nitrogen (N) given with P (Treatment *3* vs. 1) shifted the route of recycling from excreta to plant litter, a less efficient process; application of P prevented a steady decline in extractable soil P (Treatment *3* vs. 2); giving P with maximum N (Treatment *3*) increased milk yield without changing the P status of the cow, by increasing the mass of green herbage on offer.

References

Adams, S.N. and Honeysett, J.L. (1964) Some effects of soil waterlogging on the cobalt and copper status of pasture plants grown in pots. *Australian Journal of Agricultural Research* 15, 357–367.

AFRC (1991) Technical Committee on Responses to Nutrients Report No. 6: a reappraisal of the calcium and phosphorus requirements of sheep and cattle. *Nutrition Abstracts and Reviews* 61, 573–612.

Ammerman, C.B., Baker, D.H. and Lewis, A.J. (1995) In: *Bioavailability of Nutrients for Animals.* Academic Press, San Diego, pp. 383–398.

Arosemena, A., DePeters, E.J. and Fadel, J.G. (1995) Extent of variability in the nutrient composition within selected by-product feedstuffs. *Animal Feed Science and Technology* 54, 103–120.

Aschbacher, P.W., Cragle, R.G., Swanson, E.W. and Miller, J.K. (1966) Metabolism of oral iodide and 3,5-diiodosalicylic acid in the pregnant cow. *Journal of Dairy Science* 49, 1042–1045.

Beeson, K.C., Gray, L. and Adams, M.B. (1947) The absorption of mineral elements by forage plants. 1. The phosphorus, cobalt, manganese and copper content of some common grasses. *Journal of the American Society of Agronomy* 39, 356–362.

Berger, L.L. (1995) Why do we need a new NRC data base? *Animal Feed Science and Technology* 53, 99–107.

Bowen, H.J.M. and Dymond, J.A. (1955) Strontium and barium in plants and soils. *Proceedings of the Royal Society of London, B* 144, 355–368.

Bramley, R.G.V. (1990) Cadmium in New Zealand agriculture. *New Zealand Journal of Agricultural Research* 33, 505–519.

Bremner, I. and Campbell, T.K. (1980) The influence of dietary copper intake on the toxicity of cadmium. *Annals of the New York Academy of Sciences* 355, 319–332.

Bremner, I., Humphries, W.R., Phillippo, M., Walker, M.J. and Morrice, P.C. (1987) Iron-induced copper deficiency in calves: dose–response relationships and interactions with molybdenum and sulphur. *Animal Production* 45, 403–414.

Buckley, W.T. (1988) The use of stable isotopes in studies of mineral metabolism. *Proceedings of the Nutrition Society* 47, 407–416.

Burridge, J.C., Reith, J.W.S. and Berrow, M.L. (1983) Soil factors and treatments affecting trace elements in crops and herbage. In: Suttle, N.F., Gunn, R.G., Allen, W.M., Linklater, K.A. and Wiener, G. (eds) *Trace Elements in Animal Production and Veterinary Practice.* British Society of Animal Production Occasional Publication No. 7, Edinburgh, pp. 77–86.

Campbell, J.K. and Mills, C.F. (1979) The toxicity of zinc to pregnant sheep. *Environmental Research* 20, 1–13.

Cantor, A.H., Scott, M.L. and Noguchi, T. (1975) Biological availability of selenium in feedstuffs and selenium compounds for prevention of exudative diathesis in chicks. *Journal of Nutrition* 105, 96–105.

Coates, D.B., Kerridge, P.C., Miller, C.P. and Winter, W.H. (1990) Phosphorus and beef production in Australia. 7. The effect of phosphorus on the composition, yield and quality of legume based pasture and their relation to animal production. *Tropical Grasslands* 24, 209–220.

Comar, C.L. (1956) Radiocalcium studies in pregnancy. *Annals of the New York Academy of Sciences* 64, 281.

COSAC (1982) Chapter 10 Trace elements in Scottish soils and their uptake by crops, especially herbage. In: *Trace Element Deficiency in Ruminants.* Report of a Scottish Agricultural Colleges (SAC) and Research Institutes (SARI) Study Group. SAC, Edinburgh, pp. 49–50.

Davison, T.M., Orr, W.N., Doogan, V. and Moody, P. (1997) Phosphorus fertilizer for nitrogen-fertilized dairy pastures. 2. Long-term effects on milk production and a model of phosphorus flow. *Journal of Agricultural Science, Cambridge* 129, 219–231.

Dhillon, K.S. and Dhillon, S.K. (1991) Accumulation of selenium in sugarcane (*Sacharum officinarum* Linn) in seleniferous areas of Punjab, India. *Environmental Geochemistry and Health* 13, 165–170.

Dougall, H.W. and Bogdan, A.V. (1958) The chemical composition of the grasses of Kenya. *East African Agricultural Journal* 24, 17–23.

Fairweather-Tait, S.J., Johnson, A., Eagles, J., Gurt, M.I., Ganatra, S. and Kennedy, H. (1989) A double label stable isotope method for measuring calcium absorption from foods. In: Southgate, D., Johnson, I. and Fenwick, G.R. (eds) *Nutrient Availability: Chemical and Biological Aspects.* Royal Society of Chemistry Special Publication No. 72, Cambridge, pp. 45–47.

Gartner, R.J.W. and Twist, J.O. (1968) Mineral content of a variety of sorghum grain. *Australian Journal of Experimental Agriculture and Animal Husbandry* 8, 210–211.

Gladstones, J.S. and Drover, D.P. (1962) The mineral composition of lupins. 1. A survey of the copper, molybdenum and manganese contents of lupins in the south west of Western Australia. *Australian Journal of Experimental Agriculture and Animal Husbandry* 2, 46–53.

Grace, N.D. (1981) Phosphorus kinetics in the sheep. *British Journal of Nutrition* 45, 367–374.

Grace, N.D., Rounce, J.R. and Lee, J. (1996) Effect of soil ingestion on the storage of Se, vitamin B_{12}, Cu, Cd, Fe, Mn and Zn in the liver of sheep. *New Zealand Journal of Agricultural Research* 39, 325–331.

Hartmans, J. and Bosman, M.S. (1970) Differences in the copper status of grazing and housed cattle and their biochemical backgrounds. In: Mills, C.F. (ed.) *Trace Eelement Metabolism in Animals – 1.* Livingstone, Edinburgh, pp. 362–366.

Hazell, T. (1985) Minerals in foods: dietary sources, chemical forms, interactions, bioavailability. *World Review of Nutrition and Dietetics* 46, 1–123.

Healy, W.B., Rankin, P.C. and Watts, H.M. (1974) Effect of soil contamination on the element composition of herbage. *New Zealand Journal of Agricultural Research* 17, 59–61.

Heth, D.A. and Hoekstra, W.G. (1965) Zinc-65 absorption and turnover in rats. 1. A procedure to determine zinc-65 absorption and the antagonistic effect of calcium in a practical diet. *Journal of Nutrition* 85, 367–374.

Hopkins, A., Adamson, A.H. and Bowling, P.J. (1994) Response of permanent and reseeded grassland to fertilizer nitrogen. 2. Effects on concentrations of Ca, Mg, Na, K, S, P, Mn, Zn, Cu, Co and Mo in herbage at a range of sites. *Grass and Forage Science* 49, 9–20.

Johnson, J.M. and Butler, G.W. (1957) Iodine content of pasture plants. 1. Method of determination and preliminary investigations of species and strain differences. *Physiologia Plantarum* 10, 100–111.

Johnson, P.E., Lykken, G.I. and Korynta, E.D. (1991) Absorption and biological half-life in humans of intrinsic and extrinsic ^{54}Mn tracers for foods of plant origin. *Journal of Nutrition* 121, 711–717.

Jones, R.J. (1990) Phosphorus and beef production in northern Australia. 1. Phosphorus and pasture productivity – a review. *Tropical Grasslands* 24, 131–139.

Jumba, I.O., Suttle, N.F., Hunter, E.A. and Wandiga, S.O. (1996) Effects of botanical composition, soil origin and composition on mineral concentrations in dry season pastures in western Kenya. In: Appleton, J.D., Fuge, R. and McCall, G.J.H. (eds) *Environmental Geochemistry and Health.* Geological Society Special Publication No. 113, London, pp. 39–45.

Kappel, L.C., Morgan, E.B., Kilgore, L., Ingraham, R.H. and Babcock, D.K. (1983) Seasonal changes of mineral content in Southern forages. *Journal of Dairy Science* 66, 1822–1828.

Kiatoko, M., McDowell, L.R., Bertrand, J.E., Chapman, H.C., Pete, F.M., Martin, F.G. and Conrad, J.H. (1982) Evaluating the nutritional status of beef cattle herds from four soil order regions of Florida. I. Macroelements, protein, carotene, vitamins A and E, haemoglobin and haematocrits. *Journal of Animal Science* 55, 28–47.

Kirchgessner, M., Friesecke, H. and Koch, G. (1967) *Nutrition and the Composition of Milk.* Crosby Lockward, London, pp. 209–238.

Krishnamurti, C.R., Ramberg, C.F., Shariff, M.A. and Boston, R.C. (1997) A compartmental model depicting short-term exchanges in selenium metabolism in ewes fed hay containing normal or inadequate levels of selenium. *Journal of Nutrition* 127, 95–102.

Kume, S.-I. and Tanabe, S. (1993) Effect of parity on colostral mineral concentrations of Holstein cows and value of colostrum as a mineral source for newborn calves. *Journal of Dairy Science* 76, 1654–1660.

Ledgard, S.F. and Upsdell, M.P. (1991) Sulphur inputs from rainfall throughout New Zealand. *New Zealand Journal of Agricultural Research* 34, 105–111.

Leng, R.A. and Preston, T.R. (1984) Nutritional strategies for the utilisation of agroindustrial by-products by ruminants and extension of the principles and technologies to the small farmer in Asia. In: *Proceedings of the Fifth World Conference on Animal Production, Tokyo*, pp. 310–318.

Lengemann, F.W. (1969) Comparative metabolism of $Na^{125}IO_3$ and $Na^{131}I$ in lactating cows and goats. In: *Trace Mineral Studies with Isotopes in Domestic Animals.* International Atomic Energy Agency, Vienna, pp. 113–120.

Lengemann, F.W., Comar, C.L. and Wasserman, R.H. (1957) Absorption of calcium and strontium from milk and nonmilk diets. *Journal of Nutrition* 61, 571–583.

Littell, R.C., Lewis, A.J. and Henry, P.R. (1995) Statistical evaluation of bioavailability assays. In: *Bioavailability of Nutrients For Animals.* Academic Press, San Diego, pp. 5–34.

Littell, R.C., Lewis, A.J., Henry, P.R. and Ammerman, C.B. (1997) Estimation of relative bioavailability of nutrients. *Journal of Animal Science* 75, 2672–2683.

McDowell, L.R. (1992) *Minerals in Human and Animal Nutrition.* Academic Press, New York, pp. 496–511.

MAFF (1990) *UK Tables of the Nutritive Value and Chemical Composition of Foodstuffs.* In: Given, D.I. (ed.) Rowett Research Services, Aberdeen.

Minson, D.J. (1990) *Forages in Ruminant Nutrition.* Academic Press, San Diego, California, pp. 208–229.

Mitchell, R.L. (1957) The trace element content of plants. *Research, UK* 10, 357–362.

Morin, D.E., Rowan, L.L., Hurley, W.L. and Braselton, W.E. (1995) Composition of milk from llamas in the United States. *Journal of Dairy Science* 78, 1713–1720.

Odell, B.L., Burpo, C.E. and Savage, J.E. (1972) Evaluation of zinc availability in foodstuffs of plant and animal origin. *Journal of Nutrition* 102, 653–660.

Pott, F.B., Henry, P.R., Ammerman, C.B., Merritt, A.M., Madison, J.B. and Miles, R.D. (1994) Relative bioavailability of copper in a copper:lysine complex for chicks and lambs. *Animal Feed Science and Technology* 45, 193–203.

Saylor, L. and Finch, C.A. (1953) Determination of iron absorption using two isotopes of iron. *American Journal of Physiology* 172, 372–376.

Schwarz, R., Balko, A.Z. and Wien, E.M. (1982) An *in vitro* system for measuring intrinsic dietary mineral exchangeability: alternatives to intrinsic labelling. *Journal of Nutrition* 112, 497–504.

Scott, M.L. and Thompson, J.N. (1971) Selenium content of feedstuffs and effects of dietary selenium levels upon tissue selenium in chicks and poults. *Poultry Science* 50, 1742–1748.

Sherrell, C.G. (1978) A note on sodium concentrations in New Zealand pasture species. *New Zealand Journal of Experimental Agriculture* 6, 189–190.

Sherrell, C.G. and Smith, E.R. (1978) A note on the elemental concentrations of New Zealand pasture seeds. *Journal of Experimental Agriculture* 6, 191–194.

Shirley, R.L. (1978) Water as a source of minerals. In: Conrad, J.H. and McDowell, L.R. (eds) *Latin American Symposium on Mineral Nutrition Research with Grazing*

Ruminants. Animal Science Department, University of Florida, Gainsville, pp. 40–47.

Smart, M.E., Cohen, R., Christensen, D.A. and Williams, C.M. (1986) The effects of sulphate removal from the drinking water on the plasma and liver copper and zinc concentrations of beef cows and their calves. *Canadian Journal of Animal Science* 66, 669–680.

Smith, G.S., Middleton, K.R. and Edmonds, A.S. (1978) A classification of pasture and fodder plants according to their ability to translocate sodium from their roots into aerial parts. *New Zealand Journal of Experimental Agriculture* 6, 183–188.

Suleiman, A., Okine, E. and Goonewardne, L.A. (1997) Relevance of National Research Council feed composition tables in Alberta. *Canadian Journal of Animal Science* 77, 197–203.

Suttle, N.F. (1983) Assessing the mineral and trace element status of feeds. In: Robards, G.E. and Packham, R.G. (eds) *Proceedings of the Second Symposium of the International Network of Feed Information Centres, Brisbane.* Commonwealth Agricultural Bureaux, Farnham Royal, Slough, UK, pp. 211–237.

Suttle, N.F. (1985) A concept of availability and its technical implications. In: Taylor, T.G. and Jenkins, W.K. (eds) *Proceedings of the 13th International Congress of Nutrition.* John Libbey, London, pp. 232–237.

Suttle, N.F. (1991) Mineral supplementation of low quality roughages. In: *Proceedings of Symposium on Isotope and Related Techniques in Animal Production and Health.* International Atomic Energy Commission, Vienna, pp. 101–104.

Suttle, N.F. and Jones, D. (1986) Copper and disease resistance in sheep: a rare natural confirmation of interaction between a specific nutrient and infection. *Proceedings of the Nutrition Society* 45, 317–325.

Suttle, N.F., Alloway, B.J. and Thornton, I. (1975) An effect of soil ingestion on the utilization of dietary copper by sheep. *Journal of Agricultural Science, Cambridge, UK* 84, 249–254.

Suttle, N.F., Abrahams, P. and Thornton, I. (1984) The role of a soil × dietary sulphur interaction in the impairment of copper absorption by ingested soil in sheep. *Journal of Agricultural Science, Cambridge* 103, 81–86.

Thornton, I. (1974) Biogeochemical and soil ingestion studies in relation to the trace-element nutrition of livestock. In: Hoekstra, W.G., Suttie, J.W., Ganther, H.E. and Mertz, W. (eds) *Trace Element Metabolism in Animals –2.* University Park Press, Baltimore, Maryland, pp. 451–454.

Turnlund, J.R. (1989) The use of stable isotopes in mineral nutrition research. *Journal of Nutrition* 119, 7–14.

Underwood, E.J. (1977) *Trace Elements in Human and Animal Nutrition,* 4th edn. Academic Press, New York, 545 pp.

Underwood, E.J., Robinson, T.J. and Curnow, D.H. (1947) The manganese content of Western Australian cereal grains and their by-products and of other poultry feeds. *Journal of the Department of Agriculture for Western Australia* 24, 259–270.

Vaithiyanathan, S. and Singh, M. (1994) Seasonal influences on soil ingestion by sheep in an arid region. *Small Ruminant Research* 14, 103–106.

White, C.L., Robson, A.D. and Fisher, H.M. (1981) Variation in nitrogen, sulfur, selenium, cobalt, manganese, copper and zinc contents of grain from wheat and two lupin species grown in a range of Mediterranean environments. *Australian Journal of Agricultural Research* 32, 47–59.

Wilman, D. and Derrick, R.W. (1994) Concentration and availability to sheep of N, P, K, Ca, Mg and Na in chickweed, dandelion, dock, ribwort and spurrey compared with perennial ryegrass. *Journal of Agricultural Science, Cambridge* 122, 217–223.

Winks, L. (1990) Phosphorus and beef production in northern Australia. 2. Responses to phosphorus by ruminants – a review. *Tropical Grasslands* 24, 140–158.

Wright, D.E., Towers, N.R., Hamilton, P.B. and Sinclair, D.P. (1978) Intake of zinc sulphate in drinking water by grazing beef cattle. *New Zealand Journal of Agricultural Research* 21, 215–221.

The Detection and Correction of Mineral Imbalances in Animals

3

In this chapter, the general principles that govern the choice and effectiveness of procedures for the detection and correction of mineral disorders in farm animals are considered. Detection is usually based on clinical, pathological and biochemical examinations of the animals and appropriate tissues and fluids, coupled with a search for anomalous amounts and proportions of minerals in the pastures or rations. Soil mineral analyses also have some diagnostic value, since the mineral concentration of plants reflects to varying degrees the mineral content of the soil. The information obtained from any one of these sources alone is rarely conclusive. Mineral deprivation and toxicity states are frequently complicated by the presence of or clinical similarity to other nutritional disorders, parasitic infestations and bacterial, viral or protozoal infections. Differential diagnosis in such circumstances is difficult. Even when the evidence from clinical, pathological and biochemical examinations of the animal and from chemical analysis of the diet is combined and assessed, it may be impossible to define any nutritional abnormality of mineral origin, particularly when it is mild. The ultimate criterion of any mineral inadequacy, imbalance or excess is the improvement in growth, health, fertility or productivity that occurs in response to appropriate changes in the intake or utilization of the mineral or minerals in question (Phillippo, 1983). Simple but well-designed and critically conducted supplementation experiments with animals can be of the greatest value whenever and wherever mineral disorders are suspected. Furthermore, the magnitude of the response to supplementation can be used to refine the interpretation of biochemical criteria of deficiency, as Clark *et al.* (1985) did with oral cobalt and the interpretation of plasma vitamin B_{12} concentrations in lambs. The subject of dose–response trials is returned to in Chapter 19.

The Significance of Soil Mineral Status

Soils that are abnormal in a given mineral tend to produce plants that are abnormal in that mineral. On a broad geographical basis, areas where some

© CAB *International* 1999. *Mineral Nutrition of Livestock*
(E.J. Underwood and N.F. Suttle)

mineral imbalances in livestock are more or less likely to occur can be predicted by mapping techniques. One example is the use of stream sediment reconnaissance for molybdenum to predict areas of induced copper deficiency in cattle (Plant and Stevenson, 1985). However, such techniques have rarely been successfully deployed to predict health problems in livestock (Fordyce *et al.*, 1996). Relationships are far from simple or precise for several reasons: the yield of the plant, as well as its mineral concentration, is affected by soil mineral status; different species and strains of plants can vary greatly in mineral composition, even when growing on the same soil; climatic and seasonal conditions, as well as stage of growth, affect the mineral composition of plants; and the chemical form of the mineral in soil, soil pH and degree of aeration or waterlogging influence the availability of some soil minerals (see Chapter 2). The concentration of a mineral in a soil is thus an uncertain guide to its concentration in the herbage or crops on a given farm.

Attempts have been made to obtain a more satisfactory relationship between the amounts in the soil and in the plant by the use of various soil extractants, such as 0.1 M hydrochloric acid (HCl) or 2.5% acetic acid, which aim at measuring the 'plant-available' minerals in the soil. These extractants give a higher correlation with plant yields than they do with the concentrations of the mineral in the plants produced, but, for minerals such as cobalt and iodine which rarely significantly affect plant growth, the correlation between the 'plant-available' mineral in the soil and that in the plant can be high. Mitchell (1957) claimed that the incidence of cobalt deficiency in grazing sheep in Scotland could be predicted confidently by determining the cobalt soluble in 2.5% acetic acid in the soil. However, it is more usual to find that correlations between extractable soil minerals and concentrations in plants are poor (e.g. Silanpää, 1982; McClaren *et al.*, 1984; Jumba *et al.*, 1995). More sophisticated simulations of the root microenvironment are needed to accurately define the location and extent of mineral deficiencies and excesses in grazing animals by soil analysis in most areas.

The Mineral Content of Pastures and Feeds

An initial assessment of the actual or likely occurrence of a dietary mineral inadequacy or excess can be made by comparing the mineral composition of the diet with appropriate standards of adequacy or safety. Numerous dietary standards are now available, but they can vary widely according to the source (Chapter 1). Errors inherent in the analytical procedures are decreasing in importance as techniques and quality control improve, provided that the sample is not contaminated during handling, storage or grinding. Milling through steel sieves with steel blades is still a common source of error in trace element analyses, but is avoidable by the use of appropriate equipment (e.g. mills with agate-tipped blades). The use of certified standards will quickly identify most errors but not necessarily those from laboratory milling, because the standard comes ready-milled. However, the detection and diagnosis of mineral disorders of dietary origin based entirely on mineral analysis of the feed can be misleading and there will always be an intermediate band of mineral concentrations in feeds between the inadequate and adequate (i.e. a 'marginal' supply), for the following reasons.

Selection

In foraging situations, the diet sample collected may not represent the material actually eaten by the animal, because of selective grazing and soil contamination in the field. Animals show preferences for different types and parts of plants, which can vary widely in mineral concentration, even when growing together. This problem has caused particular concern to those attempting to assess dietary phosphorus status (see Chapter 5). Selective grazing could therefore greatly influence actual mineral intakes by animals, particularly where there is a mixture of pasture and browse material (Fordyce *et al.*, 1996). Oesophageal fistulation has been employed to obtain better information on what the grazing animal consumes, but contamination with saliva can complicate interpretation (Langlands, 1987). For poorly absorbed minerals, faecal analysis can give reasonable predictions of intake. These and other techniques for measuring mineral intakes by grazing animals have been reviewed by Langlands (1987).

Availability

Estimates of mineral intake take no account of differences in absorption or utilization by the animal (Chapter 2). For example, a particular dietary level of total phosphorus may be adequate for poultry if it is in inorganic or non-phytin form but inadequate when it is present largely as phytate phosphorus. The adequacy of a particular dietary concentration of calcium varies with the vitamin D status of the animal. Particular concentrations of zinc and manganese can be adequate when the diet is 'normal' in calcium or phosphorus and inadequate when it is high in those elements; and certain concentrations of copper (Cu) can be inadequate when molybdenum and sulphur intakes are high but adequate or even excessive when dietary molybdenum (Mo) and sulphur are low. The influence of the dietary Cu:Mo ratio on copper utilization in the animal can be so great that in some areas pastures containing 6 mg Cu kg^{-1} dry matter (DM) are fully satisfactory for growth, health and wool quality in sheep, whereas in others 15 mg Cu kg^{-1} DM is necessary (see Chapter 11).

Balance

The significance of overall dietary balance for the optimal absorption and utilization of mineral nutrients by animals cannot be overstressed. Balance between macroelements is important because cations, such as sodium (Na^+), potassium (K^+), calcium (Ca^{2+}) and magnesium (Mg^{2+}), are alkali-producing, whereas anions, such as chloride (Cl^-) and sulphate (SO_4^{2-}), are acid-producing. The preservation of the acid–base balance is important for all species, from poultry to the dairy cow, and will be discussed in detail in Chapter 8. The influence of dietary ratios can be especially important in assessing the safety or toxicity to animals of a potentially toxic element, such as cadmium or selenium. The toxicity of cadmium is influenced by the dietary concentrations of the divalent metals zinc and copper and the toxicity of selenium by the amounts of mercury or lead that the diet supplies.

Many other examples could be cited to show that measurement of the total concentration of a mineral in the pasture or ration cannot always detect or predict inadequacy or toxicity of that mineral in the animal.

Clinical and Pathological Changes in the Animal

Severe abnormalities

All mineral deficiencies and most excesses, in their more severe forms, are manifested by clinical and pathological disturbances in the animal, but these are rarely specific for a single element. Moreover, the nature of the clinical and pathological changes varies greatly, even within a single species or breed, with the degree and the duration of the mineral deficiency or toxicity and with the age and sex of the animal. Many of the most obvious manifestations, such as subnormal growth, inappetence, anaemia, bone abnormalities, structural defects in the skin and its appendages, impaired lactation, poor reproductive performance and a reduction in quantity and quality of egg production, occur to varying degrees with deficiencies of a wide range of mineral elements (Table 3.1). For these reasons, few illustrations of the appearance of 'deficient animals' are presented in this book. Anaemia is characteristic of iron, copper and cobalt deficiencies and is also a manifestation of molybdenum, selenium and zinc toxicities. Resistance to infection is influenced by copper, iron, selenium and zinc. Gross skeletal deformities occur as a consequence of dietary deficiencies of vitamin D, copper, manganese and zinc, as well as of calcium and phosphorus. Impaired reproduction occurs in phosphorus, copper, manganese, zinc, iodine and selenium deficiencies and in fluorine and molybdenum toxicities. The impaired reproduction may be due to a primary defect in some phase of the reproductive process itself, as seems likely in manganese deficiency in cows and goats, selenium deficiency in ewes, zinc deficiency in rams and molybdenum toxicity in bulls, or it may arise as a secondary consequence of inanition or some other dysfunction. For instance, poor lamb and calf crops in enzootic fluorosis are usually the result of impoverishment of the mother, due to fluorine-induced inappetence and deformities of the teeth and joints, which restrict mastication and grazing. The various functional and structural disorders apparent to the clinician and the pathologist may be merely the final expressions of a defect arising at one of several possible points in a chain of metabolic events. Different minerals can exert their main effects at different points in the same chain, but the end result in the animal may be the same.

Table 3.1. Examples of the sequence of pathophysiological events during the development of five mineral-responsive diseases.

	Depletion	Deficiency	Dysfunction	Disease
Calcium	Young: bone ↓	Serum ↓	Chondrodystrophy	Rickets
	Old: bone ↓	Serum ↓	Irritability ↓	Recumbency
Magnesium	Young: bone ↓	Serum ↓	Irritability ↑	Convulsion
	Old: bone ?	Serum ↓	Irritability ↑	Convulsion
Copper	Liver ↓	Serum ↓	Disulphide bonds ↓	Wool crimp ↓
Cobalt	Liver B_{12} ↓ Serum B_{12} ↓	Serum B_{12} < 350 pmol l^{-1}	MMA ↑	Inappetence
Iodine	T_4 ↓ Thyroid colloid ↓ Thyroid I ↓	T_3 ↓	BMR ↓ Thyroid hypertrophy	Goitre
Selenium	GSH-Px in erythrocyte ↓	Serum Se ↓	Serum creatine kinase ↑	Myopathy

T_4, thyroxine; BMR, basal metabolic rate.

Mild abnormalities

Mild mineral deficiencies or excesses are especially difficult to identify, because their effects are frequently indistinguishable from those resulting from semistarvation or underfeeding, protein deficiency or intestinal parasitism. A further complication arises because a depression in appetite, with resulting undernutrition, is a common early expression of many mineral deficiencies, as it is with deficiencies of other essential nutrients. For instance, cobalt deficiency in ruminants has no clinical or pathological manifestation that is entirely specific for cobalt. The appearance of even a severely cobalt-deficient animal, one of emaciation and listlessness, is indistinguishable from that of a starved animal, except that the visible mucous membranes are blanched. A sticky discharge from the eyes may be seen in a mildly deficient animal but its 'unthrifty' appearance, can only be attributed to lack of cobalt by measuring the responses to oral dosing with cobalt or injections of vitamin B_{12}. The development of enlarged thyroid glands or goitres in newborn animals has long been recognized as a sign of an absolute or conditioned dietary deficiency of iodine, but mild enlargement can occur in healthy offspring (see Chapter 12). Selenium deficiency can exacerbate iodine deficiency and some of the mild consequences hitherto ascribed to selenium *per se* may involve and be influenced by variation in iodine status (Arthur, 1992). Numerically and economically, mild abnormalities now exceed severe abnormalities in importance.

Biochemical Indicators of Deprivation in Tissues and Fluids

When taken in conjunction with clinical and pathological observations, appropriate biochemical analyses of tissues and fluids of animals are valuable aids in the early detection and differential diagnosis of mineral abnormalities in livestock (Mills, 1987). A dietary deficiency of a mineral is sooner or later reflected in subnormal concentrations of the mineral in certain of the animal's tissues and fluids, and a dietary excess of a mineral is similarly reflected in above-normal concentrations. In addition, both deficiencies and toxicities are usually accompanied by significant tissue or fluid changes in the concentrations of particular enzymes, metabolites or organic compounds with which the mineral in question is functionally associated. Many of these changes can be detected before the onset of clinically obvious signs of deficiency or excess in the animal.

Depletion, deficiency, dysfunction and disease

A general model of pathophysiological events during mineral depletion is presented in Fig. 3.1 and it provides a rational basis for the differential diagnosis of disorders due to mineral imbalances. The model divides events into four phases: firstly 'depletion', during which storage pools of the mineral are reduced; secondly, 'deficiency', during which transport pools of the mineral are reduced; thirdly, 'dysfunction', when mineral-dependent functions become rate-limiting to particular metabolic pathways; and fourthly, 'disease', during which clinical abnormalities become apparent to the naked eye. There may be variable degrees of overlap between phases; for example, deficiency

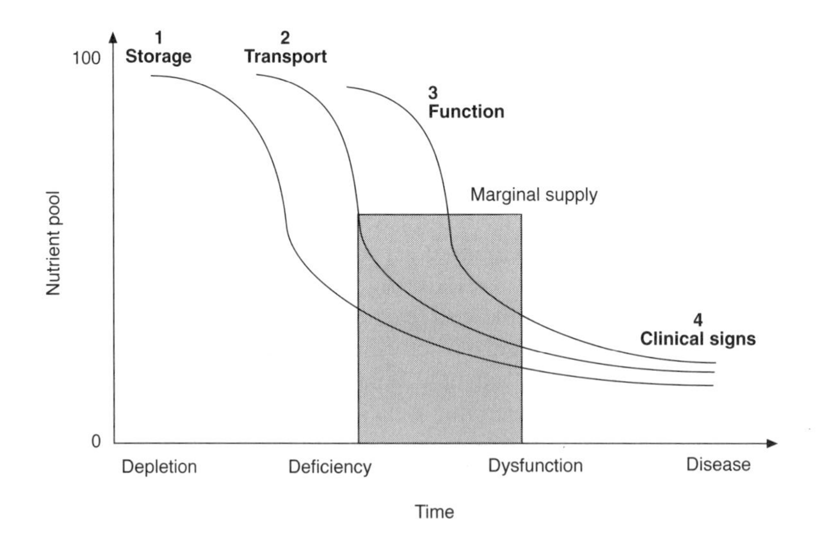

Fig. 3.1. Schematic sequence of pathophysiological changes in livestock given an inadequate mineral supply: for some elements (e.g. Ca and P), storage and function are linked by bone strength; for elements with small or slowly mobilized stores (Zn and Se), curves 1 and 2 are superimposed or interposed; there is usually a zone of marginal supply (shaded area) where mineral-dependent functions begin to fail but the animal remains outwardly healthy (upper limit on the *y* axis represents maximum attainable pool size).

may commence before reserves are entirely exhausted (e.g. iron; Chesters and Arthur, 1988). The above terminology will be adhered to throughout this book, to avoid the confusion which has arisen in the past from the varied use of words such as 'deficient' to describe diets low in minerals, healthy animals of subnormal mineral status and clinically sick animals. A diet is only 'mineral-deficient' if it is likely to induce subnormal mineral status and, if fed, does not necessarily result in clinical disease. The precise sequence of events during the development of clinical disease varies widely from mineral to mineral and slightly for a given mineral, depending on the magnitude of the deficit between daily mineral requirement and the daily supply from endogenous (e.g. body stores) and dietary sources. For example, there are limited body stores of zinc and reductions in albumin-bound, transport zinc in plasma occur soon after exposure to a diet inadequate in zinc, whereas with phosphorus mobilization of skeletal reserves can preserve serum phosphorus concentrations during long periods of depletion. Although similar mobilization of calcium will often maintain plasma calcium concentrations long after exposure to inadequate calcium supplies commenced, sudden increases in milk yield and hence calcium demand with the onset of lactation can lead to hypocalcaemia and acute clinical disease (milk-fever) while plentiful reserves remain. The rate of transition from deficiency to dysfunction and disease is therefore variable and depends partly upon the demand being placed on the critical pathway. A low selenium status may be tolerable until changes in diet and increased freedom of movement (e.g. turning out to spring pasture) present a twofold oxidative stress (Chapter 12); a low iodine status may be tolerable until a cold stress induces a thermogenic response (Chapter 11); simultaneous deficiencies of selenium and iodine increase the likelihood that one or both become dysfunctional.

What message can a particular analyte convey?

The investigator of a suspected mineral-responsive disorder is faced with a number of biochemical options when considering how to proceed with a differential diagnosis. The information provided varies with the analyte chosen but is rarely a direct measure of dysfunction and less so of disease. Historically, sample sites and substances have been selected for convenience rather than diagnostic insight. Thus, for the six elements chosen in Table 3.1, the popular biochemical criteria generally convey information about depletion and deficiency, stores and transport forms, rather than dysfunction or disease. Low values are not therefore synonymous with loss of health. The relationship with dysfunction may be tightened by lowering the threshold of normality (e.g. for vitamin B_{12}) to one at which dysfunction is more likely to occur. Arguments as to whether a marker for depletion (e.g. liver copper) or transport (serum or ceruloplasmin copper) is the superior measure of 'status' are often rehearsed but are largely irrelevant in terms of diagnosis, i.e. differentiating the diseased and dysfunctional animal from the healthy one (see Chapter 11). It follows that most commonly used biochemical criteria for animals, like those for soil and feeds, require to be interpreted using a three-tier system (Table 3.2), with a marginal band to separate the deficient from the dysfunctional individual (Fig. 3.1). Throughout this book, all biochemical criteria will be interpreted using a three-tier system, rather than a diagnostically unhelpful reference range. This system is by no means the ultimate and can be improved upon by treating an indicator of dysfunction as a continuous variable and predicting the probability of a given response (e.g. minimum economic) occurring (see Chapter 19 and Clark *et al.*, 1989).

Delineation of marginal values

The use of clinical biochemistry to diagnose or predict mineral-responsive disease is likely to remain an imprecise science. What is needed are detailed studies of the pathophysiological events associated with all economically important mineral imbalances of the type conducted by White (1996) on pooled data for zinc-deficient lambs (Fig. 3.2). The relationship between plasma or serum zinc and growth rate was fitted by a Mitscherlich equation and used to define a threshold value (6.7 μmol Zn l^{-1} plasma), 95% of the critical value or asymptote below which growth impairment became likely. This indicates the lower limit of the 'marginal' band. Once again, a familiar complication arises: the chosen value depends on the index of adequacy and, with optimal wool growth as the goal, the lower limit of marginality rises to 7.7 μmol l^{-1} (White *et al.*, 1994). Sadly, worldwide reductions in the funding of studies pertaining to the diagnosis of nutritional diseases of livestock in developed countries means that many uncertainties are unlikely to be resolved.

Criteria of dysfunction

There are two approaches which potentially offer a way out of the morass and they both relate to the recognition of dysfunction. Biochemical markers, such as methylmalonic acid (MMA) and creatine kinase (CK) (Table 3.1), indicate when particular functions have become rate-limited, MMA coenzyme A (CoA) isomerase activity in cobalt or vitamin B_{12} deficiency (Chapter 10)

Table 3.2. Interpretation of biochemical and other indices of mineral deprivation (or excess) in livestock is enhanced by the use of a marginal band between values consistent with health and those consistent with ill health.

Label	Phase[a]	Responsiveness to supplement (or antidote)
Normal	Equilibrium, depletion (or accretion)	Unlikely
Marginal	Deficiency (or excess)	Possible
	Dysfunction	
Abnormal	Disorder	Probable

[a] See Fig. 3.1 for illustration of terminology.

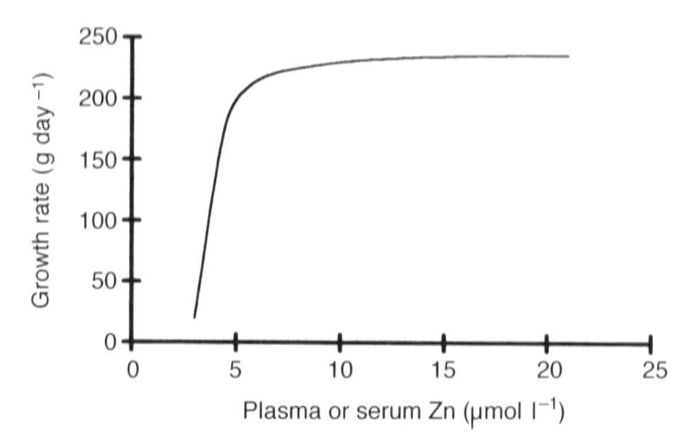

Fig. 3.2. Relationship between growth rate and plasma or serum zinc in experimentally deprived lambs (White, 1996).

and muscle membrane permeability in selenium deficiency (Chapter 15). However, a rise in plasma MMA may precede the important clinical sign of cobalt deficiency, loss of appetite (see Chapter 10). By studying relationships between markers of depletion or deficiency and dysfunction, the interpretation of each one can be clarified (Mills, 1987). It is becoming apparent that animals have internal biochemical systems for recognizing low mineral status and correcting it; mRNAs for serum ferritin – the storage protein – are switched off in iron deficiency and those for glutathione peroxidase (GPX) switched off in selenium deficiency, while those for cholecystokinin are induced in zinc deficiency (see appropriate chapters). The use of mRNAs for routine diagnosis is likely to remain prohibitively expensive, but their experimental use could resolve some long-standing arguments by letting the animal decide how mineral-deficient it is.

Physiological and developmental changes

When using the mineral composition of animal tissues and fluids to indicate the quality of mineral nutrition, it is important to avoid the confounding effects of physiological and developmental changes. The normal newborn animal often has 'abnormal' tissue or blood mineral composition when judged inappropriately by adult standards (see Chapter 11 for liver and blood copper). Most blood mineral concentrations change abruptly and briefly around parturition, and sampling at this time is only appropriate for acute disorders of calcium and magnesium metabolism (see Chapters 4 and 6).

Choice of Sample

The choice of tissue or fluid for analysis varies with the mineral under investigation. Blood, urine, saliva and hair have obvious advantages because of their accessibility without harm to the animal.

Blood

Whole blood, blood serum or plasma is widely sampled and serum is usually chosen for analysis, because it avoids the cost and possible analytical complications of adding an anticoagulant, gives a more stable (i.e. haemolysis-free) form for transportation and usually reflects the status of the transport pool of the element, i.e. one step nearer to dysfunction than storage. The use of whole blood or erythrocytes brings complications and new possibilities into the assessment of mineral status. Minerals in their functional units are often incorporated into the immature erythrocyte prior to release into the bloodstream, and the mineral content of the younger erythrocytes alone accurately reflects recent mineral nutrition. Once released, the mineral and functional status of the erythrocyte may remain unchanged, despite marked fluctuations in the mineral status of the diet or tissues. Thus erythrocyte enzymes indicative of copper and selenium status (copper–zinc superoxide dismutase (CuZnSOD) and GPX, respectively) normally show a delay of about 3 weeks before indicating a rise in copper or selenium provision, with the arrival of sufficient recently enriched cells to increase mean values for the entire erythrocyte population. This may be an advantage if animals have been treated or inadvertently gained access to a new mineral supply (during handling) prior to blood sampling, since erythrocyte enzyme activities in the subsequent blood sample would not have been influenced. However, it could be disadvantageous in masking a recent downturn in copper or selenium supply. In extremely deficient animals, erythrocytes may be released in which the enzyme lacks its functional mineral (apoenzyme). Assay of enzyme protein by enzyme-linked immunosorbent assay (ELISA) or the extent of *in vitro* reconstitution may offer new insights into the severity of depletion in such circumstances (see Chapter 16). Where the normal function of an element is expressed by the erythrocyte, e.g. stability of the membrane and/or resistance to oxidative stress, as in copper, phosphorus and selenium deficiencies, functional status may be indicated by *in vitro* haemolysis tests or staining for Heinz bodies; however, such tests would obviously be non-specific.

Secretions

Analysis of saliva provides diagnostic information on several minerals, and samples are readily procured, using an empty 20 ml syringe fitted with a semirigid polypropylene probe and using a 'trombone slide' gag to open the mouth. The composition of the saliva is particularly sensitive to a dietary deficiency of sodium; as the concentration of sodium declines, that of potassium rises and the Na:K ratio in parotid saliva can decrease tenfold or more, providing a means of detecting sodium deficiency in the animal (Morris, 1980). Preparatory mouthwashes can reduce contamination from feed and digesta without disturbing the Na:K ratio. The analysis of Na:K in muzzle secretions can also indicate sodium status (Kumar and Singh, 1981). The zinc content of saliva has been suggested as a sensitive indicator of zinc status in humans (Henkin *et al.*, 1975), although its diagnostic value is not entirely supported by other investigations (Johnson *et al.*, 1978; Greger and Sickles, 1979) and it has yet to be critically evaluated in domestic livestock.

Urine and faeces

Subnormal urinary sodium output points to a dietary deficiency of sodium, and high urinary fluoride levels suggest fluorine toxicity. However, high urinary fluorine may reflect high previous intakes of the mineral, because urinary excretion from the skeletal fluoride stores continues for some time after high intakes have ceased. Interpretation of urine analysis is improved by also measuring indices of dilution (e.g. s.g. or creatinine). Concentrations of minerals in faeces are influenced by diet and animal factors, but, after prolonged deprivation, they can contribute to a diagnosis (e.g. faecal phosphorus).

Tissues

Of the body tissues, liver and bone have proved especially useful in anticipating mineral disabilities in livestock, because they are storage organs for certain minerals and because simple sampling techniques by aspiration biopsy or trephine are available. For instance, subnormal concentrations of iron, copper and cobalt (or vitamin B_{12}) in the liver are suggestive of possible dietary deficiencies of these elements. Subnormal concentrations of calcium and phosphorus in the bones can suggest deficiencies of calcium, phosphorus or vitamin D, and high fluoride levels in bone indicate excess intakes of fluorine. Results may, however, vary widely from bone to bone, with the age of animal and with the mode of expression (fresh, dried, defatted or ashed bone basis), making interpretation difficult (see Chapter 5 in particular). With heterogeneous organs, such as the kidney, results can be affected by the area sampled; for example, selenium concentrations are two- to five-fold higher in the cortex than in the medulla (Millar and Meads, 1988), making precise description of and adherence to sampling protocols essential. Relationships between mineral concentrations in storage organs and ill health can, however, be poor because stores may be exhausted before any dysfunction or disorder arises (Fig. 3.1).

Appendages

The analysis of minerals in body appendages, such as hair, hoof, fleece or feathers, for diagnostic purposes has had a chequered history (Combs, 1987). Relationships to other indices of status, both biochemical and performance, have rarely been sufficiently precise. There are a number of reasons: exogenous surface contamination; the presence of skin secretions (e.g. suint in wool); failure to standardize on a dry-weight basis; and variable time periods over which the sample has accumulated. The value of analysing material from body appendages can clearly be increased by the rigorous washing and drying of samples. Historical profiles can be obtained by dividing hair and core samples of hoof into proximal, medial and distal portions. Alternatively, the sampling of regrowth from a recently shaved site (e.g. a liver biopsy site) can give a measure of mineral status to compare with contemporary diet, blood or tissue samples.

Functional forms and indices

Advances in assay procedures for enzymes and hormones have greatly increased the range and sensitivity of the diagnostic techniques now available. Serum assays for vitamin B_{12} rather than cobalt, triiodothyronine (T_3) rather than protein-bound iodine, ceruloplasmin rather than copper and GPX rather than selenium now provide alternative indicators of the dietary and body status of these elements. These newer assays brought with them fresh problems of standardization, which were only slowly recognized, and their use still does not invariably improve the accuracy of diagnosis. The diagnostic strength and limitations of these and similar estimations are considered later, in the chapters dealing with individual elements and their interrelations. Secondary changes in serum enzymes, arising from tissue changes or damage due to mineral deficiencies or toxicities, also have diagnostic value, but caution is still necessary.

Biochemical Indicators of Mineral Excesses

Chronic exposure to mineral excesses leads to a sequence of biochemical changes which is, in some respects, a mirror image of events during deprivation (compare Fig. 3.3 with Fig. 3.1). Firstly, there is *accretion* at storage sites; secondly, levels in transport pools may rise; thirdly, *dysfunction* may be manifested by the accumulation of abnormal metabolites or constituents in the blood, tissues or excreta; and, fourthly, clinical signs of disorder become visible. For example, marginal increases in plasma copper and a rise in serum aspartate aminotransferase (AST) precede the haemolytic crisis in chronic copper poisoning in sheep; the latter is indicative of hepatic dysfunction. However, increases in AST also occur during the development of white muscle disease, caused by selenium deficiency, because the enzyme can leak from damaged muscle as well as liver. Assays of glutamate dehydrogenase (found only in liver) and creatine kinase (found only in muscle) are of greater value than AST in distinguishing the underlying site of dysfunction. The whole sequence in Fig. 3.3 can become telescoped during acute toxicity. There will be similar marginal bands of uncertainty when assessing mineral excesses, as when deprivation is assessed.

The Prevention and Correction of Mineral Deficiencies

Successful procedures for the prevention and control of all mineral deficiencies (and many mineral toxicities) have been developed. The procedure of choice varies greatly with different elements, climatic environ-

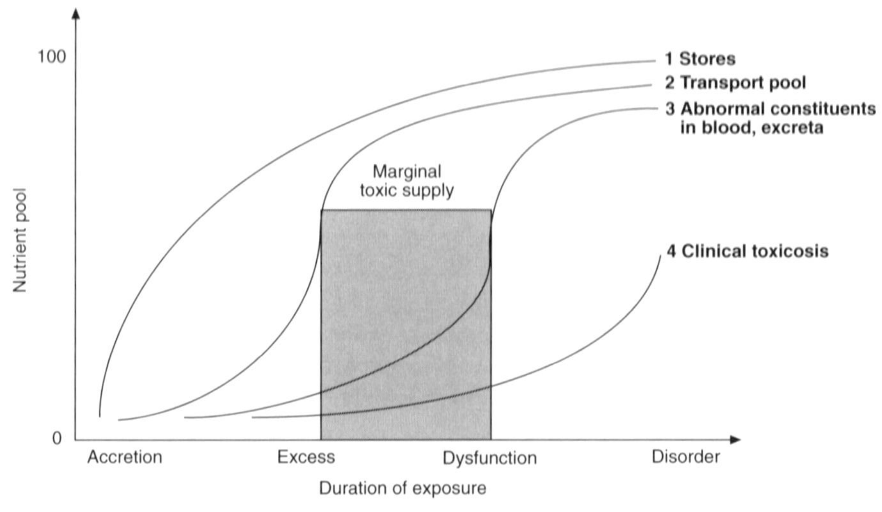

Fig. 3.3. Sequence of pathophysiological events during chronic exposure to excessive amounts of mineral: sequences become telescoped during acute exposure; rate of transition is also affected by dietary attributes (e.g. absorbability, Fig. 1.2), animal attributes (mineral may be safely disposed of in products like milk and eggs) and factors such as disease (e.g. liver toxins or pathogens).

ments, conditions of husbandry and economic circumstances. The methods available fall into three categories: indirect methods that affect the mineral composition of pastures and feeds while they are growing and continuous or discontinuous direct methods, involving administration of minerals to the animals.

Indirect methods

Correcting mineral imbalances in grazing stock by treatment of the soil often has serious economic limitations under extensive range conditions, because productivity from each unit area is often limited by inadequate or erratic rainfall or by low winter temperatures, while transport and application of the fertilizer or soil amendment are invariably costly. Furthermore, climatic effects may be dominant over soil effects in determining the mineral content of the herbage, as is pointed out in respect of phosphorus in Chapter 5. In more favourable environments, treatment of the soil is widely and successfully practised to improve both the yield and the mineral composition of herbage. Treatment with copper- or phosphorus-containing fertilizers can increase plant yield and is therefore a logical first step when pasture growth has been poor. Soil treatments with cobalt and selenium do not affect plant yields but can be a practical means of ensuring adequate mineral intakes. By incorporating small proportions of cobalt or selenium into fertilizers, such as superphosphate, used to maximize herbage yields, the costs of application are minimized and all animals can secure more mineral from the treated herbage.

Problems arise if deficiency is caused by poor availability rather than poor mineral content of the soil. For example, cobalt-containing fertilizers are ineffective on calcareous or highly alkaline soils, because of the low availability of cobalt to plants in such conditions. Soils of low available phosphorus concentration are likely to trap added fertilizer phosphorus in unavailable forms. The application of sulphur or gypsum to seleniferous soils already high in sulphate or which contain a high proportion of their selenium in organic combination will not significantly reduce selenium uptake by plants, although it can be effective in other circumstances. Treatment with copper of pastures very high in molybdenum may not raise the copper content of the herbage high enough to counteract a severe conditioned copper deficiency. The residual effects of all soil treatments and therefore the frequency with which they must be applied can vary widely from one location to another.

Continuous direct methods

Animals that are being hand-fed can best be supplemented by mixing the required minerals with the food offered. When adequate trough time or space is allowed, variation in individual consumption is not pronounced and the amounts included should not exceed average mineral requirements. The precise amounts and proportions of supplementary minerals needed will depend on the nature and degree of the deficiency and the form and intensity of production. Where individual animal productivity is high, husbandry conditions are intensive and farm labour costs generally high, it is usual to purchase mixed feeds compounded to contain the required minerals in adequate quantities and with a universally sufficient margin of safety; however, they often contain minerals that are not locally required as supplements. Where the conditions of husbandry are less intensive and farm labour is more plentiful and less costly, the same results can be achieved by home mixing and only those minerals known to be needed are added. Furthermore, home-grown roughage, rather than expensive concentrates, can be used as the carrier for the mineral supplement. In sparse grazing conditions, where little or no regular hand-feeding is practised, treatment of the water-supply may be practicable.

Free-access mineral mixtures

The most widely used method of supplementation is the provision of mineral mixtures in loose ('lick') or block form. Common salt (NaCl) is a vital ingredient of such supplements, because it is so palatable and attractive to most animals. Individuals are relied on to consume sufficient for their needs where supplies are freely available and to seek it out actively where it is less accessible. The siting of dispensers can be arranged to attract stock to parts of a grazing area that they might otherwise undergraze. Because of its palatability, salt is a valuable 'carrier' of other minerals. So long as these mineral mixtures contain 30–40% salt, they are commonly consumed in amounts sufficient to meet the needs of livestock for other minerals (McDowell *et al.*, 1993). However, individual variation in consumption from licks or blocks can be marked (Bowman and Sowell, 1997; see p. 450). In one study, 19% of the ewe flock commonly consumed nothing, while others consumed 0.4–1.4 kg day^{-1} (Ducker *et al.*, 1981). In a study of salt supplementation (presented in protected, trailer-drawn blocks) of a beef suckler herd grazing Californian rangeland, mean daily consumption per head (cow ± calf) ranged from 0 to 129 about a mean of 27 g day^{-1} over consecutive 7-day periods (Fig. 3.4; Morris *et al.*, 1980); provision is therefore 'semicontinuous', rather than 'continuous'. Seasonal variation in mineral consumption can also occur, but intakes are more uniform from granulated than from block sources of minerals (Rocks *et al.*, 1982). Provision of several, widely scattered sources can minimize competition and hence individual variation in uptake of minerals. This form of treatment cannot be relied on where the water-supplies are saline and may be unsatisfactory with an element like cobalt, which is required by the animal regularly.

Discontinuous direct methods

The oral dosing, drenching or injection of animals with mineral solutions, suspensions or pastes has the advantage that all animals receive known amounts of the required mineral at known intervals. This type of treatment is unsatisfactory where labour costs are high and animals have to be driven long distances and handled frequently and specifically for treatment. With minerals such as copper, which are readily stored in the liver to provide reserves against periods of inadequate intake, large doses given several months apart can be satisfactory. Oral dosing with selenium salts works well in selenium-deficient areas, particularly where it can conveniently be combined with oral dosing of anthelmintics. In contrast, cobalt deficiency cannot always be prevented fully if the oral doses of cobalt salts are weeks

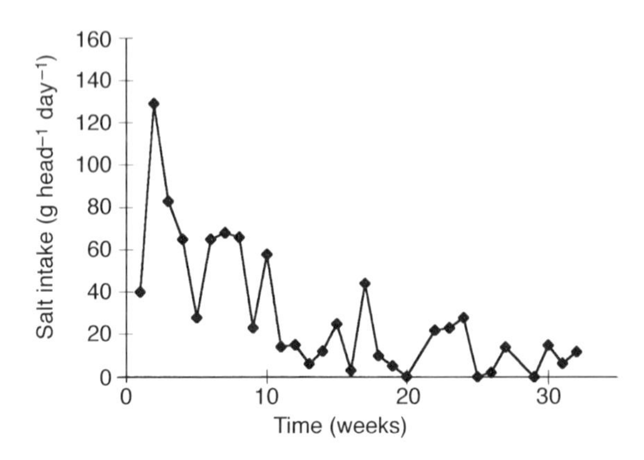

Fig. 3.4. Variation in average salt intake by a group of beef suckler cows given free access to salt from covered trailers on Californian rangeland: observations began with calving in November (data from Morris *et al.*, 1980).

apart. With iron, iodine and copper, some of the disadvantages of oral dosing can be overcome by the use of injectable organic complexes of the minerals. Such complexes are more expensive, but, when injected subcutaneously or intramuscularly, they are translocated slowly to other tissues and provide protection against a dietary deficiency of the injected element for lengthy periods. For instance, a single intramuscular injection of iron-dextran, supplying 100 mg Fe, at 2–4 days of age can control piglet anaemia (Chapter 13).

Slow-release methods

The efficacy and costs of administering minerals to individual animals can be improved by using large doses in relatively inert, slowly mobilized forms. The administration of heavy pellets (Millar and Meads, 1988), glass boluses (Millar *et al.*, 1988) or particles of the mineral (e.g. copper oxide: Chapter 11), which are retained in the gastrointestinal tract, all act as slow-release sources. The problem of cobalt deficiency can be solved by use of heavy cobalt pellets, as described in Chapter 10, which lodge in the reticulorumen and yield a steady supply of cobalt for many months, unless they are regurgitated or become coated.

Mineral sources

The choice of a mineral supplement is determined by: (i) cost per unit of the element or elements required; (ii) the chemical form in which the element is combined; (iii) its physical form, especially its fineness of division; and (iv) its freedom from harmful impurities, particularly fluorine. With calcium and phosphorus supplements, these factors can greatly influence the choice of material used, but, with the trace elements, because of the small quantities required and the relatively low costs involved, such considerations are generally of minor significance and the choice of supplement widens. Molybdate-containing salt-licks have been highly effective in preventing chronic copper poisoning in sheep and cattle in certain areas. There is a remarkable range in the composition of proprietary free-access mineral mixtures, which is influenced by factors such as cost of ingredients and colour, as well as nutritional need, and care is needed in choosing appropriate mixtures (e.g. Suttle, 1983). The mineral supplements commonly employed are given in detail in Appendix Table 2 on p. 600 and the optimum methods of treatment discussed in more detail in the following chapters.

The Misuse of Mineral Supplements

Overfeeding

The provision of extra minerals beyond the animal's needs is economically wasteful, confers no additional benefit on the animal and can be harmful. Excesses of phosphorus and magnesium can cause death from urinary calculi (Hay and Suttle, 1993), while an excess of copper soon causes toxicity in sheep. Commercial mineral mixtures often contain minerals which, although basically essential to the animal, are already present in adequate amounts in the pastures and feeds the animals will consume. There can be no automatic justification for the purchase and use of such 'shotgun' mixtures of minerals, which are designed to cover a very wide range of environments and feeding regimens and which often contain an unnecessarily wide margin of safety as an insurance against deficiency. The shotgun approach can be dangerous because of a possible disturbance of the overall dietary mineral balance and the consequent adverse effects on the absorption and utilization of certain minerals by the animal. Numerous examples of the importance of mineral balance are given in the chapters that follow. Furthermore, the surplus minerals are largely excreted and some (notably phosphorus and copper) are causing increasing concern as pollutants of the environment.

False claims

The manufacture and sale of ready-mixed mineral supplements by responsible firms is a legitimate and desirable business of considerable value to individual farmers. However, every farmer and stock-raiser should be on guard against exaggerated claims of advertising propaganda and salesmanship and should critically examine claims in the light of the particular mineral needs of stock under local conditions. Claims that the use of chelated forms of trace elements are more cost-effective than simple inorganic sources should be ignored until there is published evidence in reputable scientific journals that such benefits are consistently attainable. Such complexes must be strong enough to resist natural dietary antagonists and yet deliver the complexed element to the tissues in a usable form. There is no evidence that such complexes are superior as copper sources for ruminants (Suttle, 1994). While benefits of chelated minerals might be expected in non-ruminant diets, where phytate is such an important antagonist, these expectations have yet to be fulfilled and it may be better to attack the antagonist with phytase (see Chapters 5 and 16). Evidence that chelation alters the form or distribution of the complexed element at an early stage of digestion (e.g. in the rumen) is not proof of nutritional benefit. Evidence that chelation increases tissue element concentrations for elements whose uptake is normally regulated by need (e.g. zinc) may indicate a bypassing of the homeostatic mechanism and again provides no proof of nutritional benefit. The fact that chelation may be useful for one element (e.g. chromium, see Chapter 17) does not mean that it will be useful for others.

False concepts

Contrary to popular belief, an appetite for minerals is not a reliable measure of the animal's needs (Pamp *et al.*, 1976). The voluntary consumption of mineral mixtures or 'licks' is determined as much by palatability as by physiological need. Sodium-deficient ruminants will consume sodium sources, such as sodium bicarbonate ($NaHCO_3$), which are normally aversive, and the relish with which livestock consume mineral deposits around receding water sources may have selective nutritional benefit. Pullets approaching lay will self-select diets rich enough in calcium to meet the needs of medullary bone deposition and onset of shell calcification and the laying hen continues to show preference for calcium-rich diets on egg-laying days (Chapter 4), but these are exceptions. Experiments by Cunningham (1949, personal communication) showed that grazing sheep dosed orally with minerals to overcome deficiencies in the pastures ate as much lick as similar undosed animals. Furthermore, observations in many countries have shown that animals will regularly consume considerable quantities of licks supplying minerals in which they are not deficient and yet they will not always voluntarily consume minerals in which they are deficient.

References

Arthur, J.R. (1992) Selenium metabolism and function. *Proceedings of the Nutrition Society of Australia* 17, 91–98.

Bowman, J.G.P. and Sowell, B.F. (1997) Delivery method and supplement consumption by grazing ruminants: a review. *Journal of Animal Science* 75, 543–550.

Chesters, J.K. and Arthur, J.R. (1988) Early biochemical defects caused by dietary trace element deficiencies. *Nutrition Research Reviews* 1, 39–56.

Clark, R.G., Wright, D.F. and Millar, K.R. (1985) A proposed new approach and protocol to defining mineral deficiencies using reference curves: cobalt deficiency in young sheep used as a model. *New Zealand Veterinary Journal* 33, 1–5.

Combs, D.K. (1987) Hair analysis as an indicator of the mineral status of livestock. *Journal of Animal Science* 65, 1753–1758.

Ducker, M.J., Kendall, P.T., Hemingway, R.G. and McClelland, T.H. (1981) An evaluation of feed blocks as a means of providing supplementary nutrients to ewes grazing upland/hill pastures. *Animal Production* 33, 51–58.

Fordyce, F.M., Masara, D. and Appleton, J.D. (1996) Stream sediment, soil and forage chemistry as predictors of cattle mineral status in northeast Zimbabwe. In: Appleton, J.D., Fuge, R. and McCall, G.H.J. (eds) *Environmental Geochemistry and Health.* Geological Society Special Publication No. 113, London, pp. 23–37.

Greger, J.L. and Sickles, V.S. (1979) Saliva zinc levels: potential indicators of zinc status. *American Journal of Clinical Nutrition* 32, 1859–1866.

Hay, L. and Suttle, N.F. (1993) Urolithiasis. In: Aitken, I.D. and Martin, W.B. (eds) *Diseases of Sheep,* 2nd edn. Blackwell Scientific, London, pp. 250–253.

Henkin, R.I., Mueller, C.W. and Wolf, R.O. (1975) Estimation of zinc concentration of parotid saliva by flameless atomic absorption spectrophotometry in normal subjects and in patients with idiopathic hypogeusia. *Journal of Laboratory and Clinical Medicine* 86, 175–180.

Johnson, P.J., Freeland, J. and Ebangit, M.L. (1978) Saliva: a diagnostic tool for evaluating zinc status in man. *Federation Proceedings* 37, 253.

Jumba, I.O., Suttle, N.F., Hunter, E.A. and Wandiga, S.O. (1995) Effects of soil origin and mineral composition on the mineral composition of forages in the Mount Elgon region of Kenya. 2. Trace elements. *Tropical Grasslands* 29, 47–52.

Kumar, S. and Singh, S.P. (1981) Muzzle secretion electrolytes as a possible indicator of sodium status in buffalo (*Bubalus bubalis*) calves: effects of sodium depletion and aldosterone administration. *Australian Journal of Biological Sciences* 34, 561–568.

Langlands, J.P. (1987) Assessing the nutrient status of herbivores. In: Hacker, J.B. and Ternouth, J.H. (eds) *Proceedings of the Second International Symposium on the Nutrition of Herbivores*. Academic Press, Sydney, pp. 363–385.

McClaren, R.G., Sarift, R.S. and Quin, B.F. (1984) EDTA-extractable copper, zinc and manganese in the soils of the Canterbury Plains. *New Zealand Journal of Agricultural Research* 27, 207–217.

McDowell, L.R., Conrad, J.H. and Hembry, F.G. (1993) *Minerals for Grazing Ruminants in Tropical Regions*, 2nd edn. Animal Science Department, University of Florida, Gainsville, USA.

Millar, K.R. and Meads, W.J. (1988) Selenium levels in the blood, liver, kidney and muscle of sheep after the administration of iron/selenium pellets or soluble glass boluses. *New Zealand Veterinary Journal* 36, 8–10.

Millar, K.R., Meads, W.J., Albyt, A.T., Scahill, B.G. and Sheppard, A.D. (1988) The retention and efficacy of soluble glass boluses for providing selenium, cobalt and copper to sheep. *New Zealand Veterinary Journal* 36, 11–14.

Mills, C.F. (1987) Biochemical and physiological indicators of mineral status in animals: copper, cobalt and zinc. *Journal of Animal Science* 65, 1702–1711.

Mitchell, R.L. (1957) The trace element content of plants. *Research, UK* 10, 357–362.

Morris, J.G. (1980) Assessment of the sodium requirements of grazing beef cattle: a review. *Journal of Animal Science* 50, 145–152.

Morris, J.G., Dalmas, R.E. and Hall, J.L. (1980) Salt (sodium) supplementation of range beef cows in California. *Journal of Animal Science* 51, 722–731.

Pamp, D.E., Goodrich, R.D. and Meiske, J.C. (1976) A review of the practice of feeding minerals free choice. *World Review of Animal Production* 12, 13–17.

Phillippo, M. (1983) The role of dose–response trials in predicting trace element disorders. In: Suttle, N.F., Gunn, R.G., Allen, W.M., Linklater, K.A. and Wiener, G. (eds) *Trace Elements in Animal Production and Veterinary Practice*. British Society of Animal Production Special Publication No. 7, Edinburgh, pp. 51–60.

Plant, J.A. and Stevenson, A.G. (1985) Regional geochemistry and its role in epidemiological studies. In: Mills, C.F., Bremner, I. and Chesters, J.K. (eds) *Proceedings of the 5th International Symposium on Trace Elements in Man and Animals*. Commonwealth Agricultural Bureaux, Farnham Royal, UK, pp. 900–906.

Rocks, R.L., Wheeler, J.L. and Hedges, D.A. (1982) Labelled waters of crystallization in gypsum to measure the intake of loose and compressed mineral supplements. *Australian Journal of Experimental Agriculture and Animal Husbandry* 22, 35–42.

Silanpää, M. (1982) *Micronutrients and the Nutrient Status of Soils: a Global Study*. Food and Agricultural Organization, Rome.

Suttle, N.F. (1983) Meeting the mineral requirements of sheep. In: Haresign, E. (ed.) *Sheep Production*. Butterworths, London, pp. 167–183.

Suttle, N.F. (1994) Meeting the copper requirements of ruminants. In: Garnsworthy, P.C. and Cole, D.J.A. (eds) *Recent Advances in Animal Nutrition – 1994*. Nottingham University Press, Nottingham, pp. 173–187.

White, C.L. (1996) Understanding the mineral requirements of sheep. In: Masters,

D.G. and White, C.L. (eds) *Detection and Treatment of Mineral Nutrition Problems in Grazing Sheep.* ACIAR Monograph 37, Canberra, pp. 15–30.

White, C.L., Martin, G.B., Hynd, P.T. and Chapman, R.E. (1994) The effect of zinc deficiency on wool growth and skin and wool histology of male Merino lambs. *British Journal of Nutrition* 71, 425–435.

Calcium

4

Introduction

The discovery during the 18th century that bone consisted primarily of calcium phosphate led to the use of calcium in the prevention of rickets, a childhood disorder of bone development which had plagued humans for centuries. Similar disorders in farm livestock were quickly linked to calcium deficiency, induced by feeding pigs and laying hens diets low in calcium and prevented in calves by feeding calcium-rich diets. As animal production intensified, energy-rich, grain-based diets were increasingly fed to livestock in protected environments and the incidence of bone disorders multiplied. Unwittingly, livestock were being fed diets naturally deficient in calcium while simultaneously being 'starved' of vitamin D_3, essential to the efficient utilization of calcium, by cutting them off from sunlight. Animal breeders ensured that interest in calcium nutrition was maintained by selecting for traits which had high requirements for calcium – growth, milk yield, litter size and egg production. By encouraging reproduction while skeletal growth is still incomplete, producers have ensured that bone disorders in poultry remain commonplace. In the high-yielding dairy cow, an acute calcium deficiency still strikes many animals at calving with no sign of bone disorders and controversy still surrounds the optimum level and pattern of calcium provision for averting such problems.

Natural Sources of Calcium

Forages

Forages are generally satisfactory sources of calcium (Ca) for grazing livestock, particularly when they contain leguminous species. Minson (1990) gives the average published values as 14.2 and 10.1 g Ca kg^{-1} dry matter (DM) for temperate and tropical legumes and 3.7 and 3.8 g Ca kg^{-1} DM for the corresponding grasses. The average temperate grass sward will meet the

© CAB *International* 1999. *Mineral Nutrition of Livestock*
(E.J. Underwood and N.F. Suttle)

calcium requirements of sheep, but a contribution from legumes is needed for the dairy cow (see Tables 4.6 and 4.7). Cultivar differences in calcium content can be marked, but it is the maturity of the sward which has the more widespread influences. The leaf generally contains twice as much calcium as the stem, and pasture calcium concentrations are therefore increased by applying nitrogenous fertilizer and decrease with advancing maturity. Slowing of pasture growth by flooding or a seasonal decline in soil temperature increases herbage calcium levels. Selective grazing is likely to result in higher calcium concentrations in ingested forage than in hand-plucked samples. In the UK, conservation as silage will usually result in higher calcium concentrations than conservation as the more mature hay (6.4 vs. 5.6 g kg^{-1} DM (MAFF, 1990)). The application of lime or limestone to soils to correct soil acidity has surprisingly little effect on forage calcium concentrations, possibly because of increases in herbage yield. As far as other forage species and roughage sources are concerned, maize silage commonly contains 2.0–5.0 g Ca kg^{-1} DM (see Table 2.1) and is usually poorer in calcium than herbage, while cereal straws contain around 3 g Ca kg^{-1} DM.

Absorbability of forage calcium

With minerals such as calcium, which are absorbed according to need, the full potential of a feed as provider of absorbable calcium can only be tested under conditions were requirements (R) are barely met by intake (I). AFRC (1991) plotted all recorded absorption coefficients for calcium (A_{Ca}) in sheep against R/I, found no evidence that they could not raise A_{Ca} when necessary from any feed and suggested an average absorbability of 0.68 for all feeds, including forages (Fig. 4.1). However, much lower A_{Ca} have been reported for green-feed oats and ryegrass–white clover (0.24–0.47) given to stags with high requirements for antler growth (Muir *et al.*, 1987), for herbages (0.17–0.19) given to lactating ewes (Chrisp *et al.*, 1989a) and for lucerne hay (0.26) given to high-yielding dairy cows (Martz *et al.*, 1990). Higher values were reported for lucerne–maize silage (0.49; Martz *et al.*, 1990) and for lucerne (*Medicago sativa*) given to non-lactating goats (0.50; Freeden, 1989), which may have had the capacity to absorb even more. Thompson *et al.* (1988) estimated A_{Ca} of 0.41–0.64 for different grass species given to lambs and considered that the higher values, obtained with the most nutritious species, were closest to the maximum potential of forages. Why, then, should stags, lactating ewes and milking cows not suceed in absorbing more of their dietary calcium when their needs were maximal? The low A_{Ca} may be an indirect consequence of obligatory resorption of bone matrix during seasonal peaks in requirements, which maintains plasma calcium concentrations and lessens the perceived need for dietary calcium (Sykes and Geenty, 1986). It is noteworthy that the stags were not hypocalcaemic (Muir *et al.*, 1987), while A_{Ca} for the lactating ewes increased to 0.30 when a protein supplement was given and milk yield (i.e. need) increased (Chrisp *et al.*, 1989a), suggesting that maximal A_{Ca} had not been reached. A low A_{Ca} in some lucerne crops may be due to the presence of unavailable calcium oxalates. There is some

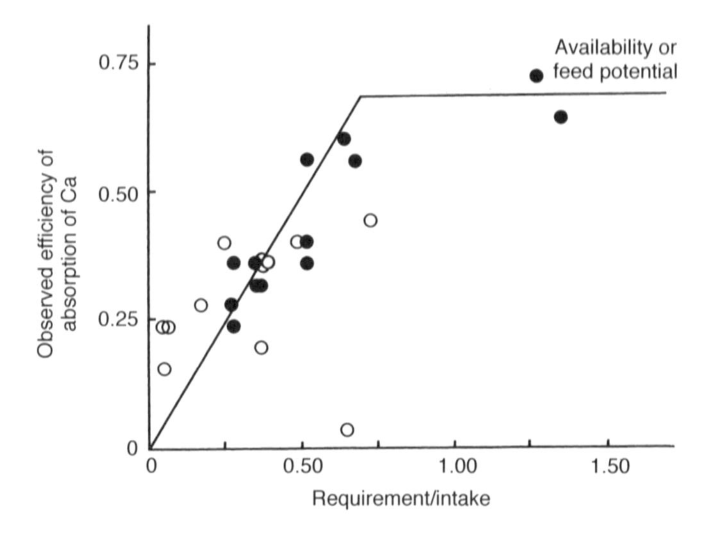

Fig. 4.1. Calcium is absorbed according to need: thus pooled data for the efficiency of calcium absorption in lambs plotted against adequacy of dietary supply (net Ca requirement: Ca intake) shows a curvilinear rise to a plateau representing the full potential of the dietary Ca source (from AFRC, 1991): •, nutritionally balanced diets; ○, nutritionally imbalanced diets.

metabolism of oxalate by rumen microbes, particularly when there has been time to adapt to high-oxalate diets, but the appearance of oxalate crystals in cattle faeces indicates that it is not always complete (Ward *et al.*, 1979). Oxalates may have greater nutritional significance in grass species, such as kikuyu (for review, see Barry and Blaney, 1987), which contain relatively little calcium (Jumba *et al.*, 1995), and in forage trees, such as the mulga (*Acacia aneura*). Horses are particularly vulnerable to the calcium-depleting effects of oxalate-rich roughages.

Concentrates

Most concentrates are low in calcium. Cereals rarely contain >1 g Ca kg^{-1} DM and cereal by-products rarely >1.5 g kg^{-1} DM. Maize typically contains only 0.2 g Ca, wheat 0.6 g and oats and barley 0.6–0.9 g Ca kg^{-1} DM (see Table 2.1). Supplementation with vegetable protein sources, with the exception of rapeseed meal (8.4 g Ca kg^{-1} DM), will not give adequate calcium levels in mixed rations because these usually contain no more than 2–4 g Ca kg^{-1} DM (MAFF, 1990). Sugar-beet pulp is an excellent supplement for grain-based diets, being rich in calcium (6.0–7.5 g Ca kg^{-1} DM) but low in phosphorus. Fish-meal and meat- and bone-meal are also good sources of calcium, levels of 50–100 g Ca kg^{-1} DM being commonplace.

Absorbability of calcium from concentrates

Cereals and vegetable protein sources are important determinants of A_{Ca} to non-ruminants, because the phytate which they contain forms unabsorbable complexes with calcium in feeds, as well as with any added inorganic calcium. Thus animals with high calcium requirements might be prevented from absorbing according to need. The only data for A_{Ca} in a foodstuff for non-ruminants, published in a review by Soares (1995), indicated a relative value for sesame seeds only 65% of that of calcium carbonate ($CaCO_3$) when given to rats. The review abounds with relative values for inorganic sources when added to various diets but gives no indication of the quantitative influence of dietary phytate on calcium availability. Hill (1984) suggested that, as a rule of thumb, 1.3 g Ca should be added per gram of phytate phosphorus (P) present in excess of 2 g kg^{-1} in rations for poultry: the calculation was based on the formula of calcium phytate. Improvements in apparent A_{Ca} have been reported when phytase was added to a maize–soybean meal diet low in phosphorus (4.3 g kg^{-1} DM) for young pigs (Pallauf *et al.*, 1992); further examples are given in Chapter 5. Correction of Pallauf *et al.*'s data for endogenous faecal calcium (FE_{Ca}) at 32 mg kg^{-1} live weight (LW) (ARC, 1981) gives an A_{Ca} coefficient of 0.58 for the basal diet and similar calculations for data from Han *et al.* (1997) give a coefficient of 0.72. With other papers indicating even higher A_{Ca} values for cereal/vegetable protein diets, with inorganic sources providing most of the calcium (e.g. Kornegay and Qian, 1996), an average absorbability of 0.70 will be used in later calculations of requirement. As far as absorbability of calcium in concentrates to ruminants is concerned, phytate is not a threat, because it is degraded in the rumen. Low absorption, associated with negative calcium balances, has been reported in normocalcaemic lactating ewes given energy-rich diets (Braithwaite, 1983b), but, as with forages, this is unlikely to represent a limitation imposed by the feed (AFRC, 1991).

Metabolism

The priority of all mammals is to maintain calcium concentrations in plasma and extracellular fluids (ECF) close to 2.5 mmol (100 mg) l^{-1} in the face of large fluctuations in demand and lesser fluctuations in supply (Hurwitz, 1996). Such constraints are necessarily relaxed in egg-laying avian species, and the processes for achieving the contrasting objectives of the dairy cow and the laying hen were reviewed by Horst (1986) and Gilbert (1983), respectively.

Absorption

Homeostasis is achieved partly by the hormonal regulation of absorption. Calcium is absorbed according to need up to the limits set by the absorbability of the mineral in the diet (Schneider *et al.*, 1985; Bronner, 1987); this is close to 90% for milk and probably rarely $< 50\%$ of the total calcium supply from solid diets, although few studies have been conducted with farm animals given marginal supplies of calcium to prove the point (AFRC, 1991). Small amounts of calcium may be absorbed from the rumen (Yano *et al.*, 1991), but the major absorptive site is the small intestine. Control of absorption is achieved by two hormones, parathyroid hormone (PTH) and the physiologically active form of vitamin D_3, 1,25-dihydroxycholecalciferol (calcitriol, 1,25-$(OH)_2D_3$). The parathyroid gland is acutely sensitive to small deviations in the ionic calcium concentration in ECF and, when concentrations fall, PTH is normally secreted (Brown, 1991) and activates vitamin D_3 (see Omdahl and DeLuca, 1973; Borle, 1974). Vitamin D_3 is hydroxylated to 25-hydroxy-D_3 (25-OHD_3) in the liver and further in the kidney to two compounds, 24,25-$(OH)_2D_3$ or 1,25-$(OH)_2D_3$. While the primary site of 1,25-$(OH)_2D_3$ synthesis is the kidney, it is now clear that other tissues, such as the bone marrow and skin, can synthesize 1,25-$(OH)_2D_3$ for local calcium needs, i.e. it is a paracrine rather than an endocrine activity (Norman and Hurwitz, 1993). In the intestinal mucosa, 1,25-$(OH)_2D_3$ acts relatively sluggishly, opening up calcium channels and facilitating calcium uptake and transfer, with the help of the calcium-binding protein, calbindin (Hurwitz, 1996). The role of calbindin in facilitating absorption of calcium according to supply and demand is illustrated by data for fast- and slow-growing chicks given diets varying in calcium (Fig. 4.2; Hurwitz *et al.*, 1995). Provision of vitamin D in a

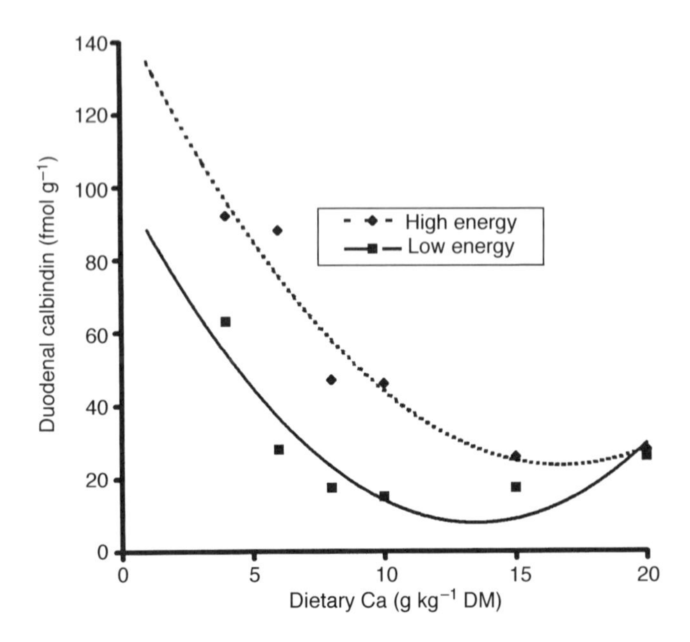

Fig. 4.2. Mechanism for absorbing calcium according to need: as dietary Ca concentrations decrease, synthesis of a potentiator of absorption (calbindin) increases; if high energy diets are fed, demand for Ca rises and the broiler chick synthesizes yet more duodenal calbindin (from Hurwitz *et al.*, 1995).

hydroxylated form overrides the normal absorptive control mechanisms, enhancing the efficiency of absorption from diets providing plentiful calcium (Braithwaite, 1983a). When the supply of calcium is excessive, the homeostatic mechanisms are reversed by the secretion of calcitonin (Beckman *et al.*, 1994). The vast quantities of calcium required by the laying hen are mostly absorbed by passive mechanisms (Gilbert, 1983).

Bone resorption

Equally important to the regulation of circulating concentrations of ionic calcium is the net flow of calcium from the enormous reserve in the skeleton. Bain and Watkins (1993) describe the situation thus: 'The skeletal morphology of the adult animal represents an elegant compromise between structural obligation and metabolic responsibility, serving the organism in support and locomotion while actively participating in the regulation of calcium homeostasis'; this view is equally true of the growing animal. The hormonal partnership which facilitates absorption also facilitates calcium mobilization from bone, but the mechanisms are more complex and involve nuclear receptors for 1,25-$(OH)_2D_3$ on cartilage- and bone-forming cells, the chrondrocytes and osteoblasts. Bone-resorbing cells (osteoclasts) respond indirectly to 1,25-$(OH)_2D_3$ via cytokines released by the osteoblasts (Norman and Hurwitz, 1993). The balance between calcium accretion and resorption can be set to mobilize around one-fifth of calcium from the skeleton in late pregnancy and lactation in species as diverse as the rat (Brommage, 1989), sheep (Braithwaite, 1983b) and dairy cow (Ramberg *et al.*, 1984); this process may be obligatory, since it is not abated by supplying more dietary calcium (AFRC, 1991) and not accompanied by hypocalcaemia (Braithwaite, 1983b). Protein depletion during pregnancy can also lead to calcium resorption, which is not preventable by supplying more calcium in the diet (Sykes and Field, 1972). Thus rapidly growing offspring can be reared successfully by dams on diets which are deficient in calcium or protein. Bone resorption matters little if the bones are fully calcified at the outset, if depletion is restricted to some 10–20% of the total bone minerals and if repletion takes place before the next period of intensive demand. Capacity for resorption decreases with age (Ramberg *et al.*, 1976).

Excretion

The modulation of excretion by faecal and urinary routes generally plays little part in calcium homeostasis and in determining risk of calcium dysfunction. The FE_{Ca} is unaffected by marked reductions in dietary calcium supply and calcium status, remaining consistently low, at about 16 mg Ca kg^{-1} LW in the ruminant (ARC, 1980). Increases in dry-matter intake from dry diets gives rise to proportional rises in FE_{Ca} in sheep (AFRC, 1991) and a similar relationship applies to grass diets (Chrisp *et al.*, 1989a, b). High FE_{Ca} values (up to 50 mg Ca kg^{-1} LW) have been reported when animals with low requirements (6.5-month-old lambs) were given a frozen, grass–clover diet rich in calcium (12.1 g Ca kg^{-1} DM; Chrisp *et al.*, 1989b). Urinary calcium excretion also tends to remain low and constant, regardless of calcium status, although it may rise significantly in the alkalotic dairy cow and in the laying hen on days when no shell is formed (Gilbert, 1983).

Functions of Calcium

Calcium is the most abundant mineral in the body and 99% is found in the skeleton. A basic function of calcium is therefore to provide a strong framework for supporting and protecting delicate organs, jointed to allow movement and malleable to allow growth.

Bone growth and mineralization

Bones grow in length by the proliferation of cartilaginous plates at the ends of bones (Vaughan, 1970). Cells towards the end of the regular columns of chondrocytes that constitute the growth or epiphyseal plate become progressively hypertrophic and degenerative. They concentrate calcium and phosphorus at their peripheries and exfoliate vesicles rich in amorphous calcium phosphate ($Ca_3(PO_4)_2$) (Wuthier, 1993). Thus a calcium-rich milieu is provided to impregnate the osteoid (organic matrix) laid down by osteoblasts. The crystalline bone mineral hydroxyapatite, $Ca_{10}(PO_4)_6(OH)_2$, accumulates in a zone of provisional calcification around the decaying chondrocytes, replacing them with an apparently disorganized and largely inorganic matrix of trabecular bone. Bones grow in width by inward deposition of concentric shells or lamellae of osteoid beneath the surface of the bone shaft or periosteum. Mammalian young are born with poorly mineralized bones and, while they suckle, they do not receive enough calcium to fully mineralize the bone growth that energy-rich milk can sustain (AFRC, 1991). After weaning, there is normally a progressive increase in bone mineralization (for data on lambs, see Field *et al.*, 1975; Spence *et al.*, 1985: for data on pigs, see Chapter 5), stimulated by increased load-bearing, thus providing added strength and reserves of both calcium and phosphorus. On average, bone contains 36% calcium and 17% phosphorus but there are wide differences between different bones (see Chapter 5). Activity increases bone growth in broilers, which also develop increasingly mineralized bones as they age (Bond *et al.*, 1991); providing a perch increases bone mineralization in hens. Avian species show unique changes in bone morphology with the onset of sexual maturity. Under the influence of oestrogen, the formation of cancellous or structural bone ceases. Instead, new medullary bone is laid down or woven on the surface of existing cortical bone and in the bone marrow to provide a labile calcium reserve to maintain plasma calcium concentrations during shell formation (Gilbert, 1983; Whitehead, 1995). Formation of medullary bone may also occur during the preparation for lactation in the dairy cow (B. Thorp, personal communication).

Non-skeletal functions

The small fraction of body calcium that lies outside the skeleton (1%) is also important to survival. Extraskeletal calcium occurs as the free ion, bound to serum proteins and complexed to organic and inorganic acids. It is the ionized calcium – 50–60% of total plasma calcium – which is the essential element for such physiological functions as nerve conduction and muscle contraction and relaxation, including that of heart muscle. Calcium can activate or stabilize some enzymes, contributes to regulation of the cell cycle and is required for normal blood clotting (Hurwitz, 1996); calcium must be present for prothrombin to form thrombin, which reacts with fibrinogen to form the blood clot, fibrin. Changes in calcium concentrations within cells, modulated by vitamin D_3 and the calcium-binding proteins, calmodulin and osteopontin, may be an important step in cell signalling (Carafoli, 1991), including the triggering of immune responses (Nonnecke *et al.*, 1993). In poultry, calcium performs the unique function of protecting the egg through the deposition of an eggshell during passage down the oviduct. The shell matrix is heavily impregnated with $CaCO_3$ and the need to furnish about 2 g calcium for every egg produced dominates calcium metabolism in the laying hen (Gilbert, 1983; Bar and Hurwitz, 1984).

Biochemical Indices of Disordered Calcium Metabolism

Calcium-responsive disorders can arise in one of two ways, either as a result of an acute increase in demand ('metabolic deprivation') or as a result of chronic dietary deprivation. The biochemical changes which arise in these two contrasting situations are illustrated in Fig. 4.3.

Acute deprivation in dairy cows

When a cow calves, she can lose 23 g of calcium in 10 kg of colostrum within 24 h and yet there is only 3 g of calcium in her entire bloodstream and she is unlikely to have eaten for several hours. All older cows, therefore, show a sharp but limited reduction in plasma calcium at parturition, which may recur at roughly 9-day intervals as they readjust to the demands of lactation (Hove, 1986). The changes which occur in association with parturition and with milk-fever in the dairy cow involve an acute hypocalcaemia and are illustrated in Fig. 4.4; those seen concurrently in circulating levels of PTH and 1,25-$(OH)_2D_3$ are more pronounced than those in calcium. There is clearly a failure of homeostatic mechanisms and it consists of a failure to produce or utilize 1,25-$(OH)_2D_3$ rather than PTH, due to a lack of target organ sensitivity to PTH, which shows the normal characteristic response to hypocalcaemia (Goff *et al.*, 1989b). The higher susceptibility of older cows may be due to a reduction in 1,25-$(OH)_2D_3$ receptors in intestine and bone (Horst *et al.*, 1990). Serum inorganic phosphorus (P_i) also declines in milk-fever, to about one-third normal (0.5 mmol l^{-1}), and serum magnesium may also be subnormal (< 0.8–1.2 mmol (20–30 mg) l^{-1}; Fenwick, 1988). Mobilization of collagenous

(a) Chronic dietary deprivation

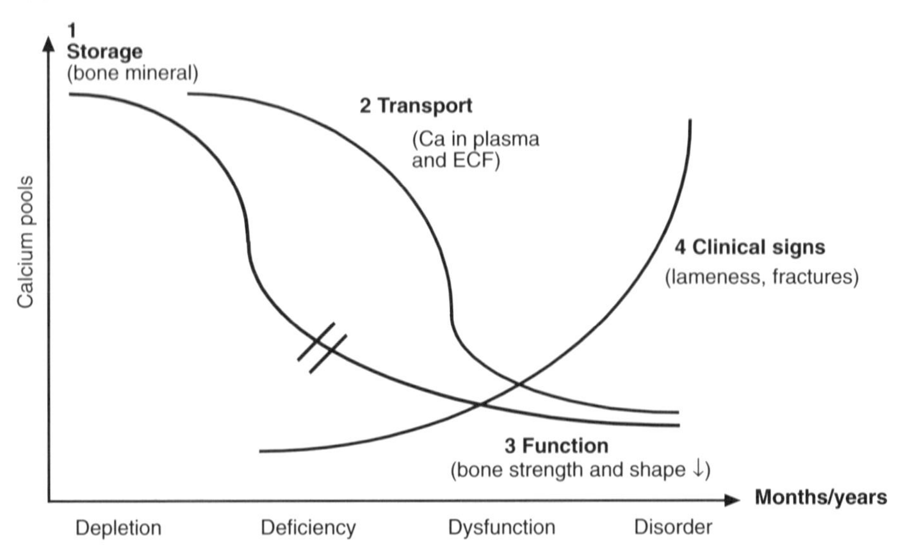

(b) Acute metabolic deprivation

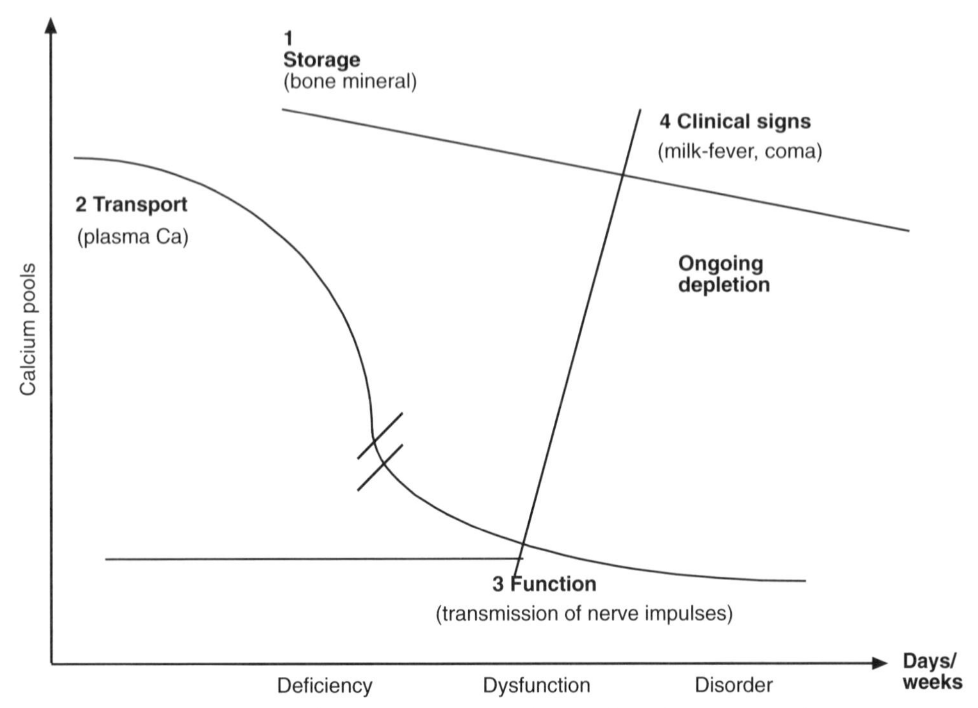

Fig. 4.3. Sequences of biochemical changes leading to clinical signs in (a) chronic dietary deprivation (i.e. skeletal disorders) and (b) acute metabolic deprivation of calcium (e.g. milk-fever).

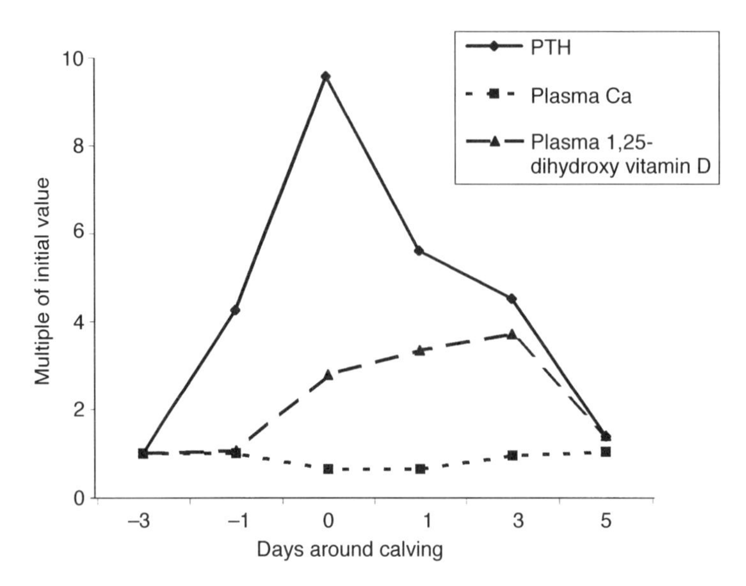

Fig. 4.4. Precipitous increases in the demand for calcium at calving prompt an early rise in plasma parathyroid hormone (PTH) concentrations and a slower rise in dihydroxylated vitamin D in an attempt to curtail life-threatening reductions in plasma Ca (values are given relative to those 3 days before calving which were approximately 60 pg ml^{-1} for PTH, 30 pg ml^{-1} for 1,25-$(OH)_2D$ and 2.4 mmol l^{-1} for Ca (data for non-relapsing cows; Goff *et al.*, 1989b).

bone matrix causes a periparturient rise in hydroxyproline in plasma and urine, collagen containing about half the hydroxyproline in the body (Black and Capen, 1971). The secretion of an additional PTH-like hormone during pregnancy and lactation may give further protection to the fetus or suckled offspring (Goff *et al.*, 1991). In ewes, the corresponding changes occur before lambing, particularly in twin-bearing ewes, because the demand for calcium for the fetus becomes greater than that for lactation (Sansom *et al.*, 1982).

Acute deprivation in laying hens

The entire skeleton of the laying hen only contains enough calcium to supply ten eggs and her daily intake of calcium from unsupplemented rations would be less than 0.03 g Ca day^{-1}. During a normal laying year, up to 30 times the hen's total body calcium is deposited in shell (Gilbert, 1983). Skeletal resorption of calcium cannot make a significant contribution over a 12-month period of lay and is stretched to its limits every day that an egg is prepared for lay. Laying hens normally have higher and more variable serum calcium values (5.0–7.5 mmol (200–300 mg) l^{-1}) than non-laying hens or chickens (2.25–3.0 mmol (90–120 mg) l^{-1}) due to the presence of a Ca-lipophosphoprotein, vitellogenin, a precursor of egg-yolk proteins (Hurwitz, 1996). During the period of shell calcification, serum calcium values decline, particularly if a low-calcium diet is given, but mobilization of bone mineral may prevent the serum calcium from falling further (see Table 4.1).

Blood changes in chronic deprivation

The serum calcium in lactating ewes confined to cereal grain rations declines to one-half or one-third of normal values (2.2–2.9 mmol l^{-1}) within a few weeks, but the rates of fall in lambs, weaners and pregnant ewes are progressively slower, due to decreasing demands for calcium (Franklin *et al.*, 1948; Fig. 4.5). A slow decline in serum calcium, with associated hormonal and metabolic changes, also occurs in response to a dietary deficiency of calcium and depletion of bone reserves in pigs (Table 4.2) and broilers. There is usually a tendency for plasma P_i to rise in chronic calcium deficiency (see Fig. 5.5) and for plasma calcium to rise in phosphorus deficiency (Fig. 4.5), necessitating separate physiological control mechanisms for the two minerals (see Chapter 5).

Table 4.1. Effect of a low-calcium diet, unsupplemented and supplemented with $CaCO_3$, on blood calcium and egg production in hens (from Buckner *et al.*, 1930).

Weeks on Ca-deficient diet	Serum Ca (mg l^{-1})		No. of eggs per hen		Weight of shell (g)		Weight of contents (g)	
	+Ca	−Ca	+Ca	−Ca	+Ca	−Ca	+Ca	−Ca
−2–0	213	209	6.3	6.9	5.0	4.8	52.3	53.2
0–2	248	152	6.6	4.7	5.4	3.6	53.9	53.5
2–4	215	161	9.0	1.5	5.7	3.8	56.7	47.4

Table 4.2. Effects of dietary calcium concentrations on selected performance, skeletal and blood indices of calcium status in the baby pig given a synthetic milk diet from 0 to 6 weeks of age (trial 2: Miller *et al.*, 1962).

Criterion	Dietary Ca (g kg^{-1} DM)				
	4	6	8	10	
Live-weight gain (g day^{-1})	170	210	180	200	
Dry-matter intake (g)	230	240	230	250	Performance
Feed/gain (g g^{-1})	1.33	1.16	1.31	1.27	
Serum inorganic P (mmol l^{-1})	3.29	3.54	3.43	3.52	
Serum Ca (mmol l^{-1})	2.60	2.60	2.72	3.00	Blood
Serum alkaline phosphatase (BL units)	9.90	8.58	8.50	9.75	
Humerus ash (%)[a]	42.2	43.3	46.1	47.1	
Femur s.g.	1.14	1.15	1.16	1.18	Bone
Femur weight (g)[b]	47.9	–	47.2	–	
Eighth rib weight (g)[b]	3.62	–	3.44	–	
Femur breaking load (kg)	57.7	65.9	65.9	84.5	

[a] Dry, fat-free basis.
[b] Trial 1 data.
s.g. = specific gravity.

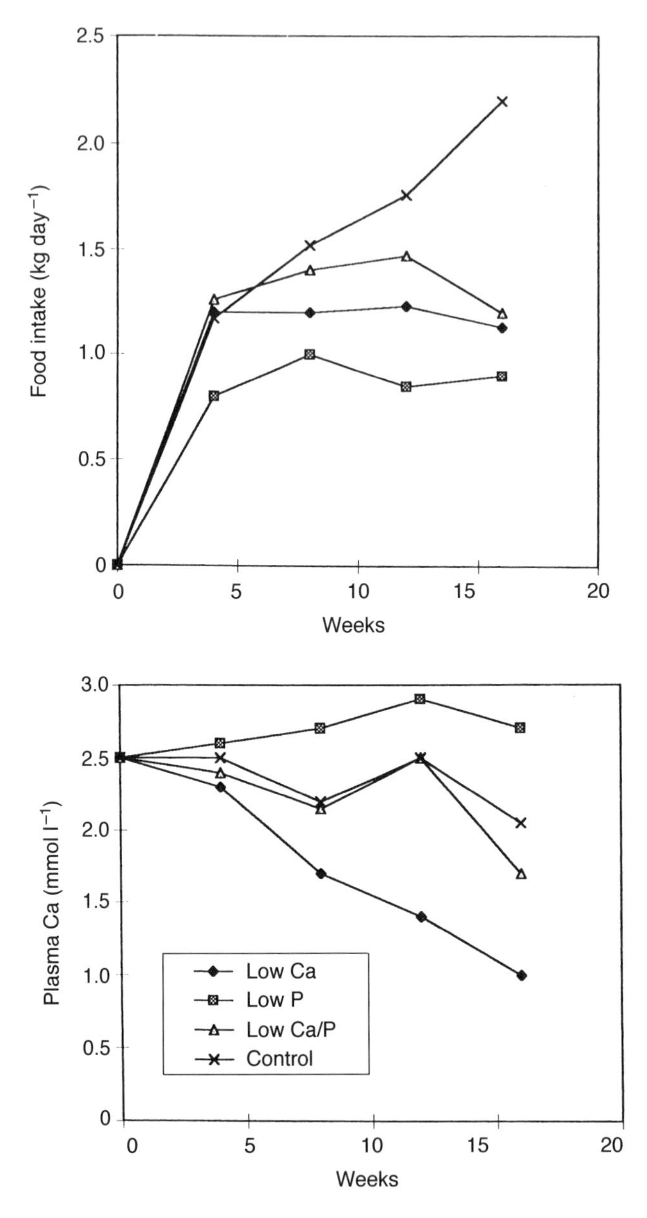

Fig. 4.5. Effects of single or combined deficiencies of calcium and phosphorus on food intake and plasma Ca in lambs: note that Ca deprivation has lesser effects on both parameters when P is also in short supply (from Field *et al.*, 1975).

Bone changes in chronic deprivation

Chronic inadequacies in dietary calcium supply are reflected by chemical, physical, histological and radiological changes, which reflect the reduction in mineralization of bones. The changes seen in growing animals are illustrated by data for baby pigs in Table 4.2 and for lambs in Table 4.3. The changes are, in most respects, like those seen when phosphorus is lacking (compare Table 4.2 with Table 5.1 on p. 116), since mineralization needs the presence of both minerals. The average Ca:P ratio of 2.1:1 can decrease or increase *slightly* in response to deficiencies of calcium or phosphorus, respectively (Table 4.3). Early establishment of the 'elegant compromise' is illustrated by the way in which the skeleton can increase substantially in size and weight in lambs on a calcium-deficient diet with little increase in bone calcium (Table 4.3). Modelling and remodelling can presumably go hand in hand to allow the scarce mineral reserve to be redistributed in a larger volume and permit growth without loss of function. Calcium deficiency does not reduce bone growth to the extent that phosphorus deprivation does (Table 4.3), because the formation of bone matrix is less impaired (also compare Tables 4.2 and 5.1). Poor mineralization of matrix, leading to a widened and weakened growth plate, is, however, more prominent when dietary calcium is lacking than when phosphorus is lacking in the growing animal. Normally, the chemical, histological, radiological and physical measures of bone quality agree well (Table 4.2) but in sexually mature birds, non-physical measures overestimate bone strength, because medullary bone lacks the strength of structural bone (Whitehead, 1995).

Withdrawal or resorption of minerals does not take place equally readily from different parts of the adult skeleton. The spongy bones, ribs, vertebrae, sternum and cancellous ends of the long bones, which are the lowest in ash, are the first to be affected and the compact shafts of the long bones and small bones of the extremities the last. In each case, the essential change is a reduction in the total mineral content of the bones, with little alteration in the proportions of the minerals in the remaining ash. More details are presented in the next chapter, since it is in the context of diagnosing phosphorus deficiency that they are most widely used. Particular mention will be made here of one bone parameter which may uniquely distinguish calcium from

Table 4.3. Effects of feeding diets low in calcium and/or phosphorus for 18 weeks on the growth and mineral content of the lamb skeleton (Field *et al.*, 1975).

	Changes in fat-free skeleton				Total mineral ratios	
Diet	Weight (kg)	Ca (g)	P (g)	Mg (g)	Ca:P	Ca:Mg
Control	+1.44	+197	+67	4.7	2.14	40.5
Low Ca	+1.00	+15	−18	2.4	2.13	25.7
Low P	+0.43	+57	−10	0.4	2.44	49.0
Low Ca/P	+0.92	+35	−14	0.8	2.30	39.4

phosphorus insufficiencies. A marked reduction in the Ca:magnesium (Mg) ratio from 40 to 26 was noted in lambs depleted of calcium, whereas phosphorus depletion tended to raise the ratio in the entire skeleton to 49 (Table 4.3).

Parasitism and bone mineralization

Studies with sheep infected experimentally with the larvae of nematodes which parasitize the gastrointestinal tract have revealed species- and dose-dependent demineralization of the skeleton of a severity that matches anything achieved with calcium- (or phosphorus-) deficient diets (Fig. 4.6; for reviews, see Sykes, 1983, 1987). Since natural infections of these parasites are commonplace in young grazing livestock, they are likely to affect susceptibility to dietary deficiencies of these elements in the field. The demineralization associated with abomasal parasitism (by *Ostertagia circumcincta*) occurs in the presence of normal plasma calcium and phosphorus concentrations and normal absorption. Loss of appetite contributes to demineralization, but comparisons with pair-fed, uninfected controls showed that there was an additional loss, which was probably caused by the mobilization of amino acids from bone matrix to compensate for gut losses of protein. Parasitism of the small intestine has adverse effects on phosphorus metabolism (see Chapter 5) and the resultant hypophosphataemia may be partly responsible for the loss of calcium from the skeleton. The systemic effects of gut parasites on bone demineralization may be triggered by similar factors to those which mobilize bone matrix during early lactation to furnish amino acids for the mammary gland and milk-protein synthesis. Parasites of the liver have far less

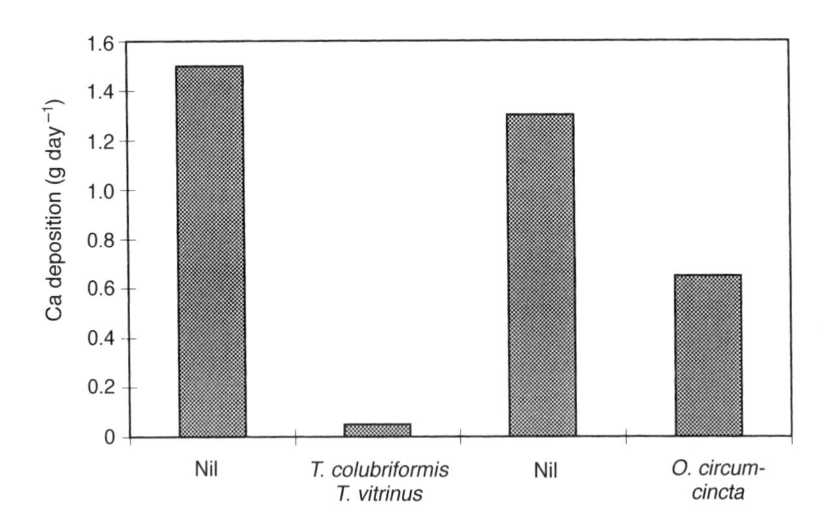

Fig. 4.6. Deposition of calcium in the skeleton of lambs is greatly reduced by parasitic infections of the abomasum (by *Ostertagia circumcincta*) or small intestine (by *Trichostrongylus vitrinus* and *T. colubriformis*); trickle infections of the nematode larvae were used and controls were pair-fed (from Sykes, 1987), Nil, non-infected.

effect on bone mineralization (Sykes, 1983). It would be surprising if debilitating infections of the gut by other species (e.g. coccidia) did not also cause demineralization of the skeleton, in which case pigs and poultry might also be adversely affected by parasite × macromineral interactions.

Clinical Manifestations

Poor growth and survival

Feeding marginally inadequate calcium levels (5.1 vs. 7.8 g Ca kg^{-1} DM) during the rearing of chicks for egg laying increased mortality from 1.9 to 10.6% and reduced body weight at 86 days from 1013 to 962 g. It is noteworthy that characteristic calcium deficiency (i.e. skeletal) abnormalities were not seen, most deaths being attributable to omphalitis (Hamilton and Cipera, 1981). The first abnormality noted when lambs were weaned on to a diet very low in calcium (0.68 g Ca kg^{-1} DM) (Field *et al.*, 1975) was a loss of appetite after 4 weeks and a subsequent proportional retardation of growth (Fig. 4.5).

Abnormalities of bones

Prolonged calcium deprivation in domestic livestock will eventually cause the following: lameness; enlarged and painful joints; bending, twisting, deformation or fractures of the pelvis and long bones; arching of the back, with posterior paralysis in pigs (Miller *et al.*, 1962); facial enlargements, involving particularly the submaxillary bones in horses; and malformations of the teeth and jaws, especially in young sheep. Stiff-gaited movements of grazing cattle ('peg-leg') hamper their ability to secure feed and water and can result in death. In prolonged calcium deficiency in growing chicks, as much as 25% of the total bone minerals can be lost, chiefly from the axial skeleton but with sufficient from the long bones to lead to fractures (Common, 1938). Abnormalities of the bones can occur at any age. Rickets is the term used to denote the skeletal changes and deformities of the growing bone of young calcium-deprived animals. *Rickets* is characterized by a uniform widening of the epiphyseal–diaphyseal cartilage and an excessive amount of osteoid or uncalcified tissue, causing enlargement of the ends of the long bones and of the rib costochondral junction. The tibial dyschondroplasia (TD) which affects broilers is histologically distinguishable from rickets by the plug of avascular cartilage which forms beneath the growth plate, but it is inducible by feeding diets low in calcium (Edwards and Veltmann, 1983). *Osteomalacia* is used to describe a condition which can affect the calcium-deprived adult; there is excessive mobilization of minerals, leaving a surfeit of matrix in bones in which the growth plate has 'closed'. Other terms, such as osteofibrosis, have been applied to the bone dystrophy occurring in young horses. The term *osteoporosis* is used to describe a situation in which bones contain less mineral than normal and proportionately less matrix, so that the degree of mineralization of the matrix remains normal. Withdrawal of calcium from the skeletal reserves takes place normally in response to the demand of egg laying and most hens end lay in an osteoporotic state (Whitehead, 1995). Matrix osteoporosis is characteristic of protein deficiency rather than simple calcium or phosphorus deficiency (Sykes and Field, 1972; Sykes and Geenty, 1986).

Teeth

Remarkably little attention has been given to the effects of calcium nutrition on tooth development. The chemical composition and histology of the outer regions of the tooth, i.e. cementum, bear a close similarity to those of bone. However, calcium in the tooth is not a metabolically active compartment. Studies at the Rowett Institute, Aberdeen, in the 1950s, which reported marked effects of calcium (and phosphorus) deprivation on bone mineralization (e.g. Duckworth *et al.*, 1962), made no comment on tooth abnormalities. The rate of cementum deposition is probably reduced by calcium (or phosphorus) deprivation during tooth development, but not necessarily out of proportion with skeletal development. Disorders of the dentition of grazing animals, such as the premature shedding of incisor teeth in sheep (broken mouth), have often been linked with inadequate mineral nutrition, but careful experimentation has failed to confirm a causal link (Spence *et al.*, 1985). Other factors, such as soil ingestion, can cause excessive tooth wear (see Chapter 2).

Milk-fever or parturient paresis

Parturient paresis is a metabolic disease of bovines associated with parturition and the initiation of lactation; older (5–10 years old) dairy cows and small breeds, such as Jersey, are particularly prone. Within 48 h of calving (usually for the third time or more), the cow becomes listless and shows muscular weakness, circulatory failure, muscular twitching, anorexia and rumen stasis; the condition develops through a second stage of drowsiness, staring, dry eyes with dilated pupils and sternal recumbency to a final stage of lateral recumbency and loss of conciousness. The basic problem is an acute one of unequal calcium supply and demand. Incidence of milk-fever is increased on diets rich in potassium (see Chapter 8). In affected cows, serum calcium falls rapidly to an average of 1.0–1.25 mmol (4–5 mg) l^{-1} or even lower. Severe hypocalcaemia is associated with decreased blood flow to peripheral tissues (Barzanji and Daniel, 1987), hypoxia (Barzanji and Daniel, 1988), hypothermia (Fenwick, 1994), hyponatraemia (Fenwick, 1988), hypomotility of the rumen and loss of appetite (Huber *et al.*, 1981); these secondary consequences may underlie the failure of some cows to respond to calcium alone. Serum magnesium is also often low and convulsions or tetany, as described in Chapter 6, may then accompany the usual signs of milk-fever. Lactating females of all species, including the nanny, sow and mare, occasionally develop milk-fever.

Hypocalcaemia in ewes

The term 'hypocalcaemia' is a biochemical rather than a diagnostic criterion, but it is used to describe a clinical condition of ewes which develops with the approach of lambing. It is characterized by restlessness, apparent blindness and, in the worst cases (55%: Mosdol and Wange, 1981), recumbency, tetany and death. Symptomatically and, to some extent, aetiologically, 'hypocalcaemia' resembles toxaemia of pregnancy (a consequence of energy deficiency), being most likely to occur in older, twin-bearing ewes exposed to a change in or shortage of feed and/or stressed by transportation (Hughes and Kershaw, 1958) or adverse weather. The two conditions, where they are separate entities, can be distinguished biochemically by the presence of low plasma calcium or raised β-hydroxybutyrate concentrations. Since loss of appetite is a feature of both conditions, each is likely to develop as a consequence of the other, because calcium and energy supplies are simultaneously decreased. To complicate matters further, hypomagnesaemia often accompanies hypocalcaemia and may explain the wide range of clinical signs associated with hypocalcaemia, which includes convulsions. The disease rarely affects ewes in their first pregnancy (Mosdol and Wange, 1981), and it is physiologically similar to milk-fever in cattle in that it is caused by disequilibrium in the supply of and demand for calcium. The onset of disorder is pre- rather than postparturient, because the peak calcium requirement of the twin-bearing ewe occurs before lambing and the onset of lactation. There is no clear association with low dietary calcium concentrations, indicating once again the equal importance of mobilizing calcium from the skeleton. Suggestions that excess dietary phosphorus predisposes to hypocalcaemia (Jonson *et al.*, 1971, 1973) might be explained by retarded bone resorption. There is disagreement on the role of excess magnesium (Pickard *et al.*, 1988), but excess potassium may predispose to the disease (see Chapter 8). Vaccination with cortisol-inducing vaccines in late pregnancy may be a risk factor and should be as early as possible before lambing (Suttle and Wadsworth, 1991). Hypocalcaemia may also occur at the end of a drought, when a sudden luxuriant growth of grass low in calcium becomes available following a period of feeding grain low in calcium (Larsen *et al.*, 1986).

Transport tetany

Clinical disease, in the form of paresis associated with hypocalcaemia, is a risk factor associated with the transfer of weaned ruminants to feedlots. Lucas *et al.* (1982) reported such an incident involving lambs and the transportation of horses can give rise to a similar condition. The problem is probably caused by a combination of factors, including lack of food and therefore dietary calcium, lack of mobilization from a poorly mineralized skeleton and increased soft-tissue uptake of calcium.

Depression of milk yield

Where cows have been fed on roughage and grains low in calcium for long periods, the skeletal reserves can be depleted to the point where milk yield is impaired throughout lactation (Fig. 4.7). An almost complete failure of milk production has been observed in sows fed on a calcium-deficient diet during the previous pregnancy. These drops in milk yield may result from the loss of appetite which can occur after severe calcium depletion (Fig. 4.5).

Reduction in egg yield and quality

The most vulnerable class of livestock are laying hens, which are often unable to satisfy their high calcium requirements by demineralizing bone. Egg yield, egg weight, hatchability and eggshell thickness are consequently reduced, shell strength being the most sensitive index of dysfunction (Hamilton and Cipera, 1981). In the early experiments of Deobald *et al.* (1936), egg production had virtually ceased by the 12th day after removal of supplementary calcium and the ash content of eggshells from some of the birds was less than 25% of normal. Gilbert (1983) found that egg production (%) increased in direct proportion to dietary calcium over the range 1.9–30.9 g Ca kg^{-1} DM, rising from 10 to 93%.

Occurrence of Calcium-responsive Disorders

Calcium responses are predominantly associated with the feeding of grain-based diets and are inevitable in pigs and poultry fed largely on low-calcium cereals and in high-yielding dairy cows given concentrate diets, unless they are appropriately supplemented.

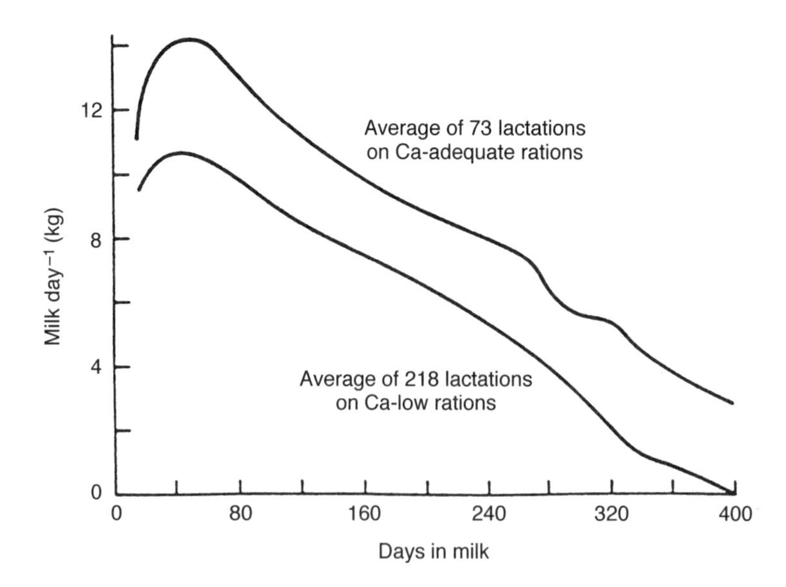

Fig. 4.7. Average daily milk yield of Jersey cows before and during the use of bone-meal as a supplement to low-Ca rations (from Becker *et al.*, 1953).

Poultry

Shell quality cannot be maintained for 1 day without calcium supplementation, and poor shell quality is a major source of loss to the poultry industry (Roland, 1986). The transitional calcium nutrition around the onset of lay is also crucial. If diets of inadequate calcium concentration are fed, the pullet overconsumes in an effort to meet her burgeoning calcium requirement, increasing the risk of fatty liver haemorrhagic syndrome (Roland, 1986). Inadequate calcium nutrition during lay will contribute to 'cage layer fatigue', but this osteoporotic condition also occurs when calcium nutrition is adequate and it is not a specific consequence of calcium dysfunction (Whitehead, 1995). Similarly, the leg weakness or TD which commonly affects young broilers and older turkey poults can be induced by diets low in calcium (7.5 g Ca kg^{-1} DM) and yet occur when supplies are apparently adequate (Table 4.4; Whitehead, 1995). Control of TD can be obtained by dietary supplementation with 1,25-$(OH)_2D_3$ (Table 4.4) but not with the unhydroxylated D_3 (i.e. normal dietary sources). It has been suggested that modern hybrids have a genetic inability to produce enough endogenous vitamin at the growth plate in what is the fastest-growing bone in the broiler's body; dietary calcium deficiency exposes this more basic disturbance (Whitehead, 1995). Bone fractures commonly occur during the transportation and slaughter of broilers and 'spent' hens; like cage layer fatigue and TD, they cannot be specifically attributed to inadequate calcium nutrition (Wilson, 1991), but it may play a part.

Grain-feeding in other species

Without supplementation, grain-based diets will cause rickets in growing pigs and leg weakness in boars, caused by a similar growth-plate lesion to that seen in TD. Calcium deprivation may also occur in horses and sheep fed largely on grain diets, designed as drought rations when little or no grazing is available (Larsen *et al.*, 1986). Severe stunting of growth, gross dental

Table 4.4. Effects of dietary calcium and 1,25-$(OH)_2D_3$ on incidence of tibial dyschondroplasia (TD) in broiler chicks (Whitehead, 1995).

	Ca	1,25-$(OH)_2D_3$ (µg kg^{-1})			
	(g kg^{-1})	0	2.0	3.5	5.0
TD incidence (%)	7.5	50	15	5	0
	10.0	10	20[a]	5	5
	12.5	15	15	0	0

[a] The complex interactions between calcium and 1,25-$(OH_2)D_3$ are due mainly to: (i) useful enhancement of calcium uptake by the vitamin at the lowest dietary calcium level; (ii) a growth-inhibiting but TD-protective hypercalcaemia at high vitamin levels; and (iii) opposing influences of high calcium and 1,25-$(OH)_2D_3$ levels on phosphorus status.

abnormalities and some deaths were observed in lambs and young, weaned sheep fed a wheat-grain-based diet without additional calcium (Franklin *et al.*, 1948). Young horses developed osteofibrosis and lost weight when fed on high-cereal concentrate rations (Groenewald, 1937). Acute calcium deprivation, in the form of milk-fever, affects 3–10% of the intensively fed dairy herds in developed countries and similar proportions of affected animals fail to respond to treatment (Littledike *et al.*, 1981). Allen and Sansom (1985) estimated that the disease cost the UK dairy industry £10 million per annum.

Grazing livestock

Calcium deprivation is much less of a problem in grazing livestock than phosphorus deficiency and McDonald (1968) found no authentic record of a primary calcium deficiency in grazing cattle or sheep. Calcium deficiency has occasionally occurred in grazing, high-producing dairy cows and has led to osteodystrophic diseases and low productivity in other livestock on acid, sandy or peaty soils in humid areas, where the available grazing is composed mainly of quick-growing grasses which contain < 2 g Ca kg^{-1} DM, as reported from parts of India, the Philippines and Guyana. The ability of an animal to absorb and use calcium depends on its supply of vitamin D_3. Grazed herbage is not normally rich in this vitamin, so that the extent of vitamin D_3 formation from dietary precursors, such as 7-dehydrocholesterol, by UV irradiation of the animal's skin becomes important. Where animals are housed during the winter or graze in areas of high latitude, few of the UV rays reach the skin and deficiencies develop at relatively high intakes of calcium. Figure 4.8 shows the marked seasonal decline in vitamin D status in sheep kept outdoors during winter in Edinburgh (Smith and Wright, 1981); a similar phenomenon was believed to have contributed to the development of rickets in young sheep in central Scotland (Bonniwell *et al.*, 1988).

Diagnosis of Calcium Disorders

Diagnosis of acute calcium-responsive disorders, such as milk-fever, 'hypocalcaemia' in ewes and transport tetany, is based heavily on clinical assessment and confirmed by a response to urgent treatment. Subsequent analysis of blood samples taken prior to treatment for serum calcium will confirm diagnosis and appropriate diagnostic ranges are given in Table 4.5. Plasma calcium levels appear to remain equally low (*c.* 1.0 mmol l^{-1}) at successive stages in the clinical progression of milk-fever (Fenwick, 1988). If blood samples are unobtainable, analysis of calcium in vitreous humour, withdrawn from the eye of the dead animal, can be informative because there is a reasonable correlation with concentrations in plasma. The analysis of ionized calcium is useful as an experimental tool (Phillippo *et al.*, 1994) but unhelpful in the diagnostic context. Chronic bone disorders in all species of livestock can have various causes and a role for calcium is readily confirmed by the

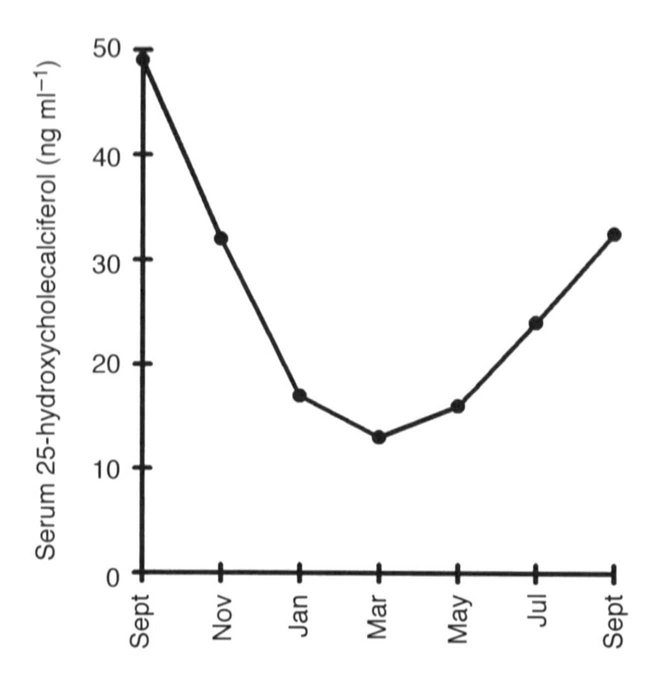

Fig. 4.8. Concentrations of the vitamin D metabolite, 25-hydroxycholecalciferol, in the serum of pregnant ewes out-wintered in Scotland, decline markedly due to lack of solar irradiation (data from Smith and Wright, 1981).

Table 4.5. Marginal bands[a] for mean total serum calcium in assessing low calcium status in various classes of livestock.

	mmol l^{-1}	mg l^{-1}
Periparturient dairy cow	1.3–2.0	50–80
Periparturient ewe	1.5–2.25	60–90
Young ruminants	1.8–2.0	70–80
Pigs	1.75–2.0	70–90
Poultry,[b] laying	3.8–5.0[c]	180–200
Poultry, non-laying	2.0–2.5	70–100
Horses	2.25–2.5	60–100

[a] Mean values within bands indicate possible and values below bands probable dysfunction in some individuals; individual values below bands indicate need for wider sampling and possibility of dysfunction.
[b] Turkeys have slightly higher ranges than those quoted, which are for fowl.
[c] Hypercalcaemia and risk of tissue damage, particularly to the kidney, is indicated by values > 3.2 mmol l^{-1} in most stock but > 6.0 mmol l^{-1} in laying hens.

presence of low serum calcium (Table 4.5). Such analyses are less helpful in ascertaining the cause of health problems and poor eggshell thickness in the laying hen, because of diurnal fluctuations in serum calcium. The initial assumption in the case of eggshell problems should be that calcium nutrition is inadequate with respect to either the forms (lack of coarse particles) or daily quantities of calcium supplied. Diagnosis is again confirmed by positive responses to remedial measures. Assessments of bone quality alone can never provide a specific diagnosis of chronic calcium deficiency, because calcium is only one of many factors that influence bone mineralization. Guidelines for the assessment of bone quality are given in the next chapter.

Prevention and Control of Calcium Disorders

There are two distinctive approaches to the control of disorders relating to calcium deficiency, depending on whether their onset is acute, as in milk-fever, or chronic, as in bone or egg-laying disorders.

Milk-fever

Parenteral treatment

The key to control and prevention of milk-fever is to restrict the postparturient fall in plasma calcium to a level of 1.75 mmol (70 mg) l^{-1}. Milk-fever may be controlled by a single intravenous infusion of 8–10 g Ca as calcium borogluconate, but some cows need repeated treatments; after three failures or 24 h, they are designated 'downer cows'. The need for repeat treatments increases with the time taken to apply the first treatment and the plasma calcium concentration at the time of that treatment. Where a veterinary practitioner is not immediately available, the herdsman should intervene, warming the bags of infusate before administration and giving a second infusion if there is not an immediate recovery (Fenwick, 1994).

Parenteral prevention is better than cure, because the cow that recovers from milk-fever is more prone to limb injuries and diseases such as mastitis and ketosis than unaffected cows. The key to prevention is to maximize the total supply of calcium absorbed from the gut and mobilized from the skeleton. Increasing the calcium content of the diet prior to calving is ineffective and, in fact, high-calcium diets may predispose cows to milk-fever (Ender *et al.*, 1956). Massive dosage with vitamin D_3 (2–30 million units a day for 5–7 days before calving) has been successful (Hibbs and Pounden, 1955), but the timing of treatment in relation to calving is critical. Doses are far lower and timing is less crucial when the 1α analogue of vitamin D_3 is used (Sachs *et al.*, 1987a, b) and an intramuscular injection of 0.25 mg of 1α-OHD_3 given within 2 h of calving averted the drop in plasma calcium after calving (Sansom *et al.*, 1977). However, the administration of 1α-OHD_3 may merely delay the onset of symptoms (Hove, 1986) because the treated cow becomes refractory to hypocalcaemia, failing to mount a PTH response (Horst, 1986).

Intramuscular injection of 0.4 g of 1,25-$(OH)_2D_3$ 5 days before predicted calving date, with reinjections every 5 days until calving, has maintained serum calcium and phosphorus during the critical period, 24 h before and 48 h after calving (Gast *et al.*, 1979). Intramuscular injections of PTH are also effective but impractical (Goff *et al.*, 1989a).

Dietary prevention

One early dietary approach was to feed diets low in calcium prior to calving to prime the homeostatic pathways, notably synthesis of PTH and hence 1,25-$(OH)_2D_3$ (Boda, 1956; Boda and Cole, 1956; Green *et al.*, 1981), but the necessary diets are not easily formulated on many farms and at some point the calcium deficit has to be made good. The most manageable and effective treatment for milk-fever is to feed acidic diets around calving (for review, see Block, 1994); they work by enhancing the sensitivity of the kidney to PTH and thus raising 1,25-$(OH)_2D_3$ concentrations in plasma (Gaynor *et al.*, 1989; Abu Damir *et al.*, 1994) and calcium absorption (Freeden *et al.*, 1988). Acidic diets are those in which the cation:anion balance – essentially (potassium (K^+) + sodium (Na^+)) – (chloride (Cl^-) + sulphate (SO_4^{2-})) – in milliequivalents (atomic weight ÷ valency) is below 100. Dietary acidity can be monitored via the pH of urine, which should be below 7.5. This line of treatment was instigated by Ender *et al.* (1971), who noted that milk-fever rarely occurred in cows given silage preserved with acids. The feeding of silage rather than hay as roughage will itself decrease the incidence of milk-fever. Large anion inputs are needed to offset the high potassium levels in the herbage of spring-calving herds, but they can be achieved with large calcium chloride ($CaCl_2$) supplements (providing 50 g Ca: Goff and Horst, 1993). The associated risks of acidosis and aspiration pneumonia can be decreased by giving $CaCl_2$ in gel or emulsion form (in polyethylene glycol) and/or as calcium propionate gel (providing 75 g Ca: Goff and Horst, 1994). The use of calcium salts to 'acidify' diets is commended because 'acidification' *per se* may be ineffective if the diet is low in calcium (Oetzel *et al.*, 1988). The acidic diets should be gradually discontinued after 21 days, because they may depress milk yield (see Chapter 7). Since the primary source of excess cations is potassium-rich forage, steps should be taken to minimize the potassium content of pastures (see Chapter 8).

Hypocalcaemia in ewes

'Hypocalcaemia' in pregnant ewes is also treated with calcium borogluconate, and response to treatment confirms the diagnosis. The subcutaneous dose is 50 ml of a 40% solution of calcium borogluconate, with added magnesium. Preventive measures have not been studied in ewes, but those which reduce the incidence of milk-fever in dairy cows should be just as effective in ewes, provided they begin well before lambing. The use of acid-preserved silage, 1α-OHD_3 injections and oral supplementation with $CaCl_2$ and/or calcium propionate are suggested, the latter having particular potential where there is a parallel risk of pregnancy toxaemia. The same treatments are indicated for transport tetany. Treatments that rely on increased synthesis of 1,25-$(OH)_2D_3$ require a supply of the unhydroxylated vitamin; this may be in short supply in early spring and require simultaneous supplementation with vitamin D_3 (Smith and Wright, 1981).

Prevention of bone and eggshell disorders

Since the clinical effects of osteodystrophy caused by calcium deprivation are mostly irreversible and weak eggshells irreparable, preventive measures must be adopted early in development. Prevention is simply a matter of seeing that calcium requirements are met from one or more of a wide range of suitable calcium sources. With housed ruminants receiving regular concentrate supplements, additional calcium is either incorporated into the whole mixed diet or into the concentrate portion of the ration, usually as calcium carbonate or dibasic calcium phosphate ($CaHPO_4$). The same salts are routinely added to free-access mineral mixtures for housed and grazing livestock; although they are rarely needed continuously as far as the grazing animal is concerned, they are essential components of all rations for non-ruminants.

Calcium supplements

There is little to choose between the common mineral sources of calcium, when they are compared with calcium carbonate (Soares, 1995). The exceptions are dolomitic limestone and soft-rock phosphate, which have relatively low availabilities (0.65 and 0.70, respectively) when fed to pigs or poultry (Dilworth *et al.*, 1964; Reid and Weber, 1976). All mineral sources of calcium are likely to have a lower than average availability when fed in high-phytate, low-phytase diets. Claims that the calcium from $CaCO_3$ is much more available to pigs than that from $CaHPO_4$ (Eeckhout *et al.*, 1995) may reflect saturation of the inhibitory effect of phytate at the high calcium levels which were achieved when adding $CaCO_3$, thus confounding the effect of source and level. Particle size can be important in the choice of calcium supplements. The high calcium needs of laying hens are commonly met by supplementation with 4% calcium as limestone, a relatively cheap and plentiful source, but replacing one-half to two-thirds with granular sources, such as oyster shell, has a sparing effect, so that 3% calcium can be sufficient in respect of shell quality (Roland *et al.*, 1974). It seems that large particles of calcium remain in the gizzard longer and provide more retainable calcium than finely ground sources (Scott *et al.*, 1971), probably by achieving synchrony with diurnal fluctuations in the calcium requirement for shell calcification (Whitehead, 1995). Particle size of $CaCO_3$ influences the phosphorus requirement of broilers (see Chapter 5). There are other reasons for discriminating between sources, however; phosphorus-free calcium sources are best in grain-based diets for ruminants and monobasic calcium phosphate (CaH_2PO_4) is unsuitable for laying hens, because it results in acidosis and poor shell quality (see Chapters 5 and 7). The acidotic calcium chloride or sulphate may be useful for manipulating acid–base balance in dairy cows to avoid milk-fever. When calcium and other minerals are mixed with the feed, it is important to ensure that they are evenly distributed. Inadequate mixing of limestone with cereals may have contributed to hypocalcaemia after drought feeding (Larsen *et al.*, 1986) and can be avoided by using molasses to stick such supplements to the grain (see Chapter 9 for detail). Growing use is being made of calcium complexes with organic acids ('calcium soaps') as energy sources in ruminant nutrition, though not as improved sources of calcium. In the poultry industry, supplementation of diets with calcium formate reduces the contamination of carcasses and eggs with pathogenic bacteria.

Calcium Requirements

The principal factors affecting the mineral requirements of farm animals, including calcium, were described in Chapter 1. Several developments in livestock husbandry have combined to steadily raise dietary calcium requirements. These include: (i) genetic improvement in the animal, resulting in faster growth and higher yields, as exemplified particularly in broiler production; (ii) the increasing use of energy-dense diets; (iii) the practice of early weaning, which emphasizes the decrease in efficiency of utilization associated with solid as opposed to milk diets; (iv) the breeding of animals at a young age, while they are still growing; and (v) the use of antibiotics, hormones and other feed additives as growth stimulants. Where these growth effects are exerted through higher feed consumption, the requirements – expressed as dietary concentrations – are little affected, but, where feed efficiency is also improved, as is usual, requirements rise correspondingly. Liberal supplies of vitamin D_3 enable the animal to make the best use of limited intakes of calcium and are important to livestock housed for long periods, particularly high-yielding dairy cows and laying hens. The following recommendations assume that supplies of vitamin D_3 are adequate.

Sheep

There have been no rigorous attempts to define the calcium requirements of sheep (or cattle) by means of feeding trials and it is therefore necessary to rely on factorial estimates of requirement. Unfortunately, these have varied considerably from one authority to another, largely due to disagreement on a realistic coefficient of absorption with which to generate gross requirements (AFRC, 1991). A major reduction in calcium requirements was first proposed by the Agricultural Research Council (ARC, 1980), on the grounds that ruminants could absorb calcium with a high efficiency when necessary: information published subsequently and reviewed on p. 68 gives no convincing reason to doubt that assertion. Minimum requirements should therefore be calculated assuming the maximum attainable efficiency of absorption of 0.68. The AFRC (1991) endorsed that assumption and their estimates are therefore commended and presented in Table 4.6. The requirements for lambs were tested by feeding at 75%, 100% and 125% of the recommended level and

Table 4.6. Requirements of sheep for dietary calcium (AFRC, 1991) and phosphorus (modified from AFRC, 1991) at the given dry-matter intakes (DMI).

	Live weight (kg)	Production level/stage	Diet quality	DMI (kg day^{-1})	Ca (g kg^{-1} DM)	P (g kg^{-1} DM)
Growing lambs	20	100 g day^{-1}	L	0.67	3.7	2.6
			H	0.40	5.7	2.8
		200 g day^{-1}	L	U	–	–
			H	0.57	7.0	3.9
	40	100 g day^{-1}	L	1.11	2.4	2.0
			H	0.66	3.4	2.0
		200 g day^{-1}	L	1.77	2.6	2.3
			H	0.93	4.0	2.4
Pregnant ewe carrying twins[a]	75	9 weeks	L	1.10	1.4[c]	1.6[c]
			H	0.71	1.6[c]	1.0[c]
		13 weeks	L	1.28	2.0	2.0
			H	0.85	2.6	1.6
		17 weeks	L	1.68	2.9	2.3
			H	1.13	3.9	2.0
		Term	L	2.37	3.2	2.2
			H	1.62	4.3	1.8
Lactating ewe nursing twins	75	2–3 kg milk day^{-1}	L	2.8–3.7	2.8 (m)	2.7
			L	2.3–3.2	3.1 (bm)[b]	3.0
			H	1.8–2.4	3.8 (m)	2.7
			H	1.5–2.1	4.3 (bm)	3.0

[a] The requirements for small ewes carrying single lambs are similar, assuming that they will eat proportionately less DM.
[b] Requirements are influenced by the ability of the diet to meet energy need and prevent loss in body weight (m); diets which allow loss of body weight at 0.1 kg day^{-1} (bm) are associated with higher requirements.
[c] Sufficient for dry ewe.
L, poorly digestible diet with '*q*' value (q = ME/GE, the metabolizability of gross energy at maintenance) 0.5; H, highly digestible diet, *q* = 0.7; U, unattainable performance.

there was every indication that they still err on the generous side (Wan Zahari *et al.*, 1990). The important features of the estimates shown in the table are as follows:

1. Requirements for growth decrease with age but increase with growth rate.
2. Requirements for pregnant ewes rise rapidly in late pregnancy to a level equal to that of the lactating ewe.
3. No figure need be met on a day-to-day basis and it would be surprising if performance suffered on any diet providing an average of 3 g Ca kg^{-1} DM throughout the year.

Cattle

In view of evidence that cattle can absorb calcium very efficiently when necessary (Van Klooster, 1976; Van Leuwen and De Visser, 1976), the AFRC (1991) estimates are again commended for general use (Table 4.7). The important features are as follows:

1. Requirements for growth decrease with age but increase with growth rate and are similar to those of lambs.
2. Requirements increase slowly during pregnancy and in high-yielding dairy cows increase by a further 33% or more with the onset of lactation (the table does not allow for secretion of calcium-rich colostrum).
3. Because there is little time for cows to replenish lost calcium reserves between lactations, the average calcium concentrations in the annual diet should be approximately 4.5 g Ca kg^{-1} DM, 50% higher than for sheep.
4. When cows 'milk off their backs' (i.e. lose body weight to sustain production), requirements in concentration terms increase substantially (by 80% in the example given).

Pigs

Empirical feeding trials have been extensively used to define the calcium requirements of pigs. ARC (1981) tabulated the results of 56 trials published between 1964 and 1976 and concluded that requirements had increased substantially (by up to 50%) since 1967, as judged by the average 'optimum requirement'. They attributed the changes to increases in pig performance on diets of increased nutrient density. There are, however, two reasons why the average 'optimum requirement' will overestimate needs for calcium. Firstly, many of the trials contained only two widely spaced treatments, one less than and one probably in excess of requirement, the upper value being taken to be 'optimal'. If the data set is restricted to trials with at least three calcium levels, the mean optimum concentration is 7.9 ± 1.7 g Ca kg^{-1} DM ($n = 18$) for pigs of 20–90 kg LW as opposed to 8.5 g Ca kg^{-1} DM for unselected data. A further source of bias is that many trials were based on the assumption that, as dietary calcium concentrations increased, those of phosphorus must be similarly increased to maintain a constant Ca : P ratio. Since inorganic phosphorus supplements lower the degradation of phytate (see Chapter 5) and phytate lowers the availability of calcium, this assumption will increase the apparent calcium requirement. High calcium 'requirements' (> 10 g kg^{-1} DM) were invariably accompanied by high phosphorus provision (≥ 10 g kg^{-1} DM). If the data set is further restricted to trials using the factorially derived minimum phosphorus requirements (5.9–8.8 g P kg^{-1} DM: Table 4.8), the mean 'optimum' calcium requirements declines further to 7.6 g kg^{-1} DM.

The factorial derivation of calcium requirements overcomes some of these difficulties but generates others, such as the choice of absorption coefficient to convert net to gross requirements (see Chapters 1 and 2). ARC (1981) assumed that the efficiency of absorption declined from 67 to 47% as pigs grew from 25 to 90 kg. However, the growth requirement declines markedly

Table 4.7. Requirements of cattle for dietary calcium (AFRC, 1991) and phosphorus (modified from AFRC, 1991) at the given dry-matter intakes (DMI).

	Live weight (kg)	Production level	Diet quality	DMI (kg day^{-1})	Ca (g kg^{-1} DM)	P (g kg^{-1} DM)
Growing cattle	100	0.5 kg day^{-1}	L	2.8	5.2	3.6
			H	1.7	8.0	3.6
		1.0 kg day^{-1}	L	4.5	6.3	4.1
			H	2.4	10.8	4.8
	300	0.5 kg day^{-1}	L	5.7	3.0	2.3
			H	3.4	4.4	1.8
		1.0 kg day^{-1}	L	8.3	3.5	2.6
			H	4.7	5.5	2.5
	500	0.5 kg day^{-1}	L	10.9	2.6	1.5
			H	6.1	3.6	1.1
		1.0 kg day^{-1}	L	11.6	2.8	2.2
			H	6.5	4.3	1.9
Pregnant cow, calf weight 40 kg at birth	600	23 weeks	L	6.3	2.1	1.6
			H	4.0	2.7	0.9
		31 weeks	L	7.2	2.3	1.8
			H	4.7	3.0	1.1
		39 weeks	L	9.1	2.7	2.0
			H	6.1	3.5	1.4
		Term	L	11.2	2.8	2.1
			H	7.5	3.6	1.6
Lactating cow	600	10 kg day^{-1}	L	12.0	2.9 (m)	2.5
			L	9.9	3.3 (bm)	2.7
		20 kg day^{-1}	H	11.4	4.6 (m)	2.5
			H	10.1	5.1 (bm)	2.7
		40 kg day^{-1}	H	19.3	4.9 (m)	2.8
			H	17.8	5.3 (bm)	3.0

L, poorly digestible diet with 'q' value 0.5; H, highly digestible diet, $q = 0.7$; m, fed to maintain bodyweight; bm, fed below maintenance.

over that weight range and the decline in reported absorptive efficiency reflects an increasingly generous calcium supply. It is more likely that pigs will maintain a high efficiency of absorption for predominantly inorganic calcium throughout life if fed to requirement for calcium and phosphorus. The requirements given in Table 4.8 use the components of net calcium requirement given by ARC (1981) but a uniformly high coefficient for A_{Ca} (0.7) (see Chapter 5) and agree well with the requirement indicated by the restricted set of feeding trials. Early calcium provision is higher and later provision slightly lower than that currently advocated by the National Research Council (NRC, 1988). It is becoming increasingly important to define and feed to minimum calcium requirements in order to make the most efficient use of phytate P.

Table 4.8. Factorially derived estimates[a] of the calcium and phosphorus requirements (g kg^{-1} DM) of growing pigs for the development of fully mineralized skeletons (modified from ARC, 1981). Corresponding NRC (1988) recommendations are given in parentheses.

Live weight (kg)	Assumed DMI (kg day^{-1})	Calcium[b]	Phosphorus[c] High phytate		Low phytate
5	0.3	11.6 (8.8)	9.2	(7.2)	9.2
25	1.05	8.6 (6.9)	8.8	(6.0)	6.3
45	1.8	6.1 (6.0)	5.9	(5.0)	4.2
90	2.8	4.5 (5.0)	4.2	(4.0)	3.0

[a] Using ARC (1981) requirements for growth (g kg^{-1} LWG) (i.e. P = 10.0 – 0.1 LW and Ca = 12.5 – 0.1 LW for birth – 50 kg; 5.0 g P and 7.5 g Ca for 50–90 kg) and maintenance (20 mg P and 32 mg Ca kg^{-1} LW).
[b] Absorption coefficients for Ca: baby pigs, 0.98; weaned pigs, 0.70 for all weights.
[c] Absorption coefficients for P: baby pigs, 0.80; weaned pigs, 0.50 for high phytate, 0.70 for low phytate diets; the lower value (0.50) is appropriate for diets which provide phytin P (availability 0.15) and non-phytin P (availability 0.8) in equal proportions and no added phytase.

Poultry

The calcium requirements for growing birds according to NRC (1994) are summarized in Table 4.9; they decline with age for all types of bird and are highest for turkey poults. Considerable variation in the estimated calcium requirements of hens for egg production and quality is apparent from reports published over the last 60 years, values ranging from 2.9 to 6.2 g day^{-1} (Roland, 1986). Expression of requirements as dietary concentrations narrowed the reported requirement range to 28–48 g Ca kg^{-1} DM. This variation stems from: differences in the basal diets with regard to phosphorus, vitamin D and other interacting nutrients; the strain and age of the birds; the sources of the supplemental calcium; the ability of hens to raise food intake to acquire sufficient calcium; exaggerated safety margins; environmental conditions, including ambient temperatures; and differences in contribution of calcium from the skeleton. Tolerance of low calcium provision early in lay was demonstrated by Hamilton and Cipera (1981), who found no reduction in eggshell strength when hens were given only 20 g Ca kg^{-1} DM during the first month of lay and 32 g Ca kg^{-1} DM thereafter. The basis for increasing need during lay is illustrated in Fig. 4.9. In addition, a decline in efficiency of calcium absorption in older hens is probably allowed for in the requirements given in Table 4.10, which continue to increase to the end of lay. Temperatures well above 20°C tend to reduce feed consumption and hence to increase mineral requirements when expressed as a proportion of the diet. The use of energy-dense diets, e.g. with added fat, increases the calcium concentration needed and the latest NRC (1994) estimates give three requirements for white-egg laying hens – 40.6, 32.5 and 27.1 g Ca kg^{-1} DM to allow for food intakes of 80, 100 and 120 g day^{-1}, respectively (i.e. decreasing

Table 4.9. Dietary requirements (g kg^{-1} DM) of calcium and phosphorus for growth in Leghorn (L) or broiler (B) chicks and turkey poults (T) (from NRC, 1994).

			Growth stage[a]		
			Early	Middle	Late
Calcium	Chick	L	9.0	8.0	20.0
		B	10.0	9.0	8.0
	Poult	T	12.0	8.5	5.5
Non-phytate phosphorus	Chick	L	4.0	3.5	3.2
		B	4.5	3.5	3.0
	Poult	T	6.0	4.2	2.8

[a] Growth stages:

	Age (weeks)			Dietary energy density (kcal ME kg^{-1} DM)		
	L	B	T	L	B	T
Early	0–6	0–3	0–4	2850	3200	2800
Middle	6–18	3–6	8–12	2850	3200	3000
Late	> 18	6–8	20–24	2900	3200	3300

ME, metabolizable energy.

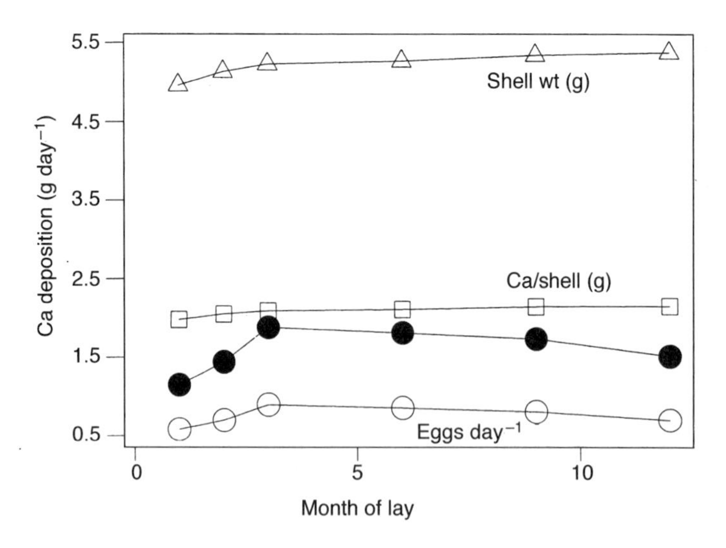

Fig. 4.9. Daily calcium deposition in eggs by laying hens (solid circle) peaks after about 3 months of lay due to the combined effects of increases in rate of egg production, the amount of shell per egg and the Ca concentration in the shell (data from Roland, 1986).

energy density). Recommended values were some 10% lower for brown-egg laying hens, which consume more food, and are much lower than those advocated by Roland (1986; Table 4.10), but they carry the proviso that they may not give maximum eggshell thickness. It has been suggested that calcium requirements should take account of the size of egg produced, increasing from 3.8 g to 4.5 g Ca day^{-1} as egg size increases from 50 to 60 g (Simons, 1986). For pullets entering lay, the optimum transitional feeding regimen for calcium is to increase levels from 10 to 30 g Ca kg^{-1} DM 1 week before the first egg is anticipated (Roland, 1986). Thereafter, the need to meet calcium requirements for peak egg production on a daily and even an hourly basis at times is in contrast to the needs of any other class of livestock for any mineral, with the possible exception of calcium for the recently calved cow. Indeed, the laying hen will voluntarily consume more of a calcium supplement on laying than on non-laying days (Gilbert, 1983).

Calcium Toxicity

Calcium is not generally regarded as a toxic element, because homeostatic mechanisms ensure that excess dietary calcium is extensively excreted in faeces. However, doubling the dietary calcium concentration of chicks to 2.05 g kg^{-1} DM caused hypercalcaemia and growth retardation, fast-growing strains being more vulnerable than slow-growing strains (Hurwitz *et al.*, 1995). The adverse nutritional consequences of feeding excess calcium are generally indirect and arise from impairments in the absorption of other elements when the digesta are enriched with calcium; thus deficiencies of phosphorus and zinc are readily induced in non-ruminants. Dietary provision of calcium soaps of fatty acids (CSFA) represents a convenient way of increasing the fat (i.e. energy) content of the diet without depressing fibre digestibility in the rumen. Significant improvements in milk yield can be obtained in dairy cows (Jenkins and Palmquist, 1984) and ewes (Sklan, 1992)

Table 4.10. Requirements of laying hens for dietary calcium and phosphorus (g day^{-1}) (Roland, 1986).

	Weeks in production			
Time in lay (weeks)	−1 to 8[a]	9–16	10–25	26–moult
Ca	3.75	3.75	4.00	4.25
P				
Total	0.70	0.70	0.60	0.50
Non-phytate	0.50	0.50	0.40	0.30

[a] It is recommended that for this first period figures be used as %DM fed without adjustment for food intake; thereafter, dietary concentrations should be adjusted for estimated food intake to give the desired daily provision.

but not without feeding (by necessity) gross excesses of calcium (*c.* 10 g Ca kg^{-1} DM) and sometimes (by choice) matching excesses of phosphorus (6 g P kg^{-1} DM). Whether or not these excesses are related to adverse effects of CSFA on conception rate in primiparous cows (Sklan *et al.*, 1994) remains to be investigated. Hypercalcaemia can cause life-threatening tissue calcification but usually occurs as a secondary consequence of phosphorus deficiency or overexposure to vitamin D_3 (Payne and Manston, 1967) and its analogues (Whitehead, 1995). Of particular interest is the tissue calcification in cattle caused by ingestion of *Solanum malacoxylon*, which contains a vitamin D analogue. The first sign of trouble is the development of osteophagia and there may be no alternative but to temporarily destock pastures infested with the water-loving plant. Factors which safely maximize the deposition of calcium in bone (heavy-boned breeds; ample dietary phosphorus; alkaline mineral supplements; grain feeding) are worth exploring as preventive measures where *S. malacoxylon* is a problem.

References

Abu Damir, H., Phillippo, M., Thorp, B.H., Milne, J.S., Dick, L. and Nevison, I.M. (1994) Effects of dietary acidity on calcium balance and mobilization, bone morphology and 1,25 dihydroxyvitamin D in prepartal dairy cows. *Research in Veterinary Science* 56, 310–318.

AFRC (1991) A reappraisal of the calcium and phosphorus requirements of sheep and cattle. Technical Committee on Responses to Nutrients, Report Number 6. *Nutrition Abstracts and Reviews (Series B)* 61, 573–612.

Allen, W.M. and Sansom, B.F. (1985) Milk-fever and calcium metabolism. *Journal of Veterinary Pharmacology and Therapeutics* 8, 19–21.

ARC (1980) *The Nutrient Requirements of Ruminant Livestock.* Commonwealth Agricultural Bureaux, Farnham Royal, Slough, UK, pp. 184–192.

ARC (1981) *The Nutrient Requirements of Pigs.* Commonwealth Agricultural Bureaux, Farnham Royal, Slough, UK, pp. 215–248.

Bain, S.D. and Watkins, B.A. (1993) Local modulation of skeletal growth and bone modelling in poultry. *Journal of Nutrition* 123, 317–322.

Bar, A. and Hurwitz, S. (1984) Egg shell quality, medullary bone ash, intestinal calcium and phosphorus absorption and calcium-binding protein in phosphorus-deficient hens. *Poultry Science* 63, 1975–1981.

Barry, T.N. and Blaney, B.J. (1987) Secondary compounds of forages. In: Hacker, J.B. and Ternouth, J.H. (eds) *The Nutrition of Herbivores.* Academic Press, Sydney, Australia, pp. 91–120.

Barzanji, A.A.H. and Daniel, R.C.W. (1987) Effect of hypocalcaemia on blood flow distribution in sheep. *Research in Veterinary Science* 42, 92–95.

Barzanji, A.A.H. and Daniel, R.C.W. (1988) The effects of hypocalcaemia on blood gas and acid–base parameters. *British Veterinary Journal* 144, 93–97.

Becker, R.B., Arnold, P.T.D., Kirk, W.G., Davis, G.K. and Kidder, R.W. (1953) *Minerals for Beef and Dairy Cattle.* Bulletin 153, Florida Agricultural Experiment Station.

Beckman, M.J., Goff, J.P., Reinhardt, T.A., Beitz, D.C. and Horst, R.L. (1994) *In vivo* regulation of rat intestinal 24-hydroxylase: potential new role of calcium. *Endocrinology* 135, 1951–1955.

Black, H.E. and Capen, C.C. (1971) Urinary and plasma hydroxyproline during pregnancy, parturition and lactation in cows with parturient hypocalcaemia. *Metabolism* 20, 337–344.

Block, E. (1994) Manipulation of dietary cation–anion balance difference on nutritionally related production diseases, productivity and metabolic responses of dairy cows. *Journal of Dairy Science* 77, 1437–1450.

Boda, J.M. (1956) Further studies on the influence of dietary calcium and phosphorus on the incidence of milk fever. *Journal of Dairy Science* 39, 66–72.

Boda, J.M. and Cole, H.H. (1956) Calcium metabolism with special reference to parturient paresis (milk fever) in dairy cattle: a review. *Journal of Dairy Science* 39, 1027–1038.

Bond, P.L., Sullivan, T.W., Douglas, J.H. and Robeson, L.G. (1991) Influence of age, sex and method of rearing on tibia length and mineral deposition in broilers. *Poultry Science* 70, 1936–1942.

Bonniwell, M.A., Smith, B.S.W., Spence, J.A., Wright, H. and Ferguson, D.A.M. (1988) Rickets associated with vitamin D deficiency in young sheep. *Veterinary Record* 122, 386–388.

Borle, A.B. (1974) Calcium and phosphate metabolism. *Annual Review of Physiology* 36, 361–390.

Braithwaite, G.D. (1983a) Effect of 1α-hydroxy cholecalciferol on calcium and phosphorus metabolism in sheep given high or low calcium diets. *Journal of Agricultural Science, Cambridge* 96, 291–299.

Braithwaite, G.D. (1983b) Calcium and phosphorus requirements of the ewe during pregnancy and lactation. 1. Calcium. *British Journal of Nutrition* 50, 711–722.

Brommage, R. (1989) Measurement of calcium and phosphorus fluxes during lactation in the rat. *Journal of Nutrition* 119, 428–438.

Bronner, F. (1987) Intestinal calcium absorption: mechanisms and applications. *Journal of Nutrition* 117, 1347–1352.

Brown, E.M. (1991) Extracellular Ca^{2+} sensing, regulation of parathyroid cell function and role of Ca^{2+} and other ions as extracellular (first) messengers. *Physiological Reviews* 71, 371–411.

Buckner, G.D., Martin, J.H. and Insko, W.M., Jr (1930) The blood calcium of laying hens varied by the calcium intake. *American Journal of Physiology* 94, 692–695.

Carafoli, E. (1991) Calcium pump of the plasma membrane. *Physiological Reviews* 71, 129–149.

Chrisp, J.S., Sykes, A.R. and Grace, N.D. (1989a) Kinetic aspects of calcium metabolism in lactating sheep offered herbages with different calcium concentrations and the effect of protein supplementation. *British Journal of Nutrition* 61, 45–58.

Chrisp, J.S., Sykes, A.R. and Grace, N.D. (1989b) Faecal endogenous loss of calcium in young sheep. *British Journal of Nutrition* 61, 59–65.

Common, R.H. (1938) Observations on the mineral metabolism of pullets. 3. *Journal of Agricultural Science, Cambridge* 28, 347–366.

Deobald, H.J., Lease, E.J., Hart, E.B. and Halpin, J.G. (1936) Studies on the calcium metabolism of laying hens. *Poultry Science* 15, 179–185.

Dilworth, B.C., Day, E.J. and Hill, J.E. (1964) Availability of calcium in feed grade phosphates to the chick. *Poultry Science* 43, 1121–1134.

Duckworth, J., Benzie, D., Cresswell, E., Hill, R. and Boyne, A.W. (1962) Studies of the dentition of sheep. III A study of the effects of vitamin D and phosphorus deficiencies in the young animal on the productivity, dentition and skeleton of Scottish Blackface ewes. *Research in Veterinary Science* 3, 408–415.

Edwards, H.M. and Veltmann, J.R. (1983) The role of calcium and phosphorus in the aetiology of tibial dyschondroplasia in young chicks. *Journal of Nutrition* 113, 1568–1575.

Eeckhout, W., de Poepe, M., Warnants, N. and Bekaert, H. (1995) An estimation of the minimal P requirements of growing–finishing pigs, as influenced by Ca level of the diet. *Animal Feed Science and Technology* 52, 29–40.

Ender, F., Dishington, I.W. and Helgebostad, A. (1956) Parturient paresis and related forms of hypocalcemic disorders induced experimentally in dairy cows. *Nordisk Veterinaermedicin* 8, 507–513.

Ender, F., Dishington, I.W. and Helgebostad, A. (1971) Calcium balance studies in dairy cows under experimental induction and prevention of hypocalcaemic parturient paresis. *Zeitschrift Tierphysiologie Tierernahrung Fultermittelkdunde* 28.

Fenwick, D.C. (1988) The relationship between certain blood cations in cows with milk-fever and both the state of conciousness and the position of the cows when attended. *Australian Veterinary Journal* 65, 374–375.

Fenwick, D.C. (1994) Limitations to the effectiveness of subcutaneous calcium solutions as a treatment for cows with milk-fever. *Veterinary Record* 134, 446–448.

Field, A.C., Suttle, N.F. and Nisbet, D.I. (1975) Effects of diets low in calcium and phosphorus on the development of growing lambs. *Journal of Agricultural Science, Cambridge* 85, 435–442.

Franklin, M.C., Reid, R.L. and Johnstone, I.L. (1948) *Studies on Dietary and Other Factors Affecting the Serum-calcium Levels of Sheep*. Bulletin, Council for Scientific and Industrial Research, Australia, No. 240.

Freeden, A.H. (1989) Effect of maturity of alfalfa (*Medicago sativa*) at harvest on calcium absorption in goats. *Canadian Journal of Animal Science* 69, 365–371.

Freeden, A.H., DePeters, E.J. and Baldwin, R.C. (1988) Effects of acid–base disturbances caused by differences in dietary fixed ion balance on kinetics of calcium metabolism in ruminants with high calcium demand. *Journal of Animal Science* 66, 174–184.

Gast, G.R., Horst, R.L., Jorgensen, N.A. and DeLuca, H.F. (1979) Potential use of 1,25-dihydroxycholecalciferol for prevention of parturient paresis. *Journal of Dairy Science* 62, 1009–1013.

Gaynor, P.J., Mueller, F.J., Miller, J.K., Ramsey, N., Goff, J.P. and Horst, R.L. (1989) Parturient hypocalcaemia in Jersey cows fed alfalfa haylage-based diets with different cation–anion ratios. *Journal of Dairy Science* 72, 2525–2531.

Gilbert, A.B. (1983) Calcium and reproductive function in the hen. *Proceedings of the Nutrition Society* 42, 195–212.

Goff, J.P. and Horst, R.L. (1993) Oral administration of calcium salts for treatment of hypocalcaemia in cattle. *Journal of Dairy Science* 76, 101–108.

Goff, J.P. and Horst, R.L. (1994) Calcium salts for treating hypocalcaemia: carrier effects, acid–base balance and oral versus rectal administration. *Journal of Dairy Science* 77, 1451–1456.

Goff, J.P., Kerli, M.E., Jr and Horst, R.L. (1989a) Periparturient hypocalcaemia in cows: prevention using intramuscular parathyroid hormone. *Journal of Dairy Science* 72, 1182–1187.

Goff, J.P., Reinhardt, T.A. and Horst, R.L. (1989b) Recurring hypocalcaemia of bovine parturient paresis is associated with failure to produce 1,25 hydroxyvitamin D. *Endocrinology* 125, 49–53.

Goff, J.P., Reinhardt, T.A., Lee, S. and Hollis, B.W. (1991) Parathyroid hormone-related peptide content of bovine milk and calf blood assessed by radioimmunoassay and bioassay. *Endocrinology* 129, 2815–2819.

Green, H.B., Horst, R.L., Beitz, D.C. and Littledike, E.T. (1981) Vitamin D metabolites in plasma of cows fed a prepartum low calcium diet for prevention of parturient hypocalcaemia. *Journal of Dairy Science* 64, 217–226.

Groenewald, J.W. (1937) Osteofibrosis in equines. *Onderstepoort Journal of Veterinary Science and Animal Industry* 9, 601–620.

Hamilton, R.M.G. and Cipera, J.D. (1981) Effects of dietary calcium levels during brooding, rearing and early egg laying period on feed intake, egg production and shell quality in white leghorn hens. *Poultry Science* 60, 349–357.

Han, Y.N., Yang, F., Zhou, A.G., Miller, E.R., Ku, P.K., Hogberg, M.G. and Lei, X.G. (1997) Supplemental phytases of microbial and cereal sources improve dietary phytate utilization by pigs from weaning through finishing. *Journal of Animal Science* 75, 1017–1025.

Hibbs, J.W. and Pounden, W.D. (1955) Studies on milk-fever in dairy cows. 4. Prevention by short-time, prepartum feeding of massive doses of vitamin D. *Journal of Dairy Science* 38, 65–72.

Hill, R. (1984) Mineral and trace element requirements of poultry. In: Haresign, W. and Cole, D.J.A. (eds) *Recent Advances in Animal Nutrition.* Butterworths, London, pp. 99–109.

Horst, R.L. (1986) Regulation of calcium and phosphorus homeostasis in the dairy cow. *Journal of Dairy Science* 69, 604–616.

Horst, R.L., Goff, J.P. and Reinhardt, T.A. (1990) Advancing age results in reduction of intestinal and bone 1,25 dihydroxyvitamin D receptor. *Endocrinology* 126, 1053–1057.

Hove, K. (1986) Cyclic changes in plasma calcium and the calcium homeostatic endocrine system of the postparturient dairy cow. *Journal of Dairy Science* 69, 2072–2082.

Huber, T.L., Wilson, R.C., Stattelman, A.J. and Goetsch, D.D. (1981) Effect of hypocalcaemia on motility of the ruminant stomach. *American Journal of Veterinary Research* 42, 1488–1490.

Hughes, L.E. and Kershaw, G.F. (1958) Metabolic disorders associated with movement of hill sheep. *Veterinary Record* 70, 77–78.

Hurwitz, S. (1996) Homeostatic control of plasma calcium concentration. *Critical Reviews in Biochemistry and Molecular Biology* 31, 41–100.

Hurwitz, S., Plavnik, I., Shapiro, A., Wax, E., Talpaz, H. and Bar, A. (1995) Calcium metabolism and requirements of chickens are affected by growth. *Journal of Nutrition* 125, 2679–2686.

Jenkins, T.C. and Palmquist, D.L. (1984) Effect of fatty acids or calcium soaps on rumen and total nutrient digestability of dairy rations. *Journal of Dairy Science* 67, 978–986.

Jonson, G., Luthman, J., Mollerberg, L. and Persson, J. (1971) Hypocalcaemia in pregnant ewes. *Nordisk Veterinaermedicin* 23, 620–627.

Jonson, G., Luthman, J., Mollerberg, L. and Persson, J. (1973) Mineral feeding flocks of sheep with cases of clinical hypocalcaemia. *Nordisk Veterinaermedicin* 25, 97–103.

Jumba, I.O., Suttle, N.F., Hunter, E.A. and Wandiga, S.O. (1995) Effects of soil origin and mineral composition and herbage species on the mineral composition of forages in the Mount Elgon region of Kenya. 1. Calcium, phosphorus, magnesium and sulphur. *Tropical Grasslands* 29, 40–46.

Kornegay, E.T. and Qian, H. (1996) Replacement of inorganic phosphorus by microbial phytase for young pigs fed on a maize–soybean-meal diet. *British Journal of Nutrition* 76, 563–578.

Larsen, J.W.A., Constable, P.D. and Napthine, D.V. (1986) Hypocalcaemia in ewes after a drought. *Australian Veterinary Journal* 63, 25–26.

Littledike, E.T., Young, J.W. and Beitz, D.C. (1981) Common metabolic diseases in cattle: ketosis, milk-fever and downer cow complex. *Journal of Dairy Science* 64, 1465–1482.

Lucas, M.J., Huffman, E.M. and Johnson, L.W. (1982) Clinical and clinicopathological features of transport tetany of feedlot lambs. *Journal of American Veterinary Medical Association* 181, 381–383.

McDonald, I.W. (1968) The nutrition of the grazing ruminant. *Nutrition Abstracts and Reviews* 38, 381–395.

MAFF (1990) *UK Tables of the Nutritive Value and Chemical Composition of Foodstuffs.* Givens, D.I. (ed.), Rowett Research Services, Aberdeen.

Martz, F.A., Belo, A.T., Weiss, M.F. and Belyca, R.L. (1990) True absorption of calcium and phosphorus from alfalfa and corn silage when fed to lactating cows. *Journal of Dairy Science* 73, 1288–1295.

Miller, E.R., Ullrey, D.C., Zutaut, G.L., Baltzer, B.V., Schmidt, D.A., Hoefer, J.A. and Luecke, R.W. (1962) Calcium requirement of the baby pig. *Journal of Nutrition* 77, 7–17.

Minson, D.J. (1990) Calcium. In: *Forage in Ruminant Nutrition.* Academic Press, New York, pp. 208–229.

Mosdol, G. and Waage, S. (1981) Hypocalcaemia in the ewe. *Nordisk Veterinaemedicin* 33, 310–326.

Muir, P.D., Sykes, A.R. and Barrell, G.K. (1987) Calcium metabolism in red deer (*Cervus elaphus*) offered herbages during antlerogenesis: kinetic and stable balance studies. *Journal of Agricultural Science, Cambridge* 109, 357–364.

Nonnecke, B.J., Franklin, S.T., Reinhardt, T.A. and Horst, R.L. (1993) *In vitro* modulation of proliferation and phenotype of resting and mitogen stimulated bovine mononuclear leukocytes by 1,25 hydroxyvitamin D3. *Veterinary Immunology and Immunopathology* 38, 75–89.

Norman, A.W. and Hurwitz, S. (1993) The role of vitamin D endocrine system in avian bone biology. *Journal of Nutrition* 123, 310–316.

NRC (1988) *Nutrient Requirements of Swine*, 9th edn. National Academy of Sciences, Washington, DC.

NRC (1994) *Nutrient Requirements of Poultry*, 9th edn. National Academy of Sciences, Washington, DC.

Oetzel, G.R., Olson, J.D., Curtis, C.R. and Fettman, M.J. (1988) Ammonium chloride and ammonium sulphate for prevention of parturient paresis in dairy cows. *Journal of Dairy Science* 71, 3302–3309.

Omdahl, J.L. and DeLuca, H.F. (1973) Regulation of vitamin D metabolism and functions. *Physiological Reviews* 53, 327–372.

Pallauf, V.J., Hohler, D., Rimbach, G. and Neusser, H. (1992) Effect of microbial phytase supplementation to a maize–soya diet on the apparent absorption of phosphorus and calcium in piglets. *Journal of Animal Physiology and Animal Nutrition* 67, 30–40.

Payne, J.M. and Manston, R. (1967) The safety of massive doses of vitamin D_3 in the prevention of milk-fever. *Veterinary Record* 81, 214–216.
Phillippo, M., Reid, G.W. and Nevison, I.M. (1994) Parturient hypocalcaemia in dairy cows: effects of dietary acidity on plasma minerals and calciotropic hormones. *Research in Veterinary Science* 56, 303–309.
Pickard, D.W., Field, B.G. and Kenworthy, E.B. (1988) Effect of magnesium content of the diet on the susceptibility of ewes to hypocalcaemia in pregnancy. *Veterinary Record* 123, 422.
Ramberg, C.F., Jr, Mayer, G.P., Kronfeld, D.S. and Potts, J.T., Jr (1976) Dietary calcium, calcium kinetics and plasma parathyroid hormone concentration in cows. *Journal of Nutrition* 106, 671–679.
Ramberg, C.F., Johson, E.K., Fargo, R.D. and Kronfeld, D.S. (1984) Calcium homeostasis in dairy cows with special reference to parturient hypocalcaemia. *American Journal of Physiology* 246, 698–704.
Reid, B.L. and Weber, C.W. (1976) Calcium availability and trace mineral composition of feed grade calcium supplements. *Poultry Science* 55, 600–605.
Roland, D.A. (1986) Calcium and phosphorus requirements of commercial Leghorns. *World Poultry Science Journal* 42, 154–165.
Roland, D.A., Sr, Sloan, D.R. and Harms, R.H. (1974) Effect of various levels of calcium with and without pellet-sized limestone on shell quality. *Poultry Science* 53, 662–666.
Sachs, M., Bar, A., Nir, O., Ochovsky, D., Machnai, B., Meir, E., Weiner, B.Z. and Mazor, Z. (1987a) Efficacy of 1α hydroxyvitamin D3 in the prevention of bovine parturient paresis. *Veterinary Record* 120, 39–42.
Sachs, M., Perlman, R. and Bar, A. (1987b) Use of 1α hydroxyvitamin D3 in the prevention of bovine parturient paresis. IX. Early and late effects of a single injection. *Journal of Dairy Science* 70, 1671–1675.
Sansom, B.F., Allen, W.M., Davies, D.C., Hoare, M.N., Stenton, J.R. and Vagg, M.R. (1977) The effects of 1α-OH-cholecalciferol on calcium, magnesium and phosphorus metabolism in dairy heifers and its potential value for the prevention of milk-fever. *Calcified Tissue Research* 22, 397–399.
Sansom, B.F., Bunch, K.J. and Dew, S.M. (1982) Changes in plasma calcium, magnesium, phosphorus and hydroxyproline concentrations in ewes from twelve weeks before until three weeks after lambing. *British Veterinary Journal* 138, 393–401.
Schneider, K.M., Ternouth, J.H., Sevilla, C.C. and Boston, R.C. (1985) A short-term study of calcium and phosphorus absorption in sheep fed on diets high and low in calcium and phosphorus. *Australian Journal of Agricultural Research* 36, 91–105.
Scott, M.I., Hull, S.J. and Mullenhoff, P.A. (1971) The calcium requirements of laying hens and effects of dietary oyster shell upon egg quality. *Poultry Science* 50, 1055–1063.
Simons, P.C.M. (1986) In: Fisher, C. and Boorman, K.N. (eds) *Nutrient Requirements of Poultry and Nutritional Research*. Butterworths, London, pp. 141–145.
Sklan, D. (1992) A note on production responses of lactating ewes to calcium soaps of fatty acids. *Animal Production* 55, 288–291.
Sklan, D., Kaim, M., Moallem, V. and Fulman, V. (1994) Effect of dietary calcium soaps on milk yield, body weight, reproductive hormones and fertility in first parity and older cows. *Journal of Dairy Science* 77, 1652–1660.
Smith, B.S.W. and Wright, H. (1981) Seasonal variation in serum 25-hydroxyvitamin D concentration in sheep. *Veterinary Record* 109, 139–141.

Soares, J.H. (1995) Calcium bioavailability. In: Ammerman, C.B., Baker, D.H. and Lewis, A.J. (eds) *Bioavailability of Nutrients for Animals.* Academic Press, New York, pp. 1195–1198.

Spence, J.A., Sykes, A.R., Atkinson, P.J. and Aitchison, G.U. (1985) Skeletal and blood biochemical characteristics of sheep during growth and breeding: a comparison of flocks with and without broken mouth. *Journal of Comparative Pathology* 95, 505–522.

Steevens, B.J., Bush, L.J., Stout, J.D. and Williams, E.I. (1971) Effects of varying amounts of calcium and phosphorus in rations for dairy cows. *Journal of Dairy Science* 54, 655–661.

Suttle, N.F. and Wadsworth, I.R. (1991) Physiological responses to vaccination in sheep. *Proceedings of the Sheep Veterinary Society* 15, 113–116.

Sykes, A.R. (1983) Effects of parasitism on metabolism in the sheep. In: Haresign, W. (ed.) *Sheep Production.* Butterworths, London, pp. 317–334.

Sykes, A.R. (1987) Endoparasites and herbivore nutrition. In: Hacker, J.B. and Ternouth, J.H. (eds) *The Nutrition of Herbivores.* Academic Press, Canberra, pp. 211–232.

Sykes, A.R. and Field, A.C. (1972) Effects of dietary deficiencies of energy, protein and calcium on the pregnant ewe. I. Body composition and mineral content of the ewes. *Journal of Agricultural Science, Cambridge* 78, 109–117.

Sykes, A.R. and Geenty, K.C. (1986) Calcium and phosphorus balances of lactating ewes at pasture. *Journal of Agricultural Science, Cambridge* 106, 369–375.

Thompson, J.K., Gelman, A. and Weddell, J.R. (1988) Mineral retention and body composition of grazing lambs. *Animal Production* 46, 53–62.

Van Klooster, A. (1976) Adaptation of calcium absorption from the small intestine of dairy cows to changes in the dietary calcium intake and at the onset of lactation. *Zietschrift Tierphysiologie Teirernahrung Futtermittelkunde* 37, 169–182.

Van Leeuwen, J.M. and De Visser, H. (1976) Dynamics of calcium metabolism in lactating cows when the calcium content of the rations is reduced. *Tijdschrift Diergeneeskunde* 101, 825–834.

Vaughan, J.M. (1970) Bone growth and modelling. In: *The Physiology of Bone.* Clarendon Press, Oxford, pp. 83–99.

Wan Zahari, M., Thompson, J.K., Scott, D. and Buchan, W. (1990) The dietary requirements of calcium and phosphorus for growing lambs. *Animal Production* 50, 301–308.

Ward, G., Harbers, L.H. and Blaha, J.J. (1979) Calcium-containing crystals in alfalfa: their fate in cattle. *Journal of Dairy Science* 62, 715–722.

Whitehead, C.C. (1995) Nutrition and skeletal disorders in broilers and layers. *Poultry International* 34, 40–48.

Wilson, J.H. (1991) Bone strength of caged layers as affected by dietary calcium and phosphorus concentrations, reconditioning and ash content. *British Poultry Science* 32, 501–508.

Wuthier, R.E. (1993) Involvement of cellular metabolism of calcium and phosphate in calcification of avian growth plate. *Journal of Nutrition* 123, 301–309.

Yano, F., Yano, H. and Breves, G. (1991) Calcium and phosphorus metabolism in ruminants. In: *Proceedings of the Seventh International Symposium on Ruminant Physiology.* Academic Press, New York, pp. 277–295.

Phosphorus

5

Introduction

An early indication that a shortage of phosphorus could have serious conseqences for livestock production came from the pioneering work of Sir Arnold Theiler (1912) early this century. He investigated two debilitating diseases of cattle and sheep grazing pastures at Armoedsvlakte in South Africa's northern Cape, diseases known locally as 'styfziekte' and 'lamziekte'. They were characterized by high mortality, poor growth and fertility, fragile bones and periodic cravings among survivors for the bones of their less fortunate predecessors. Looking to the bones for a clue as to the possible underlying nutritional deficiency, phosphorus emerged as the prime suspect and, within 20 years, a series of successful preventive measures had been developed (Theiler and Green, 1932). However, the optimal level and pattern of phosphorus supplementation for cattle at Armoedsvlakte is still the subject of investigation (De Waal *et al.*, 1996). Other pastoral areas lacking in phosphorus were subsequently identified on many continents, particularly those of the southern hemisphere. Indoor experiments were conducted to confirm and quantify the need for phosphorus and the synergism with calcium whereby the skeleton could develop while maintaining its strength. As the story unfolded, it came apparent that phosphorus had equally important roles to play in the soft as well as the hard tissues of the body and that exchanges between them influenced the development of clinical abnormalities, just as much as the dietary supply. Pigs and poultry would continually succumb to phosphorus deficiency if their grain-based rations were not routinely supplemented with phosphorus. The success of early supplementation studies with mixtures of calcium and phosphorus (bone-meal or dibasic calcium phosphate ($CaHPO_4$)) delayed recognition of the source of the problem (low phosphorus absorbability in grains), the antagonistic role of dietary calcium and the contrasting metabolism of the two elements. In all three respects, non-ruminants differ markedly from ruminants and these contrasts can best be drawn out by breaking with convention and giving phosphorus a chapter

© CAB *International* 1999. *Mineral Nutrition of Livestock*
(E.J. Underwood and N.F. Suttle)

on its own; requirements for phosphorus are, however, tabulated alongside those for calcium in the preceding chapter for convenience.

Dietary Sources of Phosphorus

Forages

The phosphorus status of forages varies widely and is influenced primarily by the phosphorus status of the soil, the stage of maturity of the plant and the climate. On average, phosphorus concentrations increase by 0.03–0.05 g kg^{-1} dry matter (DM) mg^{-1} extractable soil phosphorus (Minson, 1990; Jumba *et al.*, 1995). Temperate forages generally contain more phosphorus (P) than tropical forages (3.5 vs. 2.3 g P kg^{-1} DM) and legumes slightly more than grasses (3.2 vs. 2.7 g P kg^{-1} DM) (Minson, 1990), but there are exceptions. Tropical legumes, such as *Stylosanthes,* grow vigorously on soils that provide insufficient phosphorus for other species, but their phosphorus status remains low (often < 1.0 g P kg^{-1} DM). Distribution of phosphorus between leaf and stem is relatively uniform, but there is a marked reduction in whole-plant phosphorus concentrations as the forage matures, particularly during the dry season (see p. 127). The higher phosphorus levels reported for silages than hays in the UK (3.2 vs. 2.2 g P kg^{-1} DM: MAFF, 1990) probably reflect the earlier growth stage at which silage is harvested. Recent evidence suggests that a uniformly high proportion (0.64–0.86) of the phosphorus in dry or fresh forages is absorbable (A_P) and absorbed by both sheep (Field *et al.*, 1984; Dayrell and Ivan, 1989; Ternouth, 1989; Rajaratne *et al.*, 1994; Scott *et al.*, 1995) and cattle (Martz *et al.*, 1990; Ternouth *et al.*, 1990, 1996; Coates and Ternouth, 1992; Bortolussi *et al.*, 1994; Hendricksen *et al.*, 1994; Ternouth and Coates, 1997).

Selective grazing

The preference of grazing animals for phosphorus-fertilized pasture (Jones and Betteridge, 1994) and their ability to select herbage of higher phosphorus concentration than that present in the standing sward has long been recognized. Workers at Armoedsvlakte noted two- to threefold increases in phosphorus levels in autumn samples of selected herbage from oesophageal fistulae, compared with hand-plucked samples (Engels, 1981), and became so disillusioned about the value of herbage sampling as an index of phosphorus supply that they, like others (Karn, 1997), have stopped herbage sampling altogether (Read *et al.*, 1986a, b; De Waal *et al.*, 1996)! The problem arises when the sward contains a mixture of young, mature and senescent material and/or a mixture of species which vary in phosphorus concentration and palatability (McLean *et al.*, 1990). Selection of different parts of a given plant by the animal as opposed to the human hand is not a major factor, because distribution of phosphorus within the plant is relatively uniform. Under experimental conditions, the use of resident, oesophageally fistulated animals can be used to get a closer measure of phosphorus intake, with appropriate use of ^{32}P to correct for contamination by salivary phosphorus (Langlands, 1987). Such values sometimes show reasonable agreement with hand-plucked samples (Coates *et al.*, 1987; Coates and Ternouth, 1992). If the entire stand is virtually removed during a grazing period, the fact that selection influences phosphorus intake early on is of little consequence in terms of total phosphorus supply, and hand-plucked samples can be standardized to give samples of known maturity for particular species (Kerridge *et al.*, 1990). It is therefore recommended that herbage phosphorus concentrations be measured, wherever possible, by consistent, defined methods to record changes in the supply of available phosphorus to plants from the soil from year to year.

Absorbability of phosphorus in concentrates

Cereals contain relatively uniform and apparently adequate phosphorus concentrations (2.7–4.3 g P kg^{-1} DM) and vegetable protein sources even more (5–12 g P kg^{-1} DM). Most of this (50–80%) is present as phytate (P_p), which is well utilized by ruminants, high A_p of 0.78–0.81 having been reported for two cereal and three vegetable protein sources for sheep (Field *et al.*, 1984). Phosphorus deficiency should never arise in ruminants given significant amounts of concentrate feeds. The situation with non-ruminants has, until recently, been quite different. Comparisons with inorganic standards (P_i) of high availability (sodium and potassium phosphates) indicate that cereals and vegetable proteins generally provide phosphorus with only 20–45% of the absorbability of mineral sources for pigs and only slightly more (25–50%) for chickens (Soares, 1995). Limited evidence for turkeys indicates that they may be able to utilize plant phosphorus much more efficiently than pigs or chickens, relative values of 80% being found for maize and cottonseed meal (Andrews *et al.*, 1972). Sources such as sunflower meal, palm-kernel cake and groundnut meal were used very inefficiently by pigs (< 15% relative availability, each in single studies). Cromwell (1992) reported that availabilities of phosphorus in cottonseed meal, maize, dehulled soybean meal, oats and wheat were 0, 12, 20, 23 and 51%, respectively, and that the average relative value for a maize–soybean blend was 15%. Availability is not simply a reflection of the proportion of P_p present, however. The apparent absorption of phosphorus in pigs is much higher in wheat and barley than in maize (47 and 39 vs. 17%), despite similar proportions of P_p (70.7 and 63.6 vs. 65.6% of total P) (Jongbloed and Kemme, 1990). The superiority of wheat was related to a high phytase content and the attribute was shared by wheat middlings. To avoid uncertainties regarding availability, a convention was widely adopted whereby only non-phytate phosphorus was regarded as available to pigs and poultry. The convention is unduly pessimistic as far as absorbability of P_p is concerned, particularly for pigs. Values for A_p from published studies with pigs, obtained by correcting faecal excretion for an endogenous (FE_p) contribution of 20 mg kg^{-1} live weight (LW) (ARC, 1981), commonly fall between 0.5 and 0.7 for diets in which most of the phosphorus is provided by phytate and excesses of calcium are avoided. With growing concerns about the amounts of phosphorus which are discharged via farm wastes into the environment and the costliness of finite global supplies of phosphorus, the profligate convention no longer goes unquestioned. Animal protein sources, other than feather meal, are rich in terms of both concentration and availability of phosphorus (Jongbloed and Kemme, 1990). Meat and bone-meals and fish-meals contain > 30 g P kg^{-1} DM with high relative availabilities of > 85% for pigs and poultry (Soares, 1995).

Utilization of phytate phosphorus

The value of P_p as a source of phosphorus is influenced by many factors but particularly by the phytase activity in the feed (Pointillart *et al.*, 1987), the levels of added calcium and phosphorus and the age of the animal. Attainment of high temperatures (> 84°C) during pelleting can inactivate phytases present in the feed (Simons *et al.*, 1990). Soaking the grain can enhance the beneficial effects of added phytase (Liu *et al.*, 1997), and the high relative availabilities of phosphorus in moist maize and wheat (41–53%: Soares, 1995) may be attributable to enhanced phytase activity. Nelson *et al.* (1971) were the first to study purified phytase as a dietary supplement, using chickens. Addition of 850 U phytase released 1 g P_p (about 50% of that present) in a recent study with broiler chicks (Schoner *et al.*, 1993), and similar releases were achieved in pigs with 400 U (Hoppe *et al.*, 1993) and 246 U (Kornegay and Qian, 1996) of phytase. Small supplements of phytase (250–750 U kg^{-1} DM) can fully replace P_i supplements at all stages of pig production, from the nursery (Murry *et al.*, 1997) through growing (Liu *et al.*, 1997) to finishing (O'Quinn *et al.*, 1997) stages, with cereals as diverse as pearl millet, maize and sorghum. Bran is rich in phytase and addition of 250 U in this form can also avoid the need to add P_i to diets for pigs (Han *et al.*, 1997). Addition of phytase concurrently improves the absorption of calcium in both pigs (Kornegay and Qian, 1996; Qian *et al.*, 1996; Liu *et al.*, 1997; O'Quinn *et al.*, 1997) and poultry (Simons *et al.*, 1990), but this may be due in part to the increased need for calcium which resulted from the improved phosphorus supply from deficient basal diets. Another effective approach has been to give large supplements of 1,25-dihydroxycholecalciferol (1,25-$(OH)_2D_3$) (Edwards, 1993). Biotechnology may facilitate both approaches, recombinant phytase being successfully used in pigs (Cromwell *et al.*, 1995) and the 1α-analogue of vitamin D_3 in chicks (Biehl *et al.*, 1995; Biehl and Baker, 1997a, b); combining the two approaches gives additional benefits (Biehl *et al.*, 1995). Early assumptions that hydroxylated vitamin D_3 had increased intestinal phytase activity have not been substantiated, and the improved utilization of P_p may be a secondary effect following the increased absorption of calcium from the gut (Biehl and Baker, 1997a). Utilization of P_p can also be improved by feeding calcium strictly to minimum requirement standards. Edwards and Veltmann (1983) found that the availability of P_p to broiler chicks varied from 11 to 39% as dietary calcium and phosphorus varied (from 0.63 to 1.67% and from 0.53 to 1.01%, respectively), low utilization being associated with high concentrations of each mineral. Added calcium also reduces the efficiency of phytase additions to pig rations (Lei *et al.*, 1994).

Metabolism of Phosphorus

The control of phosphorus metabolism is necessarily quite different from that of calcium. Provided that phosphorus is present in absorbable forms in the diet, it is, like other anions, extensively absorbed, even when supplied well in excess of need, and absorption from milk is almost complete (Challa and Braithwaite, 1989). The kidney and gut are both routes for the excretion of phosphorus that is surplus to requirement, the gut (by way of saliva) being the major route for excretion in grazing ruminants or those fed roughage diets, whereas the kidney is of greater importance in monogastrics and in ruminants fed diets that promote low salivary flow rates (processed, pelleted or concentrate diets). There is no tight hormonal control of inorganic phosphorus concentrations in the bloodstream and serum values range widely above and below the renal threshold of 2–3 mmol l^{-1} in perfectly healthy animals.

Absorption in weaned ruminants

The relationship between absorbed and ingested inorganic phosphorus is linear over wide ranges of intake and the slopes indicate A_p of about 0.74 in weaned sheep (Braithwaite, 1986) and 0.83 in weaned calves (Challa *et al.*, 1989). Furthermore, these high levels of A_p can be attained with indigestible natural sources, such as straw (Ternouth, 1989) and dry, summer forage (Coates and Ternouth, 1992), though less so with bran (0.63; Field *et al.*, 1984). The P_p present in the form of inositol tetra-, penta- and hexaphosphates in straw and seeds is extensively hydrolysed in the rumen by microbial phytase and phosphate is thus released (Reid *et al.*, 1947; Nelson *et al.*, 1976; Morse *et al.*, 1992). There are marginal but quantitatively important increases in A_p in the phosphorus-deficient animal (Scott *et al.*, 1984, 1985; Coates and Ternouth, 1992) and small decreases at excessive phosphorus intakes (Braithwaite, 1984b; Challa *et al.*, 1989). Phosphorus is absorbed principally from the proximal small intestine (Bown *et al.*, 1989; Yano *et al.*, 1991).

Absorption in weaned monogastrics

Non-ruminants are dependent on the presence of phytases found either in the grains themselves or in intestinal secretions to utilize phytate phosphorus. Undegraded phytate can be extensively converted to insoluble mineral–phytate complexes before the site of phosphorus absorption is reached, thus lowering the amount absorbed. Older animals have a greater ability than younger to digest phytate. Adverse effects of calcium–phytate interactions on the absorption of manganese and zinc have been widely demonstrated (Biehl *et al.*, 1995) and are considered in later chapters.

Salivary phosphorus secretion in ruminants

Ruminants also differ from non-ruminants in that large quantities of phosphorus are secreted in saliva during the process of rumination and the major source of phosphorus flowing into the rumen is not the diet but the salivary secretions (Tomas *et al.*, 1973). The principal influence on the amount of phosphorus secreted in saliva is the P_i in plasma, the relationship between the two variables being linear over the range 1–3 mmol P_i l^{-1} and equivalent to a doubling of salivary phosphorus output in cattle (Challa and Braithwaite, 1988; Challa *et al.*, 1989). Phosphorus does not appear to be reabsorbed by the salivary gland, and parathyroid hormone (PTH) can only increase salivary phosphorus secretion when there is an ample supply (Wright *et al.*, 1984). The nature of the diet, however, has a major influence. As the roughage content of the diet increases, the partition of initially absorbed phosphorus between salivary secretion and urinary excretion shifts towards the salivary route (AFRC, 1991). The outcome and prediction of phosphorus partitioning in ruminants is of crucial importance in determining the actual or likely incidence of phosphorus deficiency and is still the subject of controversy (Ternouth, 1990; Scott *et al.*, 1995). One problem lies in the difficulty of separating the two important determinants of salivary phosphorus secretion, dry-matter intake (DMI) and phosphorus intake. Without a phosphorus-free diet, the specific effect of DMI is only obtainable with diets of different phosphorus concentration fed at several different intakes. Endogenous faecal losses do increase with DMI in the short term (AFRC, 1991), but, if the diet is deficient in phosphorus, plasma and salivary phosphorus will decline with duration of feeding and FE_p will gradually decline (Ternouth, 1989; AFRC, 1991; Coates and Ternouth, 1992) as the major homeostatic adjustment to deficiency. The physical forms of ingested roughage reported on so far do not appear to be important (Ternouth, 1989; Scott *et al.*, 1995), but they did not include long roughage. Yano *et al.* (1991) cite unpublished data which indicate a doubling of salivary phosphorus flow when the length of hay being fed to sheep was increased from 1 to 6 cm. Because absorptive efficiency for salivary phosphorus is high (70–80%: Challa *et al.*, 1989), the ruminant is able to recycle phosphorus very efficiently via saliva, and diets which stimulate secretion may improve the utilization of dietary phosphorus (Yano *et al.*, 1991).

Faecal and urinary excretion

Some phosphorus is irreversibly lost via the faeces (FE_p) by ruminants and the amount depends on the phosphorus concentration in the saliva (Braithwaite, 1984b; Challa *et al.*, 1989), the volume of saliva produced and the precise efficiency of absorption. Since the latter is subject to genetic variation in sheep, marked differences in FE_p are found within (Field *et al.*, 1984) and between (Field *et al.*, 1986) breeds of sheep. When phosphorus supply is so inadequate that absorptive efficiency rises beyond the normally high level, there is likely to be an additional conservation of salivary phosphorus by reductions in FE_p, which may sometimes be of crucial importance. Despite attempts to conserve phosphorus, FE_P becomes the predominant source of faecal phosphorus (70–90%) on diets low in phosphorus. As phosphorus intakes rise, the capacity for faecal endogenous excretion becomes saturated and most of any excess over need is excreted in urine. On pelleted, energy-dense diets, which stimulate little salivary secretion, significant urinary phosphorus excretion may occur at levels of supply which fail to allow complete mineralization of bone. Theoretically, such urinary losses should be allowed for in devising requirements for optimum bone strength (Braithwaite, 1984b). However, such diets are never likely to provide insufficient phosphorus for ruminants. Obligatory urinary losses of phosphorus are more likely to warrant attention and replacement in non-ruminants.

Bone growth and mineralization in the young

Another important mechanism for adjusting to inadequate dietary supplies of phosphorus involves the modulation of bone growth and mineralization of the skeleton. As was the case with calcium (see Chapter 4), there are differences between growing and mature animals. Weanling lambs given low-phosphorus diets can continue to increase their skeletal size without requiring more phosphorus. In Field *et al.*'s (1975) study with diets low in phosphorus, *weight of the fat-free skeleton increased by 19%* during a 4-month period, *while the phosphorus content decreased by 11%* (Table 4.3 on p. 79). The adjustment involved decreases in the amount of bone matrix in the limb bones and in the degree of matrix mineralization in cancellous bones. There may, however, be major differences between species in the extent to which such an adaptation can sustain growth. In the pig, as in other species, there is a substantial increase in the degree of mineralization of bone after weaning, bone ash content increasing from around 45% to 55% between 15 and 85 kg LW in one study (Fig. 5.1; Reinhart and Mahan, 1986). That increase was seen in groups given diets sufficiently low in phosphorus to retard growth, indicating a higher priority to provide sufficient phosphorus to mineralize bone. Presumably, it is impossible for species with such high relative growth rates to find sufficient additional phosphorus by redistributing the pool of phosphorus in the skeleton; instead, pigs deprived of phosphorus restrict growth. To describe bone mineralization in the pig as a 'sensitive criterion' of phosphorus status (Koch and Mahan, 1985) is somewhat misleading in such circumstances. The relationship between growth and bone mineralization for a given species may be partly determined by serum P_i. Data from Reinhart and Mahan (1986) show clear relationships between growth rate and serum P_i at each growth phase in the pig (Fig. 5.2), and the optimum serum P_i is over 2.5 mmol l^{-1}, a higher value than that which is considered optimal for growth in the lamb. The chick also shows a linear relationship between bone ash and serum P_i, up to values of around 2.5 mmol l^{-1} (Mohammed *et al.*, 1991).

Bone mineralization in the adult

In mature animals of all species, the quiescent remodelling process can be awakened to allow the net withdrawal of substantial amounts of phosphorus,

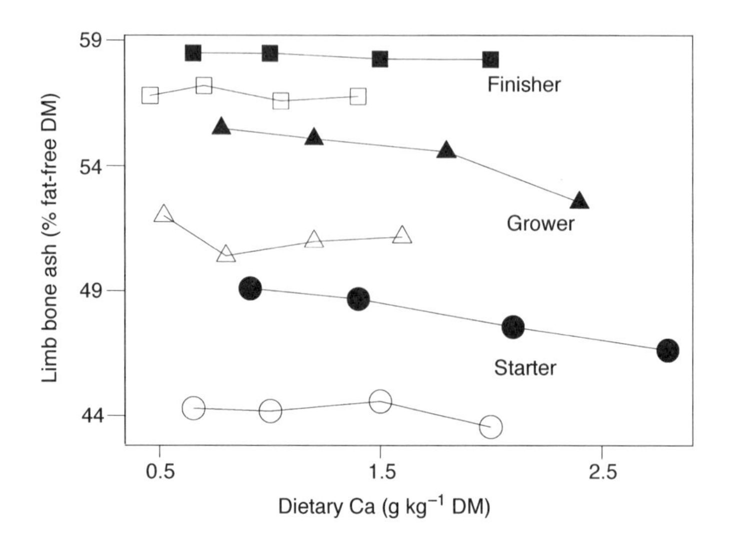

Fig. 5.1. The effects of dietary P and age on the mineralization of limb bones in young pigs: note the large increases in bone ash from the starter to the finisher stage, regardless of whether diets low (open symbol, 0.5, 0.4 and 0.3 g P kg^{-1} DM) or high in phosphorus (closed symbol, 0.7, 0.6 and 0.5 g P kg^{-1} DM for successive stages) were fed (data from Reinhart and Mahan, 1986).

and mobilization of stored phosphorus becomes more important than dietary phosphorus intake in determining animal health and production. It can be calculated that lactating cows and ewes need to mobilize only 0.7% of their skeletal reserves each day to sustain peak daily milk phosphorus outputs (Suttle, 1987). There are three distinctive features as far as phosphorus is concerned: firstly, the process does not appear to be under the same hormonal control as calcium, since PTH concentrations fall during phosphorus deficiency (Somerville *et al.*, 1985); secondly, bone resorption provides too little phosphorus relative to calcium (Ca) for milk production, since the Ca:P ratio in milk is only 1.3:1, as opposed to 2:1 in bone; and, thirdly, the utilization of stores is less efficient than for calcium, because more of the mobilized mineral will be lost via the faeces, despite the eventual decline in plasma phosphorus, salivary secretion and FE_P.

Hormonal regulation

It is arguable whether there is any specific and effective hormonal regulation of phosphorus metabolism. While PTH secretion improves the renal tubular reabsorption of phosphorus, the hypercalcaemia which usually accompanies phosphorus deficiency inhibits PTH secretion. While administration of vitamin D or its analogues often improves phosphorus absorption and retention in pigs (Pointillart *et al.*, 1986; Lei *et al.*, 1994) and poultry (Mohammed *et al.*, 1991; Biehl *et al.*, 1995), such effects may be secondary to those on calcium, for two reasons: firstly, any improvement in calcium absorption

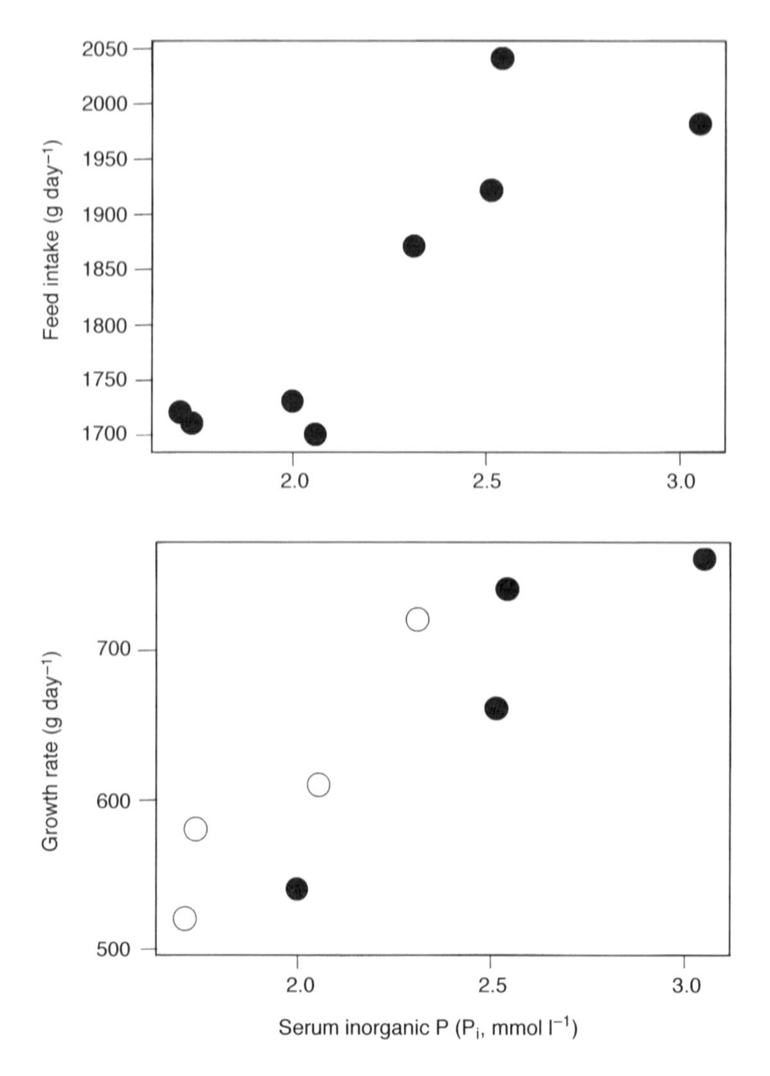

Fig. 5.2. Relationship between serum inorganic phosphorus (P_i), growth and appetite in growing pigs: note that feed intake reaches a plateau before growth rate as P_i rises, suggesting a specific effect of phosphorus deprivation on growth (data from Reinhart and Mahan, 1986); open circle, low P; closed circle, high P; dietary Ca varies.

lessens opportunities for calcium to form unabsorbable phytates in the gut; and, secondly, any improvement in retention of calcium in the skeleton must be accompanied by retention of phosphorus. There are differences between species in endocrine responses to phosphorus depletion. Increases in renal 1α-hydroxylase activity and plasma 1,25-$(OH)_2D_3$ have been reported in the phosphorus-depleted pig and chick (for review, see Littledike and Goff, 1987), but there is no such response in sheep (Breves *et al.*, 1985). One possible explanation for a lesser role of vitamin D in regulating phosphorus metabolism in ruminants is that phytate does not present an obstacle to phosphorus absorption; another is the greater reliance which ruminants can place on skeletal sources of phosphorus. Increases in intestinal calbindin can occur in the phosphorus-depleted pig without changes in the circulatory levels of 1,25-$(OH)_2D_3$ (Pointillart *et al.*, 1989), and any influence of vitamin D on phosphorus metabolism may again be secondary to that on calcium.

Gut parasites as phosphorus antagonists

The remarkable demineralization of the skeleton induced by parasitic nematode infections of the sheep's gut was outlined in Chapter 4; here, the specific effects on phosphorus metabolism will be dealt with. Infections of the small intestine (e.g. by *Trichostrongylus colubriformis*) can reduce the absorption of dietary and endogenous phosphorus by about 40% and induce hypophosphataemia, but infections of the abomasum (e.g. by *Ostertagia circumcincta*) have little or no effect (Wilson and Field, 1983). Combined infections have similar phosphorus-depleting effects, which last up to 17 weeks (Fig. 5.3; Bown *et al.*, 1989). Although food intake was reduced by 60%, this alone did not induce phosphorus deficiency, because the ryegrass/clover diet contained a respectable 3.3 g P kg^{-1} DM and was sufficient to maintain normal plasma phosphorus levels in pair-fed controls. While the sustained, dual infectious challenge was severe (generally 12,000 infectious larvae kg^{-1} herbage DM) in comparison with conditions commonly encountered on intensively grazed temperate pastures, it is likely that seasonal nematode infections often increase phosphorus requirements and may reduce the capacity to withstand later seasonal shortages in dietary phosphorus supply (Suttle, 1994).

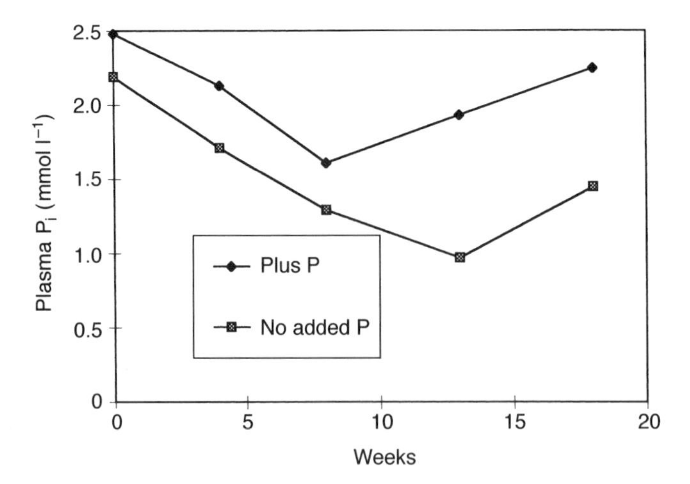

Fig. 5.3. Effects of prolonged dual infection of the abomasum and small intestine by parasitic nematodes on plasma inorganic phosphorus (P_i) concentrations in lambs (Bown *et al.*, 1989).

Functions of Phosphorus

Phosphorus is the second most abundant mineral in the animal body and about 80% is found in the bones and teeth. As with calcium, the formation and maintenance of bone are quantitatively the most important function, and the changes in bone structure and composition that result from phosphorus deprivation are in most respects the same as those described for calcium deprivation in Chapter 4. Phosphorus, however, is required for the formation of the organic bone matrix as well as the mineralization of that matrix. The 20% of phosphorus not present in the skeletal tissues is widely distributed in the fluids and soft tissues of the body, where it serves a range of essential functions. Phosphorus is a component of deoxy- and ribonucleic acids, which are essential in cell growth and differentiation; as phospholipid, it contributes to cell-membrane fluidity and integrity; as phosphate, it helps to maintain osmotic and acid–base balance; and it plays a vital role in a host of metabolic functions, including energy utilization and transfer via AMP, ADP and ATP, with implications for glucogenesis, fatty acid transport, amino acid and protein synthesis and the activity of the sodium/potassium (Na^+/K^+) pump. Disturbances of glycolytic metabolism have been noted in the erythrocytes from phosphorus-deficient cattle (Wang *et al.*, 1985). In ruminants, the requirements of the rumen and caecocolonic microflora are also important, and microbial protein synthesis may be impaired on low-phosphorus diets (Petri *et al.*, 1989; Ternouth and Sevilla, 1990b). Phosphorus is further involved in the control of appetite, in a manner not yet fully understood, and in the efficiency of feed utilization (Ternouth, 1990). Phosphorus is arguably the most potent of all the mineral elements.

Biochemical, Physical and Histological Abnormalities

The clinical sequelae of phosphorus deprivation are accompanied or preceded by biochemical changes in the blood and tissues.

Changes in bone

The chemical changes in the bones that precede or accompany the structural changes associated with phosphorus dysfunction are apparent from two early experiments with pigs and sheep. Growing pigs fed for 6 weeks on a synthetic milk diet containing 2.0 g P kg^{-1} had bones which were less well mineralized than those of comparable animals receiving phosphorus-supplemented diets (Table 5.1). In a longer experiment (14–18 months) with growing sheep fed on a moderately phosphorus-deficient diet, the ash concentration of the ribs and vertebrae was some 20% lower and that of the long bones over 8% lower than that of similar bones from sheep supplemented with phosphate (Stewart, 1934/35). Reductions occur in both the ash and organic-matter content of the bone in phosphorus-deprived pigs (Koch and Mahan, 1985) and lambs (Field *et al.*, 1975), so that ash percentage

Table 5.1. Effects of dietary phosphorus concentrations on selected performance, skeletal and blood indices of phosphorus status in the baby pig given a synthetic milk diet from 0 to 6 weeks of age (Trial 1; Miller *et al.*, 1964).

	Dietary P (g kg^{-1} DM)			
Criterion	2	4	6	
Live-weight gain (g day^{-1})	130	290	290	Performance
Dry-matter intake (g)	240	380	370	
Feed/gain (g g^{-1})	1.85	1.31	1.28	
Serum inorganic P (mmol l^{-1})	1.03	2.71	3.03	Blood
Serum Ca (mmol l^{-1})	3.20	3.05	2.58	
Serum alkaline phosphatase (BL units)	20.5	8.3	4.5	
Humerus ash (%)	33.4	44.1	47.5	Bone
Femur s.g.	1.11	1.15	1.17	
Eighth rib weight (g)	3.30	4.49	4.67	
Femur breaking load (kg)	24	61	81	

s.g., specific gravity.

underestimates the degree of deprivation. Long bones tend to be depleted more rapidly than metacarpals or metatarsals. In young calves depleted of phosphorus, the ends of the long bones showed greater reductions than the shaft, while the rib showed no change after short-term (6 week) depletion (Fig. 5.4). However, phosphorus (or calcium) is removed more extensively from the rib than from axial and limb bones after prolonged depletion (Little, 1984), a fact confirmed histologically (Field *et al.*, 1975). Careful histological comparisons of the rachitic long bones from phosphorus- and calcium-deprived chicks showed that the former could be distinguished by elongated metaphyseal vessels within the physis and by hypertrophic chondrocytes (Lacey and Huffer, 1982). Histologically diagnosed, hypophosphataemic rickets in broilers was associated with low bone ash (49.6 vs. 55.6%), low calcium + phosphorus content of ash (40.5 vs. 50.2%) and a low P:Ca ratio in the long bones (Thorp and Waddington, 1997). Histological abnormalities of the growth plate were the most sensitive index of phosphorus status in turkey poults (Fig. 1.3; Qian *et al.*, 1996). Demineralization is also accompanied by changes in physical properties indicating the strength of bone, and shear strength is now regarded as one of the more dependable and sensitive tests in the pig (Combs *et al.*, 1991c).

Changes in the blood

Early responses to a diet low in phosphorus are a fall in serum or plasma P_i (Fig. 5.5), a rise in serum alkaline phosphatase (Table 5.1) and a small rise in serum calcium (see Fig. 4.5). After a few weeks or months, serum values fall to 1.0 mmol P_i l^{-1} or below but by then the first clinical sign, severe loss of appetite, may have appeared. The onset and severity of hypophosphataemia is not simply a reflection of dietary phosphorus supply. In young lambs given

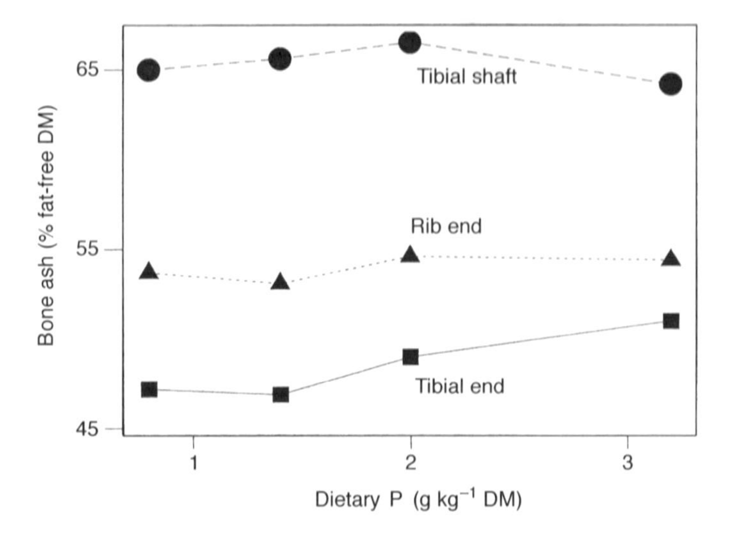

Fig. 5.4. Effects of sample site on the degree of bone demineralization recorded after short-term phosphorus depletion in calves (Miller *et al.*, 1987).

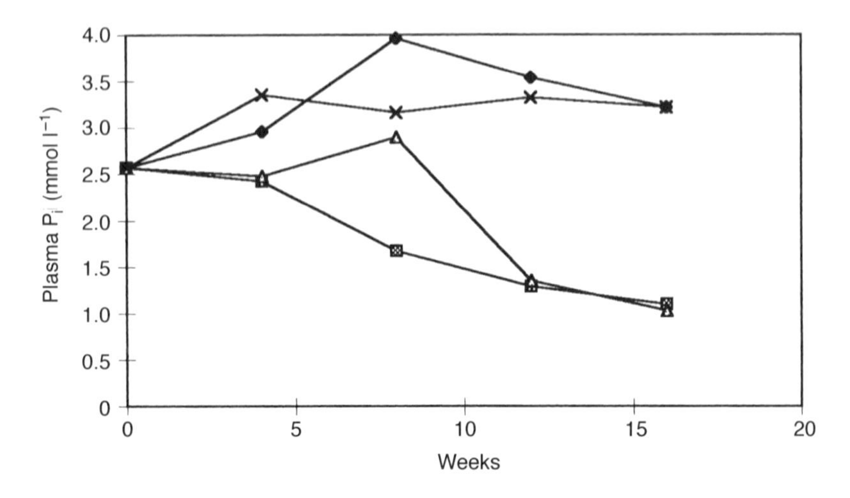

Fig. 5.5. Effects of diets low in phosphorus and/or calcium on plasma inorganic phosphorus (P_i) in lambs: note that feeding a diet low in both minerals (Δ) delays the fall in P_i compared with a diet low only in phosphorus (■), while feeding one low in calcium but adequate in phosphorus (♦) raises P_i when compared with a diet adequate in both minerals (X) (from Field *et al.*, 1975).

diets equally low in phosphorus (1.2 g P kg^{-1} DM), the decline in plasma P_i occurred 5 weeks later, when the diet was also low in calcium (0.66 vs. 4.3 g Ca kg^{-1} DM) (Fig. 5.5; Field *et al.*, 1975). Supplementary calcium does not invariably decrease phosphorus absorption from such diets (Field *et al.*, 1983, 1985), and the lowering of serum P_i may be partly due to an increase in phosphorus retention by the skeleton, a common response to calcium supplementation (Braithwaite, 1984a; Field *et al.*, 1985; Rajaratne *et al.*, 1994). Such dietary interactions do not invalidate the use of serum P_i for diagnostic purposes.

Changes in rumen phosphorus

Like levels in blood, rumen phosphorus concentrations are also influenced by factors other than dietary phosphorus intake. With a diet marginal in phosphorus (1.9 g P kg^{-1} DM), rumen P_i levels in lambs were 66% lower when the diet contained 6.8 rather than 3.5 g Ca kg^{-1} DM and accompanied by a major reduction in plasma P_i, from over 2.0 to around 1.3 mmol l^{-1} (Wan Zahari *et al.*, 1990). Rumen phosphorus had fallen towards levels associated with impaired rumen microbial activity (3 mmol l^{-1}) and were linked to a 20% reduction in growth rate on the high-calcium diet. Reductions in rumen microbial activity are usually accompanied by reductions in food intake.

Changes in saliva

Phosphorus concentrations in saliva can reflect the concentrations in plasma in some circumstances (Challa *et al.*, 1989), but they are unreliable indices of dietary phosphorus supply. Levels in saliva are about three to eight times those in plasma, but they are influenced by both salivary flow rate and phosphorus intake. At low flow rates, salivary phosphorus concentrations are much the same in both replete and phosphorus-deficient sheep, but, at high flow rates, concentrations are much lower in the phosphorus-deficient animal (Wright *et al.*, 1984). This variability makes spot saliva samples unreliable as an indicator of phosphorus status.

Changes in urine

The majority of sheep and cattle fed roughage diets excrete very little phosphorus in their urine, even at plasma P_i levels that are well above 1 mmol l^{-1}, and this situation continues despite an increase in phosphorus intake sufficient to increase plasma P_i levels to over 2 mmol l^{-1}. If they are switched to diets that promote low salivary flow (e.g. pelleted concentrates), a similar increase in phosphorus intake is matched by a proportional increase in urinary phosphorus excretion. This seems to involve a change in renal reabsorptive efficiency (proportion of filtered phosphorus which is reabsorbed), rather than a change in renal threshold. Similar renal adaptation to physiological demand (pregnancy, lactation or low phosphorus intake) has been reported in other species and is not under any known hormonal control. A low urinary phosphorus level is therefore not a good indicator of phosphorus status in the grazing situation, where deficiencies are mostly likely to arise.

Changes in faeces

Although faecal phosphorus levels reflect endogenous loss as well as intake of phosphorus, they are being used to monitor the response of grazing cattle to phosphorus supplementation on pastures deficient in both nitrogen and phosphorus (Holechek *et al.*, 1985; Grant *et al.*, 1996; Karn, 1997). Faecal phosphorus levels below 2 g kg^{-1} faecal DM, together with plasma P_i levels less than 1 mmol l^{-1}, indicate low phosphorus status.

Clinical Manifestations of Phosphorus Deprivation

A dietary deficiency of phosphorus, if sufficiently severe or prolonged, leads to abnormalities of the bones and teeth, subnormal growth, milk yield and egg production, depressed appetite, poor efficiency of feed use and the development of pica or depraved appetite; fertility may be impaired.

Abnormalities of bones and teeth

The abnormalities associated with phosphorus deprivation are almost identical to those described in Chapter 4 for calcium deficiency, since bone and tooth mineral cannot be produced if either mineral is lacking. This was illustrated by parallel studies of phosphorus and calcium requirements in baby pigs (Miller *et al.*, 1964; Tables 5.1 and 4.2), which revealed similar rachitic lesions whichever element was lacking.

Poor growth, appetite and feed utilization

Loss of appetite and subnormal growth in young animals and weight loss in mature animals are characteristic of phosphorus deprivation in all species and are illustrated by data for yearling cattle in Table 5.2 (Little, 1968) and for young pigs in Table 5.1. In newly weaned pigs deprived of phosphorus, growth can be retarded before appetite is impaired (Fig. 5.2; Reinhart and Mahan, 1986), but the primary effect in ruminants is on food intake. The anorexia may be immediate (Fig. 4.5; Field *et al.*, 1975) but usually takes several weeks or months to develop in lambs (Wan Zahari *et al.*, 1990), beef cattle (Call *et al.*, 1978, 1986; Gartner *et al.*, 1982) and dairy cows (Call *et al.*, 1986), depending perhaps on the adequacy of phosphorus supply for rumen microbes. Early reductions in appetite have been confined to situations in which semipurified diets very low in phosphorus have been fed (Field *et al.*, 1975; Miller *et al.*, 1987; Ternouth and Sevilla, 1990a); these diets stimulate little or no rumination, depriving the rumen microflora of salivary as well as dietary phosphorus. It is noteworthy that a more severe anorexia eventually develops in such circumstances (Fig. 4.5). Dry-matter intake can be closely related to serum P_i and governed by systemic as well as gut-based mechanisms (Ternouth and Sevilla, 1990a, b). In beef cattle, improvements in

Table 5.2. Effect of phosphorus supplementation on growth and feed consumption of yearling cattle fed on Townsville lucerne (*Stylosanthes humilis*) containing 10% crude protein and 0.07% phosphorus for 8 weeks (mean values ± SE) (from Little, 1968).

Rate of supplementation with phosphorus	Phosphorus intake (g day^{-1})	Weight gain (kg $week^{-1}$)	Dry-matter intake (kg day^{-1})	Blood phosphorus[a] (mg l^{-1})
Nil	3.3 ± 0.1	0.9 ± 0.2	5.2 ± 0.1	33 ± 1.0
Low	7.4 ± 0.1	1.7 ± 0.2	5.8 ± 0.4	60 ± 1.0
Medium	11.4 ± 0.1	1.7 ± 0.2	6.0 ± 0.2	63 ± 3.0
High	15.5 ± 0.3	2.7 ± 0.7	6.0 ± 0.5	67 ± 0.4

[a] Divide by 31 to give mmol l^{-1}.

food intake and growth rate following phosphorus supplementation are variable but usually occur if dietary nitrogen is > 15 g kg^{-1} DM and phosphorus < 1.5 g kg^{-1} DM (Winks, 1990). Reduced egg production in the phosphorus-deficient hen has been attributed to a reduction in food intake (Bar and Hurwitz, 1984). Feed conversion efficiency (FCE) is invariably poor in phosphorus-deprived pigs, presumably due to a disturbance in energy metabolism (Koch and Mahan, 1985; Reinhart and Mahan, 1986; Fig. 5.7), but reductions in FCE in ruminants depleted of phosphorus are rare (Ternouth, 1990) and were not found in lambs (Field *et al.*, 1975). A decline in FCE was observed in beef cows during lactation when severe phosphorus deficiency was imposed from the last 16 weeks of pregnancy and oat straw formed the bulk of the diet (Fishwick *et al.*, 1977); similar findings have been reported in beef cattle (Long *et al.*, 1957) and lactating goats (Muschen *et al.*, 1988).

Pica

The loss of appetite caused by phosphorus deprivation is often paralleled by a craving for and a consumption of abnormal materials, such as soil, wood, flesh and bones. The depraved appetite (pica) may either take a generalized form, known as allotriophagia, or it can be expressed more specifically as osteophagia (craving for bones) and as sarcophagia (craving for flesh). These forms of pica are not specific to phosphorus deficiency, since they have been observed in animals suffering from lack of sodium and potassium and also in sheep receiving insufficient energy and protein under field conditions (Underwood *et al.*, 1940). Bone-chewing in cows has been induced by exteriorizing the parotid salivary duct and feeding a low-phosphorus diet, thus preventing the recycling of phosphorus (D. Denton, 1979, personal communication). A complete blocking of the drive to eat bones was achieved within 1 h by intravenous infusion of sodium phosphate sufficient to raise serum P_i to normal. Pica, whatever its cause, can be disastrous in areas where the carcasses are infected with *Clostridium botulinum*, and toxin formation during the process of putrefaction can cause botulism and death. Mortality in cattle from this cause provided the original stimulus for the South African investigations which led to the discovery of phosphorus deficiency. Losses of sheep and cattle from botulism (toxic paralysis) have been reported from several parts of the world, due to the widespread occurrence of *C. botulinum* and of dietary phosphorus deficiency (Underwood *et al.*, 1939). The animal can be protected against botulism by vaccination, but the problem of aphosphorosis remains unless remedial measures are taken.

Reproductive disturbances

Poor lamb and calf crops have been a recurring feature of flocks and herds confined to severely phosphorus-deficient grazing, but responses to phosphorus supplements have been inconsistent. In an early study of 200 breeding cattle in South Africa, only 51% produced calves in the untreated group, compared with 80% in animals receiving a bone-meal supplement. The subnormal fertility was associated with depressed or irregular oestrus, which prevented or delayed conception; this has persisted in cattle, but sheep in the same area are not affected (Read *et al.*, 1986a, b). In the light of later evidence, it seems unlikely that infertility is a specific effect of simple phosphorus inadequacy. No adverse effects on age of puberty or pregnancy occurred in Hereford heifers fed on a low-phosphorus (1.4 g P kg^{-1}) diet for 2 years (Call *et al.*, 1978) or on conception rate during rebreeding (Call *et al.*, 1986); infertility did not develop until the seventh year and, by that time, other symptoms of deficiency had developed. Improvements in reproduction rate have rarely been recorded under Australian conditions and, where they occur, they are probably secondary to an improvement in body weight and condition at conception (Winks, 1990). Furthermore, conception rate appears to be less sensitive to phosphorus deficiency than milk yield in dairy cows, remaining unaffected by a reduction in dietary phosphorus from 3.2 to 2.4 g P kg^{-1} DM, which reduced peak milk yield significantly, by 15% (Call *et al.*, 1987); the diet contained 7.2 g Ca kg^{-1} DM. However, delayed conception, causing temporary infertility, has been observed in dairy cows in Australia mated during full lactation with slightly subnormal serum P_i (Snook, 1958). Improved fertility in dairy cows, also with slightly subnormal serum P_i, has been reported after defluorinated superphosphate was added to the drinking-water (Scharp, 1979). Both the phosphorus and the protein in the lucerne (alfalfa) pastures being grazed were apparently adequate (3.9 g P and 175 g crude protein kg^{-1} DM) but calcium was high (15.5 g kg^{-1} DM). The first-service pregnancy rate increased from 36.5 to 63.2%, the mean calving to conception interval decreased from 109 to 85 days and the number of cows culled each year for infertility decreased from 15 to 5 following supplementation. Similar 'before and after' responses were reported by Brooks *et al.* (1984) to a mixture of supplementary measures applied to cows on red-clover-dominant pastures containing 2.9 g P kg^{-1} DM but again rich in calcium (7.0–14.5 g kg^{-1} DM).

Depression of milk yield

A lactating animal responds to a dietary deficiency of phosphorus by reducing its yield of milk without changing the concentrations of minerals in the milk produced (Muschen *et al.*, 1988). In the early stages of a deficiency or where the deficiency is moderate, the animal is able to draw on its skeletal reserves to maintain yield; several successive lactations may be necessary before bone defects and other clinical signs become obvious and milk production is impaired. In parts of South Africa where phosphorus deprivation was severe and prolonged, increases in milk production of 40 to 140% were recorded from the use of bone-meal supplements. These increases could not be attributed solely to phosphorus, since the bone-meal also provided protein (Bisschop, 1964), but improvements of up to 27.5% in weaning weight to specific phosphorus supplements have subsequently been reported (Read *et al.*, 1986a, b). In an experiment with beef cows given adequate dietary protein, phosphorus deficiency imposed during late pregnancy and early lactation lowered milk yield (Fishwick *et al.*, 1977). In lactating goats, transfer to a low-phosphorus diet after 6 weeks' lactation was sufficient to suppress milk yield (Muschen *et al.*, 1988). Cows pregnant with their second calf, given diets low in phosphorus (2.4 g P kg^{-1} DM) from the seventh month of gestation and for 7.5–10 months of lactation, showed maximum depression of milk yield (32–35%) between 18 and 34 weeks of lactation and proportionate reductions in food intake (Call *et al.*, 1986). Depression of milk yield may be a secondary consequence of loss of appetite or reduced synthesis of rumen microbial protein. On the whole, improvements in milk yield or weaning weight (an index of milk yield) have rarely been found but are most likely to occur during the first lactation, when animals are still growing (Winks, 1990; Karn, 1997).

Reduction in egg yield and quality in poultry

Phosphorus deprivation is manifested by a decline in egg yield, hatchability and shell thickness, but it is much less likely to occur than calcium deficiency, because of the far smaller requirement for phosphorus. Egg production and hatchability of fertile eggs from caged hens receiving all-plant diets, containing 3.4–5.4 g P kg^{-1} DM, decreased rapidly until 0.9 g P_i kg^{-1} DM was added, but the same amount of phosphorus added as phytate in hominy was ineffective (Waldroup *et al.*, 1967). Supplementation of diets of commercial type with inorganic phosphorus may thus be necessary for maximum egg yield and quality, but the effects are essentially an induced calcium deficiency resulting from the failure to build up the skeletal reserve of calcium needed to sustain egg production.

Diagnosis of Phosphorus Deprivation in Grazing Animals

The only definite evidence that lack of phosphorus has critically restricted livestock performance comes from improvements seen when appropriate supplements have been given. The following indices provide only rough guidance.

Serum inorganic phosphorus

Although serum P_i concentrations can fall spectacularly, particularly in the young growing animal depleted of phosphorus, their diagnostic value is limited. In the absence of tight homeostatic control, further factors, such as recent feeding (+), milk yield (−) and handling stress (−), can each significantly influence serum P_i in the direction indicated (Teleni *et al.*, 1976; Forar *et al.*, 1982). Concentrations are also influenced by the site of sampling (coccygeal and mammary vein > jugular vein; Teleni *et al.*, 1976) and interval between sampling, clot formation or separation of plasma and analysis. Use of buffered trichloroacetic acid (TCA) as a protein precipitant can reduce postsampling differences (Teleni *et al.*, 1976), but results obtained with colorimetric assay kits, which require no protein precipitation, will remain vulnerable to postsampling changes. For these and other reasons, it is a mistake to interpret phosphorus levels using a single threshold, and the recognition of a marginal band for serum P_i is crucial. Serum concentrations of 1.25–1.75 mmol P_i l^{-1} are 'marginal' for ruminants because there are many instances in which values of this order have been recorded in healthy grazing livestock, which do not respond to phosphorus, and others where responses to supplementation have been obtained at the upper limit (e.g. Wadsworth *et al.*, 1990; Coates and Ternouth, 1992). Marginal serum P_i concentrations are most likely to precede clinical aphosphorosis in young, rapidly growing stock or lactating animals receiving diets of high nutritive value. Older beef cattle on dry-season pastures of low nutritive value can have mean serum P_i values around 1.0 mmol l^{-1} and not benefit from phosphorus supplementation (Engels, 1981; Wadsworth *et al.*, 1990). The equivalent marginal range for young pigs is higher (Fig. 5.2) and probably narrower than for ruminants – 2.5–2.8 mmol l^{-1} is suggested; other papers confirm that growth retardation can occur with serum P_i as high as 1.9 mmol l^{-1} (Koch and Mahan, 1985).

Bone criteria in general

Since bone contains the reserves of calcium and phosphorus and calcium cannot be removed without changes in or relating to bone phosphorus, abnormalities are only indicative of a phosphorus deficit if calcium is non-limiting (compare Table 5.1 with Table 4.2). Protein depletion can also lead to poor mineralization of bones through the resorption of bone matrix or failure to form that matrix. The adequacy of phosphorus (or calcium) nutrition can be reflected by the degree of mineralization of the skeleton in sheep (Table 4.3), cattle (Little, 1984; Williams *et al.*, 1991a), pigs (Table 5.1) and turkeys (Fig. 1.3), but there are examples of differences in production attributable to phosphorus deprivation not being reflected by significant differences in bone quality (Wan Zahari *et al.*, 1990; Coates and Ternouth, 1992). The inconsistency is to be expected, given that the primary determinants of health and performance are factors that influence satiety (both ruminal and systemic), rather than bone strength, and that the bone sample is a minute portion of a heterogeneous reserve. Having acknowledged the limitations of bone indices, it is important to choose the most informative from a bewildering array of possibilities.

In vivo bone assessment methods

Rib biopsy has been widely used since its introduction for cattle (Little and Minson, 1977) and recently applied to pigs (Combs *et al.*, 1991a). The rib biopsy is particularly useful in experimental studies of patterns of change in skeletal mineralization over time, provided that the selected site is consistent (Beighle *et al.*, 1993) and repeated biopsies of the same rib are performed dorsally to earlier biopsies after a reasonable interval (> 3 months; Little and Ratcliff, 1979). Having obtained a disc of bone by trephine (1.5 cm diameter in cattle), a number of analytical options are available, of which the best is a measure of bone density, such as weight (specific gravity), ash or phosphorus (or calcium) per unit volume. Mineral concentrations per gram of ash have been advocated, because they show minimum variation in normal animals (Beighle *et al.*, 1994). However, the mineral composition of bone is not greatly changed by depletion of phosphorus (or calcium) (Qian *et al.*, 1996) and P (or Ca) g^{-1} ash provides little or no indication of phosphorus (or calcium) status (Williams *et al.*, 1991a). Fluctuations in the mineralization of rib biopsy samples in lactating animals can be marked, particularly on phosphorus-deficient pastures (Fig. 5.6; Read *et al.*, 1986c), and they can give a different assessment of phosphorus depletion from that given by plasma P_i. Phosphorus depletion tends to increase Ca:magnesium (Mg) ratios in the skeleton of cattle (Fig. 5.6) as well as lambs (Table 4.3). Tail-bone biopsies can be used, but they contain less ash and are probably less responsive to changes in phosphorus status than rib samples. In the past, experimental stations with access to X-ray equipment have monitored cortical bone thickness and recently dual photon absorptiometry (PA), radiographic photometry (RP) and ultrasound (US) have been used to obtain *in vivo* measurements of bone mineralization (Ternouth, 1990; Williams *et al.*, 1991b). There are also biochemical tests on serum or urine which indicate the extent to which bone matrix is being broken down; they include assays of hydroxyproline (McLean *et al.*, 1990) and pyridinium X-links in urine. These biochemical measures of bone demineralization *in vivo* are prognostic rather than diagnostic tools, and they have more relevance to racehorses than commercial livestock, since avoidance of stress injury is crucial to racing performance (Price *et al.*, 1995).

Post-mortem assessment of bone mineralization

In dead animals, the X-ray, morphometric or mineral analysis of bones may be the only diagnostic option. There are, however, considerable variations between species and also within and between particular bones in the degree of mineralization at a given age (Fig. 5.4). Furthermore, most indices increase with age, irrespective of mineral nutrition (Fig. 5.1; Beighle *et al.*, 1994; Chapter 4). Mineral: or ash:volume ratios in the entire bone are the simplest and best biochemical indices of bone mineralization (Field *et al.*, 1975; Williams *et al.*, 1991a). Guideline values are given in Table 5.3, which summarizes the contribution that each parameter can make towards a diagnosis.

Occurrence of Phosphorus Deprivation

Natural shortages of phosphorus in livestock usually develop in different circumstances from those of calcium, and situations in which the two minerals are both limiting are rare.

Grazing cattle

Phosphorus deprivation is predominantly a chronic condition of grazing cattle, arising from a combination of soil and climatic effects on herbage phosphorus concentrations. The presence of soils low in plant-available phosphorus (< 10 mg kg^{-1} DM) resulted in herbage low in phosphorus (Kerridge *et al.*, 1990). As pasture productivity and milk yield rise with high nitrogen use, the critical soil phosphorus value can rise to as much as 30 mg P kg^{-1} DM (Davison *et al.*, 1997a). Acid, iron-rich soils are particularly likely

Table 5.3. Marginal bands[a] for interpreting the most diagnostically useful biochemical indices of phosphorus status in livestock.

		Bone ash[b] (% dry weight)	Plasma P_i[c] (mmol l^{-1})
Cattle	Calf	60–70 (R,d)[d]	1.3–1.9
		48–56 (R,c)	
	Cow	50–60 (R,c)	1.0–1.5
Pigs	Starter (10–25 kg)	45–48 (d)	2.6–3.2
	Grower (25–65 kg)	52–55 (d)	2.3–2.6
	Finisher (65–80 kg)	56–58 (M,d)	2.3–2.6
Sheep	Lamb	20–30 (V)	1.3–1.9
	Ewe	30–36 (T)	1.0–1.5
Broiler	Chick – < 4 weeks	40–45 (T,d)	
	Chick – > 6 weeks	45–50 (T,d)	2.0–2.2
		50–55 (Tc,d)	
Hen	Laying	–	–
Turkey	Poult	38–40 (T)	

[a] Values below bands indicate a probability of impaired health or performance; values within bands indicate possibility of future losses if phosphorus supplies are not improved.
[b] Bone ash can be converted approximately to specific mineral concentrations on the assumption that ash contains 0.18 g P, 0.36 g Ca and 0.09 g Mg kg^{-1}; ash/unit volume is roughly similar to ash/unit dry fat-free weight.
[c] Multiply by 31 to obtain units in mg l^{-1}.
[d] Source and pretreatment of bone sample: R, whole rib; M, metacarpal/tarsal; V, lumbar vertebra; T, tibia; d, defatted bone; c, cortical bone.

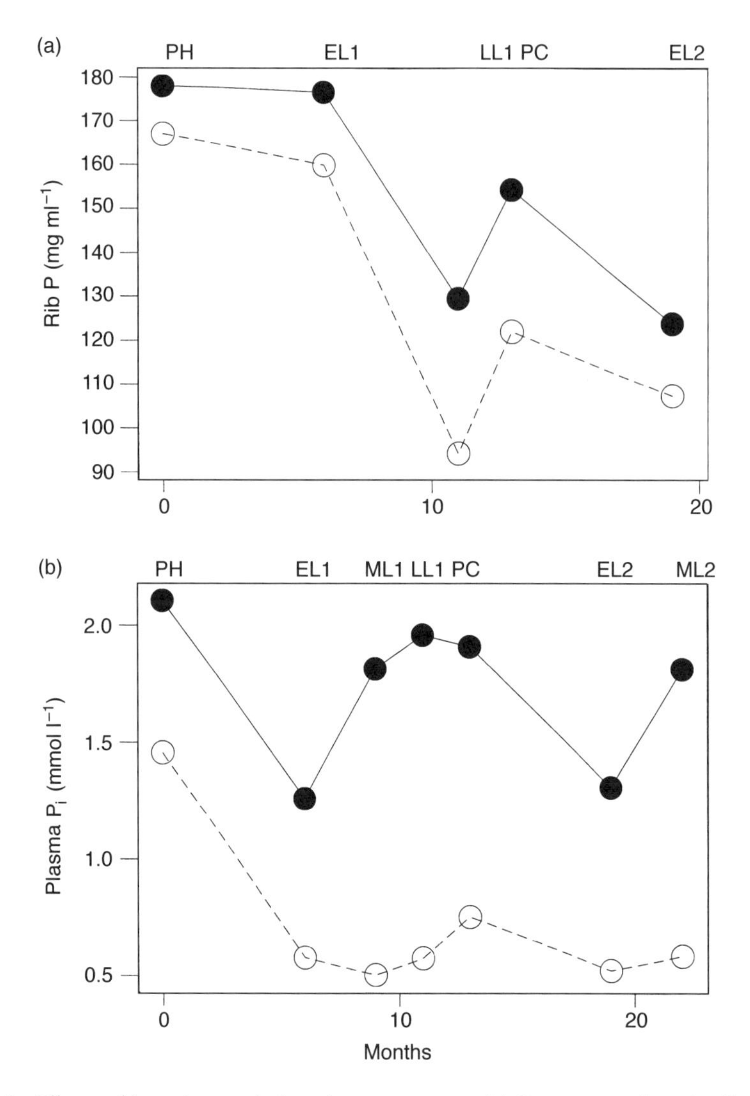

Fig. 5.6. Effects of lactation and phosphorus status on (a) P concentrations in rib biopsy; (b) plasma P_i; (c) rib magnesium (Mg); and (d) rib Ca : Mg ratio. Samples taken from two groups of beef cows, one receiving supplementary phosphorus (●) and one not (○), grazing phosphorus-deficient pasture: note that consistently low plasma P_i from the sixth month did not reflect the skeletal changes (data from Engels, 1981). PH or PC, pregnant heifer or cow; EL, ML and LL, early, mid- or late lactations; PR, pregnant.

to provide insufficient phosphorus. The occurrence of a dry period, when plants mature and seed is shed, accentuates any soil effect. In the veld country of South Africa, where the classical studies of bovine aphosphorosis

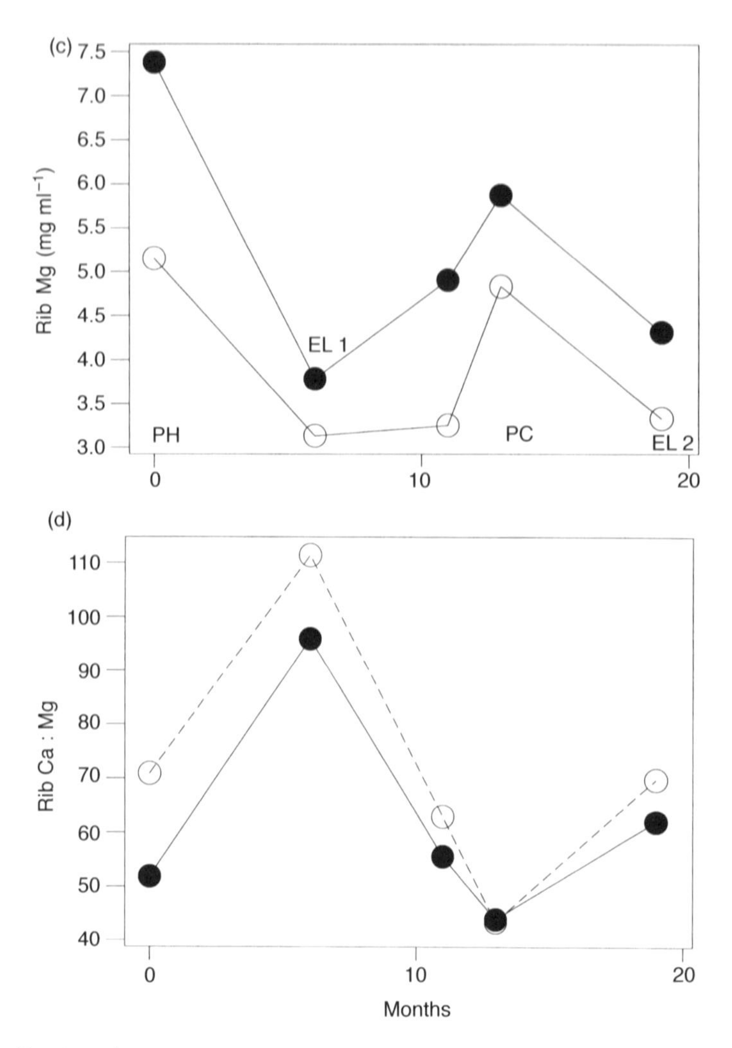

Fig. 5.6. *Continued*

were made, herbage concentrations fell typically from 1.3–1.8 g to 0.5–0.7 g P kg^{-1} DM between the wet summer and the dry winter and remained low for 6–8 months. Similarly, in the Mitchell grass (*Astrebla* spp.) pastures of northern Australia, values fell from 2.5 to as low as 0.5 g P kg^{-1} DM (Davies *et al.*, 1938). Forages from the northern Great Plains of the USA are regarded as marginal in phosphorus (Karn, 1997). The introduction of tropical legumes (e.g. *Stylosanthes* spp.), tolerant of low available soil phosphorus, lifted the constraint on herbage DM production in northern Queensland (Jones, 1990), but merely exacerbated deficiencies in livestock by producing highly digestible forage, exceedingly low in phosphorus (often < 1 g P kg^{-1} DM); the problem has required extensive research (Miller *et al.*, 1990). The protein, digestible energy and sulphur concentrations in the herbage also fall with maturity (Grant *et al.*, 1996; Karn, 1997; for sulphur, see Chapter 9), so that other deficiencies often contribute to the malnutrition of livestock in phosphorus-deficient areas. This was apparent from early studies of zebu cattle in East Africa (Lampkin *et al.*, 1961), of the savannah pastures during the dry winter season in South Africa (Van Niekerk and Jacobs, 1985) and in subtropical Australia (Cohen, 1975, 1979). Responses to phosphorus supplements are unlikely to be obtained while other nutritional deficits prevent growth (Miller *et al.*, 1990).

Other grazing livestock

Phosphorus deprivation is less common and usually less severe in grazing sheep than in grazing cattle. One reason suggested for this was that sheep (and also goats) have a higher feed consumption per unit of body weight. However, the maintenance requirement for phosphorus increases with DMI (AFRC, 1991), putting smaller ruminants at a disadvantage. There are other possible explanations: sheep (and goats) also have a smaller proportion of bone to body weight than cattle and therefore smaller growth requirements; and, because of their different methods of prehension, sheep are probably more able to select from mixed herbage those plants that are less phosphorus-deficient. The most important species difference, however, is the much shorter period of the annual reproductive cycle during which sheep are lactating (Read *et al.*, 1986a, b); this allows time for depleted skeletal reserves to be replenished. Early Australian studies failed to reveal any benefit from phosphorus supplements to sheep in some areas where a deficiency of phosphorus restricted cattle (Underwood *et al.*, 1940); a similar situation has been confirmed in South Africa (Read *et al.*, 1986a, b).

Poultry and pigs

Phosphorus deprivation would be common in poultry and pigs if inorganic phosphorus supplements were not routinely added to cereal-based diets, due to the poor absorbability of phytate phosphorus. For example, mortality was increased in broiler chicks given a maize/mixed vegetable protein diet containing 4.5 g P kg^{-1} DM, unless it was supplemented with phosphorus (Simons *et al.*, 1990). Phosphorus-responsive bone deformities (rickets) and fragilities will eventually develop, but they are usually preceded by life-threatening reductions in food intake. Nevertheless, a histological study of the proximal tibiotarsus from lame 35-day-old broiler chicks from Holland, where there is great pressure to reduce phosphorus levels in feeds, revealed a high (44%) incidence of hypophosphataemic rickets, which may increase susceptibility to tibial dyschrondroplasia (TD) (see Chapter 4) and bacterial chondronecrosis (Thorp and Waddington, 1997). Growing pigs are less vulnerable than poultry because they need less phosphorus and are normally fed lower concentrations of dietary calcium (see Chapter 4). Skeletal abnormalities are therefore again less likely to occur than in ruminants, but growth of pigs is poor on grain-based rations without phosphorus supplements.

The Prevention and Control of Phosphorus Inadequacies

Phosphorus deficiencies in grazing livestock can be prevented or overcome by direct treatment of the animal through supplementing the diet or the water-supply, or indirectly by appropriate fertilizer treatment of the soils. The choice of procedure depends on the conditions of husbandry.

Use of fertilizers

In climatically favoured and intensively farmed areas with sown pastures, phosphate applications, designed primarily to increase herbage yields, also increase herbage phosphorus concentrations (Falade, 1973). Minson (1990) calculated that, on average, pasture phosphorus was increased from 1.7 to 2.4 g kg^{-1} DM with 47 kg fertilizer P ha^{-1} applied, but the ranges in the literature surveyed were wide (increases of 0.2–3.5 to 0.5–3.9 g P kg^{-1} DM from 9–86 kg fertilizer P). Phosphorus can be applied as rock phosphate (RP), as single or triple superphosphate or combined with nitrogen (N) and potassium in complete fertilizers, but it is impossible to generalize about recommended rates (Jones, 1990). For example, lactation milk yields were increased from 3930 to 4310 and 4610 kg cow^{-1} ($P < 0.05$) in animals on tropical grass pasture when 22.5 and 45.0 kg P ha^{-1}, respectively, were applied with 300 kg N ha^{-1} (Davison *et al.*, 1997a, b). However, with 100 kg N or none, there was no response to fertilizer phosphorus. The responses were attributed to increases in the amount of green leaf on offer, rather than an increase in herbage phosphorus concentration (see also Fig. 2.6). Although herbage phosphorus increased approximately twofold (from 0.6 to 1.3 g P kg^{-1} DM) (Davison *et al.*, 1997a), no benefit was gained from providing an additional dietary phosphorus supplement (Walker *et al.*, 1997). Pastures treated with phosphatic fertilizers at rates that maximize herbage yield do not necessarily meet the requirements of grazing animals at all times and the process is inefficient on acid, iron-rich soils with a high sorption capacity. On sparse, extensive phosphorus-deficient native pastures, other methods are necessary because transport and application costs are high and herbage productivity is usually limited by climatic disadvantages.

Dietary supplementation

Phosphorus can be provided directly to grazing livestock in phosphatic salt-licks and blocks and also in the water-supply. Typical levels of provision for beef cattle are at least 5 g P $head^{-1}$ day^{-1} for growing cattle and 10 g P $head^{-1}$ day^{-1} for breeding stock (Miller *et al.*, 1990), but intakes >8 g P day^{-1} in growing cattle can retard growth (Grant *et al.*, 1996). The easiest and cheapest procedure is to provide a phosphatic lick in troughs or boxes that afford protection from rain and are situated near the watering-places. A simple 1:1 mixture of dicalcium phosphate (DCP, $CaHPO_4$) and salt, with a small proportion of molasses, is well consumed by most animals. Field experience has shown that voluntary consumption of such licks is often acceptable on a herd or flock basis, although consumption varies greatly between individual animals and at different times of the year. Individual dosing requires undesirably frequent handling. Supplementation through the drinking-water is applicable where the access to water is controlled (i.e. there are no natural water sources) but requires water-soluble sources, such as sodium phosphate (Na_2HPO_4) or ammonium polyphosphate (Hemingway and Fishwick, 1975), which are more expensive than DCP. Furthermore, the amounts ingested fluctuate with water consumption and treatment may limit water consumption, cause toxicity and corrode dispensers (Miller *et al.*, 1990). Superphosphate is the cheapest source of water-soluble phosphate (Du Toit *et al.*, 1940; Snook, 1949; Scharp, 1979). Where calcium phosphates are used in free-access mixtures, supplementary calcium is automatically supplied, although it is rarely needed by grazing livestock and may already be present at levels sufficiently high to have contributed to any aphosphorosis. In low-cost livestock production systems, where reliance is often placed on low-quality roughages, phosphorus is often lacking (Suttle, 1991) and can be cheaply provided by including 20–30% wheat or rice bran in a molasses–urea block or loose supplement (see Chapter 9 for more detail). Phosphorus supplementation is, however, pointless if underlying deficiencies of nitrogen are not corrected (Grant *et al.*, 1996) and was not beneficial in a Queensland study (Walker *et al.*, 1997).

Combined methods

Where legume species tolerant of phosphorus-deficient soils have been sown, the most efficient procedure is to combine direct and indirect methods of supplementation. Phosphorus given as fertilizer is restricted in amount to that required to maximize herbage yield (up to 10 kg P ha^{-1} $year^{-1}$) and any additional phosphorus required by livestock is given directly (Miller *et al.*, 1990).

Mineral sources

The choice of supplement depends on chemical composition, biological availability, cost, accessibility and freedom from toxic impurities and dust hazards. Differences in the acidogenic properties of monobasic and dibasic phosphates, with consequences for performance and eggshell quality, have been reported (Keshavarz, 1994), but the levels of monobasic phosphates found to be harmful were far in excess of need at 10.2 g P kg^{-1} DM. The relative availability of phosphorus in different compounds has been extensively investigated with different species of livestock (Soares, 1995). Colloidal phosphate or 'soft phosphate with colloidal clay' was ineffective and given a relative value of only 40% of that of other sources, contradicting an earlier finding with chicks (Motzok *et al.*, 1956). Relative values of 90–95% were accorded to bone-meal, defluorinated phosphate (DFP), diammonium phosphate and DCP compared with monosodium phosphate (MSP) or a similar standard given to poultry. Bone-meal (80%) and DFP (85%) were generally less available to pigs, which lack the ability to trap and grind down coarse particles in a gizzard. This apparent species difference was not confirmed by Coffey *et al.* (1994), who sought but did not find significant correlations between solubility in neutral ammonium citrate (range 60–91%) and availability for five sources of DFP; the mean values were 86 and 83%, respectively, for pigs and poultry relative to MSP. Essentially similar findings in respect of a variety of mineral phosphate supplements have been reported for sheep, for growing beef cattle and for horses (Soares, 1995). Miller *et al.* (1987) found that DFP and DCP provided phosphorus of equal availability to calves. A wider range of mineral supplements, including urea phosphate, monoammonium phosphate (Fishwick and Hemingway, 1973), two magnesium phosphates and tricalcium phosphate (Fishwick, 1978), gave similar responses in growth and phosphorus retention when tested with growing, phosphorus-deficient sheep. In sheep given coarsely ground RP, particles can be trapped in the alimentary tract, giving a falsely high estimate (0.66) of phosphorus (and presumably calcium) absorbability (Suttle *et al.*, 1989); a value two-thirds that of DCP (0.55 vs. 0.80) was considered to be more appropriate and is confirmed by Dayrell and Ivan (1989). Ground RP is also less palatable and can contain enough fluoride to cause fluorine toxicity, as discussed in Chapter 18.

Phosphorus Requirements

Various phosphorus requirements can be derived, depending on the objective. The data in Table 5.1 and Figs 5.1 and 5.7 show that pigs, for example, need less phosphorus for optimal growth than for optimal bone quality, and the situation is similar in turkey poults (Fig. 1.3). Traditionally, the most demanding criterion has provided the goal, but environmental considerations demand that optimal bone strength should only be used where strength is required for either welfare or production reasons (e.g. poultry) or where the prior establishment of a bone reserve is essential to meet subsequent peak demands (egg laying or lactation). Adequate phosphorus nutrition depends on the chemical forms in which phosphorus occurs in the diet, the vitamin D status of the diet or the animal (Pointillart *et al.*, 1986), food intake, level of performance and the dietary calcium concentration. A dietary Ca:P ratio between 1:1 and 2:1 was assumed to be ideal for growth and bone formation in all species, the upper limit approximating the ratio of the two minerals in bone. However, most livestock can radically change extreme ratios of dietary Ca:P by homeostatic control to acceptable ratios of absorbed and retained Ca:P unless and until phytate sets a ceiling on their absorption. A diet deficient in both calcium and phosphorus can have an apparently 'ideal' Ca:P ratio. The Ca:P ratio, therefore, has little or no place in defining requirements for phosphorus or calcium: each should be formulated independently. This is not to say that calcium never adversely affects the utilization of phosphorus and vice versa, rather that these interactions only have a nutritional significance when the supply of one element is limiting and the other is excessive.

Influence of calcium in ruminants

Ruminants can tolerate a wide range of Ca:P ratios when their vitamin D status is adequate and the dietary supply of each mineral is adequate. In a trial with sheep, the high Ca:P ratio of 10:1 had no adverse effect with a diet containing 2.6 g P kg^{-1} DM, but severe bone disorders arose when the diet contained only 0.8 g P kg^{-1} DM (Young *et al.*, 1966). There are other instances of the addition of calcium to phosphorus-deficient diets exacerbating or inducing a state of phosphorus deficiency in sheep (Field *et al.*, 1975; Wan Zahari *et al.*, 1990). In an experiment with calves given three levels of dietary calcium (2.7, 8.1 and 24.3 g kg^{-1} DM) and three of phosphorus (1.7, 3.4 and 6.8 g kg^{-1} DM), nine Ca:P ratios, ranging from 0.4:1 to 14.3:1, were 'tested'. Dietary ratios between 1:1 and 7:1 all gave satisfactory and similar results, but, with Ca:P ratios below 1:1 and over 7:1, growth and feed efficiency decreased significantly (Wise *et al.*, 1963). No adverse effect was apparent with heifers fed for 2 years on diets with Ca:P ratios ranging from 7:1 to 9:1 (Call *et al.*, 1978). None of these experiments critically examined the effects of the Ca:P ratio independently of calcium and phosphorus supplies. In a subsequent critical experiment, Field *et al.* (1983) showed that the Ca:P ratio *per se* was unimportant over the range 0.6–3.6, although increases in dietary calcium from 3.4 to 5.4 g kg^{-1} DM decreased the efficiency of absorption of added inorganic phosphorus by 18% at the higher of two supplementation levels. Calcium had no such effect on phosphorus absorption from the basal diet (1.5 g P kg^{-1} DM). The greater responses to phosphorus reported in South African cattle on calcareous soils than in Australian beef cattle grazing on non-calcareous soils were explained by Cohen (1979) in terms of the influence of calcium on phosphorus metabolism. It is worth noting that tropical legumes often have extreme Ca:P ratios (> 10.0: N.F. Suttle, unpublished data).

Influence of calcium in non-ruminants

Pigs and poultry appear to be less tolerant of high dietary Ca:P ratios than ruminants, but the problem is again one of sensitivity to high calcium when phosphorus supplies are marginal (Koch and Mahan, 1985; Reinhart and Mahan, 1986; Mohammed *et al.*, 1991; Lei *et al.*, 1994). Unfortunately, preoccupation with dietary Ca:P ratios has dominated the design and reporting of many later experiments, especially those dealing with the phosphorus nutrition of pigs (Koch and Mahan, 1985; Reinhart and Mahan, 1986). The extensive data already cited from one of these experiments (Figs 5.1 and 5.2; Reinhart and Mahan, 1986) can be used to illustrate the specific and dominant effect of calcium over Ca:P ratio on phosphorus metabolism (Fig. 5.7). Newly weaned pigs were taken through 'starter', 'grower' and 'finisher' phases on diets 5% below or 10% above the then phosphorus standards (NRC, 1979) for each phase (0.5 and 0.7, 0.4 and 0.6, 0.35 and 0.5 g P kg^{-1} DM, respectively), while maintaining Ca:P ratios of 1.3:1, 2:1, 3:1 or 4:1 at each phosphorus level. For a given phase, any gap between the two lines in each graph indicates a phosphorus effect, any slope a calcium effect and any difference

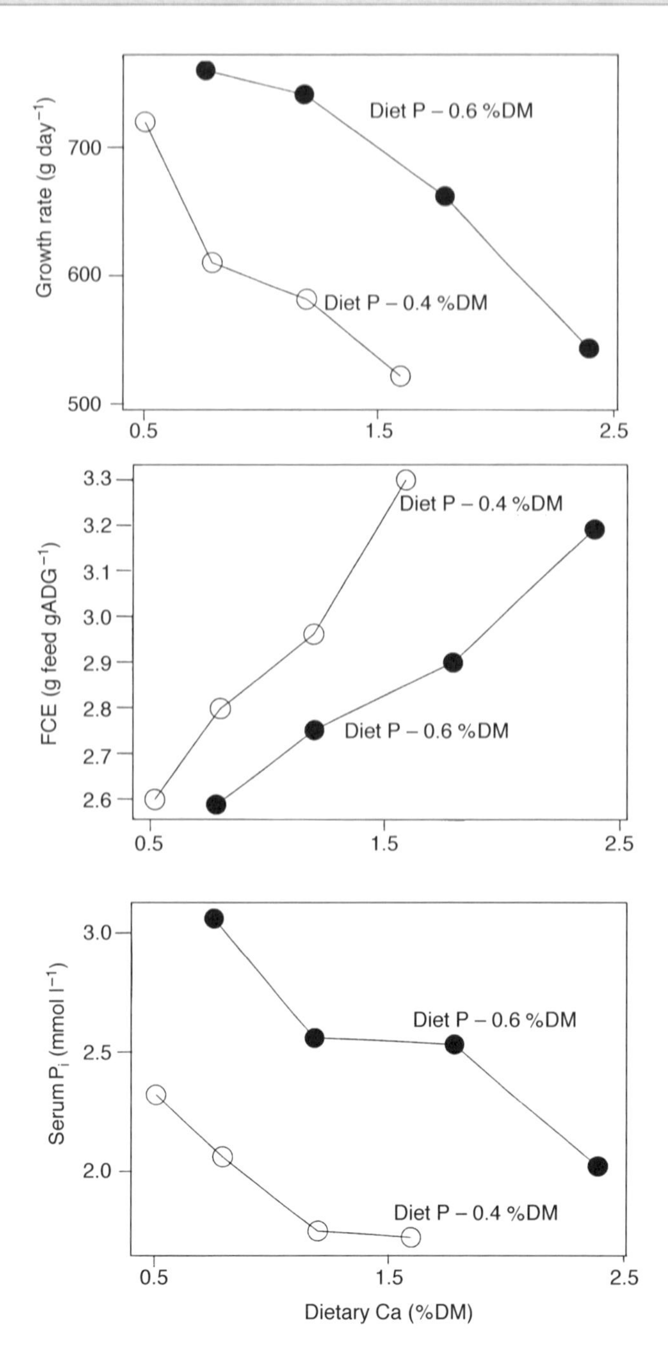

Fig. 5.7. Effects of dietary phosphorus and calcium concentrations on growth, feed conversion efficiency (FCE) and serum inorganic phosphorus (P_i) in 'grower' pigs (data from Reinhart and Mahan, 1986).

in slope a Ca × P interaction. The lower phosphorus level was inadequate throughout, particularly for growth (Fig. 5.7) and bone ash (Fig. 5.1), providing a sensitive test of both the National Research Council (NRC, 1979) standard and any further influence of dietary calcium. Growth and serum P_i (Fig. 5.7) both showed similar linear patterns of decline as dietary calcium increased. In each growth phase, *any* addition of calcium was detrimental to *each* index of phosphorus status. The most likely explanation is that the smallest addition of calcium was sufficient to reduce phosphorus absorption by a fixed amount, probably by completing the conversion of P_P in the maize–soybean meal diet (0.25–0.34 g P_P kg^{-1} DM) to non-absorbable complexes. The explanation is supported by the much smaller effects of calcium addition seen when a diet with maize starch rather than P_P-rich maize grain was used (Koch and Mahan, 1985). Similar adverse effects of dietary calcium on the absorption of plant phosphorus are seen in poultry (Mohammed *et al.*, 1991). Addition of P_i to a P_P-rich diet can dramatically increase the apparent absorption of total phosphorus (e.g. from 15 to 35%; Cromwell *et al.*, 1995: from 27.6 to 57.2%; Kornegay and Qian, 1996), but, when calcium is slavishly added in equal measure, no improvement may been seen (Simons *et al.*, 1990). If any addition of calcium has a fixed effect on phytate phosphorus absorption, irrespective of the amount of added P_i, the preservation of added Ca:added P_i ratios serves no useful nutritional purpose, as Eeckhout *et al.* (1995) concluded, but merely drives up the phosphorus requirement in pigs and growing poultry. For laying hens, any acceptable ratio must be considerably higher than 2:1, due to their far greater requirements for calcium than for phosphorus.

Phosphorus requirements for sheep

Early experiments with lambs, aged 5–6 months, indicated that 1.3 g P kg^{-1} DM was borderline, 1.0 g P kg^{-1} DM inadequate and 1.7 g P kg^{-1} DM adequate, as judged by growth and serum P_i, for growing sheep (Beeson *et al.*, 1944). These values were soon confirmed (Mitchell, 1947), but they are low when compared with the NRC (1985) recommendations for growing lambs of 1.7–3.8 g P kg^{-1} DM. The discrepancy is due to the fact that field observations deduce requirements which allow skeletal reserves to be depleted, while scientific estimates of requirement allow for the complete mineralization of bone at all times and therefore have a built-in safety factor (AFRC, 1991). The NRC (1985) estimates of requirements were 2.0–2.9 g P kg^{-1} DM for pregnant and lactating ewes. The derivation of factorial estimates of phosphorus requirements for ruminants is particularly difficult, due to the problems of defining the maintenance requirement at a level commensurate with satisfactory bone mineralization and production (Braithwaite, 1984b; Challa *et al.*, 1989; Scott *et al.*, 1995). The factorial method for estimating net requirements proposed by AFRC (1991) will be used here, since there is evidence that the faecal endogenous losses for whole-roughage diets was appropriately estimated (Ternouth, 1989; Scott *et al.*, 1995). However, a higher efficiency of phosphorus absorption (0.74 vs.

0.64) will be used to derive gross requirements, based on the mean of 11 recent estimates for roughages by previously cited authors (see 'Dietary Sources of Phosphorus') (0.74 ± 0.09 SD). The resultant requirements are given for convenience alongside calcium in Table 4.6. It must be emphasized that sheep will often consume far less phosphorus than that recommended and remain healthy, particularly when concentrates are fed; these are guidelines for feed formulation, not diagnostic criteria.

Phosphorus requirements for cattle

Studies with Holstein calves (Wise *et al.*, 1958) and Hereford bullocks (Tillman *et al.*, 1959) indicated minimum growth requirements of 2.1 g P kg^{-1} DM between 3.5 and 11 months of age. Much higher needs (up to 3.2 g P kg^{-1} DM) were suggested for young calves by Miller *et al.* (1987), but these may be peculiar to the semipurified diet used. Australian investigations with grazing beef cattle indicated that 1.8 g P kg^{-1} DM in pastures is sometimes adequate for all except lactating cows and young fast-growing animals and that 1.2 g P kg^{-1} DM is sometimes adequate for animals of 200 kg LW growing at 0.5 kg day^{-1} (Cohen, 1975). The latest NRC (1984) recommendations are similar to those given previously (NRC, 1975), ranging from 2.2 to 4.3 g P kg^{-1} DM for calves and 1.9 to 3.9 g P kg^{-1} DM for lactating cows, but there is growing evidence that they remain generous for growing and breeding beef cattle (Ternouth *et al.*, 1996; Ternouth and Coates, 1997) and for dairy cows. In earlier work, no reduction in weight gain (0.45 kg day^{-1}), feed efficiency or appetite was observed in Hereford heifers fed 66% of NRC (1975) recommendations for 2 years (1.4 as opposed to 3.6 g P kg^{-1} DM) (Call *et al.*, 1978). Similarly, feeding phosphorus at 3.2 g kg^{-1} DM (80% of NRC (1975) requirement) to lactating cows for 7.5–12 months did not reduce milk yield or induce hypophosphataemia. Again, there is a place for factorial estimates, but an important adjustment is necessary to the AFRC (1991) model; using an absorption coefficient of 0.75 rather than 0.58 to convert net to gross requirements lowers the estimates by 23% (Table 4.7). Requirements are greatest for young, rapidly growing calves and decline markedly with age. The use of high-quality feeds does not increase requirements stated as concentrations, but they do increase as milk yield increases. These trends are shared by the estimated sheep requirements (Table 4.6); again, their limitations must be stressed and they need *not* be met on a day-to-day basis. If fulfilled during periods of supplementary feeding, these recommendations should provide a skeletal reserve sufficient to sustain the herd until the next period of supplementation. No adjustment has been made to AFRC (1991) maintenance requirements, despite suggestions by Australian workers that they were grossly overestimated (Coates and Ternouth, 1992; Ternouth *et al.*, 1996; Ternouth and Coates, 1997). Their recent studies appear to confirm that FE_P is related to DMI and plasma P_i, but, if the target plasma P_i is 1.5 mmol (45 mg) l^{-1}, the FE_P values used by AFRC (1991) are close to those observed in growing and pregnant beef cattle in Queensland. Maintenance needs for lactation may be only 50% of those used by AFRC (1991), according to

Ternouth and Coates (1997), but more evidence is needed before further adjustments are made.

Phosphorus requirements for pigs

According to Miller *et al.* (1964), baby pigs up to 6–8 weeks of age required 4.0 g P kg^{-1} DM to permit normal growth and feed utilization, 5.0 g P kg^{-1} DM to maintain normal serum P_i and alkaline phosphatase values and an adequate rate of skeletal development and 6.0 g P kg^{-1} DM for maximum bone density and breaking strength (Table 5.1). Using the most sensitive criterion, optimum bone strength, empirical estimates of requirement for phosphorus still vary (ARC, 1981) and, for reasons argued in Chapter 4, averaging the literature values for feeding trials will overestimate minimum requirements. The preferred approach is that of factorial modelling, using the components of net requirement for growing pigs given by the Agricultural Research Council (ARC, 1981). The choice of absorption coefficient is critical, and the use of a single value to cover diets varying in concentration of phytate phosphorus and calcium is unrealistic. Ranges of phosphorus requirement are therefore advocated to accompany absorption coefficients ranging from 0.50 to 0.70 (Table 4.10), assuming that calcium will not be overfed (see Chapter 4). The lower requirements would apply to diets with wheat and animal protein components or added phytase and the higher requirements to maize–soybean meal diets for young pigs.

The need to minimize the amounts of phosphorus accumulating in animal wastes has introduced a new perspective into what constitutes a minimum requirement (Eeckhout *et al.*, 1995). Quotas have been imposed on farmers in the Netherlands to limit the amounts of minerals which are dispersed on land as animal wastes, and they are so restrictive that whole enterprises may have to halt production. In these circumstances, the supply of sufficient phosphorus to obtain maximum bone strength in pigs grown for meat production is a luxury. Although small deviations from NRC (1979) requirements (these were not changed much in the subsequent NRC (1988) report) caused losses of performance, phosphorus and calcium were increased or decreased together (Combs *et al.*, 1991b). Eeckhout *et al.*'s recent (1995) study of the effects of small but independent changes in phosphorus and calcium provision in wheat–soybean meal diet supports the estimates given in Table 4.10 for heavier pigs given a high-phytate diet. Since bone strength is important in boars, they should always be fed phosphorus at the higher level. Gilts should also be fed generous amounts of phosphorus (and calcium) during the first reproductive cycle. Nimmo *et al.* (1981b) found that a high proportion of gilts (30%) became lame when fed 5.5 g P (and 7.2 g Ca) kg^{-1} DM during gestation, whereas those given 8.3 g P (and 10.7 g Ca) kg^{-1} DM did not. There were associated differences in bone strength but no residual effects on offspring with common levels (5.0 g P and 7.5 g Ca kg^{-1} DM) fed during lactation (Nimmo *et al.*, 1981a).

Phosphorus requirements for poultry

Unlike calcium requirements, those for phosphorus have declined over the years. According to the NRC (1966), 6.0 g P kg^{-1} DM was adequate for growth and normal bone formation in chicks, so long as at least 4.5 g P_i kg^{-1} DM was present. The latest report (NRC, 1994) recommends 4.0 g non-phytate (np)P kg^{-1} DM for the youngest Leghorn-type chicks, decreasing to 3.0 g kg^{-1} DM by 12–18 weeks; values for the youngest broilers remain at 4.5 g kg^{-1} DM (Table 4.8). In contrast, the requirements for growth in turkeys have increased, mainly because the type of bird has changed and rate of growth has increased considerably, as has efficiency of feed use. The latest recommendation from NRC (1994) is a sliding scale, which decreases from 6.0 to 4.2 g between 4 and 8 weeks of age and to 2.8 g npP kg^{-1} DM by weeks 20–24. However, Qian *et al.*'s (1996) results for turkey poults between 1 and 21 days (Fig. 1.3) suggest that 5 g npP kg^{-1} DM was optimal for all criteria of phosphorus status. National Research Council provisions for the laying hen plummeted from 429 in 1966 to 250 mg npP day^{-1} in 1994. A more cautious approach is to reduce the phosphorus provision over a similar range as lay progresses (Table 4.9; Roland, 1986). A high rate of egg-laying increases phosphorus needs more than would be expected from the small amount of phosphorus in the egg (about 120 mg), because the increase in phosphorus catabolism associated with egg production increases endogenous losses. In two experiments with caged layers (Owings *et al.*, 1977), it was concluded that diets providing about 200 mg available phosphorus day^{-1} maintained egg production but that at least 300 mg was necessary to maintain the 'livability' of the hens, i.e. to prevent the 'cage layer fatigue syndrome' (see Chapter 4). Phosphorus requirements are further affected by environmental conditions. High temperatures reduce feed consumption and hence raise phosphorus requirements, when these are expressed as dietary proportions (NRC, 1994). A similar reasoning will, of course, apply to other minerals and nutrients. Floor-housed birds are able to recycle phosphorus through coprophagy and their requirements are therefore lower than those of caged hens (Singsen *et al.*, 1962).

Phosphorus requirements for horses

It is difficult to give precise estimates of the requirements of growing horses, because the optimal growth rates for maximum performance are unknown and the rate of growth, like that of other domestic livestock, has increased over the years. Experiments with young growing ponies (Schryver *et al.*, 1974) and with mature horses (Schryver *et al.*, 1970, 1971) indicated relatively high requirements for younger animals. Estimates were derived from the mineral composition of the growth increment, the obligatory losses of minerals and an intestinal absorption efficiency of 30–50%. From the average feed consumption observed in these experiments (H.F. Hintz, 1979, personal communication), it can be calculated that the phosphorus requirements of foals would be met by diets containing 5.0 g P kg^{-1} DM and those of mature horses from diets containing 1.5 g P kg^{-1} DM.

Phosphorus Toxicity

Phosphate is intrinsically a well-tolerated ion and, for this reason, animals can allow circulating levels of phosphate to fluctuate widely. This, coupled with ready excretion of excess phosphate via the urine, means that livestock generally tolerate excessive intakes of phosphorus. Problems arise through adverse interactions or cumulative effects with other minerals. High phosphorus intakes predispose animals to urinary calculi in sheep, this is more likely to happen when the diet also provides excess magnesium, because formation of magnesium phosphates is integral to growth of the phosphatic calculus (Suttle and Hay, 1986). Dietary excesses of phosphorus (5.5–8.3 g npP kg^{-1} DM) predispose broilers to TD (see Chapter 4), but the damage can be greatly restricted by concurrently feeding excess calcium (15–17 g Ca kg^{-1} DM) (Edwards and Veltmann, 1983). Excesses of acidogenic sources, such as monobasic phosphate, disturb the acid–base balance in laying hens (Keshavarz, 1994; see also Chapter 6). In the horse, the feeding of excess phosphorus in diets low in calcium can cause secondary hyperparathyroidism.

References

AFRC (1991) A reappraisal of the calcium and phosphorus requirements of sheep and cattle. Technical Committee on Responses to Nutrients, Report Number 6. *Nutrition Abstracts and Reviews (Series B)* 61, 573–612.

Andrews, T.L., Damron, B.L. and Harms, R.H. (1972) Utilisation of various sources of plant phosphorus by the turkey poult. *Nutrition Reports International* 6, 251–257.

ARC (1981) *The Nutrient Requirements of Pigs.* Commonwealth Agricultural Bureaux, Farnham Royal, Slough, UK, pp. 215–248.

Bar, A. and Hurwitz, S. (1984) Egg shell quality, medullary bone ash, intestinal calcium and phosphorus absorption and calcium-binding protein in phosphorus-deficient hens. *Poultry Science* 63, 1975–1981.

Beeson, W.M., Johnson, R.F., Bolin, D.W. and Hickman, C.W. (1944) The phosphorus requirement for fattening lambs. *Journal of Animal Science* 3, 63–70.

Beighle, D.E., Boyazoglu, P.A. and Hemken, R.W. (1993) Use of bovine rib bone in serial sampling for mineral analysis. *Journal of Dairy Science* 76, 1047–1072.

Beighle, D.E., Boyazoglu, P.A., Hemken, R.W. and Serumaga-Zake, P.A. (1994) Determination of calcium, phosphorus and magnesium values in rib bones from clinically normal cattle. *American Journal of Veterinary Research* 55, 85–89.

Biehl, R.R. and Baker, D.H. (1997a) 1α-Hydroxycholecalciferol does not increase the specific activity of intestinal phytase but does improve phosphorus utilisation in both caecotomised and sham-operated chicks fed cholecalciferol-adequate diets. *Journal of Nutrition* 127, 2054–2059.

Biehl, R.R. and Baker, D.H. (1997b) Utilisation of phytate and non-phytate phosphorus in chicks as affected by source and amount of vitamin D_3. *Journal of Animal Science* 75, 2986–2993.

Biehl, R.R., Baker, D.H. and DeLuca, H.F. (1995) 1α-Hydroxylated cholecalciferol compounds act additively with microbial phytase to improve phosphorus, zinc and manganese utilization in chicks fed soy-based diets. *Journal of Nutrition* 125, 2407–2416.

Bisschop, J.H.R. (1964) *Feeding Phosphates to Cattle*. Science Bulletin, Department of Agricultural Technical Services, South Africa.

Bortolussi, G., Ternouth, J.H. and McMeniman, N.P. (1994) The effects of dietary nitrogen and calcium supplementation on *Bos indicus* cattle offered low P diets. *Proceedings of the Australian Society of Animal Production* 19, 381–384.

Bown, M.D., Poppi, D.P. and Sykes, A.R. (1989) The effect of a concurrent infection of *Trichostrongylus colubriformis* and *Ostertagia circumcincta* on calcium, phosphorus and magnesium transactions along the digestive tract of lambs. *Journal of Comparative Pathology* 101, 11–20.

Braithwaite, G.D. (1984a) Changes in phosphorus metabolism in sheep in response to the increased demands for phosphorus associated with an intravenous infusion of calcium. *Journal of Agricultural Science, Cambridge* 102, 135–139.

Braithwaite, G.D. (1984b) Some observations in phosphorus homeostatis and requirements. *Journal of Agricultural Science, Cambridge* 102, 295–306.

Braithwaite, G.D. (1986) Phosphorus requirements of ewes in pregnancy and lactation. *Journal of Agricultural Science, Cambridge* 106, 271–278.

Breves, G., Ross, R. and Holler, H. (1985) Dietary phosphorus depletion in sheep: effects on plasma inorganic phosphorus, calcium, 1,25-$(OH)_2$-Vit. D and alkaline phosphatase and on gastrointestinal P and Ca balances. *Journal of Agricultural Science, Cambridge* 105, 623–629.

Brooks, H.V., Cook, T.G., Mansell, G.P. and Walker, G.A. (1984) Phosphorus deficiency in a dairy herd. *New Zealand Veterinary Journal* 32, 174–176.

Call, J.W., Butcher, J.E., Blake, J.T., Smart, R.A. and Shupe, J.L. (1978) Phosphorus influence on growth and reproduction of beef cattle. *Journal of Animal Science* 47, 216–225.

Call, J.W., Butcher, J.E., Shupe; J.L., Blake, J.K. and Olsen, A.E. (1986) Dietary phosphorus for beef cows. *American Journal of Veterinary Research* 47, 475–481.

Call, J.W., Butcher, J.E., Shupe, J.L., Lamb, R.C., Boman, R.L. and Olsen, A.E. (1987) Clinical effects of low dietary phosphorus concentrations in feed given to lactating cows. *American Journal of Veterinary Research* 48, 133–136.

Challa, J. and Braithwaite, G.D. (1988) Phosphorus and calcium metabolism in growing calves with special emphasis on phosphorus homeostasis. 2. Studies of the effect of different levels of phosphorus, infused abomasally, on phosphorus metabolism. *Journal of Agricultural Science, Cambridge* 110, 583–589.

Challa, J. and Braithwaite, G.D. (1989) Phosphorus and calcium metabolism in growing calves with special emphasis on phosphorus homeostasis. 4. Studies on milk-fed calves given different amounts of dietary phosphorus but a constant intake of calcium. *Journal of Agricultural Science, Cambridge* 113, 285–289.

Challa, J., Braithwaite, G.D. and Dhanoa, M.S. (1989) Phosphorus homeostasis in growing calves. *Journal of Agricultural Science, Cambridge* 112, 217–226.

Coates, D.B. and Ternouth, J.H. (1992) Phosphorus kinetics of cattle grazing tropical pastures and implications for the estimation of their phosphorus requirements. *Journal of Agricultural Science, Cambridge* 119, 401–409.

Coates, D.B., Schachenmann, P. and Jones, R.J. (1987) Reliability of extrusa samples collected from steers fistulated at the oesophagus to estimate the diet of resident animals in grazing experiments. *Australian Journal of Experimental Agriculture* 27, 739–745.

Coffey, R.D., Mooney, K.W., Cromwell, G.L. and Aaron, D.K. (1994) Biological availability of phosphorus in defluorinated phosphates with different phosphorus solubilities in neutral ammonium citrate for chicks and pigs. *Journal of Animal Science* 72, 2653–2660.

Cohen, R.D.H. (1975) Phosphorus for grazing beef cattle. *Australian Meat Research Committee Review* No. 23, 1–16.

Cohen, R.D.H. (1979) The calcium–phosphorus interaction and its relation to grazing ruminants. In: *Symposium on Phosphorus, Nutrition Society of Australia (Queensland Branch), 8 May, 1979.*

Combs, N.R., Kornegay, E.T., Lindemann, M.D., Notter, D.R. and Welker, F.H. (1991a) Evaluation of a bone biopsy technique for determining the calcium and phosphorus status of swine from weaning to market weight. *Journal of Animal Science* 69, 664–673.

Combs, N.R., Kornegay, E.T., Lindemann, M.D. and Notter, D.R. (1991b) Calcium and phosphorus requirements of swine from weaning to market weight. I. Development of response curves for performance. *Journal of Animal Science* 69, 673–681.

Combs, N.R., Kornegay, E.T., Lindemann, M.D., Notter, D.R., Wilson, J.H. and Mason, J.P. (1991c) Calcium and phosphorus requirements of swine from weaning to market weight. II. Development of response curves for bone criteria and comparison of bending and shear bone testing. *Journal of Animal Science* 69, 682–693.

Cromwell, G.L. (1992) The biological availability of phosphorus in feedstuffs for pigs. *Pig News Information* 13, 75.

Cromwell, G.L., Coffey, R.D., Parker, G.R., Monegue, H.J. and Randolph, J.H. (1995) Efficacy of a recombinant-derived phytase in improving the bioavailability of phosphorus in corn–soybean meal diets for pigs. *Journal of Animal Science* 73, 2000–2008.

Davies, J.G., Scott, A.E. and Kennedy, J.F. (1938) The yield and composition of a Mitchell grass pasture for a period of 12 months. *Journal of the Council for Scientific and Industrial Research, Australia* 11, 127–139.

Davison, T.M., Orr, W.N., Silver, B.A., Walker, R.G. and Duncalfe, F. (1997a) Phosphorus fertilizer for nitrogen-fertilized dairy pastures. 1. Long-term effects on pasture, soil and diet. *Journal of Agricultural Science, Cambridge* 129, 205–215.

Davison, T.M., Orr, W.N., Doogan, V. and Moody, P. (1997b) Phosphorus fertilizer for nitrogen-fertilized dairy pastures. 2. Long-term effects on milk production and a model of phosphorus flow. *Journal of Agricultural Science, Cambridge* 129, 219–231.

Dayrell, M.deS. and Ivan, M. (1989) True absorption of phosphorus in sheep fed corn silage and corn silage supplemented with dicalcium or rock phosphate. *Canadian Journal of Animal Science* 69, 181–186.

De Waal, H.O., Randall, J.H. and Keokemoer, G.J. (1996) The effect of phosphorus supplementation on body mass and reproduction of grazing beef cows supplemented with different levels of phosphorus at Armoedsvlakte. *South African Journal of Animal Science* 26, 29–36.

Du Toit, P.J., Malan, A.I., Van Der Merwe, P.K. and Louw, J.G. (1940) Mineral supplements for stock: the composition of animal licks. *Farming in South Africa* 15, 233–248.

Edwards, H.M. (1993) Dietary 1,25-dihydroxycholecalciferol supplementation increases natural phytate phosphorus utilisation in chickens. *Journal of Nutrition* 123, 567–577.

Edwards, H.M. and Veltmann, J.R. (1983) The role of calcium and phosphorus in the etiology of tibial dyschondroplasia in young chicks. *Journal of Nutrition* 113, 1568–1575.

Eeckhout, W., de Paepe, M., Warnants, N. and Bekaert, H. (1995) An estimation of the minimal P requirements for growing-finishing pigs, as influenced by the Ca level of the diet. *Animal Feed Science and Technology* 52, 29–40.

Engels, E.A.N. (1981) Mineral status and profiles (blood, bone and milk) of the grazing ruminant with special reference to calcium, phosphorus and magnesium. *South African Journal of Animal Science* 11, 171–182.

Falade, J.A. (1973) Effect of phosphorus on the growth and mineral composition of four tropical forage legumes. *Journal of the Science of Food and Agriculture* 24, 795–802.

Field, A.C., Suttle, N.F. and Nisbet, D.I. (1975) Effects of diets low in calcium and phosphorus on the development of growing lambs. *Journal of Agricultural Science, Cambridge* 85, 435–442.

Field, A.C., Kamphues, J. and Woolliams, J.A. (1983) The effect of dietary intake of calcium and phosphorus on the absorption and excretion of phosphorus in chimaera-derived sheep. *Journal of Agricultural Science, Cambridge* 101, 597–602.

Field, A.C., Woolliams, J.A., Dingwall, R.A. and Munro, C.S. (1984) Animal and dietary variation in the absorption and metabolism of phosphorus by sheep. *Journal of Agricultural Science, Cambridge* 103, 283–291.

Field, A.C., Woolliams, J.A. and Dingwall, R.A. (1985) The effect of dietary intake of calcium and dry matter on the absorption and excretion of calcium and phosphorus by growing lambs. *Journal of Agricultural Science, Cambridge* 105, 237–243.

Field, A.C., Woolliams, J.A. and Woolliams, C. (1986) The effect of breed of sire on the urinary excretion of phosphorus and magnesium in lambs. *Animal Production* 42, 349–354.

Fishwick, G. (1978) Utilisation of the phosphorus and magnesium in some calcium and magnesium phosphates by growing sheep. *New Zealand Journal of Agricultural Research* 21, 571–575.

Fishwick, G. and Hemingway, R.G. (1973) Magnesium phosphates as dietary supplements for growing sheep. *Journal of Agricultural Science, Cambridge* 81, 441–444.

Fishwick, G., Fraser, J., Hemingway, R.G., Parkins, J.J. and Ritchie, N.S. (1977) The effects of dietary phosphorus inadequacy during pregnancy and lactation on the voluntary intake and digestibility of oat straw by beef cows and the performance of their calves. *Journal of Agricultural Science, Cambridge* 88, 143–150.

Forar, F.L., Kincaid, R.L., Preston, R.L. and Hillers, J.K. (1982) Variation of inorganic phosphorus in blood plasma and milk of lactating cows. *Journal of Dairy Science* 65, 760–763.

Gartner, R.J.W., Murphy, G.M. and Hoey, W.A. (1982) Effects of induced, subclinical phosphorus deficiency on feed intake and growth in heifers. *Journal of Agricultural Science, Cambridge* 98, 23–29.

Grant, C.C., Biggs, H.C., Meisnner, H.H. and Basson, P.A. (1996) The usefulness of faecal phosphorus and nitrogen in interpreting differences in live-mass gain and the response to phosphorus supplementation in grazing cattle in arid regions. *Onderstepoort Journal of Veterinary Research* 63, 121–126.

Han, Y.N., Yang, F., Zhou, A.G., Miller, E.R., Ku, P.K., Hogberg, M.G. and Lei, X.G.

(1997) Supplemental phytases of microbial and cereal sources improve dietary phytate phosphorus utilisation by pigs from weaning through finishing. *Journal of Animal Science* 75, 1017–1025.

Hemingway, R.G. and Fishwick, G. (1975) Ammonium polyphosphate in the drinking-water as a source of phosphorus for growing sheep. *Proceedings of the Nutrition Society* 34, 78A–79A.

Hendricksen, R.E., Ternouth, J.D. and Punter, L.D. (1994) Seasonal nutrient intake and phosphorus kinetics of grazing steers in northern Australia. *Australian Journal of Agricultural Research* 45, 1817–1829.

Holechek, J.L., Galyean, M.L., Wallace, J.D. and Wofford, H. (1985) Evaluation of faecal indices for predicting phosphorus status of cattle. *Grass and Forage Science* 40, 489–492.

Hoppe, P.P., Schoner, F.-J., Wiesche, H., Schwarz, G. and Safer, S. (1993) Phosphorus equivalency of *Aspergillus niger*-phytate for piglets fed a grain–soya bean meal diet. *Journal of Animal Physiology and Animal Nutrition* 69, 1–9.

Jones, R.J. (1990) Phosphorus and beef production in northern Australia. 1. Phosphorus and pasture productivity – a review. *Tropical Grasslands* 24, 131–139.

Jones, R.J. and Betteridge, K. (1994) Effect of superphosphate and its component elements (phosphorus, sulfur and calcium), on the grazing preference of steers on a tropical grass–legume pasture grown on a low phosphorus soil. *Australian Journal of Experimental Agriculture* 34, 349–353.

Jongbloed, A.W. and Kemme, P.A. (1990) Apparent digestible phosphorus in the feeding of pigs in relation to availability, requirement and environment. 1. Digestible phosphorus in feedstuffs from plant and animal origin. *Netherlands Journal of Agricultural Science* 38, 567–575.

Jumba, I.O., Suttle, N.F., Hunter, E.A. and Wandiga, S.O. (1995) Effects of soil origin and composition and herbage species on the mineral composition of forages in the Mount Elgon region of Kenya. 1. Calcium, phosphorus, magnesium and sulphur. *Tropical Grasslands* 29, 40–46.

Karn, J.F. (1997) Phosphorus supplementation of range cows in the Northern Great Plains. *Journal of Range Management* 50, 2–9.

Kerridge, P.L., Gilbert, M.A. and Coates, D.B. (1990) Phosphorus and beef production in northern Australia. 8. The status and management of soil phosphorus in relation to beef production. *Tropical Grasslands* 24, 221–230.

Keshavarz, K. (1994) Laying hens respond differently to high dietary levels of phosphorus in monobasic and dibasic calcium phosphate. *Poultry Science* 73, 687–703.

Koch, M.E. and Mahan, D.C. (1985) Biological characteristics for assessing low phosphorus intake in growing swine. *Journal of Animal Science* 60, 699–708.

Kornegay, E.T. and Qian, H. (1996) Replacement of inorganic phosphorus by microbial phytase for young pigs fed a maize–soya bean meal diet. *British Journal of Nutrition* 76, 563–578.

Lacey, D.L. and Huffer, W.E. (1982) Studies on the pathogenesis of avian rickets. 1. Changes in epiphyseal and metaphyseal vessels in hypocalcaemic and hypophosphataemic rickets. *American Journal of Pathology* 109, 288–301.

Lampkin, G.H., Howard, D.A. and Burdin, M.L. (1961) Studies on the production of beef from zebu cattle in East Africa. 3. The value of feeding a phosphatic supplement. *Journal of Agricultural Science, Cambridge* 57, 39–47.

Langlands, J.P. (1987) Assessing the nutrient status of herbivores. In: Hacker, J.B. and

Ternouth, J.H. (eds) *The Nutrition of Herbivores.* Academic Press, Sydney, pp. 363–390.
Lei, X.G., Ku, P.K., Miller, E.R., Yokoyama, M.T. and Ullrey, D.E. (1994) Calcium level effects the efficacy of supplemental microbial phytase in corn–soybean meal diets for pigs. *Journal of Animal Science* 72, 139–143.
Little, D.A. (1968) Effect of dietary phosphate on the voluntary consumption of Townsville lucerne (*Stylosanthes humilis*) by cattle. *Proceedings of the Australian Society of Animal Production* 7, 376–380.
Little, D.A. (1984) Definition of objective criterion of body phosphorus reserves in cattle and its evaluation *in vivo. Canadian Journal of Animal Science* 64 (suppl.), 229–231.
Little, D.A. and Minson, D.J. (1977) Variation in the phosphorus content of bone samples obtained from the last three ribs of cattle. *Research in Veterinary Science* 23, 393–394.
Little, D.A. and Ratcliff, D. (1979) Phosphorus content of bovine rib: influence of earlier biopsy of the same rib. *Research in Veterinary Science* 27, 239–241.
Littledike, E.T. and Goff, J.P. (1987) Interactions of calcium, phosphorus and magnesium and vitamin D that influence their status in domestic meat animals. *Journal of Dairy Science* 70, 1727–1743.
Liu, J., Bollinger, D.W., Ledoux, D.R., Ellersieck, M.R. and Veum, T.L. (1997) Soaking increases the efficacy of supplemental microbial phytase in a low-phosphorus corn–soybean meal diet for growing pigs. *Journal of Animal Science* 75, 1292–1298.
Long, T.A., Tillman, A.D., Nelson, A.B., Gallup, W.D. and Davis, W. (1957) Availability of phosphorus in mineral supplements. *Journal of Animal Science* 16, 444–450.
McLean, R.W., Hendricksen, R.E., Coates, D.B. and Winter, W.H. (1990) Phosphorus and beef production in northern Australia. 6. Dietary attributes and their relation to cattle growth. *Tropical Grasslands* 24, 197–208.
MAFF (1990) *UK Tables of Nutritive Value and Chemical Composition of Feedstuffs.* Edited by D.I. Givens and A.R. Moss. Rowett Research Services, Aberdeen, pp. 12–78.
Martz, F.A., Belo, A.T., Weiss, M.F., Belyell, R.L. and Goff, J.P. (1990) True absorption of calcium and phosphorus from alfalfa and corn silage when fed to lactating cows. *Journal of Dairy Science* 73, 1288–1295.
Miller, C.P., Winter, W.H., Coates, D.B. and Kerridge, P.C. (1990) Phosphorus and beef production in northern Australia. 10. Strategies for phosphorus use. *Tropical Grasslands* 24, 239–249.
Miller, E.R., Ullrey, D.C., Zutaut, C.L., Baltzer, B.V., Schmidt, D.A., Hoefer, J.A. and Luecke, R.W. (1964) Phosphorus requirement of the baby pig. *Journal of Nutrition* 82, 34–40.
Miller, W.J., Neathery, M.W., Gentry, R.P., Blackmon, D.M., Crowe, C.I., Ware, G.C. and Fielding, A.J. (1987) Bioavailability of phosphorus from defluorinated and dicalcium phosphates and phosphorus requirement of calves. *Journal of Dairy Science* 70, 1885–1892.
Minson, D.J. (1990) Phosphorus. In: *Forage in Ruminant Nutrition.* Academic Press, New York, pp. 230–264.
Mitchell, H.H. (1947) The mineral requirements of farm animals. *Journal of Animal Science* 6, 365–377.
Mohammed, A., Gibney, M. and Taylor, T.G. (1991) The effects of levels of inorganic

phosphorus, calcium and cholecalciferol on the digestability of phytate-P by the chick. *British Journal of Nutrition* 66, 251–259.
Morse, D., Head, H.H. and Wilcox, C.J. (1992) Disappearance of phosphorus in phytase from concentrates *in vitro* and from rations fed to lactating dairy cows. *Journal of Dairy Science* 75, 1979–1986.
Motzok, I., Arthur, D. and Branion, H.D. (1956) Utilization of phosphorus from various phosphate supplements by chicks. *Poultry Science* 35, 627–649.
Murry, A.C., Lewis, R.D. and Amos, H.E. (1997) The effect of microbial phytase in a pearl millet–soybean meal diet on apparent digestibility and retention of nutrients, serum mineral concentration and bone mineral density of nursery pigs. *Journal of Animal Science* 75, 1284–1291.
Muschen, H., Petri, A., Breves, G. and Pfeffer, E. (1988) Response of lactating goats to low phosphorus intake. 1. Milk yield and faecal excretion of P and Ca. *Journal of Agricultural Science, Cambridge* 111, 255–263.
Nelson, T.S., Shieh, T.R., Wodzinski, R.J. and Ware, J.H. (1971) Effect of supplemental phytase on the utilisation of phytate phosphorus by chicks. *Journal of Nutrition* 101, 1289–1293.
Nelson, T.S., Daniels, L.B., Hall, J.R. and Shields, L.G. (1976) Hydrolysis of natural phytate phosphorus in the digestive tract of calves. *Journal of Animal Science* 42, 1509–1512.
Nimmo, R.D., Peo, E.R., Maser, B.D. and Lewis, A.J. (1981a) Effect of level of dietary calcium and phosphorus during growth and gestation on performance, blood and bone parameters of swine. *Journal of Animal Science* 52, 1330–1342.
Nimmo, R.D., Peo, E.R., Crenshaw, D.D., Maser, B.D. and Lewis, A.J. (1981b) Effect of level of dietary calcium and phosphorus during growth and gestation on calcium-phosphorus balance and reproductive performance of first litter sows. *Journal of Animal Science* 52, 1343–1349.
NRC (1966) *Nutrient Requirements of Poultry*, 6th edn. National Academy of Sciences, Washington, DC.
NRC (1975) *Nutrient Requirements of Beef Cattle*, 5th edn. National Academy of Sciences, Washington, DC.
NRC (1979) *Nutrient Requirements of Swine*, 8th edn. National Academy of Sciences, Washington, DC.
NRC (1984) *Nutrient Requirements of Beef Cattle*, 6th edn. National Academy of Sciences, Washington, DC.
NRC (1985) *Nutrient Requirements of Sheep*, 5th edn. National Academy of Sciences, Washington, DC.
NRC (1988) *Nutrient Requirements of Swine*, 9th edn. National Academy of Sciences, Washington, DC.
NRC (1989) *Nutrient Requirements of Horses*, 5th edn. National Academy of Sciences, Washington, DC.
NRC (1994) *Nutrient Requirements of Poultry*, 9th edn. National Academy of Sciences, Washington, DC.
O'Quinn, P.R., Knabe, D.A. and Gregg, E.J. (1997) Efficacy of Natuphos® in sorghum-based diets for finishing swine. *Journal of Animal Science* 75, 1299–1307.
Owings, W.J., Sell, J.L. and Balloun, S.L. (1977) Dietary phosphorus needs of laying hens. *Poultry Science* 56, 2056–2060.
Petri, A., Muschen, H., Breves, G., Richter, O. and Pfeffer, E. (1989) Response of lactating goats to low phosphorus intake. 2. Nitrogen transfer from rumen ammonia to rumen microbes and proportion of milk protein derived from

microbial amino-acids. *Journal of Agricultural Science, Cambridge* 111, 265–271.

Pointillart, A.L., Fontaine, N. and Thomasset, M. (1986) Effects of vitamin D on calcium regulation in vitamin D deficient pigs given a phytate-phosphorus diet. *British Journal of Nutrition* 56, 661–669.

Pointillart, A.L., Fourdin, A. and Fontaine, N. (1987) Importance of cereal phytase activity for phytate phosphorus utilisation by growing pigs fed triticale or corn. *Journal of Nutrition* 117, 907–913.

Pointillart, A.L., Fourdin, A., Bordeau, A. and Thomasset, M. (1989) Phosphorus utilisation and hormonal control of calcium metabolism in pigs fed phytic phosphorus diets containing normal or high calcium levels. *Nutrition Reports International* 40, 517–527.

Price, J.S., Jackson, B., Eastell, R., Goodship, A.E., Blumsohn, A., Wright, I., Stoneham, S., Lanyon, L.E. and Russell, R.G.G. (1995) Age-related changes in biochemical markers of bone metabolism in horses. *Equine Veterinary Journal* 27, 210–217.

Qian, H., Kornegay, E.T. and Veit, H.P. (1996) Effects of supplemental phytase and phosphorus on histological, mechanical and chemical traits of tibia and performance of turkeys fed on soyabean meal-based semi-purified diets high in phytate phosphorus. *British Journal of Nutrition* 76, 263–272.

Rajaratne, A.A.J., Scott, D. and Buchan, W. (1994) Effects of a change in phosphorus requirement on phosphorus kinetics in the sheep. *Research in Veterinary Science* 56, 262–264.

Read, M.P., Engels, E.A.N. and Smith, W.A. (1986a) Phosphorus and the grazing ruminant. 1. The effect of supplementary P on sheep at Armoedsvlakte. *South African Journal of Animal Science* 16, 1–6.

Read, M.P., Engels, E.A.N. and Smith, W.A. (1986b) Phosphorus and the grazing ruminant. 2. The effect of supplementary P on cattle at Glen and Armoedsvlakte. *South African Journal of Animal Science* 16, 7–12.

Read, M.P., Engels, E.A.N. and Smith, W.A. (1986c) Phosphorus and the grazing ruminant. 3. Rib bone samples as an indicator of the P status of cattle. *South African Journal of Animal Science* 16, 13–27.

Reid, R.L., Franklin, M.C. and Hallsworth, E.G. (1947) The utilization of phytate phosphorus by sheep. *Australian Veterinary Journal* 23, 136–140.

Reinhart, G.A. and Mahan, D.C. (1986) Effects of various calcium:phosphorus ratios at low and high dietary phosphorus for starter, grower and finishing swine. *Journal of Animal Science* 63, 457–466.

Roland, D.A. (1986) Egg shell quality. II. Calcium and phosphorus requirements of commercial Leghorns. *Worlds Poultry Science Journal* 42, 166–171.

Scharp, D.W. (1979) Effect of adding superphosphate to the drinking water on the fertility of dairy cows. *Australian Veterinary Journal* 55, 240–243.

Schoner, F.-J., Hoppe, P.P., Schwarz, G. and Wiesche, H. (1993) Comparison of microbial phytase and inorganic phosphate in male chickens: influence on performance data, mineral retention and dietary calcium. *Journal of Animal Physiology and Animal Nutrition* 69, 235–244.

Schryver, H.F., Craig, P.H. and Hintz, H.F. (1970) Calcium metabolism in ponies fed varying levels of calcium. *Journal of Nutrition* 100, 955–964.

Schryver, H.F., Hintz, H.F. and Craig, P.H. (1971) Phosphorus metabolism in ponies fed varying levels of phosphorus. *Journal of Nutrition* 101, 1257–1263.

Schryver, H.F., Hintz, H.F., Lowe, J.E., Hintz, R.L., Harper, R.B. and Reid, J.T. (1974)

Mineral composition of the whole body, liver and bone of young horses. *Journal of Nutrition* 104, 126–132.

Scott, D., McLean, A.F. and Buchan, W. (1984) The effect of variation in phosphorus intake on net intestinal phosphorus absorption, salivary phosphorus secretion and pathway of excretion in sheep fed roughage diets. *Quarterly Journal of Experimental Physiology* 69, 439–452.

Scott, D., Whitelaw, F.C., Buchan, W. and Bruce, L.A. (1985) The effect of variation in phosphorus secretion, net intestinal phosphorus absorption and faecal endogenous phosphorus excretion in sheep. *Journal of Agricultural Science, Cambridge* 105, 271–277.

Scott, D., Rajaratne, A.A.J. and Buchan, W. (1995) Factors affecting faecal endogenous phosphorus loss in the sheep. *Journal of Agricultural Science, Cambridge* 124, 145–151.

Simons, P.C.M., Versteegh, H.A.J., Jongbloed, A.W., Kemme, P.A., Slump, P., Bos, K.D., Walters, M.G.E., Beudeker, R.F. and Verschoar, G.J. (1990) Improvement of phosphorus availability by microbial phytase in broilers and pigs. *British Journal of Nutrition* 64, 525–540.

Singsen, E.P., Spandorf, A.H., Matterson, L.D., Serafin, J.A. and Tlustohowicz, J.J. (1962) Phosphorus in the nutrition of the adult hen. 1. Minimum phosphorus requirements. *Poultry Science* 41, 1401–1414.

Snook, L.C. (1949) Phosphorus deficiency in dairy cows: its prevalence in South-Western Australia and possible methods of correction. *Journal of the Department of Agriculture for Western Australia* 26, 169–177.

Snook, L.C. (1958) Phosphorus deficiency as a cause of bovine infertility. *Proceedings of the Australia and New Zealand Association for the Advancement of Science, Adelaide.*

Soares, J.H. (1995) Phosphorus bioavailability. In: Ammerman, C.B., Baker, D.H. and Lewis, A.J. (eds) *Bioavailability of Nutrients for Animals.* Academic Press, New York, pp. 257–294.

Somerville, B.A., Maunder, E., Ross, R., Care, A.D. and Brown, R.C. (1985) Effect of dietary calcium and phosphorus depletion on vitamin D metabolism and calcium binding protein in the growing pig. *Hormone and Metabolism Research* 17, 78–81.

Stewart, J. (1934/35) The effect of phosphorus deficient diets on the metabolism, blood and bones of sheep. In: *University of Cambridge Institute of Animal Pathology Fourth Report,* pp. 179–205.

Suttle, N.F. (1987) The absorption, retention and function of minor nutrients. In: Hacker, J.B. and Ternouth, J.H. (eds) *The Nutrition of Herbivores.* Academic Press, Sydney, pp. 333–361.

Suttle, N.F. (1991) Mineral supplementation of low quality roughages. In: *Proceedings of Symposium on Isotope and Related Techniques in Animal Production and Health.* International Atomic Energy Commission, Vienna, pp. 101–144.

Suttle, N.F. (1994) Seasonal infections and nutritional status. *Proceedings of the Nutrition Society* 53, 545–555.

Suttle, N.F. and Hay, L. (1986) Urolithiasis. In: Martin, W.B. and Aitken, I.D. (eds) *Diseases of Sheep,* 2nd edn. Blackwell Scientific Publications, Oxford, pp. 250–254.

Suttle, N.F., Dingwall, R.A. and Munro, C.S. (1989) Assessing the availability of dietary phosphorus for sheep. In: Southgate, D.A.T., Johnson, I.T. and Fenwick, G.R. (eds) *Nutrient Availability: Chemical and Biological Aspects.* Royal Society of Chemistry Special Publication No. 72, Cambridge, pp. 268–270.

Teleni, E., Dean, H. and Murray, R.M. (1976) Some factors affecting the measurement of blood inorganic phosphorus in cattle. *Australian Veterinary Journal* 52, 529–533.

Ternouth, J.H. (1989) Endogenous losses of phosphorus in sheep. *Journal of Agricultural Science, Cambridge* 113, 291–297.

Ternouth, J.H. (1990) Phosphorus and beef production in Northern Australia. 3. Phosphorus in cattle – a review. *Tropical Grasslands* 24, 159–169.

Ternouth, J.H. and Coates, D.B. (1997) Phosphorus homeostasis in grazing breeder cattle. *Journal of Agricultural Science, Cambridge* 128, 331–337.

Ternouth, J.H. and Sevilla, C.L. (1990a) The effect of low levels of dietary phosphorus upon dry matter intake and metabolism in lambs. *Australian Journal of Agricultural Research* 41, 175–184.

Ternouth, J.H. and Sevilla, C.L. (1990b) Dietary calcium and phosphorus repletion in lambs. *Australian Journal of Agricultural Research* 41, 413–420.

Ternouth, J.H., Bortolussi, G., Coates, D.B., Hendricksen, R.E. and McLean, R.W. (1996) The phosphorus requirements of growing cattle consuming forage diets. *Journal of Agricultural Science, Cambridge* 126, 503–510.

Theiler, A. (1912) Facts and theories about styfziekte and lamziekte. In: *Second Report of the Directorate of Veterinary Research*, pp. 7–78.

Theiler, A. and Green, H.H. (1932) Aphosphoris in ruminants. *Nutrition Abstracts and Reviews* 1, 359–385.

Thorp, B.H. and Waddington, D. (1997) Relationships between the bone pathologies, ash and mineral content of long bones in 35-day-old broiler chickens. *Research in Veterinary Science* 62, 67–73.

Tillman, A.D., Brethour, J.R. and Hansard, S.L. (1959) Comparative procedures for measuring the phosphorus requirements of cattle. *Journal of Animal Science* 18, 249–255.

Tomas, F.M., Moir, R.J. and Somers, M. (1973) Phosphorus turnover in sheep. *Australian Journal of Agricultural Research* 18, 635–645.

Underwood, E.J., Beck, A.B. and Shier, F.L. (1939) Further experiments on the incidence and control of pica in sheep in the botulism areas of Western Australia. *Australian Journal of Experimental Biology and Medical Science* 17, 249–255.

Underwood, E.J., Shier, F.L. and Beck, A.B. (1940) Experiments in the feeding of phosphorus supplements to sheep in Western Australia. *Journal of the Department of Agriculture for Western Australia* 17, 388–405.

Van Niekerk, B.D.H. and Jacobs, G.A. (1985) Protein energy and phosphorus supplementation of cattle fed low-quality roughage. *South African Journal of Animal Science* 15, 133–136.

Wadsworth, J.C., McClean, R.W., Coates, D.B. and Winter, J.H. (1990) Phosphorus and beef production in northern Australia. 5. Animal phosphorus status and diagnosis. *Tropical Grasslands* 24, 185–196.

Waldroup, P.W., Simpson, C.F., Damron, B.L. and Harms, R.H. (1967) The effectiveness of plant and inorganic phosphorus in supporting egg production in hens and hatchability and bone development in chick embryos. *Poultry Science* 46, 659–664.

Walker, R.G., Davison, T.M., Orr, W.N. and Silver, B.A. (1997) Phosphorus fertilizer for nitrogen-fertilized dairy pastures. 3. Milk responses to a dietary phosphorus supplement. *Journal of Agricultural Science, Cambridge* 129, 233–236.

Wang, X.-L., Gallagher, C.H., McClure, T.J., Reeve, V.E. and Canfield, P.J. (1985)

Bovine post-parturient haemoglobinuria: effect of inorganic phosphate on red cell metabolism. *Research in Veterinary Science* 39, 333–339.

Wan Zahari, M., Thompson, J.K., Scott, D. and Buchan, W. (1990) The dietary requirements of calcium and phosphorus for growing lambs. *Animal Production* 50, 301–308.

Williams, S.W., Lawrence, L.A., McDowell, L.R., Wilkinson, N.S., Ferguson, P.W. and Warmick, A.C. (1991a) Criteria to evaluate bone mineralisation in cattle. I. Effect of dietary phosphorus on chemical, physical and mechanical properties. *Journal of Animal Science* 69, 1232–1242.

Williams, S.W., McDowell, L.R., Lawrence, L.A., Wilkinson, N.S., Ferguson, P.W. and Warmick, A.C. (1991b) Criteria to evaluate bone mineralisation in cattle. II. Noninvasive techniques. *Journal of Animal Science* 69, 1243–1254.

Wilson, W.D. and Field, A.C. (1983) Absorption and secretion of calcium and phosphorus in the alimentary tract of lambs infected with daily doses of *Trichostrongylus colubriformis* or *Ostertagia circumcincta*. *Journal of Comparative Pathology* 93, 61–71.

Winks, L. (1990) Phosphorus and beef production in northern Australia. 2. Responses to phosphorus by ruminants – a review. *Tropical Grasslands* 24, 140–158.

Wise, M.B., Smith, S.E. and Barnes, L.L. (1958) The phosphorus requirement of calves. *Journal of Animal Science* 17, 89–99.

Wise, M.B., Ordoveza, A.L. and Barrick, E.R. (1963) Influence of variations in dietary calcium:phosphorus ratio on performance and blood constituents of calves. *Journal of Nutrition* 79, 79–84.

Wright, R.D., Blair-West, J.R., Nelson, J.F. and Tregear, G.W. (1984) Handling of phosphate by a parotid gland (ovine). *American Journal of Physiology* 246, F916–F926.

Yano, F., Yano, H. and Breves, G. (1991) Calcium and phosphorus metabolism in ruminants. In: *Proceedings of the Seventh International Symposium of Ruminant Physiology*. Academic Press, New York, pp. 277–295.

Young, V.R., Richards, W.P.C., Lofgreen, G.P. and Luick, J.R. (1966) Phosphorus depletion in sheep and the ratio of calcium to phosphorus in the diet with reference to calcium and phosphorus absorption. *British Journal of Nutrition* 20, 783–794.

Magnesium

6

Introduction

Magnesium has been known for over 70 years to be an essential element in the diet of higher animals (Leroy, 1926). A few years later, the main manifestations of magnesium deficiency in laboratory species were described and they included hyperirritability and convulsions (Kruse *et al.*, 1932). A metabolic disorder in cows, known as 'kopziekte', had been recognized in Europe for some years before Sjollema and Seekles (1930) associated the condition with subnormal serum magnesium values and termed it 'grass tetany'. Neither this name nor others, such as 'lactation tetany' or 'grass staggers', were entirely appropriate, because the disease is not limited to lactating or grazing animals and is characterized by convulsions rather than tetany. Nevertheless, a great stimulus was given to studies of the nutritional physiology of magnesium in ruminants. Comparable research with pigs and poultry came much later, largely because magnesium is abundant in most common feedstuffs relative to their requirements and natural deficiencies are unheard of. Interest in the consequences of magnesium deficiency in ruminants has been sustained by the fact that intensification of grassland production, through the increased use of nitrogenous and potassic fertilizers, has increased the incidence of grass tetany by increasing milk yields while lowering the absorbability of magnesium from the pasture. In 1979, it was estimated that grass tetany cost US producers $70m annually from deaths alone, i.e. not counting any subclinical or residual effects on health and performance (Fisher and Wilson, 1979, cited by Hurley *et al.*, 1990). Current disease statistics for the USA and the UK show that grass tetany remains a problem. The susceptibility of ruminants to magnesium deficiency hinges upon the fact that the primary site of absorption is usually the rumen (Tomas and Potter, 1976a; Field and Munro, 1977), whereas in the non-ruminant it is the small intestine. The general level of absorptive efficiency for magnesium in ruminants is roughly half that of non-ruminants (35 vs. 70% of intake) and open to a number of dietary influences, which lead to great variability and

© CAB *International* 1999. *Mineral Nutrition of Livestock*
(E.J. Underwood and N.F. Suttle)

sometimes exceedingly low values (< 20%). The factors influencing magnesium metabolism in the rumen will be reviewed at length, so that hypomagnesaemic tetany can be the better understood and controlled, and emphasis throughout will be on the magnesium nutrition of ruminants.

Sources of Magnesium

Forages

The magnesium content of herbage plants varies with the species and with the soil and climatic conditions in which the plants are grown. In temperate pastures, leguminous species are usually richer than grasses in magnesium (Mg), as they are in calcium (Thomas *et al.*, 1952; Turner *et al.*, 1978). Minson (1990) calculated mean values of 1.8 and 3.6 g Mg kg^{-1} dry matter (DM), respectively, for temperate and tropical grasses and 2.6 and 2.8 g Mg kg^{-1} DM for the corresponding legumes among 930 forage samples. Most temperate grasses were considered to be inadequate sources of magnesium for grazing livestock. Species and varietal differences among grasses may influence the incidence of hypomagnesaemic tetany, magnesium concentrations being 30% lower in 'Ajax' and S24 perennial ryegrass than in 'Augusta' (hybrid ryegrass) and 'Melle' (perennial ryegrass) grown in the UK, but they also tend to be lower in crude protein (CP) (Fig. 6.1; MAFF, 1992); the negative association between CP and serum magnesium will be discussed later. Seasonal variations in the magnesium concentration in herbage are generally small (Minson, 1990), although Allcroft (1954) found that the values in English pastures fell during the first few days of spring and Thomas *et al.* (1955) found that the magnesium concentration in a series of hay crops ranged from 0.6 to 1.5 g kg^{-1} DM. Conserved forages generally have similar species means and ranges to those of fresh forages, and recent UK data (Table 6.1; MAFF, 1992) show higher values for legume (lucerne) than grass hays and for legume (clover) than grass silages. As far as other roughages are concerned, straws are low in magnesium, while root and leafy brassica crops are similar to grasses.

Concentrates

The concentrates used in animal feeding vary widely in magnesium content. Cereal grains generally contain 1.1–1.3 g, oil-seed meals 3.0–5.8 g and fish-meals intermediate levels of 1.7–2.5 g Mg kg^{-1} DM (Table 6.1; MAFF, 1992). Animal products used as protein supplements vary with the proportion of bone they contain. Thus 'pure' meat-meal may contain only 0.4 g Mg kg^{-1} DM, while commercial meat-and-bone-meals commonly contain around 2 g but occasionally as much as 10 g Mg kg^{-1} DM. The above materials, which make up the bulk of the diets of pigs and poultry, are well supplied with magnesium compared with the estimated minimum requirements of these species.

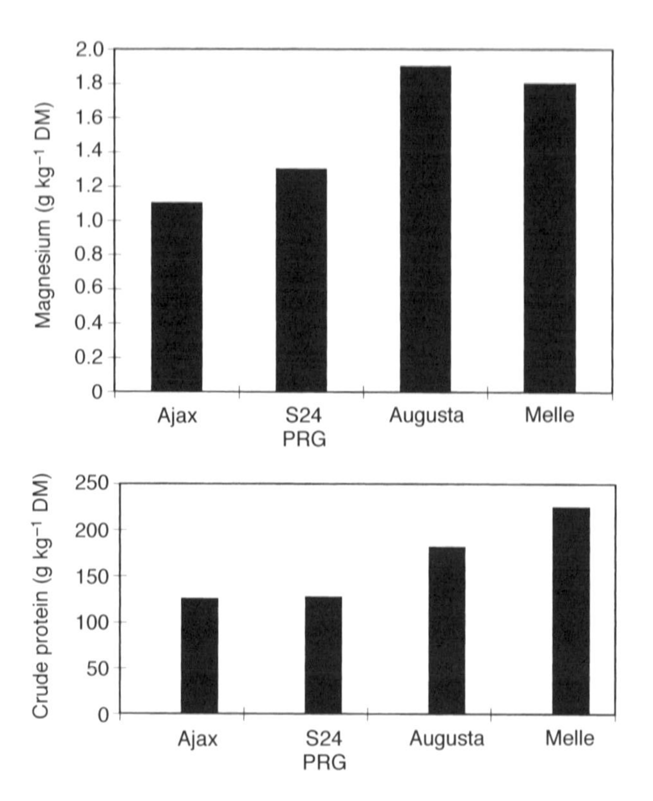

Fig. 6.1. Differences in magnesium concentrations between newer hybrid strains of perennial ryegrass: the benefits of high Mg values may partly be offset by correspondingly high crude protein (CP) levels which lower absorbability (see Fig. 6.5; data from MAFF, 1992).

Absorbability of magnesium

The nutritive value of magnesium sources is determined by absorbability (A_{Mg}) as well as by concentration, particularly for ruminants. In a recent literature review (Henry and Benz, 1995), the apparent availability of magnesium in fresh grasses or grass hays was noted to vary from −4 to +66% but little attempt was made to identify systematic sources of variation. Earlier, the Agricultural Research Council (ARC, 1980) had remained unconvinced of differences between concentrate and forage diets but took the precaution of adopting a low coefficient of absorption (0.20) in deriving requirements. Suttle (1987) subsequently suggested that the difference in absorbability between grass and hay + concentrate diets might be two- to threefold. In view of the probable impact of diet type on A_{Mg}, the linear regression coefficients relating the fractional apparent absorption of magnesium (AA_{Mg}) to potassium concentration were derived using pooled experimental data in the literature (Adediji and Suttle, 1999). Data were classified by species (sheep or cattle) and diet type (grass, conserved forage, coarse hay +

Table 6.1. Mean magnesium concentrations in livestock feeds commonly used in the UK (SD in parentheses) (MAFF, 1992).

Roughages		Concentrates		By-products	
Barley straw	0.7 (0.31)	Barley	1.2 (0.2)	Wheat bran	6.2 (2.7)
Oat straw	0.9 (0.31)	Cassava	1.1 (0.57)	Brewers' grains	1.7 (0.36)
Grass	1.6 (0.56)	Maize	1.3 (0.13)	Distillers' grains[a]	3.3 (0.34)
Kale	1.6 (0.18)	Oats, winter	1.0 (0.06)	Citrus pulp	1.7 (0.50)
White clover	2.2 (0.5)	Wheat	1.1 (0.13)	Sugar-beet pulp	1.1 (0.19)
Grass silage	1.7 (0.54)	Maize gluten	4.1 (0.70)		
Clover silage	2.3 (0.75)	Cottonseed meal	5.8 (0.43)		
Lucerne silage	1.8 (0.23)	Fish-meal, white	2.3 (0.31)		
Maize silage	2.2 (0.69)	Groundnut meal	3.5 (0.21)		
Fodder beet	1.6 (0.30)	Linseed meal	5.4 (0.09)		
Swedes	1.1 (0.07)	Maize-germ meal	2.1 (0.65)		
		Palm-kernel meal	3.0 (0.41)		
		Rapeseed meal	4.4 (0.53)		
		Soybean meal	3.0 (0.23)		
		Sunflower-seed meal	5.8 (0.49)		

[a] Barley-based.

concentrate mixtures) and the outcome is summarized in Fig. 6.2 and Table 6.2. There are enormous differences between both intercepts and slopes for sheep and cattle, particularly where forages are concerned. When potassium is low (1 g 100 g^{-1} DM), sheep (Fig. 6.2a) absorb forage magnesium roughly three times more efficiently than cattle (Fig. 6.2c). However, AA_{Mg} is much more sensitive to potassium (K) in sheep than in cattle and, when concentrations are high (5 g K 100 g^{-1} DM), coefficients for forages in the two species are equally low (*c.* 0.10). When mixed dry (roughage/concentrate (R/C)) diets low in potassium are fed, AA_{Mg} is again higher for sheep (Fig. 6.2b) than for cattle (Fig. 6.2d), but the margin is lower and AA_{Mg} is equally sensitive to increases in potassium in the two species. Thus systematic effects of diet type, dietary potassium and species on AA_{Mg} appeared to be subject to three-way interactions. The contrasts probably reflect the importance of the rumen as a site for digestion and magnesium absorption for a given species on a given diet, and hence the scope for K × Mg antagonism. Knowledge of A_{Mg} is extremely meagre for pigs and poultry, but values are usually higher than those for ruminants. Employing a combination of comparative balance and isotope dilution techniques, Güenter and Sell (1974) estimated that the true absorption of magnesium (%) in chickens was 55.7 from maize, 56.8 from wheat, 82.7 from oats, 54.5 from barley, 60.3 from soybean meal, 62 from dried skimmed milk and 42.5 from rice. Sell (1979) considered that diets containing sizeable amounts of these ingredients should meet Mg requirements for non-ruminants.

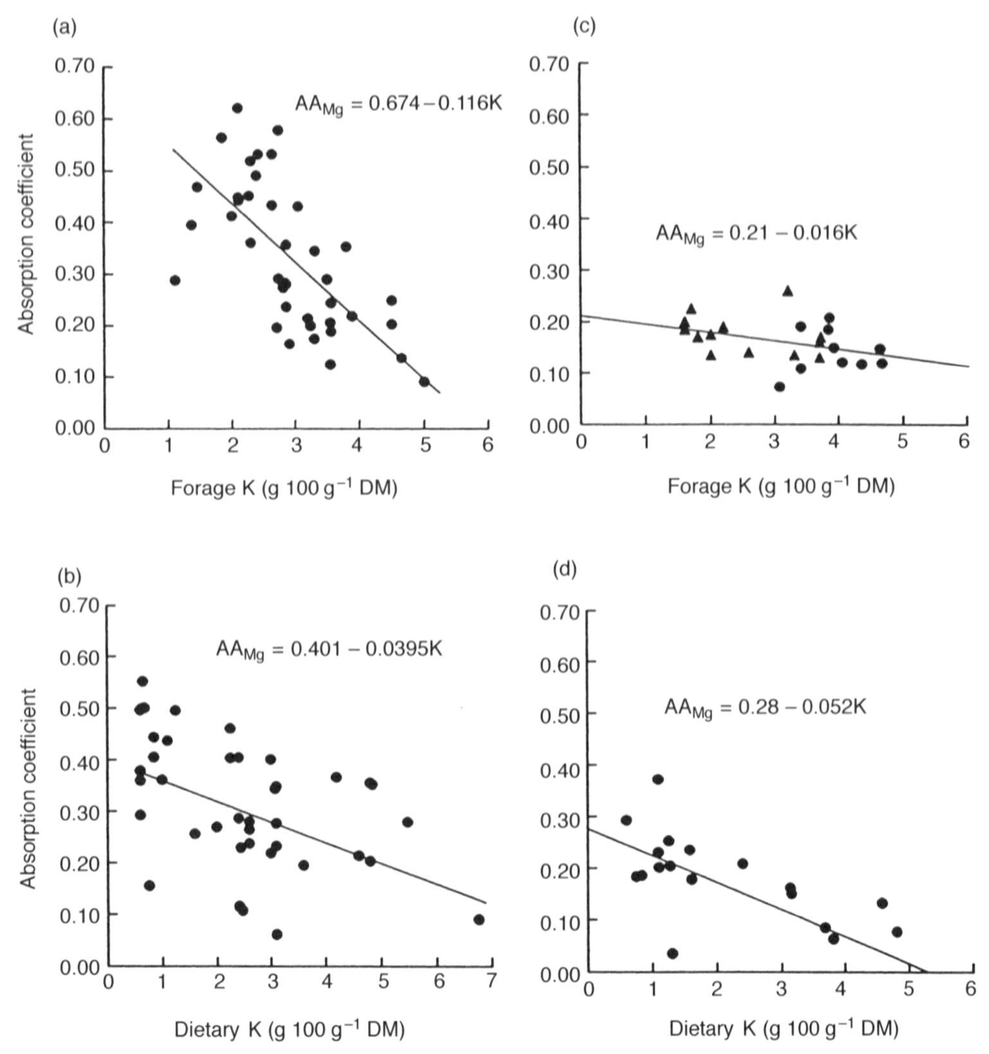

Fig. 6.2. Magnesium absorbability is inversely related to dietary potassium concentrations but the interaction is species- and diet-dependent (from Adediji and Suttle, 1999). (a) Sheep given fresh grasses; (b) sheep given roughage + concentrates; (c) cattle given hays (▲) or grasses (●); (d) cattle given hay + concentrates.

Absorption of Magnesium from the Rumen

In practice, the problems of inadequate magnesium nutrition begin with the development of a functional rumen.

Effects of potassium and sodium

Magnesium is absorbed from the rumen by two active transport processes (Dua and Care, 1995) against an electrochemical gradient (Martens, 1983;

Table 6.2. Intercepts (*c*) and coefficients (*a*) for multiple linear regression equations for the relationships between fractional AA_{Mg} and dietary potassium (g 100 g^{-1} DM) in sheep and cattle on different diets: SD are given in parentheses and r^2 gives the variation accounted for. (P = probability.)

Species	Diet	*n*	*c*	*a*K	r^2 (%)	P
Sheep	FH	57	0.67 (0.058)	−0.12 (0.019)	46.9	0.001
	CF	15	0.66 (0.131)	−0.16 (0.054)	35.2	0.002
	R/C	41	0.40 (0.034)	−0.04 (0.012)	20.1	0.002
Cattle	FH + CF	23	0.21 (0.028)	−0.016 (0.0085)	9.8	0.08
	R/C	19	0.28 (0.041)	−0.052 (0.0158)	35.6	0.004

FH, fresh herbage; CF, conserved forage; R/C, roughage/concentrate mixture.

Care *et al.*, 1984). The process is inhibited by potassium, which can increase the potential difference (p.d.) (blood positive) across the rumen epithelium from 30 to 55 mV (Scott, 1966; Martens and Blume, 1986). Infusion of potassium into the rumen decreases A_{Mg} substantially (Care *et al.*, 1984), whereas a similar quantity infused into the abomasum or ileum has no such effect (Greene *et al.*, 1983b; Wylie *et al.*, 1985). Studies with the isolated rumen showed that sodium decreased the p.d. and enhanced A_{Mg} (Martens and Blume, 1986). Furthermore, feeding sheep a diet low in sodium (Martens *et al.*, 1987) increased salivary and rumen potassium concentrations (a characteristic response to sodium depletion: see Chapter 7), facilitating a two-pronged attack on A_{Mg} from the rumen. The practical importance of these physiological interactions is twofold: firstly, they occur over ranges of sodium and potassium intakes commonly encountered by grazing ruminants (Powley *et al.*, 1977); and, secondly, they can halve A_{Mg} (Newton *et al.*, 1972; Field and Suttle, 1979; Martens *et al.*, 1987). A model of magnesium absorption in sheep which relied partly on the influence of potassium on transmural p.d. and hence net magnesium flux from the rumen achieved a remarkably good fit to observed balance data (Robson *et al.*, 1997).

Influence of pH

Rumen pH is important because it dramatically influences the solubility and therefore the absorbability of magnesium in the rumen. Ultrafilterable magnesium concentrations in rumen liquor decreased from around 6.0 mmol l^{-1} to below 0.5 mmol l^{-1} in cows as rumen pH increased from 5.6 to 7.2 (Johnson *et al.*, 1988), with values 1.0–1.5 mmol l^{-1} lower on grass than on hay + concentrate diets for a given pH. Rumen solubility of magnesium decreased from 80% to 20%, approximately, when pH was increased *in vitro* from 5 to 7 (Dalley *et al.*, 1997b). However, infusions of hydrochloric acid (HCl) and VFA can lower rumen pH substantially without increasing AA_{Mg} in sheep on a hay diet (Giduck *et al.*, 1988).

Effects of ammonium and other ions

Infusion of ammonium chloride into the rumen can lower A_{Mg} without affecting transmural p.d. (Care *et al.*, 1984), but effects are inconsistent and transient (Gabel and Martens, 1986). Transfer of ruminants from diets of hay + concentrates to grass or grass products will almost invariably raise rumen ammonia concentrations as well as pH (Johnson *et al.*, 1988; Johnson and Aubrey Jones, 1989) and both factors have long been associated with the incidence of hypomagnesaemia and tetany (Wilcox and Hoff, 1974; Fenner, 1979). However, other factors also change (notably potassium) and a causal role for ammonium (NH_4^+) has yet to be proved. Anion intake also affects rumen pH and hence magnesium uptake. Substitution of chloride by sulphate halved the A_{Mg} from the isolated rumen (Martens and Blume, 1986). Supplementing the diet of sheep with potassium as the acetate rather than the chloride enhanced the antagonism of magnesium absorption (Suttle and Field, 1967). Potassium given as the bicarbonate can raise pH by 15% (Wylie *et al.*, 1985) and may inhibit magnesium absorption more than that given as potassium chloride (KCl) (Schonewille *et al.*, 1997a). Responses to sodium and potassium will, therefore, be influenced by the accompanying anion, and the effects of employing salts, such as KCl and sodium chloride (NaCl), in experiments will not necessarily simulate those of grasses of different sodium and potassium content.

Absorption of Magnesium Beyond the Rumen

Some magnesium absorption can take place from the small and large intestine (Grace and MacRae, 1972; Tomas and Potter, 1976b; Dalley and Sykes, 1989), but it was thought to be a relatively inefficient process, either passive or active, hard to saturate and unable to fully compensate for impaired absorption from the rumen in sheep fed discontinuously (Field and Munro, 1977; Wilson, 1980). That conclusion was brought into question by evidence that magnesium from a source of low solubility in the rumen (Mg-mica) was absorbed so well from the small intestine that it equalled rumen-soluble sources, such as magnesium oxide (MgO) and magnesium hydroxide ($Mg(OH)_2$) in terms of total absorption along the entire gastrointestinal tract (Hurley *et al.*, 1990). Recent studies have shown that, when magnesium absorption from the rumen is decreased by adding potassium to the diet of sheep, fed every 2 h, there is a substantial compensatory increase in post-ruminal absorption of magnesium (Dalley *et al.*, 1997a). The compensation was believed to have occurred principally in the hind gut and was most marked at high magnesium intakes. Magnesium absorption from the hind gut may also be influenced by pH of the digesta (Dalley *et al.*, 1997b). Earlier work had shown that postruminal absorption was far more important when sheep were fed once daily, rather than continuously (Grace and MacRae, 1972). Dalley *et al.* (1997a) found that the overall reduction in magnesium absorption caused by potassium was independent of magnesium intake, whereas earlier work with discontinuously fed cattle showed the potassium effect to be proportional to magnesium intake (Field and Suttle, 1979). It is unclear whether the difference is attributable to pattern of feeding or differences in the K × Mg antagonism between species. In monogastric species, magnesium is absorbed mostly from the small and large intestine (Guenter and Sell, 1973; Partridge, 1978). Guinea-pigs can absorb magnesium from grass and lucerne hays more than twice as efficiently as lambs given the same hays (Reid *et al.*, 1978). Conditions in the intestinal lumen continue to affect absorbability, but there is no evidence that the efficiency of absorption is regulated according to need, as is the case for calcium. High intakes of calcium and phosphorus are alleged to suppress absorption of magnesium (Littledike and Goff, 1987), but there is no convincing evidence that this occurs in ruminants. In one early study with sheep (Chicco *et al.*, 1973), effects of both calcium and phosphorus were small and it could be argued that, as far as calcium (Ca) was concerned, magnesium was utilized with higher efficiency when a calcium-deficient diet (1.4 g Ca kg^{-1} DM) was fed rather than suppressed by high levels of 4.2 or 7.8 g Ca kg^{-1} DM.

Postabsorptive Metabolism

Magnesium in serum is present in two forms – either tightly bound to protein (32%) or free and ionized (68%) – and in one study the proportions remained unchanged around parturition, despite a rise in total serum magnesium from around 1.0 to 1.5 mmol l^{-1} (Riond *et al.*, 1995). Linear relationships between plasma magnesium and magnesium intakes well above requirement in ruminants (Chester-Jones *et al.*, 1989, 1990) confirm that magnesium metabolism is less rigidly controlled than calcium metabolism. Magnesium metabolism is not influenced by a specific hormone, although it may be influenced indirectly by the calcium-regulating hormones, calcitonin and parathyroid hormone (PTH), and by other hormones, including aldosterone (Littledike and Goff, 1987; Charlton and Armstrong, 1989), thyroxine and insulin (Ebel and Günther, 1980). The circulating levels of magnesium are influenced by the supply of absorbed magnesium from the gut, unavoidable losses, such as those in secreted milk, and equilibria attained with intracellular, soft-tissue and skeletal pools of magnesium. Thus, fasting causes a rapid fall in serum magnesium, while cessation of milking causes abrupt increases (Littledike and Goff, 1987). It was suggested that a shift of magnesium from the extracellular to the intracellular pool occurs when sheep consume potassium-rich grass, contributing to the development of hypomagnesaemia (Larvor, 1976), a view supported by the data of McLean *et al.* (1985).

Magnesium in the skeleton

In bone, magnesium is found largely in the hydration shell of hydroxyapatite crystals, creating a pool of exchangeable magnesium which can be mobilized when dietary supplies are inadequate (Gardner, 1973). Lability of the skeletal magnesium pool declines with age (Blaxter and McGill, 1956), but this may reflect the waning indirect influence of calcium-regulating hormones on bone resorption (see Chapter 4). Magnesium is unavoidably mobilized from the skeleton when calcium and phosphorus are withdrawn (see Table 4.3), and there is pronounced seasonal cycling of bone magnesium reserves in association with lactation (Fig. 5.6c, d). The changes in Ca:Mg ratio in rib reported by Engels (1981) were not attributable to magnesium deficiency, because plasma magnesium concentrations were normal throughout the study. The feeding of gross excesses of magnesium (24 g kg^{-1} DM) to lambs and steers increased magnesium concentrations in the rib from 3.6–4.1 to 5.8–6.4 g kg^{-1} DM (Chester-Jones *et al.*, 1989, 1990), but smaller supplements (5.5 g kg^{-1} DM) have also increased bone magnesium by 22% (Chicco *et al.*, 1973). Assumptions that magnesium retention is insignificant in mature animals may not always be safe.

Faecal endogenous losses

There was general agreement that the faeces is not used as a route of excretion for magnesium absorbed in excess of need. Faecal endogenous losses of magnesium (FE_{Mg}) were estimated to vary little from 3 mg Mg kg^{-1} live weight (LW) in ruminants by ARC (1980), who set a widely followed precedent by using this value as a constant to correct faecal magnesium excretion data and thus derive true from apparent absorption values. It has been recently argued that there is appreciable secretion of magnesium in saliva and, since absorptive efficiency is far from complete, FE_{Mg} should not be constant but increased by factors, such as potassium, which reduce absorption from the primary site, the rumen (Dua and Care, 1995). However, earlier work with sheep found no such effect of potassium (Newton *et al.*, 1972). Allsop and Rook (1979) found evidence of *decreased* FE_{Mg} in magnesium-deprived sheep, while Grace *et al.* (1985) found that 65% less magnesium was recycled to the rumen via parotid and mandibular saliva when sheep were fed fresh ryegrass, which was high in potassium and low in magnesium, rather than lucerne hay, which was low in potassium and high in magnesium (saliva flow was not greatly affected by forage type); furthermore, the amounts of magnesium secreted remained relatively small (0.85 and 0.30 mg Mg kg^{-1} LW, respectively). Potassium-loaded sheep might therefore secrete less salivary magnesium than unsupplemented sheep, because they absorb less magnesium from their diet, offsetting any tendency to reabsorb less of the magnesium which is secreted. The constancy of FE_{Mg} has yet to be disproved.

Urinary losses

Magnesium is less readily filtered at the glomerulus than most macrominerals, but sufficient is filtered and escapes tubular reabsorption, once the renal threshold of 0.92 mmol l^{-1} is exceeded, to allow urine to be the major route of excretion for excessive dietary supplies (Ebel and Gunther, 1980).

Maternal transfer of magnesium

The early magnesium nutrition of the newborn deserves comment. The magnesium concentration in main milk is low (0.3 mmol l^{-1}) and only one-tenth that of calcium, but it is maintained during depletion and represents a continuous drain on maternal reserves. Colostrum contains 0.1 g Mg kg^{-1}, two to three times more magnesium than main milk, giving a heightened maternal demand for magnesium at calving (Goff and Horst, 1997). Magnesium is absorbed very efficiently from milk by the young lamb and calf, with absorption from the large intestine making a large contribution to that high efficiency (Dillon and Scott, 1979), but values fall in calves from 87% at 2–3 weeks to 32.3% at 7–8 weeks of age (ARC, 1980). Prolonged rearing of calves on a simulated milk diet will induce hypomagnesaemic tetany (Blaxter *et al.*, 1954), as will the artificial maintenance of adult ruminants on a similar diet (Baker *et al.*, 1979). The magnesium concentration of milk cannot be raised by dietary magnesium supplements.

Genetic variation in magnesium metabolism

Magnesium metabolism is influenced by animal as well as dietary factors. Repeatable individual variation has long been identified as a potential stumbling-block to research workers, which must be minimized by appropriate experimental designs (e.g. the Latin square) if reliable results are to be obtained. Studies with identical twin cows showed that such variation was partly heritable (Field and Suttle, 1979). In a comparison of beef breeds on a common diet, Aberdeen Angus were found to have relatively low serum magnesium concentrations (Littledike *et al.*, 1995). Marked bovine breed differences in magnesium metabolism have been linked to differences in susceptibility to grass tetany (Greene *et al.*, 1986), but the small numbers representing each breed or cross ($n = 5$) and inconsistent rankings for crosses raise the need for confirmatory studies. Shockey *et al.* (1984) reckoned that sheep were 1.75 times more efficient at absorbing magnesium from their diet than cattle and they explained their findings by the higher ratio of rumen surface area:rumen contents in sheep. The regression techniques used to make the comparison were not ideal for such purposes, but evidence for a species difference in AA_{Mg} has now been confirmed (Table 6.2). The magnesium concentration required to saturate absorption from the rumen can be three times greater in cattle than in sheep (12.5 vs. 4 mmol l^{-1}; Martens, 1983): this may partially compensate for any limitation of surface area when magnesium supplements are used but leaves cattle more susceptible to the K × Mg antagonism than sheep. The validity of using sheep as a model for magnesium in cattle requires urgent study. Individuality is seen in sheep given their nutrients by abomasal infusion (Baker *et al.*, 1979) and is not merely a reflection of variation in rumen metabolism.

Function

Although most of the body magnesium (60–70%) is present in the skeleton (Todd, 1969; Fig. 6.3), magnesium ranks second to potassium in quantity in the intracellular fluids and organelles. Unlike potassium, it is largely (80%) protein-bound and associated predominantly with the microsomes (Ebel and Gunther, 1980). Magnesium also occurs in relatively low but life-sustaining concentrations in extracellular fluids, including the cerebrospinal fluid (CSF), and in the blood it is present in both plasma and erythrocytes. Magnesium is vitally involved in the metabolism of carbohydrates, lipids, nucleic acids and proteins, mostly as a catalyst of a wide array of enzymes (Wacker, 1969), facilitating the union of substrate and enzyme by first binding to one or the other (Ebel and Gunther, 1980). Magnesium is thus required for: oxidative phosphorylation, leading to ATP formation, which sustains processes such as the sodium (Na^+)/K^+ pump; pyruvate oxidation and conversion of α-oxoglutarate to succinylcoenzyme A; phosphate transfers, including those effected by alkaline phosphatase, hexokinase and deoxyribonuclease; and the β-oxidation of fatty acids (Shils, 1997). The transketolase reaction of the pentose monophosphate shunt is also magnesium-dependent (Pike and Brown, 1975). Magnesium also performs functions which do not rely on enzyme activation: the binding of magnesium to phosphate groups on ribonucleotide chains influences their folding; magnesium ions modulate neuromuscular activity and affect autonomic control in the heart; low ionic concentrations accelerate the transmission of nerve impulses; and muscle contraction depends partly on exchanges with calcium (Ebel and Gunther, 1980; Shils, 1997). Finally, magnesium affects cell membrane integrity by binding to phospholipids.

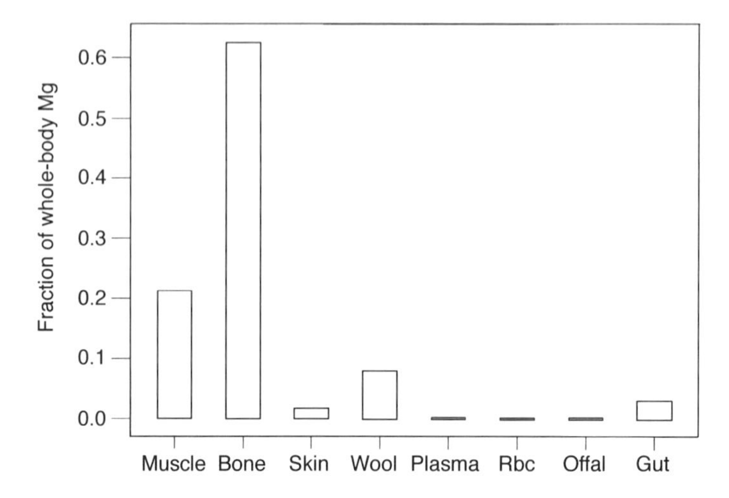

Fig. 6.3. Distribution of magnesium in the unshorn empty body of mature sheep (data from Grace, 1983).

In the light of these manifold and diverse functions, it is not surprising that magnesium deprivation in animals is manifested clinically in a wide range of disorders, which include retarded growth, hyperirritability and tetany, peripheral vasodilatation, anorexia, muscular incoordination and convulsions. Various other metabolic aberrations have been reported in the deficient animal, notably calcification of the kidneys, decline in hepatic oxidative phosphorylation, increased prostanoid synthesis, reduction in blood pressure and body temperature, decreased thiamine concentrations in the tissues (Sell, 1979; Shils, 1997) and changes in the fluidity of red-cell membranes (Tongyai *et al.*, 1989). In the case of acute magnesium dysfunction in the field, it is the physiological (i.e. neuromuscular) function which becomes rate-limiting, but, in chronic disorders, the rate-limiting function(s) remain unknown among the wide range of catalytic and structural functions performed by magnesium. Rumen microorganisms require magnesium to catalyse many of the enzymes essential to cellular function in mammals. Under experimental conditions, feeding sheep on a semipurified diet virtually devoid of magnesium rapidly impairs the cellulolytic activity of rumen microorganisms.

Biochemical Changes During Magnesium Deprivation

The hypomagnesaemia with tetany that arises suddenly in animals turned out to spring grass is not a simple dietary magnesium deficiency, because its onset is too rapid, it can occur on pastures or rations of normal or even high magnesium content and it is not accompanied by any detectable fall in skeletal magnesium: yet complete remission can be secured by a single injection of magnesium. The problem is, nevertheless, basically one of supply and demand and the failure of past models for predicting disease incidence is that they have failed to identify the 'customer' (i.e. the central nervous system (CNS), the site of dysfunction) and have oversimplified supply by ignoring appetite and mobilization from the skeleton as transient but crucial supply factors. When animals are first given a low-magnesium diet, they show a rapid fall in plasma magnesium, followed by a gradual recovery, which is usually attributed to slow mobilization of the skeletal reserve (Baker *et al.*, 1979). This is supported by evidence that skeletal mobilization of magnesium is extensive in the longer term (Gardner, 1973) and that inhibition of bone resorption in the magnesium-deprived sheep lowers plasma magnesium further (Matsui *et al.*, 1994). The oldest cows in a herd are likely to have the lowest plasma magnesium values (Fig. 6.4) because they have slower bone resorption rates than younger animals (see Chapter 4). Further complications arise from the dynamic nature of events in the pasture and the animal at times of high risk.

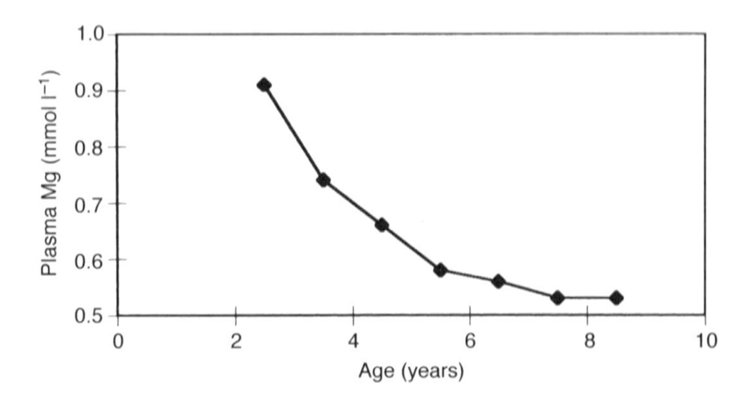

Fig. 6.4. Plasma magnesium concentrations can decrease markedly with age in a beef suckler herd but cumulative nutritional deficits may contribute to the apparent 'age' effect (from Suttle *et al.*, 1980).

Blood biochemistry

Hypomagnesaemia is most likely to develop following the onset of lactation in ewes, and those bearing twins are most vulnerable (Sansom *et al.*, 1982). Subnormal serum calcium values usually accompany the low serum magnesium and shifts in the balance between hypocalcaemia and hypomagnesaemia can influence clinical manifestations; no other inorganic blood constituent is consistently affected. Severe hypomagnesaemia inhibits the customary PTH response to mild hypocalcaemia and fasting has a similar effect (Littledike and Goff, 1987); it is therefore possible that the inappetence associated with severe hypomagnesaemia is responsible for inhibiting PTH secretion. There are parallel reductions in magnesium concentrations in other body fluids (e.g. vitreous humour, CSF), but the magnesium concentrations of the milk (Rook and Storry, 1962), soft tissues and bones of affected cows and ewes remain within normal limits. In contrast, calves chronically depleted on a low-magnesium diet reduce the magnesium reserves in their skeletal tissues and Ca:Mg ratios increase from a normal 50:1 to 150:1.

Pasture biochemistry

Despite a great volume of research in many countries, the aetiology of grass tetany remains incompletely understood. Influences of herbage constituents, such as citric and aconitic acids (Bohman *et al.*, 1969), the higher fatty acids (which cause magnesium soaps to form in the rumen) and aluminium (really a soil contaminant) have been proposed (Grunes *et al.*, 1970; Suttle, 1987). The organic acids are unlikely aetiological factors, because they are destroyed in the rumen and cannot complex magnesium or reduce its availability and the role of aluminium has been discounted (see Chapter 18). Numerous studies have demonstrated a positive relationship between the incidence of tetany and fertilizer treatment of the pastures with nitrogen (N) and potassium (Kemp *et al.*, 1961). Thus 't Hart and Kemp (1956), in a study of 3942 cows on Dutch farms, found the incidence of grass tetany to be 0.5% on pastures low in potassium; 5.2% with excess potassium; 4.3% on pastures treated with more than 50 kg N ha^{-1}; and 6.5% with excess potassium and nitrogen; such fertilizer treatments with N and K also decrease herbage magnesium concentrations (see Chapter 8). Signs of grass tetany have been induced by increasing the intake of potassium in sheep (Suttle and Field, 1969) and in cattle (Bohman *et al.*, 1969). Dutch workers brought the hypothetical influence of ammonia formation into their prediction of serum magnesium concentrations by incorporating dietary CP concentrations in a manner which treated the influences of both factors sensibly as continuous variables (Fig. 6.5), but the basis for their predictions was never shown. High nitrogen and potassium concentrations have also been implicated in the disorder which occurs when cereals are grazed as forages (Mayland *et al.*, 1976), but the two factors are confounded and potassium is by far the more influential. Pasture potassium and nitrogen concentrations rise rapidly in spring in response to rises in soil temperature, but the patterns of change in soil temperature and therefore potassium:magnesium ratios in herbage will vary greatly from week to week and year to year.

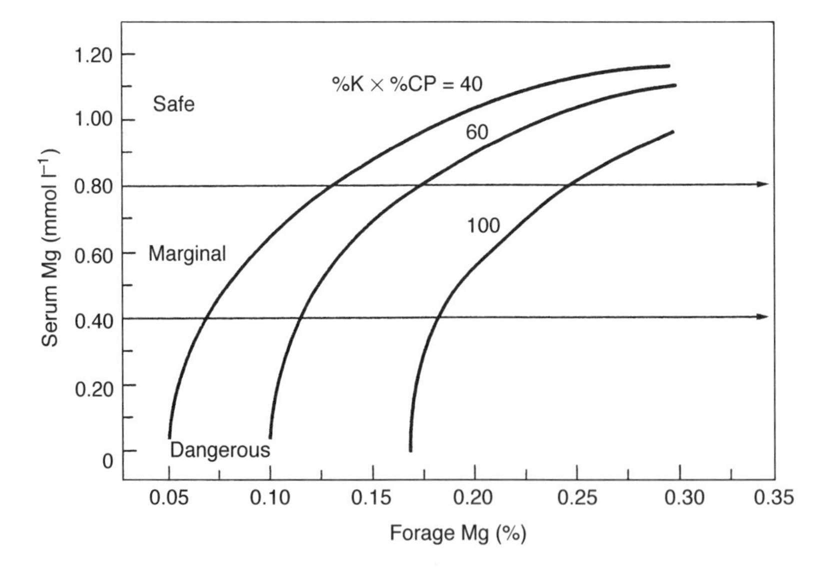

Fig. 6.5. Risk of hypomagnesaemia in the Netherlands has been predicted from the magnesium, crude protein (CP) and potassium concentrations in pasture (Netherlands Ministry of Agriculture and Fisheries, 1973): note the use of a marginal range for serum Mg within which onset of tetany is uncertain.

The roles of dietary change and anorexia

The transition from winter rations to spring grass is accompanied by marked changes in the rumen (raised pH, NH_4^+, insolubility of magnesium) and these have been linked to the development of hypomagnesaemia. The magnitude of the changes eventually diminishes (Johnson *et al.*, 1988; Johnson and Aubrey Jones, 1989) and may contribute to a gradual recovery in plasma magnesium. A change of diet can depress plasma magnesium even when the change is from grass to hay (Field, 1983). A fall in plasma magnesium can be associated with delayed rumen contractions in response to feeding and with impaired motility in the lower intestinal regions (Bueno *et al.*, 1980). Intravenous infusions of magnesium sufficient to restore normal plasma concentrations do not immediately restore normal intestinal motility. Such responses may explain the loss of appetite for roughage which can be seen in recently calved cows on low-magnesium diets (Braak *et al.*, 1986). Loss of appetite compounds the problem of low absorbability by lowering magnesium intake and plasma magnesium further (Herd, 1966). These changes are usually taking place at a time of increasing milk yield, associated with increases in the quality and quantity of digestible organic matter on offer from spring pastures. Close correlations between dietary composition and incidence of tetany are unlikely to be found in such circumstances and a further search for missing dietary factors in the aetiology of the disease is unlikely to pay dividends.

Clinical Signs of Magnesium Deprivation

Tetany in cattle

The initial signs of hypomagnesaemic tetany in cattle are those of nervous apprehension, with ears pricked, head held high and staring eyes. At this stage, the animal's movements are stiff and stilted, it staggers when walking and there is a twitching of the muscles, especially of the face and ears. Within a few hours or days, extreme excitement and violent convulsions may develop; the animal lies flat on its side, the forelegs 'pedal' periodically and the jaws work, making the teeth grate. If treatment is not given at this stage, death usually occurs during one of the convulsions or after the animal has passed into a coma. The presence of a scuffed arc of pasture by the feet of a dead animal is an important diagnostic feature. However, the subclinical stage may be followed by spontaneous recovery from hypomagnesaemia and does not invariably progress to the acute disorder.

'Hypomagnesaemia' in ewes

Preconvulsive clinical signs in sheep are less clearly defined than in cattle and can be confused with those of 'hypocalcaemia' or pregnancy toxaemia. Affected ewes breathe rapidly and their facial muscles tremble; some ewes cannot move, while others move with a stiff, awkward gait. Eventually, they collapse and show repeated tetanic spasms, with the legs rigidly extended.

The disease occurs within 4–6 weeks of lambing and can be precipitated by bad weather or flock handling.

Loss of appetite

When sheep are fed semipurified diets grossly deficient in magnesium, there is a marked loss of appetite within a few days due to rumen dysfunction (Ammerman *et al.*, 1971). With less deficient diets (0.5 g Mg kg^{-1} DM), there is eventually a marginal loss of appetite (Suttle and Field, 1969). Improvements in milk-fat yield in grazing cows given supplements of 10 g Mg day^{-1} were most noticeable in early lactation in underfed herds (Young *et al.*, 1981); this response may have been partly elicited by an improvement in appetite.

Occurrence of Magnesium-related Disorders

Hypomagnesaemia and 'tetany' in cattle

This disease occurs wherever dairying or beef production from pasture is highly developed. In Europe and North America, the disease is most common in cows turned out to graze lush pastures in the spring, but it can also occur, in combination with hypocalcaemia, when early growths of cereal crops are grazed ('wheat staggers' or 'wheat poisoning': Littledike and Goff, 1987). In New Zealand, where cows are pastured throughout the year, the disease occurs most frequently in late winter and early spring. Almost half the herds in some areas have 'subnormal' serum magnesium levels (< 0.64 mmol l^{-1}), and significant reductions in incidence of metabolic disorders were obtained by giving magnesium supplements (Young *et al.*, 1981). Out-wintered beef cattle can be affected, particularly during adverse weather conditions (Menzies *et al.*, 1994), when disturbed grazing and feeding patterns lower food and magnesium intakes. Outbreaks have also occurred in stall-fed animals and in animals subjected to intense excitement, as in transit or sale-yard conditions, giving rise to the term 'transit tetany'. The economic importance of grass tetany arises from its sudden occurrence and high death rate, the mortality among untreated clinical cases being 30% or more. Early English and Dutch studies indicated an overall incidence of 1–3% of dairy cows and the situation had probably not improved by 1982 (Whittaker and Kelly, 1982). In a New Zealand study of 477 dairy herds, the incidence of presumed grass tetany varied from 0.2 to 3.9% (Cairney, 1964), but elsewhere the incidence in individual herds can be as high as 20%. Grass tetany was reported in various parts of the USA (see extensive review by Grunes *et al.*, 1970). Subclinical disorders have been recognized in grazing cows, in which milk yield, particularly the fat component, is reduced (Wilson, 1980; Young *et al.*, 1981) and the herd is 'irritable' (Whittaker and Kelly, 1982). Moderate hypomagnesaemia is a feature of the subclinical condition and it was recently estimated to cost the small dairy industry in Northern Ireland £4m per annum (McCoy *et al.*, 1993). Incidence of both clinical and subclinical forms varies from breed to breed (Greene *et al.*, 1985), from year to year (Whittaker and Kelly, 1982) and in different seasons of the year (McCoy *et al.*, 1993, 1996; Fig. 6.6). In Northern Ireland, the incidence of hypomagnesaemia is now far higher in beef herds than in dairy herds (McCoy *et al.*, 1993), while, in the UK, incidence is higher in non-lactating than lactating members of dairy herds (Whittaker and Kelly, 1982); the explanation is that beef cows and dry dairy cows receive smaller amounts of concentrates and magnesium supplements than milking cows. Overall, hypomagnesaemictetary incidence has declined in recent years in the UK (Fig. 6.6).

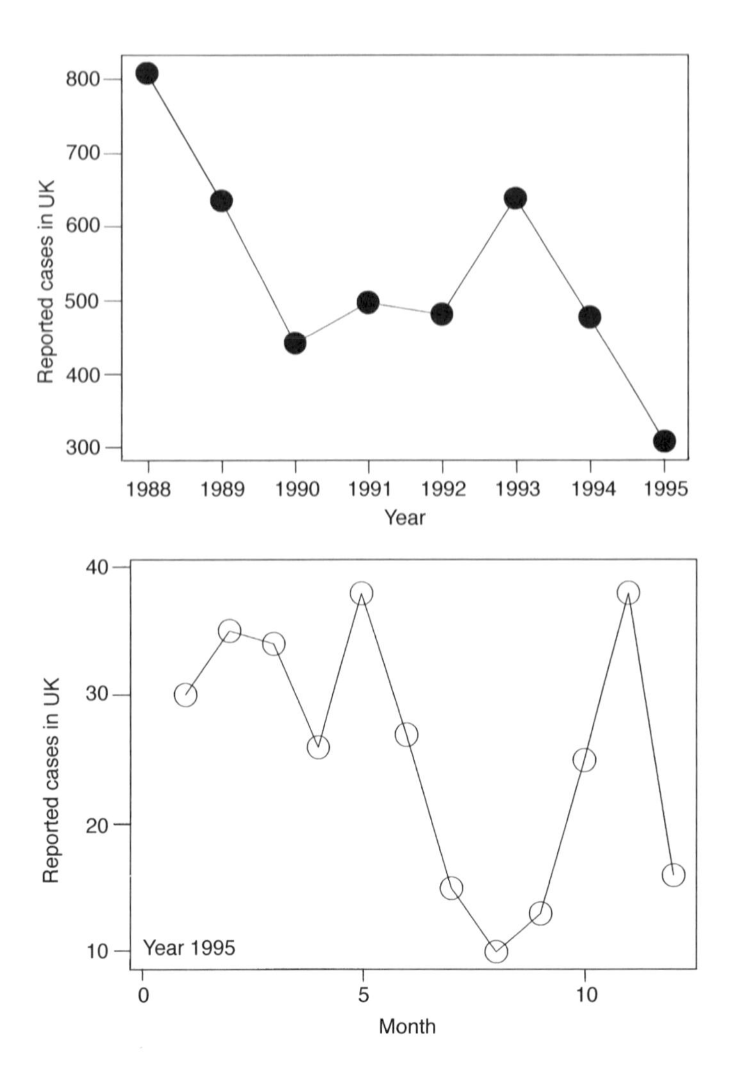

Fig. 6.6. Reported cases of hypomagnesaemic tetany in the UK have decreased in recent years and showed a seasonal fall in summer, 1995 (source: VIDA statistics, Weybridge, UK).

Hypomagnesaemia and tetany in sheep

In Australia, a high incidence of hypomagnesaemic tetany in breeding ewes has been correlated with periods of rapid winter growth of pastures (Blumer *et al.*, 1939). The disease is also recognized in the UK, where 20–40 outbreaks are reported annually, with twin-bearing ewes in the first month of lactation the most likely to be affected. Disruptive handling procedures and the grazing of cereal 'pastures' are risk factors.

Low milk-fat syndrome

Dairy cows on grain-based diets often produce milk with unacceptably low fat concentrations, and addition of calcined magnesite (MgO) to such rations can remedy the problem (Xin *et al.*, 1989). The property is shared by other minerals, some (like MgO) with buffering capacity (e.g. sodium bicarbonate ($NaHCO_3$)) and others with none (e.g. NaCl). The syndrome is therefore not a specific consequence of magnesium deprivation, and responses to MgO are unlikely to be attributable simply to added buffering capacity in the rumen. Increased rumen dilution rate and decreases in rumen propionate production may be responsible (see Chapter 8).

Diagnosis

Diagnosis of hypomagnesaemic tetany has to be instantaneous and is confirmed by response to treatment, followed by the demonstration of hypomagnesaemia in samples taken prior to treatment. Tetany usually occurs when serum magnesium falls below 0.5 mmol (12 mg) l^{-1} in cattle and below 0.2 mmol (5 mg) l^{-1} in sheep. For example, Sjollema and Seekles (1930) recorded values ranging from 1.0 to 11.6 mg Mg l^{-1} and averaging 5.0 mg l^{-1} (0.2 mmol l^{-1}) for 55 cows suffering from lactation tetany. However, it is important to note that higher values can occur in affected individuals and lower values in unaffected individuals, the difference probably being the rate at which magnesium deficiency is developing (fast and slow, respectively). Subclinical disease in cattle is associated with serum magnesium concentrations < 0.6 mmol l^{-1}, with values of 0.6–0.8 mmol l^{-1} regarded as marginal (Sutherland *et al.*, 1986; McCoy *et al.*, 1996). Similar ranges are used for sheep (Table 6.3). In the dead animal, vitreous humour from the eye can be analysed, values < 0.75 mmol l^{-1} being indicative of hypomagnesaemic tetany; the correlation with serum magnesium over the normal range is poor (McCoy and Kennedy, 1994), but serum magnesium is not a perfect marker for the disease. Pauli and Allsop (1974) preferred the analysis of magnesium in CSF, because concentrations decline more slowly than serum magnesium and may better reflect magnesium status at crucial sites; values < 0.65 mmol l^{-1} are regarded as critical, provided that the sample is blood-free. Presence of Ca:Mg ratios in rib bone $> 50:1$ indicate a probability of hypomagnesaemic tetany in sucking or milk-fed calves and lambs but are unreliable in the lactating animal (see Fig. 5.6).

Table 6.3. Marginal bands[a] for assessing the magnesium status of ruminant livestock.

	Serum	CSF	Vitreous humour	Urine Mg:creatine molar ratio	Rib (mg Mg g^{-1} ash)
	(mmol l^{-1})[b]				
Cattle	0.5–0.75	0.6–0.8	0.5–0.75	0.4–0.8	< 12 (6–12 months) < 7 (> 36 months)
Sheep	0.6–0.75	–	0.4–0.8	0.4–0.8	–

[a] When mean values for the sampled population fall below the ranges given, the flock or herd will probably benefit from supplementation; when values fall within these bands, there is a possibility of current or future responses in some individuals if status does not improve.
[b] Multiply by 24.3 for mg l^{-1}.

Prognosis

As far as prognosis is concerned, there have been no reports of the success or otherwise of using the widely quoted guidelines in Fig. 6.5. The average grass pasture in the UK containing 156 g CP and 1.6 g Mg kg^{-1} DM (MAFF, 1992) would present a marginal risk of grass tetany, while the richest in CP (300 g kg^{-1} DM) would be dangerous if they contained more than 30 g K kg^{-1} DM, according to those guidelines. Kemp and 't Hart (1957) showed that, when the ratio K $(Ca + Mg)^{-1}$ in forage was less than 2.2 (milli-equivalent basis), there were few tetany cases (0.77% of 4658 animals), whereas, with a ratio wider than 2.2, the incidence was much higher (6.66% of 1908 animals). The value of this ratio as an indicator of tetany-prone pastures has been confirmed in Scotland (Butler, 1963), New Zealand (Butler and Metson, 1967) and the USA (Ritter *et al.*, 1984), although the beneficial influence of calcium has no known nutritional explanation. The relationship in Scotland was of a linear rather than a threshold nature, disease incidence increasing by 4.5 ± 1.63 percentage units for each unitary increase in the ratio (Butler, 1963). Urinary Mg:creatinine ratios (molar) below 0.4 are considered to be 'critical' by some European workers (Braak *et al.*, 1986), while workers in New Zealand suggest that production may well improve if herds with values < 1.0 are given supplementary magnesium (Sutherland *et al.*, 1986). The prognostic value of serum magnesium has been questioned, because it is poorly correlated with urinary magnesium (Sutherland *et al.*, 1986), but this view arises from the attempt to fit linear relationships to what is bound to be a curvilinear or plateau relationship between the two variables (Fig. 6.7): corrected urinary magnesium values between 0.4 and 0.8 should be regarded as 'marginal' (Table 6.3).

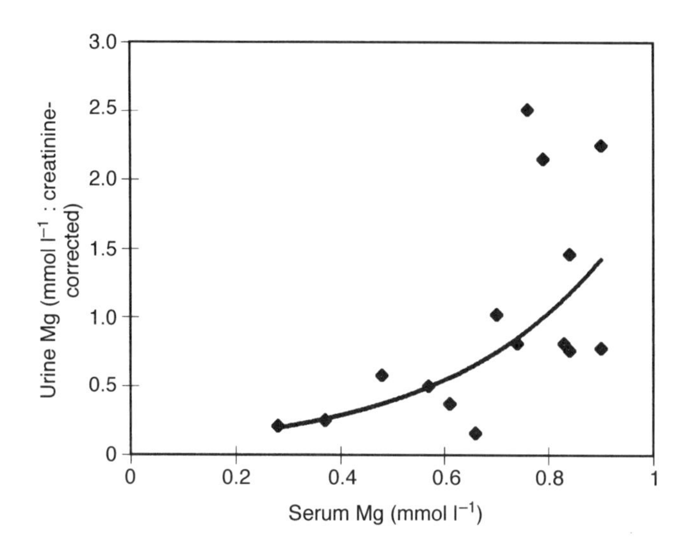

Fig. 6.7. Curvilinear relationships between two diagnostic parameters – in this case urine and plasma magnesium – will mean that they have different contributions to make: urine Mg reflects the dietary supply of Mg in excess of need, plasma Mg the degree of inadequacy (data from Sutherland *et al.*, 1986).

Treatment

Subcutaneous injection of a single dose of 200–300 ml of a 20% solution of magnesium sulphate or a similar dose of magnesium lactate restores the serum magnesium of an affected cow to near normal within about 10 min and is almost always followed by improvement or disappearance of signs of tetany. Equally rapid responses were found when artificially depleted calves were given 51 mg Mg kg^{-1} LW per rectum as magnesium chloride ($MgCl_2$) (Bacon *et al.*, 1990). Serum magnesium is likely to fall again unless the cow is immediately removed from the tetany-producing herbage and fed on hay and concentrates treated with magnesium or given oral doses of magnesium. Similarly, the affected ewe will respond quickly to intravenous magnesium, but this is invariably given with calcium, usually as 50 ml of a 250 g l^{-1} solution of calcium borogluconate, containing 25 g of magnesium hypophosphite; calcium should also be given to the recovering cow, using a dose of 500 ml to correct any hypocalcaemia.

Prevention

Several means of raising magnesium intakes to prevent losses from 'lactation tetany' have been devised, but they vary in efficacy. A recent survey (McCoy *et al.*, 1996) illustrated the way in which choice of treatment is influenced by farm circumstances in Northern Ireland. Supplementation via concentrates was preferred by most dairy farmers, but reliance on magnesium fertilizers increased as farm size increased. Beef farmers relied heavily on the magnesium block and made little or no use of fertilizers.

Pasture fertilizers

The value of applying various magnesium compounds, such as calcined magnesite (MgO), kieserite ($MgSO_4.H_2O$) or dolomitic limestone ($CaCO_3$.$MgCO_3$), to pastures to raise their magnesium concentrations has been extensively studied (Allcroft, 1961; McIntosh *et al.*, 1973). The magnesium limestone was less effective than the other sources, which increased pasture values from 2 to about 3 g Mg kg^{-1} DM. Treatment of lucerne (alfalfa) with kieserite at four rates, up to 4480 kg ha^{-1}, increased concentrations from 2 to 3.2 g Mg kg^{-1} DM at the highest application, with no effect on availability of the magnesium present (Reid *et al.*, 1979). Application rates of 1125 kg ha^{-1} for magnesite or kieserite and 5600 kg ha^{-1} for magnesium limestone are recommended, but this method of control has limitations on many soil types. Foliar dusting of pastures with fine calcined magnesite (MgO) before or during tetany-prone periods has proved effective, but effects are short-lived, regardless of the amount applied (Fig. 6.8; Kemp, 1971). Rogers (1979) recommended that at least 17 kg ha^{-1} be used and repeated after heavy rainfall

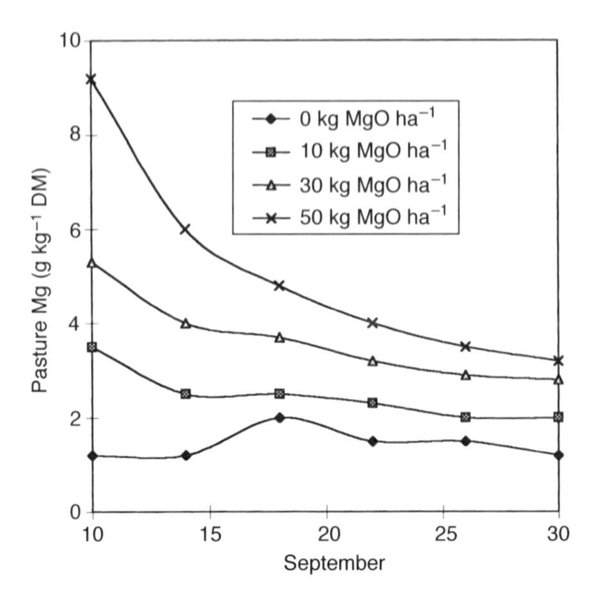

Fig. 6.8. Control of hypomagnesaemic tetany by dusting pastures with magnesite (MgO) is short-lived because the method relies on surface adherence (Kemp, 1971): repeated applications are necessary and a single treatment with a suspension of magnesium hydroxide may be more effective.

and at not more than 10-day intervals; treatments can be integrated with a 'strip-grazing' system. Spray application of magnesium hydroxide in fine suspension can produce short-term increases in herbage and serum magnesium from low application rates (5 kg Mg ha^{-1}: Parkins and Fishwick, 1999). Foliar dusting and spraying use the pasture as a short-term carrier for the dietary magnesium supplement.

Dietary supplements

For calves and dairy cows which are being fed on concentrates, the provision of sources such as MgO in the concentrate mixture can prevent the disorder. Alternatively, incorporating magnesium into mineral mixes, drenches, salt- and molasses-based, free-choice licks, adding magnesium sulphate ($MgSO_4$) to the drinking-water or even sprinkling the mineral on feeds, such as cereals, chopped roots or silage, have all been used as methods of supplementation. Magnesium must be given to all stock continuously during the tetany-susceptible period, since the disease can occur within 48 h of withholding the supplement while cows are at grass. 'Secure' prophylactic doses have been given as 50–60 g MgO or its equivalent day^{-1} (25–30 g Mg) for adult dairy cattle, 7–15 g day^{-1} for calves and 7 g day^{-1} for lactating ewes. In a study with calving beef cows, 56.8 g MgO day^{-1} maintained serum magnesium to a mean of 1.0 mmol l^{-1}, compared with 0.5 mmol l^{-1} in untreated cows grazing similar pastures (Boling *et al.*, 1979). However, these recommendations seem far too generous in the light of even the worst estimates of requirement (Tables 6.4 and 6.5) and should be halved once the immediate risk has passed. Conservative use of MgO is indicated, because the amounts previously recommended will shift acid:base balance and raise urine pH (Xin *et al.*, 1989): such changes increase the risk of milk-fever (see Chapters 4 and 8). Excessive feeding of magnesium prior to parturition did nothing to alleviate the hypomagnesaemia in beef cows associated with onset of grazing in spring (Ritter *et al.*, 1984).

In New Zealand, the method of choice is to give 10 g Mg as $MgCl_2$ daily from the milking platform to each individual, in bloat control drenches. This method assures uniformity of magnesium intake, which free-choice methods cannot give. The use of magnesium alloy bullets, lodged in the rumen, has not proved a reliable means of preventing hypomagnesaemia in dairy cows (Kemp and Todd, 1970), because they cannot always release sufficient magnesium daily throughout the risk period, but they may be slightly more effective in the beef cow, which requires less magnesium (Stuedemann *et al.*, 1984). A recent survey showed that the provision of magnesium-containing blocks to beef suckler herds in Northern Ireland did not reduce the incidence of hypomagnesaemia (McCoy *et al.*, 1996). Addition of a soluble magnesium salt, such as the chloride, sulphate or acetate, to the water-supply has been extensively studied, with mostly beneficial results (see Rogers and Poole, 1976). However, water intake by drinking varies widely between individuals and with weather and pasture conditions, and it is not easy to ensure by this means the necessary intake. The value of soluble carbohydrates, as a source

Table 6.4. Factorially derived estimates of the average magnesium requirements of grazing and housed sheep (g kg^{-1} DM).

	Live weight (kg)	Growth/ production (kg day^{-1})	Dietary requirement (g kg^{-1} DM)		DMI[b] (kg day^{-1})
			At grass[a]	Housed[a]	
Growth	20	0.1	1.0	0.50	0.50
		0.2	0.9	0.45	0.76
	30	0.1	1.0	0.50	0.67
		0.2	0.9	0.45	1.00
		0.3	0.8	0.40	U
	40	0.1	1.0	0.50	0.83
		0.2	0.9	0.45	1.23
		0.3	0.7	0.35	1.80
		Fetuses (−12 weeks to term)			
Pregnancy	40	1	0.9	0.45	0.64–0.96
		2	0.9	0.45	0.74–1.25
	75	1	1.1	0.55	1.03–1.51
		2	0.9	0.45	1.17–1.93
		Milk (kg day^{-1})			
Lactation	40	1	1.2	0.60	1.18
		2	1.2	0.60	1.90
	75	1	1.4	0.70	1.48
		2	1.3	0.65	2.18
		3	1.3	0.65	2.90

[a] Absorbability values were 0.20 for sheep grazing pasture high in potassium (> 30 g K kg^{-1} DM), 0.40 for housed sheep.
[b] Dry-matter intakes for diets of intermediate nutritive value, $q = 0.6$ (see ARC, 1980).
U, unattainable performance on such diets.

of energy which concurrently lowers rumen pH and improves magnesium absorption, has received some attention (Metson *et al.*, 1966; Giduck *et al.*, 1988; Schonewille *et al.*, 1997b) and may contribute to the efficacy of liquid mixtures of molasses and magnesium.

Genetic selection

In view of the evidence that magnesium metabolism and susceptibility to 'grass tetany' are subject to genetic variation, it would seem worthwhile to explore the possibility of selecting for resistance to the disease via indices which indicate high efficiencies of magnesium utilization. The strategy is attractive in that it targets the vulnerable minority of any population, whereas preventive supplementation methods involve treatment of the whole herd or flock, regardless of individual levels of risk. Ratios of Mg:creatinine in the urine of artificial insemination (AI) sires on a standard ration might allow the

Table 6.5. Factorially derived estimates of the average magnesium requirements of grazing and housed cattle (mg kg^{-1} DM).

	Live weight (kg)	Growth/ production (kg day^{-1})	Dietary requirement (g kg^{-1} DM)		DMI[b] (kg day^{-1})
			At grass[a]	Housed[a]	
Growth	100	0.5	1.4	0.91	2.1
		1.0	1.3	0.85	3.2
	200	0.5	1.5	0.98	3.3
		1.0	1.3	0.85	4.7
	400	0.5	1.6	1.04	5.2
		1.0	1.3	0.85	7.3
		Weeks to term			
Pregnancy	600	12	2.0	1.30	5.8
		4	1.7	1.02	7.3
		Milk (kg day^{-1})			
Lactation	600	10	2.2	1.40	9.4
		20	2.0	1.33	14.0
		30	1.9	1.26	18.8

[a] Based on absorbability coefficients of 0.15 and 0.23 for grass high in potassium (> 30 g K kg^{-1} DM) and hay + concentrates, respectively.
[b] Assuming a diet of moderate digestibility, with a *q* value of 0.6.

selection of resistant lines (Suttle, 1996). Progress can also be made by selecting for plants rich in magnesium. Thus hypomagnesaemia was controlled in sheep by grazing them on a grass cultivar high in magnesium (Moseley and Baker, 1991).

Mineral sources

The effectiveness of different chemical forms of the element have mostly been compared in ruminants (Henry and Benz, 1995). Magnesium phosphate, a calcium magnesium phosphate and magnesium ammonium phosphate are all satisfactory sources for growing sheep given dry diets (Fishwick and Hemingway, 1973) and magnesium phosphate is satisfactory for lactating cows on spring pasture (Ritchie and Fishwick, 1977). Magnesium retentions were similarly increased in growing wether sheep by magnesium oxide and two forms of magnesium phosphate (Fishwick, 1978), but the oxide and hydroxide were given low average values of 0.75 (n = 18) and 0.60 (n = 3) relative to $MgSO_4$ for sheep by Henry and Benz (1995). Commercial sources of calcined magnesites (i.e. MgO) vary widely in particle size and processing – particularly the calcination temperature – and hence in nutritive value (Xin *et al.*, 1989; Adam *et al.*, 1996). Coarse particle sizes (> 500 μm) and low calcination temperatures (< 800°C) are associated with low estimates of apparent absorption. Magnesium chloride may be better than acetate or sulphate salts, in view of the anion effects described earlier. The A_{Mg} value for magnesium chloride was nearly twice as high when added to cereals than when added to hay (0.3 vs. 0.16: Suttle, 1987) and MgO was a consistently better source when added to a maize-based rather than a dried grass diet (Adam *et al.*, 1996). A source such as magnesium-mica, which releases its magnesium postruminally, may be less susceptible to the rumen-based antagonism from potassium (Hurley *et al.*, 1990). Using weight gain and magnesium in the tibia ash as criteria, it was shown that the oxide, carbonate and sulphate are all sources of highly available magnesium for broiler chickens (McGillivray and Smidt, 1975). The A_{Mg} value for reagent grade $MgSO_4.7H_2O$ was 0.57 at 0.4 g Mg kg^{-1} DM in a semipurified diet for chickens (Guenter and Sell, 1974), confirming the contrast with ruminants.

Magnesium Requirements

The dietary magnesium requirements of livestock vary with the species and breed of the animal and its age and rate of growth or production, but mostly with the absorbability of the mineral in the diet of ruminants. The concurrent dietary levels of calcium and phosphorus influence requirements in the non-ruminant. Vitamin D does not directly affect magnesium absorption, but it may do so indirectly through effects on the rates of accretion or mobilization of bone salts (Richardson and Welt, 1965). The magnesium status of the animal has been said to influence requirements (McAleese *et al.*, 1961), because of the magnesium reserves in the skeleton. However, bone reserves provide only limited and temporary protection, particularly when the fall in absorbed magnesium is abrupt. Requirement figures used in ration formulation should therefore be sufficient to maintain the skeletal reserves in all species, although failure to provide them will not necessarily cause disease.

Sheep and cattle

The variation in absorbability of magnesium is so wide that it must be taken into account if meaningful requirements are to be calculated. The mean absorbability values in Table 6.2 have therefore been used to generate magnesium requirements for each diet type for sheep and cattle in Tables 6.4 and 6.5, respectively. The most noteworthy features of the estimates are:

- higher requirements for cattle than sheep;
- higher requirements for lactating than non-lactating animals;
- requirements do not increase with level of performance;
- requirements for housed stock are only 50% of those for grazing stock.

The use of a single absorption coefficient for magnesium in pasture for each species is an oversimplification, given the influence of continuous variables, such as pasture potassium and CP concentrations (Fig. 6.5). The intersections of each graph with the horizontal line, indicating the lower limit of marginality, suggest a two- to threefold variation in magnesium requirement (from 0.7 to 1.8 g Mg kg^{-1} DM) as potassium and CP increase from low to maximal levels.

The requirements for the average bovine in Table 6.5 cover the worst-case scenario predicted by the Dutch guidelines for potassium-rich pastures. It could be argued that a further safety factor is needed to protect the most vulnerable individual in a herd or flock, a procedure followed by ARC (1980). If potassium is added to hay + concentrate diets, the low requirements no longer apply and hypomagnesaemia can be induced with diets containing 1.7 g Mg and 40 g K kg^{-1} DM (Braak *et al.*, 1986). The requirement for cellulose digestion and food intake for a semipurified diet (30% cellulose) was 8–10 mg Mg kg^{-1} LW (Ammerman *et al.*, 1971), equivalent to 0.5–0.8 g Mg kg^{-1} DM and high enough not to be met by some low-quality roughages.

Pigs

The minimum magnesium requirement of baby pigs receiving a purified diet has been given as 325 mg kg^{-1} DM (Miller *et al.*, 1965). For optimum growth and prevention of signs of deficiency in weaner pigs, higher requirements were reported (400–500 mg kg^{-1} DM: Mayo *et al.*, 1959), but they are influenced by the protein content of the diet (Hendricks *et al.*, 1970). According to the National Research Council (NRC, 1988), 400 mg Mg kg^{-1} DM should be given to pigs of all ages. Little is known of the precise requirements of sows during pregnancy and lactation, but the experiments of Harmon *et al.* (1976) indicated that 400 mg Mg kg^{-1} DM was adequate during pregnancy and that 150 mg kg^{-1} DM was seriously inadequate for lactation. These empirical findings can be compared with factorial estimates derived by ARC (1981). Assuming an endogenous loss of 0.4 mg Mg kg^{-1} LW, tissue accretion rates declining from 460 to 380 mg Mg kg^{-1} live-weight gain (LWG) and an absorptive efficiency of 80%, they calculated that the dietary requirement fell from 410 to 160 mg Mg kg^{-1} DM as pigs grew from 5 to 90 kg LW. The figures for all but the youngest pigs are probably underestimated by the single high figure used for absorptive efficiency, which was based on studies with baby pigs: with absorption at 60% efficiency, the requirement for the fattening pig would be nearer 210 mg kg^{-1} DM.

Poultry

The magnesium requirements of poultry have been more extensively investigated than those of pigs. Growth rate probably affects requirement, because 0.2 g Mg kg^{-1} DM sufficed for early growth in White Leghorn chicks, whereas faster-growing broiler chicks needed 0.5 g kg^{-1} DM (McGillivray and Smidt, 1975). Calcium and phosphorus have small effects on the requirement of chicks (Nugara and Edwards, 1963); increasing dietary phosphorus (P) from 3 to 9 g kg^{-1} DM and dietary calcium from 6 to 12 g kg^{-1} DM raised the requirement from 0.46 to 0.59 g Mg kg^{-1} DM. Almquist (1942) concluded that chicks needed 0.4 g Mg kg^{-1} DM for growth and survival. Wu and Britton (1974) concluded that 0.25 g Mg kg^{-1} DM was sufficient to support maximum growth and prevent excessive mortality, with 6 g P kg^{-1} DM. The requirements of laying hens depend on the criteria of adequacy employed and the type of diet (Sell, 1979). To maintain normal serum magnesium values, only

0.16 g Mg kg^{-1} DM is required, while 0.26 g kg^{-1} DM is necessary to support a high rate of egg production. To obtain, in addition, a normal magnesium concentration in the eggs produced, satisfactory egg weight and a high rate of hatchability of viable chicks, the diet should contain 0.36 g Mg kg^{-1} DM (Hajj and Sell, 1969). It is unlikely that, with commercial diets, either growing chicks or laying hens would ever need supplementary magnesium for lengthy periods. A significant increase in weight gain to 3 weeks of age, but not to market weight, was observed when magnesium was added to high-energy diets (McGillivray and Smidt, 1977). Supplemental $MgSO_4$, at 0.05%, increased shell weight and thickness of eggs produced by hens on a practical ration (Bastien *et al.*, 1978), but no such benefit was observed by others (Sell, 1979). The latest recommendations by NRC (1994) of uniformly high concentrations of 0.5 g Mg kg^{-1} DM for broilers, turkey poults and laying hens (with a food intake of 100 g day^{-1}) contain a generous safety margin.

Toxicity

Lack of homeostatic mechanisms for the control of metabolism for a given mineral is usually associated with tolerance of extreme intakes, but this is barely true of magnesium as far as the range provided by natural diets is concerned. The NRC (1980) placed the tolerable limit for ruminants at 5 g Mg kg^{-1} DM and subsequent studies showed that 14 g kg^{-1} DM was mildly toxic to sheep (Chester-Jones *et al.*, 1989) and steers (Chester-Jones *et al.*, 1990). The first abnormalities were loosening of the surface (stratum corneum) of rumen papillae. Higher levels (25 and 47 g Mg kg^{-1} DM) caused severe diarrhoea and drowsiness and severe degeneration of the stratified squamous epithelium of the rumen in the steers. Early loss of appetite can occur with > 13 g Mg kg^{-1} DM (Gentry *et al.*, 1978), but this may be partly attributable to the unpalatability of the MgO used to raise magnesium intakes. Chicks also develop diarrhoea with > 30 g Mg (as MgO) kg^{-1} diet DM (Lee and Britton, 1987). Presumably, high intakes of potassium will increase the laxative effects of magnesium in ruminants by increasing the concentrations of magnesium in digesta reaching the large intestine. Linear increases in serum magnesium with increasing magnesium intakes took values above 3 mmol Mg l^{-1} after 71 days' supplementation at 47 g kg^{-1} DM in steers (Chester-Jones *et al.*, 1990); at such levels, there are likely to be loss of muscle tone and deep tendon reflexes and muscle paralysis (Ebel and Gunther, 1980). Increases in soft-tissue magnesium were, however, small and mostly non-significant. The significance of such studies relates to the practice of providing grazing animals with free-access to magnesium-rich mineral mixtures. In one instance, a mixture containing 130 g Mg kg^{-1} DM exacerbated diarrhoea in lambs (Suttle and Brebner, 1995). Given the variable individual and day-to-day intakes of such mixtures, it is quite likely that some members of a flock or herd will ingest enough magnesium to disturb gut function. Excessive magnesium intakes increase risks of urolithiasis in goats (James and

Chandron, 1975), sheep (Chester-Jones *et al.*, 1989) and calves (Chester-Jones *et al.*, 1990), but incidence may partly depend on the phosphorus concentration in the urine (see Chapter 5).

References

Adam, C.L., Hemingway, R.G. and Ritchie, N.S. (1996) Influence of manufacturing conditions on the bioavailability of magnesium in calcined magnesites measured *in vivo* and *in vitro*. *Journal of Agricultural Science, Cambridge* 127, 377–386.

Adediji, O. and Suttle, N.F. (1999) Influence of diet type, potassium and animal species on the absorption of magnesium by ruminants. *Proceedings of the Nutrition Society* 58, 31A.

Allcroft, R. (1954) Hypomagnesaemia in cattle. *Veterinary Record* 66, 517–522.

Allcroft, R. (1961) Prevention of hypomagnesaemia. *Proceedings of the British Veterinary Association.*

Allsop, T.F. and Pauli, J. (1985) Magnesium concentrations in the ventricular and lumbar cerebrospinal fluid of hypomagnesaemic cows. *Research in Veterinary Science* 38, 61–64.

Allsop, T.F. and Rook, J.A.F. (1979) The effect of diet and blood-plasma magnesium concentration on the endogenous faecal loss of magnesium in sheep. *Journal of Agricultural Science, Cambridge* 92, 403–408.

Almquist, H.J. (1942) Magnesium requirement of the chick. *Proceedings of the Society for Experimental Biology and Medicine* 49, 544–555.

Ammerman, C.B., Chicco, C.F., Moore, J.E., van Walleghem, P.A. and Arrington, L.R. (1971) Effect of dietary magnesium on voluntary food intake and rumen fermentations. *Journal of Dairy Science* 54, 1288–1293.

ARC (1980) *The Nutrient Requirements of Ruminant Livestock.* Commonwealth Agricultural Bureaux, Farnham Royal, Slough, UK, pp. 201–211.

ARC (1981) *The Nutrient Requirements of Pigs.* Commonwealth Agricultural Bureaux, Farnham Royal, Slough, UK, pp. 248–251.

Bacon, J.A., Bell, M.C., Miller, J.K., Ramsey, N. and Mueller, F.J. (1990) Effect of magnesium administration route on plasma minerals in Holstein calves receiving either adequate or insufficient magnesium in their diets. *Journal of Dairy Science* 73, 470–473.

Baker, R.M., Boston, R.C., Boyes, T.E. and Leever, D.D. (1979) Variations in the response of sheep to experimental magnesium deficiency. *Research in Veterinary Science* 26, 129–133.

Bastien, R.W., Bradley, J.W., Pennington, B.L. and Ferguson, T.M. (1978) Bone strength and egg characteristics as affected by dietary minerals. *Poultry Science* 57, 1117.

Blaxter, K.L. and McGill, R.F. (1956) Magnesium metabolism in cattle. *Veterinary Reviews and Annotations* 2, 35–55.

Blaxter, K.L., Rook, J.A.F. and MacDonald, A.M. (1954) Experimental magnesium deficiency in calves. 1. Clinical and pathological observations. *Journal of Comparative Pathology* 64, 157–175.

Blumer, C.C., Madden, F.J. and Walker, D.J. (1939) Hypocalcaemia, grass tetany or grass staggers in sheep. *Australian Veterinary Journal* 15, 2–27.

Bohman, V.R., Lesperance, A.L., Harding, G.D. and Grunes, D.L. (1969) Induction of experimental tetany in cattle. *Journal of Animal Science* 29, 99–102.

Boling, J.A., Okolo, T.O., Gay, N. and Bradley, N.W. (1979) Effect of magnesium and energy supplementation on blood constituents of fall-calving beef cows. *Journal of Animal Science* 48, 1209–1215.

Braak, A.E. van de, Klooster, A.T., van't and Malestein, A. (1986) Influence of a deficient supply of magnesium during the dry period on the rate of calcium mobilisation by dairy cows at parturition. *Research in Veterinary Science* 42, 101–108.

Bueno, L., Fioramonti, J., Geux, E. and Raissiguer, Y. (1980) Gastrointestinal hypomotility in magnesium deficient sheep. *Canadian Journal of Animal Science* 60, 293–301.

Butler, E.J. (1963) The mineral element content of spring pasture in relation to the occurrence of grass tetany and hypomagnesaemia in dairy cows. *Journal of Agricultural Science, Cambridge* 60, 329–340.

Butler, G.W. and Metson, A.J. (1967) Hypomagnesaemic tetany in relation to New Zealand dairy farming. In: *Dairyfarming Annual*, pp. 142–153.

Cairney, I.M. (1964) Grass staggers in beef cattle: results of survey of disease in Hawke's Bay. *New Zealand Journal of Agriculture* 109, 45–49.

Care, A.D., Brown, R.C., Farrar, A.R. and Pickard, D.W. (1984) Magnesium absorption from the digestive tract of sheep. *Quarterly Journal of Experimental Physiology* 69, 577–587.

Charlton, J.A. and Armstrong, D.G. (1989) The effect of an intravenous infusion of aldosterone upon magnesium metabolism in the sheep. *Quarterly Journal of Experimental Physiology* 74, 329–337.

Chester-Jones, H., Fontenot, J.P., Veit, H.P. and Webb, K.E. (1989) Physiological effects of feeding high levels of magnesium to sheep. *Journal of Animal Science* 67, 1070–1081.

Chester-Jones, H., Fontenot, J.P., Veit, H.P. and Webb, K.E. (1990) Physiological effects of feeding high levels of magnesium to steers. *Journal of Animal Science* 68, 4400–4413.

Chicco, C.F., Ammerman, C.B., Feaster, J.P. and Dunavant, B.G. (1973) Nutritional interrelationships of dietary calcium, phosphorus and magnesium in sheep. *Journal of Animal Science* 36, 986–993.

Dalley, D.E. and Sykes, A.R. (1989) Magnesium absorption from the large intestine of sheep. *Proceedings of the New Zealand Society of Animal Production* 49, 229–232.

Dalley, D.E., Isherwood, P., Sykes, A.R. and Robson, A.B. (1997a) Effect of intraruminal infusion of potassium on the site of magnesium absorption within the digestive tract of sheep. *Journal of Agricultural Science, Cambridge* 129, 99–106.

Dalley, D.E., Isherwood, P., Sykes, A.R. and Robson, A.B. (1997b) Effect of *in vitro* manipulation of pH on magnesium solubility in ruminal and caecal digesta in sheep. *Journal of Agricultural Science, Cambridge* 129, 107–112.

Dillon, J. and Scott, D. (1979) Digesta flow and mineral absorption in lambs before and after weaning. *Journal of Agricultural Science, Cambridge* 92, 289–297.

Dua, K. and Care, A.D. (1995) Impaired absorption of magnesium in the aetiology of grass tetany. *British Veterinary Journal* 151, 413–426.

Ebel, H. and Günther, T. (1980) Magnesium metabolism: a review. *Journal of Clinical Chemistry and Clinical Biochemistry* 18, 257–270.

Engels, E.A.N. (1981) Mineral status and profiles (blood, bone and milk) of the grazing ruminant with special reference to calcium, phosphorus and magnesium. *South African Journal of Animal Science* 11, 171–182.

Fenner, H. (1979) Magnesium nutrition of the ruminant. In: *Proceedings of the Second*

International Minerals Conference, St Petersburg Beach, Florida, Illinois. International Minerals and Chemical Corporation.

Field, A.C. (1983) Dietary factors influencing magnesium utilisation. In: Fontenot, J.P., Bunce, G.E., Webb, K.E., Jr and Allen, V.G. *Role of Magnesium in Animal Nutrition.* John Lee Pratt Annual Nutrition Program, Blacksburg, Virginia, USA, pp. 159–171.

Field, A.C. and Munro, C.S.M. (1977) The effect of site and quantity on the extent of absorption of Mg infused into the gastro-intestinal tract of sheep. *Journal of Agricultural Science, Cambridge* 89, 365–371.

Field, A.C. and Suttle, N.F. (1979) Effect of high and low potassium intakes on the mineral metabolism of monozygotic twin cows. *Journal of Comparative Pathology* 89, 431–439.

Field, A.C., McCallum, J.W. and Butler, E.J. (1958) Studies on magnesium in ruminant nutrition. Balance experiments on sheep with herbage from fields associated with lactation tetany and from control pastures. *British Journal of Nutrition* 12, 433–446.

Fisher, L.J., Dinn, N., Tait, R.M. and Shelford, J.A. (1994) Effect of level of potassium on the absorption and excretion of calcium and magnesium by lactating cows. *Canadian Journal of Animal Science* 74, 503–509.

Fishwick, G. (1978) Utilisation of the phosphorus and magnesium in some calcium and magnesium phosphates by growing sheep. *New Zealand Journal of Agricultural Research* 21, 571–575.

Fishwick, G. and Hemingway, R.G. (1973) Magnesium phosphates as dietary supplements for growing sheep. *Journal of Agricultural Science, Cambridge* 8, 441–444.

Gabel, G. and Martens, H. (1986) The effect of ammonia on magnesium metabolism in sheep. *Journal of Animal Physiology and Animal Nutrition* 55, 278–287.

Gardner, J.A.A. (1973) Control of serum magnesium levels in sheep. *Research in Veterinary Science* 15, 149–157.

Gentry, R.P., Miller, W.J. and Pugh, D.G. (1978) Effects of feeding high magnesium to young dairy calves. *Journal of Dairy Science* 61, 1750.

Giduck, S.A., Fontenot, J.P. and Rahmena, S. (1988) Effect of ruminal infusion of glucose, volatile fatty acids and hydrochloric acid on mineral metabolism in sheep. *Journal of Animal Science* 66, 532–542.

Goff, J.P. and Horst, R.L. (1997) Physiological changes at parturition and their relationship to metabolic disorders. *Journal of Dairy Science* 80, 1260–1268.

Grace, N.D. (1983) Amounts and distribution of mineral elements associated with fleece-free empty body weight gains in the grazing sheep. *New Zealand Journal of Agricultural Research* 26, 59–70.

Grace, N.D. and MacRae, J.C. (1972) Influence of feeding regimen and protein supplementation on the sites of net absorption of magnesium in sheep. *British Journal of Nutrition* 27, 51–55.

Grace, N.D., Carr, D.H. and Reid, C.S.W. (1985) Secretion of sodium, potassium, phosphorus, calcium and magnesium via the parotid and mandibular saliva in sheep offered chaffed lucerne hay or fresh 'Grasslands Ruanui' ryegrass. *New Zealand Journal of Agricultural Research* 28, 449–455.

Greene, L.W., Baker, J.F., Byers, F.M. and Schelling, G.T. (1985) Incidence of grass tetany in a cow herd of a five-breed diallel during four consecutive years. *Journal of Animal Science* 61 (Suppl. 1) 60.

Greene, L.W., Fontenot, J.P. and Webb, K.E. (1983a) Effect of dietary potassium on absorption of magnesium and other macroelements in sheep fed different levels of magnesium. *Journal of Animal Science* 56, 1208–1213.

Greene, L.W., Webb, K.E. and Fontenot, J.P. (1983b) Effect of potassium level on site of absorption of magnesium and other macroelements in sheep. *Journal of Animal Science* 56, 1214–1221.
Greene, L.W., Fontenot, J.P. and Webb, K.E. (1983c) Site of magnesium and other macromineral absorption in steers fed high levels of potassium. *Journal of Animal Science* 57, 503–510.
Greene, L.W., Solis, J.C., Byers, F.M. and Schelling, G.T. (1986) Apparent and true digestibility of magnesium in mature cows of five breeds and their crosses. *Journal of Animal Science* 63, 189–196.
Grunes, D.L., Stout, P.R. and Brownell, J.R. (1970) Grass tetany of ruminants. *Advances in Agronomy* 22, 331.
Guenter, W. and Sell, J.L. (1973) Magnesium absorption and secretion along the gastrointestinal tract of the chicken. *Journal of Nutrition* 103, 875–881.
Guenter, W. and Sell, J.L. (1974) A method for determining 'true' availability of magnesium from foodstuffs using chickens. *Journal of Nutrition* 104, 1446–1457.
Hajj, R.N. and Sell, J.L. (1969) Magnesium requirement of the laying hen for reproduction. *Journal of Nutrition* 97, 441–448.
Harmon, B.G., Liu, C.T., Jensen, A.H. and Baker, D.H. (1976) Dietary magnesium levels for sows during gestation and lactation. *Journal of Animal Science* 42, 860–865.
Hendricks, D.G., Miller, E.R., Ullrey, D.E., Hoefer, J.A. and Luecke, R.W. (1970) Effect of source and level of protein on mineral utilization by the baby pig. *Journal of Nutrition* 100, 235–240.
Henry, P.R. and Benz, S.A. (1995) Magnesium bioavailability. In: Ammerman, C.B., Baker, D.H. and Lewis, A.J. (eds) *Bioavailablility of Nutrients for Animals.* Academic Press, New York, pp. 239–256.
Herd, R.P. (1966) Fasting in relation to hypocalcaemia and hypomagnesaemia in lactating cows and ewes. *Australian Veterinary Journal* 42, 269–272.
Hurley, L.A., Greene, L.W., Byers, F.M. and Carstens, G.E. (1990) Site and extent of apparent magnesium absorption by lambs fed different sources of magnesium. *Journal of Animal Science* 68, 2181–2187.
James, C.S. and Chandran, K. (1975) Enquiry into the role of minerals in experimental urolithiasis in goats. *Indian Veterinary Journal* 52, 251–258.
Johnson, C.L. and Aubrey Jones, D.A. (1989) Effect of change of diet on the mineral concentration of rumen fluid, on magnesium metabolism and on water balance in sheep. *British Journal of Nutrition* 61, 583–594.
Johnson, C.L., Helliwells, S.H. and Aubrey Jones, D.A. (1988) Magnesium metabolism in the rumens of lactating dairy cows fed on spring grass. *Quarterly Journal of Experimental Physiology* 73, 23–31.
Kemp, A. (1971) *The Effects of K and N Dressings on the Mineral Supply of Grazing Animals.* Proceedings of the 1st Colloquium of the Potassium Institute, IBS, Wageningen, pp. 1–14.
Kemp, A. and 't Hart, M.L. (1957) Grass tetany in grazing milking cows. *Netherlands Journal of Agricultural Science* 5, 4–17.
Kemp, A. and Todd, J.R. (1970) Prevention of hypomagnesaemia in cows: the use of magnesium alloy bullets. *Veterinary Record* 86, 463–464.
Kemp, A., Deijs, W.B., Hemkes, O.J. and Van Es, A.J.H. (1961) Hypomagnesaemia in milking cows: intake and utilization of magnesium from herbage by lactating cows. *Netherlands Journal of Agricultural Science* 9, 134–149.
Kruse, H.D., Orent, E.R. and McCollum, E.V. (1932) Studies on magnesium deficiency

in animals. 1. Symptomatology resulting from magnesium deprivation. *Journal of Biological Chemistry* 96, 519–539.

Larvor, P. (1976) Magnesium kinetics in ewes fed normal or tetany-prone grass. *Cornell Veterinarian* 66, 413–422.

Lee, S.R. and Britton, W.M. (1987) Magnesium-induced catharsis in chicks. *Journal of Nutrition* 117, 1907–1912.

Leroy, J. (1926) Nécessite du magnésium pour la croissance de la souris. *Comptes Rendus des Séances de la Société de Biologie* 94, 341.

Littledike, E.T. and Goff, J.P. (1987) Interactions of calcium, phosphorus and magnesium and vitamin D that influence their status in domestic meat animals. *Journal of Animal Science* 65, 1727–1743.

Littledike, E.T., Wittam, J.E. and Jenkins, T.G. (1995) Effect of breed, intake and carcass composition on the status of several macro and trace minerals of adult beef cattle. *Journal of Animal Science* 73, 2113–2119.

McAleese, D.M., Bell, M.C. and Forbes, R.M. (1961) Magnesium-28 studies in lambs. *Journal of Nutrition* 74, 505–514.

McCoy, M.A. and Kennedy, D.G. (1994) Evaluation of post-mortem magnesium concentration in bovine eye fluids as a diagnostic aid for hypomagnesaemic tetany. *Veterinary Record* 135, 188–189.

McCoy, M.A., Goodall, E.A. and Kennedy, D.G. (1993) Incidence of hypomagnesaemia in dairy and suckler cows in Northern Ireland. *Veterinary Record* 132, 537.

McCoy, M.A., Goodall, E.A. and Kennedy, D.G. (1996) Incidence of bovine hypomagnesaemia in Northern Ireland and methods of supplementation. *Veterinary Record* 138, 41–43.

McGillivray, J.J. and Smidt, M.J. (1975) Biological evaluation of magnesium sources. *Poultry Science* 54, 1792–1793.

McGillivray, J.J. and Smidt, M.J. (1977) Energy level, potassium, magnesium and sulfate interaction in broiler diets. *Poultry Science* 56, 1736–1737.

McIntosh, S., Crooks, P. and Simpson, K. (1973) Sources of magnesium for grassland. *Journal of Agricultural Science, Cambridge* 81, 507–511.

McLean, A.F., Buchan, W. and Scott, D. (1985) The effect of potassium and magnesium infusion on plasma Mg concentration and Mg balance in ewes. *British Journal of Nutrition* 54, 713–718.

MAFF (1992) *Feed composition – UK Tables of Feed Composition and Nutritive Value for Ruminants*, 2nd edn. Ministry of Agriculture, Fisheries and Food Standing Committee on Tables of Feed Composition, Chalcombe Publications, Canterbury.

Martens, H. (1983) Saturation kinetics of magnesium efflux across the rumen wall in heifers. *British Journal of Nutrition* 49, 153–158.

Martens, H. and Blume, I. (1986) Effect of intraruminal sodium and potassium concentrations and of the transmural potential difference on magnesium absorption from the temporarily isolated rumen of sheep. *Quarterly Journal of Experimental Physiology* 71, 409–415.

Martens, H., Kubel, O.W., Gabel, G. and Honig, H. (1987) Effects of low sodium intake on magnesium metabolism of sheep. *Journal of Agricultural Science, Cambridge* 108, 237–243.

Matsui, T., Yans, H., Kawabata, T. and Harumoto, T. (1994) The effect of suppressing bone resorption on magnesium metabolism in sheep (*Ovis aries*). *Comparative Biochemistry and Physiology* 107A, 233–236.

Mayland, H.F., Grunes, D.L. and Lazar, V.A. (1976) Grass tetany hazard of cereal forages based upon chemical composition. *Agronomy Journal* 68, 665–667.

Mayo, R.H., Plumlee, M.P. and Beeson, W.M. (1959) Magnesium requirement of the pig. *Journal of Animal Science* 18, 264–274.

Menzies, F.D., Bryson, D.G. and McCallion, T. (1994) *Bovine Mortality Survey.* Department of Agriculture for Northern Ireland Publication, Belfast, p. 11.

Metson, A.J., Saunders, W.M.H., Collie, T.W. and Graham, V.W. (1966) Chemical composition of pastures in relation to grass tetany in beef breeding cows. *New Zealand Journal of Agricultural Research* 9, 410–436.

Miller, E.R., Ullrey, D.E., Zutaut, C.L., Baltzer, B.V., Schmidt, D.A., Hoefer, J.A. and Luecke, R.W. (1965) Magnesium requirement of the baby pig. *Journal of Nutrition* 85, 13–20.

Minson, D.J. (1990) Magnesium. In: *Forage in Ruminant Nutrition.* Academic Press, Sydney, pp. 265–290.

Moseley, G. and Baker, D.H. (1991) The efficacy of a high magnesium grass cultivar in controlling hypomagnesaemia in grazing animals. *Grass and Forage Science* 46, 375–380.

Netherlands Ministry of Agriculture and Fisheries, Committee on Mineral Nutrition (1973) *Tracing and Treating Mineral Disorders in Dairy Cattle.* Centre for Agricultural Publishing and Documentation, Wageningen, 61 pp.

Newton, G.L., Fontenot, J.P., Tucker, R.E. and Polan, C.E. (1972) Effects of high dietary potassium intake on the metabolism of magnesium by sheep. *Journal of Animal Science* 35, 440–445.

NRC (1977) *Nutrient Requirements of Poultry*, 7th edn. National Academy of Sciences, Washington, DC.

NRC (1980) *Mineral Tolerances of Domestic Animals.* National Academy of Sciences, Washington, DC.

NRC (1988) *Nutrient Requirements of Swine*, 9th edn. National Academy of Sciences, Washington, DC.

NRC (1994) *Nutrient Requirements of Poultry*, 9th edn. National Academy of Sciences, Washington, DC.

Nugara, D. and Edwards, H.M., Jr (1963) Influence of dietary Ca and P levels on the Mg requirements of the chick. *Journal of Nutrition* 80, 181–184.

Parkins, J.J. and Fishwick, G. (1999) Magnesium hydroxide as a novel herbage spray supplement for lactating cows and ewes. *Animal Science* 66 (in press).

Partridge, I.G. (1978) Studies on digestion and absorption in the intestines of growing pigs. 3. Net movements of mineral nutrients in the digestive tract. *British Journal of Nutrition* 39, 527–537.

Pauli, J.V. and Allsop, T.P. (1974) Plasma and cerebrospinal fluid magnesium, calcium and potassium in dairy cows with hypomagnesaemic tetany. *New Zealand Veterinary Journal* 22, 227–231.

Pike, R.L. and Brown, M.L. (1975) *Nutrition: an Integrated Approach*, 2nd edn. John Wiley & Sons, New York, 186 pp.

Powley, G., Care, A.D. and Johnson, C.L. (1977) Comparison of daily endogenous faecal magnesium excretion from sheep eating grass with high sodium or high potassium concentrations. *Research in Veterinary Science* 23, 43–46.

Reid, R.L., Jung, G.A., Roemig, I.J. and Kocher, R.E. (1978) Mineral utilisation in lambs and guinea pigs fed Mg-fertilized grass and legume hays. *Agronomy Journal* 70, 9–14.

Reid, R.L., Jung, G.A., Wolf, C.H. and Kocher, R.E. (1979) Effects of magnesium fertilization on mineral utilization and nutritional quality of alfalfa for lambs. *Journal of Animal Science* 48, 1191–1201.

Richardson, J.A. and Welt, L.G. (1965) The hypomagnesemia of vitamin D administration. *Proceedings of the Society for Experimental Biology and Medicine* 118, 512–514.

Riond, J.-L., Kocabagli, N., Spichiger, V.E. and Wanner, M. (1995) The concentration of ionized magnesium in serum during the peri-parturient period of non-paretic cows. *Veterinary Research Communications* 19, 195–203.

Ritchie, N.S. and Fishwick, G. (1977) Magnesium phosphate as a dietary supplement for lactating cows at spring pasture. *Journal of Agricultural Science, Cambridge* 88, 71–73.

Ritter, R.J., Boling, J.A. and Gay, N. (1984) Labile magnesium reserves in beef cows subjected to different prepasture supplementation regimes. *Journal of Animal Science* 59, 197–203.

Robson, A.B., Field, A.C. and Sykes, A.R. and McKinnon, A.E. (1997) A model for magnesium metabolism in sheep. *British Journal of Nutrition* 78, 975–992.

Rogers, P.A.M. (1979) Hypomagnesaemia and its clinical syndromes in cattle: a review. *Irish Veterinary Journal* 33, 115–124.

Rogers, P.A.M. and Poole, D.B.R. (1976) Control of hypomagnesaemia in cows: a comparison of magnesium acetate in the water supply with magnesium oxide in the feed. *Irish Veterinary Journal* 30, 129–136.

Rook, J.A.F. and Storry, J.E. (1962) Magnesium in the nutrition of ruminants. *Nutrition Abstracts and Reviews* 32, 1055–1076.

Sansom, B.F., Bunch, K.G. and Dew, S.M. (1982) Changes in plasma calcium, magnesium, phosphorus and hydroxyproline concentrations in ewes from twelve weeks before until three weeks after lambing. *British Veterinary Journal* 138, 393–401.

Schonewille, J.T., Ram, L., Van't Klooster, A.T., Wouterse, H. and Beynen, A.C. (1997a) Intrinsic potassium in grass silage and magnesium absorption in dry cows. *Livestock Production Science* 48, 99–110.

Schonewille, J.T., Ram, L., Van't Klooster, A.T., Wouterse, H. and Beynen, A.C. (1997b) Native corn starch versus either cellulose or glucose in the diet and the effects on apparent magnesium absorption in goats. *Journal of Dairy Science* 80, 1738–1743.

Scott, D. (1966) The effects of sodium depletion and potassium supplementation upon electrical potentials in the rumen of the sheep. *Quarterly Journal of Experimental Physiology* 51, 60–69.

Sell, J.L. (1979) Magnesium nutrition of poultry and swine. In: *Proceedings of the Second Annual International Minerals Conference, St Petersburg Beach, Florida, Illinois.* International Minerals and Chemical Corporation.

Shils, M.E. (1997) Magnesium. In: O'Dell, B.L. and Sunde, R.A. (eds) *Handbook of Nutritionally Essential Mineral Elements.* Marcel Dekker, New York, pp. 117–152.

Shockey, W.L., Conrad, H.R. and Reid, R.L. (1984) Relationship between magnesium intake and faecal magnesium excretion of ruminants. *Journal of Dairy Science* 67, 2594–2598.

Sjollema, B. and Seekles, L. (1930) Uber Storungen des mineralen Regulationsmechanismus bei Krankheiten des Rindes. (Ein Beitrag zur Tetaniefrage.) *Biochemische Zeitschrift* 229, 338–380.

Stuedemann, J.A., Wilkinson, S.R. and Lowrey, R.S. (1984) Efficacy of a large magnesium alloy rumen bolus in the prevention of hypomagnesaemic tetany in cows. *American Journal of Veterinary Research* 45, 698–702.

Sutherland, R.J., Bell, K.C., McSporran, K.D. and Carthew, G.W. (1986) A comparative study of diagnostic tests for the assessment of herd magnesium status in cattle. *New Zealand Veterinary Journal* 34, 133–135.

Suttle, N.F. (1987) The absorption, retention and function of minor nutrients. In: Hacker, J.B. and Ternouth, J.H. (eds) *The Nutrition of Herbivores*. Academic Press, Sydney, pp. 333–361.

Suttle, N.F. (1996) Non-dietary influences on the mineral requirements of sheep. In: Masters, D.G. and White, C.L. (eds) *Detection and Treatment of Mineral Nutrition Problems in Grazing Sheep*. ACIAR Monograph No. 37, Australian Centre for International Agricultural Research, Canberra, pp. 31–44.

Suttle, N.F. and Brebner, J. (1995) A putative role for larval nematode infection in diarrhoea of lambs which do not respond to anthelmintic drenches. *Veterinary Record* 137, 311–316.

Suttle, N.F. and Field, A.C. (1967) Studies on magnesium in ruminant nutrition. 8. Effect of increased intakes of potassium and water on the metabolism of magnesium, phosphorus, sodium, potassium and calcium in sheep. *British Journal of Nutrition* 21, 819–831.

Suttle, N.F. and Field, A.C. (1969) Studies on magnesium in ruminant nutrition. 9. Effect of potassium and magnesium intakes on development of hypomagnesaemia in sheep. *British Journal of Nutrition* 23, 81–90.

Suttle, N.F., Field, A.C., Nicolson, T.P., Mathieson, A.O., Prescott, J.H.D., Scott, N. and Johnson, W.S. (1980) Some problems in assessing the physiological and economic significance of hypocuproemia in beef suckler herds. *Veterinary Record* 106, 302–304.

't Hart, M.L. and Kemp, A. (1956) De invloed van de weersomstandigheden op het optreden van kopziekte bij rundvee. *Tijdschrift voor Diergeneeskunde* 81, 84–95.

Thomas, B., Thompson, A., Oyenuga, V.A. and Armstrong, R.H. (1952) The ash constituents of some herbage plants at different stages of maturity. *Empire Journal of Experimental Agriculture* 20, 1–22.

Thomas, B., Holmes, W.B. and Clapperton, J.L. (1955) A study of meadow hays from the Cockle Park plots. 2. Ash constituents. *Empire Journal of Experimental Agriculture* 23, 101–108.

Todd, J.R. (1969) Magnesium metabolism in ruminants: review of current knowledge. In: *Trace Mineral Studies with Isotopes in Domestic Animals*. International Atomic Energy Agency, Vienna, pp. 131–140.

Tomas, F.M. and Potter, B.J. (1976a) The site of magnesium absorption from the ruminant stomach. *British Journal of Nutrition* 36, 37–45.

Tomas, F.M. and Potter, B.J. (1976b) Interaction between site of magnesium absorption in the digestive tract of the sheep. *Australian Journal of Agricultural Research* 27, 437–446.

Tongyai, S., Raysigguer, Y., Molta, C., Gueux, E., Maurois, P. and Heaton, F.W. (1989) Mechanism of increased erythrocyte membrane fluidity during magnesium deficiency in rats. *American Journal of Physiology* 257, 270–276.

Turner, M.A., Neall, V.E. and Wilson, G.F. (1978) Survey of magnesium content of soils and pastures and incidence of grass tetany in three selected areas of Taranaki. *New Zealand Journal of Agricultural Research* 21, 583–592.

Wacker, W.E.C. (1969) The biochemistry of magnesium. *Annals of the New York Academy of Sciences* 162, 717–726.

Whittaker, D.A. and Kelly, J.M. (1982) Incidence of clinical and subclinical hypomagnesaemia in dairy cows in England and Wales. *Veterinary Record* 110, 450–451.

Wilcox, G.E. and Hoff, J.E. (1974) Grass tetany: an hypothesis concerning its relationship with ammonium nutrition of spring grasses. *Journal of Dairy Science* 57, 1085–1089.

Wilson, G.F. (1980) Effects of magnesium supplements on the digestion of forages and milk production of cows with hypomagnesaemia. *Animal Production* 31, 153–157.

Wu, C.L. and Britton, W.M. (1974) The influence of phosphorus on magnesium metabolism in the chick. *Poultry Science* 53, 1645 (abstract).

Wylie, M.J., Fontenot, J.P. and Greene, L.W. (1985) Absorption of magnesium and other macrominerals in sheep infused with potassium in different parts of the digestive tract. *Journal of Animal Science* 61, 1219–1229.

Xin, Z., Tucker, W.B. and Hemken, R.W. (1989) Effect of reactivity rate and particle size of magnesium oxide on magnesium availability, acid–base balance, mineral metabolism and milking performance of dairy cows. *Dairy Science* 72, 462–470.

Young, P.W., Rys, G. and O'Connor, M.B. (1981) Hypomagnesaemia and dairy production. *Proceedings of the New Zealand Society of Animal Production* 41, 61–67.

7 Sodium and Chlorine

Introduction

It is convenient to consider sodium and chlorine together because of their related functions and requirements in the animal body and their interactions with each other and because sodium and chlorine form common salt, the cheapest, most palatable and most freely used of all mineral supplements. The value of salt in the diet of animals and humans was recognized hundreds of years before the nature and extent of its need were established. The strong craving for salt exhibited by grazing animals was recorded by the early settlers and the regular provision of salt came to be identified with good stock husbandry in a wide range of conditions. The transition from a nomadic to an agricultural way of life, with dependence on cereals or vegetables rather than meat and milk, was only sustainable with dietary supplements of salt (Denton, 1982). The value of supplementary salt for cattle was first demonstrated experimentally by Boussingault (1847) but it was many years before Babcock (1905) reported his classical studies of the effect of salt deprivation on lactating dairy cows. A further half-century passed before Aines and Smith (1957) identified sodium rather than chlorine as the element primarily concerned. During this period, the need for supplemental salt by pigs and poultry fed on cereal-based rations was established and again sodium rather than chlorine was found to be the critically limiting element. In recent times, attention has switched to the problems of natural or commercially induced dietary excesses of sodium. Approximately one-third of the earth's land surface is affected by salinity, sodicity and aridity in various combinations (Chiy and Phillips, 1995). The exploitation of halophytic browse species and sea water for irrigation in arid coastal areas is largely a problem of managing sodium excesses (Pasternak *et al.*, 1985). Overgenerous provision of sodium for intensively reared pigs and poultry increases water consumption and excretion, creating problems of soiling and waste disposal. Unequivocal evidence of a naturally occurring dietary deficiency of chlorine is rare (Summers *et al.*, 1967), but chlorine deficiency has been established

© CAB *International* 1999. *Mineral Nutrition of Livestock*
(E.J. Underwood and N.F. Suttle)

experimentally in poultry (Leach and Nesheim, 1963) and calves (Neathery *et al.*, 1981). Although potassium is accorded its own chapter (8), some of its interactions with sodium and chloride will be dealt with here.

Dietary Sources of Sodium and Chloride

Concentrates

The sodium and chloride concentrations in some common foodstuffs are given in Table 7.1. A survey by the American Feed Industry Association (Berger, 1990) showed that sodium values were much lower than those widely quoted from the US–Canadian Tables of Feed Composition, the mean value for maize (0.07 g kg^{-1} dry matter (DM)) being only 23% of the listed value. Most grain and vegetative foodstuffs (with the exception of sunflower meal) were poor in sodium, although values were not normally distributed (Berger, 1990). Root crops, root by-products and animal products are generally much richer in sodium than cereals, while animal protein sources, such as fish-meal, meat-meal and dried skimmed milk, are higher in sodium than most plant protein sources (Table 7.1). Cereal grains generally provide more chloride (Cl) than sodium, with maize providing the least and barley the most (0.5 and 1.8 g Cl kg^{-1} DM, respectively). Cereal straws contain three- to sixfold more chloride than the grains. Vegetable protein supplements are consistently low in chloride (0.3–0.7 g kg^{-1} DM), and it is possible to compound mixtures with cereals which provide pigs and poultry with insufficient chloride requirements, but inclusion of animal by-products and grass meal would correct any deficit. Ruminants denied access to roughage could possibly receive inadequate dietary chloride intakes.

Forages

Minson (1990) reported that the distribution of pasture sodium (Na) concentrations was skewed towards low values, with 50% of samples containing < 1.5 g Na kg^{-1} DM. Low values were more common in tropical than in temperate pastures, with legumes particularly deficient, half of them containing < 0.4 g Na kg^{-1} DM. There are consistent differences among species and varieties of grass in their sodium concentration; among the more widely grazed species, the order is *Lolium perenne* > *Dactylis glomerata* > *Festuca pratensis* > *Phleum pratense*, with varietal differences most marked in the richest species (Chiy and Phillips, 1993). *Phalaris* spp. are rich in sodium, while *Pennisetum clandestinum* (kikuyu) is a natrophobic species. There are also major differences among legumes, high leaf sodium concentrations being found in natrophilic species, such as white clover, subterranean clover and trefoil, and low values in natrophobic species, such as lucerne (alfalfa) and red clover (Smith and Middleton, 1978b). Decreases in the sodium concentration of grasses (but not legumes) as they mature contribute to the reported variability. Morris *et al.* (1980) reported minimum values of 0.1–0.2 g Na kg^{-1} DM in Californian rangeland pasture in September, compared with

Table 7.1. Mean (SD) sodium concentrations (g kg^{-1} DM) in pastures and foodstuffs mostly from the UK (from MAFF, 1990) and chloride concentrations for corresponding foodstuffs from the USA* (NRC as quoted by McDowell, 1992).

	Na	Cl		Na	Cl		Na	Cl
Herbage	2.5 (2.1)	4–6	Fish-meal	11.2 (1.5)	–	Fodder beet	3.0 (1.6)	–
Grass hay	2.1 (1.7)							
Grass silage	2.7 (1.6)	–	Soybean meal (ext)	0.2 (0.1)	0.4	Swedes	0.7[a]	
Clover silage (mix)	0.8 (0.5)	–	Safflower meal	1.0 (1.2)	1.3	Turnips	2.0[a]	
Lucerne (alfalfa) silage	1.3 (1.0)	4.1	Palm-kernel cake	0.2 (0.1)	–	Cassava meal	0.6 (0.3)	–
Maize silage	0.3 (0.2)	1.8	Maize gluten	2.6 (1.4)	1.0			
Grass hay	2.1 (1.7)	4–5	Rapeseed meal	0.4 (0.3)	–	Meat- and bone-meal	8.0 (1.0)	8–12
Lucerne (alfalfa) hay	0.6 (0.1)	3.0	Cottonseed meal	0.2 (0.2)	0.5	Molasses (beet)	25.0 (8.2)	16.4
Dried grass	2.8 (1.6)	–	Linseed meal	0.7 (0.0)	0.4	Molasses (cane)	1.2 (0.9)	31.0
Dried lucerne (alfalfa)	1.3 (0.8)	4.8				Molassed beet pulp	4.4 (0.3)	
						Beet pulp	3.2 (2.1)	0.4
Barley straw	1.3 (1.4)	6.7	Maize		0.5	Brewers' grains	0.3 (0.3)	1.7
Wheat straw	0.6 (1.0)	3.2	Barley	0.3 (0.4)	1.8	Wheat feed	0.1 (0.1)	0.5
Alkali-treated barley straw	32.0(10.5)	–	Wheat	0.1 (0.1)	0.5	Distillers' grains:		
			Oats	0.2 (0.1)	1.1	Wheat	3.1 (3.9)	–
			Sorghum	0.5 (0.0)	1.0	Barley	0.3 (0.3)	–

[a] New Zealand data from Cornforth *et al.* (1978).

SD, standard deviation.

* Chloride values much higher for coastal locations.

0.5 g Na kg^{-1} DM in spring. Pastures growing within 25 km of the New Zealand coast can be three times richer in sodium (0.3 vs. 0.1 g Na kg^{-1} DM) than those growing inland, due to deposition of sodium from sea spray (Smith and Middleton, 1978a). Pasture sodium concentrations are influenced by the application of potassium (K) and nitrogen (N) fertilizers; N increases pasture sodium in a dose-dependent manner, but the concurrent application of K limits the N response, particularly at high application rates (Fig. 7.1; Kemp, 1971). Maize silage is the poorest forage source of sodium in Table 7.1. Most pastures are appreciably richer in chloride than they are in sodium, with little difference between legumes and grasses or between fresh grass and hay. For example, Thomas *et al.* (1952) reported mean values of 0.8 g Na and 4.0 g Cl kg^{-1} DM in leguminous and 1.4 g Na and 5.0 g Cl kg^{-1} DM in grass pasture species. Pastures invariably provide sufficient chloride for grazing animals.

Drinking-water

The water available to livestock varies enormously in the amount of sodium, chloride and other minerals it contains, from virtually none in some streams and reticulation sources to highly saline supplies, usually from deep wells or bores. The drinking-water can therefore constitute a valuable source of supplementary sodium in areas where intakes of sodium from feed are low, but it may provide excessive quantities of salt (and of other minerals normally

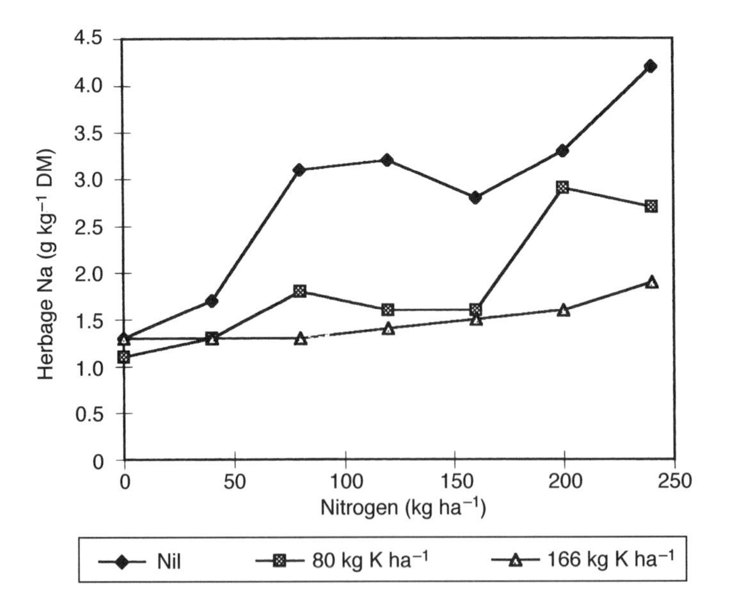

Fig. 7.1. Herbage sodium concentrations are increased by the application of nitrogenous fertilizers but if potassium is simultaneously applied the effect can be lost (data from Kemp, 1971).

present with the salt, such as sulphates and bicarbonates of sodium, calcium and magnesium). Sodium-deprived animals show a marked preference for salty water (Blair-West *et al.*, 1968; Smith and Middleton, 1978a) and they will even consume sodium bicarbonate ($NaHCO_3$), which is usually avoided (Bell, 1995). When the diet provides sufficient sodium, cattle will discriminate against water containing 12.5 g sodium chloride (NaCl) l^{-1} in favour of pure water; pygmy goats show a similar preference, while both sheep and normal goats remain indifferent (Goatcher and Church, 1970). The extent to which drinking-water meets the supplementary sodium needs of livestock will depend on the degree of deficiency, the amount of sodium in the water and the volume of water consumed. The high individual and seasonal variation in water consumption was stressed in Chapter 2. It has been calculated that cattle drinking 40 l day^{-1} would obtain their sodium requirements from water containing about 0.5 g NaCl l^{-1} (Loosli, 1978). Supplementary NaCl has been given to poultry via drinking-water (Ross, 1979).

Metabolism of Sodium and Chloride

Palatability

The presence of salt in a feed can contribute to the palatability of that feed (Grovum and Chapman, 1988), whereas the addition of salt to a feed replete with sodium can lower feed intake (Wilson, 1966; Moseley, 1980) and may be used as a means of restricting the intake of supplementary foods (De Waal *et al.*, 1989). However, sodium appetite is relative rather than absolute and can easily be affected by experience.

Absorption

Both sodium and chloride are readily absorbed, but each element can influence the absorption of the other (for a review, see Henry, 1995). Sodium uptake from the gut lumen is achieved by coupling to glucose and amino acid uptake via cotransporters and also by exchange with hydrogen ions (H^+) via an Na–H antiporter, intracellular H^+ being generated by carbonic anhydrase in the enterocytes of the gut mucosa (Harper *et al.*, 1997). Absorption of chloride from dietary and endogenous (gastric secretions) sources is achieved by exchange for another anion, bicarbonate (HCO_3^-), also generated intracellularly by carbonic anhydrase and secreted into the gut lumen.

Membrane transport

Sodium and chloride are highly labile in the body. At the cellular level, the continuous exchange of sodium and potassium via ATP-dependent Na^+–K^+ pumps provides the basis for glucose and amino acid uptake (by cotransport) (Fig. 7.2), maintaining high intracellular potassium concentrations but requiring about 50% of the cell's maintenance need for energy (Milligan and Summers, 1986). Any production increases Na^+ and K^+ transport and

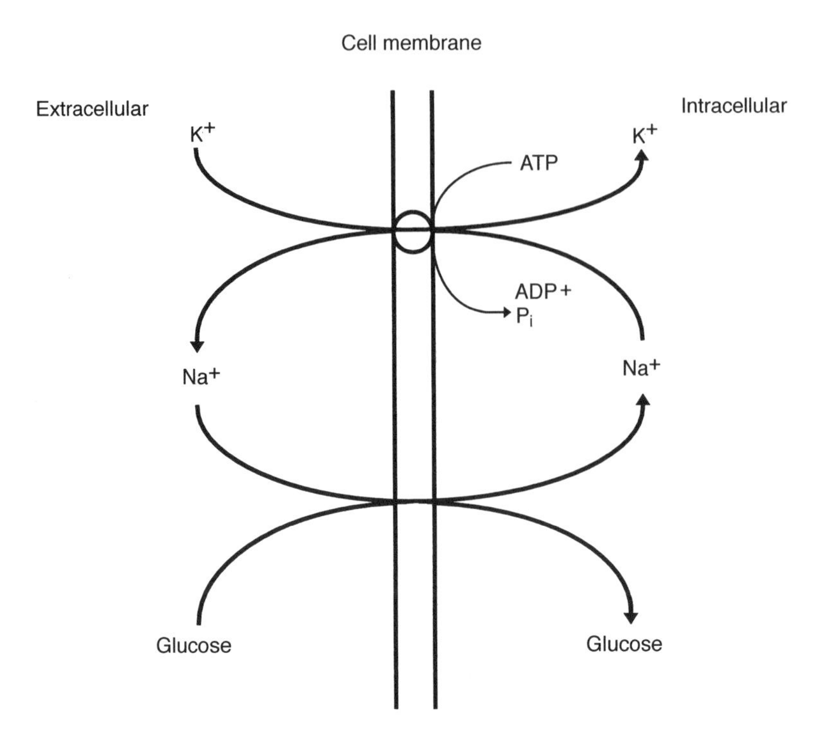

Fig. 7.2. The 'sodium pump' is vital to the maintenance of electrochemical differences across membranes and the cellular uptake of glucose through the action of a Na^+–K^+ dependent ATPase, but activity is sustained at a significant energy cost.

modelling studies suggest the associated contribution to energy expenditure rises from 18 to 23% as lamb growth increases from 90 to 230 g day^{-1}, increased metabolite transport (e.g. amino acids) in the gut mucosa being primarily responsible (Gill *et al.*, 1989). Sodium transport across membranes is also achieved by a wide variety of complementary mechanisms: by an Na–H exchanger; by electroneutral Na–K–2Cl cotransporters; by Na–Cl and Na–magnesium (Mg) exchangers; and by voltage-gated Na^+ channels. Chloride transport is almost as complex, with voltage, mechanically ('stretch') activated and Ca-activated channels and also Cl^-–HCO_3^- exchangers contributing to fluxes (Harper *et al.*, 1997).

Secretion

Much of the sodium that enters the gastrointestinal tract comes from saliva, particularly in ruminants, which daily secrete about 0.3 l kg^{-1} live weight (LW) containing sodium (150 mmol l^{-1}) as the major cation. The rumen can contain 50% of the available body sodium, providing an important reserve (Bell, 1995), but on potassium-rich diets it is displaced by potassium (Suttle and Field, 1967). In sodium deficiency, salivary sodium is replaced on a molar basis by potassium to conserve sodium (Blair-West *et al.*, 1963), in an adaptation modulated by aldosterone, a hormone secreted by the adrenal gland. All three elements (sodium, chloride and potassium) are lost via skin secretions, but there are major differences between species. In non-ruminants, including the horse, sodium is the major cation in sweat and salt concentrations in sweat can reach 4.5%. Horses, mules and donkeys sweat profusely when exercised, but the high loss of sodium balances the loss of water and provides a defence against hypernatraemia. Sodium concentrations in milk (normally 17 mmol l^{-1} in sheep, 27 mmol l^{-1} in cattle) decline only slightly during sodium depletion and there are no compensatory changes in milk potassium to compare with those seen in saliva (Morris and Peterson, 1975).

Excretion

Regulation of sodium status in the face of fluctuations in sodium intake is achieved principally by the control of reabsorption in the proximal tubule of the kidney. Mediation is achieved by active transport and changes in membrane permeability. Sodium reabsorption in the distal tubule can be impaired by excess potassium but enhanced by aldosterone, so that urinary losses become negligible when sodium intakes are low. Regulation is highly complex and beyond the scope of this chapter (see Harper *et al.*, 1997). Chloride is also reabsorbed in the kidney, but by a passive process. Dietary excesses of sodium and chloride are predominantly excreted via the kidney.

Imbalance

Risk of osmotic imbalance caused by dietary deficiencies or excesses of sodium, potassium and chloride is countered by complex physiological control mechanisms; these include the activation of renin to angiotensin I and II, which, with vasopressin, regulate aldosterone secretion, extracellular fluid (ECF) volume and blood pressure through appropriate adjustments in thirst and water balance (Bell, 1995; Michell, 1995). An increase in dietary sodium from 1 (low) to 4 (modest excess) g kg^{-1} DM increased the water intake of lactating sows from 12.4 to 13.9 l day^{-1} (Seynaeve *et al.*, 1996). Such responses are found in all species and they affect the profitability of animal production in diverse ways: in pig units, the volume of slurry is increased; in poultry units, birds produce excessively moist excreta ('wet droppings') (Mongin, 1981); in ruminants, rate of outflow of digesta from the rumen is increased, with adverse consequences for digestibility (Arieli *et al.*, 1989) but possible benefits in terms of undegraded protein outflow (Hemsley *et al.*, 1975). Effects of sodium and K/Na imbalance on magnesium metabolism in ruminants are considered in detail in Chapters 6 and 8, respectively.

Functions of Sodium and Chloride

Sodium and chloride maintain osmotic pressure, regulate acid–base equilibrium and control water metabolism in the body. Sodium is the major cation in the ECF and chloride the major anion, at concentrations of 140 and 105 mmol l^{-1}, respectively. Sodium plays a key role and Michell (1985) graphically described the physiological process as 'providing an osmotic skeleton' which was 'clothed' with an appropriate volume of water. When ion intakes increase, water intakes also increase (Wilson, 1966; Suttle and Field, 1967), to protect the gut, facilitate excretion and to 'clothe' the enlarged 'skeleton'. The 'osmotic skeleton' is sustained by the Na^+–K^+ ATPase pump in cell membranes, which actively transports sodium out of the cell, thus converting the energy of ATP into osmotic gradients along which water can flow (Fig. 7.2) and fuelling other cation-transporting mechanisms. Transmembrane potential differences are established through the activity of the pumps and these influence the uptake of other cations and are essential for excitability (i.e. response to stimuli). Amino acid and glucose uptake is dependent on sodium. Exchange of Na^+ with H^+ influences pH regulation, while that with Ca^+ influences vascular tone. The contribution of sodium to acid–base balance is covered in more detail in Chapter 8 (p. 222). There are some marked contrasts between sodium and potassium (Fig. 8.3). Sodium makes up over 90% of the bases of the serum, but little is present in the blood cells. There are significant stores of sodium in the ovine skeleton, which may contain *c.* 3–5 g Na kg^{-1}. Chlorine is found both within the cells and in the body fluids, including the gastric secretions, where it occurs as hydrochloric acid (HCl) and in the form of salts. Respiration is based on 'the chloride shift', whereby the potassium salt of oxyhaemoglobin exchanges oxygen (O_2) for carbon dioxide (CO_2) via bicarbonate in the tissue and reverses that process in the lung, where reciprocal chloride exchanges maintain the anion balance (Block, 1994).

SODIUM

Biochemical and Physiological Manifestations of Sodium Deficiency

The clinical signs of sodium deficiency, described later, occur without a significant decline in plasma or milk sodium concentrations (e.g. Seynaeve *et al.*, 1996) until the animals are *in extremis*. Urinary sodium declines rapidly to extremely low values and faecal sodium is also reduced (Jones *et al.*, 1967; Michell *et al.*, 1988), through reabsorption against a concentration gradient in the lower intestine (Bott *et al.*, 1964). Remarkable changes have been reported in the lower intestine of hens depleted of sodium (Elbrond *et al.*, 1991). After 3–4 weeks on a low-sodium diet, the microvillus surface area more than doubled, due to an increase in epithelial cell number and the surface area per cell. Greater density of open sodium channels and an increase in short-circuit current contributed to greatly increased net sodium transport. Similar studies have yet to be made in other species of farm livestock, but the gut may become at least as important as the kidney in regulating sodium balance during deficiency in sheep (Michell, 1995). Morphological changes are also seen in the adrenal, the width of the zona glomerulosa as a proportion of the cortex doubling in sodium-depleted cattle (Morris, 1980). These changes are linked to reciprocal changes in sodium and potassium concentrations in the saliva. Murphy and Plasto (1972) found that, with bullocks grazing sodium-deficient pastures, normal values were 145 and 7 mmol l^{-1} for sodium and potassium, respectively (an Na:K ratio of 20), whereas the comparable values for the deficient animals were 40 and 90 (giving an Na:K ratio of 0.45). This adaptive change in the Na:K ratio of parotid saliva is sufficiently sensitive to have been used to estimate the sodium requirements of lactating ewes (Morris and Peterson, 1975). The relationship between plasma aldosterone and salivary Na:K is curvilinear, rapid increases in hormone concentrations occurring when Na:K falls below 5.0 in sheep and goats (Fig. 7.3; McSweeney *et al.*, 1988).

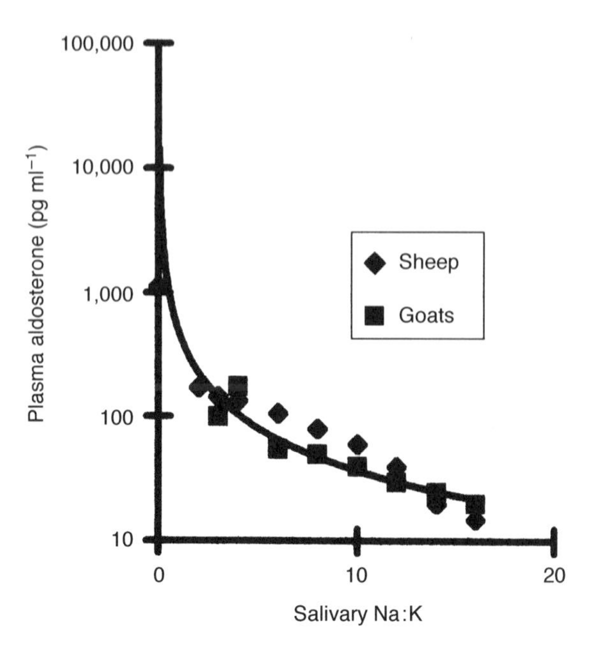

Fig. 7.3. Sodium is conserved through the action of the adrenal hormone aldosterone which facilitates the replacement of sodium by potassium in saliva and urine: low Na:K ratios in saliva are therefore correlated with plasma aldosterone and a useful indicator of sodium deprivation (data from McSweeney *et al.*, 1988).

Clinical Manifestations of Sodium Deprivation

The first sign of sodium deprivation in milking cows is a pica or craving for salt, manifested by avid licking of wood or soil and the urine or sweat of other animals. An extreme appetite for salt can occur within 2–3 weeks of deprivation. Excessive consumption of water (polydipsia) and hence high urine output has been reported (Whitlock *et al.*, 1975). After several weeks, appetite begins to decline and the animal develops a haggard appearance and rough coat, loses weight and reduces any milk yield. The decrease in milk yield is accompanied by some reduction in fat content, but the sodium concentration is only marginally reduced (Schellner *et al.*, 1971). When these events follow calving in a high-producing cow, the breakdown can be sudden and death ensues. If supplementary salt is provided before the state of collapse is reached, a dramatic recovery occurs. In growing pigs, poultry, sheep and goats, sodium deprivation is also manifested within a few weeks by inappetance, growth retardation and inefficiency of feed use, due to impairment of protein and energy metabolism, but digestibility is not affected. Laying hens on low-salt rations lose weight, are prone to cannibalism and reduce both egg production and egg size, but small differences in susceptibility exist between strains of birds (Sherwood and Marion, 1975). The feeding of 'low'-sodium rations (1.3 g Na kg^{-1} DM) has been considered as a means of inducing moult in laying hens and reduced egg production from 60 to 20% over 15 days (Ross and Herrick, 1981), despite the fact that the level attained was only marginally below requirement (see Table 7.4). Adverse effects on mammalian reproduction are not generally listed as important consequences of sodium deficiency, but this may have to change. In sodium-depleted lactating sows, the interval between weaning and oestrus increased from 6.2 to 12.6 days (P = 0.029), while the live weight of the sow and growth rate of her litter remained unaffected (Seynaeve *et al.*, 1996). Sensitivity of conception rate to sodium deprivation has also been suggested for dairy cows in Victoria, where pastures were rich in potassium and urinary sodium outputs were low (Harris *et al.*, 1986).

Occurrence of Sodium Deprivation

A critical shortage of sodium is most likely to occur in the following conditions: in rapidly growing young animals fed on cereal- or grass-meal-based diets, which are inherently low in sodium (Table 7.1); during lactation, as a consequence of sodium losses in the milk, particularly in cows; in tropical or hot, semiarid areas, where large losses of water and sodium occur in the sweat; in animals engaged in heavy or intense physical work, also through excessive sweating – this applies particularly to horses, mules and donkeys; and in animals grazing pastures on soils naturally low in sodium or pastures heavily fertilized with potash, which depresses herbage sodium contents.

When one or more of these conditions apply continuously for long periods and extra salt is not provided, sodium deprivation is inevitable.

Dietary problems

Extensive areas of sodium deprivation in livestock occur in the centres of many continents, including Saharan (Little, 1987), southern (De Waal *et al.*, 1989) and tropical Africa, Latin America (McDowell *et al.*, 1995) and the arid inland areas of Australia, where the water-table is low and the grass pastures commonly contain 0.1–0.8 g Na kg^{-1} DM (Denton *et al.*, 1961; Howard *et al.*, 1962; Murphy and Plasto, 1972). Even the grains and seeds from these parts of the world carry subnormal sodium concentrations (Chamberlain, 1955; Morris and Gartner, 1971). Temperate continents (Chiy and Phillips, 1995) and islands, such as New Zealand (Towers and Smith, 1983), can also be affected. Sherrell (1978) examined 21 New Zealand pasture species and found one-third of them to contain less than 0.5 g Na kg^{-1} DM and similar results were reported by Smith and Middleton (1978a). The introduction of productive legumes, such as lucerne, low in sodium (Table 7.1) has caused problems in New Zealand dairy cows (Joyce and Brunswick, 1975). In Scotland, the application of potassium sulphate at 210 kg ha^{-1} reduced the sodium concentrations in *L. perenne* from 2.2 to 1.2 g kg^{-1} DM (Rahman *et al.*, 1960), while, in New Zealand, increasing potassium chloride (KCl) applications from 10 to 80 kg ha^{-1} progressively reduced values to 0.4 g Na kg^{-1} DM in natrophilic pasture species, such as *D. glomerata* (McNaught, 1959). Dutch workers were particularly concerned with this problem in dairy cows (Lehr *et al.*, 1963; Fig. 7.1). Supplementary salt can improve the growth and performance of all types of cattle in such conditions (Walker, 1957; Joyce and Brunswick, 1975) and also those of high-producing pigs and poultry and milking cows fed liberally on cereal grain-based diets. With mature, non-lactating, non-working animals or with younger ruminants whose diets permit only a limited growth rate, the sodium-conserving mechanisms of the body are so effective that deprivation is likely only in hot environments with exceedingly low sodium intakes from the herbage.

Environmental and physiological problems

These have traditionally been associated with animals kept only for work, particularly equine species, which secrete sweat rich in sodium. The increased use of the lactating cow for draught purposes in developing countries will increase the risk of sodium deprivation.

Disease problems

Infections of the small and large intestine perturb mechanisms that normally contribute to sodium homeostasis. Acute bacterial infections of the gut, which cause severe diarrhoea in calves, usually necessitate electrolyte replacement therapy (Michell, 1974). Parasitic infections of the abomasum cause effluxes of sodium, which can be reabsorbed at uninfected distal sites, but, when the small intestine is also parasitized, as is often the case in young lambs and calves at pasture, sodium deficiency (Suttle *et al.*, 1996) and salt appetite may be induced on occasion, despite adequate pasture sodium levels (Suttle and Brebner, 1995). Responses to sea water have been reported in sheep that became starved and diarrhoeic during prolonged transportation by sea; lack of sodium was believed to have hindered water absorption from the gut (Black, 1997). Ideally, treatment of sodium depletion associated with diarrhoea requires the provision of glucose, bicarbonate and even potassium, as well as sodium, in a carefully balanced rehydration therapy (Michell, 1989).

The Diagnosis of Sodium Deprivation

The craving for salt, which is the earliest and most obvious sign of sodium deprivation, is non-specific; it is difficult to distinguish from other forms of pica (e.g. in deficiency of phosphorus (Chapter 5)) and some animals eat salt avidly when they are not sodium-deficient (Denton, 1982). The other manifestations of sodium deprivation are also non-specific. Since plasma sodium concentrations only fall in the terminal stages of sodium deficiency, they are of little diagnostic use (Table 7.2). Salivary Na:K ratios are more useful, but early changes reflect the operation of highly effective homeostatic mechanisms and it is necessary to recognize a band of marginal concentrations (footnote to Table 7.2). Some workers have questioned the usefulness of Na:K ratios, because they can vary widely within a group at a given sampling and fluctuate markedly from day to day and from week to week in the same animal. There are, however, reciprocal movements in plasma aldosterone (N.F. Suttle, unpublished data), which indicate the sensitivity of the Na:K ratio and the delicacy of the balances that are continuously being struck. The median Na:K ratio and the proportion of values < 2.0 are more informative than the mean salivary Na:K ratio. A simple method of collecting saliva to determine the sodium status of sheep and cattle was presented by Murphy and Connell (1970). Parotid saliva can be easily obtained using a 'trombone-slide' gag to open the mouth and a disposible 20 ml syringe fitted with a semi-rigid probe long enough to reach the area between cheek and teeth. Uniform positioning of the probe is important, because the different salivary glands produce saliva differing in Na:K ratio. Contamination by recently consumed or regurgitated food can be minimized by mouth-washing. If saliva cannot be obtained, faecal sodium may be used, values < 1 g kg^{-1} DM and Na:K < 1.0 being suggested as indicative of deficiency (Little, 1987); the dependability of faecal sodium concentrations has, however, been questioned (Morris, 1980) and would be influenced by factors such as dry-matter intake (DMI) and digestibility. Urine sodium is more dependable, concentrations < 3 mmol Na l^{-1} indicating possible responses to supplementation (Morris, 1980), but values will be influenced by urine output and therefore differ between animals on lush or dry diets, unless creatinine is used as a marker. The most certain means of diag-

Table 7.2. Suggested marginal bands[a] for sodium and chloride in serum, urine and diet as a guide to the diagnosis of dietary deprivation (D) or excess (E)[d] of each element for livestock.

			Serum[b] ($mmol\ l^{-1}$)	Urine ($mmol\ l^{-1}$)	Diet ($g\ kg^{-1}$ DM)
Sodium	Cattle[c]	D	124–135	< 3	0.5–1.0
		E	140–150	40–60	30–60
	Sheep[c]	D	140–145	1–3	0.5–1.0
		E	150–160	40–60	30–60
	Pigs	D	137–140	1–3	0.3–1.0
		E	150–180	> 11	20–40
	Poultry	D	130–145	–	0.8–1.5
		E	150–160	–	50–80
Chloride	Cattle	D	70–85	2–5	1–3
		E	> 150	>100	>80
	Sheep	D	70–85	2–5	1–3
		E	> 150	>100	>50
	Pigs	D	88–100	–	0.9–2.4
		E	> 110	–	20–49
	Poultry	D	90–115	100	0.8–2.0
		E	> 174	–	6.0–12.0

[a] Mean values below (for D) or above (for E) the given bands indicate probability of ill health from deprivation or toxicity; values within bands indicate a possibility of future disorder.
[b] If serum is unobtainable, similar interpretations can be applied to concentrations in vitreous humour.
[c] Salivary ratios of Na:K are particularly useful for diagnosing insufficiency in cattle and sheep, median values of 4.0–10.0 being regarded as 'marginal'.
[d] Excesses are less well tolerated if drinking-water is restricted.

nosing sodium deficiency is by assessing the response in appetite, appearance and productivity of the animal which follows quickly when supplementary salt is supplied.

Prevention of Sodium Deficiency

Supplementary sodium is invariably supplied in practical husbandry conditions as common salt, because of its palatability, relatively low cost and ready availability. Unfortunately, neither the cost nor the availability of this supplement is always satisfactory in tropical, developing countries, where the need may be greatest. In pig and poultry feeding, it has been common to include salt with the whole diet at about 5 g kg^{-1} DM. Half this quantity is usually adequate (see Tables 7.3 and 7.4), especially where rations contain animal protein supplements. For growing or fattening beef cattle being hand-fed on staple diets, a similar procedure is recommended, the salt being added to provide a total of 0.8–1.0 g Na kg^{-1} DM. With lactating dairy cows, mixing salt with the concentrate ration, to give at least 1.5 g Na kg^{-1} DM in the total ration, is recommended. With milking cows and ewes at pasture, it is customary to rely on the voluntary consumption of loose or block salt made continuously available; this is usually sufficient, although individual consumption can be highly variable (Morris *et al.*, 1980; Fig. 3.4) and is sometimes greatly in excess of actual needs (Suttle and Brebner, 1995; see Chapter 3). The physical form of the salt affects voluntary consumption. Smith *et al.* (1953) found that lactating cows and heifers at pasture consumed significantly more loose than block salt and sheep show similar preferences (see Chapter 3). The lower intakes of block salt were still adequate to meet the sodium needs of lactation. Where nitrogenous fertilizers are normally applied to pastures to raise herbage yields, a single spring dressing of Chilean nitrate of soda at up to 400 kg ha^{-1} can maintain sodium above 1.5 g Na kg^{-1} DM, but repeated small dressings may be more sensible. The use of a low-grade fertilizer, such as kainite, containing 22% subsidiary sodium, has been recommended. Application of NaCl (Chiy and Phillips, 1993) or sodium nitrate ($NaNO_3$) (Chiy *et al.*, 1993a) as a fertilizer at 10–60 kg Na ha^{-1} $year^{-1}$ may be doubly beneficial, increasing the pasture yield from heavily fertilized *L. perenne* pastures (Chiy and Phillips, 1993). Where the problem is one of potassium-induced sodium deficiency in spring pasture, there is clearly a need to restrict and delay potassium inputs on the farm and to restrict the recycling of potassium via slurries. In the Netherlands, potassium concentrations in the region of 50–100 g kg^{-1} DM are now frequently found. Potassium inputs should be restricted to those necessary to optimize pasture yield, and these are usually associated with pasture potassium levels of 30 g K kg^{-1} DM.

Requirements for Sodium

The comparative abundance and cheapness of common salt in most areas and its low toxicity have encouraged the use of supplementary sodium and chloride in excess of minimum requirements. This is now changing as more data become available as the energy cost to the animal (Arieli *et al.*, 1989) and cost to the environment of providing excess salt become clearer and the cost of delivering salt to sodium-deficient areas in tropical countries becomes better appreciated. Sodium, like chloride and potassium, presents no problem of low or variable availability, as occurs with other macroelements. However, the estimated requirement will depend on the criteria of adequacy used. Morris and Peterson (1975) used salivary Na:K ratios to assess the adequacy of diets containing 0.2–2.3 g Na kg^{-1} DM for lactating ewes (Fig. 7.4) and concluded that the minimum level needed to maximize sodium status was 0.80–0.87 g Na kg^{-1} DM, only 37.5% of the National Research Council (NRC, 1968) requirement at the time. Even this provision was generous in terms of animal performance, because the unsupplemented diet, containing 0.2 g kg^{-1} DM, did not reduce appetite, ewe body weight or lamb growth rate. Tolerance of sodium depletion commencing in pregnancy was demonstrated by Michell *et al.* (1988). The ability of mammals to conserve sodium and

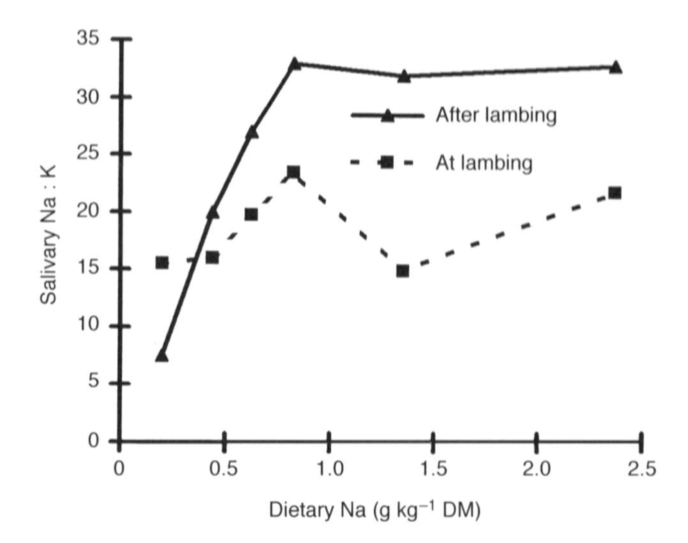

Fig. 7.4. Salivary Na:K reflects dietary sodium supply after (solid line) but not before (dashed line) lambing because of the higher sodium requirement for lactation than gestation and possibly a change in the set point for Na:K around parturition (data from Morris and Peterson, 1975).

draw upon pools in the skeleton and gastrointestinal tract (particularly in ruminants) means that requirements need not be met on a daily basis and are commonly overstated (Michell, 1989). However, substitution of potassium for sodium in the rumen on potassium-rich diets (i.e. most fertilized pastures) may increase sodium requirements and limit tolerance of deficiency. An argument for *increases* over NRC (1989) requirements for sodium (and potassium) at the onset of lactation in the dairy cow has recently been proposed (Silanikove *et al.*, 1997), to avoid the large negative balances which can occur. However, massive amounts of sodium (and potassium) were being excreted in urine at the time (1.0–2.6 g day^{-1}), suggesting either a failure of normal homeostatic mechanisms or – as seems more likely – unavoidable losses due to tissue catabolism.

Pigs

The experiments of Meyer *et al.* (1950) with weaner pigs fed on diets very low in sodium, potassium and chloride provided the first critical data on minimum requirements of these minerals. They found that for optimum growth the diet should contain 0.8–1.1 g Na kg^{-1} DM, a figure later confirmed for a conventional maize- and soybean-meal diet (Hagsten and Perry, 1976). An alternative approach to the definition of sodium and salt requirements for pigs was taken by the Agricultural Research Council (ARC, 1981). Using values of 1.0–1.5 g Na kg^{-1} weight gain as growth requirement, 56 g Na for total accumulation and hence need for pregnancy, a sodium content in sow's milk of 0.3–0.4 g kg^{-1}, obligatory endogenous losses of 1 mg kg^{-1} body weight (BW) and an efficiency of absorption of 90%, factorial estimates of sodium requirement were derived (Table 7.3). The prediction is that sodium requirements fall markedly with age, to values well below the flat rate currently recommended by NRC (1988) of 1 g Na kg^{-1} DM. Use of the lower rate would obviously decrease feed costs and counteract environmental concerns over water and salt accumulation in liquid manure pits. Excess sodium in the diet increases water consumption.

Poultry

The latest and most comprehensive national recommendations are summarized in Table 7.4. The requirements of sodium (and also of potassium and chloride) by growing birds of all types probably decline with the approach to mature body weight, but this is not always allowed for. Requirements have increased since early work showed that the needs of broiler chicks decreased from 1.3 to 0.7 g Na kg^{-1} DM between 1–3 and 7–9 weeks of age (Hurwitz *et al.*, 1974), while 0.5–0.6 g Na kg^{-1} DM was sufficient for mature, non-laying birds (Shaw and Phillips, 1953). Some recent work suggests an even higher sodium requirement for broiler chicks of 4–5 g kg^{-1} DM in the first week, declining to 3 g kg^{-1} DM by 3 weeks of age (Edwards, 1984; Britton, 1990); chloride requirements were also much higher (4–5 g kg^{-1} DM) early in life. Previous low estimates of sodium requirement were probably due in part to the presence of sodium in the drinking-water, but this cannot be relied upon. The average hen's egg contains 73 mg Na, 88 mg Cl and 82 mg K, but these additions to the daily requirement are more than met by increases in food intake, so that requirements for all three elements during lay are relatively low (Table 7.4). The use of high-energy diets greatly increases sodium and all mineral requirements when they are stated as dietary concentrations. A dietary deficiency of sodium in laying hens (0.5 g Na kg^{-1} DM) was aggravated by restriction of chlorine to 0.8 g kg^{-1} DM and was partly compensated for by increasing the dietary potassium from

Table 7.3. Sodium requirements of growing pigs (based on the lower of two estimates for sodium retention in growing tissues given by ARC (1981)).

Live weight (kg)	Growth rate (kg day^{-1})	Net requirement (g day^{-1})	Total requirement (g day^{-1})	DM intake (kg day^{-1})	Dietary Na concentration (g kg^{-1} DM)[a]
5	0.27	0.416	0.462	0.37	1.25
25	0.55	0.659	0.732	1.04	0.71
45	0.78	0.851	0.946	1.78	0.53
90	0.79	0.833	0.926	2.78	0.33

[a] These sodium requirements correspond to requirements for NaCl of 0.84–3.18 g kg^{-1} dietary DM.

Table 7.4. Dietary requirements (g kg^{-1} DM) of growing broiler (B) and Leghorn (L) chicks, turkey poults (P) and laying birds for sodium and chloride (from NRC, 1994).

	Bird type	Growth stage[a]			Lay stage[b]		
		Early	Middle	Late	High	Medium	Low
Sodium	B	2.0	1.5	1.2		–	
	L	1.5	1.5	1.5	1.9	1.5	1.3
	P	1.7	1.2	1.2		1.2	
Chloride	B	2.0	1.5	1.2		–	
	L	1.5	1.2	1.5	1.6	1.3	1.1
	P	1.5	1.4	1.2		1.2	

[a] Early = 0–3, 0–6 and 0–4 weeks; middle = 3–6, 6–12 and 8–12 weeks; late = 6–8 weeks, 18 weeks to first lay and 20–24 weeks of age for broilers, Leghorn chicks and poults, respectively.
[b] High, medium and low represent declining energy densities and food intakes of 80, 100 and 120 g day^{-1}, respectively.

7 to 12 g kg^{-1} DM (Sauveur and Mongin, 1978). On the basis of the estimated requirements given in Table 7.4, supplementation of conventional poultry rations with sodium is essential.

Sheep

The minimum sodium requirements of lambs for satisfactory growth and of lactating ewes for maintenance of body weight and milk production were estimated to be 1.0 g and 0.8 g Na kg^{-1} DM, respectively, in feeding trials (Hagsten *et al.*, 1975; Morris and Peterson, 1975). Merino wethers fed on high-grain diets needed more than 0.6 g Na kg^{-1} DM (McClymont *et al.*, 1957). Factorial estimates of sodium requirements by ARC (1980) were much higher (0.8–2.7 g kg^{-1} DM) but probably overestimated maintenance requirements (Michell, 1995), which were based on a single experiment, in which lambs were given a semipurified diet low in sodium, supplemented with sodium hydroxide (NaOH) but *without* added chloride (Devlin and Roberts, 1963). More realistic values for the obligatory losses of sodium in urine and faeces of lambs can be obtained from the data of Jones *et al.* (1967); values of 1 and 5 mg Na kg^{-1} LW for urine and faeces, respectively, were reported on a low-sodium hay which did not impair growth. Lower values of 0.3 and 1.0 mg kg^{-1} LW were recorded for urine and faecal sodium losses in healthy pregnant ewes (Michell *et al.*, 1988). Factorial estimates of sodium requirement (Table 7.5) based on these lower maintenance needs are in agreement with field experience, with needs generally < 1 g kg^{-1} DM and decreasing as productivity rises.

Cattle

The sodium requirements of beef and dairy cattle have been investigated in a wide range of conditions: in tropical and temperate climates, on good and heavily fertilized pastures and poor natural grazings and at low and high rates

Table 7.5. Estimates of the minimum dietary sodium concentrations required by sheep[a] and cattle at the given dry-matter intakes (DMI) (after ARC, 1980).

	Live weight (kg)	Growth or product	DMI (kg day^{-1})	Gross[b] Na requirement	
				g day^{-1}	g kg^{-1} DM
Lamb	20	0.1 kg day^{-1}	0.5	0.35	0.7
		0.2 kg day^{-1}	0.8	0.47	0.6
	40	0.1 kg day^{-1}	0.83	0.58	0.7
		0.2 kg day^{-1}	1.23	0.71	0.6
Ewe	75	0	0.8	1.0	1.25
Pregnant ewe	75	1 fetus – 12 weeks	1.03	1.0	1.0
		term	1.51	1.0	0.7
Lactating ewe	75	1 kg day^{-1}	1.48	1.2	0.81
		2 kg day^{-1}	2.18	1.8	0.83
		3 kg day^{-1}	2.90	2.2	0.76
Steer	200	0.5 kg day^{-1}	3.3	2.3	0.70
		1.0 kg day^{-1}	4.7	3.1	0.65
	400	0.5 kg day^{-1}	5.2	3.9	0.75
		1.0 kg day^{-1}	7.3	4.7	0.65
Cow (dry)	600	0	5.0	4.5	0.9
Cow (pregnant)	600	−12 weeks	5.8	4.3	0.9
		term	9.0	7.2	0.8
Cow (milking)	600	10 kg day^{-1}	9.4	10.3	1.1
		20 kg day^{-1}	14.0	16.8	1.2
		30 kg day^{-1}	18.8	22.6	1.2

[a] Values for sheep reduced by using a lower value for faecal endogenous loss of 5 as opposed to 20 mg Na kg^{-1} LW.
[b] Absorption coefficient = 0.91.

of growth and of milk production. All of these variables may influence requirements, so that single minimum dietary intakes are of limited use. For example, requirements increase with temperature and humidity. In unacclimatized cattle, the dribbling of saliva can result in daily losses of up to 1.4 g Na and 0.9 g Cl 100 kg^{-1} LW (Aitken, 1976). Field experiments indicate that reasonable weight gains can be expected in beef cattle on diets containing 0.6–0.8 g Na kg^{-1} DM (Morris and Gartner, 1971; Morris and Murphy, 1972), but lactating beef cows should be given slightly more (1.0 g Na kg^{-1} DM: Morris, 1980). Reductions in salivary Na:K after parturition in grazing beef cows confirmed the increased demand associated with lactation (Morris *et al.*, 1980). However, the provision of supplemental NaCl in block form did not improve calf weaning weight with pasture containing only 0.15–0.50 g Na kg^{-1} DM and the drinking-water only 10–25 mg Na l^{-1}. Substantially higher sodium requirements of 1.8 g Na kg^{-1} DM have been given for lactating dairy cows (NRC, 1989), but factorial estimates by ARC (1980) gave a maximum need which was 30% lower (Table 7.5). Although cow's milk averages about 0.5 g Na, 1.2 g Cl and 1.5 g K l^{-1}, the concentration of sodium required in the diet increases only slightly with a rise in milk yield, provided that there is an increase in feed consumption to meet demands for more energy and protein. Chiy *et al.* (1993a) have repeatedly observed increases in milk yield when the sodium concentration in grazed herbage was increased from an apparently adequate level (2 g kg^{-1} DM) to 5 g Na kg^{-1} DM by applying NaCl as a fertilizer. However, these reponses may be due to short-term effects on palatability or rumen physiology, rather than alleviation of deficiency. Sodium fertilization of a pasture can increase grazing and ruminating time and bite rate (Chiy *et al.*, 1993a). High estimates of the prevalence of sodium deficiency in dairy cows are produced when higher performance-related standards of requirement are used (Chiy and Phillips, 1995).

Goats

From a study of growth, reproduction and lactation of goats at two dietary concentrations of sodium (0.3 and 1.7 g kg^{-1} DM: Schellner, 1972/73), the lowest level was severely inadequate and the higher level appeared to be sufficient. No report on the chlorine and potassium requirements of goats is available.

Sodium and Salt Toxicities

Dietary excesses of osmotically active elements, such as sodium and chloride, can disturb body functions (e.g. induce oedema). Excesses of sodium and chloride are usually concurrent and they can occur under natural circumstances, coming from saline drinking-water or ingestion of plants growing on saline soils. Excesses can also arise from accidental or intentional human interventions (e.g. access to NaCl-containing fertilizers or mineral mixtures, failed irrigation schemes or alkali treatment of grain and roughage). Guidelines for the assessment of excesses in the diet and animal are given in Table 7.2.

Salt supplements

Daily bolus doses of 10.5 g NaCl kg^{-1} LW day^{-1} given via rumen cannulae to grazing lambs depressed growth within 4 weeks and, after 9 months, reductions of 26% in weight gain and 14% in clean wool yield were recorded (De Waal *et al.*, 1989). Protein supplementation reduced the adverse effects of NaCl dosage, raising the possibility that the mineral had reduced microbial synthesis of protein in the rumen. The practical question that arises from these findings is whether the intermittent intake of salt from free-access mixtures is ever sufficient to impair production.

Saline drinking-water

Water containing up to 5 g NaCl l^{-1} is safe for lactating cattle and up to 7 g l^{-1} is safe for non-lactating cattle and sheep (Shirley, 1978), but stock can adapt to concentrations considerably higher than these, at least in temperate climates. Where the winters were cool to mild and the pasture lush, sheep tolerated water containing 13 g NaCl l^{-1}, but with 20 g NaCl l^{-1}, feed consumption and body weight declined and some animals became weak and emaciated (Peirce, 1957, 1965). In high environmental temperatures and dry grazing conditions, the tolerable salt concentration is reduced, because of increased water consumption. Toxicity also varies with the chemical nature of the constituent salts. Sodium chloride appears to be the least harmful and magnesium salts are much more toxic than sodium salts, including sulphates and carbonates, in drinking-water. Both young and old pigs tolerate drinking-water containing 10 g NaCl l^{-1}, but 15 g l^{-1} is toxic (Heller, 1932). In experiments with laying hens given water containing 0, 2, 5 and 8 g NaCl l^{-1}, the highest concentration increased mortality and reduced egg production and body weight (Sherwood and Marion, 1975). Much higher salt intakes are tolerated when added to the diet if pure water is freely available, because the animal can compensate to some degree by increasing its intake of fresh water, thereby increasing the salt-excreting capacity of the kidneys, but when the water is itself rich in salt the animal is unable to adapt in this way. Adverse effects of saline drinking-water on the laying hen are of particular concern, because of the speed with which eggshell quality is reduced by relatively low concentrations (2 days' exposure to 0.2 g NaCl l^{-1}) and the residual effects of prolonged exposure (at least 15 weeks) (Balnave and Yoselewitz, 1985). The percentage of damaged shells increased linearly to a maximum of 8.9% as sodium concentrations increased up to 0.6 g l^{-1}, a level found in some of the local (New South Wales) underground sources.

Salt-tolerant plants

Saline soils support an unusual flora, made up of species that tolerate sodium by various means, involving avoidance (deep roots), exclusion (salt glands) and dilution (bladder cells, succulence) (Gorham, 1995). There is therefore no simple correlation between soil salinity and plant sodium concentrations. Species such as the bladder saltbush (*Atriplex vesicaria*) are essential to livestock production in many arid regions, but high ash concentrations (200–300 g kg^{-1} DM, mostly as salt) depress digestibility and lead to energy losses, associated with low digestibility and greatly increased sodium turnover (Arieli *et al.*, 1989). Tolerance of salt-rich plants and diets is highly dependent on free access to water low in salt and is high relative to the suggested limits for NaCl in the diet (15 g kg^{-1} DM for pigs, up to 25 g NaCl kg^{-1} DM for cattle) under favourable conditions (for a review, see Marai *et al.*, 1995). Toxicity is indicated clinically by anorexia and water retention and physiologically by intracellular dehydration, due to hypertonicity of the ECF, with sodium and chloride both contributing to disturbances. Control is only likely to be achieved by a variety of approaches, including the encouragement of deep-rooted (saline soil-avoiding) species, irrigation systems that limit upward migration of salt and the use of complementary feeds low in sodium (e.g. maize and lucerne).

Sodium hydroxide-treated feeds

Although the treatment of grains, straws and whole-crop silages with NaOH can improve nutrient utilization by ruminants in the short term, incidence of kidney lesions (nephritis) may increase and NH_4OH is a preferable agent.

CHLORINE

Deficiency of Chloride

While chloride deficiency is not believed to occur naturally, Coppock (1986) has questioned the assumption that sodium will always be the more limiting element and has identified diets based on maize (grain, gluten and silage) as presenting a risk of chloride deficiency. Chloride depletion might also occur in hot climates, since cattle exposed to 40°C for 7 h day^{-1} are estimated to lose slightly more chloride than sodium in sweat (1 g day^{-1} for a 200 kg steer and 1.69 g day^{-1} for a 500 kg cow: ARC, 1980). Increases in muscle chloride concentrations have been reported in 'downer cows' (see Chapter 4) and may reflect a perturbed 'chloride shift'.

Signs of Chloride Deprivation

Chloride deficiency has been produced experimentally by giving young calves a diet containing 0.63 g Cl kg^{-1} DM and removing their chloride-rich abomasal contents twice daily (Neathery *et al.*, 1981). The calves became anorexic and lethargic after 7 days, with mild polydipsia and polyuria. Severe eye defects developed after 24–46 days' depletion. Plasma chloride fell from 96 to 31 mmol l^{-1} and there was secondary alkalosis, with reductions in plasma sodium and potassium. Salivary chloride fell from 25 to 16 mmol l^{-1}. Control calves given a diet with 4.8 g Cl kg^{-1} DM grew normally, despite the daily removal of abomasal digesta. More recently, Fettman *et al.* (1984) induced symptoms of deficiency in lactating cows given a diet containing 1 g Cl kg^{-1} DM; they included pica, lowered milk yield, constipation and cardiovascular depression and were not seen with a diet containing 2.7 g Cl kg^{-1} DM. Chloride-deficient animals showed a craving for KCl as well as NaCl and the biochemical changes were similar to those found by Neathery *et al.* (1981). In chicks given a diet containing only 0.2 g Cl kg^{-1} DM, mortality increased and nervous symptoms were reported (Leach and Nesheim, 1963). Abomasal parasitism inhibits the normal secretion of HCl into the gut lumen (Coop, 1971), but the consequences of this in terms of acid:base balance and the pathogenesis of infection have not been explored. Acute challenge infections with *Haemonchus contortus* raise salivary and presumably serum chloride and perturb salivary Na:K ratios by increasing salivary potassium, indicating the possibility of acidosis and loss of intracellular potassium (N.F.

Suttle, unpublished data). Guidelines for the assessment of chloride status are given in Table 7.2.

Chloride Requirements

Chloride requirements are similar to those for sodium and a 1:1 ratio for Na:Cl is commonly proposed.

Cattle and sheep

Little is known of the chloride requirements of cattle for growth or milk production. The ARC (1980) estimates of the daily chloride requirement of beef cattle gaining 1.0 kg day^{-1} are equivalent to a dietary concentration of 0.7 g Cl kg^{-1} DM. They should be substantially higher than the 0.9–1.2 g kg^{-1} DM given for sodium in lactating dairy cows, because cow's milk contains more than twice as much chloride as sodium. Another unusual feature of chloride in relation to milk production is that the chloride concentration in milk, in contrast to most minerals, rises with advancing lactation. Lengemann *et al.* (1952) observed the following mean values for the milk of Holstein cows: first month, 1.16; fifth month, 1.29; tenth month of lactation, 1.90 g Cl l^{-1}. However, milk may represent a route of excretion for chloride that need not be sustained from the diet. The minimum chloride requirements of sheep have apparently not been studied experimentally, but this element does not present a problem in practical conditions.

Pigs and poultry

Earlier estimates for weaner pigs were that they required 1.2–1.3 g Cl kg^{-1} DM (Meyer *et al.*, 1950) and they have changed little. The latest chloride requirements for poultry (NRC, 1994; Table 7.4) are of a similar order to those of pigs and to those for sodium.

Chloride Toxicity

This is synonymous with salt toxicity and described on pp. 203–204.

References

Aines, P.D. and Smith, S.E. (1957) Sodium versus chloride for the therapy of salt-deficient dairy cows. *Journal of Dairy Science* 40, 682–688.

Aitken, F.C. (1976) *Sodium and Potassium in Nutrition of Mammals.* Technical Communication of the Commonwealth Bureau of Nutrition No. 26.

ARC (1980) *Nutrient Requirements of Ruminants.* Commonwealth Agricultural Bureaux, Farnham Royal, UK, pp. 213–216.

ARC (1981) *Nutrient Requirements of Pigs.* Commonwealth Agricultural Bureaux, Farnham Royal, UK, pp. 252–257.

Arieli, A., Naim, E., Benjamin, R.W. and Pasternak, D. (1989) The effect of feeding saltbush and sodium chloride on energy metabolism in sheep. *Animal Production* 49, 451–457.

Babcock, S.M. (1905) The addition of salt to the ration of dairy cows. In: *University of Wisconsin Experiment Station 22nd Annual Report*, p. 129.

Balnave, D. and Yoselewitz, I. (1985) The relation between sodium chloride concentration in drinking water and egg-shell damage. *British Journal of Nutrition* 58, 503–509.

Bell, F.R. (1995) Perception of sodium and sodium appetite in farm animals. In: Phillips, C.J.C. and Chiy, P.C. (eds) *Sodium in Agriculture.* Chalcombe Publications, Canterbury, UK, pp. 82–90.

Berger, L.L. (1990) Comparison of National Research Council feedstuff mineral composition data with values from commercial laboratories. In: *Proceedings of Georgia Nutrition Conference, Atlanta, 1990,* University of Georgia, Atlanta, pp. 54–62.

Black, H. (1997) Sea water in the treatment of inanition in sheep. *New Zealand Veterinary Journal* 45, 122.

Blair-West, J.R., Coghlan, J.P., Denton, D.A., Goding, J.R., Wintour, M. and Wright, R.D. (1963) The control of aldosterone secretion. In: Plincus, S. (ed.) *Recent Progress in Hormone Research.* Academic Press, New York, pp. 311–383.

Blair-West, J.R., Coghlan, J.P., Denton, D.A., Nelson, J.F., Orchard, E., Scoggins, B.A., Wright, R.D., Myers, K. and Junqueira, C.L. (1968) Physiological, morphological and behavioural adaptation to a sodium deficient environment by wild native Australian and introduced species of animals. *Nature, UK* 217, 922–928.

Block, E. (1994) Manipulation of dietary cation–anion difference on nutritionally related production diseases, productivity and metabolic responses in dairy cows. *Journal of Dairy Science* 77, 1437–1450.

Bott, E., Denton, D.A., Goding, J.R. and Sabine, J.R. (1964) Sodium deficiency and corticosteroid secretion in cattle. *Nature UK* 202, 461–463.

Boussingault, J.B. (1847) *Comptes Rendus des Séances de l'Académie des Sciences* 25, 729. Cited by McCollum, E.V. (1957) *A History of Nutrition.* Houghton Mifflin, Boston, Massachusetts.

Britton, W.M. (1990) Dietary sodium and chlorine for maximum broiler growth. In: *Proceedings of Georgia Nutrition Conference, Atlanta,* University of Georgia, Atlanta, pp. 152–157.

Chamberlain, G.T. (1955) The major and trace element composition of some East African feedingstuffs. *East African Agricultural Journal* 21, 103–107.

Chiy, P.C. and Phillips, C.J.C. (1993) Sodium fertilizer application to pasture. 1. Direct and residual effects on pasture production and composition. *Grass and Forage Science* 48, 189–202.

Chiy, P.C. and Phillips, C.J.C. (1995) Sodium in forage crops. Sodium fertilization. In: Phillips, C.J.C. and Chiy, P.C. (eds) *Sodium in Agriculture.* Chalcombe Publications, Canterbury, UK, pp. 43–81.

Chiy, P.C., Phillips, C.J.C. and Bello, M.R. (1993) Sodium fertilizer application to pasture. 2. Effects on dairy cow production and behaviour. *Grass and Forage Science* 48, 203–212.

Coop, R.L. (1971) The effect of large doses of *Haemonchus contortus* on the level of plasma pepsinogen and the concentration of electrolytes in the abomasal fluid of sheep. *Journal of Comparative Pathology* 81, 213–219.

Coppock, C.E. (1986) Mineral utilisation by the dairy cow – chlorine. *Journal of Dairy Science* 69, 595–603.

Cornforth, I.S., Stephen, R.C., Barry, T.N. and Baird, G.A. (1978) Mineral content of swedes, turnips and kale. *New Zealand Journal of Experimental Agriculture* 6, 151–156.

Devlin, T.J. and Roberts, W.K. (1963) Dietary maintenance requirement of sodium for wether lambs. *Journal of Animal Science* 22, 648–653.

Denton, D.A. (1982) *The Hunger for Salt: an Anthropological, Physiological and Medical Analysis.* Springer-Verlag, Berlin.

Denton, D.A., Goding, J.R., Wintour, M. and Wright, R.D. (1961) Adaptation of ruminant animals to variation of salt intake. In: *Proceedings of Teheran Symposium.* Arid Zone Research XIV, UNESCO, p. 3.

De Waal, H.O., Baard, M.A. and Engels, E.A.N. (1989) Effects of sodium chloride on sheep. 1. Diet composition, body mass changes and wool production in young Merino wethers grazing mature pasture. *South African Journal of Animal Science* 19, 27–42.

Edwards, H.M. (1984) Studies on the aetiology of tibial dyschondroplasia in chickens. *Journal of Nutrition* 114, 1001–1013.

Elbrond, V.S., Danzer, V., Mayhew, T.M. and Skadhouge, E. (1991) Avian lower intestine adapts to dietary salt (NaCl) depletion by increasing transepitheal sodium transport and microvillus membrane surface area. *Experimental Physiology* 76, 733–744.

Fettman, M.J., Chase, L.E., Bentinck-Smith, C., Coppock, E. and Zinn, S.A. (1984) Nutritional chloride deficiency in early lactation Holstein cows. *Journal of Dairy Science* 67, 2321–2335.

Gill, M., France, J., Summers, M., McBride, B.W. and Milligan, L.P. (1989) Simulation of the energy costs associated with protein turnover and Na^+K^+-transport in growing lambs. *Journal of Nutrition* 119, 1287–1299.

Goatcher, W.D. and Church, D.C. (1970) Taste responses in ruminants. III. Reactions of pigmy goats, normal goats, sheep and cattle to sucrose and sodium chloride. *Journal of Animal Science* 31, 364–372.

Gorham, J. (1995) Sodium content of agricultural crops. In: Phillips, C.J.C. and Chiy, P.C. (eds) *Sodium in Agriculture.* Chalcombe Publications, Canterbury, UK, pp. 17–32.

Grovum, W.L. and Chapman, H.W. (1988) Factors affecting the voluntary intake of food by sheep. 4. The effect of additives representing the primary tastes on sham intakes by oesophageally fistulated sheep. *British Journal of Nutrition* 59, 63–72.

Hagsten, I. and Perry, T.W. (1976) Evaluation of dietary salt levels for swine. 1. Effect on gain, water consumption and efficiency of feed conversion. *Journal of Animal Science* 42, 1187–1190.

Hagsten, I., Perry, T.W. and Outhouse, J.B. (1975) Salt requirements of lambs. *Journal of Animal Science* 42, 1187–1190.

Harper, M.-E., Willis, J.S. and Patrick, J. (1997) Sodium and chloride in nutrition. In: O'Dell, B.L. and Sunde, R.A. (eds) *Handbook of Nutritionally Essential Mineral Elements.* Marcel Dekker, New York, pp. 93–116.

Harris, J., Caple, I.W. and Moate, P.J. (1986) Relationships between mineral homeostasis and fertility of dairy cows grazing improved pastures. In: *Proceedings of the Sixth International Conference on Production Disease in Farm Animals, Belfast.* Veterinary Research Laboratory, Stormont, pp. 315–318.

Heller, V.G. (1932) Saline and alkaline drinking waters. *Journal of Nutrition* 5, 421–429.

Hemsley, J.A., Hogan, J.P. and Weston, R.H. (1975) Effect of high intake of sodium chloride on the utilisation of a protein concentrate by sheep. *Australian Journal of Agricultural Research* 26, 715–727.

Henry, P.R. (1995) Sodium and chlorine bioavailability. In: Ammerman, C.B., Baker, D.H. and Lewis, A.J. (eds) *Bioavailability of Nutrients for Animals.* Academic Press, New York, pp. 337–348.

Howard, D.A., Burdin, M.L. and Lampkin, G.H. (1962) Variation in the mineral and crude-protein content of pastures at Muguga in the Kenya Highlands. *Journal of Agricultural Science, Cambridge* 59, 251–256.

Hurwitz, S., Cohen, I., Bar, A. and Minkov, U. (1974) Sodium and chloride requirements of the 7–9 week-old broiler. *Poultry Science* 53, 326–331.

Jones, D.I.H., Miles, D.G. and Sinclair, K.B. (1967) Some effects of feeding sheep on low-sodium hay with and without sodium supplement. *British Journal of Nutrition* 21, 391–397.

Joyce, J.P. and Brunswick, I.C.F. (1975) Sodium supplementation of sheep and cattle. *New Zealand Journal of Experimental Agriculture* 3, 299–304.

Kemp, A. (1971) *The Effects of K and N Dressings on the Mineral Supply of Grazing Animals.* Proceedings of the 1st Colloquium of the Potassium Institute IBS, Wageningen, pp. 1–14.

Leach, R.M. and Nesheim, McC. (1963) Studies on chloride deficiency in chicks. *Journal of Nutrition* 81, 193–199.

Lehr, J.J., Grashuis, J. and Van Koetsveld, E.E. (1963) Effect of fertilization on mineral–element balance in grassland. *Netherlands Journal of Agricultural Science* 11, 23–37.

Lengemann, F.W., Aines, P.D. and Smith, S.E. (1952) The normal chloride concentration of blood plasma, milk and urine of dairy cattle. *Cornell Veterinarian* 42, 28–35.

Little, D.A. (1987) The influence of sodium supplementation on the voluntary intake and digestibility of low-sodium *Setaria sphacelatae* cv Nandi by cattle. *Journal of Agricultural Science, Cambridge* 108, 231–236.

Loosli, J.K. (1978) Sodium and chlorine requirements of ruminants. In: Conrad, J.H. and McDowell, L.R. (eds) *Latin American Symposium on Mineral Nutrition Research with Grazing Ruminants.* Animal Science Department, University of Florida, Gainsville, pp. 54–58.

McClymont, G.L., Wynne, K.N., Briggs, P.K. and Franklin, M.C. (1957) Sodium chloride supplementation of high-grain diets for fattening Merino sheep. *Australian Journal of Agricultural Research* 8, 83–90.

McDowell, L.R. (1992) *Minerals in Human and Animal Nutrition.* Academic Press, New York, pp. 496–511.

McNaught, K.J. (1959) Effect of potassium fertilizer on sodium, magnesium, and calcium in plant tissues. *New Zealand Journal of Agriculture* 99, 442.

McSweeney, C.S., Cross, R.B., Wholohan, B.T. and Murphy, M.R. (1988) Diagnosis of sodium status in small ruminants. *Australian Journal of Agricultural Research* 39, 935–942.

MAFF (1990) *UK Tables of the Nutritive Value and Chemical Composition of Feedstuffs.* Edited by D.I. Givens. Rowett Research Services, Aberdeen, UK.

Marai, I.F.M., Habeeb, A.A. and Kamal, T.H. (1995) Response of livestock to excess sodium intake. In: Phillips, C.J.C. and Chiy, P.C. (eds) *Sodium in Agriculture.* Chalcombe Publications, Canterbury, pp. 173–180.

Meyer, J.H., Grummer, R.H., Phillips, R.H. and Bohstedt, G. (1950) Sodium, chlorine, and potassium requirements of growing pigs. *Journal of Animal Science* 9, 300–306.

Michell, A.R. (1974) Body fluids and diarrhoea: dynamics of dysfunction. *Veterinary Record* 94, 311–315.
Michell, A.R. (1985) Sodium in health and disease: a comparative review with emphasis on herbivores. *Veterinary Record* 116, 653–657.
Michell, A.R. (1989) Practice tip: oral and parenteral rehydration therapy. *In Practice*, May, 96–99.
Michell, A.R. (1995) Physiological roles for sodium in mammals. In: Phillips, C.J.C. and Chiy, P.C. (eds) *Sodium in Agriculture*. Chalcombe Publications, Canterbury, UK, pp. 91–106.
Michell, A.R., Moss, P., Hill, R., Vincent, I.C. and Noakes, D.E. (1988) The effect of pregnancy and sodium intake on water and electrolyte balance in sheep. *British Veterinary Journal* 144, 147–157.
Milligan, L.P. and Summers, M. (1986) The biological basis of maintenance and its relevance to assessing responses to nutrients. *Proceedings of the Nutrition Society* 45, 185–193.
Minson, D.J. (1990) *Forage in Ruminant Nutrition*. Academic Press, New York, pp. 291–308.
Mongin, P. (1981) Recent advances in dietary anion–cation balance applications in poultry. *Proceedings of the Nutrition Society* 40, 285–294.
Morris, J.G. (1980) Assessment of sodium requirements of grazing cattle: a review. *Journal of Animal Science* 50, 145–151.
Morris, J.G. and Gartner, R.J.W. (1971) The sodium requirements of growing steers given an all-sorghum grain ration. *British Journal of Nutrition* 25, 191–205.
Morris, J.G. and Murphy, G.W. (1972) The sodium requirements of beef calves for growth. *Journal of Agricultural Science, Cambridge* 78, 105–108.
Morris, J.G. and Peterson, R.G. (1975) Sodium requirements of lactating ewes. *Journal of Nutrition* 105, 595–598.
Morris, J.G., Delmas, R.E. and Hull, J.L. (1980) Salt supplementation of range beef cows in California. *Journal of Animal Science* 51, 71–73.
Moseley, G. (1980) Effects of variation in herbage sodium levels and salt supplementation on the nutritive value of perennial ryegrass for sheep. *Grass and Forage Science* 35, 105–113.
Murphy, G.M. and Connell, J.A. (1970) A simple method of collecting saliva to determine the sodium status of cattle and sheep. *Australian Veterinary Journal* 46, 595–598.
Murphy, G.M. and Plasto, A.W. (1972) Sodium deficiency in a beef cattle herd. *Australian Veterinary Journal* 48, 129.
Neathery, M.W., Blackmon, D.M., Miller, W.J., Heinmiller, S., McGuire, S., Tarabula, J.M., Gentry, R.F. and Allen, J.C. (1981) Chloride deficiency in Holstein calves from a low chloride diet and removal of abomasal contents. *Journal of Dairy Science* 64, 2220–2233.
NRC (1968) *Nutrient Requirements of Sheep*, 3rd edn. National Academy of Sciences, Washington, DC.
NRC (1988) *Nutrient Requirements of Swine*, 9th edn. National Academy of Sciences, Washington, DC.
NRC (1989) *Nutrient Requirements of Dairy Cattle*, 6th edn. National Academy of Sciences, Washington, DC.
NRC (1994) *Nutrient Requirements of Poultry*, 9th edn. National Academy of Sciences, Washington, DC.

Pasternak, D., Danon, A., Arouson, J.A. and Benjamin, R.W. (1985) Developing the seawater agriculture concept. *Plant and Soil* 89, 337–348.
Peirce, A.W. (1957) Saline content of drinking water for livestock. *Veterinary Reviews and Annotations* 3, 37–43.
Peirce, A.W. (1965) Studies on salt tolerance of sheep. 5. The tolerance of sheep for mixtures of sodium chloride, sodium carbonate and sodium bicarbonate in the drinking water. *Australian Journal of Agricultural Research* 49, 815–823.
Rahman, H., McDonald, P. and Simpson, K. (1960) Effects of nitrogen and potassium fertilizers on the mineral status of perennial ryegrass (*Lolium perenne*). 2. Anion–cation relationships. *Journal of the Science of Food and Agriculture* 11, 429–432.
Ross, E. (1979) The effect of water sodium on the chick requirement for dietary sodium. *Poultry Science* 58, 626–630.
Ross, E. and Herrick, R.B. (1981) Forced rest induced by moult or low-salt diet and subsequent hen performance. *Poultry Science* 60, 63–67.
Sauveur, B. and Mongin, P. (1978) Interrelationships between dietary concentrations of sodium, potassium and chloride in laying hens. *British Poultry Science* 19, 475–485.
Schellner, G. (1972/73) Die Wirkung von Natriummangel und Natriumbeifutterung auf Wachstum, Milch- und Milchfettleistung und Fruchtbarkeit bei Ziegen. *Jahrbuch für Tierernahrung und Futterung* 8, 246–259.
Schellner, G., Anke, M., Ludke, H. and Henning, A. (1971) Die Abhangigkeit oer Milchleistung und Milchzusammensetzung von der Natriumversorgung. *Archiv für Experimentelle Veterinarmedizin* 25, 823–827.
Seynaeve, M., de Widle, R., Janssens, G. and de Smet, B. (1996) The influence of dietary salt level on water consumption, farrowing and reproductive performance of lactating sows. *Journal of Animal Science* 74, 1047–1055.
Shaw, R.K. and Phillips, P.H. (1953) *Lancet* 73, 176.
Sherrell, C.G. (1978) A note on sodium concentrations in New Zealand pasture species. *New Zealand Journal of Experimental Agriculture* 6, 189–190.
Sherwood, D.H. and Marion, J.E. (1975) Salt levels in feed and water for laying chickens. *Poultry Science* 54, 1816 (abstract).
Shirley, R.L. (1978) Water as a source of minerals. In: Conrad, J.H. and McDowell, L.R. (eds) *Latin American Symposium on Mineral Nutrition Research with Grazing Ruminants.* Animal Science Department, University of Florida, Gainsville, pp. 40–47.
Silanikove, N., Malz, E., Halevi, A. and Shinder, D. (1997) Metabolism of water, sodium, potassium and chloride by high yielding dairy cows at the onset of lactation. *Journal of Dairy Science* 80, 949–956.
Smith, G.S. and Middleton, K.R. (1978a) A classification of pasture and fodder species according to their ability to translocate sodium from their roots to aerial parts. *New Zealand Journal of Experimental Agriculture* 6, 183–188.
Smith, G.S. and Middleton, K.R. (1978b) Sodium and potassium content of topdressed pastures in New Zealand in relation to plant and animal nutrition. *New Zealand Journal of Experimental Agriculture* 6, 217–225.
Smith, S.E., Lengemann, F.W. and Reid, J.T. (1953) Block vs. loose salt consumption by dairy cattle. *Journal of Dairy Science* 36, 762–765.
Summers, J.D., Moran, E.T. and Pepper, W.F. (1967) A chloride deficiency in a practical diet encountered as a result of using common sulphate antibiotic potentiating procedure. *Poultry Science* 46, 1557–1560.

Suttle, N.F. and Brebner, J. (1995) A putative role for larval nematode infection in diarrhoeas which did not respond to anthelmintic drenches. *Veterinary Record* 137, 311–316.

Suttle, N.F. and Field, A.C. (1967) Studies on magnesium in ruminant nutrition. Effect of increased intakes of potassium and water on the metabolism of magnesium, phosphorus, sodium, potassium and calcium in sheep. *British Journal of Nutrition* 21, 819–831.

Suttle, N.F., Brebner, J., McClean, K. and Hoeggel, U. (1996) Failure of mineral supplementation to avert apparent sodium deficiency in lambs with abomasal parasitism. *Animal Science* 63, 103–109.

Thomas, B., Thompson, A., Oyenuga, V.A. and Armstrong, R.H. (1952) The ash constituents of some herbage plants at different stages of maturity. *Empire Journal of Experimental Agriculture* 20, 10–22.

Towers, N.R. and Smith, G.S. (1983) Sodium (Na). In: Grace, N.D. (ed.) *The Mineral Requirements of Grazing Ruminants.* Occasional Publication No. 9, New Zealand Society of Animal Production, Palmerston North, pp. 115–124.

Walker, C.A. (1957) Studies of the cattle of Northern Rhodesia. 1. The growth of steers under normal veld grazing and supplemented with salt and protein. *Journal of Agricultural Science, Cambridge* 49, 394–400.

Whitlock, R.H., Kessler, M.J. and Tasker, J.B. (1975) Salt (sodium) deficiency in dairy cattle: polyuria and polydypsia as prominent clinical features. *Cornell Veterinarian* 65, 512–526.

Wilson, A.D. (1966) The tolerence of sheep to sodium chloride in food or drinking water. *Australian Journal of Agricultural Research* 17, 503–514.

Potassium

Introduction

The essentiality of potassium in animal diets was demonstrated in experimental studies early in the 19th century. Naturally occurring potassium deficiency in livestock is rare, although potassium has long been known to be vitally involved in nerve and muscle excitability and in the water and acid–base balance of the body (Ward, 1966). The element is so abundant in common rations and pastures that, in the words of Thompson (1972), 'nutritionists have generally regarded potassium as a useful but non-critical nutrient'. The picture has changed slightly in recent years (Preston and Linser, 1985), but the nutritional problems presented by potassium commonly involve excess rather than deficiencies.

Dietary Sources of Potassium

Forages

Few natural foodstuffs show the range of mineral concentrations exhibited by forage potassium (K). In the UK, the range of 1.0–6.0 about a mean of 3.1 g K kg^{-1} dry matter (DM) most recently reported for fresh herbage (MAFF, 1992) is higher than an earlier report (Table 8.1). Similar variation is found in grass conserved as silage in the Netherlands (Fig. 8.1; Schonewille *et al.*, 1997). Potassium concentrations are influenced by the potassium status of the soil, the plant species and its state of maturity and the way the sward is managed. Cool-season grass species (e.g. *Lolium perenne*) maintain higher potassium concentrations than warm-season species (e.g. Robinson, 1985) and tropical legumes have lower levels than temperate legumes (Lanyon and Smith, 1985; Table 8.2). For a given sward, potassium concentrations can decrease markedly as the season progresses (Reid *et al.*, 1984), although not all species show the same effect (Fig. 8.2; Perdomo *et al.*, 1977). For a given sward at a given time, the potassium level attained will be determined by the potassium status of the soil.

© CAB *International* 1999. *Mineral Nutrition of Livestock*
(E.J. Underwood and N.F. Suttle)

Table 8.1. Mean (SD) potassium concentrations (g kg^{-1} DM) in pastures and foodstuffs in the UK (mostly from MAFF, 1990).

Forages		Concentrates		Roots and by-products	
Herbage	24.3 (6.6)	Fish-meal	10.2 (1.3)	Fodder beet	17.5 (4.8)
Grass silage	25.8 (6.8)	Soybean meal (ext)	25.0 (1.0)	Swedes	28.6[a]
Clover silage (mixed)	27.4 (7.6)	Safflower meal	17.1 (1.8)	Turnips	37.8[a]
Lucerne (alfalfa) silage	24.6 (4.0)	Palm-kernel cake	6.9 (1.2)	Cassava meal	8.1 (1.5)
Maize silage	12.3 (4.1)	Maize gluten	12.5 (2.7)		
Grass hay	20.7 (5.3)	Rapeseed meal	14.3 (2.2)	Meat and bone-meal	5.2 (0.6)
Lucerne (alfalfa) hay	27.3 (5.0)	Cottonseed meal	15.8 (0.8)	Molasses (beet)	49.1 (5.4)
Dried grass	26.0 (8.0)	Linseed meal	11.2 (0.13)	Molasses (cane)	38.6 (15.7)
Dried lucerne (alfalfa)	25.4 (8.3)			Molasses beet pulp	18.2 (1.9)
				Beet pulp	11.7 (7.6)
Barley straw	16.0 (6.5)	Maize	3.5 (0.22)	Brewer's grains	0.4 (–)[b]
Wheat straw	10.2 (3.7)	Barley	5.0 (0.7)	Wheat feed	13.0 (2.2)
Alkali-treated barley straw	11.6 (5.3)	Wheat	4.6 (0.4)	Distiller's dark grains: barley	10.2 (0.8)
		Oats	5.0 (0.9)		
		Sorghum	3.2 (0.1)		

[a] Data from New Zealand (Cornforth *et al.*, 1978).
[b] Extreme value of 4.5 omitted.

All pastoral systems require inputs of potassium in artificial fertilizers to maintain a potassium balance. When the pasture is grazed, extensive recycling of potassium via urination and slurry dispersal can lead to minimal fertilizer requirements. Conservation of the pasture, however, changes the picture completely, necessitating major inputs of fertilizer potassium. The best practice is to use, whenever possible, mixtures of nitrogen and phosphorus without potassium for the first fertilizer application in spring, when herbage potassium is maximal, and to restrict subsequent applications to the minimum needed to reach the critical concentration for maximum plant yield (Table 8.2).

Other feeds

Most feeds show less variation in potassium concentration than forage, but there is considerable overlap between types (Table 8.1). Energy sources are generally low (3–5 g K kg^{-1} DM) and brewer's grains exceptionally low (0.4 g K kg^{-1} DM) in potassium. Protein sources generally contain 10–20 g K kg^{-1} DM, and it is hard to compound a balanced ration from the ingredients given in Table 8.1 without meeting all the requirements of ruminants and non-ruminants. Over-reliance on cereals, cereal by-products and non-protein (also non-potassium) nitrogen sources might, however, create a potassium-deficient diet if alkali-treated straw is used as the source of roughage; treatment with ammonia or sodium hydroxide (NaOH) appears to reduce potassium concentrations by about 25% (Table 8.1). At the other extreme, root crops, molasses, molasses by-products (beet pulp, bagasse), cottonseed meal and soybean meal can add significant amounts of potassium

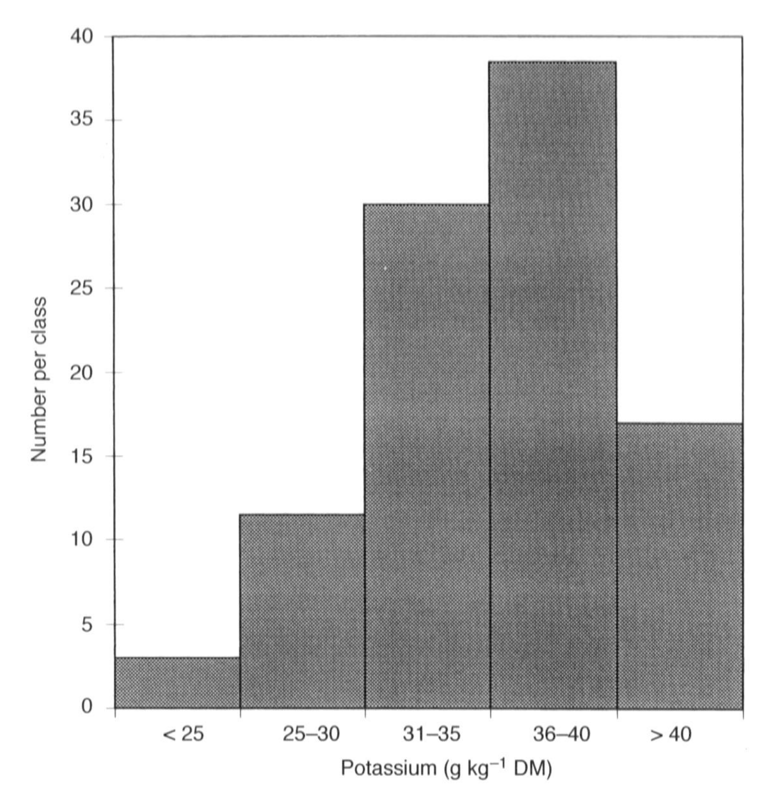

Fig. 8.1. Frequency distribution of potassium concentrations in silages in the Netherlands: the median value exceeds that needed for optimum silage yield and reflects excessive inputs of K from artificial fertilizer and recycled excreta (data from Schonewille *et al.*, 1997).

Table 8.2. Critical concentrations for potassium (g kg^{-1} DM) below which plant yield may be reduced in forage species (from Lanyon and Smith, 1985; Robinson, 1985).

Climate	Grass		Legume	
Cool temperate	*Phleum pratensis*	16–20	*Trifolium repens*	10
	Dactylis glomerata	23–35	*Medicago sativa*	12
	Lolium multiflorum	26–30	*Trifolium fragiferum*	10
	Lolium perenne	26–30		
	Festuca elatior	25–28		
Tropical	*Digitaria decumbens*	12–14	*Stylosanthes humilis*	6.0
	Sorghum bilcolor sudanensis	15–18	*Centrosema brasilianum*	11.2[a]
	Cynodon dactylon	15–17	*Lotononis bainsii*	9.0
			Lotononis capitata	9.8–12.2[a]
			Centrosema pubescens	7.5[a]

[a] In the dry season, critical values are < 7.

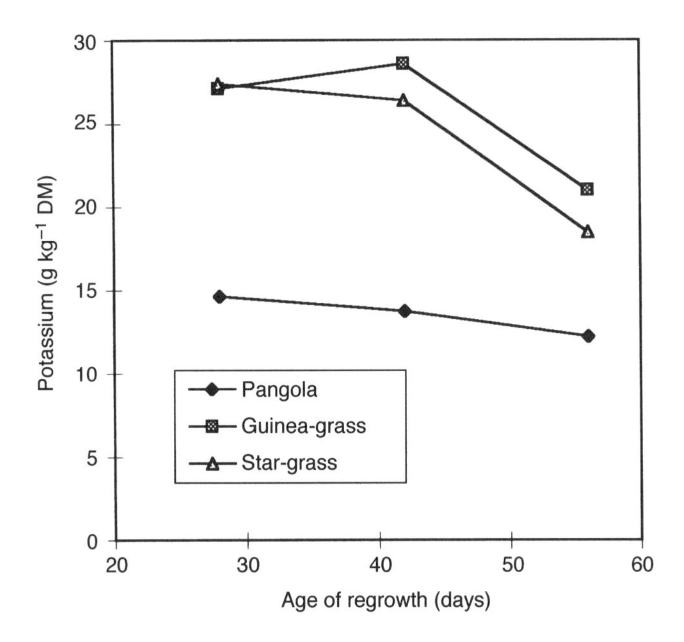

Fig. 8.2. Potassium concentrations in herbage decline as the sward matures (data for tropical grasses from Perdomo *et al.*, 1977).

to supplements for dairy cows and should be avoided where there is difficulty in restricting potassium accumulation in the pasture.

Absorbability

Dietary sources of potassium are highly soluble and almost completely absorbed. Availability can be measured from urinary responses to dietary supplements (Combs *et al.*, 1975), and one study estimated that potassium, given as maize and soybean meal, was 90–95% and 97%, respectively, as available to young pigs as potassium acetate (Combs and Miller, 1985). Apparent absorption of potassium in four tropical grasses cut at three stages of regrowth varied little about a high mean of 86% (Perdomo *et al.*, 1977).

Metabolism of Potassium

Absorption

Potassium absorption occurs principally in the small intestine in non-ruminants by unregulated processes, but in the ruminant extensive absorption occurs from the rumen. Entry into the bloodstream occurs largely via conductance channels in the basolateral membrane of the gut mucosa.

Membrane transport

There are more mechanisms for transporting potassium across membranes than there are for any other element, reflecting the difficulty but essentiality of maintaining high intracellular concentrations of potassium. In addition to the familiar sodium (Na^+)K^+ ATPase pump and cotransporters (see Chapter 7), there are hydrogen (H^+)K^+ ATPases and six types of potassium channel, each distinctively regulated (Peterson, 1997). Short-term adjustments to fluctuating potassium supply can be made through changes in the net flux of potassium into cells, under the influence of insulin (Lindeman and Pederson, 1983). Depletion of cell potassium may be partially offset by uptake of H^+, but only at the expense of intracellular acidosis.

Excretion

The major problem faced by ruminants in dealing with potassium relates to excess, rather than deficiency. Adaptation to potassium loading is believed to involve splanchnic sensors, which provide early warning of the ingestion of potentially lethal amounts (Rabinowitz, 1988). Response to the sensors involves an increase in Na^+K^+ ATPase activity and the number of pumps in the basolateral membrane of both the distal renal tubule and colon, leading to increased potassium excretion by both the urinary and faecal routes (Hayslett and Binder, 1982). Aldosterone modulates the renal but not the colonic response. The need for regulation lies in the cytotoxicity of high circulatory levels of potassium, values > 6 mmol l^{-1} serum being sufficient to cause heart failure. Regulation of body potassium status is achieved principally by the kidney, where tubular reabsorption is restricted during overload under the influence of aldosterone (Kem and Trachewsky, 1983; see also Chapter 7). Michell (1995) has suggested that aldosterone may often be more important to the grazing ruminant by controlling potassium excess than by the more widely studied alleviation of sodium deficiency. However, the mechanisms are not as effective as those for sodium, and plasma potassium rises in response to increases in dietary potassium (Combs *et al.*, 1985). Early suggestions that rises in plasma potassium increased sodium appetite were later questioned (Michell, 1978).

Secretion

In ruminants, potassium is the major cation in sweat, due possibly to the high K:Na ratio in their natural diet, grass (Bell, 1995); losses increase with environmental temperature and are greater in *Bos indicus* than in *Bos taurus* (Johnson, 1970) at a given temperature, despite a lower sweating rate. Potassium is also the major cation secreted in milk (36 mmol l^{-1} in sheep and cattle); concentrations do not reflect high dietary potassium intake (Sasser *et al.*, 1966), but they decline slightly during severe potassium deprivation (Pradhan and Hemken, 1969).

Functions of Potassium

Potassium is the major intracellular ion in tissues and is usually present at concentrations of 100–160 mmol l^{-1}, which are 25–30 times greater than those in plasma (Ward, 1966). The established gradients create an electrical potential, which is essential for the maintenance of responsiveness to stimuli and muscle tone. Potassium inevitably contributes to the regulation of the acid–base balance and participates in respiration, via the chloride shift (see Chapter 7). All soft tissues are much richer in potassium than in sodium, making potassium the third most abundant mineral in the body, at about 3.0 g kg^{-1} live weight (LW), ahead of sodium, at 1.2 g kg^{-1} LW (Fig. 8.3). The highest potassium concentrations are found in muscle (*c.* 4 g K kg^{-1}: ARC, 1980) and it is possible to estimate lean body mass by measuring the activity of the naturally occurring potassium radioisotope, ^{40}K, in a whole-body monitor (Ward, 1966). Many enzymes have specific or facilitative requirements for potassium and the element influences many intracellular reactions involving phosphate with effects on enzyme activities and muscle contraction (Ussing, 1960; Thompson, 1972).

Signs of Potassium Deprivation

The clinical and biochemical manifestations of potassium deprivation in farm animals are not as well documented as they are for sodium, a reflection of

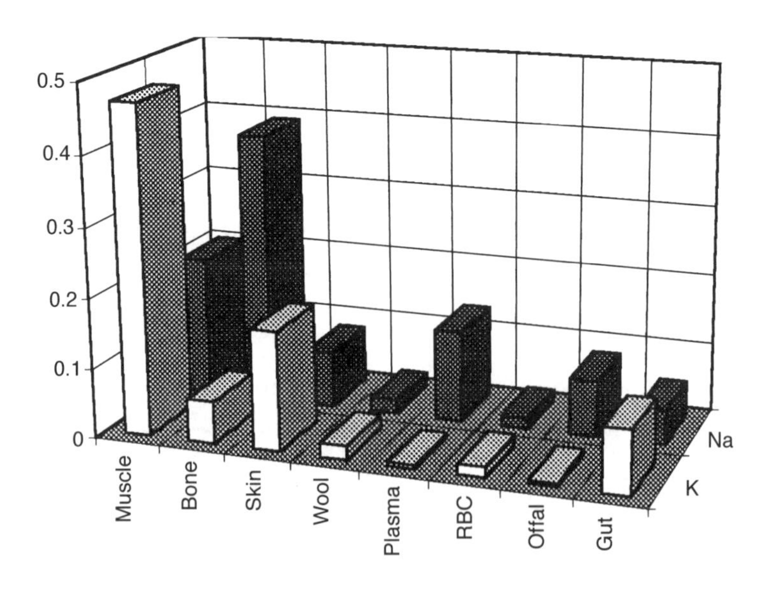

Fig. 8.3. Contrasting distributions of sodium and potassium in the unshorn, empty body of young sheep (data from Grace, 1983).

the ample supplies of this mineral in most foodstuffs. Reduced appetite was one of the first signs shown when growing pigs (Hughes and Ittner, 1942) and feeder lambs (Campbell and Roberts, 1965) were given semipurified, low-potassium diets. Poor growth, muscular weakness, stiffness, paralysis and intracellular acidosis have also been reported. These disturbances are to be expected, as the animal body contains virtually no reserve of potassium. It seems that the poor growth derives primarily from inappetence and partly from an impairment of protein metabolism, rather than from any adverse effect on digestibility. When lactating dairy cows were given diets, based on beet pulp, brewers' grains and maize, containing only 0.6 or 1.5 g K kg^{-1} DM, severe anorexia developed within 4 weeks (Pradhan and Hemken, 1969). Milk yield fell and pica, loss of coat condition and decreased pliability of the hide were evident when comparisons were made with periods when the same cows had been given 8 g K kg^{-1} DM. Potassium deprivation led to a greater reduction in milk potassium than in plasma potassium and there was a small compensatory rise in milk sodium. Haematocrit increased from 35.3 to 38.1 on the low-potassium diets. In a study involving less severe depletion, improved recovery of body weight (+92.9 vs. +15.7 kg) was observed in dairy cows after peak lactation when the diet contained 6.6 rather than 4.5 g K kg^{-1} DM, but there was no accompanying improvement in milk yield, despite an improved appetite (Dennis *et al.*, 1976). In all species there are often reductions in serum potassium below the normal range of 4–5 mmol l^{-1} when dietary potassium supplies are inadequate (Table 8.3) but in dairy calves, marginally deprived of potassium, loss of appetite and reduced weight gain occurred, with plasma potassium unaffected at 6.1–6.6 mmol l^{-1} (Weil *et al.*, 1988). In beef cattle, urinary creatinine:potassium ratios (molar) below 2.1 indicate the likelihood of a negative balance (Karn and Clanton, 1977).

Table 8.3. Suggested marginal bands[a] for potassium in serum, urine and diet as a guide to the diagnosis of dietary deprivation (D) or excess (E) in livestock.

		Serum (mmol l^{-1})	Urine (mmol l^{-1})	Diet (g kg^{-1} DM)
Cattle	D	< 2.5	< 19	< 8
	E	6–10	> 120	30–40
Sheep	D	2.4–4.0	10–20	< 8
	E	5–10	> 150	30–40
Pigs	D	2.5–3.5	3–5	3–6
	E	5–10	>150	> 18
Poultry	D	3.0–5.5	–	1.0–3.5
	E	> 10	–	> 9

[a] Mean values below (for D) or above (for E) the given bands indicate probability of ill health from deprivation or toxicity; values within bands indicate a possibility of future disorder.

Natural Occurrence of Deficiency

Preston and Linser (1985) reported that 'the potassium supplementation of finishing diets for beef is now a common practice to improve gains and mortality'. They also allude to improvements in milk production in dairy cows and the allevation of stiffness and excessive irritability in swine when potassium was added to maize-based diets, hitherto thought to provide adequate amounts of the element. However, they cite no published scientific papers to support such claims. Potassium concentrations decline in dry swards, due to a combination of ageing and leaching effects, so that values of 0.9–5.7 g K kg^{-1} DM can occur in winter range pasture (Preston and Linser, 1985). Responses to potassium supplementation have been obtained under such rangeland conditions when urea was also given, but they varied from year to year (Karn and Clanton, 1977). Responses are more likely to occur if cereal/urea supplements are deployed to alleviate the more serious constraints of digestible energy and degradable nitrogen in such pastures. The inclusion of potassium in supplements fed during droughts may be beneficial. Unusual conditions, such as acidosis, stress and diarrhoea, may lead to potassium depletion (Preston and Linser, 1985).

Potassium Requirements

Pigs and poultry

Reinvestigation of the potassium requirement of weanling cross-bred pigs placed it at 2.6–3.3 g K kg^{-1} DM, using a purified diet supplemented with potassium acetate or bicarbonate (Combs *et al.*, 1985); the figure is close to that originally proposed by Meyer *et al.* (1950). The potassium requirements for growth in poultry appear to vary according to dietary conditions, being increased by dietary phosphorus (Gillis, 1948, 1950), chloride (Nesheim *et al.*, 1964), energy (Leach *et al.*, 1959) and protein supplies, but the latter differences probably reflect the higher growth rate achieved by particular supplemented diets. The requirement was stated as 4–6 g K kg^{-1} DM for chickens (Thompson, 1972) and was similar for turkey poults (Supplee, 1965). The latest recommendations (Table 8.4) distinguish between different types and ages of bird, with particularly high needs for the young turkey poult (7 g K kg^{-1} DM) and a low requirement for the approach of pullets to lay (2.5 g K kg^{-1} DM). Leach (1974) found that 1.0 g K kg^{-1} DM did not support optimum shell thickness, but optimum dietary concentrations will be influenced by interactions with sodium and chloride (Sauveur and Mongin, 1978).

Cattle and sheep

Precise data from feeding trials on the requirements of potassium for growth or milk production are still not available. Thompson (1972) recommended 6–8 g K kg^{-1} for growing beef cattle and 8–10 g K kg^{-1} DM for lactating dairy

Table 8.4. Dietary requirements (g kg^{-1} DM) of growing broiler (B) and Leghorn (L) chicks, turkey poults (P) and laying birds for potassium (from NRC, 1994).

	Bird type	Growth stage[a]			Lay stage[b]		
		Early	Middle	Late	High	Medium	Low
Potassium	B	3.0 (5.0)[c]	3.0	3.0		–	
	L	2.5 (4.0)[c]	2.5	2.5	1.9	1.5	1.3
	P	7.0	5.0	4.0		6.0	

[a] Early = 0–3, 0–6 and 0–4 weeks; middle = 3–6, 6–12 and 8–12 weeks; late = 6–8, 18 weeks to first lay and 20–24 weeks of age for broilers, Leghorn chicks and poults, respectively.
[b] High, medium and low represent declining energy densities and food intakes of 80, 100 and 120 g day^{-1}, respectively.
[c] A factorial model would predict higher potassium requirements for early growth in chicks: suggested values given in parentheses.

animals, and recent National Research Council (NRC) recommendations are similar. Thompson (1972) pointed out that the need for potassium for milk production was considerably greater than that of calcium and phosphorus, because of the higher concentration in milk (1.5 g K l^{-1}) and that the concentrates given to high-producing cows often provide appreciably less than the estimated requirements at the peak of lactation. Such recommendations raise the possibility that lack of potassium can limit milk production in some management systems. However, tentative factorial estimates by the Agricultural Research Council (ARC, 1980) indicated that, while the potassium requirement increases with milk yield, values were lower than those previously recommended, at 6.4–7.4 g K kg^{-1} DM (Table 8.5). Similarly, ARC (1980) estimates for growing cattle were lower, declining markedly as animals grew. Subsequent feeding trials with calves of 80 kg LW gaining 0.74 kg day^{-1} suggested a requirement of between 3.4 and 5.8 g K kg^{-1} DM (Weil *et al.*, 1988), while steers of 300 kg LW gaining 1.3 kg day^{-1} did not benefit when their dietary potassium was increased from 5.5 to 10 g kg^{-1} DM (Brink *et al.*, 1984); both studies support the lower ARC (1980) estimates. Comparison of ARC (1980) maintenance requirements with those calculated by Karn and Clanton (1977: 3 g K 100 kg^{-1} LW) gives no indication of underestimation. Thompson (1972) gave 7–8 g K kg^{-1} DM as the requirement for sheep, based on two experiments with feedlot 'finishing lambs' given graded amounts of potassium as the carbonate. In the first of these (Telle *et al.*, 1964), the best weight gains and feed efficiency were obtained at 6.2 g K kg^{-1} DM, the highest level employed; in the second trial (Campbell and Roberts, 1965), the best performance was observed at 7 g K kg^{-1} DM, also the highest level used. However, ARC (1980) concluded from the same data that growing sheep required only 3–5 g K kg^{-1} DM and their factorial estimates agreed with the lower limit of 3 g K kg^{-1} DM (Table 8.5).

Table 8.5. Examples of dietary requirements for potassium of cattle and sheep (after ARC, 1980).

Animal	Weight gain or milk yield (kg day^{-1})	Assumed DM intake (kg day^{-1})	Faecal loss (g day^{-1})	Inevitable loss in urine and through skin, saliva, etc. (g day^{-1})	Production requirement (g day^{-1})	Total[a] dietary requirement (g day^{-1})	Total[a] dietary requirement (g kg^{-1} DM)
50 kg calf	0.5	0.5	1.3	1.9	1.0	4.2	8.4
250 kg bullock	0.5	5	13.0	9.5	1.0	23.5	4.7
600 kg cow, weeks 30–40 of pregnancy	0	8	20.8	22.7	2.8	46.3	5.8
600 kg cow	10	10	26.0	22.7	15	63.7	6.4
600 kg cow	30	14	36.4	22.7	45	104.1	7.4
40 kg sheep	0.2	1	1.0	1.6	0.4	3.0	3.0

[a] Estimates did not incorporate an absorption coefficient or obligatory faecal loss component, but, with absorption almost complete (*c.* 90%), this should not matter.

Requirements for Acid: Base Balance

Optimal production requires the avoidance of acid:base imbalance (i.e. of acidosis and alkalosis). The major determinants of equilibrium are the cations Na^+ and K^+ and the anions chloride (Cl^-) and sulphate (SO_4^{2-}). Imbalance is measured by estimating cation–anion differences (CAD).

Calculations and interpretations

Some investigators include only the fixed ions in calculating CAD in terms of milliequivalents (me) (millimolecular weights ÷ valencies), excluding metabolized ions, such as SO_4^{2-}. Forages have high CAD (> 100 me kg^{-1} DM), cereals low CAD (< 20 me kg^{-1} DM) and protein supplements high (e.g. soybean meal) or low (e.g. fish-meal) CAD (NRC, 1989). Experimental evidence of the importance of CAD is conflicting, due partly to the practice of adding cations and anions together as salts and ignoring the contribution of ions not incorporated in the formula (e.g. calcium (Ca^{2+}) from calcium chloride ($CaCl_2$), ammonium (NH_4^+) from ammonium chloride (NH_4Cl) and bicarbonate (HCO_3^-) or carbonate (CO_3^{2-}) from sodium and potassium salts) and those which do not conform (e.g. alkaline anions, such as HCO_3^-) (Block, 1994). Some authorities suggest that differences in ion absorption should be allowed for, but the choice of absorption coefficients for Ca and SO_4 raises problems. Improvements in dry-matter intake and milk protein and milk-fat yields have been obtained in dairy cows given a maize–silage diet, theoretically adequate in sodium (3.1 g kg^{-1} DM), potassium (8.6 g kg^{-1} DM) and chloride (3.2 g kg^{-1} DM), by adding these elements, and 'optimum' CAD were defined at 300–500 me kg^{-1} DM day^{-1} (Sanchez *et al.*, 1994). Mongin

(1981) suggested the use of dietary electrolyte balance (DEB), defined as Na + K − Cl, and diets formulated to contain 250 me DEB kg^{-1} DM were recommended for optimum growth in chicks; similar recommendations were made for pigs, but cows have a wider optimum range (200–375 me kg^{-1} DM: Tucker *et al.*, 1988; West *et al.*, 1991). Interpretation of feeding trials is, however, difficult. For example, Ross *et al.* (1994) added 0.9 g NH_4Cl kg^{-1} DM to a basal concentrate diet for finishing steers to provide a DEB of zero, and the major 'benefits' from increasing DEB were associated with withdrawal of NH_4Cl, hardly a practical starting-point.

Local effects in the rumen

Particular attention has been given to the possible benefits of using 'buffering agents', such as sodium bicarbonate ($NaHCO_3$), to counteract the low rumen pH and milk-fat output which often accompany high levels of concentrate feeding in the dairy cow. Responses have been highly variable, with beneficial outcomes usually accompanied by improvements in food intake (Erdmann, 1988). Russell and Chow (1993) proposed that, since carbonate supplements had little effect on buffering capacity in the rumen, they were more likely to improve milk-fat yields by increasing rumen dilution rate and hence the flow of undegraded starch from the rumen, raising the acetate:propionate ratio in rumen fluid. Improvements in milk-fat yields have been reported in grazing cows given $NaHCO_3$ with their concentrate supplement (Chiy *et al.*, 1993a). Responses in milk-fat and lactose yields were also obtained from sodium applied in sodium nitrate ($NaNO_3$) as a pasture fertilizer and attributed to increases in rumen pH and ruminal outflow rate (Chiy *et al.*, 1993b). Improvements in milk-fat synthesis and acetate:propionate ratios in rumen liquor have been reported in cattle (Erdmann, 1988) and sheep (Moseley, 1980; Arieli *et al.*, 1989) given sodium chloride (NaCl), which will also increase rumen dilution rate. Increases in ruminal outflow rate will not be beneficial if rumen microbes make less use of degradable energy and protein supplies from the diet, and this may explain the variable responses to dietary buffers in the literature. The treatment of straw with NaOH to improve digestibility results in very high sodium concentrations (Table 7.1) and increases in rumen pH when the treated material is fed, but neutralization with acid has surprisingly little benefit (Wamberg *et al.*, 1985). Buffering rumen VFA with NaOH rather than sodium salts influences gastric secretion in sheep (Margan, 1988). Potassium salts may affect rumen metabolism in similar ways.

Heat stress

Heat stress can induce respiratory alkalosis, associated with panting, and supplementation with sodium bicarbonate or potassium chloride (KCl), but not potassium carbonate, can increase milk yield in the heat-stressed dairy cow (West *et al.*, 1992). In heat-stressed broiler chicks, the feeding of potassium or ammonium chlorides in the drinking-water (1.5 and 2.0 g l^{-1}, respectively) with a diet adequate in potassium (7.3 g K kg^{-1} DM) improved growth and feed conversion efficiency (Teeter and Smith, 1986); higher levels or mixtures of the two salts were, however, detrimental. Later work (Deyhim and Teeter, 1994) confirmed the benefits of drinking-water supplemented with 0.067 mol KCl or NaCl l^{-1} on growth and water consumption, but there was no restoration of lowered plasma sodium or potassium and aldosterone levels remained high; it was concluded that the osmotic stress of the heat-stressed broiler had not been alleviated. Where there is evidence of production responses to sodium or effective replacement of sodium by potassium in diets apparently adequate in sodium (e.g. in poultry), these responses probably reflect the optimization of acid–base balance, rather than responses to sodium or potassium *per se*.

Natural Occurrence of Excess

In addition to the lethal but rare direct effects of potassium toxicity, described later, moderate excesses can create imbalances, which predispose ruminants to calcium- and magnesium-responsive disorders.

Milk-fever

The importance of the acid–base balance in determining incidence of milk-fever and the preventive use of acidogenic supplements has long been known and was described in detail in Chapter 4. However, emphasis has only recently been placed on the causative role of potassium as the major source of excess bases where fresh or conserved grass is the staple feed (Horst *et al.*, 1997). Moderate additions of potassium to the diet increased the incidence of hypocalcaemia and milk-fever, particularly when dietary calcium was low (Table 8.6; Goff and Horst, 1997). It was suggested that reduction of potassium concentrations in forage could make an important contribution to control of the disorder. The underlying control mechanism is an increased contribution of calcium from bone to the exchangeable pool at a time of increased demand for lactation.

Hypomagnesaemic tetany

Increased lactational demand also contributes to the incidence of hypomagnesaemic tetany but the antagonistic role of potassium arises through an antagonism of magnesium absorption from the rumen (see Chapter 6). The antagonism may be mediated, at least in part, by aldosterone because intravenous infusion of the hormone lowers magnesium absorption from the rumen and concentrations in the plasma, while exerting similar influences on rumen and plasma potassium (Charlton and Armstrong, 1989). The antagonism appears to be expressed more strongly when the diet consists of fresh herbage or forage + concentrates than when forage alone is given (Fig. 6.2). It would appear that potassium levels at the upper end of the normal range for herbage (40–50 g kg^{-1} DM) are associated with a doubling of magnesium requirement. Reductions in forage potassium may therefore have a twofold benefit in reducing incidence of two of the major metabolic diseases of dairy cows.

Table 8.6. Influence of dietary potassium on the incidence of milk-fever (C, %) in a dairy herd and on the success of treatment (T, number per case) at two dietary calcium levels (from Goff and Horst, 1997).

Dietary K ($g\ kg^{-1}$ DM)		Dietary Ca ($g\ kg^{-1}$ DM) 5	15
11	C	0	20
	T	0	1.0
21	C	36	66
	T	2.25	1.5
31	C	80	23
	T	2.0	1.25

Potassium Toxicity

Potassium is less well tolerated than sodium under acute challenge (Neathery *et al.*, 1979). Calves discriminate against diets containing 20 g K as KCl kg^{-1} DM when given a choice, but, when given no choice, appetite was not depressed (Neathery *et al.*, 1980). However, appetite and growth declined with 60 g K kg^{-1} DM, a level sometimes encountered in herbage under intensive grassland management and found in molasses (Wythes *et al.*, 1978). The major direct effect of excess potassium is a disturbance of the acid:base balance, hyperkalaemia and cardiac arrest (Neathery *et al.*, 1979), but chronic exposure may lead indirectly to ill health through the induction of magnesium deficiency in spring, when pasture potassium concentrations are maximal (see Chapter 6).

References

ARC (1980) *Nutrient Requirements of Ruminants.* Commonwealth Agricultural Bureaux, Farnham Royal, UK, pp. 211–212.

Arieli, A., Naim, E., Benjamin, R.W. and Pasternak, D. (1989) The effect of feeding saltbush and sodium chloride on energy metabolism in sheep. *Animal Production* 49, 451–457.

Bell, F.R. (1995) Perception of sodium and sodium appetite in farm animals. In: Phillips, C.J.C. and Chiy, P.C. (eds) *Sodium in Agriculture.* Chalcombe Publications, Canterbury, UK, pp. 82–90.

Block, E.R. (1994) Manipulation of dietary cation–anion difference on nutritionally related production diseases, productivity and metabolic responses of dairy cows. *Journal of Dairy Science* 77, 1437–1450.

Brink, D.R., Turgeon, O.A., Harmon, D.L., Steele, R.T., Mader, T.L. and Britton, R.A. (1984) Effects of additional limestone of various types on feedlot performance of beef cattle fed high corn diets differing in processing method and potassium level. *Journal of Animal Science* 59, 791–798.

Campbell, L.D. and Roberts, W.K (1965) The requirements and role of potassium in ovine nutrition. *Canadian Journal of Animal Science* 45, 147–156.

Charlton, J.A. and Armstrong, D.G. (1989) The effect of an intravenous infusion of aldosterone upon magnesium metabolism in sheep. *Quarterly Journal of Experimental Physiology* 74, 329–337.

Chiy, P.C., Phillips, C.J.C. and Bello, M.R. (1993a) Sodium fertilizer application to pasture. 2. Effects on dairy cow production and behaviour. *Grass and Forage Science* 48, 203–212.

Chiy, P.C., Phillips, C.J.C. and Omed, H.M. (1993b) Sodium fertilizer application to pasture. 3. Rumen dynamics. *Grass and Forage Science* 48, 249–259.

Combs, N.R. and Miller, E.R. (1985) Determination of potassium availability in K_2CO_3, $KHCO_3$, corn and soybean meal for the young pig. *Journal of Animal Science* 60, 715–719.

Combs, N.R., Miller, E.R. and Ku, P.K. (1985) Development of an assay to determine the bioavailability of potassium in feedstuffs for the young pig. *Journal of Animal Science* 60, 709–714.

Cornforth, I.S., Stephen, R.C., Barry, T.N. and Baird, G.A. (1978) Mineral content of swedes, turnips and kale. *New Zealand Journal of Experimental Agriculture* 6, 151–156.

Deyhim, F. and Teeter, R.G. (1994) Effect of heat stress and drinking water salt supplements on plasma electrolytes and aldosterone concentrations in broiler chickens. *International Journal of Biometeorology* 38, 216–217.

Dennis, R.J., Hemken, R.W. and Jacobson, D.R. (1976) Effect of dietary potassium percent for lactating cows. *Journal of Dairy Science* 59, 324–328.

Erdmann, R.A. (1988) Dietary buffering requirements of the lactating dairy cow: a review. *Journal of Dairy Science* 71, 3246–3266.

Gillis, M.B. (1948) Potassium requirement of the chick. *Journal of Nutrition* 36, 351–357.

Gillis, M.B. (1950) Further studies on the role of potassium in growth and bone formation. *Journal of Nutrition* 42, 45–57.

Goff, J.P. and Horst, R.L. (1997) Effect of dietary potassium and sodium but not calcium on the incidence of milk fever in dairy cows. *Journal of Dairy Science* 80, 176–186.

Grace, N.D. (1983) Amounts and distribution of mineral elements associated with fleece-free empty body weight gains in the grazing sheep. *New Zealand Journal of Agricultural Research* 26, 59–70.

Hayslett, J.P. and Binder, H.J. (1982) Mechanism of potassium adaptation. *American Journal of Physiology* 243, F103–F112.

Horst, R.L., Goff, J.P., Reinhardt, T.A. and Buxton, D.R. (1997) Strategies for preventing milk fever. *Journal of Dairy Science* 80, 1269–1280.

Hughes, E.H. and Ittner, N.R. (1942) The potassium requirement of growing pigs. *Journal of Agricultural Research* 64, 189–192.

Johnson, K.G. (1970) Sweating rate and the electrolyte content of skin secretions of *Bos taurus* and *Bos indicus* cross-bred cows. *Journal of Agricultural Science, Cambridge* 75, 397–402.

Karn, J.F. and Clanton, D.C. (1977) Potassium in range supplements. *Journal of Animal Science* 45, 1426–1434.

Kem, D.C. and Trachewsky, D. (1983) Potassium metabolism. In: Whang, R. (ed.)

Potassium: Its Biological Significance. CRC Press, Boca Raton, Florida, pp. 25–35.

Lanyon, L.E. and Smith, F.W. (1985) Potassium nutrition of alfalfa and other forage legumes: temperate and tropical. In: Munson, R.D. (ed.) *Potassium in Agriculture.* American Society of Agronomy, Madison, Wisconsin, pp. 861–894.

Leach, R.M., Jr (1974) Studies on the potassium requirements of the laying hen. *Journal of Nutrition* 104, 684–686.

Leach, R.M., Jr, Dam, R., Zeigler, T.R. and Norris, L.C. (1959) The effect of protein and energy on the potassium requirement of the chick. *Journal of Nutrition* 68, 89–100.

Lindeman, R.D. and Pederson, J.A. (1983) Hypokalaemia. In: Whang, R. (ed.) *Potassium: Its Biological Significance.* CRC Press, Boca Raton, Florida, pp. 45–75.

MAFF (1990) *UK Tables of the Nutrient Value and Chemical Composition of Foodstuffs.* Givens, D.I. (ed.), Rowett Research Services, Aberdeen.

MAFF (1992) *Feed composition – UK Tables of Feed Composition and Nutritive Value for Ruminants,* 2nd edn. Ministry of Agriculture, Fisheries and Food Standing Committee on Tables of Feed Composition, Chalcombe Publications, Canterbury, UK.

Margan, D.E. (1988) Stimulation of abomasal flow in sheep with buffer infusions for abomasum and effects of same on certain aspects of intestinal function. *Australian Journal of Agricultural Research* 39, 1121–1134.

Meyer, J.H., Grummer, R.H., Phillips, R.H. and Bohstedt, G. (1950) Sodium, chlorine, and potassium requirements of growing pigs. *Journal of Animal Science* 9, 300–306.

Michell, A.R. (1978) Plasma potassium and sodium appetite: the effect of potassium infusion in sheep. *British Veterinary Journal* 134, 217–224.

Michell, A.R. (1995) Physiological roles for sodium in mammals. In: Phillips, C.J.C. and Chiy, P.C. (eds) *Sodium in Agriculture.* Chalcombe Publications, Canterbury, UK, pp. 91–106.

Miller, E.R. (1995) Potassium bioavailability. In: Ammerman, C.B., Baker, D.H. and Lewis, A.J. (eds) *Bioavailability of Nutrients for Animals.* Academic Press, New York, pp. 295–302.

Mongin, P. (1981) Recent advances in dietary anion–cation balance applications in poultry. *Proceedings of the Nutrition Society* 40, 285–294.

Moseley, G. (1980) Effects of variation in herbage sodium levels and salt supplementation on the nutritive value of perennial ryegrass for sheep. *Grass and Forage Science* 35, 105–113.

Neathery, M.W., Pugh, D.G., Miller, W.J., Whitlock, R.H., Gentry, R.F. and Allen, J.C. (1979) Potassium toxicity and acid:base balance from large oral doses of potassium to young calves. *Journal of Dairy Science* 62, 1758–1765.

Neathery, M.W., Pugh, D.G., Miller, W.J., Gentry, R.F. and Whitlock, R.H. (1980) Effects of sources and amounts of potassium on feed palatability and on potassium toxicity in dairy calves. *Journal of Dairy Science* 63, 82–85.

Nesheim, M.C., Leach, R.M., Jr, Zeigler, T.R. and Serafin, J.A. (1964) Interrelationships between dietary levels of sodium, chlorine and potassium. *Journal of Nutrition* 84, 361–366.

NRC (1989) *Nutrient Requirements of Dairy Cattle,* 6th edn. National Academy of Sciences, Washington, DC.

NRC (1994) *Nutrient Requirements of Poultry,* 9th edn. National Academy of Sciences, Washington, DC.

Perdomo, J.T., Shirley, R.L. and Chicco, C.F. (1977) Availability of nutrient minerals in four tropical forages fed freshly chopped to sheep. *Journal of Animal Science* 45, 1114–1119.

Peterson, L.N. (1997) Potassium in nutrition. In: O'Dell, B.L. and Sunde, R.A. (eds) *Handbook of Nutritionally Essential Mineral Elements*. Marcel Dekker, New York, pp. 153–183.

Pradhan, K. and Hemken, R.W. (1969) Potassium depletion in lactating dairy cows. *Journal of Dairy Science* 51, 1377–1381.

Preston, R.L. and Linser, J.R. (1985) Potassium in animal nutrition. In: Munson, R.D. (ed.) *Potassium in Agriculture*. American Society of Agronomy, Madison, Wisconsin, pp. 595–617.

Rabinowitz, L. (1988) Model of homeostatic regulation of potassium excretion in sheep. *American Journal of Physiology* 254, R381–R388.

Reid, R.L., Baker, B.S. and Vona, L.C. (1984) Effects of magnesium sulphate supplementation and fertilization on quality and mineral utilisation of timothy hays by sheep. *Journal of Animal Science* 59, 1403–1410.

Robinson, D.L. (1985) Potassium nutrition of forage grasses. In: Munson, R.D. (ed.) *Potassium in Agriculture*. American Society of Agronomy, Madison, Wisconsin, pp. 895–903.

Ross, J.G., Spears, J.W. and Garlich, J.D. (1994) Dietary electrolyte balance effects on performance and metabolic characteristics in finishing steers. *Journal of Animal Science* 72, 1600–1607.

Russell, J.B. and Chow, J.M. (1993) Another theory for the action of ruminal buffer salts: decreased starch fermentation and propionate production. *Journal of Dairy Science* 76, 826–830.

Sanchez, W.K., Beede, D.K. and Cornell, J.A. (1994) Interactions of sodium, potassium and chloride on lactation acid–base status and mineral concentrations. *Journal of Dairy Science* 77, 1661.

Sasser, L.B., Ward, G.M. and Johnson, J.E. (1966) Variations in potassium concentration of cow's milk. *Journal of Dairy Science* 49, 893–895.

Sauveur, B. and Mongin, P. (1978) Interrelationships between dietary concentrations of sodium, potassium and chloride in laying hens. *British Poultry Science* 19, 475–485.

Schonewille, J.T., Ram, L., Van't Klooster, A.T., Wonterse, H. and Beynen, A.C. (1997) Intrinsic potassium in grass silage and magnesium absorption in dry cows. *Livestock Production Science* 48, 99–110.

Supplee, W.C. (1965) Observations on the requirement of young turkeys for dietary potassium. *Poultry Science* 44, 1142–1144.

Teeter, R.G. and Smith, M.O. (1986) High chronic ambient temperature stress effects on acid–base balance and their response to supplemental ammonium chloride, potassium chloride and potassium carbonate. *Poultry Science* 65, 1777–1781.

Telle, P.P., Preston, R.L., Kintner, L.D. and Pfander, W.H. (1964) Definition of the ovine potassium requirement. *Journal of Animal Science* 23, 59–66.

Thompson, D.J. (1972) *Potassium in Animal Nutrition*. International Minerals and Chemical Corporation, Libertyville, Illinois.

Tucker, W.B., Harrison, G.A. and Hemken, R.W. (1988) Influence of dietary cation–anion balance on milk, blood, urine and rumen fluid in lactating dairy cows. *Journal of Dairy Science* 71, 346–354.

Ussing, H.H. (1960) The biochemistry of potassium. In: *Proceedings of the 6th Congress*. International Potash Institute, Amsterdam.

Wamberg, S., Engel, D. and Stigsen, P. (1985) Acid–base balance in ruminating calves

given sodium-hydroxide treated straw. *British Journal of Nutrition* 54, 655–657.
Ward, G.M. (1966) Potassium metabolism of domestic ruminants: a review. *Journal of Dairy Science* 49, 268–276.
Weil, A.B., Tucker, W.B. and Hemken, R.W. (1988) Potassium requirement of dairy calves. *Journal of Dairy Science* 71, 1868–1872.
West, J.W., Mullinix, B.G. and Sandifer, T.G. (1991) Changing dietary electrolyte balance for dairy cows in cool and hot environments. *Journal of Dairy Science* 74, 1662–1674.
West, J.W., Haydon, K.D., Mullinix, B.G. and Sandifer, T.G. (1992) Dietary cation–anion balance and cation source effects on production and acid–base status of heat-stressed dairy cattle. *Journal of Dairy Science* 75, 2776–2786.
Wythes, J.R., Wainwright, D.H. and Blight, G.W. (1978) Nutrient composition of Queensland molasses. *Australian Journal of Experimental Agriculture and Animal Husbandry* 18, 629–634.

9 Sulphur

Introduction

Recognition that sulphur was an important dietary component stemmed from the discovery in the 1930s that the essential amino acid, methionine, contained one atom of sulphur per molecule (McCollum, 1956). Subsequent investigations of the composition of a range of tissue proteins showed that sulphur (S) was an integral part of the majority, comprising 0.5–2.0% by weight. Other S-amino acids, such as cystine, cysteine, cystathione and taurine, contributed to protein sulphur, but each could be derived from methionine and they were not essential constituents of the diet. Strictly speaking, sulphur is only an essential nutrient for plants and microbes, because only they can synthesize S-amino acids and hence proteins from degradable inorganic sulphur sources. Mammals digest plant proteins and recombine the S-amino acids to form their own unique tissue proteins. Non-ruminants require particular balances among their dietary amino acids and methionine is often the second most limiting behind lysine. Since manipulation of dietary sulphur can make only a slender contribution to S-amino acid supply in the non-ruminant (either by sparing cystine (Lovett *et al.*, 1986) or via microbial synthesis of S-amino acids in the large intestine), those interested in this aspect of monogastric nutrition should consult texts on protein nutrition and the review of amino acid availability by Henry and Ammerman (1995). Ruminants are quite different; they possess a substantial microbial population in the fore-stomach or rumen, which can incorporate degradable inorganic sulphur sources into microbial proteins (Kandylis, 1984a). Following digestion in the small intestine, microbial protein provides a high-quality amino acid supply, ideal for all purposes except the growth of wool, hair and mohair. Furthermore, since sulphur is much cheaper to provide than dietary protein, efficient production requires the pragmatic, if not essential, use of dietary sulphur in ruminant nutrition.

© CAB *International* 1999. *Mineral Nutrition of Livestock*
(E.J. Underwood and N.F. Suttle)

Sulphur Sources for Ruminants

Sulphur in forages

The sulphur concentrations found in pasture and conserved forages range widely, from 0.5 to > 5.0 g S kg^{-1} dry matter (DM) (Table 9.1), depending mainly on the availability of soil sulphur, nitrogen and phosphorus and the maturity of the sward. In the UK, the range in fresh grass is narrower (1.2–4.0 g S kg^{-1} DM), and the mean (2.2) is adequate for plant and animal growth (MAFF, 1990). In many other countries, sulphur deficiency limits forage production. Areas of sulphur-deficient soils, crops and forages have been delineated in Australia and the south-east (Kamprath and Jones, 1986), north-east, mid-west (Hoeft and Fox, 1986), west (Rasmussen and Kresge, 1986) and north (Beaton *et al.*, 1971) of the USA. Many regions of China are also deficient in sulphur (Qi *et al.*, 1994), and the need for regular inputs to maintain the productivity of clover-based pastures in Australasia (Goh and Nguyen, 1997) and maize in Nigeria (Ojeniyi and Kayode, 1993) has recently

Table 9.1. Reported values for sulphur concentrations and total sulphur:nitrogen (S:N) ratios in some common feedstuffs for ruminants.

	Replication	S (g kg^{-1} DM)	S:N[a]	Reference
Grasses	137	2.2	0.088	MAFF (1990)
Dried grass	8	3.5	0.12	MAFF (1990)
Stylo hay	2	1.25	0.071	Bird (1974)
Lucerne hay	2	3.6	0.095	Qi *et al.* (1994)
Spear grass hay	2	0.5, 0.8	0.055, 0.205	Kennedy and Siebert (1972a, b)
Maize silage	1			Buttrey *et al.* (1986)
Sorghum silage	1	1.0	0.067	Ahmad *et al.* (1995)
Barley straw	4	2.0	0.33	MAFF (1990)
Wheat straw	2	1.0	0.32	Kennedy (1974)
Barley	10	1.5	0.073	MAFF (1990)
Maize	5	1.6	0.10	MAFF (1990)
Distillers' dried grain	5	3.7	0.084	MAFF (1990)
Swedes	?	4.8	0.179	Cornforth *et al.* (1978)
Turnips	?	6.1	0.233	Cornforth *et al.* (1978)
Kales	?	9.0	0.313	Cornforth *et al.* (1978)
Feather meal	5	18.1	0.127	MAFF (1990)
Rapeseed meal	5	16.9	0.26	MAFF (1990)
Linseed meal	5	4.1	0.065	MAFF (1990)
Cottonseed meal	5	5.0	0.083	MAFF (1990)
Soybean meal	5	4.6	0.058	MAFF (1990)
Safflower meal	1	2.2	0.029	Qi *et al.* (1994)
Fish-meal	1	4.9	0.046	Qi *et al.* (1994)
Molasses (cane)	1	7.3	0.811	Bogdanovic (1983)

[a] This ratio only improves the assessment of nutritive value for ruminants if the S and N have similar rates and extents of degradability in the rumen.

been confirmed. Plant species vary in their sulphur requirement, legumes showing optimal growth with around 2 g S kg^{-1} DM in their tissues, while maize tolerates 1.4 g S kg^{-1} DM and sugar cane < 1 g S kg^{-1} DM. Any area away from the coast and industrial activity, where the annual rainfall is moderate (> 500 mm) and the soils derive from weathered volcanic material, is likely to have leached soils that are low in sulphur and swards that respond to supplementation. There is little published information on seasonal variation in sulphur concentrations, but sulphur and protein are highly correlated and protein concentration declines at a rate of 1–3 g kg^{-1} DM day^{-1}, depending on species and growth conditions (Minson, 1990), with approach to maturity: this is equivalent to a decline of 0.1–0.3 g S kg^{-1} DM $week^{-1}$ of sward growth or regrowth after cutting. In the Mt Elgon region of west Kenya, a recent survey found that 25% of the dry-season forages contained < 1 g S kg^{-1} DM (Jumba *et al.*, 1996). Sulphur deficiency is rare in arid regions (< 250 mm rainfall p.a.), because of the upward movement of sulphates through the soil profile. Application of sulphate fertilizer increases herbage sulphur, even when baseline levels are high, mostly by increasing the sulphate (i.e. non-protein) component (Spears *et al.*, 1985). Pastures in coastal or industrial areas are enriched with sulphur from sea spray (3–5 kg S ha^{-1}: Ledgard and Upsdell, 1991) or emissions (*c.* 8 kg S ha^{-1} $year^{-1}$: McClaren, 1975) and unlikely to be deficient.

Sulphur in other feedstuffs

Sulphur concentrations in crops and feedstuffs vary widely (Table 9.1). Cereal grains tend to be low in sulphur (*c.* 1 g kg^{-1} DM: Todd, 1972), cereal straws slightly richer (1.4–4.0 g S kg^{-1} DM: Suttle, 1991), protein supplements rich (2.2–4.9 g S kg^{-1} DM: Qi *et al.*, 1994) and brassica crops too rich (4.8–9.0 g S kg^{-1} DM: Cornforth *et al.*, 1978) in sulphur. Leafy brassicas such as kale and the leaves of root crops, such as swedes and turnips, contain the most sulphur: rapeseed meal is a uniquely S-rich crop by-product. Among other by-products, molasses is one of the best sources of sulphur, containing 3–7 g S kg^{-1} DM; animal by-products are generally moderate sources, but feather meal is the richest known source at 18.1 g S kg^{-1} DM (Table 9.1).

Sulphur in the drinking-water

Surface- and groundwater supplies can contain as much as 1.5 g sulphate S l^{-1} and can therefore make significant contributions to sulphur intake. In regions such as the state of Nevada, 23% of samples in one early survey contained > 250 mg sulphate (SO_4) l^{-1} (Miller *et al.*, 1953, cited by Weeth and Hunter, 1971). In Saskatchewan, Canada, water from deep aquifers contained 500 mg S as sulphates l^{-1} and was sufficient to induce hypocuprosis (Smart *et al.*, 1986; see also Chapter 11).

Nutritive value of sulphur in feeds

The value of feeds as sources of sulphur for ruminants depends entirely on the coavailability of other factors needed for microbial protein synthesis, and their evaluation presents unique problems. Nutritive value is more closely related to S:nitrogen (N) ratios than to sulphur concentration *per se*. Although sulphur and nitrogen concentrations are well correlated within feeds, there are wide differences in S:N ratios between feeds (Table 9.1), ratios being fourfold higher in brassicas and cereal straws at the high extreme (0.2–0.3) than in cereals and vegetable protein sources at the low extreme (0.05–0.07).

Metabolism

Influence of rumen microbiota

The many strains of rumen bacteria all require sulphur, but they procure it by different pathways, some being capable of degrading inorganic sources of sulphur to sulphide and incorporating it into S-amino acids (dissimilatory) while others (assimilatory) utilize only organic sulphur (Kandylis, 1984a; Durand and Komisarczuk, 1988; Henry and Ammerman, 1995). Those dependent on organic sulphur require degradable rather than non-degradable sulphur sources. Studies with radio-labelled $^{32}SO_4$ indicated that not all SO_4^{2-} passed through the rumen sulphide pool and the fraction 'escaping' increased from 45 to 55% when molybdenum was added to the sheep's diet (Gawthorne and Nader, 1976). Approximately 50% of bacterial organic sulphur was derived from sulphide in sheep given grass or lucerne hay (Kennedy and Milligan, 1978). Dietary supply is supplemented by sulphur secreted in saliva, itself a mixture of inorganic and organic sulphur and more liberally supplied by cattle than by sheep (Kennedy and Siebert, 1972b; Bird, 1974). The proportion of total sulphur flow into the rumen which is 'captured' as rumen microbial protein (Fig. 9.1) varies widely and is

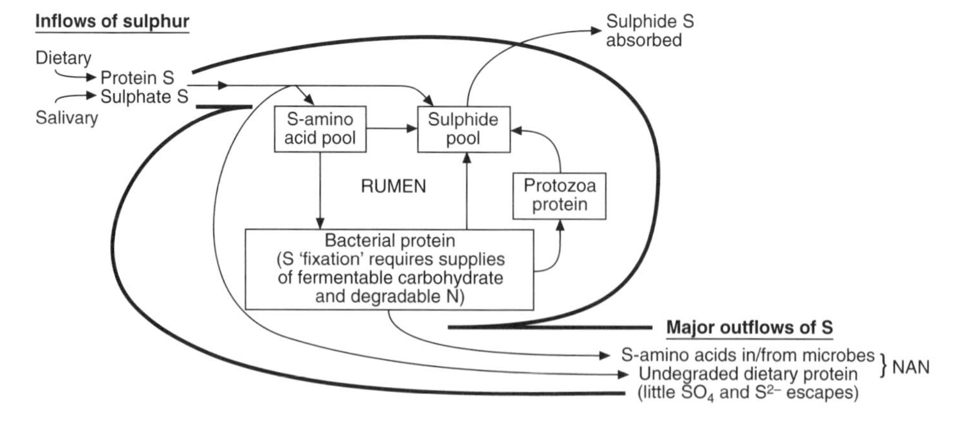

Fig. 9.1. Illustration of the primary exchanges involving sulphur (S) in the rumen and of the involvement of energy and nitrogen (N) supply in determining the most important S-containing product, non-ammonia nitrogen (NAN), entering the duodenum.

determined by factors such as the sulphur source (methionine being less degraded than other S-amino acids (Bird, 1972a)) and by the coavailability of other substrates (chiefly degradable nitrogen), i.e. by 'rumen synchrony' (Moir, 1970; Beever, 1996). Optimal microbial synthesis and sulphur capture occur when fermentable energy, degradable sulphur, nitrogen and phosphorus are supplied at rates which match the synthetic capacity of the rumen microbial biomass. Excess degradable sulphur is rapidly absorbed from the rumen as sulphide, which has little or no nutritional value but is potentially toxic. The rumen protozoa feed on rumen bacteria, returning sulphur to the sulphide pool as they do so (Kandylis, 1984a). The efficiency of rumen microbial protein synthesis can be increased by defaunating the rumen (i.e. removing the protozoa), a process which should lower rumen sulphide concentrations (Hegarty *et al.*, 1994). Similar ends may be achieved by transferring genes which encode enzymes for alternative S-fixing pathways from common species of bacterium into species which inhabit the rumen (Ward and Nancarrow, 1991). The considerable involvement of rumen sulphide in copper metabolism is discussed further in Chapter 11. Anaerobic fungi may play a significant role in the structural degradation of fibre in the rumen; fungal activity is dependent on dietary sulphur supply and contributes significantly to the synthesis of S-amino acids (Weston *et al.*, 1988).

Postruminal events

Sulphur in the digesta leaves the rumen in several forms – undegraded dietary protein, microbial and fungal protein (Fig. 9.1), sulphide and sulphate – and the proportions are determined by dietary composition and the efficiency of microbial capture (Bray and Till, 1975). From that point, sulphur metabolism in the ruminant is similar to that of the monogastric animal. Sulphate is absorbed by active transport, while the S-amino acids have first to be liberated by proteolytic digestion in the small intestine. Further degradation of SO_4^{2-} to S^{2-} and incorporation into bacterial protein can occur in the hind gut but probably makes little contribution to sulphur ulitization (Kandylis, 1984a). Methionine is demethylated in the tissues to provide methyl groups (CH_3) for the elaboration of carbon chains, leaving homocysteine, from which the non-essential S-amino acids cystathione and cysteine are formed. Tissue protein synthesis can now proceed. The major sites of cysteine accretion or secretion are the muscle and the mammary gland, and the products, such as casein, contain S:N in a ratio of about 0.067. Wool, hair and mohair are much richer in sulphur, containing 2.7–5.4% sulphur, mostly as cysteine, and with an S:N ratio of 0.2. Daily wool production rates are, however, low and insignificant in the context of total sulphur requirements, except in the mature animal, with only a maintenance requirement for sulphur, and those kept solely for fleece production.

Excretion

Sulphur is excreted principally via the urine as sulphate derived from the oxidation of sulphide and the catabolism of sulphur-containing, organic molecules in the tissues. Sulphate filtered at the glomerulus at rates which exceed the tubular reabsorption rate is rapidly excreted in urine. Sulphate competes with molybdate (Mo) for renal tubular reabsorption (Bishara and Bray, 1978); the macronutrient (S) can greatly impede the micronutrient (Mo), but molybdate can only slightly influence the urinary excretion of sulphate. Dick (1956) demonstrated a remarkable enhancement of urinary molybdenum excretion and decreased retention in molybdenum-loaded sheep when sulphate was subsequently added to the diet. When sulphur and molybdenum are simultaneously added to the diet, sulphur can *decrease* urinary molybdenum excretion by decreasing molybdate absorption and increasing retention (Grace and Suttle, 1979; see also Chapter 17, molybdenum section). The sulphur excreted via faeces is largely in organic forms.

Homeostasis

Homeostatic regulation of sulphate metabolism is of little significance in the monogastric animal, because conserved sulphate cannot be used to synthesize S-amino acids. In the ruminant, conservation might theoretically be effective because sulphur secreted in saliva can be incorporated into microbial protein. However, salivary sulphur secretion decreases as sulphur intakes and plasma sulphate concentrations decline, and there is no evidence that the capacity of the salivary glands to extract and secrete sulphur increases during sulphur deficiency. During S-amino acid deficiency, the catabolism of tissue protein will allow redistribution of the essential S-amino acids from the more expendable protein pools (e.g. bone matrix in the pregnant or lactating female) to less expendable sites and processes (fetus and mammary gland). Wool growth is more sensitive to a reduced protein supply than fetal growth, maternal weight or milk production (Masters *et al.*, 1996), and reductions in wool growth during sulphur deprivation might allow the synthesis of far more protein elsewhere, because other proteins contain far fewer S-amino acids.

Functions

The functions of sulphur are as diverse as the proteins of which they are part. Sulphur is frequently present as highly reactive sulphydryl (SH) groups or disulphide bonds, maintaining the spatial configuration of elaborate polypeptide chains and providing the site of attachment for prosthetic groups and the binding to substrates that are essential to the activity of many enzymes. Cysteine-rich molecules, such as metallothionein, play a vital role in protecting animals from excesses of copper, cadmium and zinc, while others influence selenium transport and protect tissues from selenium toxicity. Glutathione may facilitate the uptake of copper by the liver (see Chapter 11). Glutathione also protects tissues from oxidants by interconverting between the reduced (GSH) and oxidized state (GS–SG) and may thus protect erythrocytes from lead toxicity. Sulphur is present as sulphate in the chrondroitin sulphate of connective tissue and in the natural anticoagulant heparin. Sulphur is relatively abundant in the keratin-rich appendages – hoof, horn, feathers and above all wool fibre and mohair. Hormones such as insulin and oxytocin contain sulphur as do the vitamins, thiamine and biotin. However, the rate-limiting function in the sulphur-deprived ruminant is related to none of the functions listed above; instead, it concerns the failure of fermentation and microbial protein synthesis in the rumen.

Clinical and Other Signs of Sulphur Deprivation

The symptoms of sulphur deprivation in ruminants are not specific but are shared by any nutritional deficiency or factor which depresses rumen microbial activity. On fibrous diets, there is often an early depression of appetite and digestibility, due to a failure to digest cellulose in sheep (Kennedy and Siebert, 1972a, b; Hegarty *et al.*, 1994) and cattle (Kennedy, 1974). Sheep deprived of sulphur spend more time ruminating and have an increased rumen liquor volume (Weston *et al.*, 1988). Eventually, growth is retarded, wool or hair is shed, there is profuse salivation and lacrimation and the eyes become cloudy; eventually, the emaciated animal dies (Thomas *et al.*, 1951; Kincaid, 1988). Examination of rumen fluid reveals changes in biochemical (Whanger, 1972) and microbial (Weston *et al.*, 1988; Morrison *et al.*, 1990) composition. Hypoalbuminaemia is inevitable, but blood urea will be high if the intake of degradable dietary nitrogen is high. Sulphate concentrations in plasma generally reflect sulphur intake when other influential factors are non-limiting for microbial protein synthesis, but, if another factor, such as nitrogen, is limiting, serum sulphate values for a given sulphur intake are increased (Kennedy and Siebert, 1972b). Low plasma sulphate levels of around 10 mg S l^{-1} are found on poor, dry-season pasture but they rise to 20–40 mg S l^{-1} with the onset of rains and the growth of green herbage (Kennedy and Siebert, 1972a; White *et al.*, 1997). Concentrations of methionine in plasma and liver fell when calves were given a diet sufficiently low in sulphur (0.4 g S kg^{-1} DM) to retard growth (Slyter *et al.*, 1988). If dietary sulphur supply alone is limiting rumen microbial synthesis, provision of additional sulphur should lower blood urea levels and urinary nitrogen excretion. It has recently been suggested that diets low in sulphur enhance the tissue accumulation of cadmium by reducing rumen sulphide concentrations and limiting the formation of insoluble and unabsorbable cadmium sulphide (Smith and White, 1997).

Occurrence of Sulphur Deprivation

Plant factors

If crop husbandry is ideal, sulphur deficiency will not occur in livestock grazing legume-rich swards, because the animals need less sulphur than the plants. However, livestock given sugar cane or maize by-products (including silage) might need supplementation. Some species of tropical grasses, such as spear grass (*Heteropogon contortus*) are often low in sulphur (*c.* 0.5 g S kg^{-1} DM: Kennedy and Siebert, 1972a, b; Morrison *et al.*, 1990). Sulphur fertilization of ryegrass pasture (Jones *et al.*, 1982) or forage crops (Buttrey *et al.*, 1986; Ahmad *et al.*, 1995) or direct supplementation of diets based on sulphur-deficient forages can improve animal production. In all pastoral systems, risks of sulphur deprivation (but also nitrogen and phosphorus deficiencies) in livestock increase as the herbage matures. Improvements in wool production and live-weight gain are commonly found when sulphur supplements are provided for sheep grazing dry-season pastures in Western Australia (Mata *et al.*, 1997; White *et al.*, 1997). Mature forage is often harvested for winter use as a maintenance feed, and attempts to improve its nutritive value by supplementation with rumen-degradable energy and nitrogen sources will only succeed if sulphur is also added (Leng, 1990; Suttle, 1991). The converse is also true: responses will only be obtained with sulphur if the energy, nitrogen and phosphorus requirements of the microbes are met (Suttle, 1991). Marked improvements in production can be obtained by the balanced supplementation of low-quality roughages with molasses–urea mixtures, but they must contain added sulphur. In tropical areas of Queensland and Western Australia, it has been suggested that seasonal supplements of sulphur and nitrogen will be needed to obtain the full production advantages from the use of defaunated sheep (Bird and Leng, 1985; Hegarty *et al.*, 1994). If heavy reliance is placed on cereals for animal production and non-protein nitrogen sources are used to promote microbial protein synthesis, this will only work if sulphur is also given. Where cereals are used for drought feeding, it is recommended that the supplement contains 10 g urea and 1.5 g calcium sulphate ($CaSO_4$) kg^{-1} DM (after Bogdanovic, 1983).

Animal factors

High wool-producing species, such as Merino sheep and angora goats, are more susceptible to sulphur deprivation than cattle because of high requirements for fleece growth (Bird, 1974; Kennedy, 1974; Qi *et al.*, 1994). Widespread deficiencies of sulphur have been predicted even for cattle on the basis of recommended S:N ratios of 0.1 for beef and 0.08 for dairy animals (Qi *et al.*, 1994). However, the ratio of total S:total N is a questionable basis for such calculations and the ideal ratio should be closer to 0.07 for rumen degradable S:N for all ruminants. Blends of feeds for productive animals are always likely to have S:N ratios > 0.067 unless urea is used as the sole source of supplementary nitrogen.

Diagnosis of Sulphur Deprivation

Since the problem of sulphur deprivation in ruminants relates to the well-being of the rumen microflora, the best diagnosis may be afforded by obtaining samples of rumen fluid by stomach tube and determining whether or not they contain sufficient sulphide for unrestricted microbial protein synthesis. The critical level was thought to be as low as 1 μg sulphide S ml^{-1} (Bray and Till, 1975) and this was supported by Hegarty *et al.* (1991), but other recent studies suggest that it may be between 1.6 and 3.8 μg S ml^{-1} (Weston *et al.*, 1988). Unlike total sulphur, sulphide is easily determined by acid displacement of hydrogen sulphide (H_2S) and titration (Kennedy and Siebert, 1972a). Samples of rumen fluid can be obtained by means of a simple vacuum source (50 ml disposable syringe fitted with a two-way valve) and interchangeable tubes for receiving the samples. Ease of sampling depends on the consistency of the digesta and relevance of the result on the continuity of feeding; since rumen sulphide increases after a meal, three samples should be taken, one preprandial, one postprandial (1–2 h) and another after about 5 h. Low serum SO_4-sulphur (< 10 mg l^{-1}) is also indicative of sulphur deprivation. Australian workers have recently examined blood and liver GSH concentrations as indicators of sulphur status (Mata *et al.*, 1997; White *et al.*, 1997). Blood GSH showed a similar marked decline to that of plasma SO_4 during the dry season (Fig. 9.2), reaching levels that would be regarded as subnormal (340; normal > 500 mg l^{-1}). However, the decline was not diminished by dietary sulphate supplements (White *et al.*, 1997) and only partially reduced by 'protected methionine' (Mata *et al.*, 1997). It was concluded that GSH synthesis was restricted by a general lack of protein and was therefore an unreliable measure of sulphur status. However, it should be regarded as a useful complement to serum SO_4, emphasizing the limitations of sulphur supplements given in isolation when nitrogen is lacking. The surest diagnosis is provided by responses in appetite or growth (of wool or body) to the administration of sulphur. Supplementation must be continuous and will be ineffective if other factors are limiting rumen microbial activity; it may often be necessary to provide additional fermentable carbohydrate and degradable nitrogen (urea) with the sulphur supplement (flowers of sulphur or gypsum ($CaSO_4$)).

Prevention and Control of Sulphur Deprivation

Fertilizers

Many phosphorus (P) fertilizers used to be relied upon to provide sufficient adventitious sulphur to avoid the need for specific sulphur sources, but the S:P ratio in phosphorus fertilizers has markedly declined and so has the sulphur status of pastures and crops. As with phosphorus (see Chapter 5), the primary goal must be to ensure that shortage of sulphur does not limit pasture or crop growth. Sophisticated pasture growth models have been

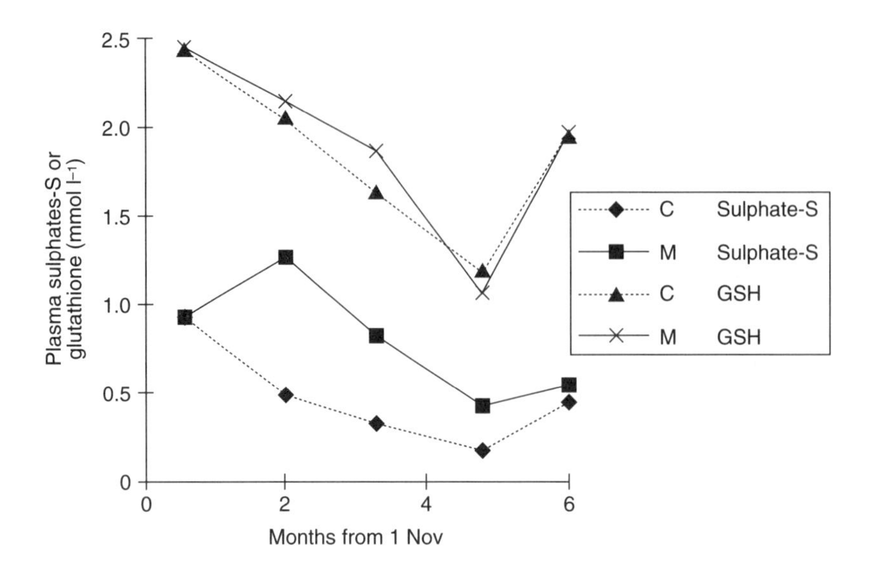

Fig. 9.2. Seasonal fluctuations in two biochemical indices of sulphur status, plasma sulphate (SO_4) and blood glutathione (GSH) in two groups of sheep grazing in Western Australia: one group (■, ×) received a sulphur-rich mineral the other (♦, ▲) did not (data from White *et al.*, 1997).

developed, based on two soil tests, which measure available SO_4^{2-} and the organic sulphur component from which SO_4^{2-} is 'mineralized' (Goh and Nguyen, 1997). Thus, it was estimated that 6.7–13.5 kg S ha^{-1} $year^{-1}$ was required to replace the net loss of sulphur from the soil SO_4^{2-} pool, stemming from removal of the element in animal products, leaching and the 'export' of excreta. Using a simpler yardstick (soil extractable SO_4^{2-} < 12 mg kg^{-1}), biennial applications of 10 kg S ha^{-1} were recommended for maize grown in the rain-forest–savannah transition zone in Nigeria (Ojeniyi and Kayode, 1993). Sulphur in fertilizer form is most usually distributed as $CaSO_4$ or potassium sulphate (K_2SO_4), and the responses in herbage sulphur-dependent upon application rate (Fig. 9.3).

Direct supplements

Lack of sulphur in the diet can readily be corrected by the inclusion of elemental sulphur, sodium sulphate (Na_2SO_4) or $CaSO_4$, but improvements in production will be limited if any concomitant lack of rumen-degradable nitrogen (Kennedy and Siebert, 1972a, b; Bird, 1974) or energy is not alleviated. It must be emphasized that poor-quality pasture or roughage is usually lacking in all three nutrients and possibly a fourth, phosphorus. Nevertheless, provision of sulphur alone may be life-saving. Inclusion of $CaSO_4$ in a molasses–urea–bran block yields an ideal supplement for stock grazing sulphur- and nitrogen-deficient pastures, with the proportions of $CaSO_4$ and urea linked to provide 0.067 g S g^{-1} N. One procedure is to heat

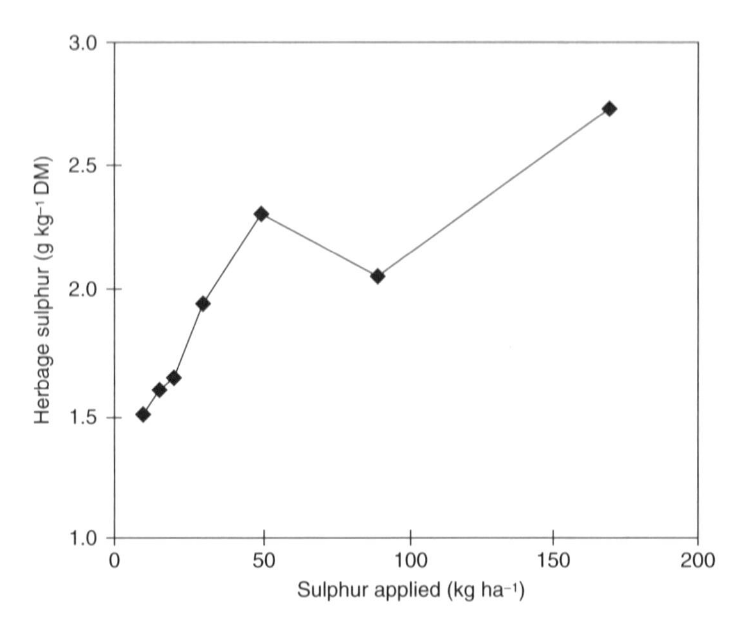

Fig. 9.3. Responses in herbage sulphur concentration to different application rates of fertilizer S as $CaSO_4$ in south-east Scotland (McLaren *et al.*, 1975).

the molasses to 90°C for 20 min, cool to 60°C and then add (kg kg^{-1}) urea, 0.1; $CaSO_4$, 0.015; magnesium oxide (MgO), 0.01; calcium hydroxide ($Ca(OH)_2$), 0.1; sodium chloride (NaCl), 0.05; wheat bran, 0.23; and trace elements as needed. These ingredients are then mixed thoroughly, poured into moulds and allowed to solidify for 24 h (modified from Habib *et al.*, 1991). However, supplements for stall-fed animals should be spread over the hand-fed fodder or feeds in the trough, and molasses may not always be necessary. Pastured sheep offered 175 g $head^{-1}$ day^{-1} of a mineral mixture containing 46 g S kg^{-1} while grazing consumed enough (167 g day^{-1}) to provide 1.15 g S day^{-1}, and probably contribute to improved performance (White *et al.*, 1997).

Nutritive value of inorganic sources

The value of mineral supplements as sources of sulphur for ruminants depends entirely on the coavailability of the other factors needed for microbial protein synthesis, and their assessment, like that of feeds, presents unique problems. While the literature contains many comparative assessments of different sulphur sources (Henry and Ammerman, 1995), those based on sulphur balance give little or no indication of ability to promote rumen microbial synthesis. Techniques based on *in vitro* cultures of rumen microbes get closer to true nutritional value to the extent that they adequately simulate ruminal events *in vivo*, but they have given inconsistent results (Henry and Ammerman, 1995). Assessments based on animal responses to supplementation have generally been more consistent. For example, elemental sulphur is accorded low nutritive values of 0–36% by *in vitro* methods, 28–69% by balance methods and 73–102% by sheep growth tests, when compared with Na_2SO_4 or methionine (Henry and Ammerman, 1995); the latter range is the most dependable because there is clear evidence that elemental sulphur is of value to ruminants (Slyter *et al.*, 1988) and potentiates the same rumen-based antagonisms as other sulphur sources (see Chapter 11 for copper (Cu) × Mo × S). There is no reason on past evidence to differentiate between sources in terms of nutrient value g^{-1} S for ruminants.

Methionine analogues

Much attention is being given to the possible advantages of feeding sulphur in the form of methionine analogues (e.g. malyl and methylmalyl methionine), which are not degraded by rumen microbes. Where rumen microbial protein synthesis cannot provide sufficient methionine to meet the needs of the animal and methionine is the most limiting amino acid, it may be more economical to feed such 'protected' sulphur sources than to feed sulphur in expensive, non-degradable protein sources, such as fish-meal. However, the first priority must be to maximize microbial protein synthesis, using inorganic sulphur supplements, if necessary. All claims for benefits from feeding costly methionine analogues should be evaluated against the background of what was or what might have been achieved with cheaper, degradable sulphur and nitrogen sources. If the supply of sulphur is so poor that it limits rumen microbial activity, the provision of 'protected' forms might be disadvantageous. The major use of methionine analogues is for types of animal and production with high methionine requirements (e.g. wool production from merinos (Coetzee *et al.*, 1995)). In one study, it was estimated that an additional 100 g day^{-1} of lupin seeds would have been needed to provide the improvements in wool and body growth provided by 3 g protected methionine day^{-1} (Mata *et al.*, 1997). Responses to methionine analogues in high-yielding dairy cows in a recent study were confined to improvements in milk-fat yield (Overton *et al.*, 1996); previous work had failed to detect any improvement in yield, even when methionine was limiting (Casper *et al.*, 1987). When fed at high levels of 12–24 g day^{-1} to sheep, methionine analogues reduce appetite (Kandylis, 1984a).

Sulphur Requirements of Ruminants

Needs of the rumen microbiota

The definition of requirements of ruminants for sulphur poses unique problems, which need a unique solution. Like those of monogastric animals, requirements are best stated as needs for protein and S-amino acids, with the differences that the supply is that which leaves the rumen rather than that ingested and supply can be manipulated, within certain constraints, by adding sulphur to the diet. The first attempt to define requirements in terms of rumen microbial protein synthesis was made by the Agricultural Research Council (ARC, 1980). This assumed that any sulphur and nitrogen naturally present in feeds would be appropriately balanced for microbial synthesis. If additional degradable nitrogen was required, it was recommended that sulphur should also be supplied in the S:N ratio found in microbial protein, 0.067. This may overestimate sulphur needs, because lower ratios of 0.054 and 0.046 have been reported for mixed rumen bacteria and protozoa, respectively (Harrison and McAllan, 1980). The optimum S:N ratio used by ARC (1980) was similar to that originally proposed by Loosli (1952) for a totally different reason, namely that animal tissues and products mostly contained S:N in a ratio close to 0.067. The coincidence is important, because it means that the efficiencies of utilization of microbial protein for diverse purposes, such as growth and milk production, will be broadly similar.

The efficiency of protein use for fleece growth, however, is singularly low. Much of the early work on sulphur nutrition was conducted by Moir and his associates in Western Australia with mature male merino sheep, and they concluded that, when low-quality roughages were supplemented with urea, total S:N ratios should not fall below 0.1 (e.g. Moir, 1970; Bird, 1974), a higher ratio than that found in microbial protein. There are three possible explanations which are consistent with the overriding needs of rumen microbes: firstly, the high proportion of sulphur requirement going to wool synthesis created a surplus of sulphur-free amino acids from which nitrogen was recycled to the rumen and reincorporated, with additional dietary sulphur, into microbial protein; secondly, the sulphur provided by the basal diet (*c.* 1 g kg^{-1} DM, predominantly from oat hulls or wheat straw) was inefficiently utilized because of poor or slow degradability; and, thirdly, the diets provided too much urea nitrogen (usually 20–30 g kg^{-1} DM – attainment of the ideal S:N ratio is only necessary up to the optimum level of provision for degradable nitrogen). The fact that cattle needed less sulphur than sheep to digest the same urea-supplemented, low-quality diet (Bird, 1974; Kennedy and Siebert, 1972a) suggests that differences in S:N requirement (and hence opportunity for recycling) are involved. Anaerobic fungi may be particularly sensitive to the ruminal supply of sulphur, and the proportion of S-amino acids in protein leaving the rumen can be increased by sulphur supplements, which increase the fungal population, rich in S-amino acids (Weston *et al.*, 1988).

Formulation of sulphur needs

The ARC (1980) approach harnessed the calculation of sulphur requirements to a complex formula for estimating protein requirements but, since such formulae are now part and parcel of ration formulation by computer, they present no major obstacle. Application of the formula resulted in minimum requirements ranging from 1.1 to 1.6 g S kg^{-1} DM and were in keeping with feeding trials with sheep and dairy cattle, which showed little or no response to added sulphur with basal diets containing 1.0–1.5 g S kg^{-1} DM (ARC, 1980). The responses to sulphur supplementation obtained recently in sheep grazing pastures containing 0.95–1.18 g S kg^{-1} DM also support the formula, although *total* S:N ratios of 0.08–0.09 appeared adequate (Mata *et al.*, 1997). The lowest sulphur requirement is for the maintenance of cattle (Bird, 1974). The ARC (1980) protein system was developed by AFRC (1993) into one which divided dietary crude protein into three components, the quickly, slowly and non-rumen-degradable protein, but no provision was made for sulphur. At the present state of knowledge, all that can be done is to provide quickly degradable sulphur (as Na_2SO_4) in a ratio of 0.067 to the conventional supplementary source of quickly available nitrogen (urea). Further information on the degradability of feed and synthetic sources of sulphur is urgently needed to match the data on protein classification. The supply of both nitrogen and sulphur in slowly degradable forms (e.g. biuret and $CaSO_4$) should theoretically improve microbial protein synthesis by up to 25%, compared with that attained with urea and Na_2SO_4, by improving microbial capture and thus minimizing losses as ammonia and sulphide.

Requirements for wool and mohair production

Recent evidence from studies with angora goats, in which responses in production and metabolic parameters were used to define optimum sulphur intakes (Qi *et al.*, 1992, 1993), appears to be at odds with the ARC (1980) model and the modification now suggested, indicating high optimum S:N ratios of 0.1–0.4. Furthermore, it has been argued that other ruminants may also benefit from higher S:N ratios than those recommended by ARC (1980) (Qi *et al.*, 1994). There is no doubt that protein and therefore sulphur requirements are high for mohair and wool growth, because the efficiency of use of microbial (or undegraded dietary) protein for fleece growth is only 26%, compared with around 80% for most other purposes. Furthermore, merino flocks selected for high fleece production show greater responses to supplements of S-amino acids than those selected for low fleece weight (Williams *et al.*, 1972). Optimal dietary sulphur concentrations for wool and fleece growth may well be nearly twice those for other purposes (approximately 2.5 vs. 1.25 g S kg^{-1} DM). However, these additional needs are unlikely to be met by increasing degradable S:N ratios above 0.067. Indeed, the optimal supplements in the experiments of Qi *et al.* (1992, 1993) provided degradable S:N ratios of between 0.047 and 0.088 for mature goats and between 0.061 and 0.11 for growing kids. Discrepancies arise when these optima are translated inappropriately to ratios of total dietary S:N for fibre-rich basal diets, high in urea nitrogen (66% of total nitrogen) but containing no added sulphur. For all production purposes, sulphur requirements should be calculated on the basis of common rumen microbial needs, not total dietary sulphur concentrations or total S:total N ratios.

Sulphur Toxicity

Direct effects

The margin between the desirable and harmful concentration of sulphur in the ruminant diet is surprisingly small, at two- to threefold, reductions in appetite and growth rate being found in cattle and sheep given diets with 3–4 g S kg^{-1} DM (for review, see Kandylis, 1984b). This low tolerance is most commonly ascribed to sulphide toxicity (Bird, 1972b), and the first site to be affected is the rumen, where sulphide is generated and motility may be impaired. The variable tolerances to different amounts and sources of sulphur in the literature (Kandylis, 1984b) partly reflect differences in the rate of ingestion of degradable sulphur, the rate of sulphide absorption across the rumen wall (which is pH-dependent) and the rate of sulphide capture by rumen microbes. Simultaneous addition of urea can lessen the depression of appetite and digestibility caused by sulphur alone (Bird, 1974). However, the performance of steers on a concentrate diet was adversely affected when ammonium sulphate ($(NH_4)_2SO_4$) was used to raise dietary sulphur from 2.0 to 2.5 g kg^{-1} DM, a very modest increase (Zinn *et al.*, 1997). Cattle reduced both water and food consumption when offered drinking-water containing 2.83 g SO_4 as Na_2SO_4 l^{-1} for 30 days (Weeth and Hunter, 1971); the treatment was equivalent to an increase in dietary sulphur to 8.4 g kg^{-1} DM. There is evidence that the rumen microflora adapt to high sulphate intakes and gradually increase their dissimilatory capacity, thus increasing rumen sulphide concentrations (Cummings *et al.*, 1995a, b). Absorbed sulphide may have harmful effects on the central nervous system (CNS), whether from drinking-water (Smart *et al.*, 1986) or dietary sulphur supplements (Raisbeck, 1982; Cummings *et al.*, 1995a, b), contributing to the development of polioencephalomalacia (PEM). This disorder developed in calves and lambs when a dietary acidifier, included to reduce risk of urolithiasis, was changed from ammonium bicarbonate to ammonium sulphate (Jeffrey *et al.*, 1994). Losses from a clinically similar PEM in lambs were associated with a high sulphur concentration (4.1 g S kg^{-1} DM) in a concentrate fed *ad libitum* (Low *et al.*, 1996). The disorders differed in fine pathology detail and in responsiveness to vitamin B_1 from those associated with thiamine deficiency, the common cause of PEM. These adverse effects are in complete contrast to those found in non-ruminants, in which undegraded sulphate entering the small and large intestine causes an osmotically driven diarrhoea.

Indirect effects

Chronic exposure to high sulphur intakes in the diet or drinking-water can have adverse indirect effects on ruminants, particularly by inducing copper deficiency, with or without the assistance of molybdenum (see Chapter 11). As intakes of inorganic or organic sulphur rise, concentrations of sulphide in digesta reaching the omasum increase, while those of soluble copper in strained rumen fluid decrease (Bird, 1970). Sulphates can be removed from 'saline' water supplies by reverse osmosis (Smart *et al.*, 1986).

Toxic sulphur compounds

Some of the sulphur in crops and forages is present in potent organic forms. The most widely studied is S-methylcysteine sulphoxide (SMCO), a free amino acid resembling methionine, which commonly causes a haemolytic anaemia in ruminants consuming large quantities of brassicas, such as kale (Barry *et al.*, 1981; Barry and Manley, 1985) or rape (Suttle *et al.*, 1987) (for review, see Whittle *et al.*, 1976). Conversion of SMCO to dimethyl disulphoxide in the rumen plays a key role in pathogenesis (Smith, 1978). The levels of SMCO can be reduced by plant breeding and limiting the available sulphur supply from the soil (McDonald *et al.*, 1981). Oral dosing with methionine provides partial protection from kale anaemia, while stimulating live-weight gain and wool growth; it was suggested that methionine reduced the normally rapid and complete degradation of SMCO in the rumen (Barry and Manley, 1985).

References

AFRC (1993) *Energy and Protein Requirements of Ruminants: an Advisory Manual Prepared by the AFRC Technical Committee on Responses to Nutrients.* CAB International, Wallingford, UK.

Ahmad, M.R., Allen, V.G., Fontenot, J.P. and Hawkins, G.W. (1995) Effect of sulphur fertilization on chemical composition, ensiling characteristics and utilisation by lambs on sorghum silage. *Journal of Animal Science* 73, 1803–1810.

ARC (1980) *The Nutrient Requirements of Ruminant Livestock.* Commonwealth Agricultural Bureaux, Farnham Royal, UK, pp. 166–168.

Barry, T.N. and Manley, T.R. (1985) Responses to oral methionine supplementation in sheep fed on kale (*Brassica oleracea*) diets containing S-methyl-L-cysteine sulphoxide. *British Journal of Nutrition* 54, 753–761.

Barry, T.N., Reid, T.C., Miller, K.R. and Sadler, W.A. (1981) Nutritional evaluation of kale (*Brassica oleracea*) diets 2. *Journal of Agricultural Science, Cambridge* 96, 269–282.

Beaton, J.D., Tisdale, S.L. and Platou, J. (1971) *Crop Responses to Sulphur in North America.* Technical Bulletin 18, Sulphur Institute, Washington, DC, pp. 1–10.

Beever, D.E. (1996) Meeting the protein requirements of ruminant livestock. *South African Journal of Animal Science* 26, 20–26.

Bird, P.R. (1970) Sulphur metabolism and excretion studies in ruminants. III. The effect of sulphur intake on the availability of copper in sheep. *Proceedings of the Australian Society of Animal Production* 8, 212–218.

Bird, P.R. (1972a) Sulphur metabolism and excretion studies in ruminants. V. Ruminal desulphuration of methionine and cysteine. *Australian Journal of Biological Sciences* 25, 185–193.

Bird, P.R. (1972b) Sulphur metabolism and excretion studies in ruminants. X. Sulphide toxicity in sheep. *Australian Journal of Biological Sciences* 25, 1087–1098.

Bird, P.R. (1974) Sulphur metabolism and excretion studies in ruminants. XIII. Intake and utilisation of wheat straw by sheep and cattle. *Australian Journal of Agricultural Sciences* 25, 631–642.

Bird, S.H. and Leng, R.A. (1985) Productivity responses to eliminating protozoa from the rumen of sheep. In: Leng, R.A. *et al.* (eds) *Reviews in Rural Science* 6. University of New England Publishing Unit, Armidale, New South Wales, pp. 109–117.

Bishara, H.N. and Bray, A.C. (1978) Competition between molybdate and sulphate for renal tubular reabsorption in sheep. *Proceedings of the Australian Society of Animal Production* 12, 123.

Bogdanovic, B. (1983) A note on supplementing whole wheat grain with molasses, urea, minerals and vitamins. *Animal Production* 37, 459–460.

Bray, A.C. and Till, A.R. (1975) Metabolism of sulphur in the gastro-intestinal tract. In: McDonald, I.W. and Warner, A.C.I. (eds) *Digestion and Metabolism in the Ruminant.* University of New England Publishing Unit, Armidale, New South Wales, pp. 243–260.

Buttrey, S.A., Allen, V.G., Fontenot, J.P. and Renau, R.B. (1986) Effect of sulfur fertilization on chemical composition, ensiling characteristics and utilisation of corn silage by lambs. *Journal of Animal Science* 63, 1236–1245.

Casper, D.P., Schingoethe, D.J., Yang, C.-M.J. and Mueller, C.R. (1987) Protected methionine supplementation with extruded blend of soybeans and soybean meal for dairy cows. *Journal of Dairy Science* 70, 321–330.

Coetzee, J., de Wet, P.J. and Burger, W.J. (1995) Effects of infused methionine, lysine and rumen-protected methionine derivatives on nitrogen retention and wool growth of Merino wethers. *South African Journal of Animal Science* 25, 87–94.

Cornforth, I.S., Stephen, R.C., Barry, T.N. and Baird, G.A. (1978) Mineral content of swedes, turnips and kale. *New Zealand Journal of Experimental Agriculture* 6, 151–156.

Cummings, B.A., Caldwell, D.R., Gould, D.H. and Hawar, D.W. (1995a) Identity and interactions of rumen microbes associated with sulphur-induced polioencephalomalacia in cattle. *American Journal of Veterinary Research* 56, 1384–1389.

Cummings, B.A., Gould, D.H., Caldwell, D.R. and Hawar, D.W. (1995b) Ruminal microbial alterations associated with sulphide generation in steers with dietary sulphate-induced polioencephalomalacia. *American Journal of Veterinary Research* 56, 1390–1395.

Dick, A.T. (1956) Molybdenum in animal nutrition. *Soil Science* 81, 229–258.

Durand, M. and Komisarczuk, K. (1988) Influence of major minerals on rumen microbiota. *Journal of Nutrition* 118, 249–260.

Gawthorne, J.M. and Nader, C.J. (1976) The effect of molybdenum on the conversion of sulphate to sulphide and microbial protein sulphur in the rumen of sheep. *British Journal of Nutrition* 35, 11–23.

Goh, K.M. and Nguyen, M.L. (1997) Estimating net annual soil sulphur mineralisation in New Zealand grazed pastures using mass balance models. *Australian Journal of Agricultural Research* 48, 477–484.

Grace, N.D. and Suttle, N.F. (1979) Some effects of sulphur intake on molybdenum metabolism in sheep. *British Journal of Nutrition* 41, 125–136.

Habib, C., Basit Ali Shah, S., Wahidullah, Jabbar, G. and Ghufranullah (1991) The importance of urea–molasses blocks and by-pass protein in animal production: the situation in Pakistan. In: *Proceedings of Symposium on Isotope and Related Techniques in Animal Production and Health.* IAEA, Vienna, pp. 133–144.

Harrison, D.G. and McAllan, A.B. (1980) Factors affecting microbial growth yields in the reticulo-rumen. In: Ruckebusch, Y. and Thivend, P. (eds) *Digestive Physiology and Metabolism in Ruminants.* MTP Press, Lancaster, pp. 205–226.

Hegarty, R.S., Nolan, J.V. and Leng, R.A. (1991) Sulphur availability and microbial fermentation in the fauna free rumen. *Archiv für Animal Nutrition, Berlin* 41, 725–736.

Hegarty, R.S., Nolan, J.V. and Leng, R.A. (1994) The effects of protozoa and of supplementation with nitrogen and sulphur on digestion and microbial metabolism in the rumen of sheep. *Australian Journal of Agricultural Research* 45, 1215–1227.

Henry, P.R. and Ammerman, C.B. (1995) Sulfur bioavailability. In: Ammerman, C.B., Baker, D.H. and Lewis, A.J. (eds) *Bioavailability of Nutrients For Animals.* Academic Press, New York, pp. 349–366.

Hoeft, R.G. and Fox, R.H. (1986) Plant response to S in the Midwest and Northeastern United States. In Tabatabai, M.A. (ed.) *Sulfur in Agriculture.* American Society of Agronomy, Madison, Wisconsin, pp. 345–356.

Jeffrey, M., Duff, J.P., Higgins, R.J., Simpson, V.R., Jackman, R., Jones, T.O., Machie, S.C. and Livesey, C.T. (1994) Polioencephalomalacia associated with the ingestion of ammonium sulphate by sheep and cattle. *Veterinary Record* 134, 343–348.

Jones, M.B., Rendig, V.V., Torell, D.T. and Inouye, T.S. (1982) Forage quality for sheep and chemical composition associated with sulfur fertilization on a sulfur deficient site. *Agronomy Journal* 74, 775–780.

Jumba, I.O., Suttle, N.F. and Wandiga, S.O. (1996) Mineral composition of tropical forages in the Mount Elgon region of Kenya. *Tropical Agriculture* 73, 108–112.

Kamprath, E.J. and Jones, U.S. (1986) Plant response to sulfur in southeastern United States. In: Tabatabai, M.A. (ed.) *Sulfur in Agriculture.* American Society of Agronomy, Madison, Wisconsin, pp. 323–343.

Kandylis, K. (1984a) The role of sulphur in ruminant nutrition: a review. *Livestock Production Science* 11, 611–624.

Kandylis, K. (1984b) Toxicity of sulphur in ruminants: review. *Journal of Dairy Science* 67, 2179–2187.

Kennedy, P.M. (1974) The utilisation and excretion of sulphur in cattle fed on tropical roughages. *Australian Journal of Agricultural Research* 25, 1015–1022.

Kennedy, P.M. and Milligan, L.P. (1978) Quantitative aspects of the transformations of sulphur in sheep. *British Journal of Nutrition* 39, 65–84.

Kennedy, P.M. and Siebert, B.D. (1972a) The utilisation of spear grass (*Heteropogon contortus*). II. The influence of sulphur on energy intake and rumen and blood parameters in sheep and cattle. *Australian Journal of Agricultural Research* 23, 45–46.

Kennedy, P.M. and Siebert, B.D. (1972b) The utilisation of spear grass (*Heteropogon contortus*). III. The influence of the level of dietary sulphur on the utilisation of spear grass by sheep. *Australian Journal of Agricultural Research* 24, 143–152.

Kincaid, R. (1988) Macro elements for ruminants. In: Church, D.C. (ed.) *The Ruminant Animal – Digestive Physiology and Nutrition.* Prentice Hall, New Jersey, pp. 326–330.

Langlands, J.P. and Sutherland, H.A.M. (1973) Sulphur as a nutrient for Merino sheep. I. Storage of sulphur in tissues and wool and its secretion in milk. *British Journal of Nutrition* 30, 529–535.

Ledgard, S.F. and Upsdell, M.P. (1991) Sulphur inputs from rainfall throughout New Zealand. *New Zealand Journal of Agricultural Research* 34, 105–111.

Leng, R.A. (1990) Factors affecting the utilisation of 'poor-quality' forages by ruminants particularly under tropical conditions. *Nutrition Research Reviews* 3, 277–303.

Loosli, J.K. (1952) *Feed Age* 2, 44–46.

Lovett, T.D., Coffey, M.T., Miles, R.D. and Combs, G.E. (1986) Methionine, choline and sulphate interrelationships in the diet of weanling pigs. *Journal of Animal Science* 63, 467–471.

Low, J.C., Scott, P.R., Howie, F., Lewis, M., Fitzsimons, J. and Spence, J.A. (1996) Sulphur-induced polioencephalomalacia in sheep. *Veterinary Record* 138, 327–329.

McClaren, R.G. (1975) Marginal sulphur supplies for grassland herbage in south east Scotland. *Journal of Agricultural Science, Cambridge* 85, 571–573.

McCollum, E.V. (1956) *A History of Nutrition.* Houghton Miffin, Boston, Massachusetts.

McDonald, R.L., Manley, T.R., Barry, T.W., Forss, D.A. and Sinclair, A.G. (1981) Nutritional evaluation of kale (*Brassica oleracea*) diets. 3: Changes in plant composition induced by soil fertility practices with special reference to SMCO and glucosinolates. *Journal of Agricultural Science, Cambridge* 97, 13–23.

MAFF (1990) *UK Tables of the Nutritive Value and Chemical Composition of Foodstuffs.* Givens, D.I. (ed.), Rowett Research Services, Aberdeen, UK.

Masters, D.G., Stewart, C.A., Mata, G. and Adams, N.R. (1996) Responses in wool and liveweight when different sources of dietary protein are given to pregnant and lactating ewes. *Animal Science* 62, 497–506.

Mata, G., Masters, D.G., Chamberlain, N.L. and Young, P. (1997) Production and glutathione responses to rumen-protected methionine in young sheep grazing dry pastures over summer and autumn. *Australian Journal of Agricultural Research* 48, 1111–1120.

Minson, D.J. (1990) *Forage in Ruminant Nutrition.* Academic Press, Sydney, pp. 178–188.

Moir, R.J. (1970) Implications of the N:S ratio and differential recycling. In: Muth, O.H. and Oldfield, J.E. (eds) *Sulphur in Nutrition – Symposium Proceedings.* AVI Publishing Company, Westport, Connecticut, pp. 165–170.

Morrison, M., Murray, R.M. and Boniface, A.N. (1990) Nutrient metabolism and rumen microorganisms in sheep fed a poor quality tropical grass hay supplemented with sulphate. *Journal of Agricultural Science* 115, 269–275.

Ojeniyi, S.O. and Kayode, G.O. (1993) Response of maize to copper and sulphur in tropical regions. *Journal of Agricultural Science, Cambridge* 120, 295–299.

Overton, T.R., LaCount, D.W., Cicela, T.M. and Clark, J.H. (1996) Evaluation of a ruminally protected methionine product for lactating dairy cows. *Journal of Dairy Science* 79, 631–638.

Qi, K., Lu, C.D., Owens, F.N. and Lupton, C.J. (1992) Sulphate supplementation of Angora goats: metabolic and mohair responses. *Journal of Animal Science* 70, 2828–2837.

Qi, K., Lu, C.D. and Owens, F.N. (1993) Sulphate supplementation of growing goats: effects on performance, acid–base balance and nutrient digestibilities. *Journal of Animal Science* 71, 1579–1587.

Qi, K., Owens, F.N. and Lu, C.D. (1994) Effects of sulphur deficiency on performance of fiber-producing sheep and goats: a review. *Small Ruminant Research* 14, 115–126.

Raisbeck, M.F. (1982) Is polioencephalomalacia associated with high sulphate diets? *Journal of the American Veterinary Medical Association* 180, 1303–1305.

Rasmussen, P.E. and Kresge, P.O. (1986) Plant responses to S in the western United States. In: Tabatabai, M.A. (ed.) *Sulfur in Agriculture.* American Society of Agronomy, Madison, Wisconsin, pp. 357–374.

Slyter, L.L., Chalupa, W. and Oltjen, R.R. (1988) Response to elemental sulphur by calves and sheep fed purified diets. *Journal of Animal Science* 66, 1016–1027.

Smart, M.E., Cohen, R., Christensen, D.A. and Williams, C.M. (1986) The effects of sulphate removal from the drinking water on the plasma and liver copper and zinc concentrations of beef cows and their calves. *Canadian Journal of Animal Science* 66, 669–680.

Smith, G.M. and White, C.L. (1997) A molybdenum–sulphur–cadmium interaction in sheep. *Australian Journal of Agricultural Research* 48, 147–154.

Smith, R.H. (1978) S-methyl cysteine sulphoxide: the kale anaemia factor (?). *Veterinary Science Communications* 2, 47–61.

Spears, J.W., Burns, J.C. and Hatch, P.A. (1985) Sulphur fertilization of cool season grasses and effect on utilisation of minerals, nitrogen and fiber by steers. *Journal of Dairy Science* 68, 347–355.

Suttle, N.F. (1991) Mineral supplementation of low quality roughages. In: *Isotope and Related Techniques in Animal Production and Health.* International Atomic Energy Agency, Vienna, pp. 101–114.

Suttle, N.F., Jones, D.G., Woolliams, C. and Woolliams, J.A. (1987) Heinz body anaemia in lambs with deficiencies of copper and selenium. *British Journal of Nutrition* 58, 539–548.

Thomas, W.E., Loosli, J.K., Williams, H.H. and Maynard, L.A. (1951) The utilisation of inorganic sulphates and urea nitrogen by lambs. *Journal of Nutrition* 43, 515–523.

Todd, J.R. (1972) Copper, molybdenum and sulphur contents of oats and barley in relation to chronic copper poisoning in housed sheep. *Journal of Agricultural Science, Cambridge* 79, 191–195.

Ward, K.A. and Nancarrow, C.D. (1991) The genetic engineering of production traits in domestic animals. *Experientia* 47, 913–921.

Weeth, H.J. and Hunter, J.E. (1971) Drinking of sulphate water by cattle. *Journal of Animal Science* 32, 277–281.

Weston, R.H., Lindsay, J.R., Purser, D.B., Gordon, G.L.R. and Davis, P. (1988) Feed intake and digestion responses in sheep to the addition of inorganic sulfur to a herbage diet of low sulfur content. *Australian Journal of Agricultural Research* 39, 1107–1119.

Whanger, P.D. (1972) Sulfur in ruminant nutrition. *World Review of Nutrition and Diet* 15, 225–237.

White, C.L., Kumagai, H. and Barnes, M.J. (1997) The sulphur and selenium status of pregnant ewes grazing Mediterranean pastures. *Australian Journal of Agricultural Research* 48, 1081–1087.

Whittle, P.J., Smith, R.H. and McIntosh, A. (1976) Estimation of S-methyl sulphoxide (kale anaemia factor) and its distribution among brassica and forage crops. *Journal of Science in Food and Agriculture* 27, 633–642.

Williams, A.J., Robards, G.E. and Saville, D.G. (1972) Metabolism of cystine by Merino sheep genetically different in wool production. II. The responses in wool growth to abomasal infusion of L-cystine and DL-methionine. *Australian Journal of Biological Science* 25, 1269–1276.

Zinn, R.A., Aloarez, E., Mendez, M., Montano, M., Ramirez, E. and Shen, Y. (1997) Influence of dietary sulphur level in growth, performance and digestive function in feedlot cattle. *Journal of Animal Science* 75, 1723–1728.

10 Cobalt

Introduction

Cobalt was first shown to be an essential nutrient for sheep and cattle as an outcome of Australian investigations of two naturally occurring diseases, 'coast disease' of sheep (Lines, 1935; Marston, 1935) and 'wasting disease', or enzootic marasmus, of cattle (Underwood and Filmer, 1935). The findings were confirmed in sheep grazing similar calcareous soils of aeolian origin on the west coast of Scotland in the same decade (for review, see Suttle, 1988). Progress in understanding the mode of action of cobalt in the animal organism was slow until 1948, when two groups of workers independently discovered that the anti-pernicious anaemia factor, subsequently designated vitamin B_{12}, contained cobalt (Rickes *et al.*, 1948; Smith, 1948). Within 3 years of that discovery, remission of all signs of cobalt deficiency in lambs was secured with parenteral injections of the vitamin (Smith *et al.*, 1951). Vitamin B_{12} has the general formula $C_{63}H_{88}N_{14}O_{14}PCo$, has a relative molecular mass of 1355 and contains 4.4% of cobalt. Cobalt deficiency in ruminants is therefore a vitamin B_{12} deficiency, brought about by the inability of the rumen microorganisms, when dietary cobalt is inadequate, to synthesize sufficient vitamin B_{12}. Non-ruminant animals cannot incorporate cobalt into its physiologically active form and must obtain their vitamin B_{12} either preformed in the food or indirectly, by ingesting faeces which has been enriched by the synthetic activity of microbes in the large intestine. Clear evidence of cobalt deficiency, as distinct from vitamin B_{12} deficiency, in non-ruminant animals has not yet appeared and this chapter will therefore deal almost exclusively with ruminants. Significant growth responses have been demonstrated in some leguminous plants and some non-legumes grown on cobalt-deficient soils treated with cobalt (Young, 1979).

© CAB *International* 1999. *Mineral Nutrition of Livestock*
(E.J. Underwood and N.F. Suttle)

Sources of Cobalt

Forages

All common pasture plants and forages contain cobalt in concentrations that vary widely with the species and soil conditions, though not with advancing maturity (Minson, 1990). In ten samples of mixed pasture herbage from Scotland, Mitchell (1957) found the cobalt concentrations to range from 0.02 to 0.22 (mean 0.09) mg kg^{-1} dry matter (DM). Legumes are usually richer in cobalt (Co) than grasses grown in the same conditions and in a study of red clover and ryegrass from 15 sites in Scotland (Mitchell, 1945), mean concentrations were 0.35 and 0.18 mg kg^{-1} DM, respectively. Lower values have been reported for Japanese-grown pasture species, but the difference in favour of legumes was greater (Hayakawa, 1962) (mean 0.12 and 0.03 mg Co kg^{-1} DM). The advantage of legumes can be lost if soils are cobalt-deficient (Minson, 1990). Twofold variation among grass species grown under similar conditions has been reported, with *Phleum pratense* and *Dactylis glomerata* poor sources compared with *Lolium perenne* (Minson, 1990). The cobalt concentrations in pasture grasses are reduced as the soil pH rises, as shown in Chapter 2 (Fig. 2.5). Waterlogging, in contrast, greatly increases herbage cobalt concentrations.

Ingested soil

Soil contamination can greatly increase herbage cobalt concentrations, particularly in spring and autumn. This ingested cobalt is partially available for ruminal synthesis of vitamin B_{12} (Brebner *et al.*, 1987) and is believed to alleviate cobalt deficiency at times (Clark *et al.*, 1989). Washing herbage will often reduce measured cobalt concentrations but does not necessarily improve the assessment of cobalt supply from the pasture. Grass species can differ in the degree to which they are contaminated with soil and in their ability to absorb cobalt from the soil (Jumba *et al.*, 1995).

Concentrates

Data on cobalt in grains and other concentrates are meagre. Cereal grains are poor sources of cobalt, with concentrations usually within the range 0.01–0.06 mg kg^{-1} DM, and are used to produce cobalt-deficient diets (Field *et al.*, 1988; Kennedy *et al.*, 1991b, 1992). Like most trace elements, cobalt is highly concentrated in the outer bran layers, so that the concentrations in bran and pollard are substantially greater than they are in whole grain. Feeds of animal origin, other than liver meal, are mostly poor sources of cobalt. This applies particularly to milk and milk products. The average cobalt concentration of normal cow's milk varies from 0.5 to 0.9 μg l^{-1} (Kirchgessner, 1959), but it is possible to increase these values several-fold by heavy supplementation with cobalt salts. Unfortunately, any cobalt not present as vitamin B_{12} in milk will not be incorporated into vitamin B_{12}, because stimulation of the oesophageal groove reflex will direct the cobalt straight to the abomasum, bypassing the rumen. Milk, however, is an important source of vitamin B_{12} for the preruminant, and concentrations of vitamin B_{12} in the milk can be increased substantially by supplementing the mother with cobalt (Hart and Andrews, 1959; Skerman and O'Halloran, 1962; Quirk and Norton, 1988).

Metabolism

Rumen synthesis of vitamin B_{12}

The ruminant makes extremely inefficient use of its dietary cobalt. In the first place, rumen microbes partition cobalt between active (cobalamins) and physiologically inactive vitamin B_{12}-like compounds (corrinoids) that the ruminant can neither absorb nor use (Gawthorne, 1970). The production of true vitamin B_{12} from cobalt accounted for about 15% of total vitamin B_{12} production in the deficient sheep and only about 3% in the cobalt-sufficient sheep (Smith and Marston, 1970a). Increasing the roughage content and total intake of the diet increases vitamin B_{12} production in the rumen (Sutton and Elliot, 1972; Hedrich *et al.*, 1973). Ruminal synthesis of vitamin B_{12} responds within hours to changes in dietary cobalt supply, but the efficiency of capture decreases as cobalt intake increases. Using area under the plasma vitamin B_{12} vs. time-after-dosing curve as a measure of vitamin synthesis by deficient lambs, synthesis increased in proportion to the square root of cobalt dose over the range 1–32 mg in one study (Fig. 10.1).

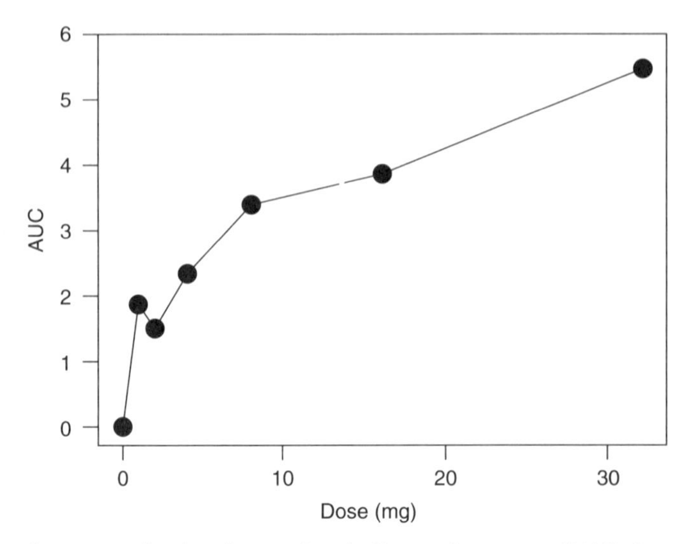

Fig. 10.1. The area under the plasma vitamin B_{12} vs. time curve (AUC) shows a curvilinear relationship to the amount of cobalt supplement when Co-deprived lambs are repleted with single doses of cobalt and kept on a low Co diet (data from Suttle *et al.*, 1989): limitations to vitamin B_{12} synthesis by rumen microbes are implied.

Absorption

Dietary corrinoids become bound to R proteins from sources such as saliva, but the complexes are normally digested by pancreatic enzymes. Cobalamins (Cbl) synthesized in the rumen are released during subsequent digestion and selectively bound by intrinsic factor (IF), produced by parietal cells in the abomasum in ruminants (McKay and McLeay, 1981). Some 10,000–24,000 international units (IU) of IF is secreted in gastric juice daily in sheep, presumably under the influence of factors that stimulate gastric secretion, such as eating. In other species, secretion of dihydroxy bile acids may inhibit the binding of Cbl to IF (Teo *et al.*, 1981), but bile salts facilitate binding of the Cbl:vitamin B_{12} complex to receptors in the ileal mucosal brush border, which complete the selective absorption of Cbl (Smith, 1997). Surprisingly little is known about cobalt, Cbl or analogue absorption by ruminants, but early evidence suggested that cobalt was more slowly and less completely absorbed in ruminants than in monogastric species (Rothery *et al.*, 1953). Of the vitamin B_{12} that is produced in the rumen, about half is lost during passage through the alimentary tract and only 3–5% is absorbed by Co-deficient sheep (Kercher and Smith, 1955; Marston, 1970). Appreciably higher apparent absorption is attained at higher dietary cobalt intakes (Hedrich *et al.*, 1973). Features of vitamin B_{12} absorption found in non-ruminant species, such as an increase in absorptive capacity in pregnancy (Nexo and Olsen, 1982) and in the colostrum-fed neonate (Ford *et al.*, 1975), may well also occur in ruminants.

Mucosal transfer and transport

The Cbl:IF complex enters the enterocyte by receptor-mediated endocytosis. The processes by which Cbl is then released, bound to carrier and presented to the bloodstream are poorly understood (Smith, 1997). The transport of absorbed cobalamins from the serosal surface is achieved by plasma proteins called transcobalamins (TC), and mammalian species show important differences in the distribution and binding properties of the three major transport proteins, TC0, TCI and TCII. There is usually excess vitamin B_{12} binding capacity in plasma and it increases during infection and inflammation, suggesting an increased cellular demand for the vitamin. This has been confirmed *in vitro* by greatly increased cobalamin uptake by stimulated lymphocytes (Quadros *et al.*, 1976) and decreased candidacidal activity in neutrophils from cobalt-deficient cattle (MacPherson *et al.*, 1987).

Storage

Although it is customary to refer to excess supplies of vitamin B_{12} as being 'stored' in the liver, the early work of Marston (1970) gave little evidence of this. There was no significant increase in liver vitamin B_{12} concentrations after 36 weeks of daily oral treatment with cobalt at 10 and 100 times a level regarded as 'marginal' (0.1 mg day^{-1}). Plasma vitamin B_{12} was slightly more responsive (Fig. 10.2), and Marston (1970) concluded that the poor capacity of sheep to store vitamin B_{12} in liver was 'a limiting factor' in the aetiology of

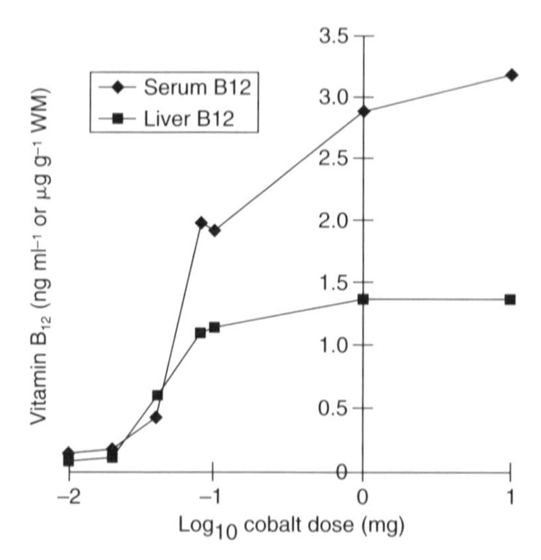

Fig. 10.2. When cobalt-deprived sheep are continuously repleted with diets providing 0.1 to 10 mg Co day^{-1}, vitamin B_{12} concentrations reach an earlier plateau in liver than in serum, suggesting limits to hepatic storage capacity (data from Marston, 1970).

cobalt-responsive disorders. Vitamin B_{12} resembles the other water-soluble vitamins in being poorly stored. Placental transfer is not marked and liver vitamin B_{12} concentrations in the ovine fetus were less than half of those found in the mother (Grace *et al.*, 1986).

Excretion

Little is known about the excretion of vitamin B_{12} in ruminants. In humans, the primary route can be via biliary secretion to the faeces, with negligible catabolism or urinary loss. In dairy cows on high cobalt intakes (17.8–31.0 mg day^{-1}), Walker and Elliot (1972) reported marked increases in urinary vitamin B_{12} to peak values of around 12 ng ml^{-1} during lactation, higher than those found in milk (3–7 ng ml^{-1}) and serum (0.6–0.8 ng ml^{-1}). The predominance of urinary excretion is also characteristic of a water-soluble vitamin. Urinary losses probably fall to negligible levels during cobalt deficiency and may be useful in assessing vitamin B_{12} status.

Functions

Synthesis by rumen or other microbes of vitamin B_{12} and similar molecules with a corrin ring as their core (corrinoids) yields one of the most complex groups of non-polymeric, natural products known. Attachments of the dimethylbenzimidazole side-chain and adenosyl or methyl ligands occur at later stages of synthesis and determine function in both microbes and mammals.

The essentiality of cobalt for mammals is linked to two distinct forms of vitamin B_{12}, with contrasting coenzyme functions. As methylcobalamin (MeCbl), cobalt assists a number of methyltransferase enzymes by acting as a donor of methyl groups and is thus involved in one-carbon metabolism, i.e. the build-up of carbon chains (Fig. 10.3). Methylcobalamin is important for microbes as well as mammals and is needed for methane, acetate and methionine synthesis by rumen bacteria (Poston and Stadman, 1975). In mammals, MeCbl enables methionine synthase to supply methyl groups to a wider range of molecules, including formate, noradrenaline, myelin and phosphatidyl ethanolamine (PE). Failure of methylation in vitamin B_{12}-deficient sheep also

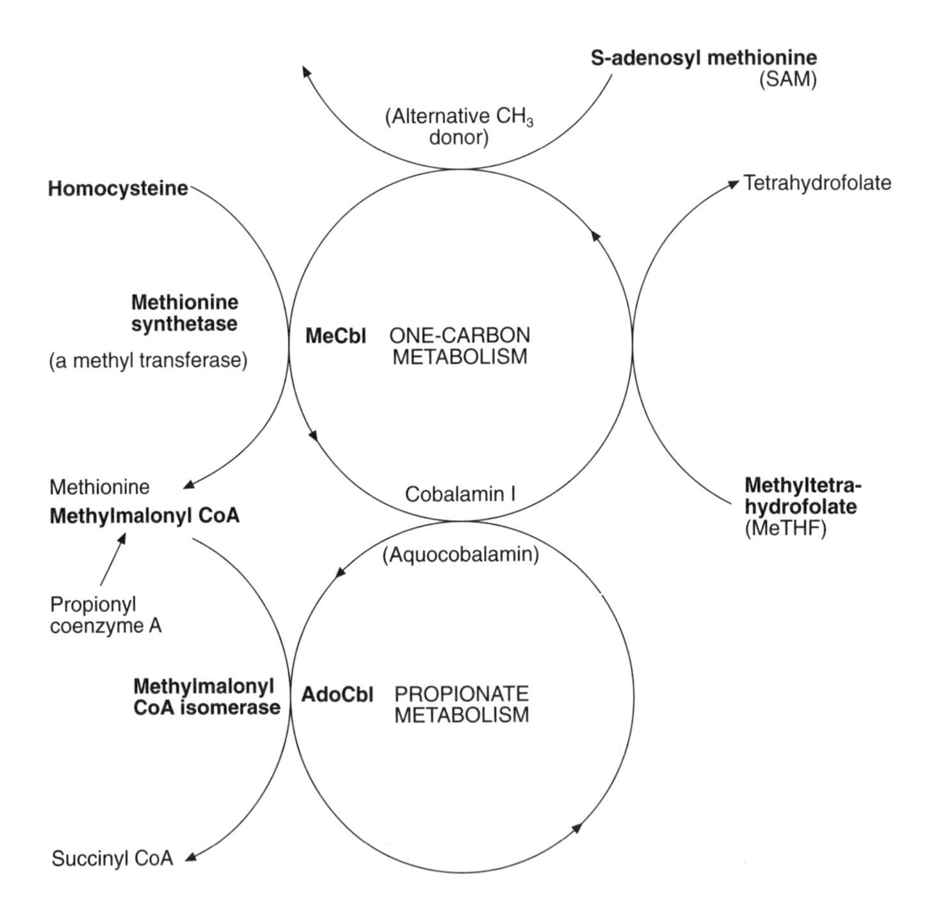

Fig. 10.3. Vitamin B_{12} (cobalamin) has two distinct and yet linked coenzyme functions in one – carbon and propionate metabolism as methyl- and adenosylcobalamin (MeCbl and AdoCbl), respectively: cobalt depletion can lead to accumulation of the substrates for each function in mammalian tissues and fluids i.e. homocysteine, MeTHF and SAM when MeCbl is limiting, MMA when AdoCbl is limiting; in rumen microbes, AdoCbl catalyses the reverse reaction and succinate accumulates (see Fig. 10.5).

inhibits folate uptake by the liver (Gawthorne and Smith, 1974; Fig. 10.3). As adenosylcobalamin (AdoCbl), cobalt influences energy metabolism, facilitating the formation of glucose by assisting methylmalonyl-coenzymeA (CoA) mutase to form succinate from propionate, chiefly in the liver (Fig. 10.3). Microbes are again equally dependent on the coenzyme, but for opposite reasons; in the propionic acid bacteria, formation of propionate from succinate is catalysed by the same mutase (Babior, 1975). The enzyme leucine 2,3-mutase is also AdoCbl-dependent and a breakdown in this pathway has been implicated in pernicious anaemia in humans (Poston, 1980). The identification of rate-limiting pathways has, however, long been a matter of controversy (Chanarin *et al.*, 1981; Scott, 1992) and it has been suggested that the myelopathy which accompanies pernicious anaemia may be attributable to MeCbl rather than AdoCbl deficiency (Small and Carnegie, 1981; Scott, 1992). There is probably interconversion of the coenzymes, as well as recycling at the tissue level and via biliary secretion. Lesions of the central nervous system (CNS) were not reported in the early studies of cobalt deficiency in ruminants, but they have been described in lambs born of deficient ewes (Fell, 1981). The rate-limiting function which affects appetite in ruminants is discussed in detail later.

Biochemical Manifestations of Cobalt Deprivation

Depletion and deficiency

The clinical and pathological signs of cobalt deprivation are preceded by characteristic biochemical changes in the tissues and fluids of the body (Somers and Gawthorne, 1969; Fig. 10.4). As soon as depletion begins, the concentrations of cobalt and of vitamin B_{12} fall in rumen fluid. Vitamin B_{12} values in blood serum also show an early decline, because they measure vitamin which is in transit, and this is largely a reflection of the adequacy of current rumen synthesis. Serum vitamin B_{12} declines before liver vitamin B_{12}, confirming that the latter does not perform like an active storage pool. The lag before liver vitamin B_{12} concentrations decline partly reflects increased efficiencies of synthesis and capture of the absorbed vitamin. Seasonal fluctuations in serum vitamin B_{12} are followed by similar fluctuations in liver vitamin B_{12} (Millar *et al.*, 1984), but the two parameters can show a significant curvilinear relationship at a given time when there has been no recent change in cobalt intake (Marston, 1970; Field *et al.*, 1988). The pattern is, however, the opposite of that linking plasma and liver copper concentrations (Chapter 11), with serum vitamin B_{12} continuing to increase in response to generous dietary supplies of cobalt after liver vitamin B_{12} values have reached a plateau. Lack of simple relationships between plasma and liver vitamin B_{12} concentrations within a flock at a given time or with indices of dysfunction has prompted some to question the diagnostic value of serum vitamin B_{12} (Millar and Penrose, 1980; Millar *et al.*, 1984) and others to defend it (Sutherland, 1980).

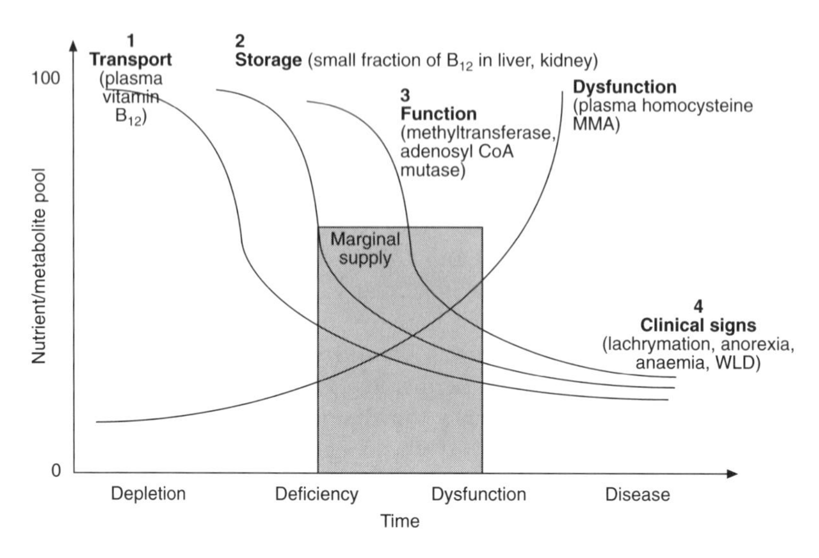

Fig. 10.4. Sequence of biochemical changes preceding the appearance of clinical symptoms in ruminants deprived of cobalt and hence vitamin B_{12}: MMA, methylmalonic acid; WLD, white liver disease. See also Fig. 3.1 and related text.

Failure of propionate metabolism

The marked failure of appetite that is such a conspicuous feature of vitamin B_{12} deprivation in ruminants is not nearly so obvious in monogastric species and this is probably due to differences in their primary sources of energy. The main source of energy to ruminants is not glucose but volatile fatty acids, produced by fermentation in the rumen. A breakdown in propionate metabolism at the point in the metabolic pathway where methylmalonyl-CoA is converted to succinyl-CoA with the help of AdoCbl might seriously affect the cobalt-deprived ruminant (Fig. 10.3; Marston *et al.*, 1961). Rumen synthesis and absorption of propionic and other fatty acids were thought to proceed normally, while the rate of clearance of propionate from the blood fell and the intermediary methylmalonyl-CoA accumulated (Marston *et al.*, 1972), causing a marked increase in urinary excretion of methylmalonic acid (MMA) (Gawthorne, 1968). Acetate clearance was similarly impaired (Somers, 1969) and attributed to an inhibitory effect of the raised blood propionate values on acetate metabolism. Accumulation of propionate has by far the greater inhibitory effect on appetite (Farningham and Whyte, 1993) and there is an inverse relationship between appetite and propionate clearance in cobalt-deficient sheep (Marston *et al.*, 1972). This long-accepted dysfunction has been challenged recently by workers in Belfast, whose results refocus attention on vitamin B_{12} requirements of the rumen microbes (Kennedy *et al.*, 1991b). Massive increases in succinate concentrations in rumen liquor within 2 weeks of transfer to a barley-based diet, low in cobalt, suggested inhibition of propionate-producing microbial species such as *Selenomonas ruminantium*.

More recent work shows that succinate accumulates within 2 days of transfer to a diet very low in cobalt (0.02 mg kg^{-1} DM) but not with 0.04 mg Co kg^{-1} DM present (Kennedy *et al.*, 1996). Early studies of natural cobalt deficiency in Florida had reported changes in rumen microbial composition (Gall *et al.*, 1949), and attempts to study vitamin B_{12} metabolism in cultures of rumen microorganisms were frustrated by instability when cobalt-deficient substrates were fed (McDonald and Suttle, 1986). The accumulation of succinate in the rumen was accompanied by a steady rise in plasma succinate (Fig. 10.5) and confirmed in cobalt-deficient sheep at pasture (Kennedy *et al.*, 1991b). Since succinate can enter the tricarboxylic acid cycle as a glucose precursor, impairment of the conversion of glycogenic propionate to succinate because of AdoCbl deficiency might be metabolically unimportant in terms of glucogenesis. Absorption of succinate does not perturb plasma MMA values (Kennedy *et al.*, 1996).

Failure of methylation

The Belfast group also investigated the extent to which lack of MeCbl causes dysfunction in methylation in cobalt-deprived lambs (Kennedy *et al.*, 1992). Impaired methionine metabolism became apparent as a rise in plasma homocysteine after 14 weeks (Fig. 10.6) and by slaughter after 28 weeks, methionine synthetase activity was reduced in liver, kidney and spinal cord, though not in brain. Increases in sulphur (S)-adenosyl methionine (SAM) in liver confirmed the reduction in methylation and the accompanying increase

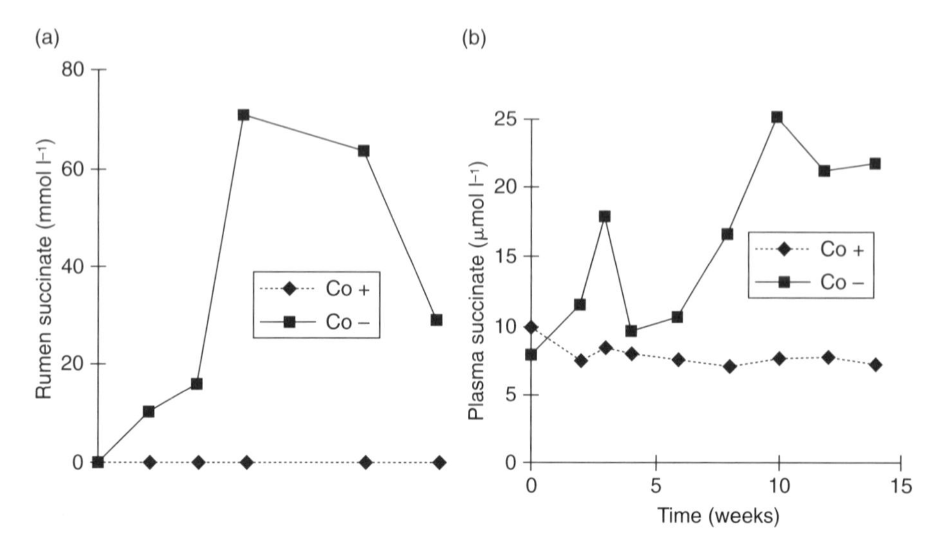

Fig. 10.5. Lambs given a barley-based diet very low in cobalt show marked early rises in rumen succinate (a) followed by increases in plasma succinate (b), a glucose precursor: this may partially overcome any constraint upon glucogenesis due to impaired propionate metabolism (see Fig. 10.3) in the liver (from Kennedy *et al.*, 1991).

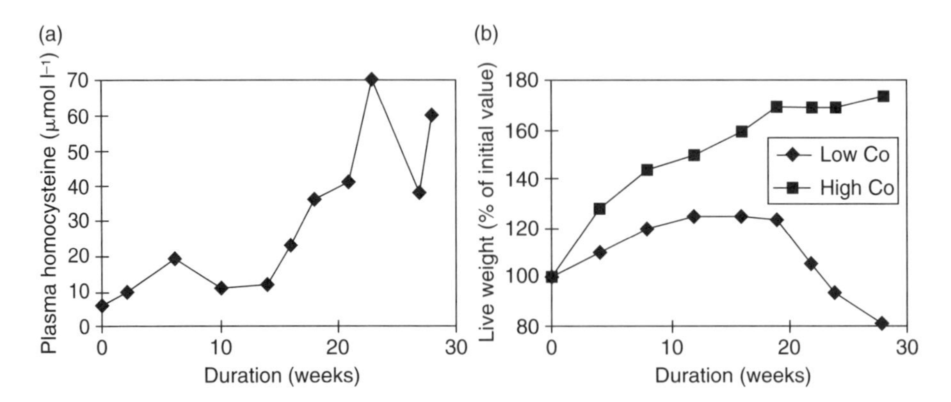

Fig. 10.6. Increases in plasma homocysteine (a) in cobalt-deprived lambs confirm an eventual dysfunction in one-carbon metabolism (see Fig. 10.3) but the change comes long after anorexia has retarded growth (b) (data from Kennedy *et al.*, 1992); methylcobalamin deficiency may not therefore be the first rate-limiting function.

in SAM:S-adenosyl homocysteine (SAH) (see Fig. 10.3) was probably responsible for a 'feedback' reduction in apoenzyme (i.e. methionine synthetase) concentration. While reductions in enzyme capacity do not always result in dysfunction if there is spare capacity, in this instance accumulation of phosphatidyl choline (PC) in liver and brain indicated a functional disturbance of PE methyltransferase. These results confirmed and extended those of Gawthorne and Smith (1974), who first reported the reduction in methyltransferase activity in the cobalt-deficient sheep, and Gawthorne (1968), who reported a rise in formiminoglutamic acid (FIGLU) in urine, indicating dysfunction in methylation during folate metabolism, with secondary effects on histidine (precursor of FIGLU). The importance of MeCbl as a methyl donor in folate metabolism was reviewed recently by Scott (1992). The MeCbl pathway might be particularly important in ruminants, because the major dietary sources of methyl groups for non-ruminants, such as choline and betaine, are degraded in the rumen (Kennedy *et al.*, 1992). Among ruminants, sheep should be more vulnerable to MeCbl deficiency than cattle, because of the high requirement of S-amino acids for wool synthesis (see also Chapter 9). Sheep are more susceptible to cobalt deficiency than cattle, and the accumulation of fat in the vitamin B_{12}-deprived liver seen in sheep (ovine white liver disease (OWLD)), but not in cattle, could be related to a methyl-group deficiency affecting hepatic lipid metabolism (Suttle, 1988). Thus, demand for a trace-element-dependent function might influence the clinical manifestation of the disorder (Chapter 3).

Abnormal lipid metabolism

Major defects in lipid metabolism, involving both Cbl-dependent pathways, have been proposed to explain the pathogenesis of OWLD on barley diets (Kennedy *et al.*, 1994b), and they are indicated by the biochemical abnormalities recorded in Table 10.1. Firstly, MMACoA accumulates, due to low activity of AdoCbl-deprived MMACoA mutase. The MMACoA is an inhibitor of the β-oxidation of free fatty acids (FFA), which the cobalt-deprived lamb mobilizes from fat depots to offset loss of appetite; FFA therefore accumulate in the liver. The normal metabolic response to FFA accumulation would be an increase in hepatic synthesis of triglycerides and their export from the liver as very low-density lipoprotein (VLDL). However, VLDL assembly requires MeCbl and methionine synthetase, whose activity is also low (Table 10.1). Triglycerides therefore accumulate, providing a pool of readily peroxidizable, unsaturated fats. Kennedy *et al.* (1994b) propose that a further metabolic anomaly of the cobalt-deprived lamb, accumulation of homocysteine (plasma concentrations were increased from 37.8 in controls to 90.4 μmol l^{-1}), initiates a chain of peroxidation, which leads to accumulation of the oxidation product, lipofuscein, depletion of the antioxidant, vitamin E, and damage to mitochondrial structure (Kennedy *et al.*, 1997). Impairment of β-oxidation of FFA may contribute to the demise of other, leaner sheep deprived of cobalt on diets low in digestible energy, without visible evidence of fat accumulation in the liver.

The accumulation of MMA in the cobalt-deficient lamb leads to another biochemical anomaly. The MMA is misincorporated into branched-chain fatty acids (BCFA), which are not normally found in large quantities in body fat (Duncan *et al.*, 1981; Kennedy *et al.*, 1994a). The abnormality is not thought to have any clinical significance. However, BCFA occur to a lesser extent naturally when grass rather than barley is fed, due to lower rumen production of propionate (Wahle *et al.*, 1979) the relative importance of the AdoCbl- and MeCbl-dependent pathways may be different for barley- and grass-fed sheep. It will be important for the abnormal biochemistry found in OWLD in barley-fed lambs to be confirmed in forage-fed lambs. The demand placed upon a micronutrient-dependent pathway may again determine whether or not that pathway becomes rate-limiting (Chapter 3).

Table 10.1. Biochemical markers of cobalt deprivation and abnormal lipid and vitamin E metabolism in the livers of lambs with ovine white liver disease (OWLD) and in cobalt-sufficient controls (Kennedy *et al.*, 1994b).

Marker[a]	Control	OWLD
Vitamin B_{12} (pmol g^{-1})	396 ± 9.7	15 ± 4.1
Vitamin E (nmol g^{-1})	13.1 ± 0.94	5.9 ± 0.56
MMACoA mutase (U g^{-1})	30.8 ± 2.2	4.4 ± 0.6
Methionine synthase (U g^{-1})	869 ± 102	105 ± 26
SAM:SAH	5.4 ± 0.76	2.5 ± 0.68
PC:PE	1.5 ± 0.04	1.1 ± 0.09
Total lipid (g kg^{-1} FW)	50.8 ± 8.0	165 ± 13.9
Triglyceride fatty acids (mg kg^{-1} FW)	5.2 ± 1.00	36.8 ± 14.9
Free fatty acids (mg g^{-1} FW)	2.9 ± 0.80	10.5 ± 1.80

[a] See text for abbreviations; data are means ± SEM for three (OWLD) or four (control) lambs.

The rate-limiting function

While failure of methylation as a consequence of cobalt deficiency has been clearly confirmed in sheep, AdoCbl may yet be the first limiting of the two vitamin B_{12}-dependent pathways. It can be seen from Fig. 10.6 that growth in the cobalt-deficient lambs had ceased before plasma homocysteine began to rise. Furthermore, appetite had been depressed from the outset and the biochemical changes found in the tissues may have been secondary to the anorexia. Such rapid effects on appetite point to a rumen-based dysfunction. When Price (1991) compared the onset of disturbances in Ado- and MeCbl function in the same sheep given cobalt-deficient hay, the former was the first to show dysfunction by 10–16 weeks (Fig. 10.7). Furthermore, MeCbl deficiency had to be exposed by intravenous loading with histidine and

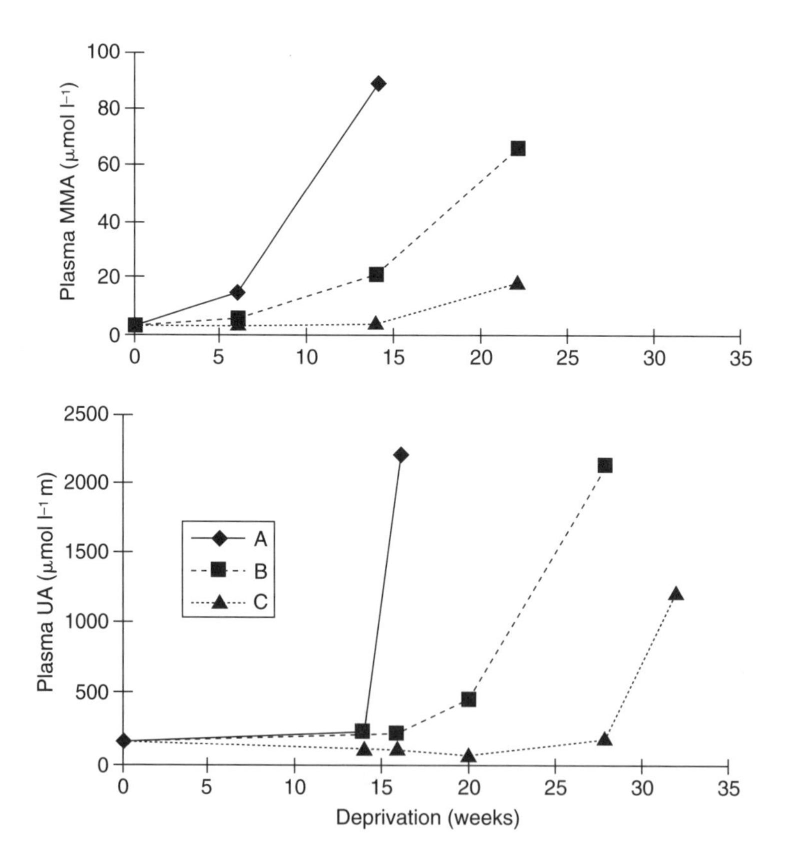

Fig. 10.7. Biochemical evidence of adenosylcobalamin dysfunction, in the form of increases in plasma methylmalonic acid (MMA) precedes that of methylcobalamin dysfunction, in the form of increases in urocanate (UA), in three cobalt-deprived sheep; appetite decreased after 9, 19 and 22 weeks in Sheep A, B and C, respectively (data from Price, 1991).

detection of urocanate (a precursor of FIGLU) in plasma, whereas AdoCbl deficiency was detected by a spontaneous rise in plasma MMA. Marston *et al.*'s (1972) explanation for the anorexia of cobalt deficiency may therefore still be essentially correct, since blood propionate levels in their cobalt-deficient sheep were raised nearly sevenfold and circulating propionate is an important determinant of appetite. Factors other than tissue depletion of AdoCbl may, however, have contributed to the reduced clearance of propionate, including the increased ruminal supply of an alternative glucose precursor, succinate, from the rumen. Cattle subjected to severe cobalt deprivation show little or no evidence of MeCbl or AdoCbl dysfunction, in keeping with their reputation of tolerance when compared with sheep (Kennedy *et al.*, 1995). Increases in the specific activity of nicotinamide adenine dinucleotide (NADH-) and NADphosphate (NADPH)-linked aquacobalamin reductases in the liver during vitamin B_{12} deficiency indicate inherent capacity to synthesize more coenzymes of the vitamin or their precursor (Cbl 1), compensating for shortages in the dietary supply (Watanabe *et al.*, 1991).

Clinical Manifestations

Anorexia and anaemia

When sheep, cattle or goats are confined to cobalt-deficient diets, there is eventually a gradual loss of appetite and failure of growth or weight loss; this is succeeded by extreme inappetence, rapid muscular wasting (marasmus), pica or depraved appetite and severe anaemia, culminating in death. If the deficit is less severe, only the young animal will be affected, exhibiting a vague unthriftiness. The appearance of a severely cobalt-deficient animal is one of extreme emaciation and listlessness, a condition difficult to distinguish from that of a starved animal, except that the visible mucous membranes are blanched and the skin is usually pale and fragile. There is a general absence of body fat, but the liver may be fatty, the spleen is haemosiderized and there is hypoplasia of erythrogenic tissue in the bone marrow and an anaemia of the normocytic, normochromic type, unlike the megaloblastic anaemia seen in humans. Inappetence and marasmus usually precede anaemia, and appetite and weight generally respond more quickly than the anaemia to cobalt feeding or vitamin B_{12} injections.

Fatty liver

Cobalt deficiency is also responsible for OWLD, a disease first described in New Zealand (Sutherland *et al.*, 1979) and which affects lambs at times of prolific pasture growth. The disease has since been identified in Northern Ireland, the Netherlands and Norway (for review, see Suttle, 1988) and produced experimentally by feeding lambs on a diet lacking only cobalt (Kennedy *et al.*, 1994b, 1997). The OWLD progresses from mild fatty infiltration of hepatocytes to a point where bile-ducts proliferate, enzymes leak into the bloodstream from damaged liver cells and plasma bilirubin levels rise (Kennedy *et al.*, 1997). Affected lambs are anorexic, anaemic and ill-thriven in appearance. The disease is sometimes accompanied by polioencaphalomalacia and is preventable by cobalt or vitamin B_{12} supplementation (Sutherland *et al.*, 1979; McLoughlin *et al.*, 1986). A few cases of white liver disease (WLD) have been described in angora goats in New Zealand (Black *et al.*, 1988).

Perinatal mortality

If the ewe is cobalt-deficient during early pregnancy, she may give birth to fewer lambs than normal (Duncan *et al.*, 1981; Quirk and Norton, 1987) and those that are born may be stillborn or slow to suck and therefore less likely to survive than lambs from cobalt-supplemented ewes (Fisher and MacPherson, 1991). An important feature of the latter study was that the poor reproductive performance was not attributable to anorexia, because ewe live weight and body condition were not affected. Depression of milk yield can reduce the growth rate of surviving lambs (Quirk and Norton, 1987); new-born calves are less affected, although their subsequent growth can be depressed (Quirk and Norton, 1988).

Disease susceptibility

Disease resistance may decline during cobalt deficiency. Ferguson *et al.* (1989) reported increased susceptibility to infection by the abomasal parasite *Ostertagia circumcincta* in the cobalt-deficient lamb, while MacPherson *et al.* (1987) suggested that immune responses to a similar parasite of cattle (*Ostertagia ostertagii*) were compromised in cobalt deficiency.

Infertility

While infertility is always likely to arise as a secondary consequence of debilitating conditions, such as severe cobalt deprivation, a recent report suggests that infertility may occur in the absence of ill-thrift in beef cows (Judson *et al.*, 1997). At one stage, five out of nine unsupplemented cows had failed to conceive, whereas, in each of three cobalt-supplemented groups of 9–11 cows, only two or three had not conceived.

Occurrence of Cobalt Deprivation in Ruminants

Cobalt deprivation occurs in grazing ruminants in particular areas in many countries, with varying degrees of severity. Horses and other non-ruminant species confined to the same areas remain healthy and grow normally; deer (Clark *et al.*, 1986) and goats (Clark *et al.*, 1987), like cattle, are less vulnerable than sheep. The vulnerability of sheep may be attributable to their high requirement of S-amino acids for wool growth (see Chapter 9). Deprivation arises on well-drained soils of diverse geological origin, including coarse volcanic soils; leached podsolized sands; sandy loams derived from granites; calcareous, wind-blown shell sands; and ironstone gravels. High levels of soil manganese (Mn) depress cobalt uptake by pasture to such an extent that, with more than 1 g Mn kg^{-1}, the application of cobalt fertilizers is unlikely to increase herbage cobalt (Minson, 1990). Cobalt deprivation may also be induced by heavy liming, which reduces the amount of plant-available soil cobalt (Poole *et al.*, 1972; Fig. 2.5). Thus, cobalt deficiency occurs under a wide range of climatic conditions, from the tropical regions of Brazil to the cool, temperate regions of Ireland and Scandinavia. Disorders occur more on long than on short pasture, because less soil is ingested (Andrews *et al.*, 1958). Soil is always richer in cobalt than the overlying pasture, providing the rumen microbes with available cobalt, and dosing with soil was used as an early remedy for cobalt deficiency in Scotland (Suttle, 1988). The more severe consequences of cobalt deficiency in sheep and cattle were given a variety of local names, such as 'bush-sickness' in New Zealand, 'coast disease' in South Australia, 'wasting disease' in Western Australia, 'salt sickness' in Florida, 'Nakuruitis' in Kenya, 'pining' in Great Britain, 'Lecksucht' in the Netherlands and Germany and 'Grand Traverse disease' in Michigan; all describe an enzootic marasmus or severe wasting condition, which can affect ruminants of all ages and types at all times and cause high mortality. Much larger areas exist in which deprivation is mild or marginal and subclinical disorder is only apparent in some seasons (Lee, 1951) in the more susceptible types of stock, such as young male lambs (Shallow *et al.*, 1989). Subclinical unthriftiness may go undetected, because it can readily be confused with the effects of parasitism or underfeeding.

Diagnosis of Cobalt-responsive Disorders

Cobalt deprivation in ruminants, in its milder forms, is impossible to diagnose clinically, because the only evidence may be a state of unthriftiness. The only sure way of establishing a critical lack of dietary cobalt is by measuring the response in temperament, appetite and weight that follows cobalt feeding or vitamin B_{12} injection. However, a flexible framework for the interpretation of biochemical indices of vitamin B_{12} status is presented in Table 10.2, based on the following assessments of individual criteria.

Dietary cobalt

An indication of the need for supplementary cobalt can be obtained from the cobalt content of the pasture or diet. If this is below 0.07 mg Co kg^{-1} DM and is maintained thus for some months, dysfunction is possible, particularly in sheep. If cobalt intake is sufficiently low to cause the cobalt concentration in the rumen fluid to fall below about 0.5 ng ml^{-1}, vitamin B_{12} synthesis by the microorganisms is inhibited (Smith and Marston, 1970b). A more detailed assessment is given in the section dealing with dietary cobalt requirements.

Table 10.2. Marginal bands[a] for the most common biochemical indices used to assess the mean[b] cobalt and vitamin B_{12} status of groups of ruminants.

		Marginal status	Interpretative problems
Diet	Bovines	0.04–0.06 mg Co kg^{-1} DM[c]	Contamination by soil Co
	Other ruminants	0.05–0.07 mg Co kg^{-1} DM[c]	
Serum B_{12}	Bovines: S	30–60 pmol l^{-1}[d]	Prone to gross underestimation and lab-to-lab variation
	W	40–80 pmol l^{-1}	
	Other ruminants: S	230–350 pmol l^{-1}	
	W	336–500 pmol l^{-1}	
Liver B_{12}	All species	280–340 nmol kg^{-1} FW	No binding problems
Milk B_{12}	Bovine	250–500 pmol l^{-1}	No binding problems
Serum MMA	All species of weaned ruminant	5–10 μmol l^{-1}	Insensitive in dam to needs of offspring

[a] Mean values within band denote possibility of sufficient individuals benefiting to justify supplementation for all.
[b] Individual values below the lower limit (or above the upper limit in the case of MMA) are suggestive of production-limiting dysfunction.
[c] Multiply by 17.0 for μmol.
[d] Multiply by 1.355 for ng l^{-1}.
S, suckling; W, weaned.

Liver cobalt

The concentration of cobalt in the livers of sheep and cattle is sufficiently responsive to changes in cobalt intake to assist in diagnosing disorders in the field, provided that care is taken during analysis. The assay of cobalt in liver by atomic absorption spectrophotometry is susceptible to interference from high iron concentrations, but steps can be taken to eliminate it (Gelman, 1976). McNaught (1948) suggested that values of 0.04–0.06 mg Co kg^{-1} DM or less in the livers of sheep and cattle indicate cobalt deficiency and that values of 0.08–0.12 mg kg^{-1} DM or more indicate a satisfactory cobalt status. Similar criteria for sheep were proposed by Andrews *et al.* (1959) and have recently been confirmed in cattle (Mitsioulis *et al.*, 1995). Studies of liver vitamin B_{12} concentrations of animals in various stages of cobalt deficiency and of the proportion of the total liver cobalt present as the vitamin indicate that liver vitamin B_{12} concentrations are the more sensitive and reliable criterion of adequacy. The relationship between liver cobalt and liver vitamin B_{12} is curvilinear with a wide scatter at high liver cobalt concentrations, but the two correlate linearly and well over the lower ranges, which have diagnostic significance (Suttle, 1995).

Liver vitamin B_{12}

The responsiveness of liver vitamin B_{12} concentrations to suboptimal cobalt intakes (Fig. 10.2) makes it a useful diagnostic aid. Liver biopsy samples can be obtained from the live animal, but considerations of welfare and cost limit their use in practice. Early guidelines on diagnostic interpretation were that values > 140 were 'normal', 81–140 'borderline', 52–74 'low' and

< 52 nmol kg^{-1} fresh weight (FW) 'very low' in sheep (Andrews *et al.*, 1959) (multiply by 0.00136 to obtain values in μg g^{-1} or mg kg^{-1}). On an individual basis, Marston (1970) reckoned that appetite fell when liver vitamin B_{12} was < 74 nmol kg^{-1}, and Millar and Lorentz (1979) reported a similar threshold below which biochemical dysfunction (raised MMA in urine) occurred. These observations are consistent with Andrews *et al.*'s (1959) proposals. Assessment of growth responses to cobalt supplements in lambs in New Zealand indicated that, on a flock basis, a much higher mean threshold value of 375 nmol kg^{-1} FW was required to avoid growth retardation in summer (Clark *et al.*, 1989). In view of changes in methodology and the need to predict flock responses from a random sample, Clark *et al.*'s (1989) standards should form the basis of diagnosis, along the lines given in Table 10.2. The diagnostic value of liver vitamin B_{12}, and still more of kidney vitamin B_{12}, is reduced if the cobalt deficiency is coexistent with other diseases or conditions resulting in loss of appetite (Andrews and Hart, 1962), because starvation tends to increase tissue concentrations (Millar *et al.*, 1984). Any fatty infiltration will lead to underestimation of effective vitamin B_{12} concentrations in liver.

Serum vitamin B_{12} methodology

Serum vitamin B_{12} determinations have obvious advantages over liver or kidney determinations, because of the ease and frequency with which blood samples can be obtained. However, the results obtained for a given sample can vary widely from laboratory to laboratory (Table 10.3; Millar and Penrose, 1980; Schulz, 1987b), making diagnosis unreliable. In the report by Schulz, the methods were all based on the principle of radioisotope dilution (RID), in which vitamin B_{12} is first freed from plasma binders and then allowed to compete with a fixed amount of added ^{57}Co-cyanocobalamin for an exogenous binder, IF, of varying purity. Problems arise if unsaturated binders in the plasma are not removed during the initial extraction (non-specific binding) or if bound vitamin B_{12} is not fully liberated (residual binding). A combination of these factors probably led to the twofold variation in assay results between laboratories ($P < 0.001$), shown in Table 10.3. The analytical problem is most pronounced in bovine samples, because the high-molecular-weight binders, TC0 and TC1, contribute significantly to vitamin B_{12}-binding capacity (Polak *et al.*, 1979; Schulz, 1987a, b; Price, 1991), particularly in the young calf (Schulz, 1987a). Seasonal fluctuations in plasma-binding properties have been reported in cattle (Millar *et al.*, 1984), and binding is also likely to be influenced by infection. Price and Wood (1993) suggested that vitamin B_{12} concentrations in bovine sera will have been seriously underestimated by RID methods. Furthermore, any residual binding will also affect microbiological assays, because they utilize similar extraction procedures to RID methods. The fact that reported vitamin B_{12} values for bovine samples are generally low compared with ovine samples and far less responsive to cobalt supplementation (e.g. Givens and Simpson, 1983; Judson *et al.*, 1997) indicate that underestimations will have been commonplace. Suggestions that up to 50%

Table 10.3. Comparison of methods for measuring vitamin B_{12} (pmol l^{-1}) in ovine and bovine sera (Schulz, 1987b).

	Sample	Laboratories 1	2	3	Commercial kits A	B	C	LSD (P = 0.05)
Ovine	1	292	307	225	202	195	225	15
	2	816	1184	951	719	554	876	30
	3	1918	2375	2292	1813	1386		90
Bovine	1	195	367	172	127	165	742	22
	2	577	689	532	569	487	1131	37
	3	1910	2075	2375	1723	1513		90

of the vitamin B_{12} in bovine plasma is in the form of analogues (Halpin *et al.*, 1984) are unreliable because the calculations rely on differences between two unreliable methods. Babidge and Babidge's (unpublished data) studies suggest that, with extraction at 100°C and a consistently high pH (9–12), non-specific and residual binding in bovine sera can be overcome.

Interpretation of serum vitamin B_{12}

When the results of cobalt supplementation trials with lambs in New Zealand were collated, the correlations between the size of the mean growth response and the mean serum or liver vitamin B_{12} concentrations in mid-trial were equally good (Clark *et al.*, 1989). Bands of serum vitamin B_{12} concentration were derived which indicated the probability of growth being limited by deprivation (see Fig. 19.2), and they are used in Table 10.2. Thresholds are again higher than those given earlier (Andrews and Stephenson, 1966), but there is a simple common explanation. Within all flocks, there is variation in vitamin B_{12} status and cobalt-responsiveness. Individuals with below-average vitamin B_{12} status will contribute most to the mean growth response. The earlier guidelines were influenced by the association between *individual* vitamin B_{12} status and clinical condition. The need to distinguish biochemical criteria for individual and flock diagnosis applies to all minerals, as does the need to sample large numbers (at least ten and preferably one-tenth) to characterize a population. Suckled ruminants can thrive with lower serum vitamin B_{12} concentrations than weaned ruminants. This probably reflects two things: firstly, a faster outflow of the vitamin from the plasma in rapidly growing animals with a high efficiency of feed utilization; and, secondly, the low demand for propionate metabolism until grass becomes the major source of nutrients. The relationship between serum vitamin B_{12} levels and an indicator of dysfunction (serum MMA) changes as lambs age, higher levels being associated with normal serum MMA in individual 4-month- as opposed to 2.5-month-old lambs (650 vs. 450 pmol B_{12} l^{-1}; MacPherson *et al.*, 1999). Appropriate diagnostic levels for the very young lamb and calf are suggested in Table 10.2, although the latter may increase as analytical methods improve.

Vitamin B_{12} in milk

Analysis of vitamin B_{12} in milk has a particular advantage in assessing the status of cattle, because there appear to be fewer analytical problems than for serum. Concentrations of vitamin B_{12} in milk can show large responses to supplementation when none are apparent in the plasma, and values correlate well with liver vitamin B_{12} concentrations (Judson *et al.*, 1997).

Indicators of metabolic dysfunction – methylmalonic acid

The indicators of AdoCbl and MeCbl deficiency – abnormally high MMA and FIGLU – have both been advocated for diagnostic purposes. Early work focused on urinary excretion of these metabolites, and both MMA (Millar and Lorentz, 1979; Quirk and Norton, 1987, 1988) and FIGLU (Russel *et al.*, 1975; Quirk and Norton, 1987, 1988) were raised in lambs or calves whose growth was retarded by cobalt deprivation. The urinary MMA method was not specific (Elliot *et al.*, 1979) and has been superseded by the determination of serum MMA (Rice *et al.*, 1987). Increases in serum MMA provide an early warning, because values rise before appetite declines (Fisher and MacPherson, 1991; Price, 1991; Fig. 10.7). The relationship between serum MMA and serum vitamin B_{12} is curvilinear, with increases in serum MMA occurring in most individuals with < 220 pmol vitamin B_{12} l^{-1} serum (Rice *et al.*, 1987). Individual serum MMA values of 5–20 μmol l^{-1} and group mean values of 5–10 μmol l^{-1} indicate marginal cobalt status in growing or mature stock with fully functioning rumens. In young sheep and calves dependent largely on milk, growth can be retarded before anomalies in MMA appear (Quirk and Norton, 1987, 1988). Furthermore, the pregnant and lactating ewe may show no abnormal MMA metabolism and yet fail to provide her offspring with enough vitamin B_{12} for optimal growth (Quirk and Norton, 1987). Methylmalonic acid has little or no diagnostic value in the marginally deficient nursing mother or her suckled offspring.

Other indicators of dysfunction

Suckling calves and lambs both show increases in FIGLU excretion when growth is retarded by cobalt deprivation, but MMA gradually becomes the dominant abnormality and AdoCbl dysfunction probably provides the larger metabolic constraint as they grow (Quirk and Norton, 1987, 1988). With the ewe and heifer, there is again evidence of failure to give priority to the offspring, since dams not excreting FIGLU in their urine can rear lambs which do so (Quirk and Norton, 1987, 1988). Other biochemical abnormalities which have been associated with Co deprivation in sheep include high plasma aspartate aminotransferase (AAT) and pyruvate values, together with low plasma glucose ascorbic acid and thiamine concentrations (MacPherson and Moon, 1974; MacPherson *et al.*, 1976); most are not consistent or specific features of the disorder, however (Hannan *et al.*, 1980). The high pyruvate may indicate an induced thiamine deficiency and may be significant, given the impairment of thiamine absorption in folate deficiency (Howard *et al.*, 1974), the link between vitamin B_{12} and folate metabolism (Fig. 10.3) and the appearance of polioencephalomalacia, a thiamine-responsive condition, in conjunction with the vitamin B_{12}-responsive OWLD. The low plasma ascorbic acid values (1 mg l^{-1} compared with a normal 4–8 mg l^{-1}) may be a factor in the increased susceptibility to infection exhibited by cobalt-deficient sheep. No biochemical marker in a suckling lamb or calf will indicate the full impact of a dietary cobalt deficiency which is severe enough to reduce milk yield in the mother.

Prevention and Control

Cobalt insufficiency in ruminants can be cured or prevented by a variety of methods, choice being dictated by local circumstances and cost–benefit analysis.

Continuous indirect supplementation

In deficient areas, adequate cobalt intakes can usually be ensured by including cobalt salts or oxide ores in the fertilizers used; in some areas, the added cobalt will also increase yields of the legume component of the pasture. The amount and the frequency of application vary with the soil type and terrain and with the husbandry practices employed. In the hill country of New Zealand, a single application of 1.5 kg cobalt sulphate ($CoSO_4$) ha^{-1} every 3–4 years was effective. On more accessible, sandy soils, where phosphatic fertilizers were applied every year or two, less $CoSO_4$ sufficed. Single dressings of 0.3 kg and 0.6 kg $CoSO_4$ ha^{-1} raised the mean cobalt concentrations in a grass/legume pasture from 0.04 to 0.19 and to 0.39 mg Co kg^{-1} DM, respectively, in the first year; by the second year, values had fallen to 0.10 and 0.16 mg Co kg^{-1} DM, respectively, but were still adequate (Rossiter *et al.*, 1948). On highly alkaline soils, such as calcareous shell sands and heavily limed soils, and on soils high in manganese oxides, which fix cobalt in an unavailable form (Adams *et al.*, 1969), the uptake of cobalt by plants is so low that fertilizer treatment is unreliable. Cobalt fertilizers have become very expensive and uneconomic in many areas in recent years; one way of using them sparingly is to treat strips of pasture only (MacPherson, 1983).

Continuous direct supplementation

Direct administration of cobalt to animals can be achieved by: incorporating the element (as a salt or oxide) into the mineral supplements or rations; supplementation of the drinking-water; or providing cobalt-containing salt-licks for voluntary consumption. With stall-fed animals, the first of these procedures is the simplest and most economical. The inclusion of cobalt in commercial mineral mixtures and prepared rations is common practice, even when there is no evidence of cobalt deficiency. A mineral mix containing 40 mg Co kg^{-1}, included at 25 kg t^{-1} in a concentrate which constitutes at least one-tenth of the total ration, will provide sufficient cobalt for ruminants.

The provision of salt-licks containing about 0.1% cobalt is a satisfactory procedure in deficient range conditions, provided that the lick is consumed uniformly at vital times of the year.

Discontinuous parenteral supplementation

Intramusuclar injections of vitamin B_{12} at the rate of 100 μg each week or of 150 μg every second week rapidly alleviated all signs of deficiency in lambs and were just as effective as cobalt administered orally at the rate of 7 mg week^{-1} (Andrews and Anderson, 1954). A single injection of hydroxycobalamin, 1 mg, protected lambs for 14 weeks in one study (Hannan *et al.*, 1980). However, 1 mg vitamin B_{12} given every 6 weeks was not sufficient to maintain normal serum vitamin B_{12} concentrations in lambs (MacPherson *et al.*, 1999) and 2 mg month^{-1} is recommended for older lambs. Doses of 6 mg vitamin B_{12} 50 kg^{-1} live weight (LW) every 6 weeks has improved the growth of beef calves (Judson *et al.*, 1982). The subcutaneous injection of 2 mg of the vitamin as a cyanocobalamin–tannin complex achieved a sustained increase in serum vitamin B_{12} without improving liver stores, when compared with an aqueous source (Judson *et al.*, 1988b). Such treatments have little value in practice, because of their high cost relative to that of orally administered cobalt. Oral administration of the vitamin is much less effective than vitamin B_{12} injection, because of poor absorption by ruminants, doses of 500 μg day^{-1} giving similar responses in cobalt-deficient lambs to those obtained by injecting 500 μg every 2 weeks. Parenteral injections of cobalt salts are not effective, because insufficient cobalt reaches the rumen, where microbial synthesis of vitamin B_{12} takes place, increases in tissue cobalt concentrations being of no value to the deficient animal.

Discontinuous oral supplementation

Oral dosing or drenching with dilute cobalt solutions is effective provided that the doses are given frequently. Dosing sheep twice each week with 2 mg Co or once each week with 7 mg Co and dosing cattle with five to ten times those amounts, depending on their size and age, is effective even in severely cobalt-deficient areas. Larger oral doses given at longer intervals (up to 4 weeks) will prevent obvious signs of deficiency but may be suboptimal. Manufacturers of anthelmintic drenches commonly add cobalt to their products, but the amounts provided range widely (for lambs) and give short-lived increases in plasma vitamin B_{12} concentrations (Field *et al.*, 1988). They cannot be relied upon to alleviate a severe deficiency when given every month as part of a worm control programme. Cumulative responses in plasma vitamin B_{12} over 14 days were increased by 40% when doses were increased from 8 to 32 mg Co (Fig. 10.1; Suttle *et al.*, 1989). Ewes should receive cobalt supplements throughout late pregnancy to avoid loss of appetite and risk of pregnancy toxaemia. Supplementation of the ewe will also increase the amounts of vitamin B_{12} secreted in milk, but the lamb should receive cobalt supplements from 6–8 weeks of age so that it can begin to synthesize its own vitamin B_{12}. Cobalt supplementation of the pregnant heifer can produce massive increases in milk concentrations of vitamin B_{12} (Quirk and Norton, 1988).

Slow-release methods

Supplementary cobalt ideally should be supplied continuously, because the vitamin is not stored extensively. Cobalt concentrations in the rumen liquor must be maintained above the critical level of 5 μg Co l^{-1}, and this can be achieved by the use of cobalt-containing pellets or bullets. Early types consisted of cobaltic oxide (60%) and finely divided ferruginous clay and ideally had a specific gravity of 4.5–5.0. When introduced into the oesophagus with an appropriate 'gun', the pellet lodges in the rumino-reticulum and remains there for months or years, slowly dissolving and supplying cobalt to the rumen bacteria (Dewey *et al.*, 1969). The efficiency of such pellets in sheep and cattle has been established under a wide range of conditions, including changes of formulation which halved their cobalt content (Judson *et al.*, 1997). In young lambs and calves, in which the rumen is not fully developed, and in lactating cows, in which feed intake is large and rumination frequent, a few animals regurgitate the pellets (Millar and Andrews, 1964; Poole and Connolly, 1967; Millar *et al.*, 1984). A more serious problem is the development of an impervious coat, consisting mostly of calcium phosphate, on the surface of the pellet. This can be minimized by introducing a small steel screw with the pellet or by administering two pellets together, which abrades the coating. Australian experience indicates that a single treatment can supply sufficient cobalt to sheep for more than 5 years (Lee and Marston, 1969) and in beef cows for up to 19 months (Judson *et al.*, 1997). More recently, a soluble glass bolus has been introduced, which is not susceptible to coating and also yields a steady supply of cobalt (Judson *et al.*, 1988a; Zervas, 1988).

Cobalt sources

Inorganic sources of cobalt must be partially soluble in the rumen to be of nutritional value to ruminants when used as food supplements. While successful as a constituent of heavy pellets, because of the large doses given, cobaltous and cobaltic oxides have lower nutritive value than soluble inorganic salts (cobalt carbonate ($CoCO_3$) and $CoSO_4$) per unit of cobalt, when assessed by comparative increases in liver cobalt (Ammerman *et al.*, 1982). This would appear to be at variance with an earlier finding that cobalt oxide (Co_3O_4) and $CoSO_4$ were equally effective in raising serum and liver vitamin B_{12} concentrations (Andrews *et al.*, 1966). However, the amounts of cobalt given in the latter study (300 mg) would have far exceeded the capacity for ruminal synthesis (see Fig. 10.1), thus putting the more soluble $CoSO_4$ at a disadvantage. Chelates of cobalt, such as cobalt ethylenediaminetetra-acetic acid (CoEDTA), are no more effective in stimulating microbial vitamin B_{12} synthesis *in vitro* than cobalt chloride ($CoCl_2$) (N.F. Suttle, unpublished data), while CoEDTA and $CoSO_4$ gave similar responses in liver and serum vitamin B_{12} when given orally to grazing lambs (Millar and Albyt, 1984).

Cobalt Requirements

Ruminants

Early field experience indicated that species differences in cobalt requirements were small and that 0.07 mg Co kg^{-1} DM was just adequate for growth and health in both sheep (ARC, 1980) and cattle (Winter *et al.*, 1977). There are, however, reports of growing sheep (Millar and Albyt, 1984) and cattle (Clark *et al.*, 1986) showing no benefit from cobalt or vitamin B_{12} supplementation on pastures containing 0.06 mg Co kg^{-1} DM. Even lower levels may be tolerated by mature stock (0.03–0.05 mg Co kg^{-1} DM), but requirements increase with the onset of lactation (Quirk and Norton, 1987, 1988), possibly because vitamin B_{12} is needed to utilize body reserves of other nutrients. Higher requirements suggested for growing lambs (namely 0.11 mg Co kg^{-1} DM: Andrews *et al.*, 1958) would probably maintain maximal liver vitamin B_{12} concentrations (ARC, 1980). Dietary concentrations of 0.05–0.08 should be regarded as marginal for sheep and 0.04–0.06 mg kg^{-1} DM marginal for cattle, deer and goats. More precise estimates, applicable under all grazing conditions, are difficult to give, because of seasonal changes in herbage cobalt concentrations and the extent of soil contamination of the herbage consumed.

Protection from toxins

An abnormal substance that appears to affect cobalt requirements occurs in the perennial grass, *Phalaris tuberosa*. Consumption of this grass is responsible, in restricted areas, for the disease in sheep and cattle known as '*Phalaris* staggers' or 'Ronpha staggers'. This disease can be prevented, but not cured, by regular oral dosing with cobalt salts (50 μg Co twice daily or 28 mg Co $week^{-1}$: Lee *et al.*, 1957), by the use of cobalt pellets (Lee and Kuchel, 1953) or by treatment of pastures with a cobalt-containing fertilizer. *Phalaris tuberosa* contains a neurotoxin which is responsible for the staggers syndrome but at high rumen cobalt concentrations the neurotoxin is inactivated or its absorption is reduced. The action of cobalt in this respect differs from its action in controlling deficiency, because administration of vitamin B_{12} is ineffective against *Phalaris* staggers. The amount of cobalt required for protection rises with increasing toxic potential of the pastures. In some areas, the staggers syndrome does not develop because the cobalt intake from soil and herbage is sufficient to detoxify the neurotoxin. Cobalt supplements have also provided protection against annual ryegrass toxicity in sheep. The pathogenic agents are corynetoxins produced by a bacterium, *Clavibacter toxicus*, which is carried into the seed head by an invading nematode, *Anguina agrostis*. Cobalt supplements delay, but do not prevent, the onset of neurological signs, irrespective of the amount of toxin (0.15 or 0.30 mg kg^{-1} LW day^{-1}) or cobalt (4 or 16 mg day^{-1}) given (Davies *et al.*, 1995).

Non-ruminants

Cobalt deficiency *per se* has never been clearly demonstrated in any monogastric species, and horses thrive on pastures containing insufficient cobalt for sheep and cattle. Non-ruminant animals have a much lower dietary requirement for vitamin B_{12} than do ruminants, due to a more efficient absorption of the vitamin and possibly to lower tissue requirements. That vitamin B_{12} performs the same functions in non-ruminants as in ruminants is demonstrated by the fact that administration of a methyl transferase inhibitor (nitrous oxide) to piglets induces symptoms of vitamin B_{12} deficiency (Kennedy *et al.*, 1991a), while the vitamin B_{12}-deficient rat excretes FIGLU and MMA in its urine to a degree which is dependent on metabolic demand (Batra *et al.*, 1979). Pigs and poultry consuming rations derived entirely from plant sources, and therefore containing little or no vitamin B_{12}, require supplements of vitamin B_{12}. Gastrointestinal synthesis, unlike that of ruminants, is inadequate to meet the vitamin B_{12} needs of growing pigs and poultry in particular, even in the presence of ample cobalt. Additional supplies of the vitamin may be obtained by the consumption of litter, refuse or faeces in which bacterial fermentation has occurred. If these extraneous sources of vitamin B_{12} are denied, marked growth responses can be obtained from supplementary vitamin B_{12} or from the addition of feeds of animal or microbial origin that are rich in this vitamin. No such response can usually be obtained from cobalt supplements. The dietary requirements of growing pigs for vitamin B_{12} have been put at 10–18 μg kg^{-1} DM (ARC, 1981), while those of poultry are lower, ranging from 3–10 μg kg^{-1} DM (NRC, 1994).

Cobalt Toxicity

Cobalt is commonly stated to be of low toxicity to all species, but this is only true if toxicity is assessed as a multiple of minimum requirement, when there is a 100-fold margin of safety. In terms of dietary concentrations, cobalt is surpassed only by copper, selenium and iodine among trace elements as a threat to health and field cases of suspected cobalt toxicity in ruminants have been reported (Dickson and Bond, 1974). There were no distinctive pathological features, but liver cobalt concentrations were very high (20–69 mg kg^{-1} DM). The Agricultural Research Council (ARC, 1980) summarized the available evidence and – noting that > 4 mg kg^{-1} LW was toxic to sheep and > 1 mg kg^{-1} LW to young cattle – recommended that dietary levels for ruminants should not exceed 30 mg Co kg^{-1} DM. Tolerance may be higher when the cobalt is given continuously in the diet, rather than as a drench, or when a non-ruminant is exposed. Huck and Clawson (1976) found that pigs tolerated 200 mg Co kg^{-1} DM. Higher levels of 400 and 600 mg Co kg^{-1} DM caused anorexia, stiffness, incoordination and muscular tremor, but symptoms were alleviated by supplements of methionine or a combination of iron + manganese + zinc. Anaemia and decreases in tissue iron indicated a cobalt × iron antagonism. In day-old chicks given 125, 250 or 500 mg Co kg^{-1} DM for

14 days, the lowest level reduced feed intake, weight gain and gain:feed ratios, while the two higher levels caused pancreatic fibrosis, hepatic necrosis and muscle lesions (Diaz *et al.*, 1994). In the chick embryo, cobalt was the third most toxic of eight elements examined, surpassed only by cadmium and arsenic (Gilani and Alibhai, 1990).

References

Adams, S.N., Honeysett, J.L., Tiller, K.G. and Norrish, K. (1969) Factors controlling the increase of cobalt in plants following the addition of a cobalt fertilizer. *Australian Journal of Soil Research* 7, 29.

Ammerman, C.B., Henry, P.R. and Loggins, P.R. (1982) Cobalt bioavailability in sheep. *Journal of Animal Science* 55 (suppl. 1), 403.

Andrews, E.D. and Anderson, J.P. (1954) Responses of cobalt-deficient lambs to cobalt and to vitamin B_{12}. *New Zealand Journal of Science and Technology* A35, 483–488.

Andrews, E.D. and Hart, L.I. (1962) A comparison of vitamin B_{12} concentrations in livers and kidneys from cobalt-treated and mildly cobalt-deficient lambs. *New Zealand Journal of Agricultural Research* 5, 403–408.

Andrews, E.D. and Stephenson, B.J. (1966) Vitamin B_{12} in the blood of grazing cobalt-deficient sheep. *New Zealand Journal of Agricultural Research* 9, 491–507.

Andrews, E.D., Stephenson, B.J., Anderson, J.P. and Faithful, W.C. (1958) The effect of length of pastures on cobalt-deficiency disease in lambs. *New Zealand Journal of Agricultural Research* 31, 125–139.

Andrews, E.D., Hart, L.I. and Stephenson, B.J. (1959) A comparison of the vitamin B_{12} and cobalt contents of livers from normal lambs cobalt-dosed lambs and others with a recent history of mild cobalt deficiency disease. *New Zealand Journal of Agricultural Research* 2, 274–282.

Andrews, E.D., Stephenson, B.J., Isaccs, C.E. and Register, R.H. (1966) The effects of large doses of soluble and insoluble forms of cobalt given at monthly intervals on cobalt deficiency disease in lambs. *New Zealand Veterinary Journal* 14, 191.

ARC (1980) *The Nutrient Requirements of Ruminant Livestock.* Commonwealth Agricultural Bureaux, Farnham Royal, Slough, UK, pp. 240–243.

ARC (1981) *The Nutrient Requirements of Pigs.* Commonwealth Agricultural Bureaux, Farnham Royal, Slough, UK, p. 201.

Babior, B.M. (1975) Cobamides as cofactors: adenosylcobamide-dependent reactions. In: Babior, D.M. (ed.) *Cobalamin: Biochemistry and Pathophysiology.* John Wiley & Sons, London, pp. 141–212.

Batra, K.K., Watson, J.E. and Stokstad, E.L.R. (1979) Effect of thyroid powder on urinary excretion of formiminoglutamic acid and methyl malonic acid. *Proceedings of the Society for Experimental Biology and Medicine* 161, 589–594.

Black, H., Hulton, J.B., Sutherland, R.J. and James, M.P. (1988) White liver disease in goats. *New Zealand Veterinary Journal* 36, 15–17.

Brebner, J., Suttle, N.F. and Thornton, I. (1987) Assessing the availability of ingested soil cobalt for the synthesis of vitamin B_{12} in the ovine rumen. *Proceedings of the Nutrition Society* 46, 766A.

Chanarin, I., Deacon, R., Perry, J. and Lumb, M. (1981) How vitamin B_{12} acts. *British Journal of Haematology* 47, 487–491.

Clark, R.G., Burbage, J., Marshall, J.McD., Valler, T. and Wallace, D. (1986) Absence of vitamin B_{12} weight gain response in two trials with growing red deer (*Cervus elaphus*). *New Zealand Veterinary Journal* 34, 199–201.

Clark, R.G., Mantelman, L. and Verkerk, G.A. (1987) Failure to obtain weight gain response to vitamin B_{12} treatment in young goats grazing pasture that was cobalt-deficient for sheep. *New Zealand Veterinary Journal* 35, 38–39.

Clark, R.G., Wright, D.F., Millar, K.R. and Rowland, J.D. (1989) Reference curves to diagnose cobalt deficiency in sheep using liver and serum vitamin B_{12} levels. *New Zealand Veterinary Journal* 37, 1–11.

Davies, S.C., White, C.L. and Williams, I.H. (1995) Increased tolerance to annual ryegrass toxicity in sheep given a supplement of cobalt. *Australian Veterinary Journal* 72, 221–224.

Dewey, D.W., Lee, H.J. and Marston, M.R. (1969) Efficacy of cobalt pellets for providing cobalt for penned sheep. *Australian Journal of Agricultural Research* 20, 1109–1116.

Diaz, G.J., Julian, R.J. and Squires, E.J. (1994) Lesions in broiler chickens following experimental intoxication with cobalt. *Avian Diseases* 38, 308–316.

Dickson, J. and Bond, M.P. (1974) Cobalt toxicity in cattle. *Australian Veterinary Journal* 50, 236.

Duncan, W.R.H., Morrison, E.R. and Garton, G.A. (1981) Effects of cobalt deficiency in pregnant and post-parturient ewes and their lambs. *British Journal of Nutrition* 46, 337–343.

Elliot, J.M., Haluska, M., Peters, J.P. and Barton, E.P. (1979) MMA in ruminant urine – a re-evaluation. *Journal of Dairy Science* 62, 785–787.

Farningham, D.A.H. and Whyte, C.C. (1993) The role of propionate and acetate in the control of food intake in sheep. *British Journal of Nutrition* 70, 37–46.

Fell, B.F. (1981) Pathological consequences of copper deficiency and cobalt deficiency. *Philosophical Transactions of the Royal Society of London, Series B* 294, 153–169.

Ferguson, E.G.W., Mitchell, G.B.B. and MacPherson, A. (1989) Cobalt deficiency and *Ostertagia circumcincta* infection in lambs. *Veterinary Record* 124, 20.

Field, A.C., Suttle, N.F., Brebner, J. and Gunn, G.W. (1988) An assessment of the efficacy and safety of selenium and cobalt included in an anthelmintic for sheep. *Veterinary Record* 123, 97–100.

Fisher, G.E.J. and MacPherson, A. (1991) Effect of cobalt deficiency in the pregnant ewe on reproductive performance and lamb viability. *Research in Veterinary Science* 50, 319–327.

Ford, J.E., Scott, K.J., Sanson, B.F. and Taylor, P.J. (1975) Some observations on the possible nutritional significance of vitamin B_{12}- and folate-binding proteins in milk: absorption of [^{58}Co] cyano-cobalamin by suckling piglets. *British Journal of Nutrition* 34, 469–492.

Gall, L.S., Smith, S.E., Becker, D.E., Stark, C.N. and Loosli, J.K. (1949) Rumen bacteria in cobalt deficient sheep. *Science* 109, 468–469.

Gawthorne, J.M. (1968) The excretion of methylmalonic and formiminoglutamic acids during the induction and remission of vitamin B_{12} deficiency in sheep. *Australian Journal of Biological Sciences* 21, 789–794.

Gawthorne, J.M. (1970) The effect of cobalt intake on the cobamide and cobinamide composition of the rumen contents and blood plasma of sheep. *Australian Journal of Experimental Biology and Medical Science* 48, 285–292.

Gawthorne, J.M. and Smith, R.M. (1974) Folic acid metabolism in vitamin B_{12}-deficient sheep: effects of injected methionine on methotrexate transport and the activity of enzymes associated with folate metabolism in liver. *Biochemical Journal* 142, 119–126.

Gelman, A.L. (1976) A note on the determination of cobalt in animal liver. *Journal of Science in Food and Agriculture* 27, 520.

Gilani, S.H. and Alibhai, Y. (1990) Teratogenicity of elements to chick embryos. *Journal of Toxicology and Environmental Health* 30, 23–31.

Givens, D.I. and Simpson, V.R. (1983) Serum vitamin B_{12} concentrations in growing cattle and their relationship with growth rate and cobalt bullet therapy. In: Suttle, N.F., Gunn, R.G., Allen, W.M., Linklater, K.A. and Wiener, G. (eds) *Occasional Publication No. 7.* British Society of Animal Production, Edinburgh, UK, pp. 145–146.

Grace, N.D., Clark, R.G. and Mortleman, L. (1986) Hepatic storage of vitamin B_{12} by the pregnant ewe and foetus during the third trimester. *New Zealand Journal of Agricultural Research* 29, 231–232.

Halpin, C.C., Harris, D.J. and Caple, I.W. (1984) Contribution of cobalamin analogues to plasma vitamin B_{12} concentrations in cattle. *Research in Veterinary Science* 37, 249–251.

Hannan, R.J., Judson, J.G., Reuter, D.J., McLaren, L.D. and McFarlane, J.D. (1980) Effect of vitamin B_{12b} injections on the growth of young Merino sheep. *Australian Journal of Agricultural Research* 31, 347–355.

Hart, L.I. and Andrews, E.D. (1959) Effect of cobaltic oxide pellets on the vitamin B_{12} content of ewes milk. *Nature (London)* 184, 1242–1243.

Hayakawa, T. (1962) Amounts of trace elements contained in grass produced in Japan. *National Institute of Animal Health Quarterly* 2, 172–181.

Hedrich, M.F., Elliot, J.M. and Lowe, J.E. (1973) Response in vitamin B_{12} production and absorption to increasing cobalt intake in the sheep. *Journal of Nutrition* 103, 1646–1651.

Howard, L., Wagner, C. and Schenker, S. (1974) Malabsorption of thiamine in folate-deficient rats. *Journal of Nutrition* 104, 1024–1032.

Huck, D.W. and Clawson, A.J. (1976) Excess dietary cobalt in pigs. *Journal of Animal Science* 43, 1231–1246.

Judson, G.J., McFarlane, J.D., Riley, M.J., Milne, M.L. and Horne, A.C. (1982) Vitamin B_{12} and copper supplementation in beef calves. *Australian Veterinary Journal* 58, 249–252.

Judson, G.J., Brown, T.H., Kempe, B.R. and Turnbull, R.K. (1988a) Trace element and vitamin B_{12} status of sheep given an oral dose of one, two or four soluble glass pellets containing copper, selenium and cobalt. *Australian Journal of Experimental Agriculture* 28, 299–305.

Judson, G.J., Shallow, M. and Ellis, N.J.S. (1988b) Evaluation of a depot vitamin B_{12} supplement for lambs. In: *Trace Elements in New Zealand: Environmental, Human and Animal Health.* Proceedings of New Zealand Trace Element Group Conference, Lincoln College, Canterbury, pp. 225–229.

Judson, G.J., McFarlane, J.D., Mitsioulis, A. and Zviedrans, P. (1997) Vitamin B_{12} responses to cobalt pellets in beef cows. *Australian Veterinary Journal* 75, 660–662.

Jumba, I.O., Suttle, N.F., Hunter, E.A. and Wandiga, S.O. (1995) Effects of soil origin and mineral composition and herbage species on the mineral composition of forages in the Mount Elgon region of Kenya. 2. Trace elements. *Tropical Grasslands* 29, 47–52.

Kennedy, D.G., Molloy, A.M., Kennedy, S., Scott, J.M., Blanchflower, W.J. and Weir, D.G. (1991a) Biochemical changes induced by nitrous oxide in the pig. In: Momcilovic, B. (ed.) *Proceedings of the Seventh International Symposium on Trace Elements in Man and Animals, Dubrovnik.* IMI, Zagreb, pp. 17-17–17-18.

Kennedy, D.G., Young, P.B., McCaughey, W.J., Kennedy, S. and Blanchflower, W.J. (1991b) Rumen succinate production may ameliorate the effects of cobalt–vitamin B_{12} deficiency on methylmalonyl CoA mutase in sheep. *Journal of Nutrition* 121, 1236–1242.

Kennedy, D.G., Blanchflower, W.J., Scott, J.M., Weir, D.G., Molloy, A.M., Kennedy, S. and Young, P.B. (1992) Cobalt–vitamin B_{12} deficiency decreases methionine synthase activity and phospholipid methylation in sheep. *Journal of Nutrition* 122, 1384–1390.

Kennedy, D.G., Kennedy, S., Blanchflower, W.J., Scott, J.M., Weir, D.G., Molloy, A.M. and Young, P.B. (1994a) Cobalt–vitamin B_{12} deficiency causes accumulation of odd-numbered, branched chain fatty acids in the tissues of sheep. *British Journal of Nutrition* 71, 67–76.

Kennedy, D.G., Young, P.B., Blanchflower, W.J., Scott, J.M., Weir, D.G., Molloy, A.M. and Kennedy, S. (1994b) Cobalt–vitamin B_{12} deficiency causes lipid accumulation, lipid peroxidation and decreased α tocopherol concentrations in the liver of sheep. *International Journal of Vitamin Nutrition Research* 64, 270–276.

Kennedy, D.G., Young, P.B., Kennedy, S., Scott, J.M., Molloy, A.M., Weir, D.G. and Price, J. (1995) Cobalt–vitamin B_{12} deficiency and the activity of methyl malonyl CoA mutase and methionine synthase in cattle. *International Journal for Vitamin and Nutrition Reseach* 65, 241–247.

Kennedy, D.G., Kennedy, S. and Young, P.B. (1996) Effects of low concentrations of dietary cobalt on rumen succinate concentrations in sheep. *International Journal for Vitamin and Nutrition Reseach* 66, 86–92.

Kennedy, S., McConnell, S., Anderson, D.G., Kennedy, D.G., Young, P.B. and Blanchflower, W.J. (1997) Histopathologic and ultrastructural alterations of white liver disease in sheep experimentally depleted of cobalt. *Veterinary Pathology* 34, 575–584.

Kercher, C.J. and Smith, S.E. (1955) The response of cobalt-deficient lambs to orally administered vitamin B_{12}. *Journal of Animal Science* 14, 458–464.

Kirchgessner, M. (1959) Wechselbeziehungen zwischen Spurenelementen in Futtermitteln und tierischen Substanzen sowie Abhangigkeitsverhaltnisse zwischen einzelnen Elementen bei der Retention. 5. Die Wechselwirkungen zwischen verschiedenen Elementen in der Colostral- und normalen Milch. *Zeitschrift für Tierphysiologie Tierernahrung und Futtermittelkunde* 14, 270–277.

Lee, H.J. (1951) Cobalt and copper deficiencies effecting sheep in South Australia. Part 1. Symptoms and distribution. *Journal of Agricultural Science, South Australia* 54, 475–490.

Lee, H.J. and Kuchel, R.E. (1953) The aetiology of *Phalaris* staggers in sheep. 1. Preliminary observations on the preventive role of cobalt. *Australian Journal of Agricultural Research* 4, 88–99.

Lee, H.J. and Marston. H.R. (1969) The requirement for cobalt of sheep grazed on cobalt-deficient pastures. *Australian Journal of Agricultural Research* 20, 905–918.

Lee, H.J., Kuchel, R.E., Good, B.F. and Trowbridge, R.F. (1957) The aetiology of *Phalaris* staggers in sheep. III. The preventive effect of various oral dose rates of cobalt. *Australian Journal of Agricultural Research* 8, 494–501.

Lines, E.W. (1935) The effect of the ingestion of minute quantities of cobalt by sheep affected with 'coast disease': a preliminary note. *Journal of the Council for Scientific and Industrial Research, Australia* 8, 117–119.

McDonald, P. and Suttle, N.F. (1986) Abnormal fermentation in continuous cultures of rumen microorganisms given cobalt deficient hay or barley as the food substrate. *British Journal of Nutrition* 56, 369–378.

McKay, E.J. and McLeary, L.M. (1981) Location and secretion of gastric intrinsic factor in the sheep. *Research in Veterinary Science* 30, 261–265.

McLoughlin, M.F., Rice, D.A. and McMurray, C.H. (1986) Hepatic lesions associated with vitamin B_{12} deficiency. In: *Proceedings of Sixth International Conference on Production Disease in Farm Animals, Belfast.* Veterinary Research Laboratory, Stormont, Northern Ireland, pp. 104–107.

McNaught, K.J. (1948) Cobalt, copper and iron in the liver in relation to cobalt deficiency ailment. *New Zealand Journal of Science and Technology* A30, 26–43.

MacPherson, A. (1983) Oral treatment of trace element deficiencies in ruminant livestock. In: Suttle, N.F., Gunn, R.G., Allen, W.M., Linklater, K.A. and Wiener, G. (eds) *Occasional Publication No. 7.* British Society of Animal Production, Edinburgh, pp. 93–103.

MacPherson, A. and Moon, F.E. (1974) Effects of long-term maintenance of sheep on a low-cobalt diet as assessed by clinical condition and biochemical parameters. In: Hoekstra, W.G., Suttie, J.W., Ganther, H.E. and Mertz, W. (eds) *Trace Element Metabolism in Animals – 2.* University Park Press, Baltimore, pp. 624–627.

MacPherson, A., Moon, F.E. and Voss, R.C. (1976) Biochemical aspects of cobalt deficiency in sheep with special reference to vitamin status and a possible involvement in the aetiology of cerebrocortical necrosis. *British Veterinary Journal* 132, 294–308.

MacPherson, A., Gray, D., Mitchell, G.B.B. and Taylor, C.N. (1987) *Ostertagia* infection and neutrophil function in cobalt-deficient and cobalt-supplemented cattle. *British Veterinary Journal* 143, 348–355.

MacPherson, A., Suttle, N.F., Linklater, K.A. and Rice, D.A. (1999) An assessment of the extent to which trace element deficiencies limit lamb growth on improved Scottish hill pastures. 2. Cobalt. *Journal of Agricultural Science, Cambridge* (in press).

Marston, H.R. (1935) Problems associated with 'coast disease' in South Australia. *Journal of the Council for Scientific and Industrial Research, Australia* 8, 111–116.

Marston, H.R. (1970) The requirement of sheep for cobalt or vitamin B_{12}. *British Journal of Nutrition* 24, 615–633.

Marston, H.R., Allen, S.H. and Smith, R.M. (1961) Primary metabolic defect supervening on vitamin B_{12} deficiency in the sheep. *Nature, UK* 190, 1085–1091.

Marston, H.R., Allen, S.H. and Smith, R.M. (1972) Production within the rumen and removal from the blood-stream of volatile fatty acids in sheep given a diet deficient in cobalt. *British Journal of Nutrition* 27, 147–157.

Millar, K.R. and Albyt, A.T. (1984) A comparison of vitamin B_{12} levels in the liver and serum of sheep receiving treatments used to correct cobalt deficiency. *New Zealand Veterinary Journal* 32, 105–108.

Millar, K.R. and Andrews, E.D. (1964) A method of preparing and detecting radioactive cobaltic oxide pellets and an assessment of their retention by sheep. *New Zealand Veterinary Journal* 12, 9–12.

Millar, K.R. and Lorentz, P.P. (1979) Urinary methyl malonic acid as an indicator of the vitamin B_{12} status of grazing sheep. *New Zealand Veterinary Journal* 27, 90–92.

Millar, K.R. and Penrose, M.E. (1980) A comparison of vitamin B_{12} levels in the livers and sera of sheep measured by microbiological and radioassay methods. *New Zealand Veterinary Journal* 28, 97–99.

Millar, K.R., Albyt, A.T. and Bond, G.C. (1984) Measurement of vitamin B_{12} in the livers and sera of sheep and cattle and an investigation of factors influencing serum vitamin B_{12} levels in sheep. *New Zealand Veterinary Journal* 32, 65–70.

Minson, D.J. (1990) *Forage in Ruminant Nutrition.* Academic Press, New York, pp. 382–395.

Mitchell, R.L. (1945) Cobalt and nickel in soils and plants. *Soil Science* 60, 63.

Mitchell, R.L. (1957) The trace element content of plants. *Research* 10, 357–362.

Mitsioulis, A., Bansemer, P.C. and Koh, T.-S. (1995) Relationship between vitamin B_{12} and cobalt concentrations in bovine liver. *Australian Veterinary Journal* 72, 70.

Nexo, E. and Olsen, H. (1982) Intrinsic factor, transcobalamin and haptocorrin. In: Dolphin, D.D. (ed.) *B_{12}* – Volume 2. *Biochemistry and Medicine.* John Wiley & Sons, New York, pp. 57–86.

Polak, D.M., Elliot, J.M. and Haluska, M. (1979) Vitamin B_{12} binding proteins in bovine serum. *Journal of Dairy Science* 62, 697–701.

Poole, D.B.R. and Connolly, J.F. (1967) Some observations on the use of the cobalt heavy pellet in sheep. *Irish Journal of Agricultural Research* 6, 281–284.

Poole, D.B.R., Fleming, G.A. and Kiely, J. (1972) Cobalt deficiency in Ireland – soil, plant and animal. *Irish Veterinary Journal* 26, 109–117.

Poston, J.M. (1980) Cobalamin-dependent formation of leucine and β-leucine by rat and human tissue. *Journal of Biological Chemistry* 255, 10067–10072.

Poston, J.M. and Stadman, T.C. (1975) Cobamides as cofactors: methyl cobamides and the synthesis of methionine, methane and acetate. In: Babior, B.M. (ed.) *Cobalamin: Biochemistry and Pathophysiology.* John Wiley & Sons, New York, pp. 111–140.

Price, J. (1991) The relative sensitivity of vitamin B_{12}-deficient propionate and 1-carbon metabolism to low cobalt intake in sheep. In: Momcilovic, B. (ed.) *Proceedings of Seventh International Symposium on Trace Elements in Man and Animals, Dubrovnik.* IMI, Zagreb, pp. 27-14–27-15.

Price, J. and Wood, S.G. (1993) Recent developments in the assay of plasma vitamin B_{12} in cattle. In: Anke, M., Meissner, D. and Mills, C.F. (eds) *Proceedings of the Eighth International Symposium on Trace Elements in Man and Animals.* Media Touristik Gersdorf, Germany, pp. 317–318.

Quadros, E.V., Matthews, D.M., Hoffbrand, A.V. and Linnell, J.C. (1976) Synthesis of cobalamin coenzymes by human lymphocytes *in vitro* and the effects of folates and metabolic inhibitors. *Blood* 48, 609–619.

Quirk, M.F. and Norton, B.W. (1987) The relationship between the cobalt nutrition of ewes and the vitamin B_{12} status of ewes and their lambs. *Australian Journal of Agricultural Research* 38, 1071–1082.

Quirk, M.F. and Norton, B.W. (1988) Detection of cobalt deficiency in lactating heifers and their calves. *Journal of Agricultural Science, Cambridge* 110, 465–470.

Rice, D.A., McLoughlin, M., Blanchflower, W.J., Goodall, E.A. and McMurray, C.H. (1987) Methyl malonic acid as an indicator of vitamin B_{12} deficiency in sheep. *Veterinary Record* 121, 472–473.

Rickes, E.L., Brink, N.G., Koniuszky, F.R., Wood, T.R. and Folkers, K. (1948) Vitamin B_{12}, a cobalt complex. *Science* 108, 134.

Rossiter, R.C., Curnow, D.H. and Underwood, E.J. (1948) The effect of cobalt sulphate on the cobalt content of subterranean clover (*Trifolium subterraneum* L. var. Dwalganup) at three stages of growth. *Journal of the Australian Institute of Agricultural Science* 14, 9–14.

Rothery, P., Bell, J.M. and Spinks, J.W.T. (1953) Cobalt and vitamin B_{12} in sheep. 1. Distribution of radiocobalt in tissues and ingesta. *Journal of Nutrition* 49, 173–181.

Russel, A.J.F., Whitelaw, A., Moberley, P. and Fawcett, A.R. (1975) Investigation into diagnosis and treatment of cobalt deficiency in lambs. *Veterinary Record* 96, 194–198.

Schulz, W.J. (1987a) Unsaturated vitamin B_{12} binding capacity in human and ruminant blood serum – a comparison of techniques including a new technique by high performance liquid chromatography. *Veterinary Clinical Pathology* 16, 67–72.

Schulz, W.J. (1987b) A comparison of commercial kit methods for assay of vitamin B_{12} in ruminant blood. *Veterinary Clinical Pathology* 16, 102–106.

Scott, J.M. (1992) Folate–vitamin B_{12} interrelationships in the central nervous system. *Proceedings of the Nutrition Society* 51, 219–224.

Shallow, M., Ellis, N.J.S. and Judson, G.J. (1989) Sex-related responses to vitamin B_{12} and trace element supplementation in prime lambs. *Australian Veterinary Journal* 66, 250–251.

Skerman, K.D. and O'Halloran, M.W. (1962) The effect of cobalt bullet treatment of Hereford cows on the birth weight and growth rate of their calves. *Australian Veterinary Journal* 38, 98–102.

Small, D.H. and Carnegie, P.R. (1981) Myelopathy associated with vitamin B_{12} deficiency: new approaches to an old problem. *Trends in Neurosciences* 4, X–XI.

Smith, E.L. (1948) Presence of cobalt in the anti-pernicious anaemia factor. *Nature, UK* 162, 144–145.

Smith, R.M. (1997) Cobalt. In: O'Dell, B.L. and Sunde, R.A. (eds) *Handbook of Nutritionally Essential Mineral Elements.* Marcel Dekker, New York, pp. 357–388.

Smith, R.M. and Marston, H.R. (1970a) Production, absorption, distribution and excretion of vitamin B_{12} in sheep. *British Journal of Nutrition* 24, 857–877.

Smith, R.M. and Marston, H.R. (1970b) Some metabolic aspects of vitamin B_{12} deficiency in sheep. *British Journal of Nutrition* 24, 879–891.

Smith, S.E., Koch, B.A. and Turk, K.L. (1951) The response of cobalt-deficient lambs to liver extract and vitamin B_{12}. *Journal of Nutrition* 144, 455–464.

Somers, M. (1969) Volatile fatty-acid clearance studies in relation to vitamin B_{12} deficiency in sheep. *Australian Journal of Experimental Biology and Medical Science* 47, 219–225.

Somers, M. and Gawthorne, J.M. (1969) The effect of dietary cobalt intake on the plasma vitamin B_{12} concentration of sheep. *Australian Journal of Experimental Biology and Medical Science* 47, 227–233.

Sutherland, R.J. (1980) On the application of serum vitamin B_{12} radioassay to the diagnosis of cobalt deficiency in sheep. *New Zealand Veterinary Journal* 28, 169–170.

Sutherland, R.J., Cordes, D.O. and Carthew, G.C. (1979) Ovine white liver disease: a hepatic dysfunction associated with vitamin B_{12} deficiency. *New Zealand Veterinary Journal* 27, 227–232.

Suttle, N.F. (1988) The role of comparative pathology in the study of copper and cobalt deficiencies in ruminants. *Journal of Comparative Pathology* 99, 241–258.

Suttle, N.F. (1995) Relationship between vitamin B_{12} and cobalt concentrations in bovine liver. *Australian Veterinary Journal* 72, 278.

Suttle, N.F., Brebner, J., Munro, C.S. and Herbert, E. (1989) Towards an optimum dose of cobalt in anthelmintics in lambs. *Proceedings of the Nutrition Society* 48, 87A.

Sutton, A.L. and Elliot, J.M. (1972) Effect of ratio of roughage to concentrate and level of feed intake on ovine ruminal vitamin B_{12} production. *Journal of Nutrition* 102, 1341–1346.

Teo, N.H., Scott, J.M., Read, B., Neale, G. and Weir, D.G. (1981) Bile acid inhibition of vitamin B_{12} binding by intrinsic factor *in vitro*. *Gut* 22, 270–276.

Underwood, E.J. and Filmer, J.F. (1935) The determination of the biologically potent element (cobalt) in limonite. *Australian Veterinary Journal* 11, 84–92.

Wahle, K.W.J., Duncan, W.R.H. and Garton, G.A. (1979) Propionate metabolism in different species of ruminant. *Annales de Recherche Vétérinaire* 10, 362–364.

Walker, C.K. and Elliot, J.M. (1972) Lactational trends in vitamin B_{12} status on conventional and restricted roughage rations. *Journal of Dairy Science* 55, 474–478.

Watanabe, F., Nakano, Y., Tachikake, N., Saido, H., Tamura, Y. and Yamanaka, H. (1991) Vitamin B_{12} deficiency increases the specific activities of rat liver NADH- and NADPH-linked aquocobalamin reductase isoenzymes involved in coenzyme synthesis. *Journal of Nutrition* 121, 1948–1954.

Winter, W.J., Siebert, B.D. and Kuchel, R.E. (1977) Cobalt and copper therapy of cattle grazing improved pasture in northern Cape York peninsula. *Australian Journal of Experimental Agriculture and Animal Husbandry* 17, 10–15.

Young, R.S. (1979) *Cobalt in Biology and Biochemistry*. Academic Press, New York, pp. 50–86.

Zervas, G. (1988) Use of soluble glass boluses containing Cu, Co and Se in the prevention of trace element deficiencies in goats. *Journal of Agricultural Science, Cambridge* 110, 155–158.

Copper

11

Introduction

Copper was first shown to be essential for growth and haemoglobin formation in rats by Hart *et al.* (1928). This important discovery was soon followed by experimental evidence that copper is essential for growth and for the prevention of a wide range of clinical and pathological disorders in all types of farm animal. A number of metalloenzymes containing copper were later identified in the cells and tissues. While the early experimental investigations were proceeding, a number of naturally occurring diseases of sheep and cattle were associated with a dietary deficiency of copper and found to respond to copper therapy, such as 'salt-sick' of cattle in Florida (Neal *et al.*, 1931) and 'lecksucht' of sheep and cattle in the Netherlands (Sjollema, 1933). An ataxic disease of newborn lambs occurring in parts of Western Australia was attributed to copper deficiency in the ewe during pregnancy (Bennetts and Chapman, 1937). Subsequently, extensive copper-deficiency areas, affecting both crops and livestock, were discovered throughout the world. The importance of an interaction between copper and molybdenum became evident when it was discovered that the severe scouring disease ('teart') of cattle, caused by the ingestion of *excessive* amounts of molybdenum from the herbage in parts of England, could be controlled by massive doses of copper (Ferguson *et al.*, 1938, 1943). Concurrently, investigation of another 'area' problem was proceeding, that of chronic copper poisoning of sheep and cattle in eastern Australia; the disorder was associated with abnormally *low* molybdenum contents in the herbage and was controllable by supplementary molybdenum (Dick and Bull, 1945). The limiting effect of molybdenum on copper retention in the animal was then demonstrated and, in a classical series of experiments, Dick (1952, 1953, 1954, 1956) showed that this effect appeared to require the presence of inorganic sulphate. The findings, indicating a three-way interaction between copper, molybdenum and sulphur, gave a great stimulus to further studies (Suttle, 1975, 1991). Copper-responsive disorders rarely occur naturally in non-ruminants and the emphasis throughout this chapter will be placed on the more vulnerable ruminant animal.

© CAB *International* 1999. *Mineral Nutrition of Livestock*
(E.J. Underwood and N.F. Suttle)

Sources of Copper for Ruminants

Absorbability

The ability of a feed to meet the copper requirements of ruminants or pose a risk of copper poisoning depends more on the absorbability (A_{Cu}) than on the concentration of copper that it contains (Suttle, 1983a). Feed sources differ widely in A_{Cu}, for reasons which are not completely understood (Table 11.1); fresh grass is a poor copper source, while brassicas and cereals are both excellent sources of copper for sheep, despite the high sulphur content of brassicas. The high value given for brassicas, however, is inconsistent with the development of hypocupraemia in cattle given kale (Barry *et al.*, 1981). Little is known about the forms in which copper occurs in feeds, apart from the early studies of forages by Mills (1954, 1956a, b). However, the variations in A_{Cu} within and between feedstuffs for ruminants is determined largely by events in the rumen, notably the synchronicity of release of copper and its potential antagonists, molybdate, sulphide and iron (Fe^{2+}) from the diet (Suttle, 1991).

The copper × sulphur and copper × molybdenum × sulphur interactions

The effects of molybdenum and sulphur on A_{Cu} in grasses, hays and silages, as determined by plasma copper repletion rates in hypocupraemic sheep, have been described by prediction equations and are illustrated in Fig. 11.1.

Table 11.1. Estimates of the absorbability of copper (A_{Cu}, %) in natural foodstuffs of low molybdenum content (< 2 mg kg^{-1} DM) to Scottish Blackface ewes.

	A_{Cu} (mean ± SD)	Number of estimates
Grazed herbage (July)	2.5 ± 1.09	7
Grazed herbage (Sept./Oct.)	1.4 ± 0.86	6
Silage	4.9 ± 3.2	7
Hay	7.3 ± 1.8	5
Root brassicas	6.7 ± 0.9	2
Cereals	9.1 ± 0.97	3
Leafy brassicas	12.8 ± 3.2	5

SD, standard deviation.

Fig. 11.1. (*Opposite*) The adequacy of fresh and conserved grass as sources of copper (Cu) for ruminants is determined largely by the extent to which sulphur (S) and molybdenum (Mo) combine to reduce Cu absorption (A_{Cu}) but the interactions vary with forage type: (a) in silages, Mo has a small and little studied effect on A_{Cu} but S reduces A_{Cu} in a logarithmic manner; (b) in hays, the inhibitory effect of Mo is detectable but less than that of S, not greatly influenced by interactions with S and A_{Cu} remains relatively high; (c) in fresh grass, A_{Cu} starts low and is further greatly reduced by small increments in Mo and S (data from Suttle, 1983a,b obtained using the technique outlined by Suttle, 1986). See Table 11.6 for effects on requirements.

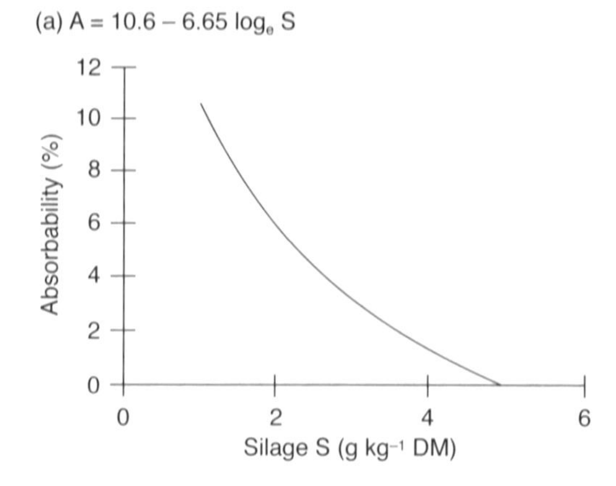

(a) A = 10.6 – 6.65 log$_e$ S
Absorbability (%)
0
2
4
6
8
10
12
0
2
4
6
Silage S (g kg^{-1} DM)

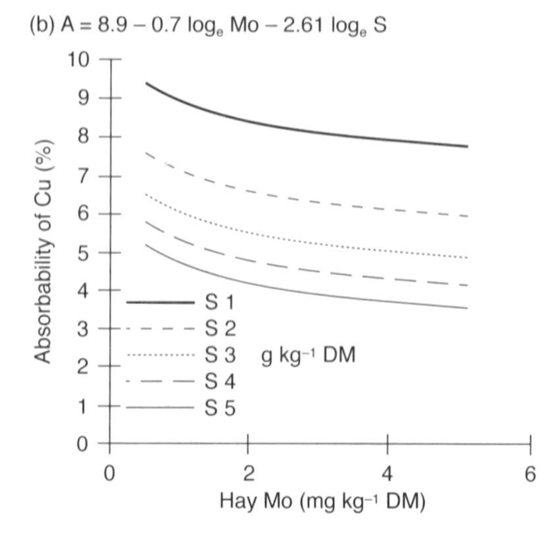

(b) A = 8.9 – 0.7 log$_e$ Mo – 2.61 log$_e$ S
Absorbability of Cu (%)
0
1
2
3
4
5
6
7
8
9
10
S 1
S 2
S 3
S 4
S 5
g kg^{-1} DM
0
2
4
6
Hay Mo (mg kg^{-1} DM)

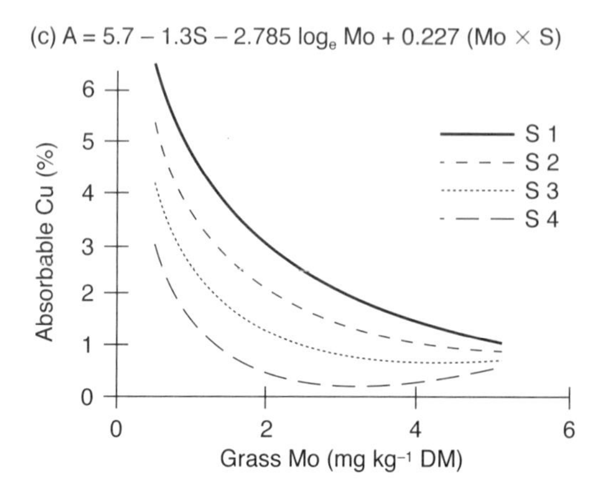

(c) A = 5.7 – 1.3S – 2.785 log$_e$ Mo + 0.227 (Mo × S)
Absorbable Cu (%)
0
1
2
3
4
5
6
S 1
S 2
S 3
S 4
0
2
4
6
Grass Mo (mg kg^{-1} DM)

Profound effects are predicted for both antagonists within commonly encountered ranges, the earliest, small rises in herbage molybdenum and sulphur having the largest inhibitory effects on A_{Cu} (see Chapter 9 for copper (Cu) × sulphur (S)); it is arguable that a Cu × S interaction will rarely be fully independent of a molybdenum influence. The outcome of the antagonisms varies between forages, sulphur *per se* having an enhanced influence in silages (Fig. 11.1a) and both antagonists having reduced influence in hays (Fig. 11.1b), when compared with fresh grass (Fig. 11.1c). In long-term grazing studies employing a liver copper repletion technique, Langlands *et al.* (1981) found the effects of molybdenum and sulphur to be far smaller, but reasons for this discrepancy have been discussed (Suttle, 1983b) and the effect of molybdenum (Mo) on hepatic copper retention can be just as dramatic as that indicated by plasma repletion (Suttle, 1977). The herbage equation (Fig. 11.1c) helped to explain the reduction in copper status in lambs grazing improved pastures (Whitelaw *et al.*, 1979; Woolliams, C. *et al.*, 1986; Woolliams, J.A. *et al.*, 1986) and the hypocuprosis seen in cattle transferred from winter feeds to spring pasture, despite an increase in dietary copper concentration (Jarvis and Austin, 1983). Both the Cu × S (Suttle, 1974; Bremner *et al.*, 1987; Fig. 11.2) and Cu × Mo × S (Suttle, 1977) interactions also affect A_{Cu} from concentrate-type diets.

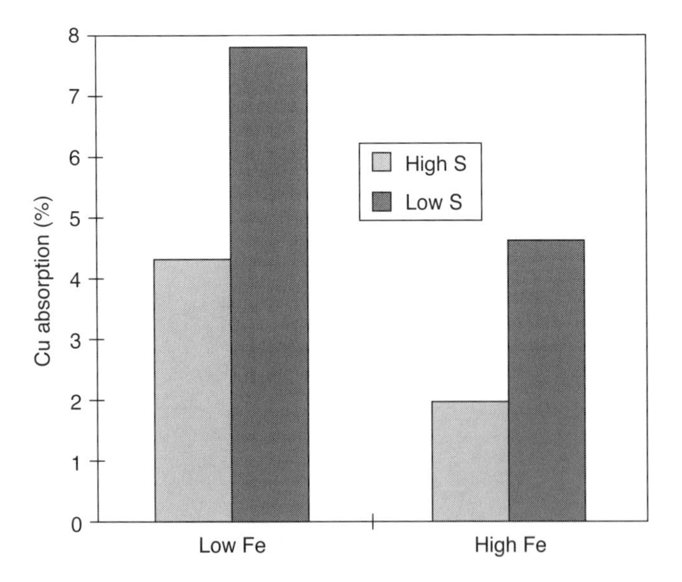

Fig. 11.2. The absorbability of copper (A_{Cu}) in diets for weaned ruminants is significantly affected by interactions between copper and iron which are partly dependent on sulphur (S): effects on A_{Cu} cannot yet be predicted in the manner used for Cu × Mo × S interactions (Fig. 11.1) and are not additive to those of Mo × S but with 800 mg added Fe kg^{-1} DM the effect is to roughly halve A_{Cu} (see Table 11.6).

The copper × iron interaction

Antagonisms between copper and iron also influence copper absorption. Jarvis and Austin (1983) invoked an Fe × Cu antagonism to further explain the recovery from hypocuprosis in summer with no change in A_{Cu}, as predicted in Fig. 11.1c. However, the low value for A_{Cu} in autumn compared with summer pasture (Table 11.1) could not be explained in terms of autumn increases in herbage iron (Suttle, 1983). Supplementation of a semipurified diet for sheep with 800 mg Fe as ferric oxide (Fe_2O_3) kg^{-1} dry matter (DM), to simulate the effects of ingesting soil iron, lowered A_{Cu} significantly, from 6 to 4% (Suttle and Peter, 1985); similar inhibitions were obtained with iron as ferrous sulphate ($FeSO_4$) (Suttle *et al.*, 1984) (Fig. 11.2), although the effect with the low-sulphur diet may have been due in part to the significant contribution of sulphur from the supplement (*c.* 0.5 g kg^{-1} DM). The accelerated depletion of liver copper reserves in weaned, iron-supplemented calves (Humphries *et al.*, 1983) probably reflects inhibition of copper absorption, and the interactions in both sheep (Suttle *et al.*, 1984) and cattle (Bremner *et al.*, 1987) are in part dependent on sulphur. Soil iron probably inhibits copper absorption by two mechanisms: the trapping of sulphide as FeS by soluble iron in the rumen (Suttle *et al.*, 1984) and the adsorption of copper by insoluble iron compounds (Suttle and Peter, 1985). Bovine copper status has been lowered by as little as 250 mg Fe kg^{-1} DM when saccharated ferrous carbonate provided the iron (Bremner *et al.*, 1987). Contamination of pastures and silages with soil iron is commonplace and should be taken into account when assessing their value as copper sources, although there is no evidence that the Cu × Fe interaction can add to the effect of the Cu × Mo × S interaction in cattle (Humphries *et al.*, 1983; Phillippo *et al.*, 1987a, b) or sheep (N.F. Suttle, unpublished data) to produce further reductions in A_{Cu}. The Cu × Fe interaction is not manifested in the preruminant calf (Bremner *et al.*, 1987) and may not be seen in sheep on concentrate diets low in sulphur (Rosa *et al.*, 1986).

Other influences upon absorbability

Inhibitory effects of dietary calcium on copper absorption were the subject of early reports and speculation but were not evident in a study of liver copper stores in cattle (Huber and Price, 1971) or in sheep, using a plasma repletion technique (N.F. Suttle, unpublished data). Inhibitory effects of zinc on copper metabolism have been reported in sheep (Bremner *et al.*, 1976) but are unlikely to be seen over the normal range of forage zinc concentrations. Small supplements of cadmium can reduce the placental transfer of copper in ewes and cows, and this interaction may be important where herbage cadmium levels have been increased by the use of superphosphates high in cadmium (see Chapter 18). Recent studies have indicated that intestinal parasitism lowers copper status in sheep (Suttle, 1996), and the low autumn A_{Cu} values in Table 11.1 may partly reflect an autumn rise in larval nematode infection on pasture. Similar problems associated with the spring hatch of nematode eggs may lower the value of spring pasture as a copper source.

Pot-ale syrup, a by-product from the distillery industry, contains high concentrations of copper in a poorly absorbable form (Suttle *et al.*, 1996).

Composition of forages

The copper, iron, molybdenum and sulphur contents of pastures and forages vary with the species, strain and maturity of the plant, the soil conditions and the fertilizers used (McFarlane *et al.*, 1990; Table 11.2). High iron concentrations reflect soil contamination and are found on soils prone to waterlogging (e.g. red gums and rendzinas, Table 11.2). Further details of sulphur and iron concentrations are given in Chapters 9 and 13, respectively. Temperate grasses tend to be lower in copper than legumes grown in the same conditions (4.7 vs. 7.8 mg Cu kg^{-1} DM), but under tropical conditions the position is reversed (grasses 7.8 vs. legumes 3.9 mg Cu kg^{-1} DM: Minson, 1990). There is little published information on molybdenum in tropical legumes, but modest increases may be all that are needed to induce hypocuprosis; on average, 3.9 mg Mo kg^{-1} DM would give a Cu:Mo ratio of 1.0. Differences within grass and legume species can be high. Beeson *et al.* (1947) found the copper concentrations in 17 grass species grown together on a sandy loam soil to range from 4.5 to 21.1 mg kg^{-1} DM. Jumba *et al.* (1995) reported much smaller but significant differences between species in west Kenya after adjustment for soil effects, *P. clandestinum* (kikuyu) having the highest and *C. gayana* (Rhodes grass) the lowest concentration. However, when available copper concentrations were predicted, the lusher kikuyu lost much of its advantage (Table 11.3), due mostly to the influence of sulphur.

Copper is unevenly distributed in temperate grasses, the leaves containing 35% higher concentrations than stems; there is thus a tendency for values in the whole plant to decline during the growing season (Minson, 1990), and the fall can be marked at some locations (Jarvis and Austin, 1983). The low copper levels found in hays and silages in Alberta (4.3 ± 2.45 and 5.0 ± 1.98 (SD) mg kg^{-1} DM, respectively, (Suleiman *et al.*, 1997) may partly reflect the

Table 11.2. The mean and range (in parentheses) of copper, molybdenum, sulphur and iron concentrations in subterranean and/or strawberry clover tops sampled on two to four occasions from sites of the major soil types in the south-east of South Australia

Soil type	No. of sites	Cu (mg kg^{-1})	Mo (mg kg^{-1})	S (g kg^{-1})	Fe (mg kg^{-1})
Sand/clay	24	6.6 (3.0–14.6)	1.4 (0.1–4.1)	2.5 (1.5–4.0)	160 (45–346)
Red gum	22	9.5[a] (4.1–15.9)	1.8 (0.1–5.4)	2.6 (1.5–4.3)	520[a] (81–2300)
Groundwater rendzina	20	8.5 (4.2–12.8)	1.6 (0.2–5.4)	2.5 (1.9–5.1)	510[a] (200–1000)
Deep sand	16	7.3 (4.0–14.0)	1.1 (0.1–3.8)	2.8 (2.1–4.6)	140 (119–154)
Peat	4	7.2 (4.4–11.4)	8.3[a] (4.7–16.2)	3.6[a] (2.8–4.5)	110 (48–320)
Calcareous sand	11	5.3 (1.9–9.5)	10.1[a] (1.6–21.8)	2.5 (1.7–3.4)	100 (70–130)

Mean values followed by [a] differ significantly ($P < 0.05$) from others: each is comprised of five to ten samples.

Table 11.3. Species differences in available copper concentraions in four grasses calculated by prediction equations (Suttle, 1983a) from Cu, Mo and S concentrations in the pastures (from Jumba *et al.*, 1995).

	Herbage concentration			Available copper ($mg\ kg^{-1}$ DM)	
	Cu	Mo	S	Grass[a]	Hay[a]
Species	($mg\ kg^{-1}$ DM)		($g\ kg^{-1}$ DM)	equation	equation
P. purpureum (Napier grass)	4.1	0.85	1.1	0.135	0.303
S. sphacelata (setaria)	3.9	0.66	1.4	0.141	0.308
C. gayana (Rhodes grass)	3.5	0.64	2.0	0.107	0.251
P. clandestinum (kikuyu grass)	5.7	1.50	1.7	0.127	0.383

[a] See Fig. 11.1.

influence of advancing maturity but are offset by concurrent decreases in sulphur and increases in availability (Table 11.1, Figs 11.1 and 11.2), when compared with grasses.

Forage copper is not influenced by soil pH, but the position of molybdenum is quite different (see Fig. 2.2). The highest herbage molybdenum concentrations occur on alkaline soils and on soils high in organic matter; the latter also give rise to high sulphur levels in herbage (Table 11.2). On granitic soils, liming can significantly raise molybdenum levels, markedly decreasing the all-important Cu:Mo ratio of herbage (Whitelaw *et al.*, 1979, 1983). In a study of fodders and grains grown in British Columbia, this ratio ranged from 0.1 to 52.7 in individual samples (Miltimore and Mason, 1971). Low molybdenum concentrations in maize silage grown in the molybdeniferous 'teart' area of Somerset (A. Adamson, personal communication) raise the possibility of important differences between forage crops, which might be exploited to control molybdenum-induced disorders.

Composition of concentrate feeds

Species differences among the graminaceous grains in copper and antagonist concentrations are relatively small and the normal concentrations are low compared with most other feeds. A study of grain grown in Northern Ireland revealed concentrations ranging from 1.5 to 8.4 mg Cu kg^{-1} DM, with averages of 3.9 and 4.9 mg kg^{-1} DM for oats and barley, respectively (Todd, 1972). The corresponding averages for molybdenum were 0.25 and 0.30 mg kg^{-1} DM, while sulphur concentrations ranged from 0.8 to 1.5 g kg^{-1} DM. The copper concentrations in leguminous and oilseed meals are high and usually range from 15 to 35 mg kg^{-1} DM, but molybdenum concentrations are also higher than in cereals, ranging from 1 to 4 mg kg^{-1} DM. The feeding of high levels of palm-kernel cake, containing 25–40 mg Cu kg^{-1} DM, has caused copper toxicity in sheep (Chooi *et al.*, 1988). Feeds of animal origin, other than liver meal and meals from Crustacea and shellfish, are poor to moderate sources of copper. Meat meal and fish-meal typically contain

5–15 mg Cu kg^{-1} DM, compared with 80–100 mg kg^{-1} DM in liver meal. Dairy milk and milk products are inherently low in copper (*c.* 1 mg Cu kg^{-1} DM) but can be contaminated with copper during processing and storage. Feeds of animal origin are also relatively poor sources of molybdenum, unless they come from animals fed on high-molybdenum, low-sulphur diets. The molybdenum concentration in milk is highly dependent on dietary intake, in contrast to copper, and can be raised several-fold above the normal of about 0.06 mg Mo l^{-1} in ewes or cows. However, molybdenum in milk will bypass the rumen and probably have little effect on the copper status of the suckling. Molybdenum-rich dairy products should present no risk of induced copper deficiency to non-ruminant consumers.

Sources of Copper for Non-ruminants

Variation in the 'availability' of copper between and among natural organic and inorganic copper sources for non-ruminants has been extensively studied, despite the lack of evidence that low availability would ever cause natural copper deficiencies. Poultry have been studied more extensively than pigs, and it is clear that natural foodstuffs, such as maize gluten meal, cottonseed meal, groundnut hulls and soybean meal, have low values (48, 41, 44 and 38%, respectively) relative to copper sulphate ($CuSO_4$) (Baker and Ammerman, 1995). Low availability may not be attributable to antagonism from phytate, because phytate supplements can enhance copper retention (Leach *et al.*, 1990). Certain by-products contained copper of very low availability (feather meal < 1%, meat- and bone-meal 4–28%), but others (poultry waste, 67%) were good sources. Unfortunately, the absolute value for $CuSO_4$ remains undefined.

Metabolism of Copper

Maternal influences

Provision for the newborn is made through the accumulation of high liver copper concentrations in the fetus. Provided that copper supply is adequate during pregnancy, copper concentrations in the newborn calf, lamb, pig and foal are likely to be in the region of 3–6 mmol kg^{-1} DM (Egan and Murrin, 1973; Underwood, 1977). Restriction of the maternal supply of copper can reduce neonatal reserves drastically at birth, hastening their subsequent depletion and revealing effects of maternal breed (compare Wiener *et al.*, 1984a and 1984b). Priority is clearly given to the fetus, because cows with low liver copper concentrations (< 0.39 mmol kg^{-1} DM) can carry calves with 5.2 mmol Cu kg^{-1} DM in their livers (Gooneratne and Christensen, 1989), but, with maternal values below 0.26 mmol kg^{-1} DM, fetal values declined by up to 50%. Copper secretion in milk is reduced when the dietary copper supply is inadequate (Beck, 1941a; Whitelaw *et al.*, 1983) but cannot normally be increased by supplementing an adequate diet with copper. There are wide differences between species in normal milk copper concentrations, the cow and goat secreting milk with low copper levels (0.15 mg or 2 μmol Cu l^{-1}) and sows with a high level (0.75 mg or 10 μmol Cu l^{-1}). The unique position of the pig reflects the need to sustain large, rapidly growing litters during a brief lactation.

Absorption

Absorption in the newborn of all species can proceed by pinocytosis – the engulfment of large, proteinaceous complexes; this, coupled with high copper concentrations in colostrum, ensures a plentiful early copper supply. The precise mechanism of absorption in later life is not clearly understood but has two components, one active and saturable, the other passive and unsaturable (Bronner and Yost, 1985); it is unlikely that ruminants and non-ruminants differ in this respect. The vulnerability of ruminants to copper deficiency is due to digestive processes in the rumen, which degrade organic and inorganic sulphur sources to sulphide (Suttle, 1974), while failing to digest 30–50% of the organic matter. The rumen protozoa are particularly important as generators of sulphide (see Fig. 9.1) and their removal in Caesarean-derived, specific pathogen-free flocks (Ivan, 1988) or by the administration of antiprotozoal ionophores (van Ryssen and Barrowman, 1987) increases copper absorption. Much of the copper released during rumen digestion is likely to be precipitated as copper sulphide and remain unabsorbed (Bird, 1970), while that released during post-ruminal digestion may become partially bound to undigested constituents. Whereas the young milk-fed lamb can absorb 70–85% of the copper ingested, the weaned lamb absorbs less than 10% (Suttle, 1974b). When the diet is enriched with molybdenum as well as sulphur, thiomolybdates are formed, which not only complex copper but leave it firmly bound to particulate matter (Allen and Gawthorne, 1987); this reduces absorbability still further (Price and Chesters, 1985), until as little as 1% may be absorbed. These complex events were reviewed by Suttle (1991). In contrast, non-ruminants given digestible diets absorb about 60% of moderate copper intakes.

Transport and cellular uptake

Binding to metallothionein in the gut mucosa is an important means of restricting the further translocation of copper (Cousins, 1985) and may contribute to the adaptation of sheep (Woolliams *et al.*, 1983) and other species to excessive copper intakes. Transfer of absorbed copper in the portal bloodstream is probably facilitated largely by binding to albumin. At the liver, copper is taken up by a two-stage process involving binding first to glutathione and then to metallothionein, before being partitioned between biliary secretion, synthesis of ceruloplasmin (the major transport protein) and storage (Bremner, 1993). Increases in the hepatic activity of enzymes involved in glutathione synthesis during copper deficiency (Chen *et al.*, 1995) may enhance the efficiency of copper uptake by the liver, as does exposure to excess selenium (Hartman and van Ryssen, 1997). Elsewhere, uptake of copper is probably achieved primarily by ceruloplasmin receptors in cell membranes (McArdle, 1992; Saenko *et al.*, 1994), although albumin and amino acids, such as histidine, facilitate copper uptake *in vitro*. The relative importance of these processes may depend on species and copper status. Ceruloplasmin normally constitutes a higher proportion of total plasma copper in non-ruminants than in ruminants (95 vs. 80%) but is depleted during deficiency and can approach undetectable levels in outwardly healthy cattle (Humphries *et al.*, 1983). Avian species are exceptional, having very little ceruloplasmin in the bloodstream.

Homeostasis

Adjustment to fluctuations in copper supply is achieved predominantly by hepatic storage and biliary copper secretion, but apportionment of copper trapped by the liver varies widely between species; those faced with endemic risks of copper deficiency (i.e. ruminants) avidly store excess copper, while species at no risk (i.e. non-ruminants) lavishly excrete excess copper via the bile and maintain low liver copper levels (Fig. 11.3). There are also differences among classes, cattle limiting hepatic storage sooner than sheep by means of biliary secretion (Phillippo and Graca, 1983). During depletion, rates of decline in liver copper are positively correlated with liver copper concentrations in ruminants (Woolliams *et al.*, 1983; Freudenberger *et al.*, 1987), confirming homeostatic control via biliary secretion. In addition to coping with excesses, biliary secretion affords opportunities for enterohepatic recycling of copper, given the moderate efficiency of copper absorption from

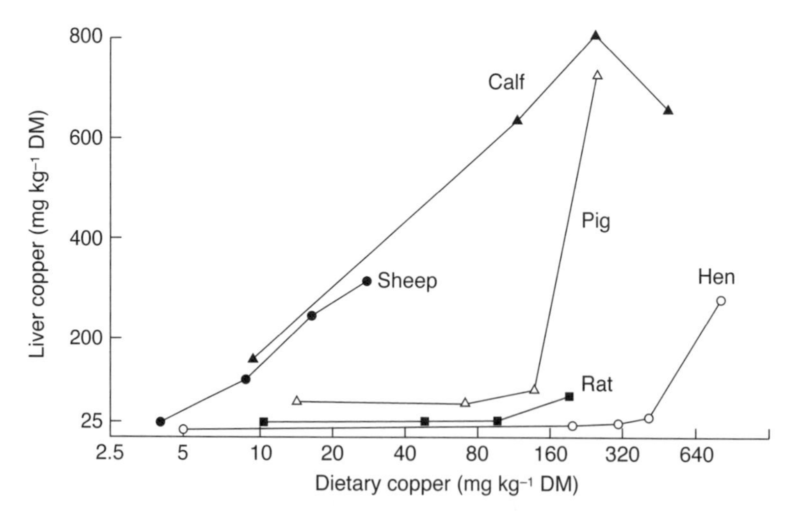

Fig. 11.3. Species differ in the extent to which they store excess dietary copper in their livers: ruminant species, for which the risk of copper deficiency is ever-present, store copper avidly while non-ruminant species, which are rarely at risk, do not.

the intestine in simple-stomached animals and the partial absorbability of biliary copper (Farrer and Mistilis, 1967). Any enhancement of absorptive efficiency during deficiency should enhance the recycling of copper. The extent to which copper is recycled by ruminants and the vulnerability of that recycling to impairment by thiomolybdates, leading to enhanced losses, remains to be thoroughly investigated, but Smith *et al.* (1968) reported a 2.5-fold increase in faecal endogenous losses of copper in sheep given large molybdenum (25 mg kg^{-1} DM) and sulphur (4.5 g kg^{-1} DM) supplements. Enhancement of endogenous loss by dietary molybdenum in sheep is also indicated by marked increases in the rate of liver copper depletion when plotted against liver copper concentration (Woolliams *et al.*, 1983; Freudenberger *et al.*, 1987), but deer showed the opposite response (Freudenberger *et al.*, 1987). Urinary excretion of copper is normally small, constant and unaffected by copper intake in all species, although it is increased in sheep by exposure to molybdenum (Smith *et al.*, 1968; Marcilese *et al.*, 1970).

Extrahepatic tissue distribution of copper

Widely different copper concentrations are maintained in different organs, heart and kidney being particularly enriched (3.1 and 3.6 mg Cu kg^{-1} (fresh weight) FW, respectively in sheep; Grace, 1983). Because the major contributors to carcass weight (muscle, fat and bone) contain the least copper, maximal copper concentrations in the whole carcass, excluding the liver, are relatively low, at 1.2 mg Cu kg^{-1} FW in sheep (Grace, 1983) and 0.8 mg kg^{-1} FW in the bovine, declining further in copper deficiency (Simpson *et al.*, 1981). Muscle provides the major extrahepatic pool of copper and relatively large amounts are deposited in the fleece (Fig. 11.4). Copper deprivation leads to widespread reductions in tissue copper concentrations but most noticeably in kidney and least in brain (Suttle and Angus, 1976). Non-ruminant species, such as the rat and chick, have higher carcass copper concentrations (4.8 and 1.7 mg kg^{-1} live weight (LW), respectively: Suttle, 1987a) than ruminants.

Genetic variation

Studies of genetic variation in copper metabolism among farmed livestock have been largely confined to the sheep, although Reetz *et al.* (1975) reported high heritabilities of 0.35–0.52 for plasma copper in pigs. The situation in livestock differs from the much studied but exotic heritable disorders of copper metabolism in humans and laboratory animals in affecting large populations (Wiener, 1987). Sheep breeds differ substantially in the liver copper concentrations they attain when given copper in excess (Woolliams *et al.*, 1982; van der Berg *et al.*, 1983), high-retaining breeds, such as the Texel, being particularly vulnerable to toxicity. Conversely, when the copper supply is inadequate, breeds characterized by poor retention, such as the Scottish Blackface, develop hypocuprosis (Woolliams, J.A. *et al.*, 1986). The differences are largely attributable to differences in the efficiency of

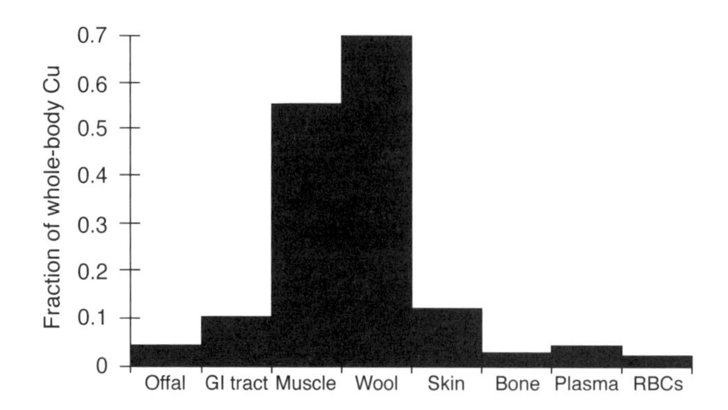

Fig. 11.4. Distribution of copper in the shorn, empty body of mature sheep: the relatively large amounts removed in the fleece are also shown (data from Grace, 1983); GI, gastrointestinal.

absorption (Wiener *et al.*, 1978; Woolliams *et al.*, 1983) but may also involve differences in the partitioning of absorbed copper (Woolliams *et al.*, 1982, 1983).

Comparable variation was looked for among six beef breeds crossed with the Friesian and given diets containing 10 or 20 mg Cu kg^{-1} DM, but it was not found (J.A. Woolliams and N.F. Suttle, unpublished data). Littledike *et al.* (1995) reported significantly higher liver copper concentrations in Limousin than eight other breeds after long-term feeding trials and suggested a high efficiency of copper absorption for the breed. However, Limousin also had the smallest livers and the highest liver zinc concentrations, suggesting an adaptation to small liver size rather than different efficiencies of overall copper utilization. Jerseys accumulated copper in their livers slightly faster than Holsteins when given the same copper-rich diet *ad libitum*, but they consumed more of the diet per unit of metabolic live weight (Du *et al.*, 1996). Differences between small groups ($n = 4$) of Aberdeen Angus and Simmental in biliary copper secretion (Gooneratne *et al.*, 1994) may be indicative of breed differences in copper metabolism in cattle. Goats retain less copper in their livers than sheep when exposed to excesses, presumably because they share with cattle a propensity for biliary copper secretion (Zervas *et al.*, 1990).

Genetic differences in copper metabolism are variably influenced by exposure to antagonists. Molybdenum supplementation has obliterated apparent breed differences in faecal endogenous copper (FE_{Cu}) loss in depleting sheep (Woolliams *et al.*, 1983). In the repleting but hypocupraemic state, molybdenum and sulphur had similar inhibitory effects on copper absorption in cattle to those reported for sheep (Suttle, 1978), but effects of these antagonists on rates of hepatic copper depletion in individuals with large liver reserves of copper seemed much greater in another bovine study (Simpson *et al.*, 1981). Since cattle excrete more copper in bile than sheep, molybdenum supplements may impair recycling and raise FE_{Cu} more in cattle, causing greater increases in endogenous losses. Deer showed no signs of abnormal plasma copper distribution when given molybdenum-supplemented silage, but sheep given the same diet did so (Freudenberger *et al.*, 1987). This was consistent with Mason *et al.*'s (1984) study, which indicated a relatively low ruminal production of thiomolybdates in deer.

Functions of Copper

Copper is surpassed only by zinc in the number of enzymes which it activates (Table 11.4), making it difficult to determine the precise reason why particular abnormalities arise in livestock deprived of copper. The essentiality of copper for reproduction, bone development, growth, connective-tissue development and pigmentation of the skin appendages (hair, fleece and

Table 11.4. Some copper-dependent enzymes found in mammalian tissues,[a] their functions and possible consequences of a marked reduction in activity.

Enzyme	Nomenclature	Functions	Abbreviation	Pathognomic significance
Ceruloplasmin (ferroxidase)	EC 1.16.3.1	$Fe^{2+} \rightarrow Fe^{3+}$, hence Fe transport[b]	CP	Anaemia
Cytochrome *c* oxidase	EC 1.9.3.1	Terminal electron transfer-respiratory chain	CCO	Anoxia (neuronal degeneration; cardiac hypertrophy)
Dopamine-ß-monoxygenase	EC 1.14.17.1	Catecholamine metabolism	DBM	Behaviour?
Lysyl oxidase	EC 1.4.3.13	Desmosine X-linkages in connective tissue	LO	Aortic rupture; joint disorders; osteoporosis
Peptidylglycine α-amidating monooxygenase	EC 1.14.17.3	Elaboration of numerous biogenic molecules, e.g. gastrin	PAM	Appetite?
Copper–zinc superoxide dismutase	EC 1.15.1.1	Dismutation of O_2^- to H_2O_2	CuZnSOD	Lipid peroxidation
Tyrosinase	EC 1.14.18.1	Tyrosine → melanin		Depigmentation

[a] Most of these enzymes or closely related molecules (e.g. ceruloplasmin-like, intracellular, Cu-transporters) are found in most tissues, though in varying concentrations and excesses, and variously supported by Cu-independent enzymes with overlapping (e.g. MnSOD) or linked (e.g. glutathione peroxidase) functions. Reductions in activity are therefore not invariably indicative of dysfunction.
[b] Ceruloplasmin may also contribute to blood clotting and antioxidant defences (Saenko *et al.*, 1994).

feathers) are all unquestioned, but only the latter functions have proved specific enzyme connections, pigmentation with tyrosinase and connective-tissue development with lysyl oxidase (for review, see Suttle, 1987a). Few other copper enzymes have been proved to become rate-limiting, even in studies with laboratory animals. The situation may become clarified by new techniques that follow the way in which the animal tries to avert the consequences of copper deprivation. Induction of mRNAs for ferritin and fetuin in the copper-deficient rat point to early priorities for controlling disturbances in iron metabolism and insulin function (Wang *et al.*, 1996). Meanwhile, long-established physiological functions are still lacking an unequivocal biochemical explanation and here focus will be placed on one of them (erythropoeisis) before dealing with some of the newer functions for the element.

Erythropoiesis

The reason why copper is essential for erythropoiesis is still a matter for controversy. Copper was thought to be involved in the absorption of iron from the intestinal mucosa, its mobilization from the tissues and its utilization in haemoglobin synthesis, rather than in haem biosynthesis. These functions could be accomplished by ceruloplasmin (CP), a copper-containing, α_2-globulin with ferroxidase activity in the plasma, facilitating iron transport through the formation of Fe (III) transferrin (Frieden, 1971). Others have since claimed that the anaemia of copper deficiency is caused by a breakdown in intracellular iron metabolism in the liver (Williams *et al.*, 1983). There are many inconsistencies in the patterns of tissue iron accumulation at different sites and in different experiments, which suggest that impairment of iron transport is a simplistic explanation for the erythropoietic function of copper (Suttle, 1987a). Ceruloplasmin also promotes the incorporation of iron into the storage protein, ferritin (Saenko *et al.*, 1994).

Protection from oxidants

Copper may protect tissues from oxidant stress via two distinct pathways, one involving impaired iron metabolism, the other a copper–zinc superoxide dismutase enzyme (CuZnSOD). Hepatic activity of the iron-containing, haem enzyme catalase is reduced in copper deficiency, despite a rise in tissue iron concentrations (Taylor *et al.*, 1988). Since iron promotes free-radical generation and catalase protects tissues from hydrogen peroxide (H_2O_2) and hydroperoxide damage, this is potentially a two-edged, pathogenic sword (Golden and Ramdath, 1987). Hepatic defences against another free radical, superoxide, may have been reduced in Taylor *et al.*'s (1988) study by a fall in CuZnSOD, which accompanies liver copper depletion in all species. Hepatic enlargement is a feature of copper deficiency in small laboratory animals but has not been reported in farm animals. Ceruloplasmin may contribute to antioxidant defences by scavenging free iron and free radicals (Saenko *et al.*, 1994). Scope for interactions with other nutrients with antioxidant properties, such as manganese, selenium and vitamin E, is thus provided (Fig. 16.2).

Heart development

Cardiac enlargement is a prominent feature of copper deficiency in male laboratory animals and is occasionally seen in cattle (Mills *et al.*, 1976). Of the newer copper-dependent enzymes, dopamine-β-monooxygenase (DBM) has been shown to become rate-limiting in the hearts of copper-deficient mice (Gross and Prohaska, 1990), and catecholamine metabolism may be critically impaired in both the heart and brain during copper deficiency. Defects in myofibrillar protein synthesis have been reported in copper-deficient rats and linked to low cytochrome *c* oxidase (CCO) concentrations (Chao *et al.*, 1994). This ubiquitous and important enzyme, responsible for the terminal electron transfer in the respiratory chain, may be involved in other dysfunctions associated with copper deficiency.

Development of the central nervous system (CNS)

One of the earliest recognized pathological consequences of copper deficiency in livestock, sway-back in young lambs, has been linked to lack of CCO activity in the neuron (Fell *et al.*, 1965). Decreases in another copper-dependent enzyme, peptidylglycine-α-amidating monooxygenase (PAM), have been reported in the brain of newborn rats from copper-depleted dams. They are of particular interest, given the multiplicity of biogenic molecules dependent on PAM (they include the appetite-regulating hormones gastrin and cholecystokinin) and the slow recovery of enzyme activity following copper repletion (Prohaska and Bailey, 1995).

Immunocompetence

Copper appears to be essential for the normal functioning of the immune system in ruminants, just as it does in small laboratory animals, although the bulk of evidence relates to *in vitro* tests showing impaired blastogenesis (Suttle and Jones, 1986) or phagocytic killing (Boyne and Arthur, 1981; Jones and Suttle, 1981; Xin *et al.*, 1991) in single-cell type cultures. More convincing evidence comes from increased leucocyte counts in the copper-deficient heifer (Arthington *et al.*, 1996; Gengelbach *et al.*, 1997) and from a field study of molybdenum-induced copper deficiency in sheep (Woolliams, C. *et al.*, 1986), which is discussed in more detail later. In severely deficient small laboratory animals, copper deficiency can affect the numbers of cells mediating immunity, increasing mast cells (i.e. non-specific immune cells) in muscle (Schuschke *et al.*, 1994) and decreasing some subpopulations of T cells (i.e. specific immune cells) (Mulhern and Koller, 1988). The greatly increased metabolism of cells of the immune system during infection and challenge might make them vulnerable to depletion of the energy-generating CCO or protective CuZnSOD. Impaired responses of splenic lymphocytes and neutrophils to *in vitro* challenges have been reported in marginally depleted male rats showing *no* reduction in liver copper or plasma ceruloplasmin concentrations (Hopkins and Failla, 1995), but this raises questions about the usefulness of the tests, unless the apparent 'defects' are associated with impaired immunocompetence *in vivo*. Both non-immune and immune responses to inflammatory stimuli can be aggravated by copper deficiency in mice (Jones, 1984), but protection against common pathogens in the copper-deficient calf (Stabel *et al.*, 1993) and heifer (Arthington *et al.*, 1996; Gengelbach *et al.*, 1997) was not compromised. Molybdenum-induced hypocupraemia has lowered the antibody response to *Brucella abortus* antigen in cattle (Cerone *et al.*, 1995).

Other functions

Studies with copper-depleted cattle have demonstrated a breakdown of basement membranes in the acinar cells of the pancreas as one of the earliest pathological lesions; failure of glycosylation or sulphation was suggested and there was a reduction of CCO activity in pancreatic tissue (Fell *et al.*, 1985). Glucose intolerance has been reported in copper-deficient rats (Hassell *et al.*, 1983) and may be explained by inhibition of insulin receptors (Wang *et al.*, 1996). Changes in lipid and lipoprotein metabolism also occur in the copper-deficient rat (Lei, 1991). Sequence homology between ceruloplasmin and blood-clotting factors indicates a possible role in haemostasis (Saenko *et al.*, 1994), which is supported by the fact that there is less ceruloplasmin in serum than in plasma from blood samples taken from the same animal.

Biochemical Consequences of Copper Deprivation

Changes in liver, blood and appendages

The sequence of changes seen in livestock deprived of copper is illustrated in Fig. 11.5. The liver is the main storage organ for copper, so that the first biochemical change to be seen is a decline in liver copper concentrations. This provides a useful index of the copper status of the previous diet and the degree of *depletion* but not necessarily the functional copper status at critical sites. Liver copper values must therefore be used with caution as diagnostic aids. When liver reserves approach exhaustion or cannot be mobilized fast enough to meet the deficit in dietary copper supply, CP synthesis decreases and plasma copper concentrations fall below normal, indicating a state of *deficiency* (Beck, 1941b). Blood copper concentrations show a slower decline, because it is some time before erythrocytes become depleted of CuZnSOD, their major copper constituent, and accumulate in sufficient numbers to add to the effect on plasma copper. Reductions in cuproenzyme activities ensue at various rates at different sites, indicating progress towards a state of local *dysfunction*, but these have rarely been defined. A breakdown in the conversion of tyrosine to melanin in the hair or wool follicle probably explains the early failure in pigmentation, since this conversion is catalysed by a copper-containing polyphenyl oxidase, tyrosinase (Holstein *et al.*, 1979).

Changes in connective tissue and skin

Events concerning connective tissue provide a further rare example of specific biochemical dysfunction being linked unequivocally to a pathological

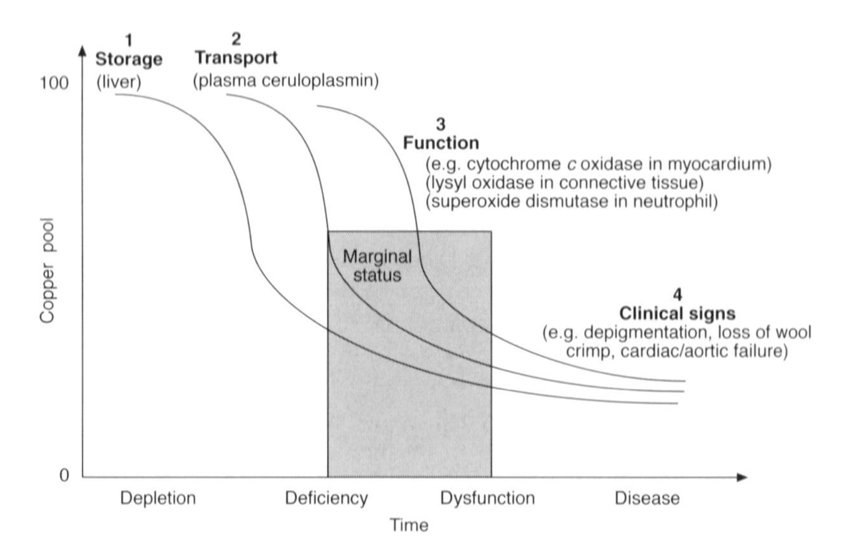

Fig. 11.5. Sequence of biochemical changes leading to the appearance of clinical signs of copper deprivation: see also Fig. 3.1 and related text.

consequence of copper deprivation, aortic rupture: the elastin content of the aorta is reduced and contains more lysine and less desmosine and isodesmosine (the key cross-linkage groups in elastin); the amine or lysyl oxidase activities of the plasma, liver and aorta are subnormal, so that the copper-deprived animal has a diminished capacity to form desmosine from lysine; and reduced elasticity is responsible for rupture of the aorta (Hill *et al.*, 1967). A more detailed account of the relationship of copper to elastin structure and function has been given by O'Dell (1976). Similar biochemical abnormalities are believed to be responsible for defective shell formation in the laying hen (Baumgartner *et al.*, 1978) and thinning of the air–blood capillary network in the avian lung (Buckingham *et al.*, 1981). Turkey poults can succumb to aortic rupture under commercial rearing conditions unless supplementary copper is given (Guenthner *et al.*, 1978). However, reductions in metallo-enzyme activity are not synonymous with dysfunction. In a recent study with copper-deprived rats, a 50–60% reduction in the activity of lysyl oxidase in the skin was not accompanied by adverse changes in collagen composition (Rucker *et al.*, 1996).

Changes in bone and joints

There is unlikely to be a single primary biochemical lesion affecting the bones and joints of copper-deprived animals; reduction in the activity of lysyl oxidase could diminish the stability and strength of bone collagen and impair mineralization of the cartilage, with consequent reductions in bone strength (Opsahl *et al.*, 1982). Markedly subnormal amine oxidase activity has been demonstrated in the bones of copper-deficient chicks (Rucker *et al.*, 1969, 1975). However, depletion of CCO may compromise osteoblastic activity.

Changes in the central nervous system

The biochemical defects responsible for the pathological lesions in swayback are also incompletely understood. Activities of CCO are subnormal in the brain of ataxic lambs (Howell and Davison, 1959) and are at their lowest in the affected neurons of the grey matter of the spinal cord (Barlow, 1963b; Fell *et al.*, 1965). However, the depletion of CCO activity in the mitochondria of the cerebral cortex of ataxic lambs is insufficient to constitute a respiratory constraint (Smith *et al.*, 1976). Furthermore, the catecholamine neurotransmitters, dopamine and noradrenaline, as well as amine oxidase and CuZnSOD, are significantly lower in the brain stem of ataxic lambs than in copper-treated controls (O'Dell *et al.*, 1976). Impaired biosynthesis of the catecholamines involving the copper-dependent enzyme DBM or free-radical damage may therefore contribute to the development of ataxia.

Changes in cardiac tissue

Studies have largely been confined to laboratory animals (e.g. Gross and Prohaska, 1990), but depletion of CCO activity has been described in the hearts of copper-deficient calves in association with abnormal mitochondrial ultrastructure (Leigh, 1975). Whether or not CCO was the rate-limiting enzyme could not be ascertained. Other enzymes are also likely to be depleted, including CuZnSOD, but recent work with rats suggests that compensatory increases in manganese superoxide dismutase (MnSOD) activity could limit the functional importance of this particular cuproenzyme (Lai *et al.*, 1994).

Changes associated with reproduction

Impaired pulsatile release of luteinizing hormone (LH) was detected in infertile heifers following molybdenum-induced copper deficiency (Phillippo *et al.*, 1987b) but not in those with an equally severe hypocupraemia induced by iron, suggesting that infertility was attributable to molybdenum excess, rather than copper deprivation. Other workers have failed to impair LH release by molybdenum supplementation (Xin *et al.*, 1993), using rations with wider Cu:Mo ratios than those used in the previous study (1.9 vs. 0.8) and which did not induce hypocupraemia; this suggests that copper deprivation *is* an important prerequisite for molybdenum-induced infertility. Xin *et al.* (1993) assayed copper levels at sites of suspected influence on LH release (hypothalamus and anterior pituitary) but found no evidence of depletion. The principle of assessing copper status at the site(s) of functional importance was sound, but similar studies need to be conducted in females affected by molybdenum-induced infertility.

Biochemical Consequences of Molybdenum Exposure in Ruminants

At high molybdenum intakes (> 8 mg Mo kg^{-1} DM) and low Cu:Mo ratios (< 1) in sheep, thiomolybdates leave the rumen in absorbable forms (Price *et al.*, 1987) and have been detected in the plasma, where they circulate as albumin-bound, trichloroacetic acid (TCA)–insoluble copper (TCAICu)/Mo-complexes (Dick *et al.*, 1975; Smith and Wright, 1975a, b), with a much slower clearance rate than normal albumin-bound copper (Mason *et al.*, 1986). The anomalously bound copper has been estimated by difference between the values of plasma treated or not treated with TCA. Cattle and deer must be exposed to lower dietary Cu:Mo ratios than sheep before TCAICu is detected (Freudenberger *et al.*, 1987; Wang *et al.*, 1988). Initially, it was assumed that TCAICu was poorly utilized by tissues, offering opportunities for further inhibition of copper metabolism at the cellular level. This concept has been brought into question by evidence that copper supplements can slowly increase ESOD, despite inducing early increases in TCAICu (Suttle and Small, 1993). By giving rats drinking-water supplemented with molybdate, enormous increases in TCAICu can be sustained in plasma for long periods, with little or no depletion of ceruloplasmin or tissue superoxide dismutase (SOD) activity, retardation of growth or anaemia (N. Sangwan and N.F. Suttle, unpublished data). The Cu–Mo–albumin complexes are not excretable, but they are slowly hydrolysed (Mason, 1986). Daily release of copper from the greatly increased albumin-bound copper pool may therefore be sufficient to sustain the supply of copper to the tissues.

When lambs with high initial liver copper reserves were continuously supplemented with molybdenum (3.0 mg kg^{-1} DM) as tetrathiomolybdate (TTM) via a diet initially low and latterly high in copper to give Cu:Mo ratios of 1.1 and 5.1, respectively, TCAICu and molybdenum initially accumulated in plasma but disappeared during repletion (Fig. 11.6) while retention of copper in liver was constantly impaired (Suttle and Field, 1983). These findings suggest that the primary consequence of exposure to molybdenum is the formation of unabsorbable copper:TTM complexes in the gut. A 'flooding' parenteral dose of TTM will temporarily inactivate ESOD (Suttle *et al.*, 1992a) and the antagonist is also capable of removing copper from metallothionein *in vitro* (Allen and Gawthorne, 1988). Serial injections of TTM have retarded growth and impaired fertility in lambs (Moffor and Rodway, 1991). In practical nutritional contexts, however, thiomolybdates are slowly and continuously absorbed and may be effectively detoxified by albumin-binding without adverse pathological consequences (Lannon and Mason, 1986). Of greater practical importance is the possibility of acid-insoluble, unabsorbable complexes forming in the abomasum when parasitic infection causes leakage of plasma into an acid, thiomolybdate-rich lumen (Ortolani *et al.*, 1993). Furthermore, exposure to molybdenum may enhance the inflammatory response whereby sheep expel gut parasites (Suttle *et al.*, 1992b).

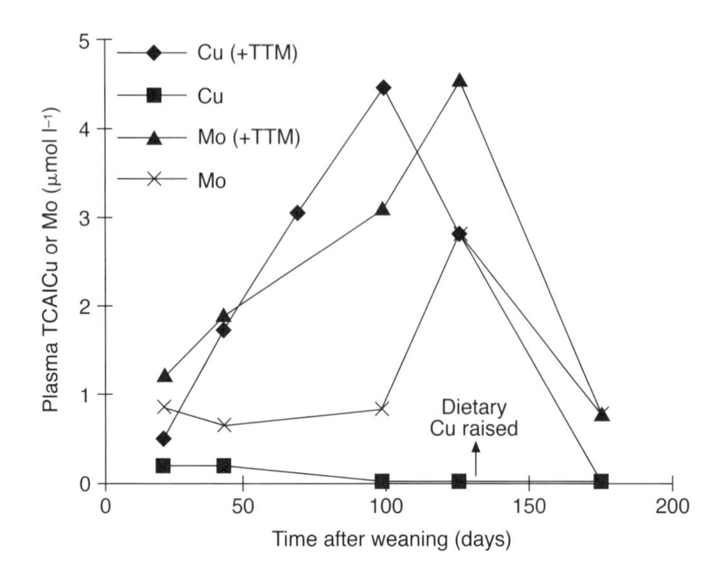

Fig. 11.6. The biochemical consequences of exposing lambs to tetrathiomolybdate (TTM, 5.4 mg Mo kg^{-1} DM, ♦ and ▲) via the diet depend on the copper (Cu) also provided: initially, when Cu is low (3.5 mg kg^{-1} DM), excess TTM is absorbed, causing TCA insoluble Cu (TCAICu) and Mo (▲) to accumulate in plasma; when Cu is raised after 130 days (to 15.5 mg kg^{-1} DM), TCAICu disappears and plasma Mo is markedly reduced due to insoluble Cu : TM complex formation in the gut (from Suttle and Field, 1983).

Clinical Manifestations of Copper Deprivation

The relative susceptibility of different copper-dependent processes and therefore the manifestations of deficiency vary with the animal species and the stage of developement at which critical dysfunctions develop.

Anaemia

Anaemia can be a feature of copper inadequacy in all species but usually develops only where deficiency has been severe or prolonged (Beck, 1941a; Baumgartner *et al.*, 1978; Suttle and Angus, 1978). In these circumstances, the blood copper falls to values as low as 1.5–3 µmol l^{-1} and normal haemopoiesis cannot be sustained. In pigs and lambs, the anaemia is of the hypochromic, microcytic type and indistinguishable from that of iron deficiency; in chicks, it is normocytic and normochromic, while in cows and ewes limited evidence indicates that the anaemia is hypochromic and macrocytic. Haemolysis may contribute to the development of anaemia. Signs of oxidative stress in the form of Heinz bodies in erythrocytes have been reported in the copper-deficient lamb (Suttle *et al.*, 1987a), and low activities of erythrocyte CuZnSOD have been implicated in the incidence of haemolytic, Heinz-body anaemia in hypocupraemic cattle given kale (Barry *et al.*, 1981). Furthermore, the survival time of erythrocytes is shorter than normal in the copper-deficient pig (Bush *et al.*, 1956).

Bone disorders

Abnormalities in bone development vary widely within and between species, but this is hardly surprising. The disturbances of endochondral ossification which give rise to uneven bone growth (osteochondrosis) can only affect growing animals, and bone morphology will be influenced by rate of growth, body-weight distribution, movement and even rate of hoof growth. 'Beading' of the ribs in sheep and cattle may be seen, due to overgrowth of chostrochondral junctions. A generalized osteoporosis and low incidence of spontaneous bone fractures can occur in grazing cattle and sheep (Cunningham, 1950; Whitelaw *et al.*, 1979). In young pigs rendered severely deficient, osteoblastic activity became depressed, the epiphyseal plate widened and gross bone disorders developed with fractures and severe deformities, including marked bowing of the forelegs (Baxter and Van Wyk, 1953; Baxter *et al.*, 1953). Osteoporosis and a reduction in osteoblastic activity can occur in young housed lambs born of copper-deficient ewes but without morphological bone changes (Suttle *et al.*, 1972). Widening of the epiphyses of the lower limb bones is a common manifestation in growing cattle (Fig. 11.7c).

Connective-tissue disorders

Osteochondrosis in young, copper-deficient, farmed red deer is accompanied by gross defects in the articular cartilages (Thompson *et al.*, 1994). These probably arise through impaired collagen and elastin development, as do the subperiosteal haemorrhages and imperfect tendon attachments seen in lambs on molybdenum-rich pastures (Hogan *et al.*, 1971; Pitt *et al.*, 1980). The abnormal gait, variously described as 'pigeon-toed', 'stiff-legged' or 'bunny-hopping', associated with copper deficiency in ruminants is, therefore, not simply a bone disorder but a combination of bone and connective-tissue disorders. In some instances, the connective-tissue defect can be predominant, lesions in the ligamentum nuchae supporting the neck and scapulae (Fell *et al.*, 1975) leading to dislocation of the scapulae, which form a 'hump', while the head droops (B. Ruksan, personal communication). In foals, angular deformities at the pasterns can respond to copper supplementation and they probably indicate disordered connective-tissue development (Barton *et al.*, 1991). Hens deprived of copper lay eggs abnormal in size and shape, either lacking shells or with wrinkled, rough shells, due to malformation of the shell membranes (Baumgartner *et al.*, 1978).

Neonatal ataxia

A nervous disorder of lambs, characterized by incoordination of movement and high mortality, has been recognized for many years in different parts of the world and given a variety of local names, 'swayback' being the most

(a)

(b)

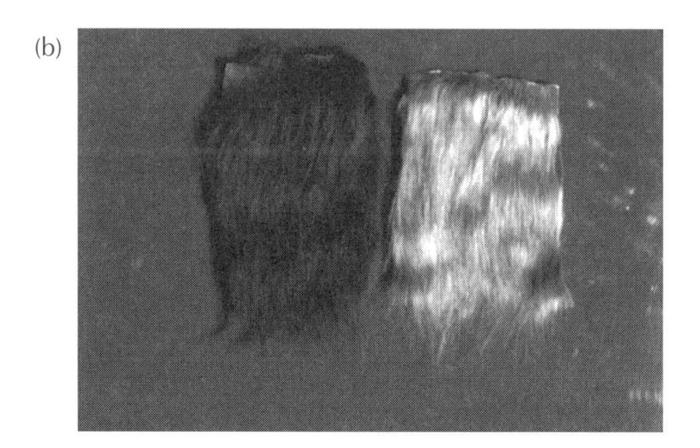

(c)

Fig. 11.7. Clinical symptoms of copper deprivation in ruminants include loss of hair or fleece colour, giving a 'spectacled' appearance (a) and 'bleached' coat (b) in cattle, and skeletal abnormalities, including widening of the epiphyses (c).

commonly used. Neonatal ataxia could be prevented by giving copper supplements to the ewe (Bennetts and Chapman, 1937), but it was some time before the disease was produced in lambs experimentally, by feeding ewes either a semipurified diet low in copper (Lewis *et al.*, 1967; Suttle *et al.*, 1970) or diets high in molybdenum and sulphur (Fell *et al.*, 1965; Suttle and Field, 1969). Three types of ataxia occur in lambs: the common form, which is manifested at birth, with primary lesions in the cerebrum; a delayed form, with primary lesions in the spinal cord, which does not appear for several weeks; and an atypical late form, in which the primary lesion is a cerebral oedema, where the lamb may stand transfixed, head quivering and apparently blind (Roberts *et al.*, 1966). With neonatal ataxia, some lambs are paralysed or ataxic when born and soon die. Calves reared in the same conditions are not affected by swayback. In the goat kid, the delayed form of ataxia is predominant (Hedger *et al.*, 1964) and cerebellar hypoplasia is an additional feature (Wouda *et al.*, 1986). Uncoordinated movements of the hind limbs, a stiff and staggering gait and swaying of the hindquarters become evident as the delayed form develops (Fig. 11.7), but cases may only become apparent when the flock is moved. Ataxia is caused by myelin aplasia rather than myelin degeneration (Barlow, 1963a) and is associated with degeneration of the motor neurons of the brain and spinal cord (Mills and Williams, 1962; Howell *et al.*, 1964; Cancilla and Barlow, 1968; Wouda *et al.*, 1986). The different manifestations within and between species nevertheless reflect the developmental state of the CNS and particularly the rate of myelination at the time copper deficiency strikes, sheep having a high rate of fetal brain myelination in mid-pregnancy and lambs and kids a phase of rapid spinal-cord myelination a few weeks after birth, while calves undergo a slow, regular myelination of the CNS (Suttle, 1987a). The lesions are irreversible and may commence as early as 6 weeks before birth and continue to develop afterwards.

Cardiovascular disorders

Cardiac lesions associated with copper deficiency first became evident in a disease of cattle occurring in Western Australia, known locally as 'falling disease' (Bennetts and Hall, 1939). The essential lesion is a slow and progressive degeneration of the myocardium, with replacement fibrosis and accumulation of iron. The sudden deaths are believed to be due to acute heart failure, usually after mild exercise or excitement. Liver copper and blood copper are extremely low in affected animals (32 μmol kg^{-1} DM and 1.6 μmol l^{-1}, respectively) and reflect the abnormally low pasture copper values (1–3 mg kg^{-1} DM). 'Falling disease' does not occur in sheep or horses grazing the same copper-deficient pastures and occurs only rarely elsewhere (Davis, 1950). A derangement in the elastic tissues of the aorta of copper-deprived chicks and pigs, with ruptures of the major blood-vessels and high mortality, was first reported under experimental conditions some 20 years ago (Carnes *et al.*, 1961; O'Dell *et al.*, 1961) and can occur under commercial conditions in turkey poults (Guenthner *et al.*, 1978).

Depigmentation

Achromotrichia is the earliest clinical sign of copper deprivation in all species, provided the breed has a pigmented coat. Lack of pigment production in black-woolled sheep and greying of black or bleaching of brown hair in cattle, especially round the eyes, can occur at dietary copper intakes sufficient to prevent other signs of deficiency. The pigmentation process in the sheep is so sensitive to changes in copper intake that alternating bands of unpigmented and pigmented wool fibres can be produced, as copper is withheld from and then replaced in the diet (Underwood, 1977). Inclusion of a few black-fleeced individuals in a flock is an effective way of monitoring the development of dysfunction. Even on fairly high copper intakes, the pigmentation process can be blocked within 2 days by raising the molybdenum and sulphur intakes sufficiently. In turkey poults reared on a low-copper milk diet, depigmentation of the feathers was noted (Savage *et al.*, 1966). Loss of coat colour is not a specific sign of copper deficiency and has been noted in many vitamin deficiencies and in cobalt deficiency (see Chapter 10). Furthermore, when cattle shed their winter coat, clinical appearances can be deceptively like those of animals lacking copper; the 'foxy-brown' coat discoloration seen in Friesian cattle is not a sign of copper deficiency (Mee, 1991).

Defective keratinization of wool and hair

Changes in the growth and physical appearance of the hair of cattle, reflected in a thin, wavy, harsh coat, commonly occur. In sheep, the characteristic crimp in the wool becomes less distinct, until the fibres emerge as almost straight, hair-like growths, to which the terms 'stringy' and 'steely' wool have been applied. The tensile strength of such wool is reduced and the elastic properties are abnormal. A spectacular restoration of the crimp and physical properties can be achieved by copper supplementation (Underwood, 1977). The abnormalities described are most obvious in the wool of merino sheep, which is normally heavily crimped, but they have been demonstrated experimentally in British breeds (Lee, 1956). 'Steely wool' was still a problem in South Australia long after the initial link with copper was established (Hannam and Reuter, 1977). The characteristic physical properties of wool, including crimp, are dependent on the presence of disulphide groups that provide the cross-linkages or bonding of keratin and on the alignment or orientation of the long-chain fibrillae in the fibre. Both of these are adversely affected in copper deficiency (Marston, 1946), but the precise biochemical involvement of copper is unknown. Nutritional stresses of various kinds can lead to temporary loss of wool strength and crimp, which is not therefore a specific sign of copper dysfunction.

Scouring or diarrhoea

A copper-responsive diarrhoea has been observed in cattle in several parts of the world, including the Netherlands, where the name 'leschucht' or 'scouring disease' was given to it. This condition can develop in simple copper deficiency, as in 'falling disease' areas, and in experimental deficiency (Fell *et al.*, 1975; Mills *et al.*, 1976), but is more common in molybdenum-induced disorders, such as 'peat scours' areas of New Zealand. At the very high molybdenum concentrations associated with 'teart scours' in Somerset, UK (> 20 mg kg^{-1} DM: Ferguson *et al.*, 1938, 1943), the diarrhoea occurs immediately upon turnout and before liver or blood copper concentrations reach subnormal levels; localized, thiomolybdate-induced copper depletion of the intestinal mucosa has been suggested to explain this anomaly (Suttle, 1991). Horses grazing similar pastures do not scour, and diarrhoea is not a feature of simple copper deprivation in pigs or sheep, but it can be induced by molybdenum-rich diets in sheep (Suttle and Field, 1968; Hogan *et al.*, 1971). Scouring has been observed in goats maintained on the Dutch pastures that induce scouring in cattle. The reasons for these species differences in susceptibility and the highly variable individual incidence of the diarrhoea in cattle (Jamieson and Allcroft, 1950) are unknown. Histochemical and ultrastructural changes were observed in the small intestine of young Friesian cattle showing clinical signs of copper deficiency, but no clear relationship between these changes and diarrhoea was apparent (Fell *et al.*, 1975) and such changes were not a feature of other clinically affected calves (Suttle and Angus, 1978).

Infertility

Low fertility, associated with delayed or depressed oestrus, occurs in cattle grazing copper-deficient pastures in several widely separated areas, and infertility, associated in some cases with aborted, small, dead fetuses, has been reported in experimental copper deprivation in ewes (Howell and Hall, 1970). However, the relationship between low copper status and infertility in the field is inconsistent. Phillippo *et al.* (1982) found that most beef suckler herds could tolerate hypocupraemia and not show impaired fertility, with the exception of one herd grazing pasture of marginally raised molybdenum concentration. The importance of molybdenum excess rather than copper deprivation was apparently confirmed in experiments, with penned heifers given a diet of marginal copper content (4 mg Cu kg^{-1} DM), with molybdenum (5 mg kg^{-1} DM) or iron (500–800 mg kg^{-1} DM) added; infertility only occurred in molybdenum-supplemented groups, despite the fact that liver and plasma copper concentrations in all treated groups seemed equally low (Humphries *et al.*, 1983; Phillippo *et al.*, 1987b). Improvements in conception rate following the parenteral administration of copper are, however, claimed in normocupraemic herds with (Ingraham *et al.*, 1987) or without (Hunter, 1977) synergism from magnesium. These may represent pharmacological effects of injected copper mediated via the pituitary, given new information from the pig (Zhou *et al.*, 1994b), discussed later. In hens fed on a severely copper-deficient diet, egg production and hatchability are markedly reduced and, after incubation, the embryos from these hens exhibit anaemia, retarded development and a high incidence of haemorrhage (Savage, 1968); such problems do not occur on natural diets.

Susceptibility to infection

Long-term studies of genetic and other causes of mortality on a Scottish hill farm revealed a marked increase in mortality when the pastures were improved by liming and reseeding. Losses were higher in breeds with low efficiencies of copper utilization and were reduced by copper supplementation or genetic selection for high copper status (Suttle and Jones, 1986; Woolliams, C. *et al.*, 1986). Microbial infections were major causes of death and, since there were parallel changes in incidence of a proved manifestation of copper deficiency – swayback, susceptibility to infection must have been increased by copper deficiency. Subsequent work indicated that perinatal mortality was increased unless copper supplements were given in late pregnancy, but the precise role of infection was not established (Suttle *et al.*, 1987b). Febrile responses to experimental viral challenge in heifers with molybdenum-induced copper deficiency remained unaffected (Arthington *et al.*, 1996) or were inconsistently related to copper status (Gengelbach *et al.*, 1997).

Subclinical Consequences of Copper Deprivation

Growth retardation is a common feature of copper deprivation in ruminants, though a late and mild feature in non-ruminants. In both field (Phillippo, 1983) and experiment (Humphries *et al.*, 1983; Phillippo *et al.*, 1987a) and in sheep (Whitelaw *et al.*, 1979, 1983; Woolliams, J.A. *et al.*, 1986) as well as cattle (Thornton *et al.*, 1972a, b), growth retardation has been associated predominantly with the consumption of pasture or diets of low Cu:Mo ratio (< 3.0 and often < 1.0). In the Aberdeen studies, growth retardation – like infertility – was only a feature of molybdenum-induced copper deficiency (Humphries *et al.*, 1983; Phillippo *et al.*, 1987a) (Fig. 11.8). These findings have been confirmed in the USA (Gengelbach *et al.*, 1994) and again the role of copper seems questionable. However, it has been argued that molybdenum produced a localized copper deficiency, which was not reflected in liver or plasma copper concentrations (Suttle, 1988). Close comparison of the copper-deprived groups in Fig. 11.8 shows that those given molybdenum, with or without iron, had lower liver copper and blood CuZnSOD concentrations than the unretarded group depleted by iron alone. Recent data (Gengelbach *et al.*, 1997) support this view: late reductions in CuZnSOD activity in the neutrophils of calves from molybdenum-supplemented heifers were found, but no reduction was seen in equally hypocupraemic offspring from iron-supplemented heifers. If growth retardation and infertility were caused by molybdenosis rather than induced hypocuprosis in these studies, molybdenum would become one of the most toxic elements known. Copper deficiency *per se* can retard ruminant growth (Mills *et al.*, 1976), just as it produces all other clinical abnormalities associated with moderate exposure to molybdenum, and exposure to excess iron in the field induces copper-responsive disorders (see next section). It is hard to

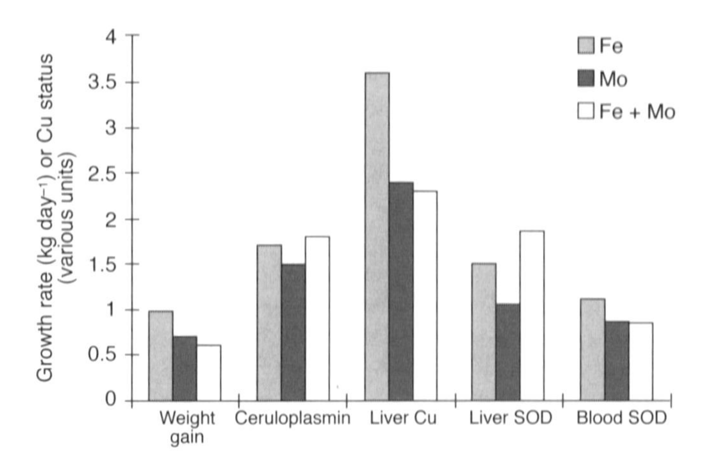

Fig. 11.8. Difficulties of assessing copper status of three groups of severely copper-deficient heifers, given iron (Fe), molybdenum (Mo) or Fe + Mo as copper antagonists; only groups given Mo grew poorly, although all had equally low plasma Cu and ceruloplasmin concentrations (Humphries *et al.*, 1983): poor growth was, however, accompanied by lower liver Cu and blood superoxide dismutase activity (SOD) and suggestions that copper deprivation of Mo-supplemented groups was not involved are unfounded.

believe that such low levels of molybdenum (5 mg Mo kg^{-1} DM) could produce an identical combination of symptoms, including growth retardation, without compromising copper-dependent functions. The number of biochemical lesions that might be invoked to explain growth retardation are almost as numerous as the functions of copper (Table 11.4), but the low milk yield of beef cows on pastures high in molybdenum is believed to have been responsible for the poor growth of their suckled calves (Wittenberg and Devlin, 1987).

Occurrence of Copper-responsive Disorders

Copper deprivation causes problems in herbivores but not in poultry or pigs fed natural feedstuffs. A copper-responsive anaemia has been reported in pigs fattened on swill which had been stored in galvanized bins (Pritchard *et al.*, 1985). The swill had a high zinc concentration of 2580–5100 mg kg^{-1} DM and only 3.2–4.7 mg Cu kg^{-1} DM. However, other swills being fed in the locality but uncontaminated with zinc had little more copper (3.5–9.0 mg kg^{-1} DM) and caused no problems. It was concluded that swills should be supplemented with copper and not stored in galvanized bins. Disorders are most prevalent in ruminants and are influenced by soil, plant seasonal and genetic factors.

Soil factors

Clinical copper deficiency or hypocuprosis occurs naturally in grazing livestock in many parts of the world under a wide range of soil conditions. In some areas, it occurs as a dual deficiency of copper and cobalt, particularly on coastal, calcareous sands or leached soils of granitic origin, as in 'salt-sick' of cattle in Florida and 'coast disease' of sheep in southern Australia (McFarlane *et al.*, 1990). Copper deficiency also occurs as a single deficiency arising from inherently low soil and herbage copper contents (Beck, 1951), as in 'falling disease' of cattle and neonatal ataxia (swayback) of lambs in Western Australia. Soils rich in organic matter have low contents of plant-available copper (Haynes, 1997). However, most of the copper deficiencies in livestock which occur naturally are 'conditioned' by the presence of dietary factors that interfere with the absorption or utilization of copper by the animal (Phillippo, 1983). In southern Australia, associations between bovine hypocupraemia and iron-rich soils (rendzinas) and pastures have been reported (McFarlane *et al.*, 1990). In parts of England, iron contamination from soil can elicit a mild copper deficiency in cattle in spring (Jarvis and Austin, 1983). Irrigation of pasture with iron-rich bore water accelerated the depletion of liver copper and induced diarrhoea in grazing cattle (Campbell *et al.*, 1974). The ingestion of sulphate-rich drinking-water from deep aquifers has caused growth retardation in heifers and neonatal mortality in their off-spring (Smart *et al.*, 1986) (see also Chapter 9). By far the most influential factor, however, is soil molybdenum status. On peat soils of the Canterbury Plains, New Zealand (Cunningham, 1950), and of southern Australia (McFarlane *et al.*, 1990), the alluvial plains of Argentina, the lower lias clay and marine shales of England (Thornton and Alloway, 1974) and north-eastern Manitoba (Boila *et al.*, 1984a, b), the grey soils of north-eastern Saskatchewan (Gooneratne and Christensen, 1989) and the humic peat soils of Europe (Szalay *et al.*, 1975), the herbage takes up sufficient molybdenum to induce clinical abnormalities.

Plant factors

The influence of molybdenum is so strong that areas of high risk of hypocuprosis can be delineated by soil or stream sediment maps of molybdenum concentration (Thornton and Alloway, 1974; Boila *et al.*, 1984b). The introduction of new farming methods, however, can require maps to be redrawn. The improvement of Scottish hill pastures by liming and reseeding can inadvertently induce hypocuprosis in lambs by raising herbage molybdenum and sulphur concentrations (Whitelaw *et al.*, 1979, 1983; Woolliams, C. *et al.*, 1986; Woolliams, J.A. *et al.*, 1986) but in practice the disorder is rare (Suttle *et al.*, 1999). Without such changes to lower copper absorbability, pasture improvement *per se* does not increase the risk of hypocuprosis (Hannam and Reuter, 1977). Outbreaks of swayback on some coastal pastures in Greece have been attributed to the ingestion of sulphate-rich, halophytic plants by the ewes (Spais, 1959). Heinz-body anaemia and growth retardation, both responsive to copper supplementation, have been reported in cattle consuming a sulphur-rich brassica (kale) crop (Barry *et al.*, 1981).

Seasonal factors

Seasonal factors affecting copper antagonists are also a major consideration. Herbage iron concentrations are maximal in spring and autumn in temperate pastures. Herbage molybdenum concentrations increase as the season progresses and can double during the grazing season in Scotland (Suttle *et al.*, 1999). Sulphur concentrations decline as herbage matures (see Chapter 9) and there is evidence that the availability of copper is higher in mature than in fresh green herbage (Hartmans and Bosman, 1970). Induced hypocuproses in grazing animals are therefore essentially 'green-sward' problems and rarely seen during dry summers. In parts of southern Australia (Hannam and Reuter, 1977) and the UK (Bain *et al.*, 1986), bovine hypocupraemia is associated with years or months of heavy rainfall. 'Swayback' of lambs is more prevalent after mild than after severe winters, possibly due to less supplementary feeding and more soil ingestion reducing the supply of absorbable copper (Suttle *et al.*, 1984). When the seasonal supply of available copper fluctuates, the occurrence of hypocuprosis will depend on the sufficiency of stores built up during times of generous supply. It is therefore impossible to define or predict the occurrence of hypocuprosis on the basis of the soil or herbage copper concentration alone.

Genetic factors

The genetic differences in copper metabolism referred to earlier can be translated into breed differences in susceptibility to copper deprivation among sheep, the Scottish Blackface being particularly and consistently susceptible in an 'improved' hill-pasture environment (Woolliams, C. *et al.*, 1986; Woolliams, J.A. *et al.*, 1986). Comparable differences between breeds of cattle have yet to be proved and are complicated by differences in natural coat colour and hence propensity to display the earliest sign of the disorder.

The Diagnosis and Prognosis of Copper-responsive Disorders

The diagnosis of copper deprivation in grazing animals rests upon three conditions: firstly, the presence of clinical symptoms or subclinical loss of production; secondly, biochemical evidence of subnormal tissue or blood copper concentrations; and, thirdly, an improvement after treatment with copper when compared with untreated contemporaries. Omission of any one component of this diagnostic profile will leave the diagnosis insecure – with the exception of swayback. A general diagnostic framework for the most vulnerable species is presented in Table 11.5: the reasons for selecting particular criteria and setting the given limits are based on the following analysis of the literature. Measurements of soil copper status are generally unhelpful (see Chapter 3) and the results of pot trials, involving recent additions of copper fertilizer (Haynes, 1997), give misleadingly good predictions of plant copper concentration.

Table 11.5. Marginal bands[a] for copper:antagonist concentrations in the diet and copper concentrations in the bloodstream and soft tissues of ruminants as aids to the diagnosis and prognosis of copper-responsive disorders in ruminants on fresh herbage (H)- or roughage (R)-based diets.

Criterion[b]	Diet	Sheep (S) and cattle (C)	Deer (D) and goats (G)	Interpretative limits
Diet Cu:Mo	H	1.0–3.0	0.5–2.0	Diet S > 2 g kg^{-1} DM and
	R	0.5–2.0	0.3–1.2	diet Mo < 8 (mg kg^{-1} DM) for S, D, G, < 15 for C
Diet Fe:Cu	H or R	50–100	50–100	
Diet Cu	H	6–8	6–8	Diet Mo < 1.5
(mg kg^{-1} DM)	R	4–6	4–6	(mg kg^{-1} DM)
Liver Cu (μmol kg^{-1} DM)[c,d]		100–300	122–244	Weaned or Adult[f]
Serum Cu (μmol l^{-1})[d]		3–9	3–9	Diet Mo < 8 for S, D, G, < 15 for C: no acute-
Blood Cu (μmol l^{-1})[d]		6–10	6–10	phase response; age > 1 week
Hair or wool Cu[d,e]		31–62 (S)	93–141 (D)	Clean sample, recent
(μmol kg^{-1} DM)		62–124 (C)	47–78 (G)	growth

[a] Values *below* the marginal band indicate high probability of current or future dysfunction and impairment of health or production; values *above* the band indicate minimal likelihood of copper supplementation being beneficial. As the *number of criteria* falling within the marginal band increases, so does the likelihood of a significant proportion of the population benefiting from extra copper.
[b] Alternative criteria, such as ceruloplasmin and erythrocyte CuZnSOD activities, are not included because they introduce additional problems of interpretation or standardization without adding significantly to the precision of diagnosis.
[c] Divide by 3.0 to obtain values on fresh-weight basis.
[d] Multiply by 0.064 to obtain values in mg kg^{-1} or mg l^{-1}.
[e] From Puls (1994).
[f] The newborn require much higher bands, e.g. 790–3150 for calves.

Dietary copper and its antagonists

Determination of copper in the diet or pasture has *no* diagnostic value in ruminants *unless* other elements with which copper interacts are determined also. The commonest approach has been to assay additionally for molybdenum and to predict risk from the Cu:Mo ratio. Although it is clear that the Cu:Mo ratio (mg:mg) is an important determinant of both copper deficiency and toxicity, there is little agreement on a critical Cu:Mo ratio. According to a Canadian study, a Cu:Mo dietary ratio lower than 2:1 can result in conditioned copper deficiency in cattle (Miltimore and Mason, 1971), while the results of an English study suggest that a ratio closer to 4:1 is necessary to avoid swayback in sheep (Alloway, 1973). There are several reasons why the Cu:Mo ratio will not give a precise indication of risk of hypocuprosis: the current ratio may not reflect the previous supply of absorbable copper and hence liver copper stores; the ratio does not allow for the influence of sulphur (Cu:Mo ratios become less important as herbage sulphur increases); the effect of molybdenum is different for grass, hay and silage (see Fig. 11.1) and will probably vary with season in the standing sward; the inhibitory effect of molybdenum on copper absorbability may be reversed at high molybdenum concentrations (> 8 mg Mo kg^{-1} DM in sheep: Suttle, 1983); and other antagonists, notably iron, may also be influential. Dietary Cu:Mo ratios, like all other diagnostic aids, require flexible interpretation, taking Cu:Mo ratios < 1.0 (or Fe:Cu ratios > 100) to indicate a high risk of disorder and ratios of 1.0–3.0 (or Fe:Cu between 50 and 100) to indicate a marginal risk of past or future problems. The omission of sulphur is merely pragmatic; it is the most difficult and costly of the interactants to assay and, in the green swards on which disorders chiefly occur, sufficient sulphur is usually present to allow expression of both antagonisms. However, sulphur intakes from drinking-water must be allowed for in some areas (Smart *et al.*, 1986). Limitations remain because the effects of iron and molybdenum cannot simply be combined, since the two antagonists do not act additively (Fig. 11.8). The use of threshold dietary copper concentrations to delineate when high Cu:Mo ratios are likely to cause problems (Boila *et al.*, 1984a) is illogical.

Liver copper

In early studies with weaned sheep and cattle, liver copper values of 20–25 mg (315–395 μmol) kg^{-1} DM or less were taken as indicative of 'deficiency', simple or conditioned. Later evidence suggested that the threshold value used to differentiate dysfunction from normal animals could be 75% lower (Smith and Coup, 1973; Suttle, 1987a). It has been suggested that Holsteins have a lower threshold for liver copper than Jerseys for the maintenance of normal plasma copper concentrations (Du *et al.*, 1996). In the Aberdeen experiments (Fig. 11.8; Humphries *et al.*, 1983; Phillippo *et al.*, 1987a, b), copper-responsive disorders could only be differentiated by liver values below 3 mg Cu kg^{-1} DM. A three-tier classification, with values < 100 μmol kg^{-1} DM indicating a high risk and 100–300 μmol (6–20 mg) kg^{-1} DM a possible risk of past or future disorder in the field, is recommended (Table 11.5). Note that higher thresholds are desirable for newborn lambs and calves.

Plasma and serum copper

The normal ranges are wide and vary between species. The norms for plasma copper are 9–15 μmol l^{-1} for ruminants (Suttle, 1994) and 3–6 μmol l^{-1} in poultry (Beck, 1956); for other non-ruminants, such as the horse and pig, a relatively high normal range of 16.5–20.0 μmol l^{-1} has recently been suggested (Suttle *et al.*, 1997). In all species, values are influenced by age, the type of sample, pregnancy and disease. Values in newborn mammals are normally around 50% of adult values but increase during the first week (McMurray, 1980), due to an increase in the major component, CP. In samples from adult ruminants, some CP is lost during clotting and serum values are 10–20% lower than those in the corresponding plasma (McMurray, 1980; Paynter, 1987; Suttle, 1994). Infectious disease and other factors that stimulate the immune system, notably vaccination, increase values substantially by inducing CP synthesis, even in initially hypocupraemic animals (Suttle, 1994), and have probably contributed to the bad reputation of serum copper as a diagnostic aid compared with liver copper (e.g. Xin *et al.*, 1993; Vermunt and West, 1994). However, hypocupraemia remains a useful component of the diagnostic profile, indicating that the liver reserve is sufficiently exhausted for synthesis of CP to be compromised. Clinical judgement and other biochemical markers of acute-phase reactions (low plasma zinc; fibrinogen > 1 g l^{-1}) can be used to identify anomalies associated with non-specific immune responses, although the latter may be affected by molybdenum (Arthington *et al.*, 1996). Serum or plasma values of 3–4.5 μmol (0.2–0.3 mg Cu) l^{-1} are common in clinically affected sheep (Whitelaw *et al.*, 1979, 1983; Woolliams, J.A. *et al.*, 1986) and cattle (McFarlane *et al.*, 1991) grazing deficient pastures, but there are many instances of cattle tolerating mean values between 3 and 9 μmol l^{-1} (Smith and Coup, 1973; Suttle *et al.*, 1980; Givens *et al.*, 1981; Humphries *et al.*, 1983; McFarlane *et al.*, 1991), and these should be regarded as appropriate marginal limits in the manner indicated for liver copper (Table 11.5). The use of a marginal band lessens the risk of infection or sample choice (plasma or serum) and preparation (water dilution or TCA treatment) taking on crucial diagnostic significance, by marginally changing the measured copper concentration (Suttle, 1994); TCA treatment gives higher values than aqueous dilution, provided there is no TCAICu present. Determination of TCAICu indicates the extent of severe molybdenum exposure but not necessarily a systemic impairment of copper metabolism (see p. 301), and it rarely contributes to diagnoses in cattle (Paynter, 1987).

Ceruloplasmin

Todd (1970) advocated CP estimations on serum as a diagnostic aid, because of the stability of the enzyme, the small size of the serum sample required and the technical convenience of the assay. This technique offers little advantage over serum copper in specificity of diagnosis in ruminants (see Fig. 11.8), as some 80–90% of copper occurs as CP and correlations between the two parameters usually have r values of 0.8–0.9. Assays for CP are helpful in situations where exposure to excess molybdenum has caused TCAICu to accumulate in the plasma, but this rarely occurs in cattle or sheep under natural grazing conditions. Enzyme assays for CP introduce problems of quality control and standardization, and this may explain the poor correlations (r = 0.55) found by Telfer *et al.* (1996), who advocate the use of CP:Cu ratios in the diagnosis of induced copper disorders in ruminants. There is no peer-reviewed evidence that CP:Cu ratios have any diagnostic value, and comparisons of bovine sera from high-molybdenum and low-molybdenum regions in Scotland revealed no anomaly in CP:Cu ratio (N. Suttle and J. Small, unpublished results).

Whole-blood copper

Whole-blood copper has rarely been routinely used for diagnostic purposes other than in England, where the adoption of a single high threshold of normality – 12.6 μmol l^{-1} – probably led to gross overestimates of the incidence of hypocuprosis (Suttle, 1993). As with serum, animals with 'subnormal' values by this standard can fail to benefit from supplementation (Givens *et al.*, 1981). From the study of Koh and Judson (1987), who measured plasma, whole-blood and liver copper in the same individuals over long periods, the marginal range for whole blood is 6–10 μmol Cu l^{-1}.

Erythrocyte copper and copper–zinc superoxide dismutase

Theoretically, ESOD activities have a distinctive diagnostic contribution to make, since they decline at a slow linear rate during deficiency, compared with the rapid exponential decline in plasma or serum copper (Fig. 11.9; Suttle and McMurray, 1983). Low values confirm a prolonged deficiency and high values a copper overload (N.F. Suttle and J.N. Small, unpublished data) and were better related to growth rate than CP in the Aberdeen studies (Fig. 11.8). However, the conventional units of measurement are not quantitative, one unit being the activity needed to impair rate of 'superoxidation' by 50%. Since superoxidation rates vary between methods, so the molecular CuZnSOD equivalence of one unit varies. Furthermore, red blood cells (RBCs) must be washed to remove plasma inhibitors and heavily diluted to minimize interference from RBC constituents, which probably include haemoglobin. Tentative diagnostic ranges for one particular kit assay method have

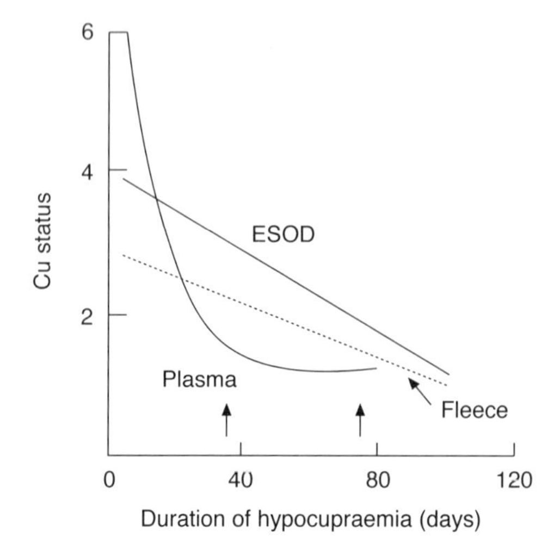

Fig. 11.9. Different time courses for changes in three indices of copper status in sheep given a low Cu diet: erythrocyte superoxide dismutase (ESOD), plasma Cu and fleece Cu; arrows indicate the onset of clinical signs and units vary (Suttle and McMurray, 1983).

been proposed (Herbert *et al.*, 1991). Since ESOD activity is highly correlated with erythrocyte copper, measurement of RBC copper will be just as useful and less problematical; neither parameter should be affected by infection (Arthington *et al.*, 1996).

Neutrophil copper and copper–zinc superoxide dismutase

Another blood constituent with a distinctive prognostic contribution to make is the neutrophil. Steers given a diet containing 10.3–17.7 mg Cu and 1 or 10 mg Mo kg^{-1} DM showed reductions in neutrophil copper from 13.0 to 5.1 pg 10^{-7} and neutrophil SOD from 0.65 to 0.31 international units (IU) 10^{-6} cells after 8 months of molybdenum-induced depletion (Xin *et al.*, 1991). At that stage, the mean liver copper concentration was marginal (285 μmol kg^{-1} DM), the animals were clinically normal and the plasma copper concentration was normal, but the *in vitro* killing capacity of the neutrophil was reduced from 27.7 to 17.3%. In a recent study in which mild exposure to molybdenum (5 mg kg^{-1} DM) induced clinical symptoms (Gengelbach *et al.*, 1997), treated calves were eventually distinguishable in terms of low neutrophil CuZnSOD activity. In some situations of molybdenum-induced deficiency, decreases in leucocyte copper may give a warning of impending dysfunction which the conventional blood criteria do not give.

Hair and fleece copper

Concentrations of copper in the hair and fleece have been widely measured without becoming established as reliable diagnostic aids. Values can reflect suboptimal copper intake in both sheep and cattle (Suttle and Angus, 1978; Suttle and McMurray, 1983) and represent the sample of choice from a grossly deteriorated carcass. Diagnostic interpretation is given in Table 11.5 but may be influenced by other nutritional deficiencies that influence hair or wool growth.

Other criteria

Plasma monoamine oxidase and liver cytochrome oxidase activities provide further indices of chronic copper depletion, but Mills *et al.* (1976) were obliged to conclude from their critical studies with copper-deficient calves that 'existing biochemical techniques are of limited value for predicting the speed or extent to which an individual will develop overt signs of deficiency'. Nothing has changed and liver CuZnSOD appears to be no more specific than other tests (Fig. 11.8).

Prevention and Control of Copper Deprivation

A variety of methods for providing livestock with sufficient copper have been devised, but they differ in efficacy and suitability.

Indirect supplementation

The application of copper-containing fertilizers can increase plant yields and raise the copper content of herbage and crops, but adequate levels are often much higher for the grazing animals (78 mg Cu kg^{-1} DM), see Table 11.7, than for the plants (≤ 4 mg Cu kg^{-1} DM: McFarlane, 1989; Haynes, 1997). The amounts of copper required vary with the soil type and climatic conditions. Early Australian experience indicated that a single dressing of 5–7 kg ha^{-1} of copper sulphate was usually sufficient for 3–4 years, except on calcareous soils (Underwood *et al.*, 1943), but longer residual effects (≥ 23 years) have been reported on sandy soils. Fertilizer treatment is unsuitable on soils high in organic matter, because the copper becomes 'fixed' in unavailable humic acid complexes, and is not advised when induced hypocuprosis occurs on pastures high in copper (McFarlane *et al.*, 1990).

Direct continuous supplementation

Under range conditions, where fertilizer treatments are uneconomic, deficiency can be prevented by the provision of salt-licks containing 0.5–1.9% copper. With housed cattle, supplementary copper is normally supplied by incorporation into mineral mixtures or concentrate, in the same way that other minerals are supplied. Iron should never be added to mineral supplements for grazing livestock, even as Fe_2O_3 to give colour. Copper can be dissolved in water and sprinkled over the forage in amounts calculated to add 5–10 mg Cu kg^{-1} total DM. Additions of 2–5 mg Cu l^{-1} to the drinking-water using a proportioning device, here controlled both hypocupraemia and hypocuprosis induced by exposure to molybdenum in cattle (Humphries, 1980). There is no necessity for supplementing dry rations for sheep, and great care has to be taken to avoid oversupply of copper from the natural feed constituents (Buck, 1970; Suttle, 1977). The only justification for copper supplementation with growing pigs and poultry rests on a growth-stimulating effect, unrelated to deficiency in the conventional sense (Braude, 1945, 1967).

Discontinuous supplementation

Discontinuous methods can be effective, because all grazing animals have the ability to store copper in their livers during periods of excess intake and to draw on those stores when intakes are inadequate. Drenching with $CuSO_4$ at monthly or longer intervals is often satisfactory. However, where molybdenum contents in the herbage are sufficiently high to induce scouring (5 mg Mo kg^{-1} DM or higher), daily copper supplementation may be necessary. Although the lesions of swayback are irreversible, they can develop over a sufficiently long period after birth to justify treatment of all lambs in a flock, following confirmation of the first case (Suttle *et al.*, 1970; Lewis *et al.*, 1981). An oral dose rate of 1 mg Cu kg^{-1} LW may well suffice for a milk-fed animal and be dangerous to exceed. Subcutaneous or intramuscular injections of a slowly translocated form of copper constitute satisfactory means of treatment, even at intervals as long as 3 months (Sutherland *et al.*, 1955). Copper glycine, copper–calcium ethylenediaminetetra-acetic acid (CuCaEDTA) and Dicuprene have been used in doses of 30–40 mg Cu for sheep and 120–240 mg for cattle (Cunningham, 1959; Camargo *et al.*, 1962). Water-soluble complexes, such as the hydroxyquinoline sulphonate and heptonate, are available and cause less tissue reaction, because they are rapidly translocated from the injection site, but the former poses a greater risk of acute toxicity. Methionate complexes are the least toxic but cause the largest reactions at the site of injection, where part of the dose can become encapsulated (Suttle, 1981a, b).

Slow release oral supplementation

The bolus administration of copper oxide 'needles' (rod-like particles of high specific gravity) into the oesophagus with a tube or balling-gun avoids tissue reactions (Dewey, 1977; Deland *et al.*, 1979). The particles are retained in the abomasum, where copper is released over a period of several weeks. The optimum dose rate is 0.1 g kg^{-1} LW (Suttle, 1987a, b) and liver copper stores can remain increased for many months after a single dose (McFarlane *et al.*, 1991). There have been reports of much shorter efficacy in the field, including situations in which diarrhoea is a problem. In two groups of lambs, given the same dose at the same time but grazing different fields, protection from hypocupraemia was afforded for 56 days to one, badly scouring group, and for over 150 days to the other, less affected group (Suttle and Brebner, 1995). Abomasal parasitism was implicated in the light of evidence from New Zealand that less copper is released from copper oxide particles if abomasal parasitism has raised abomasal pH (Bang *et al.*, 1990). To maximize the efficacy of these particles, parasite infections and other causes of diarrhoea should be controlled prior to dosing, which should precede turnout by several weeks to maximize retention. Copper oxide is equally successful when given in coarse powder form (Cavanagh and Judson, 1994) and is also available for cattle in a multimineral and vitamin bolus (Parkins *et al.*, 1994). Most recently, a soluble glass bolus has been introduced which releases copper, along with cobalt and selenium, for long periods (Koh and Judson, 1987; Givens *et al.*, 1988; McFarlane *et al.*, 1991).

Organic versus inorganic copper supplements

Given the generally low absorbability of copper in feeds for ruminants, there is considerable potential for improvement by agents which protect copper from antagonisms initiated in the rumen. Unfortunately, this potential has yet to be realized. Extravagant claims have been made for Cu–protein hydrolysate mixtures ('metallosates') and specific amino acid complexes, such as Cu_2:lysine; neither source had any advantage over $CuSO_4$ in four studies with cattle (metallosate: Wittenberg *et al.*, 1990: Cu_2:lysine: Kegley and Spears, 1993; Ward *et al.*, 1993; Du *et al.*, 1996) or one with sheep (Cu_2:lysine: Suttle and Brebner, 1996). In two studies involving three experiments and four metallosates, two of the products showed 'advantages' when compared with $CuSO_4$ in calves (Kincaid *et al.*, 1986; Ward *et al.*, 1996), but in neither case was proper account taken (i.e. covariance analysis) of large differences in initial copper status. Use of CuEDTA-type complexes has also failed to enhance availability in cattle (Miltimore *et al.*, 1978) and sheep (MacPherson and Hemingway, 1968; Suttle, 1994). Neither Cu methionine nor Cu lysine shows consistently superior availability over $CuSO_4$ in studies with pigs or poultry (Baker and Ammerman, 1995). There are no major differences among inorganic copper salts, except for cuprous chloride, which has shown a marked (44%) advantage over $CuSO_4$ in poultry (Baker and Ammerman, 1995); this might be exploitable if growth stimulation could be obtained at lower levels of supplementation and hence at lower environmental cost. Insoluble copper compounds, such as CuO and CuS, are poorly available compared with $CuSO_4$ on a mg-for-mg basis, but the large doses of inert CuO which can be safely given in particulate form make it an ideal slow-release source for ruminants, as discussed earlier.

Genetic selection

The existence of genetic variation in copper metabolism could be used more to prevent deficiencies and lessen the dependence on supplementation where problems are endemic. Selection of ram lambs for four generations on the basis of plasma copper has eliminated hypocuprosis in an inbred, vulnerable 'breed' (Woolliams, C. *et al.*, 1986; Woolliams, J.A. *et al.*, 1986). Alternatively, the use of rams of a 'tolerant' breed, such as the Texel, for crossing with the Scottish Blackface should reduce hypocuprosis problems on hill pastures.

Minimizing antagonisms

Given the prominent influence of antagonisms from iron and molybdenum on the occurrence of copper-responsive disorders, their prevention can be achieved in part by minimizing exposure to such antagonists. By avoiding the excessive use of lime and improving soil drainage, the antagonism from molybdenum can be reduced. By avoiding overgrazing and 'poaching', antagonisms from soil iron are minimized. Iron should never be used in mineral supplements for grazing or mature housed ruminants.

Copper Requirements

Ruminants

Requirements are so powerfully influenced by interactions with iron, molybdenum and sulphur that fourfold variation in the ability of feeds to provide absorbable copper must be allowed for. Net requirements can be predicted by modifying the factorial model used by the ARC (1980), reducing the requirements for growth and maintenance (Simpson *et al.*, 1981; Suttle, 1987a). The resultant estimates of gross requirements are given in Table 11.6. Predictions for mixtures of roughage and concentrate are speculative (Suttle, 1994). Lactation in cows does not raise requirements significantly.

Table 11.6. Factorial estimates of the dietary copper requirements of sheep and cattle given three types of diet with contrasting copper absorbabilities[a] (A_{Cu}).

	Liveweight (LW) (kg)	Growth rate or milk yield (kg day^{-1})	Net requirement[b] (mg day^{-1})	Food intake[c] (kg DM day^{-1})	Gross requirement (A_{Cu}) (mg kg^{-1} DM)		
					0.06	0.03	0.015
Lamb	20	0.1	0.13	0.5	4.3	8.6	17.2
		0.2	0.18	0.7	4.3	8.6	17.2
Ewe (16-week twin fetuses)	75	0	0.63	1.5	7.0	14.0	21.0[d]
Lactating ewe	75	1	0.52	1.5	5.8	11.6	23.2
		3	1.24	2.9	7.1	14.2	28.4
Calf	100	0.5	0.65	3.0	3.6	7.2	14.4
		1.0	0.90	3.6	4.2	8.4	16.8
Cow (40-week fetus)	500	0	4.1	9.1	7.5	15.0	30.0[d]
Lactating	500	20	4.0	15.2	4.4	8.8	17.6
		40	6.0	19.3	5.2	10.4	20.8

[a] Approximate A_{Cu} values for particular feed classes are as follows; 0.06 for roughage + concentrates; 0.03 for normal green swards; 0.015 for molybdenum-rich swards (> 2 mg Mo kg^{-1} DM); intermediate values are needed for brown, dry season swards (0.04) and iron-rich (> 800 mg Fe kg^{-1} DM) green swards (0.02).
[b] Components of factorial model (see p.11): M = 4 μg kg^{-1} LW, G = 0.5 mg kg^{-1} LW gain, L = 0.1 and 0.22 mg l^{-1} for cow and sheep milk, respectively, and F = 4 mg Cu kg^{-1} clean wool growth.
[c] Real food intakes will vary according to diet quality (see Table 16.6 on p. 500).
[d] Requirements for late pregnancy allow for build up of a large fetal reserve of copper.

Horses

The copper requirements of horses are ill-defined, but improvements in angular limb deformities in 2–4-month-old foals when the dietary copper concentration was raised from 8 to 25 mg kg^{-1} DM suggest a relatively high requirement for a non-ruminant species (Barton *et al.*, 1991). Molybdenum does not impair copper metabolism or raise copper requirements in horses (Strickland *et al.*, 1987).

Pigs

Detailed calculations are superfluous for non-ruminants, because variation between feeds in copper absorbability is small and dietary provision normally generous. A purified diet supplying 6 mg Cu kg^{-1} DM fully meets the requirements of baby pigs (Okonkwo *et al.*, 1979). The ARC (1981) considered that 4 mg Cu kg^{-1} DM was sufficient for growing pigs, and all normal diets composed of cereals with the usual protein supplements supply at least that level. The additional copper requirement of 4–6 mg day^{-1} which can be calculated for lactation (5–7 kg milk day^{-1} containing 0.75 mg Cu kg^{-1}) would be met by the additional food intake (3 kg day^{-1}) over that for late pregnancy if the diet contained only 4 mg Cu kg^{-1} DM and absorptive efficiency was 50%.

Poultry

Definitive data on the minimum copper requirements of chicks for growth or of hens for egg production have not been reported. The NRC (1994) give the following requirements in mg Cu kg^{-1} of feed: starting chicks (0–8 weeks), 5; growing Leghorn-type chickens (8–18 weeks), 4; broilers (all ages), 8; starting turkey poults, 8; growing turkeys, 6; breeding turkeys, 6. No requirements were given for laying or breeding hens, but their previous low recommendations (3–4 mg Cu kg^{-1} DM) have not been disputed. Poultry rations composed of normal feeds invariably contain 8 mg Cu kg^{-1} DM or more and should therefore not require supplementation.

Growth Promotion

Pigs

The mechanisms responsible for the growth stimulation and improved feed conversion efficiency induced in weaned pigs by high copper supplements (125–250 mg Cu kg^{-1} DM: Braude, 1967) are still incompletely understood. The sulphate is more effective than the sulphide or oxide and it is the copper and not the sulphate radical that is effective. Kirchgessner and Giessler (1961) and Kirchgessner *et al.* (1976) demonstrated an improvement in the digestibility of proteins by young pigs. Addition of copper sulphate to diets containing rapeseed meals improves growth, partly by detoxifying goitrogens (see Chapter 12). Birth and weaning weights can be increased by 9% by similarly supplementing the sow's diet with copper (Cromwell *et al.*, 1993). Recent studies have shown that improved appetite makes a major contribution to growth stimulation (Zhou *et al.*, 1994a), but a systemic component was indicated when intravenously administered copper histidinate also stimulated growth (Zhou *et al.*, 1994b). Release of the mRNA for growth hormone was stimulated by both the oral and intravenous routes of copper supplementation, which significantly increased brain copper concentrations. Copper lysine stimulated growth more than did copper sulphate, in keeping with most previous studies (Zhou *et al.*, 1994a), but this may reflect a higher palatability and not necessarily a higher availability of copper from Cu lysine for growth stimulation. Copper supplementation of pig rations at 250 or even 125 mg kg^{-1} DM is not always a safe procedure. It can reduce iron stores in the liver (Bradley *et al.*, 1983) and lead to anaemia, copper toxicosis or zinc deficiency, unless the diets contain sufficient iron and zinc. Concurrent supplementation with iron and zinc, each at 150 mg kg^{-1} diet, protected against the adverse effects of copper at 425 mg Cu kg^{-1} DM in one early study (Suttle and Mills, 1966a). Supplementation of pig rations can pose environmental problems, because of the high copper content of the animal wastes produced. Chronic copper poisoning in sheep grazing herbage treated with the liquid manure from pigs fed with copper supplements has been reported from the Netherlands (van Ulsen, 1972) and the USA (Kerr and McGavin, 1991). The copper in such slurry was found to be as available to sheep as that in $CuSO_4$ (Suttle and Price, 1976), but, once washed into the soil, it causes little or no rise in herbage or cereal copper concentrations.

Poultry

The effects of large copper supplements in poultry rations are highly variable, of doubtful economic significance and affected by the nature of the basal diet. Thus Jenkins *et al.* (1970) obtained significant increases in chick growth with 250 mg Cu kg^{-1} DM in a diet composed mainly of wheat and fish-meal, with added tallow, but observed either no growth response or a growth depression when a maize and soybean diet was similarly supplemented. Growth has occasionally been retarded by copper supplementation through negative interactions with other feed additives (Guenthner *et al.*, 1978). In turkey poults, growth has been retarded by 204 mg Cu kg^{-1} DM in a short-term (10-day) study (Ward *et al.*, 1994) and by 120 mg Cu kg^{-1} DM in a longer-term (24-week) study (Kashani *et al.*, 1986). The brief addition of copper to the drinking-water as well as diet is often practised but not recommended (Ward *et al.*, 1994).

Copper Poisoning

Species differences in susceptibility to copper overload are high: pigs are highly tolerant, cattle and goats less tolerant and sheep extremely intolerant of copper excess.

Occurrence in sheep

Chronic copper poisoning has frequently been encountered in housed lambs, milk sheep and pedigree rams receiving large amounts of concentrates. Complete sheep diets containing over 15 mg Cu kg^{-1} air-dry feed can cause copper poisoning (Hartmans, 1975), and it is difficult for feed compounders to keep values consistently below this limit, particularly when the feeds have been stored or processed in contact with copper-supplemented pig rations or when copper-rich constituents, such as certain distillery by-products (Suttle *et al.*, 1996) or palm-kernel cake, are used. Addition of monensin (van Ryssen and Barrowman, 1987) and possibly selenium (Hartmans and van Ryssen, 1997) to sheep feeds predisposes to copper poisoning. During the period 1970–1990, the number of reported cases of copper poisoning in sheep in the

UK rose steadily to about 100 per annum, due partly to the growing popularity of susceptible breeds, such as the Texel. Problems can also arise from excessive consumption of copper-containing salt-licks or the unwise use of any copper supplement, oral or parenteral. Chronic copper poisoning rarely occurs in grazing sheep under natural conditions, except in susceptible breeds, such as the North Ronaldsay (Wiener *et al.*, 1978), or when copper fungicides have been applied in orchards and contaminate the underlying pasture. In parts of Australia and elsewhere, normal copper intakes together with very low levels of molybdenum (0.1–0.2 mg kg^{-1} DM) or the consumption of plants such as *Heliotropium europaeum,* containing hepatotoxic alkaloids (heliotrine and lasiocarpine), can cause copper poisoning. The latter disease is known variously as 'toxaemic jaundice' or 'yellows'.

Pathogenesis in sheep

Whatever the cause, chronic copper poisoning is characterized by two phases: during the first (prehaemolytic), which is clinically 'silent', copper accumulates in the liver over a period of weeks until liver concentrations of 1000 mg Cu kg^{-1} DM or more are reached; during the second (haemolytic), copper is released from the principal hepatic storage site, in lysosomes. Consequently, blood copper values rise tenfold or more and a haemolytic crisis is precipitated, with haemoglobinuria, haemoglobinaemia and jaundice preceding death. Significant rises in serum enzymes from the damaged liver occur during the prehaemolytic phase, but serum copper concentrations barely increase. Serum aspartate aminotransferase was the first enzyme to be proposed for the early detection of chronic copper poisoning (MacPherson and Hemingway, 1969; Thompson and Todd, 1974), but glutamate dehydrogenase is more specific and just as sensitive. Increases in gamma glutamyl transferase indicate more long-standing damage to the bile-ducts. The haemolytic crisis is accompanied by a marked rise in serum acid phosphatase and creatine kinase, as well as copper. The detailed pathology of the disease has been reviewed by Howell and Gooneratne (1987). Given that sheep can show histological (King and Bremner, 1979) or biochemical (Woolliams *et al.*, 1982) evidence of liver damage at liver concentrations as low as 350 mg Cu kg^{-1} DM and widely ranging concentrations at the time of haemolytic crisis, a marginally toxic range of 350–1000 mg Cu kg^{-1} DM should replace the commonly used single threshold of 750 mg Cu kg^{-1} DM when assessing copper overload. Interpretation of liver enzyme and serum copper data also benefits from the introduction of marginal bands, as shown in Table 11.7.

Treatment and prevention in sheep

Copper poisoning can be treated and prevented by exploiting the known antagonists of copper. In the first application of this principle, the provision of molybdate-containing salt-licks eliminated mortality from toxaemic jaundice (Dick and Bull, 1945). Accumulation of copper from concentrates can also be greatly reduced by the addition of iron (Abdellatif, 1968), molybdenum, sulphur (Suttle, 1977) or zinc supplements (Bremner *et al.*,

Table 11.7. Marginal bands[a] for copper:antagonist concentrations in the bloodstream and tissues and serum enzyme activities (indicative of liver or muscle damage) in the diagnosis and prognosis of chronic copper-poisoning (CCP) in livestock.

	Sheep(S)	Cattle(C)	Goats(G)	Pigs	Poultry	Interpretative limits
Diet Cu:Mo						Low values for S, C and G apply with Cu : Mo > 10
Diet Cu (mg kg^{-1} DM)	12–36	100–300	30–100	250–750	200–400	
Liver Cu (mmol kg^{-1} DM)[b]	6.4–16.0	6.4–16.0	6.4–16.0	32.0–48.0	1.0–2.0	
Kidney Cu[c] (mmol kg^{-1} DM)	0.6–0.8	0.6–0.8	0.6–0.8	10.0–16.0		
Kidney Fe (mmol kg^{-1} DM)	16–18	16–18	16–18			
Serum Cu (μmol l^{-1})	20–25	20–25	20–25	40–60		No acute-phase response
Serum GLDH (IU l^{-1} at 30°C)	12–51	12–51				Consider other possible causes, e.g. liver fluke
Serum GGT (IU l^{-1} at 30°C)	45–100	45–100				

[a] Values below lower limit rule out CCP; values above upper limit strongly suggestive of CCP; marginal values for several criteria indicate probable CCP.
[b] Multiply by 63.4 to obtain units in mg kg^{-1} DM and 21.1 for mg kg^{-1} FW.
[c] High kidney but normal liver Cu indicates acute toxicity; high kidney Cu with normal kidney Fe indicates hepatic crisis without haemolytic crisis.
GLDH, glutamate dehydrogenase; GGT, gamma glutamyl transferase.

1976), but puzzling interactions can occur when zinc and iron are given together (Rosa *et al.*, 1986). It is common for mixtures of antagonists to be added by feed compounders to reduce the risks of toxicity, although there is no evidence that fully additive effects will be obtained. Addition of sulphur may cause problems of sulphur toxicity (see Chapter 9). More recently, the mediator of the Cu–Mo–S antagonism – tetrathiomolybdate – has been successfully used to prevent and treat chronic copper toxicity (Gooneratne *et al.*, 1981). The most convenient treatment is a course of three subcutaneous injections of 3.7 mg ammonium tetrathiomolybdate (ATTM) given on alternate days and it is effective in some sheep already in haemolytic crisis (Humphries *et al.*, 1988). Increases in total plasma copper occur after ATTM treatment, due to accumulation of TCAICu, and the success of treatment can *only* be assessed by serum enzyme assays (Suttle *et al.*, 1992a; N.F. Suttle *et al.*, unpublished data). There is also a temporary fall in blood SOD after ATTM treatment (Suttle *et al.*, 1992a), but this may have no important physiological consequences. In addition to treating subclinically affected animals, all those at risk should be transferred to a diet of lower available copper content; this can be achieved by the use of whole grain (which is invariably low in copper) or the addition of gypsum (15 g kg^{-1} DM) and/or sodium molybdate (19.2 mg kg^{-1} DM) to the forage. With vulnerable breeds or dairy sheep at recurrent risk while housed, the grazing of strips of pasture fertilized with sodium molybdate (0.27 kg ha^{-1}) during summer may lessen the accumulation of copper over the years. As with the prevention of deficiency, appropriate genetic selection could lower the susceptibility of vulnerable breeds to copper poisoning. Selection of Texel sires for low plasma copper has resulted in lower liver copper concentrations in their offspring (J.A. Woolliams, personal communication) and could lessen the vulnerability of this breed to copper poisoning when intensively fattened. Similar selection within dairy sheep would lessen their vulnerability.

Occurrence in cattle and goats

Cattle are at greatest risk of developing copper poisoning prior to weaning when given copper-rich milk substitutes from which copper is well absorbed (Shand and Lewis, 1957). Exposure to 200 mg Cu kg^{-1} DM from 3 to 45 days of age was harmful to calves, but 50 mg Cu kg^{-1} DM was tolerated (Jenkins and Hidiroglou, 1989). After weaning, copper levels up to 900 mg kg^{-1} DM can be fed for several months without causing hepatotoxicity (Felsman *et al.*, 1973), but occasionally copper toxicosis has been reported (for a recent review of cases, see Tremblay and Baird, 1991). Some incidents of chronic copper toxicity have involved massive pollution from mining activities (Gummow *et al.*, 1991). Like sheep, cattle can succumb to acute copper poisoning when given excessive amounts by injection (Mylrea and Byrne, 1974), and doses should always be carefully scaled according to the range of live weights in the group to be treated. Cases of sudden death in intensively reared bulls, with high kidney copper concentrations, moderate liver copper concentrations and no evidence of haemolysis, raised the possibility of a hepatic crisis brought on by consumption of a copper-rich mineral mixture (C. Low, personal communication), but this has still to be confirmed. Goats have tolerated 60 mg added Cu kg^{-1} DM for 137 days, whereas lambs succumbed to copper poisoning; goats must therefore be classed as a copper-tolerant species (Zervas *et al.*, 1990), although they become more susceptible when repeatedly injected with selenium and vitamin E (Hussein, 1985).

Occurrence in pigs and poultry

Copper poisoning in non-ruminants follows a similar two-phase pattern to that seen in sheep, except that the haemolytic crisis is less pronounced and growth retardation more prominent. The disorder occurs in a minority of pigs given massive copper supplements (250 mg Cu kg^{-1} DM) as growth stimulants to rations which are not suitably balanced with other minerals, and they can be protected by adding zinc and iron to the diet in broadly equivalent concentrations (Suttle and Mills, 1966a, b). Sows have been given similar supplements throughout six gestations without causing copper toxicity, although liver copper levels increased to 1899 mg Cu kg^{-1} DM (Cromwell *et al.*, 1993). Erosion of the gizzard is an additional feature of copper poisoning in poultry and occurs at dietary concentrations around 500 mg Cu kg^{-1} DM (Christmas and Harms, 1979; Jensen and Maurice, 1979). Methionine supplementation lessens the growth retardation caused by excess copper but does not prevent gizzard erosion (Kashani *et al.*, 1986). It would be interesting to examine the toxicity to the gizzard of a relatively insoluble copper salt, such as cupric carbonate. The toxicity of copper to chicks is increased when the diet is low in calcium and based on casein–gelatin rather than maize–soybean (Leach *et al.*, 1990) or marginal in selenium (Jensen, 1975). In the latter study, mortality was associated with exudative diathesis (see Chapter 15) and prevented by raising the dietary selenium level from 0.2 to 0.7 mg kg^{-1} DM (the copper level was 800 mg kg^{-1} DM).

References

Abdellatif, A.M.M. (1968) Conditioned hypocuprosis: some effects of diet on copper storage in ruminants. *Verslagen Van Landboukundige Onderzoekingen* 709, 43–67.

Allen, J.D. and Gawthorne, J.M. (1987) Involvement of the solid phase rumen digesta in the interaction between copper, molybdenum and sulphur in sheep. *British Journal of Nutrition* 58, 265–276.

Allen, J.D. and Gawthorne, J.M. (1988) Interactions between proteins, thiomolybdates and copper. In: *Proceedings of the 6th Symposium on Trace Elements in Man and Animals.* Plenum, New York, pp. 315–316.

Alloway, B.J. (1973) Copper and molybdenum in sway-back pastures. *Journal of Agricultural Science, Cambridge* 80, 521–524.

ARC (1980) *The Nutrient Requirements of Ruminant Livestock.* Commonwealth Agricultural Bureaux, Farnham Royal, UK, pp. 221–234.

ARC (1981) *The Nutrient Requirements of Pigs.* Commonwealth Agricultural Bureaux, Farnham Royal, UK, pp. 273–275.

Arthington, J.D., Corah, L.R. and Blecha, F. (1996) The effect of molybdenum-induced copper deficiency on acute phase protein concentrations, superoxide dismutase activity, leukocyte numbers and lymphocyte proliferation in beef heifers inoculated with bovine herpesvirus-I. *Journal of Animal Science* 74, 211–217.

Bain, M.S., Spence, J.B. and Jones, P.C. (1986) An investigation of bovine serum copper levels in Lincolnshire and South Humberside. *Veterinary Record* 119, 593–595.

Baker, D.H. and Ammerman, C.B. (1995) Copper bioavailability. In: Ammerman, C.B., Baker, D.H. and Lewis, A.J. (eds) *Bioavailability of Nutrients for Animals.* Academic Press, New York, pp. 127–156.

Bang, K.S., Familton, A.S. and Sykes, A.R. (1990) Effect of ostertagiasis on copper status in sheep: a study involving use of copper oxide wire particles. *Research in Veterinary Science* 49, 306–314.

Barlow, R.M. (1963a) Further observations on swayback. 1. Transitional pathology. *Journal of Comparative Pathology and Therapeutics* 73, 51–60.

Barlow, R.M. (1963b) Further observations on swayback. 2. Histochemical localisation of cytochrome oxidase activity in the central nervous system. *Journal of Comparative Pathology and Therapeutics* 73, 61–67.

Barry, T.N., Reid, T.C., Millar, K.R. and Sadler, W.A. (1981) Nutritional evaluation of kale (*Brassica oleracea*) diets. 2. Copper deficiency, thyroid function and selenium status in young cattle and sheep fed kale for prolonged periods. *Journal of Agricultural Science, Cambridge* 96, 269–282.

Barton, J., Hurtig, M. and Green, S. (1991) The role of copper in developmental orthopaedic disease in foals. In: *Highlights of Agricultural and Food Research in Ontario*. University of Guelph, Guelph, pp. 14–18.

Baumgartner, S., Brown, D.J., Salevsky, E., Jr and Leach, R.M., Jr (1978) Copper deficiency in the laying hen. *Journal of Nutrition* 108, 804–811.

Baxter, J.H. and Van Wyk, J.J. (1953) A bone disorder associated with copper deficiency. 1. Gross morphological, roentgenological and chemical observations. *Bulletin of the Johns Hopkins Hospital* 93, 1–23.

Baxter, J.H., Van Wyk, J.J. and Follis, R.H., Jr (1953) A bone disorder associated with copper deficiency. 2. Histological and chemical studies on the bones. *Bulletin of the Johns Hopkins Hospital* 93, 25–39.

Beck, A.B. (1941a) Studies on the copper content of the milk of sheep and of cows. *Australian Journal of Experimental Biology and Medical Science* 19, 145–150.

Beck, A.B. (1941b) Studies on the blood copper of sheep and of cows. *Australian Journal of Experimental Biology and Medical Science* 19, 249–254.

Beck, A.B. (1951) *A Survey of the Copper Content of Western Australian Pastures.* Leaflet No. 678, Department of Agriculture (Western Australia), Perth.

Beck, A.B. (1956) The copper content of the liver and blood of some vertebrates. *Australian Journal of Zoology* 4, 1–18.

Beeson, K.C., Gray, L. and Adams, M.B. (1947) The absorption of mineral elements by forage plants. 1. The phosphorus, cobalt, manganese, and copper content of some common grasses. *Journal of the American Society of Agronomy* 39, 356–362.

Bennetts, H.W. and Chapman, F.E. (1937) Copper deficiency in sheep in Western Australia: a preliminary account of the aetiology of enzootic ataxia of lambs and on anaemia of ewes. *Australian Veterinary Journal* 13, 138–149.

Bennetts, H.W. and Hall, H.T.B. (1939) 'Falling disease' of cattle in the south-west of Western Australia. *Australian Veterinary Journal* 15, 152–159.

Bird, P.R. (1970) Sulphur metabolism and excretion studies in ruminants. III. The effect of sulphur intake on the availability of copper in sheep. *Proceedings of the Australian Society of Animal Production* 8, 212–218.

Boila, R.J., Devlin, T.J., Drydale. R.A. and Lillie, L.E. (1984a) Geographical variation in the copper and molybdenum contents of forages grown in Northwestern Manitoba. *Canadian Journal of Animal Science* 64, 899–918.

Boila, R.J., Devlin, T.J., Drydale, R.A. and Lillie, L.E. (1984b) The severity of hypocupraemia in selected herds of beef cattle in Northwestern Manitoba. *Canadian Journal of Animal Science* 64, 919–936.

Boyne, R. and Arthur, J.R. (1981) Effect of selenium and copper deficiency on neutrophil function in cattle. *Journal of Comparative Pathology* 91, 271–276.

Bradley, B.D., Graber, G., Condon, R.J. and Frobish, L.T. (1983) Effects of graded levels of dietary copper on copper and iron concentrations in swine tissues. *Journal of Animal Science* 56, 625–630.

Braude, R. (1945) Some observations on the need for copper in the diet of fattening pigs. *Journal of Agricultural Science* 35, 163–167.

Braude, R. (1967) Copper as a stimulant in pig feeding. *World Review of Animal Production* 3, 69–82.

Bremner, I. (1993) Metallothionein in copper deficiency and toxicity. In: Anke, M., Meissner, D. and Mills, C.F. (eds) *Proceedings of the Eighth International Symposium on Trace Elements in Man and Animals.* Verlag Media Touristik, Gersdorf, pp. 507–515.

Bremner, I., Young, B.W. and Mills, C.F. (1976) Protective effect of zinc supplementation against copper toxicosis in sheep. *British Journal of Nutrition* 36, 551–561.

Bremner, I., Humphries, W.R., Phillippo, M., Walker, M.J. and Morrice, P.C. (1987) Iron-induced copper deficiency in calves: dose–response relationships and interactions with molybdenum and sulphur. *Animal Production* 45, 403–414.

Bronner, F. and Yost, J.H. (1985) Saturable and non-saturable copper and calcium transport in mouse duodenum. *American Journal of Physiology* 249, G108–G112.

Buck, W.B. (1970) Diagnosis of feed-related toxicoses. *Journal of the American Veterinary Medical Association* 156, 1434–1443.

Buckingham, K., Heng-Khou, C.S., Dubick, M., Lefevre, M., Cross, C., Julian, L. and Rucker, R. (1981) Copper deficiency and elastin metabolism in avian lung. *Proceedings of the Society for Experimental Biology and Medicine* 166, 310–319.

Bush, J.A., Jensen, W.N., Athens, J.W., Ashenbrucker, H., Cartwright, G.E. and Wintrobe, M.M. (1956) Studies on copper metabolism. 19. The kinetics of iron metabolism and erythrocyte life-span in copper-deficient swine. *Journal of Experimental Medicine* 103, 701–712.

Camargo, W.V. de A., Lee, H.J. and Dewey, D.W. (1962) The suitability of some copper preparations for parenteral copper therapy in sheep. *Proceedings of the Australian Society of Animal Production* 4, 12–17.

Campbell, A.G., Coup, M.R., Bishop, W.H. and Wright, D.E. (1974) Effect of elevated iron intake on the copper status of grazing cattle. *New Zealand Journal of Agricultural Research* 17, 393–399.

Cancilla, P. and Barlow, R.M. (1968) Structural changes in the central nervous system in swayback (enzootic ataxia) of lambs. IV. Electron microscopy of the white matter lesions of the spinal cord. *Acta Neuropathologica* 11, 294–301.

Carnes, W.H., Shields, G.S., Cartwright, G.E. and Wintrobe, M.M. (1961) Vascular lesions in copper deficient swine. *Federation Proceedings* 20, 118.

Cavanagh, N.A. and Judson, G.J. (1994) Copper oxide powder as a copper supplement for sheep. *Journal of Trace Elements and Electrolytes in Health and Disease* 8, 183–188.

Cerone, S., Sansinea, A. and Nestor, A. (1995) Copper deficiency alters the immune response of bovine. *Nutrition Research* 15, 1333–1341.

Chao, J.C.J., Medeiros, D.M., Davidson, J. and Shiry, L. (1994) Low levels of ATP synthase and cytochrome *c* oxidase subunit peptide from hearts of copper-deficient rats are not altered by the administration of dimethyl sulphoxide. *Journal of Nutrition* 24, 789–903.

Chen, Y., Saari, J.T. and Kang, Y.J. (1995) Expression of γ-glutamyl cysteine synthetase in the liver of copper deficient rats. *Proceedings of the Society for Experimental Biology and Medicine* 210, 102–106.

Chooi, K.F., Hutagalung, R.I. and Wan Mohammed, W.E. (1988) Copper toxicity in sheep fed oil palm products. *Australian Veterinary Journal* 66, 156–157.

Christmas, R.B. and Harms, R.H. (1979) The effect of supplemental copper and methionine on the performance of turkey poults. *Poultry Science* 58, 382–384.

Cousins, R.J. (1985) Absorption, transport and hepatic metabolism of copper and zinc with special reference to metallothionein and caeruloplasmin. *Physiological Reviews* 65, 238–309.

Cromwell, G.L., Monegue, H.J. and Stahly, T.S. (1993) Long-term effects of feeding a high copper diet to sows during gestation and lactation. *Journal of Animal Science* 71, 2996–3002.
Cunningham, I.J. (1950) Copper and molybdenum in relation to diseases of cattle and sheep in New Zealand. In: McElroy, W.D. and Glass, B. (eds) *Copper Metabolism: a Symposium on Animal, Plant and Soil Relationships.* Johns Hopkins Press, Baltimore, Maryland, pp. 246–273.
Cunningham, I.J. (1959) Parenteral administration of copper to sheep. *New Zealand Veterinary Journal* 7, 15–17.
Davis, G.K. (1950) The influence of copper on the metabolism of phosphorus and molybdenum. In: McElroy, W.D. and Glass, B. (eds) *Copper Metabolism: a Symposium on Animal, Plant and Soil Relationships.* Johns Hopkins Press, Baltimore, Maryland, pp. 216–229.
Deland, M.P.B., Cunningham, P., Milne, M.L. and Dewey, D.W. (1979) Copper administration to young calves: oral dosing with copper oxide compared with subcutaneous copper glycinate injection. *Australian Veterinary Journal* 55, 493–494.
Dewey, D.W. (1977) An effective method for the administration of trace amounts of copper to ruminants. *Search, Australia* 8, 326–327.
Dick, A.T. (1952) The effect of diet and of molybdenum on copper metabolism in sheep. *Australian Veterinary Journal* 28, 30–33.
Dick, A.T. (1953) The control of copper storage in the liver of sheep by inorganic sulphate and molybdenum. *Australian Veterinary Journal* 29, 233–239.
Dick, A.T. (1954) Studies on the assimilation and storage of copper in crossbred sheep. *Australian Journal of Agricultural Research* 5, 511–544.
Dick, A.T. (1956) Molybdenum in animal nutrition. *Soil Science* 81, 229–258.
Dick, A.T. and Bull, L.B. (1945) Some preliminary observations on the effect of molybdenum on copper metabolism in herbivorous animals. *Australian Veterinary Journal* 21, 70–72.
Dick, A.T., Dewey, D.W. and Gawthorne, J.M. (1975) Thiomolybdates and the copper–molybdenum–sulphur interaction in ruminant nutrition. *Journal of Agricultural Science, Cambridge* 85, 567–568.
Du, Z., Hemken, R.W. and Harmon, R.J. (1996) Copper metabolism of Holstein and Jersey cows and heifers fed diets high in cupric sulphate or copper proteinate. *Journal of Dairy Science* 79, 1873–1880.
Egan, D.A. and Murrin, M.P. (1973) Copper concentrations and distribution in the livers of equine fetuses, neonates and foals. *Research in Veterinary Science* 15, 147–148.
Farrer, P. and Mistilis, S.P. (1967) Absorption of exogenous and endogenous biliary copper in the rat. *Nature, London* 213, 291–292.
Fell, B.F., Mills, C.F. and Boyne, R. (1965) Cytochrome oxidase deficiency in the motor neurones of copper-deficient lambs: a histochemical study. *Research in Veterinary Science* 6, 170–177.
Fell, B.F., Dinsdale, D. and Mills, C.F. (1975) Changes in enterocyte mitochondria associated with deficiency of copper in cattle. *Research in Veterinary Science* 18, 274–281.
Fell, B.F., Farmer, L.J., Farquharson, C., Bremner, I. and Graca, D.S. (1985) Observations on the pancreas of cattle deficient in copper. *Journal of Comparative Pathology* 95, 573–590.
Felsman, R.J., Wise, M.B., Harvey, R.W. and Barrick, E.R. (1973) Effect of graded levels

of copper sulphate and antibiotic on performance and certain blood constituents of calves. *Journal of Animal Science* 36, 157-160.

Ferguson, W.S., Lewis, A.H. and Watson, S.J. (1938) Action of molybdenum in nutrition of milking cattle. *Nature, UK* 141, 553.

Ferguson, W.S., Lewis, A.H. and Watson, S.J. (1943) The teart pastures of Somerset. 1. The cause and cure of teartness. *Journal of Agricultural Science* 33, 44–51.

Freudenberger, D.O., Familton, A.S. and Sykes, A.R. (1987) Comparative aspects of copper metabolism in silage-fed sheep and deer (*Cervus elaphus*). *Journal of Agricultural Science, Cambridge* 108, 1–7.

Frieden, E. (1971) Caeruloplasmin, a link between copper and iron metabolism. *Advances in Chemistry Series* 100, 292–321.

Gengelbach, G.P., Ward, J.D. and Spears, J.W. (1994) Effects of dietary copper, iron and molybdenum on growth and copper status of beef cows and calves. *Journal of Animal Science* 72, 2722–2727.

Gengelbach, G.P., Ward, J.D. and Spears, J.W. (1997) Effect of copper deficiency and copper deficiency coupled with high dietary iron or molybdenum on phagocytic function and response of calves to a respiratory disease challenge. *Journal of Animal Science* 75, 1112–1118.

Givens, D.I., Hopkins, J.R., Brown, M.E. and Walsh, W.A. (1981) The effect of copper therapy on the growth rate and blood composition of young growing cattle. *Journal of Agricultural Science, Cambridge* 97, 497–505.

Givens, D.I., Zervas, G., Simpson, V.R. and Telfer, S.B. (1988) Use of soluble rumen boluses to provide a supplement of copper for suckled calves. *Journal of Agricultural Science, Cambridge* 110, 199–204.

Golden, M.H. and Ramdath, D. (1987) Free radicals in the pathogenesis of kwashiorkor. *Proceedings of the Nutrition Society* 46, 53–68.

Gooneratne, S.R. and Christensen, D.A. (1989) A survey of maternal copper status and foetal tissue copper concentrations in Saskatchewan bovine. *Canadian Journal of Animal Science* 69, 141–150.

Gooneratne, S.R., Howell, J.McC. and Gawthorne, J.M. (1981) Intravenous administration of thiomolybdate for the prevention of chronic copper poisoning in sheep. *British Journal of Nutrition* 46, 457–467.

Gooneratne, S.R., Symonds, H.W., Bailey, J.V. and Christensen, D.A. (1994) Effects of dietary copper, molybdenum and sulphur on biliary copper and zinc secretion in Simmental and Angus cattle. *Canadian Journal of Animal Science* 74, 315–325.

Grace, N.D. (1983) Amounts and distribution of mineral elements associated with the fleece-free empty body weight gains of the grazing sheep. *New Zealand Journal of Agricultural Research* 26, 59–70.

Gross, A.M. and Prohaska, J.R. (1990) Copper-deficient mice have higher cardiac epinephric turnover. *Journal of Nutrition* 120, 88–96.

Guenthner, E., Carlson, C.W. and Emerick, R.J. (1978) Copper salts for growth stimulation and alleviation of aortic rupture losses in turkeys. *Poultry Science* 57, 1313–1324.

Gummow, B., Botha, C.J., Basson, C.J. and Bastianello, S.S. (1991) Copper toxicity in ruminants: air pollution as a possible cause. *Onderstepoort Journal of Veterinary Research* 58, 33–39.

Hannam, R.J. and Reuter, D.J. (1977) The occurrence of steely wool in South Australia, 1972–75. *Agricultural Record* 4, 26–29.

Hart, E.B., Steenbock, H., Waddell, J. and Elvehjem, C.A. (1928) Iron in nutrition. 7. Copper as a supplement to iron for hemoglobin building in the rat. *Journal of Biological Chemistry* 77, 797–812.

Hartman, F. and van Ryssen, J.B.J. (1997) Metabolism of selenium and copper in sheep with and without sodium bicarbonate supplementation. *Journal of Agricultural Science, Cambridge* 128, 357–364.

Hartmans, J. (1975) The frequency of occurrence of copper poisoning and the role of sheep concentrates in it merits enquiry. *Tijdschrift Diergeneeskunde* 100, 379–382.

Hartmans, J. and Bosman, M.S.M. (1970) Differences in the copper status of grazing and housed cattle and their biochemical backgrounds. In: Mills, C.F. (ed.) *Trace Element Metabolism in Animals.* E. & S. Livingstone, Edinburgh, pp. 362–366.

Hassell, C.A., Marchello, J.A. and Lei, K.Y. (1983) Impaired glucose tolerance in copper-deficient rats. *Journal of Nutrition* 113, 1081–1083.

Haynes, R.J. (1997) Micronutrient status of a group of soils in Canterbury, New Zealand, as measured by extraction with EDTA, DTPA and HCl, and its relationship with plant response to applied Cu and Zn. *Journal of Agricultural Science, Cambridge* 129, 325–333.

Hedger, R.S., Howard, D.A. and Burdin, M.L. (1964) The occurrence in goats and sheep in Kenya of a disease closely similar to swayback. *Veterinary Record* 76, 493–497.

Herbert, E., Small, J.N.W., Jones, D.G. and Suttle, N.F. (1991) Evaluation of superoxide dismutase assays for the routine diagnostic assessment of copper status in blood samples. In: Momcilovic, B. (ed.) *Proceedings of the Seventh International Symposium on Trace Elements in Man and Animals, Dubrovnik.* IMI, Zagreb, pp. 5-15–5-16.

Hill, C.H., Starcher, B. and Kim, C. (1967) Role of copper in the formation of elastin. *Federation Proceedings* 26, 129–133.

Hogan, K.G., Money, D.F.L., White, D.A. and Walker, R. (1971) Weight responses of young lambs to copper and connective tissue lesions associated with grazing pasture of high molybdenum content. *New Zealand Journal of Agricultural Research* 14, 687–701.

Holstein, T.J., Fung, R.Q., Quevedo, W.C. and Bienieki, T.C. (1979) Effect of altered copper metabolism induced by mottled alleles and diet on mouse tyrosinase. *Proceedings of the Society for Experimental Biology and Medicine* 162, 264–268.

Hopkins, R.G. and Failla, M.L. (1995) Chronic intake of a marginally low copper diet impairs *in vitro* activities of lymphocytes and neutrophils from male rats despite minimal impact on conventional indicators of copper status. *Journal of Nutrition* 125, 2658–2668.

Howell, J.M. and Davison, A.N. (1959) The copper content and cytochrome oxidase activity of tissues from normal and swayback lambs. *Biochemical Journal* 72, 365–368.

Howell, J.M. and Gooneratne, S.R. (1987) The pathology of copper toxicity. In: Howell, J.McC. and Gawthorne, J.M. (eds) *Copper in Animals and Man,* Vol. II. CRC Press, Boca Raton, Florida, pp. 53–78.

Howell, J.M. and Hall, G.A. (1970) Infertility associated with experimental copper deficiency in sheep, guinea-pigs and rats. In: Mills, C.F. (ed.) *Trace Element Metabolism in Animals.* E. & S. Livingstone, Edinburgh, pp. 106–109.

Howell, J.M., Davison, A.N. and Oxberry, J. (1964) Biochemical and neuropathological changes in swayback. *Research in Veterinary Science* 5, 376–384.

Huber, J.T. and Price, N.O. (1971) Influence of high dietary calcium and phosphorus and Ca:P ratio on liver copper and iron stores in lactating cows. *Journal of Dairy Science* 54, 429–432.

Humphries, W.R. (1980) Control of hypocupraemia in cattle by addition of copper to water supplies. *Veterinary Record* 106, 359–362.

Humphries, W.R., Phillippo, M., Young, B.W. and Bremner, I. (1983) The influence of dietary iron and molybdenum on copper metabolism in calves. *British Journal of Nutrition* 49, 77–86.

Humphries, W.R., Morrice, P.C. and Bremner, I. (1988) A convenient method for the treatment of chronic copper poisoning in sheep using subcutaneous ammonium tetrathiomolybdate. *Veterinary Record* 123, 51–53.

Hunter, A.P. (1977) Some nutritional factors affecting the fertility of dairy cattle. *New Zealand Veterinary Journal* 25, 305–307.

Hussein, K.S.M. (1985) Copper toxicity in small ruminants – interaction of selenium and the role of metallothionein. PhD thesis, Swedish University of Agricultural Sciences, Uppsala.

Ingraham, R.H., Kappel, L.C., Morgan, E.B. and Srikandakumar, A. (1987) Correction of subnormal fertility with copper and magnesium supplementation. *Journal of Dairy Science* 70, 167–180.

Ivan, M.M. (1988) The effect of faunation of rumen on solubility and liver content of copper in sheep fed low or high copper diets. *Journal of Animal Science* 66, 1498–1501.

Jamieson, S. and Allcroft, R. (1950) Copper pine of calves. *British Journal of Nutrition* 4, 16–31.

Jarvis, S.C. and Austin, A.R. (1983) Soil and plant factors limiting the availability of copper to a beef suckler herd. *Journal of Agricultural Science, Cambridge* 101, 39–46.

Jenkins, K.J. and Hidiroglou, M. (1989) Tolerance of the calf for excess copper in milk replacer. *Journal of Dairy Science* 72, 150–156.

Jenkins, N.K., Morris, T.R. and Valamotis, D. (1970) The effect of diet and copper supplementation on chick growth. *British Poultry Science* 11, 241–248.

Jensen, L.S. (1975) Precipitation of a selenium deficiency by high dietary levels of copper and zinc (in fowls). *Proceedings of the Society for Experimental Biology and Medicine* 149, 113–116.

Jensen, L.S. and Maurice, V. (1979) Influence of sulfur amino acids on copper toxicity in chicks. *Journal of Nutrition* 109, 91–97.

Jones, D.G. (1984) Effects of dietary copper depletion on acute and delayed inflammatory responses in mice. *Journal of Comparative Pathology* 37, 205–210.

Jones, D.G. and Suttle, N.F. (1981) Some effects of copper deficiency on leucocyte function in sheep and cattle. *Research in Veterinary Science* 31, 151–156.

Jumba, I.O., Suttle, N.F., Hunter, E.A. and Wandiga, S.O. (1995) Effects of soil origin and herbage species on mineral composition of forages in the Mount Elgon region of Kenya. 2. Trace elements. *Tropical Grasslands* 29, 47–52.

Kashani, A.B., Samie, H., Emerick, R.J. and Carlson, C.W. (1986) Effect of copper with three levels of sulfur-containing amino acids in diets for turkeys. *Poultry Science* 65, 1754–1759.

Kegley, E.B. and Spears, J.W. (1993) Bioavailability of feed grade copper sources (oxide, sulphate or lysine) in growing cattle. *Journal of Animal Science* 72, 2728–2734.

Kerr, L.A. and McGavin, H.D. (1991) Chronic copper poisoning in sheep grazing pastures fertilized with swine manure. *Journal of the American Veterinary Medical Association* 198, 99–101.

Kincaid, R.L., Blauwiekel, R.M. and Gonrath, J.D. (1986) Supplementation of copper as copper sulphate and copper proteinate for growing calves fed forages containing molybdenum. *Journal of Dairy Science* 69, 160–163.

King, T.P. and Bremner, I. (1979) Autophagy and apoptosis in liver during the prehaemolytic phase of chronic copper poisoning in sheep. *Journal of Comparative Pathology* 89, 515–530.

Kirchgessner, M. and Giessler, H. (1961) Der Einfluss einer CuSO4-Zulage auf den N-Ansatz wachsender Schweine. *Zeitschrift für Tierphysiologie, Tierernahrung und Futtermittelkunde* 16, 297–300.

Kirchgessner, M., Beyer, M.G. and Steinhart, H. (1976) Activation of pepsin (EC3.4.4.1) by heavy-metal ions including a contribution to the mode of action of copper sulphate in pig nutrition. *British Journal of Nutrition* 36, 15–22.

Koh, T.-S. and Judson, G.J. (1987) Copper and selenium deficiency in cattle: an evaluation of methods of oral therapy and an observation of a copper–selenium interaction. *Veterinary Record Communications* 11, 133–148.

Lai, C.-C., Huang, W.-H., Askari, A., Wang, Y., Sarvazyan, N., Klevay, L.M. and Chin, T.H. (1994) Differential regulation of superoxide dismutase in copper deficient rat organs. *Free Radical Biology and Medicine* 16, 613–620.

Langlands, J.P., Bowles, J.E., Donald, G.E., Smith, A.J. and Paull, D.R. (1981) Copper status of sheep grazing pastures fertilized with sulphur and molybdenum. *Australian Journal of Agricultural Research* 32, 479–486.

Lannon, B. and Mason, J. (1986) The inhibition of bovine ceruloplasmin oxidase activity *in vivo* and *in vitro*: a reversible interaction. *Journal of Inorganic Biochemistry* 26, 107–115.

Leach, R.M., Jr, Rosenblum, C.I., Amman, M.J. and Burdette, J. (1990) Broiler chicks fed low-calcium diets. 2. Increased sensitivity to copper toxicity. *Poultry Science* 69, 1905–1910.

Lee, H.J. (1956) The influence of copper deficiency on the fleece of British breeds of sheep. *Journal of Agricultural Science, Cambridge* 47, 218–244.

Lei, R.Y. (1991) Dietary copper: cholesterol and lipoprotein metabolism. *Annual Reviews in Nutrition* 11, 265–283.

Leigh, L.C. (1975) Changes in the ultrastructure of cardiac muscle in steers deprived of copper. *Research in Veterinary Science* 18, 282–287.

Lewis, G., Terlecki, S. and Allcroft, R. (1967) The occurrence of swayback in the lambs of ewes fed a semi-purified diet of low copper content. *Veterinary Record* 81, 415–416.

Lewis, G., Terlecki, S., Parker, B.N.J. and Don, P.L. (1981) Prevention of delayed swayback in lambs. In: Howell, J.McC., Gawthorne, J.M. and White, C.L. (eds) *Proceedings of the Second International Symposium on Trace Element Metabolism in Man and Animals*. Australian Academy of Science, Canberra, pp. 291–293.

Littledike, E.T., Wittam, T.E. and Jenkins, T.G. (1995) Effect of breed, intake and carcass composition on the status of several macro and trace minerals of adult beef cattle. *Journal of Animal Science* 73, 2113–2119.

McArdle, H. (1992) The transport of iron and copper across the cell membrane: different mechanisms for different metals? *Proceedings of the Nutrition Society* 51, 199–209.

McFarlane, J.D. (1989) The effect of copper supply on vegetative and seed yield of pasture legumes and the field calibration of a test for detecting copper deficiency. I. Subterranean clover. *Australian Journal of Agricultural Research* 40, 817–832.

McFarlane, J.D., Judson, J.D. and Gouzos, J. (1990) Copper deficiency in ruminants in the South East of Australia. *Australian Journal of Experimental Agriculture* 30, 187–193.

McFarlane, J.D., Judson, J.D., Turnbull, R.K. and Kempe, B.R. (1991) An evaluation of copper-containing soluble glass pellets, copper oxide particles and injectable copper as supplements for cattle and sheep. *Australian Journal of Experimental Agriculture* 31, 165–174.

McMurray, C.H. (1980) Copper deficiency in ruminants. In: *Biological Roles of Copper.* Ciba Foundation 79 (New Series), Elsevier, New York, pp. 183–207.

MacPherson, A. and Hemingway, R.G. (1968) Effects of liming and various forms of oral copper supplementation on the copper status of grazing sheep. *Journal of the Science of Food and Agriculture* 19, 53–58.

MacPherson, A. and Hemingway, R.G. (1969) The relative merit of various blood analyses and liver function tests in giving an early diagnosis of chronic copper poisoning in sheep. *British Veterinary Journal* 125, 213–221.

Marcilese, N.A., Ammerman, C.B., Valsecchi, R.M., Dienavant, B.G. and Davis, G.K. (1970) Effect of dietary molybdenum and sulphate upon urinary excretion of copper in sheep. *Journal of Nutrition* 100, 1399–1405.

Marston, H.R. (1946) Nutrition and wool production. In: *Proceedings of a Symposium on Fibrous Proteins.* Society of Dyers and Colourists, Leeds, UK.

Mason, J. (1986) Thiomolybdates: mediators of molybdenum toxicity and enzyme inhibitors. *Toxicology* 42, 99–109.

Mason, J., Williams, S., Harrington, R. and Sheahan, B. (1984) Some preliminary studies on the metabolism of 99Mo-labelled compounds in deer. *Irish Veterinary Journal* 38, 171–175.

Mason, J., Woods, M. and Poole, D.B.R. (1986) Accumulation of copper and albumin *in vivo* after intravenous trithiomolybdate administration. *British Veterinary Journal* 41, 108–113.

Mee, J.F. (1991) Coat colour and copper deficiency in cattle. *Veterinary Record* 129, 536.

Mills, C.F. (1954) Copper complexes in grassland herbage. *Biochemical Journal* 57, 603–610.

Mills, C.F. (1956a) Studies of the copper compounds in aqueous extracts of herbage. *Biochemical Journal* 63, 187–190.

Mills, C.F. (1956b) The dietary availability of copper in the form of naturally occurring organic complexes. *Biochemical Journal* 63, 190–193.

Mills, C.F. and Williams, R.B. (1962) Copper concentration and cytochrome-oxidase and ribonuclease activities in the brains of copper-deficient lambs. *Biochemical Journal* 85, 629–632.

Mills, C.F., Dalgarno, A.C. and Wenham, G. (1976) Biochemical and pathological changes in tissues of Friesian cattle during the experimental induction of copper deficiency. *British Journal of Nutrition* 35, 309–311.

Miltimore, J.E. and Mason, J.L. (1971) Copper to molybdenum ratio and molybdenum and copper concentrations in ruminant feeds. *Canadian Journal of Animal Science* 51, 193–200.

Miltimore, J.E., Kalmin, C.M. and Clapp, J.B. (1978) Copper storage in the livers of cattle supplemented with injected copper and with copper sulphate and chelated copper. *Canadian Journal of Animal Science* 58, 525–530.

Minson, D.J. (1990) Copper. In: *Forage in Ruminal Nutrition.* Academic Press, Sydney, pp. 316–324.

Moffor, F.M. and Rodway, R.G. (1991) The effect of tetrathiomolybdate on growth rate and onset of puberty in ewes lambs. *British Veterinary Record* 147, 421–431.

Mulhern, S.A. and Koller, L.D. (1988) Severe or marginal copper deficiency results in a graded reduction of the immune status in mice. *Journal of Nutrition* 118, 1041–1047.

Mylrea, P.J. and Byrne, D.T. (1974) An outbreak of acute copper poisoning in cattle. *Australian Veterinary Journal* 50, 169–172.

Neal, W.M., Becker, R.B. and Shealy, A.L. (1931) A natural copper deficiency in cattle rations. *Science* 74, 418–419.

NRC (1994) *Nutrient Requirements of Poultry*, 9th edn. National Academy of Sciences, Washington, DC, pp. 20–36.

O'Dell, B.L. (1976) Biochemistry and physiology of copper in vertebrates. In: Prasad, A.S. (ed.) *Trace Elements in Human Health and Disease*, Vol. I. Academic Press, New York, pp. 391–413.

O'Dell, B.L., Hardwick, B.C., Reynolds, G. and Savage, J.E. (1961) Connective tissue defect in the chick resulting from copper deficiency. *Proceedings of the Society for Experimental Biology and Medicine* 108, 402–405.

O'Dell, B.L., Smith, R.M. and King, R.A. (1976) Effect of copper status on brain neurotransmitter metabolism in the lamb. *Journal of Neurochemistry* 26, 451–455.

Okonkwo, A.C., Ku, P.K., Miller, E.R., Keahey, K.K. and Ullrey, D.E. (1979) Copper requirement of baby pigs fed purified diets. *Journal of Nutrition* 109, 939–948.

Opsahl, W., Zeronian, H., Ellison, M., Lewis, D., Rucker, R.B. and Riggins, R.S. (1982) Role of copper in collagen cross-linking and its influence on selected mechanical properties of chick bone and tendon. *Journal of Nutrition* 112, 708–716.

Ortolani, E., Knox, D.P., Jackson, F., Coop, R.L. and Suttle, N.F. (1993) Abomasal parasitism lowers liver copper status and influences the Cu × Mo × S antagonism in lambs. In: Anke, M., Meissner, D. and Mills, C.F. (eds) *Proceedings of the Eighth International Symposium on Trace Elements in Man and Animals, Dresden.* Verlag Media Touristik, Gersdorf, GDR, pp. 331–332.

Parkins, J.J., Hemingway, R.G., Lawson, D.C. and Ritchie, N.S. (1994) The effectiveness of copper oxide powder as a component of a sustained-release multi-trace element and vitamin rumen bolus system for cattle. *British Veterinary Journal* 150, 547–553.

Paynter, D.J. (1987) The diagnosis of copper insufficiency. In: Howell, J.McC. and Gawthorne, J.M. (eds) *Copper in Animals and Man*, Vol. I. CRC Press, Boca Raton, Florida, pp. 101–119.

Phillippo, M. (1983) The role of dose response trials in predicting trace element deficiency disorders. Suttle, N.F., Gunn, R.G., Allen, W.M., Winklater, K.A. and Wiener, G. (eds). In: *Trace Elements in Animal Production and Veterinary Practice*. Occasional Publication of the British Society of Animal Production No. 7, Edinburgh, pp. 51–59.

Phillippo, M. and Graca, D.S. (1983) Biliary copper secretion in cattle. *Proceedings of the Nutrition Society* 42, 46A.

Phillippo, M., Humphries, W.R., Lawrence, C.B. and Price, J. (1982) Investigation of the effect of copper therapy on fertility in beef suckler herds. *Journal of Agricultural Science, Cambridge* 99, 359–364.

Phillippo, M., Humphries, W.R. and Garthwaite, P.H. (1987a) The effect of dietary molybdenum and iron on copper status and growth in cattle. *Journal of Agricultural Science, Cambridge* 109, 315–320.

Phillippo, M., Humphries, W.R., Atkinson, T., Henderson, G.D. and Garthwaite, P.H. (1987b) The effect of dietary molybdenum and iron on copper status, puberty, fertility and oestrus cycles in cattle. *Journal of Agricultural Science, Cambridge* 109, 321–336.

Pitt, M., Fraser, J. and Thurley, D.C. (1980) Molybdenum toxicity in the sheep: epiphysiolysis, exostoses and biochemical changes. *Journal of Comparative Pathology* 90, 567–576.

Price, J. and Chesters, J.K. (1985) A new bioassay for assessment of copper bioavailability and its application in a study of the effect of molybdenum on the distribution of available copper in ruminant digesta. *British Journal of Nutrition* 53, 323–336.

Price, J., Will, M.A., Paschaleris, G. and Chesters, J.K. (1987) Identification of thiomolybdates in digesta and plasma from sheep after administration of ^{99}Mo-labelled compounds into the rumen. *British Journal of Nutrition* 58, 127–138.

Pritchard, G.C., Lewis, G., Wells, G.A.H. and Stopforth, A. (1985) Zinc toxicity, copper deficiency and anaemia in swill-fed pigs. *Veterinary Record* 117, 545–548.

Prohaska, J.R. and Bailey, W.R. (1995) Alterations of rat peptidylglycine α-amidating monooxygenase and other cuproenzyme activities following perinatal copper deficiency. *Proceedings of the Society for Experimental Biology and Medicine* 210, 107–116.

Puls, R. (1994) *Mineral Levels in Animal Health*, 2nd edn. Diagnostic data, Sherpa International, Clearbook, BC, pp. 83–109.

Reetz, I., Wegner, W. and Feder, H. (1975) Statistics heritability and correlations between several characters of the circulatory system in female German Landrace fatteners. II. Degree of heritability and gene frequencies. *Zentrabl Veterinar Reche* A22, 741–750.

Roberts, H.E., Williams, B.M. and Harvard, A. (1966) Cerebral oedema in lambs associated with hypocuprosis and its relationship to swayback. II. Histopathological findings. *Journal of Comparative Pathology* 76, 285–290.

Rosa, I.V., Ammerman, C.B. and Henry, P.R. (1986) Interrelationships of dietary copper, zinc and iron on performance and tissue mineral concentration in sheep. *Nutrition Reports International* 34, 893–902.

Rucker, R.B., Parker, H.E. and Rogler, J.C. (1969) Effect of copper deficiency on chick bone collagen and selected bone enzymes. *Journal of Nutrition* 98, 57–63.

Rucker, R.B., Riggins, R.S., Laughlin, R., Chan, M.M., Chen, M. and Tom, K. (1975) Effect of nutritional copper deficiency on the biomechanical properties of bone and arterial elastin metabolism in the chick. *Journal of Nutrition* 105, 1062–1070.

Rucker, R.B., Romero-Chapman, N., Wong, T., Lee, J., Steinberg, F.M., McGee, C., Clegg, M.S., Reiser, K., Kosonen, T., Uriu-Hare, J.Y., Murphy, J. and Keen, C.L. (1996) Modulation of lysyl oxidase by dietary copper in the rat. *Journal of Nutrition* 126, 51–60.

Saenko, E.L., Yaroplov, A.I. and Harris, E.D. (1994) Biological functions of caeruloplasmin expressed through copper-binding sites. *Journal of Trace Elements in Experimental Medicine* 7, 69–88.

Savage, J.E. (1968) Trace minerals and avian reproduction. *Federation Proceedings* 27, 927–931.

Savage, J.E., Bird, D.W., Reynolds, G. and O'Dell, B.L. (1966) Comparison of copper deficiency and lathyrism in turkey poults. *Journal of Nutrition* 88, 15–25.

Schuschke, D.A., Saari, J.T., West, C.A. and Miller, F.N. (1994) Dietary copper deficiency increases the mast cell population of the rat. *Proceedings of the Society for Experimental Biology and Medicine* 207, 274–277.

Shand, A. and Lewis, G. (1957) Chronic copper poisoning in young calves. *Veterinary Record* 69, 618–620.

Simpson, A.M., Mills, C.F. and McDonald, I. (1981) Tissue copper retention or loss in young growing cattle. In: Howell, J.McC., Gawthorne, J.M. and White, C.L. (eds) *Proceedings of the Fourth International Symposium on Trace Element Metabolism in Man and Animals.* Australian Academy of Sciences, Canberra, pp. 133–136.

Sjollema, B. (1933) Kupfermangel als Ursache von Krankheiten bei Pflanzen und Tieren. *Biochemische Zeitschrift* 267, 151–156.

Smart, M.E., Cohen, R., Christensen, D.A. and Williams, C.M. (1986) The effects of sulphate removal from the drinking water on the plasma and liver copper and zinc concentrations of beef cows and their calves. *Canadian Journal of Animal Science* 66, 669–680.

Smith, B. and Coup, M.R. (1973) Hypocuprosis: a clinical investigation of dairy herds in Northland. *New Zealand Veterinary Journal* 21, 252–258.

Smith, B.S.W. and Wright, H. (1975a) Effect of dietary molybdenum on copper metabolism: evidence for the involvement of molybdenum in abnormal binding of Cu to plasma proteins. *Clinica Chimica Acta* 62, 55–62.

Smith, B.S.W. and Wright, H. (1975b) Copper:molybdenum interactions: effects of dietary molybdenum on the binding of copper to plasma proteins in sheep. *Journal of Comparative Pathology* 85, 299–305.

Smith, B.S.W., Field, A.C. and Suttle, N.F. (1968) Effect of intake of copper, molybdenum and sulphate on copper metabolism in the sheep. III. Studies with radioactive copper in male castrated sheep. *Journal of Comparative Pathology* 78, 449–461.

Smith, R.M., Osborne-White, W.S. and O'Dell, B.L. (1976) Cytochromes in brain mitochondria from lambs with enzootic ataxia. *Journal of Neurochemistry* 26, 1145–1148.

Spais, A.G. (1959) *Askeri veteriner dergisi* 135, 161.

Stabel, J.R., Spears, J.W. and Brown, T.T. (1993) Effect of copper deficiency on tissue, blood characteristics and immune function of calves challenged with Infectious Bovine Rhinotracheitis Virus and *Pasteurella haemolytica*. *Journal of Animal Science* 71, 1247–1255.

Strickland, K., Smith, F., Woods, M. and Mason, J. (1987) Dietary molybdenum as a putative copper antagonist in the horse. *Equine Veterinary Journal* 19, 50–54.

Suleiman, A., Okine, E. and Goonewardene, L.A. (1997) Relevance of National Research Council feed composition tables in Alberta. *Canadian Journal of Animal Science* 77, 197–203.

Sutherland, A.K., Moule, G.R. and Harvey, J.M. (1955) On the toxicity of copper aminoacetate injection and copper sulphate drench for sheep. *Australian Veterinary Journal* 31, 141–145.

Suttle, N.F. (1974) Effects of organic and inorganic sulphur on the availability of dietary copper to sheep. *British Journal of Nutrition* 32, 559–568.

Suttle, N.F. (1975a) Effects of age and weaning on the apparent availability of dietary copper to young lambs. *Journal of Agricultural Science, Cambridge* 84, 255–261.

Suttle, N.F. (1975b) Trace element interactions in animals. In: Nicholas, D.J.D. and Egan, A.R. (eds) *Trace Elements in Soil–Plant–Animal Systems*. Academic Press, New York, pp. 271–289.

Suttle, N.F. (1977) Reducing the potential toxicity of concentrates to sheep by the use of molybdenum and sulphur supplements. *Animal Feed Science and Technology* 2, 235–246.

Suttle, N.F. (1978) Determining the copper requirements of cattle by means of an intravenous repletion technique. In: Kirchgessner, M. (ed.) *Proceedings of the Third International Symposium on Trace Element Metabolism in Man and Animals*. Arbeitskreis für Terernährungsforschung, pp. 473–480.

Suttle, N.F. (1981a) Effectiveness of orally administered cupric oxide needles in alleviating hypocupraemia in sheep and cattle. *Veterinary Record* 108, 417–420.

Suttle, N.F. (1981b) Comparison between parenterally administered copper complexes of their ability to alleviate hypocupraemia in sheep and cattle. *Veterinary Record* 109, 304–307.

Suttle, N.F. (1983a) Assessing the mineral and trace element status of feeds. In: Robards, G.E. and Packham, R.G. (eds) *Proceedings of the Second Symposium of the International Network of Feed Information Centres.* Commonwealth Agricultural Bureaux, Farnham Royal, UK, pp. 211–237.

Suttle, N.F. (1983b) Effects of molybdenum concentration in fresh herbage, hay and semi-purified diets on the copper metabolism of sheep. *Journal of Agricultural Science, Cambridge* 100, 651–656.

Suttle, N.F. (1987a) The nutritional requirement for copper in animals and man. In: Howell, J.McC. and Gawthorne, J.M. (eds) *Copper in Animals and Man*, Vol. I. CRC Press, Boca Raton, Florida, pp. 21–44.

Suttle, N.F. (1987b) Safety and effectiveness of cupric oxide particles for increasing liver copper stores in sheep. *Research in Veterinary Science* 42, 219–223.

Suttle, N.F. (1988) The role of comparative pathology in the study of copper and cobalt deficiencies in ruminants. *Journal of Comparative Pathology* 99, 241–258.

Suttle, N.F. (1991) The interactions between copper, molybdenum and sulphur in ruminant nutrition. *Annual Review of Nutrition* 11, 121–140.

Suttle, N.F. (1993) Overestimation of copper deficiency. *Veterinary Record* 133, 123–124.

Suttle, N.F. (1994) Meeting the copper requirements of ruminants. In: Garnsworthy, P.C. and Cole, D.J.A. (eds) *Recent Advances in Animal Nutrition.* Nottingham University Press, Nottingham, pp. 173–188.

Suttle, N.F. (1996) Non-dietary influences on mineral requirements of sheep. In: Masters, D.G. and White, C.L. (eds) *Detection and Treatment of Mineral Nutrition Problems in Sheep.* ACIAR Monograph 37, Canberra, pp. 31–44.

Suttle, N.F. and Angus, K.W. (1976) Experimental copper deficiency in the calf. *Journal of Comparative Pathology* 86, 595–608.

Suttle, N.F. and Angus, K.W. (1978) Effects of experimental copper deficiency on the skeleton of the calf. *Journal of Comparative Pathology* 88, 135–145.

Suttle, N.F. and Brebner, J. (1995) A putative role for larval nematode infection in diarrhoeas of lambs which do not respond to anthelmintic drenches. *Veterinary Record* 137, 311–316.

Suttle, N.F. and Brebner, J. (1996) A comparison of the availability of copper in copper : lysine and copper sulphate for sheep. *Animal Science* 62, 690.

Suttle, N.F. and Field, A.C. (1968) Effect of intake of copper, molybdenum and sulphate on copper metabolism in sheep. 1. Clinical condition and distribution of copper in blood of the pregnant ewe. *Journal of Comparative Pathology* 78, 351–362.

Suttle, N.F. and Field, A.C. (1969) Effect of intake of copper, molybdenum and sulphate on copper metabolism in sheep. 4. Production of congenital and delayed swayback. *Journal of Comparative Pathology* 79, 453–464.

Suttle, N.F. and Field, A.C. (1983) Effects of dietary supplements of thiomolybdates on copper and molybdenum metabolism in sheep. *Journal of Comparative Pathology* 93, 379–389.

Suttle, N.F. and Jones, D.G. (1986) Copper and disease resistance in sheep: a rare natural confirmation of interaction between a specific nutrient and infection. *Proceedings of the Nutrition Society* 45, 317–325.

Suttle, N.F. and McMurray, C.H. (1983) Use of erythrocyte copper:zinc superoxide dismutase activity and hair or fleece concentrations in the diagnosis of hypocuprosis in ruminants. *Research in Veterinary Science* 35, 47–52.

Suttle, N.F. and Mills, C.F. (1966a) Studies of the toxicity of copper to pigs. 1. Effects of oral supplements of zinc and iron salts on the development of copper toxicosis. *British Journal of Nutrition* 20, 135–148.

Suttle, N.F. and Mills, C.F. (1966b) Studies of the toxicity of copper to pigs. 2. Effect of protein source and other dietary components on the response to high and moderate intakes of copper. *British Journal of Nutrition* 20, 149–161.

Suttle, N.F. and Peter, D.W. (1985) Rumen sulphide metabolism as a major determinant of the availability of copper to ruminants. In: Mills, C.F., Bremner, I. and Chesters, J.K. (eds) *Proceedings of the Fifth International Symposium on Trace Elements in Man and Animals, Aberdeen.* Commonwealth Agricultural Bureaux, Farnham Royal, UK, pp. 367–370.

Suttle, N.F. and Price, J. (1976) The potential toxicity of copper-rich animal excreta to sheep. *Animal Production* 23, 233–241.

Suttle, N.F. and Small, J.N. (1993) Evidence for delayed availability of copper in supplementation trials with lambs on molybdenum-rich pastures. In: Anke, M., Meissner, D. and Mills, C.F. (eds) *Proceedings of the Eighth International Symposium on Trace Elements in Man and Animals.* Verlag Media Touristik, Gersdoff, pp. 651–655.

Suttle, N.F., Field, A.C. and Barlow, R.M. (1970) Experimental copper deficiency in sheep. *Journal of Comparative Pathology* 80, 151–162.

Suttle, N.F., Angus, K.W., Nisbet, D.I. and Field, A.C. (1972) Osteoporosis in copper-depleted lambs. *Journal of Comparative Pathology* 82, 93–97.

Suttle, N.F., Field, A.C., Nicolson, T.B., Mathieson, A.O., Prescott, J.H.D., Scott, N. and Johnson, W.S. (1980) Some problems in assessing the physiological and economic importance of hypocupraemia in beef suckler herds. *Veterinary Record* 106, 302–304.

Suttle, N.F., Abrahams, P. and Thornton, I. (1984) The role of a soil × dietary sulphur interaction in the impairment of copper absorption by soil ingestion in sheep. *Journal of Agricultural Science, Cambridge* 103, 81–86.

Suttle, N.F., Jones, D.G., Woolliams, C. and Woolliams, J.A. (1987a) Heinz body anaemia in lambs with deficiencies of copper or selenium. *British Journal of Nutrition* 58, 539–548.

Suttle, N.F., Jones, D.G., Woolliams, J.A. and Woolliams, C. (1987b) Copper supplementation during pregnancy can reduce perinatal mortality and improve early growth in lambs. *Proceedings of the Nutrition Society* 46, 68A.

Suttle, N.F., Brebner, J., Small, J.N. and McLean, K. (1992a) Inhibition of ovine erythrocyte superoxide dismutase activity (ESOD; EC1.14.1.1) *in vivo* by parenteral ammonium tetrathiomolybdate. *Proceedings of the Nutrition Society* 51, 145A.

Suttle, N.F., Knox, D.P., Angus, K.W., Jackson, F. and Coop, R.L. (1992b) The effects of dietary molybdenum on nematode and host during *Haemonchus contortus* infection in lambs. *Research in Veterinary Science* 52, 230–235.

Suttle, N.F., Small, J.N.W., Jones, D.G. and Watkins, K.L. (1997) Do horses and other non-ruminants require different standards of normality from ruminants when assessing copper status. In: Fischer, P.W.F., L'Abbé, M.R., Cockell, K.A. and Gibson, R.S. (eds) *Proceedings of the Ninth International Symposium on Trace Elements in Man and Animals.* NRC Research Press, Ottawa, Canada, pp. 134–136.

Suttle, N.F., Brebner, J. and Pass, R. (1996) A comparison of the availability of copper in four whisky distillery by-products with that in copper sulphate for lambs. *Animal Science* 62, 689–690.

Suttle, N.F., MacPherson, A., Phillips, P. and Wright, C.C. (1999) The influence of trace elements status on the pre-weaning growth of lambs on improved hill pastures in Scotland. I. Copper, molybdenum, sulphur and iron. *Journal of Agricultural Science, Cambridge* (in press).

Szalay, A., Samsoni, Z. and Szilagyi, M. (1975) Manganese and copper deficiency of plants as a characteristic defect of lowmoor peat soils. *Zeitschrift für Pflanzenernahrung und Bodenkunde* 4/5, 447–458.

Taylor, C.G., Bettger, W.J. and Bray, T.M. (1988) Effect of dietary zinc or copper deficiency on the primary free radical defence system in rats. *Journal of Nutrition* 118, 613–621.

Telfer, S.B., MacKenzie, A.M., Illingworth, D.V. and Jackson, D.W. (1996) The use of caeruloplasmin activities and plasma copper concentrations as indicators of copper status in cattle. In: *Proceedings of the XIX World Buiatrics Conference, Edinburgh*, Vol. II. British Cattle Veterinary Association, Frampton-on-Severn, pp. 402–404.

Thompson, K.G., Audige, L., Arthur, D.G., Juhan, A.F., Orr, M.B., McSporran, K.D. and Wilson, P.R. (1994) Osteochondrosis associated with copper deficiency in young farmed red deer and wapiti × red deer hybrids. *New Zealand Veterinary Journal* 42, 137–143.

Thompson, R.H. and Todd, J.R. (1974) Muscle damage in chronic copper poisoning of sheep. *Research in Veterinary Science* 16, 97–99.

Thornton, I. and Alloway, B.J. (1974) Geochemical aspects of the soil–plant–animal relationship in the development of trace element deficiency and excess. *Proceedings of the Nutrition Society* 33, 257–266.

Thornton, I., Kershaw, G.F. and Davies, M.K. (1972a) An investigation into copper deficiency in cattle in the southern Pennines. 1. Identification of suspect areas using geochemical reconnaissance followed by blood copper surveys. *Journal of Agricultural Science, Cambridge* 78, 157–163.

Thornton, I., Kershaw, G.F. and Davies, M.K. (1972b) An investigation into copper deficiency in cattle in the southern Pennines. 2. Response to copper supplementation. *Journal of Agricultural Science, Cambridge* 78, 165–171.

Todd, J.R. (1970) A survey of the copper status of cattle using copper oxidase (caeruloplasmin) activity of blood serum. In: Mills, C.F. (ed.) *Trace Element Metabolism in Animals*. E. & S. Livingstone, Edinburgh, pp. 448–451.

Todd, J.R. (1972) Copper, molybdenum and sulphur contents of oats and barley in relation to chronic copper poisoning in housed sheep. *Journal of Agricultural Science, Cambridge* 79, 191–195.

Tremblay, R.R.M. and Baird, J.D. (1991) Chronic copper poisoning in two Holstein cows. *Cornell Veterinarian* 81, 205–213.

Underwood, E.J. (1977) *Trace Elements in Human and Animal Nutrition*, 4th edn. Academic Press, New York, 545 pp.

Underwood, E.J., Robinson, T.J. and Curnow, D.H. (1943) The influence of topdressing with copper sulphate on the copper content and the yield of mixed pasture at Gingin. *Journal of the Department of Agriculture of Western Australia* 20, 80–87.

van der Berg, R., Levels, F.H.R. and van der Schee, W. (1983) Breed differences in sheep with respect to the accumulation of copper in the liver. *Veterinary Quarterly* 5, 26–31.

van Ryssen, J.B.J. and Barrowman, P.R. (1987) Effect of ionophores on the accumulation of copper in the livers of sheep. *Animal Production* 44, 255–261.

van Ulsen, F.W. (1972) Schapen, varkens en koper. *Tijdschrift voor Diergeneeskunde* 97, 735–738.

Vermunt, J.J. and West, D.M. (1994) Predicting copper status in beef cattle using serum copper concentrations. *New Zealand Veterinary Journal* 42, 194–195.

Wang, Y.R., Wu, J.Y.J., Reaves, S.K. and Lei, K.Y. (1996) Enhanced expression of hepatic genes in copper-deficient rats detected by the messenger RNA differential display method. *Journal of Nutrition* 126, 1772–1781.

Wang, Z.Y., Poole, D.B.R. and Mason, J. (1988) The effects of supplementation of the diet of young steers with Mo and S on the intracellular distribution of copper in liver and on copper fractions in blood. *British Veterinary Journal* 144, 543–551.

Ward, J.D., Spears, J.W. and Kegley, E.B. (1993) Effect of copper level and source (copper lysine vs copper sulphate) on copper status, performance and immune responses in growing steers fed diets with or without supplemental molybdenum and sulphur. *Journal of Animal Science* 71, 2748–2755.

Ward, T.L., Watkins, K.L. and Southern, L.L. (1994) Interactive effects of dietary copper and water copper levels on growth, water uptake and plasma and liver copper concentrations of poults. *Poultry Science* 73, 1306–1311.

Ward, J.D., Spears, J.W. and Kegley, E.B. (1996) Bioavailability of copper proteinate and copper carbonate relative to copper sulphate in cattle. *Journal of Dairy Science* 79, 127–132.

Whitelaw, A., Armstrong, R.H., Evans, C.C. and Fawcett, A.R. (1979) A study of the effects of copper deficiency in Scottish Blackface lambs on improved hill pasture. *Veterinary Record* 104, 445–460.

Whitelaw, A., Russel, A.J.F., Armstrong, R.H., Evans, C.C. and Fawcett, A.R. (1983) Studies in the prophylaxis of induced copper deficiency in sheep grazing reseeded hill pastures. *Animal Production* 37, 441–448.

Wiener, G. (1987) The genetics of copper metabolism in man and animals. In: Howell, J.McC. and Gawthorne, J.M. (eds) *Copper in Man and Animals*, Vol. I. CRC Press, Boca Raton, Florida, pp. 45–61.

Wiener, G., Suttle, N.F., Field, A.C., Herbert, J.G. and Woolliams, J.A. (1978) Breed differences in copper metabolism in sheep. *Journal of Agricultural Science, Cambridge* 91, 433–441.

Wiener, G., Wilmut, I., Woolliams, C. and Woolliams, J.A. (1984a) The role of the breed of dam and the breed of lamb in determining the copper status of the lamb. 1. Under a dietary regime low in copper. *Animal Production* 39, 207–217.

Wiener, G., Wilmut, I., Woolliams, C. and Woolliams, J.A. (1984b) The role of the breed of dam and the breed of lamb in determining the copper status of the lamb. 2. Under a dietary regime moderately high in copper. *Animal Production* 39, 219–227.

Williams, D.M., Kennedy, F.S. and Green, B.G. (1983) Hepatic iron accumulation in copper-deficient rats. *British Journal of Nutrition* 50, 653–660.

Wittenberg, K.M. and Devlin, T.J. (1987) Effects of dietary molybdenum on productivity and metabolic parameters of lactating beef cows and their offspring. *Canadian Journal of Animal Science* 67, 1055–1066.

Wittenberg, K.M., Boila, R.J. and Shariff, M.A. (1990) Comparison of copper sulphate and copper proteinate as copper sources for copper-depleted steers fed high molybdenum diets. *Canadian Journal of Animal Science* 70, 895–904.

Woolliams, C., Suttle, N.F., Woolliams, J.A., Jones, D.G. and Wiener, G. (1986) Studies on lambs from lines genetically selected for low or high plasma copper status. 1. Differences in mortality. *Animal Production* 43, 293–301.

Woolliams, J.A., Suttle, N.F., Wiener, G., Field, A.C. and Woolliams, C. (1982) The effect of breed of sire on the accumulation of copper in lambs, with particular reference to copper toxicity. *Animal Production* 35, 299–307.

Woolliams, J.A., Suttle, N.F., Wiener, G., Field, A.C. and Woolliams, C. (1983) The long-term accumulation and depletion of copper in the liver of different breeds of sheep fed diets of different copper content. *Journal of Agricultural Science, Cambridge* 100, 441–449.

Woolliams, J.A., Woolliams, C., Suttle, N.F., Jones, D.G. and Wiener, G. (1986) Studies on lambs from lines genetically selected for low or high plasma copper status. 2. Incidence of hypocuprosis on improved hill pasture. *Animal Production* 43, 303–317.

Wouda, W., Borst, G.H.A. and Gruys, E. (1986) Delayed swayback in goat kids, a study of 23 cases. *The Veterinary Quarterly* 8, 45–56.

Xin, Z., Waterman, D.F., Hemken, R.W. and Harmon, R.J. (1991) Effects of copper status on neutrophil function, superoxide dismutase and copper distribution in steers. *Journal of Dairy Science* 74, 3078–3085.

Xin, Z., Silvia, W.J., Waterman, D.F., Hemken, R.W. and Tucker, W.B. (1993) Effect of copper status on luteinizing hormone secretion in dairy steers. *Journal of Dairy Science* 76, 437–444.

Zervas, G., Nikolau, E. and Mantzios, A. (1990) Comparative study of chronic copper poisoning in lambs and young goats. *Animal Production* 50, 497–506.

Zhou, W., Kornegay, E.T., van Laar, H., Swinkels, J.W.G.M., Wong, E.A. and Lindemann, M.D. (1994a) The role of feed consumption and feed efficiency in copper-stimulated growth. *Journal of Animal Science* 72, 2385–2394.

Zhou, W., Kornegay, E.T., Lindemann, M.D., Swinkels, J.W.G.M., Welten, M.K. and Wong, E.A. (1994b) Stimulation of growth by intravenous injection of copper in weanling pigs. *Journal of Animal Science* 72, 2395–2403.

Iodine 12

Introduction

Iodine is unique among mineral elements in that a deficiency leads to a clinical abnormality – enlargement of the thyroid gland in the neck or 'goitre', as it is called – which is easily recognized and specific for that deficiency. Attempts to treat goitre go back thousands of years, but a relationship between iodine and goitre in humans and animals only emerged in the 19th century with discoveries that salts of iodine could be used successfully in the treatment of human goitre and that the occurrence of endemic goitre was inversely correlated with concentrations of iodine in soils, foods and waters in particular areas in Europe. The manifestations of endemic goitre were found to be similar to those of deficient thyroid function in animals, both of which responded to administration of thyroid gland extracts. Iodine was then shown to be a normal constituent of the animal body, highly concentrated in the thyroid gland and diminished in concentration in goitrous thyroids. Within 20 years of the last of these discoveries, the active principle of the thyroid gland was isolated, identified as tetraiodothyronine (T_4), and named thyroxine (see Harington, 1953). Subsequently, triiodothyronine (T_3), which has three to four times the potency of thyroxine, was shown to circulate in low concentrations in the blood. Extensive goitrous areas were discovered in every continent (Fig. 12.1), affecting humans and farm stock, associated primarily with an environmental deficiency of iodine and gradually controlled by raising individual iodine intakes, usually through the medium of iodized salt. More recently, attention has focused on the adverse effects of iodine deficiency on the development of the central nervous system, which can occur in humans and farm animals (Hetzel, 1991), and the role of selenium in iodine metabolism (Beckett and Arthur, 1994) (see Chapter 15). Iodine deficiency is no longer synonymous with goitre or goitre with uncomplicated iodine deficiency (Hetzel and Welby, 1997).

The factors that limit the capacity of the human thyroid to maintain structure and function apply equally to farm animals and respond similarly to

© CAB *International* 1999. *Mineral Nutrition of Livestock*
(E.J. Underwood and N.F. Suttle)

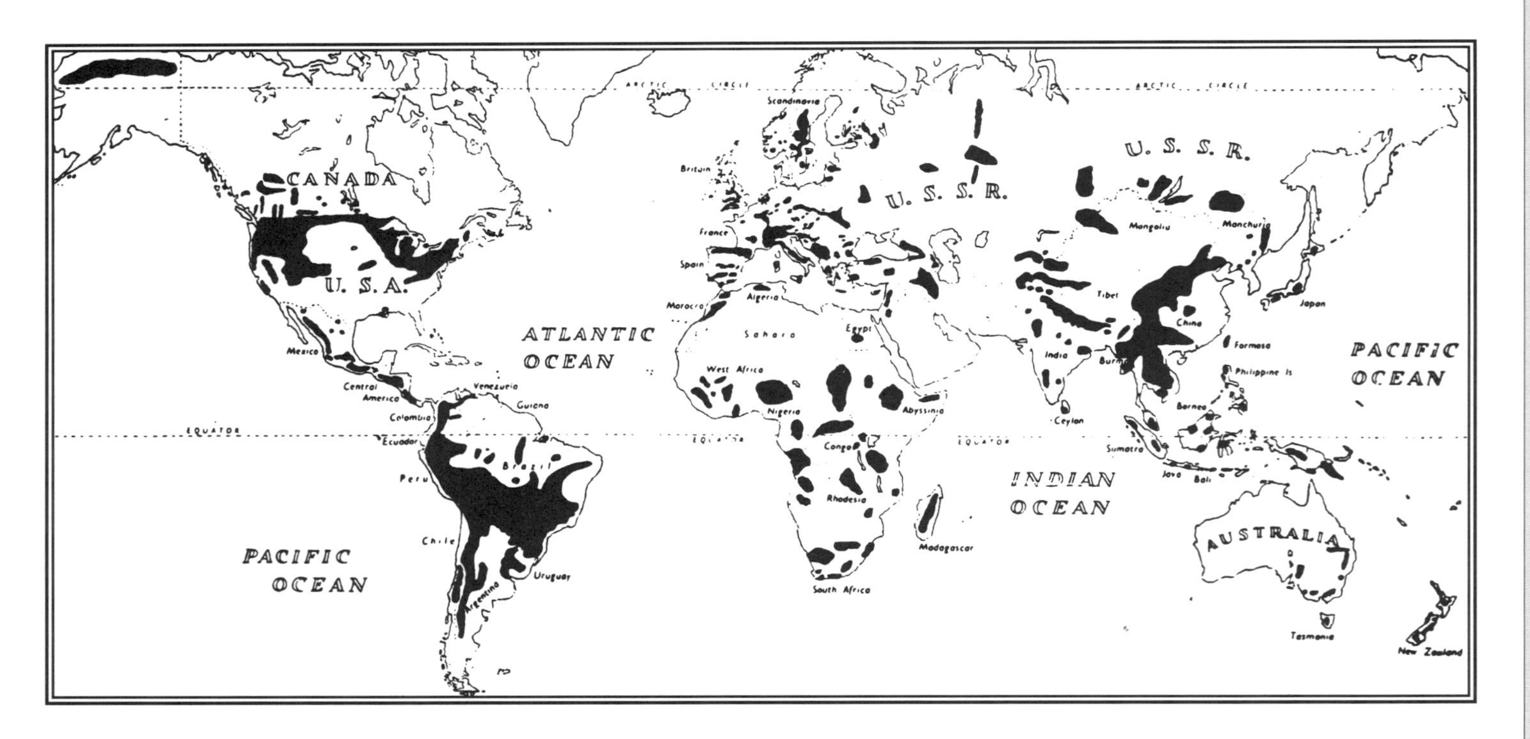

Fig. 12.1. Areas of the world where goitre was endemic (map by courtesy of the Iodine Education Bureau, London).

supplemental iodine. These factors are: (i) a gross environmental deficiency of iodine; (ii) the presence of dietary constituents, called goitrogens, which interfere with thyroid hormone synthesis by limiting the capacity of the gland either to 'trap' iodine or to incorporate this iodine into thyroactive substances; (iii) the dietary supply of other trace elements, such as selenium and iron, which influence iodine metabolism; and (iv) environmental factors, such as cold stress, which increase basal metabolic rate (BMR). Thyroid activity in farm stock influences productive functions, such as milk yield, egg production, wool growth and reproductive performance. Thyroactive iodinated proteins were investigated as a means of improving animal production in extreme wartime circumstances unrelated to deficient dietary iodine supplies, and their use in animal nutrition is not considered further.

Sources of Iodine

Forages

Plants contain iodine in highly variable concentrations, due to species and strain differences, climatic and seasonal conditions and, to a lesser extent, the type of soil and the fertilizer treatment received; interactions between these factors are also important. Soils high in iodine, such as boulder-clays and alluvial soils, generally produce plants richer in the element than iodine-low soils, such as those derived from granites (Groppel and Anke, 1986), but attempts to correlate total soil with plant iodine have not always been successful. Plant species differ widely in their ability to absorb and retain iodine from the soil (Butler and Glenday, 1962), and the botanical composition of pastures can greatly influence iodine intakes. The mean iodine concentrations in cyanogenetic and non-cyanogenetic strains of white clover, all grown in the same area in New Zealand, were 0.2 mg kg^{-1} and 0.04 mg kg^{-1} dry matter (DM), respectively (Johns, 1956). The proportionality, if it can be relied upon, means that the contribution of the two strains to iodine requirements of livestock might be of the same order. Tenfold variation in the levels of iodine in other pasture species have been reported (Table 12.1), but the extent of species variation is dependent on seasonal effects. Rapid seasonal declines in iodine concentrations were noted for five different plant species in the German Democratic Republic (GDR), but the effect was proportionally lower in leguminous than in graminaceous species (Table 12.2; Groppel and Anke, 1986). In a study of Dutch pastures, iodine concentrations in white clover and grasses ranged from 0.16 to 0.18 and from 0.06 to 0.14 mg kg^{-1} DM, respectively (Hartmans, 1974). The mean concentrations in Welsh pasture grasses were 0.2–0.3 mg kg^{-1} DM (Alderman and Jones, 1967), and similar levels have been reported recently for Northern Ireland (McCoy *et al.*, 1997). Proximity to the sea also has a major influence on plant iodine concentrations, as illustrated in Fig. 12.2. Dilution of the iodine deposited from marine sources over winter may contribute to the marked seasonal decline in iodine concentrations in spring pastures (Table 12.2). The only

Table 12.1. Iodine content of New Zealand pasture species (from Johnson and Butler, 1957).

	Herbage iodine (μg kg^{-1} DM)		
	Silt loam	Sandy loam	Hill soil
Total soil iodine (μg kg^{-1} DM)	8530	3000	1760
Perennial ryegrass	1500	1600	1350
Short-rotation ryegrass	150	168	230
Cocksfoot	175	258	225
Paspalum dilatatum	1280	1280	1700
White clover	500	725	800

Table 12.2. Effects of plant species and sampling time on plant iodine concentrations (μg kg^{-1} DM) in the German Democratic Republic (Groppel and Anke, 1986).

	24 April	16 June
Meadow fescue	184 ± 24	20 ± 4
Green rye	305 ± 108	43 ± 20
Green wheat	215 ± 53	18 ± 4
Acre red clover	294 ± 34	103 ± 37
Lucerne	358 ± 81	149 ± 37

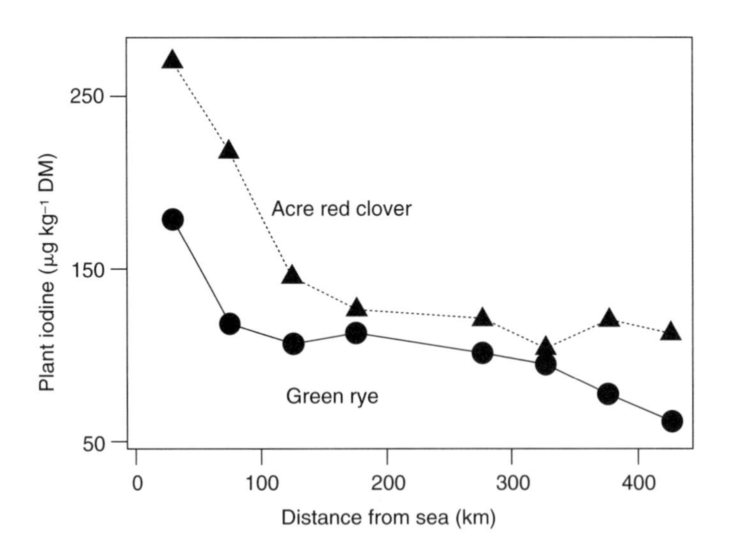

Fig. 12.2. Iodine concentrations in crops and grasses are greatly influenced by marine deposition and therefore decrease with distance from the sea (data from Groppel and Anke, 1986).

artificial fertilizer naturally high in iodine is Chilean nitrate of soda, and applications in the amounts used to supply nitrogen can double or treble plant iodine concentrations. Heavy applications of seaweed can increase the iodine content of animal fodders 10–100 times (Gurevich, 1964), but nitrogenous fertilizers generally reduce herbage iodine concentrations by increasing plant growth (Alderman and Jones, 1967).

Other foodstuffs

Animal feedstuffs vary widely in iodine concentration; the wide differences between feed types, first reported in 1952, were essentially confirmed 30 years later (Table 12.3). Cereals and oil-seed meals are poor sources of iodine and fish-meals exceedingly rich sources. The iodine content of other animal protein sources is moderate but will greatly depend on dietary iodine provision. The concentration of iodine in seaweed may be as high as 4–6 g kg^{-1} DM and enormous increases in the iodine content of milk and eggs can be obtained by feeding cows or hens with large quantities of seaweed.

Water

In many parts of the world, the iodine content of the drinking-water has been inversely correlated with the incidence of endemic goitre in humans. A similar correlation no doubt exists in farm animals where they subsist predominantly on feeds and waters of local origin. However, the water itself does not normally contribute a significant proportion of total daily iodine intakes, over 90% of which comes from the food. The correlation exists because the iodine content of the water tends to reflect the iodine content of the rocks and soils and hence of the edible plants of an area. Thus, iodine concentrations in drinking-water, like those in pastures and crops, decrease with distance from the sea but within a low range; in one study in East

Table 12.3. Confirmation from the German Democratic Republic (GDR) (Groppel and Anke, 1986) of the wide differences in iodine concentrations (μg kg^{-1} DM) between animal foodstuffs reported by the Chilean Iodine Education Bureau (CIEB) in 1952.

CIEB		GDR			
Feedstuff	Range	Feedstuff	Mean	*n*	SD
Hays and straws	100–200	Hay	136	13	61
		Cereal straw	368	5	283
Cereal grains	40–90	Maize	44	9	18
		Barley	95	69	69
Oil-seed meals	100–200	Rape-seed (ext.)	67	7	30
		Soybean (ext.)	97	8	33
Meat-meals	100–200				
Milk products	200–400	Dried skimmed milk	376	8	196
Fish-meals	800–8000	Fish-meal	6688	16	2879

SD, standard deviation.

Germany, values fell from 7.6 ± 4.4 standard deviation (SD) to 1.1 ± 0.9 μg l^{-1} as distance increased from < 50 to > 400 km (Groppel and Anke, 1986).

Milk

The iodine content of milk is extremely variable in all species, due to the ease with which this element passes the mammary barrier. Iodine concentrations in bovine milk increase linearly with intake over a wide range (Fig. 12.3) and the slope indicates recovery of about 30% of the supplement in milk with yields of 30 l day^{-1}. Marked peaks in iodine concentrations are found in bovine milk in winter in the UK and have been attributed to the overfeeding of iodine in concentrates given to the housed dairy cow (Phillips *et al.*, 1988). The iodine content of milk is influenced also by stage of lactation. Colostrum is two to three times as rich in iodine as milk taken in full lactation and there is a fall in concentration towards the end of lactation. The iodine content of cow's milk can also be increased by exogenous iodine from iodophors used to maintain dairy hygiene. Iodine differs from most mineral nutrients in milk in the extent to which it is retained by the fat. Such milk products as dried skimmed milk, buttermilk and whey usually contain slightly lower concentrations of iodine than the milk from which they are derived.

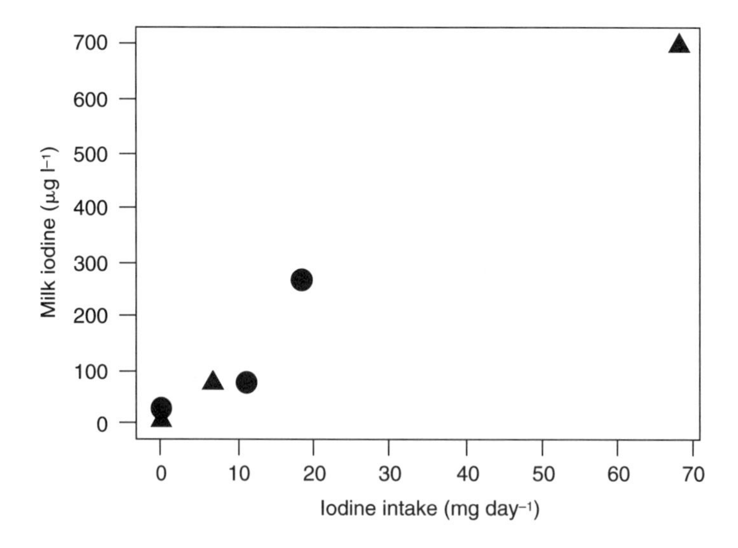

Fig. 12.3. Concentrations of iodine in milk increase linearly with increases in dietary iodine intake in the dairy cow (data from Kirchgessner, 1959 (circle) and Hemken *et al.* 1972 (triangle)).

Sources of Organic Goitrogens

The contents of goitrogens or goitrogen precursors in the diet are probably more important determinants of iodine status than dietary iodine concentration and they occur more widely than was once thought. Cruciferous plants contain variable amounts of potential goitrogens of the 'thiouracil' type (e.g. progoitrin), whereas most brassicas and white clover contain cyanogenetic glycosides that are goitrogenic because of the release of hydrocyanic acid (HCN) after structural damage to plant cells and conversion into thiocyanate in the animal. The greatest limitation to the use of cassava as a livestock feed is its content of the cyanogenetic glycosides, linamarin and lotoaustralin (Khajarern *et al.*, 1977). Toxicity of cassava can arise from the liberation of HCN when linamarin is hydrolysed by linamarinase within the plant. However, the fresh cassava root contains < 500 mg HCN kg^{-1} and levels in both leaf and root can be reduced by appropriate processing and drying. The substances which become goitrogens often have useful fungicidal or insecticidal properties in the plant and, in the early years of plant breeding, selection for disease and pest resistance also increased goitrogenicity. The pendulum has now swung and strains are being selected for 'low' goitrogen content. This was particularly necessary in the case of oil-seed rape (*Brassica campestris*), which was increasingly cultivated as a source of edible oil, generating large quantities of rape-seed meal, a valuable source of nutrients for livestock (Bell, 1984). Rape-seed meals can contribute both types of goitrogen; they contain over 0.1 mmol glucosinolates kg^{-1} and enzymatic degradation during processing and digestion produces isothiocyanates (ITC), thiocyanate and goitrin (an oxazolidinethione (OT)) (Emanuelson, 1994). Canadian workers have led the way in producing low-glucosinate (LG) cultivars, containing < 0.03 mmol glucosinolate kg^{-1}, for a product termed Canola meal; such products will have minimal effects on iodine status. Winter varieties grown in middle and western Europe are a different proposition and rape-seed meals derived from them require treatment to lower glucosinolate content. The concentration of goitrogen precursors is not reduced by minimizing the available sulphate concentration in the soil but may be reduced when nitrogen is applied at some sites (Alderman and Stranks, 1967; McDonald *et al.*, 1981).

Metabolism of Iodine

Absorption, transport, storage and excretion

Like all the anionic elements, iodine is absorbed very efficiently from the gastrointestinal tract and this enables any iodine secreted prior to absorptive sites to be extensively recycled. Whereas phosphorus is recycled via saliva in ruminants, iodine is recycled via secretion into the abomasum (Miller *et al.*, 1974b). Unheated soybeans contain a heat-labile factor which can induce goitre by impairing the intestinal recycling of iodine (Hemken, 1960).

Absorbed iodine is transported in the bloodstream loosely bound to plasma proteins. Active uptake by a sodium (Na):potassium (K)-dependent ATPase in the thyroid gland results in the capture of up to 90% of iodine which passes through that organ (Hetzel and Welby, 1997). Approximately 80% of iodine in the mammalian body is found in the thyroid gland, but some iodine accumulates in soft tissues, such as muscle and liver, when excess iodine is consumed (Downer *et al.*, 1981). Iodine captured by the thyroid is combined with tyrosine to form diiodotyrosine (T_2) and two molecules of this compound are used to form T_4, the physiologically inactive transport form of the hormone. The thyroid stores T_4 in the colloidal form as thyroglobulin, an iodinated glycoprotein, and, during iodine depletion and deficiency, the follicular cells are quickly denuded of colloid. A small proportion of the extrathyroidal iodine occurs as free iodine, which, like chloride, readily permeates all tissues. Recycling of thyroidal iodine occurs via the iodide pool. Excess dietary iodine is excreted predominantly via urine, but in lactating animals significant amounts can be secreted in milk (Fig. 12.3; Miller *et al.*, 1965; Hillman and Curtis, 1980). The iodine lost by both routes is in the form of iodide.

Regulation of thyroid hormone function

The efficiency with which iodine is captured by the thyroid gland varies according to need and is determined by secretion of two hormones with thyrotrophin-releasing and thyroid-stimulating properties (TRH and TSH, respectively), which together determine T_4 secretion rate (TSR); TSH is produced by the anterior pituitary and TRH by the hypothalamus, with feedback control by the circulating levels of free T_4 and T_3, the minor fractions of predominantly protein-bound pools (Beckett and Arthur, 1994). Activation of T_4 is achieved by three deiodinase (ID) enzymes (types I, II and III) which are selenium-dependent (Arthur *et al.*, 1990; Arthur, 1997); thiouracil inhibits type I only (Hetzel and Welby, 1997). An important feature of the formation of physiologically active T_3 is that it can occur peripherally at the point of need and is not confined to the thyroid gland. In animals of normal iodine status, 80% or more of the T_3 can be formed extrathyroidally, principally in the liver and kidney (Ingar, 1985) but also in the skin (Hetzel and Welby, 1997); that proportion falls markedly in iodine deficiency. There are major differences between ruminants and non-ruminants, the ID1 enzyme being particularly important in brown adipose tissue (BAT) in the newborn ruminant and serving as a source of T_3 for other tissues (Nicol *et al.*, 1994). Concentrations of T_4 in ovine fetal plasma are normally about twice those of the dam, but T_3 is barely detectable. A saturation gradient for globulin-bound T_4 may facilitate transfer of T_4 across the bovine placenta.

Impairment of thyroid hormone function

Dietary factors other than the supply of absorbable iodine can have a profound influence on iodine metabolism. Selenium deficiency impairs the formation of T_3 from T_4 (see Chapter 15) and can give rise to secondary or induced iodine deficiency. Activity of ID2 in BAT increased when weanling rats were exposed to low temperatures (4°C for 18 h), but the increase was slowed and eventually greatly diminished by selenium deprivation (Arthur *et al.*, 1992). Ingestion of thiouracil-type goitrogens produces a similar effect by inhibiting iodide oxidation (Spiegel *et al.*, 1993); the antagonism is reflected in high T_4:T_3 ratios and stimulation of TRH and TSH release. Inhibition of the $T_4 \rightarrow T_3$ conversion by goitrogens may occur outside the thyroid. These adverse effects cannot be overcome by providing more iodine, but those of the cyanogenetic goitrogens which competitively inhibit uptake of iodine by the thyroid are iodine-responsive.

Metabolism of Goitrogens

Goitrogens were believed to present less of a danger to ruminants than non-ruminants, because they were destroyed in the rumen, but this is no longer a safe assumption (Hill, 1991; Emanuellson, 1994). Thiocyanate and goitrin excretion in cow's milk rises when rape-seed meal is fed, a clear indication of persistence. However, thiocyanate excretion is accompanied by a reduction in milk iodide and relatively small increases in milk iodine when iodine supplements are fed. This apparently paradoxical response may arise from inhibition of iodine uptake by the mammary gland, conserving iodine in the lactating animal (Iwarsson, 1973) but only at the expense of any suckled offspring. Increases in TSH secretion in cows on high-glucosinolate diets (Laarveld *et al.*, 1981) indicate that, qualitatively, a net impairment of thyroxine synthesis can occur, but the magnitude of the effect on iodine requirements may be less in lactating than non-lactating animals, especially in sheep, which secrete far more iodine in milk than cows. The browse legume *Leucaena leucocephala* contains mimosine, which is degraded to a goitrogen in the rumen and thus causes thyroid enlargement in steers in some locations, though not on the island of Hawaii. It was duly found that the local cattle possessed a rumen microorganism which detoxified the goitrogen and that transfer of a rumen inoculum from Hawaii to other locations prevented the induction of iodine deficiency in goats (Jones and Megarrity, 1983).

Function of Iodine

Iodine has only one known but very important function as a constituent of the thyroid hormones, particularly T_3, which control oxidation rate and protein synthesis in all cells. Thyroid hormones control development of the fetus, particularly its brain, heart, lungs and wool follicles (Erenberg *et al.*, 1974; Hopkins, 1975). They also set the BMR and play active roles in digestion (Miller *et al.*, 1974a), thermoregulation, intermediary metabolism, growth, muscle function, immune defence, circulation and the seasonality of reproduction (Follett and Potts, 1990). As pace-setter, T_3 influences the needs for all other nutrients, but only through carefully orchestrated interactions with other endocrine glands and hormones (Ingar, 1985).

Biochemical Manifestations of Iodine Deprivation

A simplified sequence of biochemical events during iodine deprivation is given in Fig. 12.4. In the initial depletion phase, iodine concentrations in the thyroid decline; although the organ is only small (*c.* 4 g dry weight in the adult cow), the amounts of iodine (I) stored can be substantial (8–16 mg) relative to minimum daily need (0.3 mg I day^{-1} in heifers; McCoy *et al.*, 1997). During the deficiency phase, the forms of iodine present in the thyroid gland change disproportionally in an attempt to maintain iodine homeostasis and to use the limited supplies more efficiently; the changes include a preferential synthesis of T_3 over T_4 and an increase in the ratio of monoiodotyrosine (T_1) to T_2. The extent of these changes is evident from Table 12.4, in which the thyroid glands of goats from an endemic goitre area are compared with those from goats from an area with abundant environmental iodine (Karmarkar *et al.*, 1974). Adaptive changes in selenoenzyme activities confirm their involvement in thyroid function (see Chapter 15). Ten- to twelvefold increases in ID1 activity have been reported in the thyroids of heifers and their newborn calves deprived of iodine (Zagrodzki *et al.*, 1998). Accompanying increases in cytosolic glutathione peroxidase (CGPXI) activity

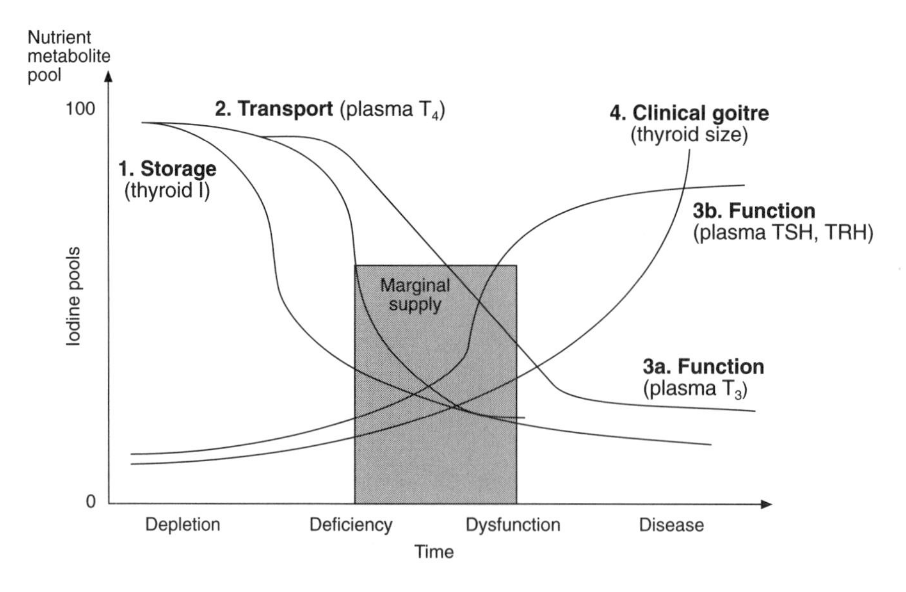

Fig. 12.4. General scheme for the sequence of biochemical changes in livestock deprived of iodine; where thiouracil-type goitrogens are involved, impaired conversion of T_4 to T_3 in and beyond the thyroid changes the picture.

Table 12.4. Distribution of labelled iodine (I) and the iodoamino acids mono-iodotyrosine (T_1), di-iodotyrosine (T_2), trio-iodothyronine (T_3) and tetra-iodothyronine (T_4) in the thyroid glands of goats from goitre and control areas, expressed as a percentage of total radioactivity (from Karmarkar *et al.* (1974)).

Area	*n*	I	T_1	T_2	T_3	T_4
Goitre	10	16.3 ± 2.42	39.1 ± 2.42	35.9 ± 1.71	6.2 ± 0.75	2.4 ± 0.43
Control	10	6.4 ± 0.98	30.8 ± 0.89	53.8 ± 1.26	0.9 ± 0.10	4.3 ± 0.47

in the thyroid probably reflected increased peroxide generation in a hyperactive gland. A study with pigs given rape-seed meal has shown that responses to iodine deficiency outside the thyroid gland may be just as important as those within the gland (Spiegel *et al.*, 1993). The presence of cyanogenetic goitrogens in the rape-seed meal was confirmed by a rise in serum thiocyanate and a sixfold rise in thyroid weight, but, in addition, liver size doubled. Although serum T_4 decreased, T_3 concentrations were maintained and T_4:T_3 ratios therefore decreased; this was attributed to the hepatomegaly, which increased T_3 synthesis from T_4 and preserved growth rate. Adaptations of this nature begin to explain the diagnostic limitations of measurements of thyroid size and serum T_4, discussed later. Where the problem is primarily one of impaired hormogenesis due to selenium deprivation, T_4:T_3 ratios in serum increase but exposure to thiouracil goitrogens decreases T_4:T_3 and reduces ID1 activity in liver and kidney in sheep (Villar *et al.*, 1998). Subclinical histological abnormalities in the thyroids of iodine-depleted heifers and their offspring are accompanied by reductions in iodine concentrations in the thyroid and plasma but not by decreases in plasma T_4 (McCoy *et al.*, 1997). The approach of dysfunction is indicated by rises in TSH and TRH in the bloodstream. Clinically significant goitre is accompanied by low plasma T_4 concentrations (e.g. Schöne *et al.*, 1988).

Clinical Manifestations of Thyroid Dysfunction

Goitre

The obvious clinical manifestation of iodine deficiency is an enlarged thyroid gland, the degree of enlargement increasing with the degree and duration of iodine deprivation. Thyroid enlargement represents an attempt to compensate for deficient production of thyroid hormones. This and other manifestations of iodine deficiency occur whether the impaired hormonogenesis arises from an absolute or from a conditioned iodine deficiency, whether it is brought about by the action of goitrogens or by a constitutional disability in the animal. Goitre is manifested predominantly in the newborn animal, which is usually delivered by a clinically normal dam.

Impaired brain development

The discovery that iodine deficiency impaired brain development in humans led to iodine deficiency disorders (IDD) replacing 'goitre' as the description of the syndrome. Studies with pregnant sheep given an exceedingly deficient diet (5–10 μg I kg^{-1} DM) showed that deficiency can lead to impaired brain maturation in the fetus, lambs being stillborn and without wool (Potter *et al.*, 1981, 1982). The brain abnormalities were reversed by giving iodine during the third trimester. Whether such changes occur in livestock in the field and are ever irreversible is questionable. Nevertheless, goitre should not be the only consequence of iodine deficiency to concern livestock nutritionists.

Impaired reproduction

Fetal development may be arrested at any stage during thyroid dysfunction, leading to early death and resorption, abortion, stillbirth or the birth of weak or hairless young of subnormal birth weights, but such problems are not easily linked unequivocally to iodine deficiency (Smyth *et al.*, 1992). Neonatal mortality occurs in lambs from ewes fed on goitrogenic kale (Andrews and Sinclair, 1962). The need for thyroid hormones to produce lung surfactants may be an important factor determining viability of the newborn (Erenberg *et al.*, 1974). Infertility or sterility and poor conception rates, due to irregular or suppressed oestrus, are allegedly features of thyroid dysfunction in cows and were attributed to the enhanced losses of iodine at peak lactation, when mating usually occurs (Hemken, 1960). Male fertility is also affected, a decline in libido and a deterioration in semen quality being characteristic of iodine deficiency in rams, bulls and stallions. In the Brown Leghorn male deprived of iodine, the testes remain small and without spermatozoa. Thyroidectomy reduces egg production in hens, but the seasonal cycle of egg production in poultry is related in part to a natural seasonal variation in thyroid activity and is not necessarily associated with fluctuations in dietary iodine. In fact, poultry can withstand a considerable degree of iodine deficiency without any marked loss of production. Thus Rogler (1958) maintained hens on an iodine-deficient diet for 35 weeks without affecting hatchability or embryo weight, but after 2 years there was a decrease in hatchability, prolongation of hatching time and retarded embryo development.

Postnatal mortality and growth retardation

In mildly iodine-deficient areas where the occurrence of clinically obvious goitre is low and seasonally variable, animal productivity can still be seriously impaired. In a flock of Polwarth ewes in Tasmania showing only minimal thyroid enlargement, a high lamb mortality of 36% was reported (King, 1976). Marked improvements in the birth weights, survival rates and growth rates of merino lambs were obtained when ewes in Queensland were given extra iodine (Knights *et al.*, 1979). These findings, together with those of Andrewartha *et al.* (1980) in eastern Australia and McGowan (1983) in New Zealand, suggest the possibility of significant production responses to iodine supplementation in areas where thyroid enlargement is slight. Given the importance of thermogenesis from BAT to survival in animals born into a cold environment, iodine deficiency may increase susceptibility to cold stress (Arthur *et al.*, 1992). There is, however, little convincing evidence as yet that iodine deprivation impairs survival in the marginally depleted calf (McCoy *et al.*, 1995, 1997; Mee *et al.*, 1995). Addition of goitrogens to a pig diet low in iodine severely restricted growth and stunted skeletal development (Sihombing *et al.*, 1974), while inclusion of rape-seed meal in a natural diet for weanling pigs reduced growth rate by 50% (Lüdke and Schöne, 1988).

Disorders of the integument

Changes in the skin and its outgrowths – hair, fur, wool and feathers – are common features of thyroid dysfunction. Pigs and calves born to iodine-deprived mothers are frequently hairless and have thick, pulpy skins due to subcutaneous oedema. This was one of the first signs of deficiency in farm animals to be recognized in goitrous regions and has been reproduced experimentally in piglets (Sihombling *et al.*, 1974). Myxoedema also develops when weanling pigs are given goitrogen-rich diets (Lüdke and Schöne, 1988). Less severe deficiency is reflected in minor changes, such as rough, dry skin, harsh coat, scanty wool and hairiness of the fleece. Reduced quantity and quality of wool growth have been widely associated with goitre in sheep. Thyroid insufficiency in the young lamb permanently impairs the quality of the adult fleece, since the normal development of the wool-producing secondary follicles requires thyroid activity in excess of that needed for general body growth (Ferguson *et al.*, 1956). Thyroidectomy of the ram reduced wool growth, but replacement therapy only needed to raise T_4 levels in the bloodstream to 30% of normal to restore yield (Maddocks *et al.*, 1985). Supplementation of the ewe with iodine has increased wool production by 6% (Statham and Koen, 1981). Administration of the goitrogen, methyl-thiouracil, to cashmere goats reduced the proportion of active secondary hair follicles and delayed the onset of moult (Rhind and McMillen, 1996). Thyroid insufficiency can be very conspicuous in some types of bird, particularly the male, in which the comb decreases in size, moulting is inhibited and the characteristic plumage is lost.

Low milk yield

Reductions in milk yield are an important feature of iodine deficiency in the dairy cow and can probably occur in all mammals. The hypothyroid state induced by the prolonged feeding of goitrogen-rich foods is accompanied by loss of appetite, leading to impaired growth and depressed milk yield (Hill, 1991). The earlier in postnatal life that maternal exposure to goitrogens begins, the greater is the impact on lamb growth. The increases in milk yield that were obtained by feeding iodinated casein to dairy cows were not due to the correction of iodine deficiency and were rarely sustained (Hemken, 1960).

Occurrence of Iodine-related Disorders

Geochemical factors

The occurrence of iodine-related disorders is traditionally indicated by maps showing where goitre has occurred, one of the most widespread of all mineral deficiency diseases (see Fig. 12.1), but important local variations in goitre incidence can be found within small countries (Statham and Bray, 1975; Wilson, 1975), which cannot be shown on such large-scale maps. A deficiency of plant-available soil iodine is the primary reason for the occurrence of goitre and other manifestations of iodine deficiency in most areas. In regions that have been subjected to comparatively recent (Pleistocene) glaciation, the low iodine status of the soils can be explained by the fact that the iodine-rich surface soils were removed and the subsequent soil-forming period has been too short to allow replenishment with airborne iodine (Hetzel and Welby, 1997). In other goitre regions, the existence of iodine-low soils can generally be attributed to small or negligible accumulation of iodine of marine origin, due to their long distance from the sea, low rainfall or both (Fuge, 1996). The importance of these factors over long periods of time was illustrated by the calculations of Hercus *et al.* (1925), who estimated that 22–50 mg of iodine per acre fell annually in the rain on the Atlantic coastal plain, compared with only 0.7 mg $acre^{-1}$ in the inland Great Lakes region of North America, where goitre was endemic. A coastal location is less beneficial when it does not face the prevailing wind. Soils low in organic matter (e.g. sandy soils) tend to be low in iodine, because they trap little of any deposited iodine.

Plant-related factors

In areas of marginally low soil iodine, the incidence and severity of goitre are influenced further by the botanical composition of the herbage, problems being more likely to occur on grass than on mixed or leguminous swards. Goitre incidence is also influenced by climatic and seasonal conditions, which affect the grazing animal directly and indirectly, through changes in pasture productivity and opportunity for marine or soil contamination of the herbage consumed. The significance of soil contamination is illustrated by the Tasmanian experience, in which goitre incidence in lambs was higher in 'good' than in 'poor' seasons (Statham and Bray, 1975). The higher herbage production of 'good', high-rainfall seasons provided less opportunity for contamination with soil and therefore less soil ingestion by animals than occurred in poor seasons of relatively low pasture productivity. Since most topsoils are very much richer in iodine than the plants which grow on them, soil ingestion can greatly increase iodine intakes. Of equal importance is the four–fivefold reduction in iodine concentration which can accompany the advancing maturity of a sward (Alderman and Jones, 1967). The iodine status of grazing animals thus tends to decline as the grazing season progresses.

Secondary iodine disorders

The occurrence of goitre can depend on dietary selenium supply (see Chapter 15) and the extent to which goitrogen-containing plants occur and are consumed. Where the animal's diet is composed largely of highly goitrogenic feeds, such as kale or cabbage, the incidence of goitre can be high. Although other leafy and root brassicas contain sufficient goitrogen precursors to increase thyroid weight, the general view is that growth rate in ruminants will not be retarded when they are consumed, thyroid enlargement compensating for any antagonism (Fitzgerald, 1983). Rape-seed meals contain both types of goitrogen and their effects will be additive. Inclusion of as little as 80 g kg^{-1} DM of a high-glucosinolate meal in pig rations will induce hypothyroidism and suppress growth, effects which iodine supplementation can only partially alleviate (see Fig. 12.5; Schöne *et al.*, 1988). The feeding of cassava can induce iodine deficiency but diets containing 500 mg HCN kg^{-1} DM have been fed to sows throughout gestation and lactation without adverse effects on sow or litter; bitter varieties pose the greatest risk of inducing iodine deficiency.

Diagnosis of Thyroid Dysfunction

Diagnosis of iodine deprivation has been based on morphological, histological or biochemical indices of iodine status, but response to iodine supplementation is often the most reliable index, unless thiouracil-type goitrogens are involved. Interpretation is improved by the adoption of marginal ranges and the use of more than one index (Table 12.5).

Thyroid morphology

Severe goitre can readily be diagnosed on clinical evidence alone, as the enlarged goitrous thyroid can be visible to the eye or detected by palpation. Diagnosis of less severe forms of goitre requires determinations of the weight, histological structure and iodine concentration of the thyroid. The studies of Sinclair and Andrews (1961) indicated that the thyroid status of newborn lambs is normal if the fresh glands weigh less than 1.3 g, doubtful if weights range between 1.3 and 2.8 g and hyperplastic if the weights exceed 2.8 g. An early source (Wilson, 1975) indicated that corresponding values for calves were < 10, 10–13 and > 13 g, respectively, but similar variation in size may occur in histologically normal glands (McCoy *et al.*, 1997), which is not attributable to variation in birth weight; wider diagnostic ranges are needed (Table 12.5).

Thyroid histology

A sequence of histological abnormalities can be detected in the thyroid gland during iodine deprivation; depletion of colloid is followed first by hyperplasia and then by hypertrophy of the cuboidal epithelia lining the thyroid follicles. Hyperplasia has been suggested as a more reliable index of deprivation than

Table 12.5. Indices of marginal iodine status[a] in livestock.

	Condition	Pigs	Poultry	Sheep	Cattle
Diet I ($mg\ kg^{-1}$ DM)	Summer	0.04–0.08	0.06–0.08	0.10–0.15	0.075–0.10
	Winter	NA	NA	0.20–0.30	0.15–0.20
	Goitrogens[b]	0.14–0.28	0.18–0.30	0.75–1.00	0.75–1.00
Thyroid weight ($gFW\ kg^{-1}$ LW)	Newborn	0.2–0.4	0.06–0.10	0.4–0.9	0.5–1.0
Thyroid I ($g\ kg^{-1}$ DM)	All ages	1.2–1.5	1.5–1.5	1.2–2.0	1.2–2.0
Serum PBI ($\mu g\ l^{-1}$)	Newborn	–	30–40	30–40	30–40
	Adult	20–33	–	–	10–25[c]
Serum T_4 ($nmol\ l^{-1}$)	D/NB	–	–[d]	1.0–1.5	2.0–2.5
	Adult	20–30	5–9	20–30	25–50[c]
Serum T_3 ($nmol\ l^{-1}$)	Adult	0.6–0.8	–	1.0–1.7	2.0–2.5
Milk or egg I ($\mu g\ l^{-1}$ or kg^{-1})	Adult	30–40	40–80	70–100	30–50

[a] Mean values falling within the ranges given indicate a possibility of response to iodine supplementation (or goitrogen removal) if sustained.
[b] Three- to fivefold increases in iodine requirement are caused by the ingestion of diets rich in cyanogenetic goitrogens.
[c] Range is normal for 1st 4 weeks after calving.
[d] Values normally increase during lay.
NA, not available, but requirements may be raised in the free-ranging animal; D/NB, Ratio of T_4 for dam/newborn.
Conversion factors: T_4 divided by 1.287 and T_3 by 1.546 for $\mu g\ l^{-1}$
I multiplied by 0.788 for $\mu mol\ kg^{-1}$ DM in diet, $nmol\ l^{-1}$ in serum or milk.

thyroid weight (Smyth *et al.*, 1992), but both can be abnormal in healthy calves (Mee *et al.*, 1995; McCoy *et al.*, 1997) and the terms 'colloid goitre' and 'hyperplastic goitre' should not be used to implicate iodine deprivation as an unequivocal cause of neonatal mortality.

Thyroid iodine

In simple goitre, the amount of iodine in the gland is reduced and the concentration in the enlarged gland is reduced even more significantly, below the normal level for mammals of 2–5 g kg^{-1} DM. Early in this century, Marine and Williams (1908) found that hyperplastic changes characteristic of goitre were regularly found when the iodine concentration fell below 1 g kg^{-1} DM. A critical iodine concentration of 1.2 g kg^{-1} DM, below which the gland cannot function properly, was reported for several farm species (Andrews *et al.*, 1948; Sinclair and Andrews, 1958), but lower values have since been found in histologically normal glands (Wilson, 1975; McCoy *et al.*, 1995, 1997).

Iodine in the bloodstream

The serum-precipitable iodine (SPI), protein-bound iodine (PBI) or butanol-extractable iodine (BEI) of the serum normally consists largely of thyroxine (T_4) and each is sensitive to gross changes in thyroid activity, but individual variability among healthy animals of the same species or breed is high. The PBI values of 30–40 μg l^{-1} have been taken as 'normal' for adult sheep and cattle, but a value of 39 μg l^{-1} in periparturient ewes was reported in an outbreak of neonatal goitre (Statham and Koen, 1981). Lower normal PBI values have been recorded in studies with the domestic fowl and with horses (Irvine, 1967). Misleadingly high PBI values are found after the administration of diiodosalicylic acid (DIS), which becomes strongly bound to serum proteins while remaining poorly available for the synthesis of thyroid hormones (Miller and Ammerman, 1995). The measurement of inorganic iodine in serum has come into vogue as a simple measure of current dietary iodine supply, but it does not measure thyroid hormone function (Mee *et al.*, 1995).

Thyroxine assays

Serum T_4 values can reflect the thyroid and iodine status of domestic animals, although this relationship is better documented for sheep than for other species. Thus Walton and Humphrey (1979) determined by radioimmunoassay the plasma thyroxine of: (i) groups of ewes from the severely iodine-deficient highlands of Papua New Guinea, with clinical evidence of goitre in their lambs; (ii) similar ewes previously given an injection of iodized oil; and (iii) ewes from a coastal flock with no history of goitre. The mean and standard deviations for the three groups were: (i) 18.4 ± 14.2; (ii) 128.2 ± 8.0; and (iii) 71.9 ± 14.5 nmol T_4 l^{-1}. However, serum T_4 values have generally been less satisfactory in domestic animals than in humans for assessing hypothyroidism, because of the many factors that can affect thyroid output. For example, serum T_4 concentrations in sheep are usually higher in winter than in early autumn or summer (Sutherland and Irvine, 1974; Andrewartha *et al.*, 1980) and are depressed by intestinal parasitism (Prichard *et al.*, 1974). In cattle, serum T_4 values fall to 20–40 μmol during early lactation (Puls, 1994). The age of the animal can also be important. Serum thyroxine concentrations in normal newborn lambs are high but decrease after birth, and it is some 8 weeks before they reach those of ewes (Andrewartha *et al.*, 1980). Similarly, T_4 values in newborn calves from heifers given adequate dietary iodine are three times higher than those of their dams (McCoy *et al.*, 1997). In lactating ewes, a critical mean plasma T_4 concentration of 50 nmol l^{-1} has been suggested below which thyroid insufficiency can be assumed (Wallace *et al.*, 1978). In the newborn lamb, a ratio of < 1 in lamb:ewe T_4 concentrations in serum may indicate deficiency in the offspring (Andrewartha *et al.*, 1980), but the data for beef cattle (McCoy *et al.*, 1997) indicate a higher threshold of < 2.5.

Triiodothyronine assays

Where the cause of hypothyroidism is impaired conversion of T_4 to T_3, as with selenium deficiency or thiouracil-type goitrogens, T_4 will be particularly misleading and T_3 measurements may give a better measure of dysfunction. For example, Wichtel *et al.* (1996) found significantly lower ($P < 0.04$) T_3 concentrations in a dairy herd which responded to selenium than in an unresponsive herd (1.81 vs. 2.06 nmol l^{-1}); T_4 concentrations were high in untreated heifers (66 and 99 nmol l^{-1}) but were reduced by selenium. However, T_4:T_3 ratios may not reflect functional T_3 status at the tissue level, and measurement of T_3 does not overcome all the difficulties of assessing iodine status, because concentrations in the bloodstream are reduced by feed and water restriction (Blum and Kunz, 1981), heat (Christopherson *et al.*, 1979) and other stressors, parasitic infection and protein (Ash *et al.*, 1985) and iron deficiencies (Beard *et al.*, 1989). Furthermore, non-ruminants show different patterns of response to ruminants (Spiegel *et al.*, 1993).

Iodine in urine and milk

The rates of iodine excretion in urine and milk can provide further useful diagnostic criteria in simple iodine deficiency. Iodine is excreted mainly in the urine, so that iodine intakes can be related to urinary iodine excretion. The iodine in milk is also extremely responsive to changes in dietary iodine intakes, and milk iodine determinations have been proposed as a means of establishing the iodine and goitre status of an area. Alderman and Stranks (1967) demonstrated a linear relationship between the iodine content of bulked cow's milk and the daily iodine intake of cows from 17 herds and suggested that, with milk iodine values less than 25 µg l^{-1}, dietary iodine supply may be suboptimal. Higher thresholds (possibly > 50 µg l^{-1}: McCoy *et al.*, 1997) would be needed for colostrum, which is much richer in iodine than main milk. Ewe's milk iodine can also be a good indicator of dietary intake, and concentrations below 80 µg l^{-1} point to a possible dysfunctional state. From field observations, Mason (1976) related average ewe-milk iodine values to the incidence of goitre in the lambs, determined by palpable thyroid enlargement (Table 12.6). Where the thyroid dysfunction is secondary to the influence of selenium or goitrogens of the thiouracil or cyanogenetic type, secretion of iodine in urine will probably be deceptively high.

Prevention and Control

Disorders caused by simple iodine deficiency or deficiency induced by cyanogenetic goitrogens can be prevented and cured by iodine supplementation; those caused by thiouracil-type goitrogens or selenium deficiency require other measures.

Continuous methods

With stall-fed or hand-fed animals, supplementation is accomplished either by the use of iodized salt-licks, by the feeding of kelp (seaweed) or by incor-

Table 12.6. Relation of ewe-milk iodine to thyroid enlargement in lambs (from Mason, 1976).

Percentage of lambs with enlarged thyroids	Ewe milk iodine (μg l^{-1})
23–69	23–64
12–22	71–98
0–4	98–111

porating iodine into the mineral mixtures or concentrates, where these are regularly given for other reasons. With the exception of DIS, the commonly used iodine salts have a uniformly high availability to ruminants and non-ruminants (Miller and Ammerman, 1995). For cattle, DIS has only 15% of the value of potassium iodide (KI), but it may be more available to non-ruminants. Potassium iodide or calcium iodate ($Ca(IO_3)_2$) is most commonly used for these purposes, but salt-licks and mixtures containing the iodide are subject to considerable losses of iodine by volatilization and by leaching, particularly in hot, humid climates, unless suitably stabilized (see Kelly, 1953). Cuprous iodide is stable and readily absorbed, despite a low solubility, and is favoured by some workers. Orthoperiodates (e.g. $Ca_5(IO_6)_2$) are also stable and not toxic at the levels required in salt-licks or mineral mixtures (0.1 g I kg^{-1}). Consumption of all free-access sources can be extremely variable (see Chapter 3). Herbage iodine concentrations can be increased by the use of fertilizers containing iodine. This method is inefficient, because of low uptake by the herbage, although added iodate is much better absorbed than iodide, especially where liming is practised (Whitehead, 1975).

Discontinuous methods

Dosing or drenching is effective in controlling the incidence of endemic goitre but is time-consuming and costly in labour, except where the animals are regularly handled for other reasons. Two oral doses of 280 mg KI or 360 mg potassium iodate (KIO_3), after the third and fourth months of pregnancy, have prevented neonatal mortality and associated goitre in lambs from ewes wintered on goitrogenic kale. A single 1 ml intramuscular injection of an iodized poppy-seed oil, containing 40% by weight of bound iodine and administered to ewes 7–9 weeks before lambing, is equally effective (Sinclair and Andrews, 1958, 1961; Statham and Koen, 1981) and can improve survival to weaning (McGowan, 1983). Statham and Koen (1981) reported an improvement in wool yield in the year after treatment, and control of goitre in the lamb lasted for 2 years. The recommended dose of iodized poppy oil for cows is only 4 ml, i.e. not proportional to body weight *vis-à-vis* sheep. Slow-release intraruminal devices have been developed, one of which lasts for several years, by employing a simple diffusion principle and an iodide-filled, polyethylene sack, which lodges in the rumen (Mason and Laby, 1978). Rigid capsules can achieve a similar purpose (Siebert and Hunter, 1982), and efficacy has been demonstrated (Ellis and Coverdale, 1982).

Countering thiouracil-type goitrogens

Soaking rape-seed meal in copper sulphate ($CuSO_4$) solution (25 g l^{-1}) and drying (Lüdke and Schöne, 1988) or supplementation of pig diets with 250 mg Cu kg^{-1} DM (Schöne *et al.*, 1988) can eliminate both types of goitrogen, ITC and OT, restoring normal growth and thyroid size in weaned pigs, which iodine supplements alone could not do (Schöne *et al.*, 1988; Fig. 12.5). The rape-seed meal goitrogens were predominantly of the thiouracil-type (11.9 g OT and 3.8 g ITC kg^{-1} DM).

Iodine Requirements

General considerations

The minimum dietary iodine requirements of farm animals cannot be given with any accuracy, partly because the criteria of adequacy employed have not been consistent. Requirements for growth are not necessarily identical to those for reproduction and lactation or for maintenance of thyroid structure and function. Over 40 years ago, Levine *et al.* (1933), on the basis of their work with rats, linked the need for iodine to energy consumption and stated that 'until precise data are obtained for different species of animals ... 20–40 µg per 1000 calories of ration is considered as the minimum iodine requirement for farm animals'. Mitchell and McClure (1937) suggested that the requirement for iodine would be more properly related to heat production

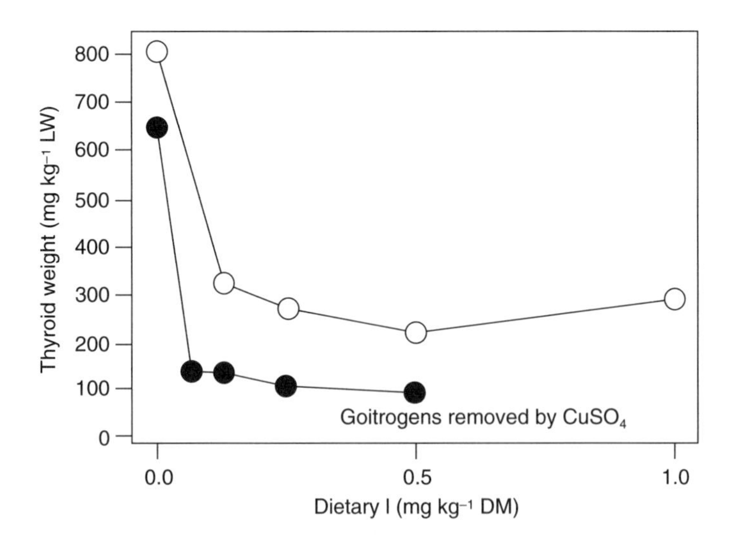

Fig. 12.5. Thyroid enlargement in pigs given diets containing some thiouracil-type goitrogens (as rape-seed meal) cannot be fully controlled by iodine supplementation but their removal by treatment with copper sulphate allows full recovery (data from Schöne *et al.*, 1988).

than to energy intake in farm animals. On this basis, the minimum iodine requirements of different classes of stock were calculated (see Table 12.3) and they compared well with the minimum iodine consumption figures obtained by Orr and Leitch (1929) for poultry, pigs and sheep in areas known to be free from goitre, and therefore presumably adequate for these species. Orr and Leitch's figures for cows were substantially higher than those in Table 12.7. The minimum amounts of iodine required per day can be readily converted into dietary concentrations, assuming a normal daily DM consumption, and for poultry, pigs and sheep requirements lie between 0.05 and 0.10 mg I kg^{-1} DM. On a similar basis, a ration containing 0.03–0.06 mg I kg^{-1} DM would suffice for milking cows. Critical levels for all species are likely to vary with environmental temperature and the efficiency of energy utilization. The Agricultural Research Council (ARC, 1980) adopted a physiological approach by estimating requirements for ruminants from the amount of iodine needed to sustain TSR, with an assumed capture efficiency of 33%. This approach will overestimate the need for iodine, because it ignores the fact that a proportion of the iodine contained in the thyroid hormones is returned to the body's iodide pool for reutilization after catabolism of the hormone. In view of these complexities, a sensible approach to defining iodine requirements would be to relate the animal's physiological assessment of hypothyroidism, i.e. TRH or TSH induction, to dietary iodine supply. Until this is done, pragmatism must suffice (Table 12.5).

Non-ruminants

Highly variable iodine requirements have been reported for poultry when determined by feeding trial. Thus Wilgus *et al.* (1953) placed the minimum needs of growing and breeding birds at close to 1 mg kg^{-1} DM and those of laying birds at 0.2–1.0 mg kg^{-1} DM. These are higher than the estimates of 0.03–0.15 mg kg^{-1} DM given by Godfrey *et al.* (1953) and of 0.075 mg kg^{-1} DM for normal growth of chicks and 0.3 mg kg^{-1} DM for completely normal thyroid structure given by Creek *et al.* (1954). A dietary intake of 0.3 mg I kg^{-1} DM was adequate for growth and normal thyroid development in pheasants and quail (Scott *et al.*, 1960). The average hen's egg contains some 4–10 μg of iodine, most of which is in the yolk. These amounts are reduced

Table 12.7. Minimum iodine requirements of farm animals (from Mitchell and McClure, 1937).

Animal	Weight (kg)	Heat production (kcal)	Iodine requirement (μg day^{-1})
Poultry	2.3	225	5–9
Sheep	50	2,500	50–100
Pigs	68	4,000	80–160
Milking cow (16 l day^{-1})	454	20,000	400–800

in goitre areas or in conditions of prolonged iodine deficiency but can be increased as much as 100-fold by feeding the hen with large amounts of iodine. The remarkable ability of the hen to adjust the iodine concentration of her eggs to the iodine content of the diet makes it difficult to relate her iodine requirements to rate of egg production. Experiments with growing pigs indicate that, on cereal and soybean-meal rations, the iodine requirement to avoid thyroid hyperplasia lies between 0.09 and 0.15 mg kg^{-1} DM (Cromwell *et al.*, 1975; Schöne *et al.*, 1988). Sihombing *et al.* (1974) showed that the growth of 6–7-week-old pigs was only slightly and inconsistently compromised when the diet contained 0.03–0.04 mg I kg^{-1} DM, after prior depletion of iodine reserves. However, thyroid weight was invariably and markedly reduced by adding 0.1 mg I kg^{-1} DM. For the pig and other species, requirements based on minimum thyroid weight or iodine uptake contain a safety margin, in terms of animal health and performance.

Ruminants

The ARC (1980) estimates for ruminants (TSR-based) allow for a substantial effect of low environmental temperature on iodine requirement. When converted to dietary concentrations, the mean requirements were 0.11 mg I kg^{-1} DM in summer and 0.54 mg I kg^{-1} DM in winter for sheep, with a similar winter value for cattle (0.52). An effect of environmental temperature on iodine uptake by the thyroid gland was demonstrated by Lengemann (1979) in lactating goats, albeit with small numbers (two), using a before-and-after design. Thyroid retention of ^{125}I was about tenfold higher (at 10% of the dose) 50 days after oral dosage when animals were kept at 5°C rather than 33°C. At the lower temperature, less ^{125}I was excreted in milk (2.6 vs. 16.8% of dose), but more was lost in urine (71.2 vs. 52.5%), and the quantitative impact in terms of requirement remained unclear. Alternative estimates of requirement, derived from the relationship between dietary iodine concentrations and incidence of disorder, suggest a lesser effect of low temperatures. Mee *et al.* (1995) reported a significant reduction in thyroid weight in perinatal calves, from 22.9 to 15.8 g, when the pregnant cow's iodine intake from a grass silage diet, containing 0.15–0.21 mg I kg^{-1} DM, was increased by about 25 mg $head^{-1}$ day^{-1}, and McCoy *et al.* (1997) reported a 120% increase in thyroid weight with only 0.06 mg I kg^{-1} DM. However, survival of the calves born in winter (February–March: Mee *et al.*, 1995) and indoors (McCoy *et al.*, 1997) was not improved by iodine supplementation. This suggests that the iodine supply was marginal and unlikely to be five times higher than the summer requirement, as suggested by ARC (1980). Herbage levels of 0.18–0.27 mg I kg^{-1} DM can sustain normal growth in cattle (Wichtel *et al.*, 1996) and wool growth in sheep in summer in New Zealand (Barry *et al.*, 1983).

Influence of Goitrogens on Iodine Requirements

Organic goitrogens

The dominant influence of organic goitrogens on iodine requirements has rarely been quantified. The ARC (1980) suggested an increase in iodine supply for ruminants to 2 mg kg^{-1} DM when the presence of substantial quantities of goitrogens is suspected. That an increase is necessary is shown by the linear increase in thyroid weight (mg kg^{-1} body weight) from 67.4 to 120.8 as the glucosinolate concentration in the lambs' diet (from rape-seed meal) increased up to 17.5 mmol kg^{-1}: the concentrate component of the diet contained 0.8 mg added I kg^{-1} DM (Hill *et al.*, 1990). Incorporation of rape-seed meal (80 g kg^{-1} DM of a high-glucosinolate type) increased the iodine requirement of weaned pigs from around 0.1 to 0.5 mg I kg^{-1} DM (Schöne *et al.*, 1988). Addition of 50 mmol KCNS kg^{-1} DM to the diet of young pigs induced goitre and a precipitous fall in serum PBI (from 45 down to 5 μg l^{-1} in 21 days), but the challenge was severe and values recovered almost as quickly when the goitrogen was removed from the low-iodine diet (0.04 mg I kg^{-1} DM) (Sihombing *et al.*, 1974). Administration of 6 mmol KCNS day^{-1} by vaginal pessary for the last two-thirds of pregnancy in ewes lowered their plasma T_3 and T_4 concentrations by 11% ($P < 0.05$) without affecting values in the newborn lamb or subsequent lamb growth (Donald *et al.*, 1993). Until further information becomes available, the ARC (1980) recommendation of 2 mg I kg^{-1} DM for thiocyanate-rich or thiocyanate-generating diets should be maintained as a safe allowance for ruminants. The problem presented by thiouracil-type goitrogens in the diet of non-ruminants cannot be remedied by increasing iodine supply and is compounded by the possibility of impairment of T_3 synthesis at sites other than the thyroid gland.

Inorganic goitrogens

Iodine requirements may be affected by gross dietary imbalance involving several minerals. High arsenic and high fluoride intakes have tentatively been incriminated as contributing factors in the incidence of human goitre in certain parts of the world, and an association between the incidence of goitre and the consumption of hard water high in calcium and magnesium has been mooted. However, evidence for an antagonism between fluorine and iodine in livestock is unconvincing (see Chapter 18). Antithyroid activity has been demonstrated for several cations and anions when ingested in excess, notably bivalent cobalt, rubidium, sodium chloride and especially perchlorate. These may act by increasing urinary excretion or by interfering with iodine uptake by the thyroid; like all potential goitrogens, they are of particular significance when dietary iodine levels are marginal. In endemic goitre areas in parts of central Europe, it has been suggested that either a deficiency or an excess of cobalt can accentuate the incidence of goitre in farm animals (Novikova, 1964), but an effect of cobalt deficiency was not evident in a recent study (Mburu *et al.*, 1994).

Iodine Toxicity

Iodine is a cumulative, chronic poison and the reports of tolerable doses for all sources naturally vary. As with many minerals given in excess, the smaller livestock species appear to be more tolerant than large animals, but this is partly because large-animal experiments usually last longer than those with pigs and poultry. Nevertheless, tolerances have been placed at 300–400 mg I kg^{-1} DM for pigs and poultry and 50 mg I kg^{-1} DM for sheep and cattle (NRC, 1980). Exposure to excess iodine paradoxically results in hypothyroidism, because of feedback inhibition of T_3 synthesis. However, there is a wide margin between the required (< 0.5 mg I kg^{-1} DM) and the tolerated concentration (< 50 mg I kg^{-1} DM) for species other than the horse. The horse appears to be exceptionally vulnerable to iodine toxicity, goitre being reported in mares and their foals given large amounts of seaweed (McDowell, 1992). The tolerance of equines has been placed as low as 5 mg I kg^{-1} DM. Beef cattle given diets containing 100–200 mg I kg^{-1} DM for 3–4 months grew poorly and developed respiratory problems; the minimum safe concentration ranges were considered to be 25–50 mg I kg^{-1} DM (Newton *et al.*, 1974). The only natural dietary circumstances which present a risk of iodine toxicity are where large proportions of seaweed or its derivatives are fed; they can contain 4–6 g I kg^{-1}.

Toxicoses have been caused by misuse of ethylenediamine dihydroiodide (EDDI) in the oral treatment of foot-rot and soft-tissue lumpy jaw, particularly in the USA. The signs of toxicity were associated with iodine intakes (as EDDI) of 74–402 mg day^{-1} and were similar to those for other sources of iodine; they include nasal and lacrymal discharge, conjunctivitis, coughing, bronchopneumonia, hair loss and dermatitis (Hillman and Curtis, 1980). Metabolic rate is elevated and indicated by raised body temperature and heart rate. Blood glucose and urea concentrations are raised and immunosuppression has been reported (Haggard *et al.*, 1980). When EDDI was incorporated into a free-access salt mixture (1.25 g I kg^{-1}) for a herd of beef cows at pasture, consumption of the mixture ranged from 20 to 70 g $animal^{-1}$ $unit^{-1}$ day^{-1} and provided sufficient iodine (24–88 mg day^{-1}) to reduce the incidence of foot-rot without harming calf performance; mean serum iodine did not exceed 0.46 mg l^{-1} (Maas *et al.*, 1984). Total serum iodine increases logarithmically as the dose increases, reaching about 4 mg I l^{-1} when 0.77 mg EDDI kg^{-1} live weight (LW) was given daily for 28 days (Maas *et al.*, 1989) and providing a rough index of exposure. Milk iodine concentrations can exceed 2 mg l^{-1} in the dairy cow and also confirm iodosis (Hillman and Curtis, 1980). Thyroid hormone concentrations present a confusing picture, sometimes remaining normal (Hillman and Curtis, 1980), but they are decreased in the newborn calf (Fish and Swanson, 1983). Tolerances of bolus doses of EDDI equivalent to 100 mg I kg^{-1} dry feed have been reported after 13 months' exposure. Onset of lactation reduces serum iodine levels and the risk of toxicity, because extra food is provided, and much of the excess iodine can be secreted in milk (Fish and Swanson, 1983). In the dairy cow, neonatal losses caused by iodosis did not present typical morphological or histological evidence of goitre (Fish and Swanson, 1983). High iodine intakes which are tolerable by the dairy cow may increase iodine levels in milk sufficiently to increase the risk of thyrotoxicosis in humans (Phillips *et al.*, 1988).

References

Alderman, G. and Jones, D.I.H. (1967) The iodine content of pastures. *Journal of the Science of Food and Agriculture* 18, 197–199.

Alderman, G. and Stranks, M.H. (1967) The iodine content of bulk herd milk in summer in relation to estimated dietary iodine intake of cows. *Journal of the Science of Food and Agriculture* 18, 151–153.

Andrewartha, K.A., Caple, I.W., Davies, W.D. and McDonald, J.W. (1980) Observations on serum thyroxine concentrations in lambs and ewes to assess iodine nutrition. *Australian Veterinary Journal* 56, 18–21.

Andrews, E.D. and Sinclair, D.P. (1962) Goitre and neonatal mortality in lambs. *Proceedings of the New Zealand Society of Animal Production* 22, 123–132.

Andrews, F.N., Shrewsbury, C.L., Harper, C., Vestal, C.M. and Doyle, L.P. (1948) Iodine deficiency in newborn sheep and swine. *Journal of Animal Science* 7, 298–310.

ARC (1980) *The Nutrient Requirements of Ruminant Livestock.* Commonwealth Agricultural Bureaux, Farnham Royal, UK, pp. 251–256.

Arthur, J.R. (1997) Selenium biochemistry and function. In: Fischer, P.W.F., L'Abbé, M.R., Cockell, K.A. and Gibson, R.S. (eds) *Proceedings of the Ninth International Symposium on Trace Elements in Man and Animals (TEMA 9).* NRC Research Press, Ottawa, Canada, pp. 1–5.

Arthur, J.R., Nicol, F. and Beckett, G.J. (1990) Hepatic iodothyronine deiodinase: the role of selenium. *Biochemical Journal* 272, 537–540.

Arthur, J.R., Nicol, F., Guo, Y. and Trayhurn, P. (1992) Progressive effects of selenium deficiency on the acute, cold-induced stimulation of type II deiodinase activity in rat brown adipose tissue. *Proceedings of the Nutrition Society* 51, 63A.

Ash, C.P.J., Crompton, D.W.T. and Lunn, P.G. (1985) Endocrine responses of protein-malnourished rats infected with *Nippostrongylus brasiliensis* (Nematoda). *Parasitology* 91, 359–368.

Barry, T.N., Duncan, S.J., Sadler, W.A., Millar, K.R. and Sheppard, A.D. (1983) Iodine metabolism and thyroid hormone relationships in growing sheep fed kale (*Brassica oleracea*) and ryegrass (*Lolium perenne*) – white clover (*Trifolium repens*) fresh forage diets. *British Journal of Nutrition* 49, 241–254.

Beard, J., Tobin, B. and Green, W. (1989) Evidence for thyroid hormone deficiency in iron-deficient, anaemic rats. *Journal of Nutrition* 119, 772–778.

Beckett, G.J. and Arthur, J. (1994) The iodothyronine deiodinases and 5′ deiodination. *Baillière's Clinical Endocrinology and Metabolism* 8, 285–304.

Bell, J.M. (1984) Nutrients and toxicants in rapeseed meal: a review. *Journal of Animal Science* 58, 996–1010.

Blum, J.W. and Kunz, P. (1981) The effects of fasting on thyroid hormone levels and kinetics of reverse tri-iodothyronine in cattle. *Acta Endocrinologica* 98, 234–239.

Butler, G.W. and Glenday, A.C. (1962) Iodine content of pasture plants. 2. Inheritance of leaf iodine content of perennial ryegrass (*Lolium perenne* L.). *Australian Journal of Biological Sciences* 15, 183–187.

Chilean Iodine Educational Bureau (1952) *Iodine Content of Foods.* London, 183 pp.
Christopherson, R.J., Gonyon, H.W. and Thompson, J.R. (1979) Effects of temperature and feed intake on plasma concentrations of thyroid hormones in beef cattle. *Canadian Journal of Animal Science* 59, 655–661.
Creek, R.D., Parker, H.E., Hauge. S.M., Andrews, F.N. and Carrick. C.W. (1954) The iodine requirements of young chickens. *Poultry Science* 33, 1052.
Cromwell, G.L., Sihombing, D.T.H. and Hays, V.W. (1975) Effects of iodine level on performance and thyroid traits of growing pigs. *Journal of Animal Science* 41, 813–818.
Donald, G.E., Langlands, J.P., Bowles, J.E. and Smith, A.J. (1993) Subclinical selenium insufficiency. 4. Effects of selenium, iodine and thiocyanate supplementation of grazing ewes on their selenium and iodine status and on the status and growth of their lambs. *Australian Journal of Experimental Agriculture* 33, 411–416.
Downer, J.V., Hemken, R.W., Fox, J.D. and Bull, L.J. (1981) Effect of dietary iodine on tissue iodine content of the bovine. *Journal of Animal Science* 52, 413–417.
Ellis, K.J. and Coverdale, O.R. (1982) The effects on new-born lambs of administering iodine to pregnant ewes. *Proceedings of the Australian Society of Animal Production* 14, 660.
Emanuelson, M. (1994) Problems associated with feeding rapeseed meal to dairy cows. In: Garnsworthy, P.C. and Cole, D.J.A. (eds) *Recent Advances in Animal Nutrition – 1994.* Nottingham University Press, Nottingham, pp. 189–214.
Erenberg, A., Omori, K., Menkes, J.N., Oh, W. and Fisher, D.A. (1974) Growth and development of the thyroidectomised ovine fetus. *Pediatrics Research* 8, 783–789.
Ferguson, K.A., Schinckel, P.G., Carter, H.B. and Clarke, W.H. (1956) The influence of the thyroid on wool follicle development in the lamb. *Australian Journal of Biological Sciences* 9, 575–585.
Fish, R.E. and Swanson, E.W. (1983) Effects of excessive iodide administered in the dry period on thyroid function and health of dairy cows and their calves in the periparturtient period. *Journal of Animal Science* 56, 162–172.
Fitzgerald, S. (1983) The use of forage crops for store lamb fattening. In: Haresign, W. (ed.) *Sheep Production.* Butterworths, London, pp. 239–286.
Follett, B.K. and Potts, C. (1990) Hypothyroidisan effects reproductive refractoriness and the seasonal oestrus period in Welsh Mountain ewes. *Journal of Endrocrinology* 127, 103–109.
Fuge, R. (1996) Geochemistry of iodine in relation to iodine deficiency diseases. In: Appleton, J.D., Fuge, R.D. and McCall, G.J.H. (eds) *Environmental Geochemistry and Health.* Geological Society Special Publication No. 113, London, pp. 201–212.
Godfrey, P.R., Carrick, C.W. and Quackenbush, F.W. (1953) Iodine nutrition of chicks. *Poultry Science* 32, 394–396.
Groppel, B. and Anke, M. (1986) Iodine content of feedstuffs, plants and drinking water in the GDR. In: Anke, M., Boumann, W., Braunich, H., Bruckner, B. and Groppel, B. (eds) *Spurenelement Symposium Proceedings,* Vol. 5. *Iodine.* Friedrich Schiller University, Jena, pp. 19–28.
Gurevich, G.P. (1964) Soil fertilization with coastal iodine sources as a prophylactic measure against endemic goiter. *Federation Proceedings* 23, T 511–T 514.
Haggard, D.L., Stowe, H.D., Conner, G.H. and Johnson, D.W. (1980) Immunological effects of experimental iodine toxicosis in young cattle. *American Journal of Veterinary Research* 41, 539–543.
Harington, C.R. (1953) *The Thyroid Gland, its Chemistry and Physiology.* Oxford University Press, London.

Hartmans, J. (1974) Factors affecting the herbage iodine content. *Netherlands Journal of Agricultural Science* 22, 195–206.

Hemken, R.W. (1960) Iodine. *Journal of Dairy Science* 53, 1138–1143.

Hemken, R.W., Vandersall, J.H., Oskarsson, M.A. and Fryman, L.R. (1972) Iodine intake related to milk iodine and performance of dairy cattle. *Journal of Dairy Science* 55, 931–934.

Hercus, C.E., Benson, W.N. and Carter, C.L. (1925) Endemic goitre in New Zealand, and its relation to the soil-iodine. *Journal of Hygiene* 24, 321–402.

Hetzel, B.S. (1991) The international public health significance of iodine deficiency. In: Momcilovic, B. (ed.) *Proceedings of the Seventh International Symposium on Trace Elements in Man and Animals, Dubrovnik.* IMI, Zagreb, 7-1–7-3.

Hetzel, B.S. and Welby, M.C. (1997) Iodine. In: O'Dell, B.L. and Sunde, R.A. (eds) *Handbook of Nutritionally Essential Mineral Elements.* Marcel Dekker, New York, pp. 557–582.

Hill, R. (1991) Rapeseed meal in the diet of ruminants. *Nutrition Abstracts and Reviews, Series B* 61, 139–155.

Hillman, D. and Curtis, A.R. (1980) Chronic iodine toxicity in dairy cattle: blood chemistry, leucocytes and milk iodine. *Journal of Dairy Science* 63, 55–63.

Hopkins, P.S. (1975) The development of the foetal ruminant. In: McDonald, I.W. and Warner, A.C.I. (eds) *Metabolism and Digestion in the Ruminant. Proceedings of the IVth International Symposium on Ruminant Physiology*, 7th edn. University of New England Publishing Unit, Armidale, pp. 1–14.

Ingar, S.H. (1985) The thyroid gland. In: Wilson, J.D. and Foster, D.W. (eds) *Williams Textbook of Endocrinology.* Saunders, Philadelphia, pp. 682–815.

Irvine, C.H.G. (1967) Protein-bound iodine in the horse. *American Journal of Veterinary Research* 28, 1687–1692.

Iwarsson, K. (1973) Rapeseed meal as a protein supplement for dairy cows. 1. The influence of certain blood and milk parameters. *Acta Veterinaria Scandinavica* 14, 570–594.

Johns, A.T. (1956) The influence of high-production pasture on animal health. In: *Proceedings of the 7th International Grassland Congress, Palmerston North, New Zealand*, pp. 251–261.

Johnson, J.M. and Butler, G.W. (1957) Iodine content of pasture plants. 1. Method of determination and preliminary investigation of species. *Physiologia Plantarum* 10, 100–111.

Jones, R.J. and Megarity, R.G. (1983) Comparative responses of goats fed on *Leucena leucephala* in Australia and Hawaii. *Australian Journal of Agricultural Research* 34, 781–790.

Karmarkar, M.G., Deo, M.G., Kochupillai, N. and Ramalingaswami, V. (1974) Pathophysiology of Himalayan endemic goiter. *American Journal of Clinical Nutrition* 27, 96–103.

Kelly, F.C. (1953) Studies on the stability of iodine compounds in iodized salt. *Bulletin of the World Health Organization* 9, 217–230.

Khajarern, S., Khajarern, J.M., Kitpanit, N. and Muller, Z.O. (1977) Cassava in the nutrition of swine. In: Nestel, B. and Graham, M. (eds) *Cassava as Animal Feed. Proceedings of a Workshop Held at the University of Guelph.* International Development Research Centre, Ottawa, pp. 56–64.

King, C.F. (1976) Ovine congenital goitre associated with minimal thyroid enlargement. *Australian Journal of Experimental Agriculture and Animal Husbandry* 16, 651–655.

Kirchgessner, M. (1959) Wechselbeziehungen zwischen Spurenelementen in Futtermitteln und tierischen Substanzen sowie Abhangigkeitsverhaltnisse zwischen einzelnen Elementen bei der Retention. 5. Die Wechselwirkungen zwischen verschiedenen Elementen in der Colostral- und normalen Milch. *Zeitschrift für Tierphysiologie Tierernahrung und Futtermittelkunde* 14, 270–277.

Knights, G.I., O'Rourke, P.K. and Hopkins, P.S. (1979) Effects of iodine supplementation of pregnant and lactating ewes on the growth and maturation of their offspring. *Australian Journal of Experimental Agriculture and Animal Husbandry* 19, 19–22.

Laarveld, B., Brockman, R.P. and Christensen, D.A. (1981) The effects of Tower and Midas rapeseed meals on milk production and concentrations of goitrogens and iodine in milk. *Canadian Journal of Animal Science* 61, 131–139.

Lengemann, F.W. (1979) Effect of low and high ambient temperatures on metabolism of radioiodine by the lactating goat. *Journal of Dairy Science* 62, 412–415.

Levine, H., Remington, R.E. and von Kolnitz, H. (1933) Studies on the relation of diet to goiter. 2. The iodine requirement of the rat. *Journal of Nutrition* 6, 347–354.

Lüdke, H. and Schöne, F. (1988) Copper and iodine in pigs diets with high glucosinolate rapeseed meal. I. Performance and thyroid hormone status. *Animal Feed Science and Technology* 22, 33–43.

MAFF (1990) *UK Tables of the Nutritive Value and Chemical Composition of Foodstuffs.* Givens, D.I. (ed.), Rowett Research Services, Aberdeen, UK.

Maas, J., Davis, L.E., Hempsteed, C., Berg, J.N. and Hoffman, K.A. (1984) Efficacy of ethylenediamine dihydriodide in the prevention of naturally occurring foot rot in cattle. *American Journal of Veterinary Research* 45, 2347–2350.

Maas, J., Berg, J.N. and Petersen, R.G. (1989) Serum distribution of iodine after oral administration of ethylenediamine dihydriodide in cattle. *American Journal of Veterinary Research* 50, 1758–1759.

McCoy, M.A., Smyth, J.A., Ellis, W.A. and Kennedy, D.G. (1995) Parenteral iodine and selenium supplementation in stillbirth perinatal weak calf syndrome. *Veterinary Record* 136, 124–126.

McCoy, M.A., Smyth, J.A., Ellis, W.A., Arthur, J.R. and Kennedy, D.G. (1997) Experimental reproduction of iodine deficiency in cattle. *Veterinary Record* 141, 544–547.

McDonald, R.C., Manley, T.R., Barry, T.N., Forss, D.A. and Sinclair, A.G. (1981) Nutritional evaluation of kale (*Brassica oleracea*) diets. 3. Changes in plant composition induced by fertility practices, with special reference to SMCO and glucosinolate concentrations. *Journal of Agricultural Science, Cambridge* 97, 13–23.

McDowell, L.R. (1992) Iodine. In: *Minerals in Animal and Human Nutrition.* Academic Press, New York, pp. 224–245.

McGowan, A.C. (1983) The use of 'Lipiodol' for sub-clinical iodine deficiency in livestock. *Proceedings of the New Zealand Society of Animal Production* 43, 135–136.

Maddocks, S., Chandrasekhar, Y. and Setchell, B.P. (1985) Effect on wool growth of thyroxine replacement in thyroidectomised lambs. *Australian Journal of Biological Science* 38, 405–410.

Marine, D. and Williams, W.W. (1908) The relation of iodine to the structure of the thyroid gland. *Archives of Internal Medicine* 1, 349–384.

Mason, R.W. (1976) Milk iodine content as an estimate of the dietary iodine status of sheep. *British Veterinary Journal* 132, 374–379.

Mason, R.W. and Laby, R. (1978) Prevention of ovine congenital goitre using iodine-releasing intramuscular devices: preliminary results. *Australian Journal of Experimental Agriculture and Animal Husbandry* 18, 653–657.

Mburu, J.N., Kamau, J.M.Z. and Badamana, M.S. (1994) Thyroid hormones and metabolic rate during induction of vitamin B12 deficiency in goats. *New Zealand Veterinary Journal* 42, 187–189.

Mee, J.F., Rogers, P.A.M. and O'Farrell, K.J. (1995) Effect of feeding a mineral–vitamin supplement before calving on the calving performance of a trace element deficient dairy herd. *Veterinary Record* 137, 508–512.

Miller, E.R. and Ammerman, C.B. (1995) Iodine bioavailability. In: Ammerman, C.B., Baker, D.H. and Lewis, A.J. (eds) *Bioavailability of Nutrients for Animals.* Academic Press, New York, pp. 157–168.

Miller, J.K., Swanson, E.W. and Hansen, S.M. (1965) Effects of feeding potassium iodide, 3,5 diodosalicylic acid or L-thyroxine on iodine metabolism of lactating dairy cows. *Journal of Dairy Science* 48, 888–894.

Miller, J.K., Swanson, E.W., Lyke, W.A., Moss, B.R. and Byrne, W.F. (1974a) Effect of thyroid status on digestive tract fill and flow rate of undigested residues in cattle. *Journal of Dairy Science* 57, 193–197.

Miller, J.K., Swanson, E.W., Spalding, G.E., Lyke, W.A. and Hall, R.F. (1974b) The role of the abomasum in recycling of iodine in the bovine. In: Hoekstra, W.G., Suttie, J.W., Ganther, H.E. and Mertz, W. (eds) *Proceedings of the Second International Symposium on Trace Elements in Man and Animals.* University Park Press, Baltimore, pp. 638–640.

Mitchell, H.H. and McClure, F.J. (1937) *Mineral Nutrition of Farm Animals.* Bulletin No. 99, National Research Council, USA.

Newton, G.L., Barrick, E.R., Harvey, R.W. and Wise, M.B. (1974) Iodine toxicity, physiological effects of elevated dietary iodine on calves. *Journal of Animal Science* 38, 449–455.

Nicol, F., Lefrane, H., Arthur, J.R. and Trayhurn, P. (1994) Characterisation and postnatal development of 5′-deiodinase activity in goat perirenal fat. *American Journal of Physiology* 267, R144–R149.

Novikova, E.P. (1964) Effects of different amounts of dietary cobalt on iodine content of rat thyroid gland. *Federation Proceedings* 23, T 459–T 460.

NRC (1980) *Mineral Tolerances of Domestic Animals.* National Academy of Sciences, Washington, DC.

Orr, J.B. and Leitch, I. (1929) *Iodine in Nutrition. A Review of Existing Information.* Special Report Series, Medical Research Council, UK, No. 123, 108 pp.

Phillips, D.I.W., Nelson, M., Barker, D.J.P., Morris, J.A. and Wood, T.J. (1988) Iodine in milk and the incidence of thyrotoxicosis in England. *Clinical Endocrinology* 28, 61–66.

Potter, B.J., McIntosh, G.H. and Hetzel, B.S. (1981) The effect of iodine deficiency on foetal brain development in the sheep. In: Hetzel, B.S. and Smith, R.M. (eds) *Fetal Brain Disorders – Recent Approaches to the Problem of Mental Deficiency.* Elsevier/North Holland Biomedical Press, Amsterdam, pp. 119–147.

Potter, B.J., Mano, M.T., Belling, G.B., McIntosh, G.H., Hua, C., Cragy, B.G., Marshall, J., Wellby, M.L. and Hetzel, B.S. (1982) Retarded fetal brain development resulting from severe iodine deficiency in sheep. *Applied Neurobiology* 8, 303–313.

Prichard, R.K., Hennessy, D.R. and Griffiths, D.A. (1974) Endocrine responses of sheep to infection with *Trichostrongylus colubriformis. Research in Veterinary Science* 17, 182–187.

Puls, R. (1994) *Mineral Levels in Animal Health*, 2nd edn. Diagnostic data, Sherpa International, Clearbrook BC, pp. 83–109.

Rhind, S.M. and McMillen, S.R. (1996) Effects of methylthiouracil treatment on the growth and moult of cashmere fibre in goats. *Animal Science* 62, 513–520.

Rogler, J.C. (1958) Effects of iodine on hatchability and embryonic development. *Dissertation Abstracts* 18, 1925–1926.

Schöne, F., Lüdke, H., Hennig, A. and Jahreis, G. (1988) Copper and iodine in pig diets with high glucosinolate rapeseed meal. II. Influence of iodine supplements for rations with rapeseed meal untreated or treated with copper ions on performance and thyroid hormone status of growing pigs. *Animal Feed Science and Technology* 22, 45–59.

Scott, M.L., van Tienhoven, A., Holm, E.R. and Reynolds, R.E. (1960) Studies on the sodium, chlorine and iodine requirements of young pheasants and quail. *Journal of Nutrition* 71, 282–288.

Siebert, B.D. and Hunter, R.A. (1982) In: Hacker, J.B. (ed.), *Nutrition Limits to Animal Production from Pastures.* Commonwealth Agricultural Bureaux, Farnham Royal, Slough, England, pp. 409–425.

Sihombing, D.T.H., Cromwell, G.L. and Hays, V.W. (1974) Effects of protein source, goitrogens and iodine level on performance and thyroid status of pigs. *Journal of Animal Science* 39, 1106–1112.

Sinclair, D.P. and Andrews, E.D. (1958) Prevention of goitre in newborn lambs from kale-fed ewes. *New Zealand Veterinary Journal* 6, 87–95.

Sinclair, D.P. and Andrews, E.D. (1961) Deaths due to goitre in newborn lambs prevented by iodized poppy-seed oil. *New Zealand Veterinary Journal* 9, 96–100.

Smyth, J.A., McNamee, P.T., Kennedy, D.G., McCullough, S.J., Logan, E.F. and Ellis, W.A. (1992) Stillbirth/perinatal weak calf syndrome: preliminary pathological, microbiological and biochemical findings. *Veterinary Record* 130, 237–240.

Spiegl, C., Bestetti, G.E., Rossi, G.L. and Blum, J.W. (1993) Normal circulating triiodothyronine concentrations are maintained despite severe hypothyroidism in growing pigs fed rapeseed presscake meal. *Journal of Nutrition* 123, 1554–1561.

Statham, M. and Bray, A.C. (1975) Congenital goitre in sheep in southern Tasmania. *Australian Journal of Agricultural Research* 26, 751–768.

Statham, M. and Koen, T.B. (1981) Control of goitre in lambs by injection of ewes with iodized poppy seed oil. *Australian Journal of Experimental Agriculture and Animal Husbandry* 22, 29–34.

Sutherland, R.L. and Irvine, C.H.G. (1974) Effect of season and pregnancy on total plasma thyroxine concentrations in sheep. *American Journal of Veterinary Research* 35, 311–312.

Villar, D., Rhind, S.J., Dicks, P., McMillen, S.R., Nicol, F. and Arthur, J.R. (1998) Effect of propylthiouracil-induced hypothyroidism on thyroid hormone profiles and tissue deiodinase activity in cashmere goats. *Small Ruminant Research* 29, 317–324.

Wallace, A.L.C., Gleeson, A.R., Hopkins, P.S., Mason, R.W. and White, R.R. (1978) Plasma thyroxine concentrations in grazing sheep in several areas of Australia. *Australian Journal of Biological Science* 31, 39–41.

Walton, E.A. and Humphrey, J.D. (1979) Endemic goitre of sheep in the highlands of Papua New Guinea. *Australian Veterinary Journal* 55, 43–44.

Whitehead, D.C. (1975) Uptake by perennial ryegrass of iodide, elemental iodine and iodate added to soil as influenced by various amendments. *Journal of the Science of Food and Agriculture* 26, 361–367.

Wichtel, J.J., Craigie, A.L., Freeman, D.A., Varela-Alvarez, H. and Williamson, N.B. (1996) Effect of selenium and iodine supplementation on growth rate and on thryoid and somatotropic function in dairy calves at pasture. *Journal of Dairy Science* 79, 1865–1872.

Wilgus, H.S., Jr, Gassner, F.X., Patton, A.R. and Harshfield, G.S. (1953) *The Iodine Requirements of Chickens.* Technical Bulletin No. 49, Colorado Agricultural Experiment Station.

Wilson, J.G. (1975) Hypothyroidism in ruminants with special reference to foetal goitre. *Veterinary Record* 97, 161–164.

Zagrodski, P., Nicol, F., McCoy, M.A., Smyth, J.A., Kennedy, D.G., Beckett, G.J. and Arthur J.R. (1998) Iodine deficiency in cattle: compensatory changes in thyroidal selenoenzymes. *Research in Veterinary Science* 64, 209–211.

Iron

13

Introduction

Iron is by far the most abundant trace element in the body and its value as a dietary constituent has been appreciated for over 2000 years. A connection between dietary iron supply and disorder of the blood was proposed in the 16th century, but its physiological basis was not discerned until much later, when horse haemoglobin was shown to contain 0.335% of iron (Zinoffsky, 1886) and the finding was confirmed in a range of animal species. Stockman (1893) ended a long controversy by showing that haemoglobin concentrations in anaemic women could be increased rapidly by supplementation with iron salts as diverse as ferrous citrate and ferrous sulphide. For farm livestock, iron supply is rarely critical, other than in the first few weeks of life, when rapid expansion of red-cell mass imposes great demands upon animal and diet to deliver sufficient iron to erythropoietic tissue.

Dietary Sources of Iron

Most plant materials used in the feeding of farm animals contain large, though variable, concentrations of iron, depending on the plant species, the type of soil on which the plants grow and the degree of contamination by soil.

Forages

In a detailed investigation of the mineral content of English pasture species, the mean concentrations of iron (FE) (mg kg^{-1} dry matter (DM)) were 306 for legumes, 264 for grasses and 358 for 'herbs' (Thomas *et al.*, 1952), but means can be deceptive. New Zealand pastures can range from 111 to 3850 mg Fe kg^{-1} DM (mean 581 ± 163) (Campbell *et al.*, 1974) and pastures in southern Australia from 70 to 2300 mg Fe kg^{-1} DM (see Table 11.2). High values are attributable to soil contamination, which is most likely to occur on soils prone

© CAB *International* 1999. *Mineral Nutrition of Livestock*
(E.J. Underwood and N.F. Suttle)

to waterlogging. Lower values, between 90 and 110 mg Fe kg^{-1} DM, have been reported for Egyptian green-cut clover and clover hay (Abou-Hussein *et al.*, 1970) and values below 30 mg Fe kg^{-1} DM for some grasses on poor sandy soils in Australia (E.J. Underwood, unpublished data).

Grains and seeds

Most cereal grains contain 30–60 mg Fe kg^{-1} DM and species differences appear to be small, although 10 and 20 mg kg^{-1} DM have been recorded for Egyptian-grown maize and barley, respectively (Abou-Hussein *et al.*, 1970). Leguminous seeds and the oil-seed meals are invariably richer in iron than the cereal grains; oil-seed meals commonly contain 100–200 mg Fe kg^{-1} DM. Iron concentrations in foodstuffs as they reach the farm may be much higher than those recorded at experimental sites; in the UK, values (mg Fe kg^{-1} DM) of 100 are found in barley and maize, 120 in oats, 140 in wheat, 480 in maize-gluten feed, 220 in wheat feed and 2600 in rice-bran meal (MAFF, 1990). It was once thought that phytate-rich foods, such as cereals and legume seeds, would provide iron of low availability through the formation of unavailable ferric–phytate complexes. That the preformed tetra-complex is poorly available is beyond doubt (Bremner *et al.*, 1976; Morris and Ellis, 1976). However, wheat is of equal value as an iron source to ferric chloride, partly because it contains iron already complexed as the monoferric phytate, which is well absorbed (Morris and Ellis, 1976), and partly because the seed-coat is rich in phytase. Welch and Van Campen (1975) found that iron was more available from mature, phytate-rich soybeans than from immature soybeans, which are low in phytate.

Animal sources

Feeds of animal origin, other than milk and milk products, are rich sources of iron, and Liebscher (1958, cited by Kolb, 1993) gave the following representative figures: blood-meal, 3108; fish-meal, 381; meat-meal, 439; dried skimmed milk 52 mg Fe kg^{-1} DM. Where the first three of these are included in cereal rations as protein supplements, overall iron intakes will be substantially increased. Dried skimmed milk, whey and buttermilk powders vary greatly in iron content as a consequence of variable opportunities for contamination during processing and storage. However, the iron content of bovine milk is inherently so low that milk products used as feeds are low in this element compared with most farm feeds. Virtually nothing is known of the physiological availability of the iron in these and other farm feeds to herbivorous animals.

Absorbability

The absorbability of iron (A_{Fe}) is affected primarily by the chemical form of the iron ingested and the amounts and proportions in the diet of other metals and compounds with which iron interacts. The highest A_{Fe} values are accorded to milk and are partly attributable to the fact that iron is present partly as a highly absorbable glycoprotein, lactoferrin, and accompanied by two enhancers of iron absorption, citrate and lactose (Lonnerdal, 1988). Lactoferrin-bound iron may not be absorbed well by all species; in the newborn calf, for example, dosage with iron-saturated lactoferrin failed to prevent a postnatal fall in blood haemoglobin, whereas the same amount of iron as ferrous sulphate ($FeSO_4$) was effective (Kume and Tanabe, 1994). Absorption may become limited by the presence of unabsorbable forms of iron in the diet and weaning on to anything other than a meat-rich diet will be accompanied by a fall in the efficiency of iron absorption. Haem forms of iron are better absorbed than the inorganic or non-haem complexes that provide the main sources of the metal to domestic livestock. The absorption of non-haem forms of iron is greatly influenced by various constituents of the diet, such as ascorbic acid, which promote absorption. Of farmed livestock, only the chicken has been used to assess the availability of iron in natural foodstuffs; plant sources generally contain iron with only 20–60% of the availability of $FeSO_4$, but the reasons for the low and variable values are unclear (Henry and Miller, 1995).

Metabolism of Iron

The advent of radioactive isotopes of iron and the application of molecular-biology techniques have greatly facilitated investigations of iron absorption, transport, storage, excretion and intermediary metabolism. Animals have a limited capacity to excrete iron (Kreutzer and Kirchgessner, 1991), so that retention is largely controlled by absorption.

Absorption

Iron is absorbed according to need and is therefore affected by factors such as the iron status of the body and age. Iron absorption is greater in deficient than in iron-sufficient animals, because iron metabolism is regulated predominantly at the gut level, where the efficiency of absorption is controlled by the iron status of the mucosa (Underwood, 1977). The primary site of absorption is the duodenum. The sensitivity of mucosal control was demonstrated by Fairweather-Tait and Wright (1984) in rats. Using ^{59}Fe retention following a standard test meal (wheat flour) as the measure of absorption, they found values to fall from 60% to 9% of the dose as the iron concentration of the previous diet increased from 8 to 1270 mg kg^{-1} DM (Fig. 13.1). Feeding the lowest iron level for 24 h was sufficient to fully enhance absorption. Sensitivity was related to the migration from the crypts of mucosal cells with increased iron-binding properties in their brush-border membranes. Enhanced capacity to reduce the ubiquitous ferric iron to its absorbable ferrous state does not play a role in enhancing iron absorption (Wein and van Campen, 1994). Iron – like most minerals – is absorbed with high efficiency in milk-fed animals and this is partly attributable to the low iron concentration in milk (< 1 mg kg^{-1} DM in all species) and the high growth rates and hence requirements created by such highly digestible, energy-rich diets.

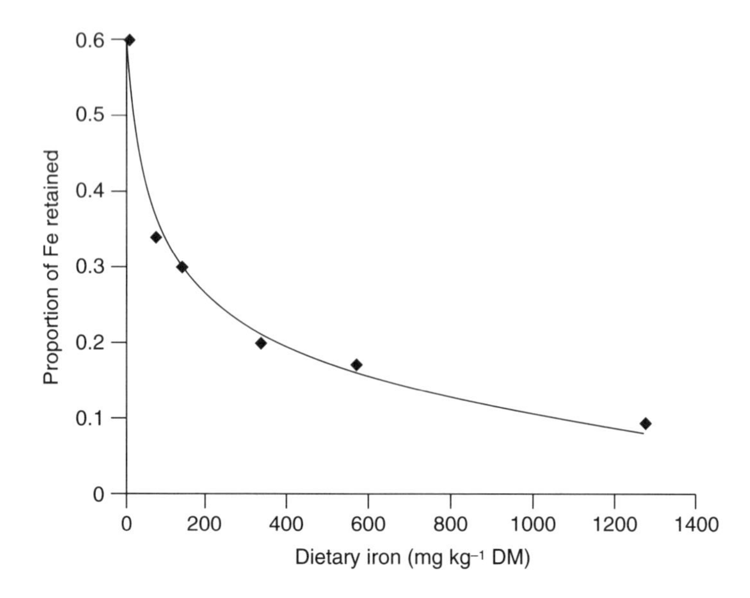

Fig. 13.1. Iron is absorbed according to need and only brief exposure of rats to excess dietary iron causes marked reductions in iron retention (data for whole body retention of ^{65}Fe from Fairweather-Tait and Wright, 1984).

Transport

Absorbed iron is delivered as ferric iron at the serosal surface, where it becomes bound to transferrin (T_f), a non-haem glycoprotein which binds two atoms of ferric iron per mole. Circulating T_f generates the total iron-binding capacity (TIBC) of the plasma and the degree of unsaturation (UIBC) reflects the proportion of iron-free apotransferrin present. Copper plays a key role in iron utilization, but the mechanism remains unclear. Ceruloplasmin, the major copper protein of the blood plasma, can function as a ferroxidase and was thus believed to facilitate the release of iron from ferritin in the intestinal mucosal and hepatic parenchymal cells and also the binding of iron to T_f (Frieden, 1971; Frieden and Hsieh, 1976), but these assumptions have been questioned (see Chapter 11). The principal ferroxidase in the mucosa and liver is the molybdeno-enzyme, xanthine dehydrogenase (see Chapter 17).

Storage

Transferrin is involved not only in the transport of absorbed iron to the tissues but also in the redistribution of storage iron and the recycling of iron from aged erythrocytes via the reticuloendothelial system (Liebold and Guo, 1992). Receptors for transferrin (T_fR) in the cell membrane carry iron by endocytosis to the cell interior. There, regulatory iron-responsive binding proteins (IREBP), sensitive to free iron concentrations, receive iron from T_f

and partition it between functional (haemoglobin synthesis) and storage (ferritin synthesis) pathways. Ferritin is the main iron storage compound of the body and its concentration in the tissues, together with that of haemosiderin, reflects the iron status of the animal. Ferritin is a non-haem protein (globulin) compound, containing up to 20% iron, which occurs widely throughout the body and particularly in the liver. A high positive correlation between serum ferritin concentrations and body iron stores exists in humans, so that its estimation is a useful indicator of iron stores (Baynes, 1997). Haemosiderin is the predominant storage form when a high iron status is attained; it contains 35% iron, primarily in a colloidal form as ferric hydroxide, and probably results from the aggregation of ferritin molecules and the subsequent denaturation of their protein constituent (Kent and Bahu, 1979). Soluble T_fR are detectable in serum and concentrations increase during iron deficiency (Baynes, 1997).

Functions of Iron

Approximately 60% of body iron is present as haemoglobin, a complex of the protoporphyrin haem and globin. The haem molecule contains one atom of iron in the centre of its ring structure and there are four rings in each haemoglobin molecule. As oxyhaemoglobin it carries oxygen from the lung to the tissues via the arterial blood, and it returns carrying carbon dioxide as carboxyhaemoglobin via the venous circulation. Haemoglobin is packaged in erythrocytes and allows the tissues to 'breathe'. Myoglobin is a simpler, less abundant Fe-porphyrin found in muscle, where its higher affinity for oxygen completes the transfer of oxygen from haemoglobin into the cell. For many years, nutritional interest in iron was focused on its role in haemoglobin formation and oxygen transport. A broader concept of the physiological significance of iron developed when the presence of iron in the haemoprotein cytochrome enzymes and the role of these enzymes in the oxidative mechanisms of all cells were established. The discovery of iron-containing flavoprotein enzymes further extended the relationship of iron to basic biochemical processes in the tissues. The ability of iron to change between the divalent and trivalent state allows the cytochromes *a*, *b* and *c* – of which iron is a part – to participate in the electron transfer chain. By activating or assisting enzymes, such as succinate dehydrogenase, iron is involved at every stage of the tricarboxylic acid (Krebs) cycle. Iron-containing catalase and peroxidases remove potentially dangerous products of metabolism, while iron-activated hydroxylases influence connective-tissue development (O'Dell, 1981). The severely iron-deficient rat shows evidence of thyroid hormone deficiency (Beard *et al.*, 1989) and decreased resistance to disease (Weinberg, 1984). Inhibition or impairment of some processes in the animal through lack of dietary iron can occur long before haemoglobin formation becomes adversely affected.

Biochemical and Physiological Manifestations of Iron Deprivation

The clinical manifestations of iron deprivation are preceded by depletion of storage iron, i.e. ferritin and haemosiderin, in the liver, kidneys and spleen, a fall in serum iron and raised TIBC and UIBC in the serum (Planas and de Castro, 1960). Serum ferritin is also reduced, but metallothionein concentrations in the blood cells increase, particularly in the reticulocytes, reflecting the increase in erythropoietic activity (Robertson *et al.*, 1989). The sequence of changes and their rates of development are fitted well by the general scheme outlined in Fig. 13.2. The initial period of *depletion* will be determined by the size of the initial liver reserve of iron, which can vary from 20 to 540 mg in the newborn calf (Charpentier *et al.*, 1966, cited by ARC, 1980). In a depletion experiment with a milk diet containing only 5 mg Fe kg^{-1} DM, the blood iron pool actually increased by 750 mg and muscle iron by 250 mg in 3 months, liver and diet contributing the necessary iron in roughly equal measure. The period of *deficiency* is marked by reductions in serum iron; haemoglobin and myoglobin concentrations begin to fall below normal without retarding growth in calves (ARC, 1980), lambs (Bassett *et al.*, 1995) and pigs (ARC, 1981). There is spare capacity to be exhausted before *dysfunction and disorder* occur. When growth rate was depressed in 'moderately' deficient calves given a diet containing 20 mg Fe kg^{-1} DM (Hostettler-Allen *et al.*, 1993), decreased activity of Fe-dependent enzymes was believed to have

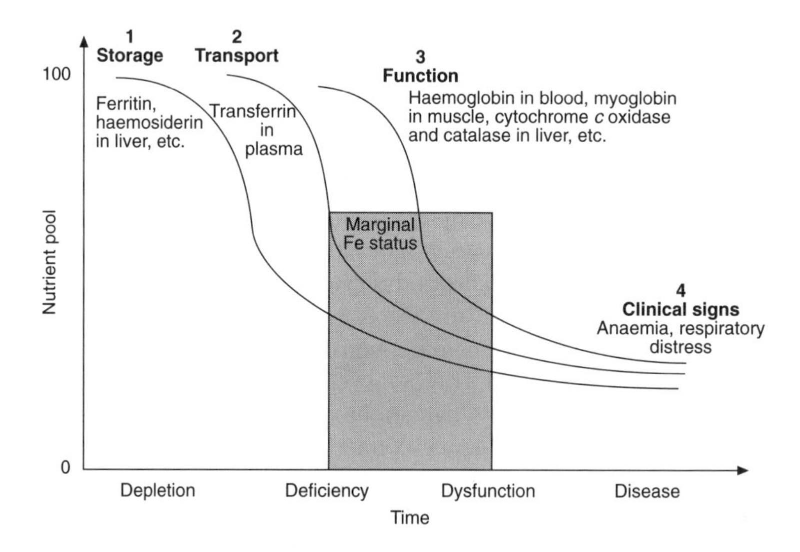

Fig. 13.2. Sequence of biochemical changes leading to the appearance of clinical signs of iron deprivation: see also Fig. 3.1 and related text.

increased anaerobic glycolysis and lactate–glucose recycling, leading to inefficient glucose utilization, i.e. *dysfunction*. In iron-deficient pigs and calves, there is a greater reduction in the catalase activity of the blood than in haemoglobin (Grassmann and Kirchgessner, 1973).

Species differences

The nature and extent of biochemical changes which accompany iron deprivation vary between animal species. In iron-deficient pigs myoglobin concentrations and cytochrome *c* oxidase activities are much reduced (Gubler *et al.*, 1957), but in calves myoglobin has priority over haemoglobin for scarce supplies of iron (ARC, 1980). In iron-deficient chicks, haemoglobin values are reduced earlier than muscle myoglobin or liver cytochrome *c* and succinic dehydrogenase activities (Davis *et al.*, 1962). The functional significance of these biochemical changes for a given species is questionable, because they precede growth retardation (ARC, 1980) and their diagnostic value is clearly limited. Increases in serum T_fR indicate reductions in the functional pool of iron in small laboratory animals (Baynes, 1997), but they have yet to be studied in farm livestock.

Clinical Manifestations of Iron Deprivation

Prolonged iron deprivation is characterized clinically by loss of appetite, poor growth, lethargy, blanching of the visible mucous membranes, increased respiration rate and, when severe, high mortality. These signs are preceded by and largely caused by the development of a progressive hypochromic, microcytic anaemia. Unseen lies a normoblastic, hyperplastic bone marrow, containing little haemosiderin. The early onset of anaemia provides a point of contrast with the late development of anaemia in cobalt and copper deprivation (see Chapters 10 and 11). Evidence of hypothyroidism in severely iron-deficient rats (haematocrit 16% as against a normal 41%) and failure to thermoregulate normally when cold-stressed (Beard *et al.*, 1989) raises the possibility that low environmental temperatures may exacerbate the clinical effects of iron deprivation in young livestock reared in cold conditions.

Cognitive dysfunction

There is currently great interest in the possibility that iron-deficiency anaemia in children is accompanied by loss of cognitive function (Pollitt, 1993). Sustained iron supplementation has improved psychomotor function in particular in the majority of trials, indicating a reversible dysfunction in mildly anaemic children. However, the spontaneous improvement in iron status in the developing farm animal would probably prevent any long-term cognitive dysfunction.

Susceptibility to infection

Decreased resistance to infection in iron deprivation has been described and the complex underlying mechanisms extensively and critically reviewed (Strauss, 1978; Weinberg, 1984). Rats with iron-deficiency anaemia are more susceptible to oral challenge with *Salmonella* and *Streptococcus pneumoniae* (Chandra, 1976), and iron-deficient piglets exhibit greater susceptibility to the endotoxin of *Escherichia coli* (Osborne and Davis, 1968) than normal healthy animals. However, the degree of anaemia established in such experiments exceeded that likely to be found on farms and mild iron deficiency can improve resistance to infection (Brock and Mainou-Fowler, 1986).

Occurrence of Iron Deprivation

Iron deprivation is of limited practical significance in farm livestock. Heavy egg production in poultry is sometimes accompanied by anaemia, but this is not associated with a lack of dietary iron. Primary iron deprivation has never been demonstrated unequivocally in grazing animals, a reflection of the normally high iron content of pastures and forages and of the opportunities for their contamination with soil and dust. The iron content of soils is commonly many times that of the plants they support. Furthermore, lactation, which is so demanding of other mineral nutrients, imposes relatively little extra demand for iron because of the low iron content of milk. Confinement increases the possibility of iron deficiency in young suckling animals or animals reared largely on a diet of milk or milk products, when extraneous sources of iron are minimal.

Piglet anaemia

This condition occurs wherever pigs are housed in concrete pens and denied access to soil or pasture, unless preventive measures are taken. It does not occur in less restricted conditions, where the foraging normal to the species is possible. Piglet anaemia is an uncomplicated iron deficiency (McGowan and Crichton, 1924; Hart *et al.*, 1930). Sometimes referred to as 'thumps', because of the laboured spasmodic breathing, the condition usually develops within 2–4 weeks of birth, by which time haemoglobin may have fallen from the normal adult value (110–120 g l^{-1}) to as low as 30–40 g l^{-1}. Mortality is high, but survivors begin a slow spontaneous recovery at about 6–7 weeks of age, by which time they have usually started to eat appreciable amounts of the sow's food. Anaemic piglets are listless and flabby; their appetite is poor and their growth subnormal; their skins become wrinkly, their hair is rough and the incidence of diarrhoea is high. Post-mortem examination shows the heart to be dilated, with an excess of pericardial fluid, and the lungs to be oedematous (Seamer, 1956). Several features of the nutritional physiology of iron peculiar to pigs combine to produce piglet anaemia. Newborn piglets absorb colostrum proteins intact into the blood from the small intestine by the process of pinocytosis (Leece, 1973). In this way, passive immunity is acquired, but plasma volume increases markedly during the first 12 h of nursing, without significant change in red-cell volume (Talbot and Swenson, 1970; Furugouri and Tohara, 1971). In other words, the piglet develops a physiological anaemia caused by a rapid increase in blood volume within the first day and a half of life. This poor start to life is compounded by an iron deficit brought about by a combination of the following factors: (i) unusually low body iron stores at birth, compared with the newborn of most other species; (ii) absence of the polycythaemia of birth common to other animal species; (iii) low levels of iron in sow's milk; (iv) very rapid early growth rate compared with that of the lamb or calf; and (v) large litter size. If adequately fed, each piglet can reach four times its birth weight at the end of 3 weeks and ten times its birth weight at the end of 8 weeks. Such a rapid growth rate requires the retention of 7–11 mg Fe day^{-1}, whereas only about 1 mg day^{-1} is obtained from milk alone (Venn *et al.*, 1947).

Lamb and calf anaemia

Anaemia can also develop in the young, preruminant lamb and calf, but the condition is far less prevalent and less severe than in piglets. Transient mild anaemia in the young, rapidly growing suckling has been reported in naturally reared calves (30% < 90 g haemoglobin l^{-1}: Hibbs *et al.*, 1963) and in lambs reared indoors (Ullrey *et al.*, 1965; Green *et al.*, 1993, 1997). Twin calves are more likely to develop anaemia than single calves, because they compete for a limited maternal supply of iron (Kume and Tanabe, 1994), and the same may be true of twin lambs. These anaemias can also be prevented by administering iron (Carleson *et al.*, 1961; Ullrey *et al.*, 1965; Kume and Tanabe, 1994). Weight gain was not improved in one study with lambs, even when the anaemia in untreated lambs (minimum haemoglobin value 81 g l^{-1}) lasted for 4 weeks (Bassett *et al.*, 1995), but it was improved by 1 kg in another larger study with 525 lambs (Green *et al.*, 1997) and also increased in the calves studied by Carleson *et al.* (1961). The variable growth responses of anaemic populations to iron supplementation probably reflect the small proportion of clinically affected individuals (e.g. 4–14%: Green *et al.*, 1993) and is attributable to variable recovery of appetite. Severe, debilitating anaemia has been produced experimentally in calves by rearing them exclusively on cow's milk (Blaxter *et al.*, 1957), but a corresponding field study revealed only a mild anaemia, which responded to iron supplementation.

Veal production

The development of anaemia is an inevitable and necessary feature of veal production, in which energy-rich milk substitutes low in iron are fed to calves to produce pale meat. Tolerance of anaemia in confined animals is demonstrated by the fact that haemoglobin values can fall to 50% of normal while rapid growth rates are sustained. The practice raises serious welfare questions, as well as adding to the difficulty of fixing iron requirements. Other intensive feeding systems in which processed materials and synthetic feeds, such as urea, replace a substantial proportion of natural feeds may lower iron status.

Copper-supplemented pigs

High dietary levels of copper (250 mg kg^{-1} DM) given to weaned pigs as a growth stimulant can reduce iron absorption sufficiently to induce a secondary iron deficiency, followed by anaemia, unless iron is also added (Suttle and Mills, 1966; Gipp *et al.*, 1974; Bradley *et al.*, 1983).

Infections as a cause of anaemia

Microbial pathogens also need iron and they have iron-binding proteins in their outer membranes to ensure that they get it. During infection, iron is redistributed by the host in an attempt to deplete the pathogen of iron (Weinberg, 1984), and antibodies to the iron-regulated outer membrane proteins of pathogens such as *Pasteurella haemolytica* are produced (Confer *et al.*, 1995). The UIBC of lactoferrin in milk increases during infections such as mastitis and has long been attributed bactericidal properties associated with iron deprivation; these have recently been confirmed *in vivo* (Teraguchi *et al.*, 1995). Parenteral administration of complexed iron has enhanced survival in chicks infected with fowl typhoid (*S. gallinarum*) (Smith and Hill, 1978). The redistribution of iron by the host may cause a secondary anaemia by depriving the erythropoietic tissues of iron. Parasitic infestation involving severe blood loss can also produce a secondary iron-deficiency anaemia. Such effects have been reported for various parasites, including *Bunostomum* and trichostrongylidae, as discussed by Kolb (1963). Infection by a blood-sucking parasite of the ovine abomasum (*Haemonchus contortus*) caused losses of up to 24 ml erythrocytes per day in lambs (Abbot *et al.*, 1988).

Diagnosis of Iron Disorders

Risk of inadequacy is initially assessed by measuring blood haemoglobin and/or the highly correlated haematocrit (packed cell volume (PCV)). The normal haemoglobin values in the blood of farm animals are well established and commonly given as 110–120 g l^{-1} in pigs, poultry and cattle and 100–110 g l^{-1} in sheep, goats and horses. However, individual variability is high, and low values are not necessarily indicative of iron deficiency. Furthermore, there are consistent changes with age; in healthy pigs, haemoglobin values declined from 125 to 85 g l^{-1} between birth and 6–8 weeks of age, despite iron supplementation, but gradually increased to 135 g l^{-1} after 5–6 months (Miller, 1961). Sheep show a lesser fall and incomplete recovery of birth values (Ullrey *et al.*, 1965). Sex differences are insignificant except in poultry, where slightly higher haemoglobin values occur in cocks than in hens and in non-laying than in laying hens. In all species, sustained iron deficiency eventually leads to the release of small, new erythrocytes with less than the normal concentration of haemoglobin. This is reflected by a low mean cell haemoglobin concentration (MCHC, calculated by dividing the haemoglobin reading by the PCV) and low mean cell volume (MCV, calculated by dividing PCV by erythrocyte count). Iron deficiency is thus confirmed by the presence of hypochromic, microcytic anaemia. Reductions in serum iron and increases in TIBC and UIBC in serum accompany the development of anaemia due to iron deprivation; individual variability in normal values is large in farm animals, but differences among species are small. The means and extreme values for sheep and cattle, reported by Underwood and Morgan (1963) and others are given in Table 13.1. Serum iron values in poultry vary with age and egg-laying; they were reported to average 2.25 mg Fe l^{-1} (range 1.73–2.90) in immature pullets and to rise until the laying period was reached, when values from 6.7 to as high as 9.4 mg Fe l^{-1} were observed (Ramsay and Campbell, 1954). The distinction between deficiency and dysfunction is blurred, because some reduction in haemoglobin and haematocrit can be tolerated (e.g. in piglets: Schrama *et al.*, 1997), but values 25% or more below the normal ranges clearly indicate anaemia and values 50–60% below normal induce clinically obvious pallor in animals (e.g. in lambs: Green *et al.*, 1993). Diagnostic guidelines for these and other blood indices of iron status are given in Table 13.2. Serum ferritin is of limited diagnostic value, because values become minimal before anaemia develops. Iron deficiency would also be indicated by low levels of iron in the liver, and a marginal band of 150–250 mg kg^{-1} DM is tentatively proposed to separate the deficient from the normal. It is important to distinguish anaemia caused by dietary iron deprivation from that associated with infection. Anaemia of infection would be indicated by fever, raised blood levels of acute-phase proteins or faecal excretion of parasite oocysts or eggs. Abnormally high plasma copper values, coupled with coccidial oocysts in faeces, are suggestive of infection in the lamb anaemia reported by Green *et al.* (1993).

Treatment of Iron Deprivation

Injection of iron by the intramuscular route is widely practised (Patterson and Allen, 1972), using an iron–dextran or a dextrin:ferric oxide compound of

Table 13.1. Normal serum iron, total iron-bonding capacity (TIBC) and percentage unsaturation (UIBC) in livestock (from Underwood and Kaneko, 1993).

	Serum iron (μmol l^{-1})	TIBC (μmol l^{-1})	UIBC (%)
Cattle	17.4 ± 5.2	40.8 ± 10.0	57
Sheep	34.6 ± 1.25	59.8 ± 3.2	42
Pigs	21.7 ± 5.9	56.8 ± 6.8	62

Table 13.2. Marginal bands for assessing the risk of iron deprivation (D) or excess (E)[a] from biochemical criteria for domestic livestock.

		Diet Fe (mg kg^{-1} DM)	Liver Fe (mmol kg^{-1} DM)[b]	Serum Fe (μmol l^{-1})[b]	Serum ferritin (μg l^{-1})
Cattle	D	40–60	1.79–2.68	8.9–17.9	10–30
	E	1000–4000	> 17.9	10.7–32.2	> 80
Pigs	Dn	60–80	0.54–0.89	2.7–10.7	5–10
	Da	30–50	1.79–2.68	14.3–17.9	35–55
	E	1000–2500	> 17.9	> 26.7	–
Poultry	D	35–45	1.79–2.68	–29.0	–
	E	200–2000	17.9–107.2	> 53.6	–
Sheep	D	30–50	1.25–1.79	–29	–
	E	600–1200	> 17.9	> 39	–

[a] Tolerable E values will be far higher if contaminant soil iron contributes largely to the dietary iron level.
[b] Multiply by 55.9 to obtain liver values in mg kg^{-1} DM and serum values in μg l^{-1}.
Dn, neonate; Da, adult.

high stability to minimize muscle damage. A dose of 2 ml iron–dextran (200 mg Fe) at 3 days of age restores haemoglobin values in piglets to those of birth and maintains them throughout the nursing period. A smaller dose (100 mg) may delay the recovery in haemoglobin without significantly reducing growth rate, appetite or antibody responses to vaccination (Schrama *et al.*, 1997). Anaemia in the neonatal lamb (Ullrey *et al.*, 1965) and calf (Carleston *et al.*, 1961) is controlled in a similar manner to piglet anaemia, intramuscular doses of 200 mg Fe being used for lambs (Green *et al.*, 1993, 1997) and 500 mg Fe for calves (Hibbs *et al.*, 1963). Daily oral supplementation with 20–40 mg Fe as $FeSO_4$ is also effective in newborn calves (Kume and Tanabe, 1994).

Prevention of Iron Deficiency

Oral route

The earliest and simplest preventive procedure for piglet anaemia was to provide a small amount of soil or pasture sods which the piglets could consume regularly. Supplementary iron can be given orally but must be given within 2–4 days of birth and usually again at 10–14 days. Iron given to the sow during pregnancy and lactation cannot be relied upon to significantly increase the iron stores of the piglet at birth or the iron content of the sow's milk (ARC, 1981), but supplementation of the sow's diet with 2 g Fe kg^{-1} DM as $FeSO_4$ can control piglet anaemia when the offspring have access to the mother's faeces (Gleed and Sansom, 1982).

Inorganic sources

Inorganic salts of iron vary little in availability, provided they are soluble. The belief that ferric salts were intrinsically less available than ferrous salts became untenable when it was realized that some ferric salts were highly insoluble. Ferric ammonium citrate compares favourably with the standard $FeSO_4$ in availability studies (relative values 115 vs. 110) and was infinitely superior to ferrous carbonate (given a value of 2) for chicks (Fritz *et al.*, 1970). Occasionally, sources such as ferrous carbonate have a higher relative value (Henry and Miller, 1995). The ferric oxide (Fe_2O_3) used as a colouring agent in mineral mixes is among the poorest of inorganic sources capable of impairing copper absorption (Suttle and Peter, 1985), and should not be used. Many of the inorganic compounds used as calcium, phosphorus and magnesium supplements will markedly and incidentally increase the overall iron contents of the supplemented rations, because of their frequent heavy contamination with iron, indicated by a rust-coloured appearance. The availability of iron presented in materials such as steamed bone-meal and soft rock phosphate is exceedingly low, while di- and monocalcium phosphates contain iron with only 62% of the value of that in $FeSO_4$ (Deming and Czarnecki-Maulden, 1989).

Organic chelates

Chelated iron proteinate given to sows during gestation and lactation (at the rate of 55 mg Fe kg^{-1} DM), with no iron supplementation of the piglets, promoted piglet growth and haemoglobin formation (Ashmead, 1979). Similar treatment of sows with an amino acid iron chelate increased the liver iron stores of the piglets and iron concentrations in the sow's milk (Brady *et al.*, 1976), but such results can be partially attributed to consumption by the piglet of iron-rich sow's faeces and may occur whatever iron source is used. There is no convincing evidence that provision of iron in chelated forms enhances bioavailability (Fox *et al.*, 1997).

Iron Requirements

Requirements for iron vary according to age and the criterion of adequacy used. In calves reared for veal production, for example, iron supply is deliberately kept well below the requirement for maximum myoglobin formation and below that for maximum haemoglobin formation. Mild anaemia (haemoglobin 70 g l^{-1}) was not associated with impaired oxygen transport or consumption and did not increase heart rate during unaccustomed exercise (Bremner *et al.*, 1976). Iron-deficiency anaemia would be more debilitating in working than in non-working livestock but is less likely to occur in the working environment, which provides more opportunities to ingest contaminant iron. Suboptimal requirements for pigs have also been considered (ARC, 1981) but should be rejected in view of the possibility that iron-deficient, mildly anaemic animals may absorb increased amounts of potentially toxic metals, such as lead and cadmium (see Chapter 18). Recommendations should therefore aim to maintain normal haemoglobin concentrations in all species and contain a marginal band to allow for variations in availability and minimize the premature diagnosis of dietary iron deprivation (Table 13.2). Since the growth requirement for iron is predominantly for haemoglobin and increase in red-cell mass constitutes a progressively smaller component of weight gain as animals grow, need in terms of dietary iron concentration will decline with age (ARC, 1981) and spontaneous recoveries in haemoglobin occur on diets initially low enough in iron to induce anaemia (Fig. 13.3; Bremner *et al.*, 1976; Bassett *et al.*, 1995).

Sheep and cattle

Calves fed exclusively on a milk diet develop a severe iron-deficiency anaemia in 8–10 weeks. Cow's milk in full lactation averages about 0.5 mg Fe kg^{-1} FW, which is equivalent to only 4 mg kg^{-1} DM (colostrum contains some three times this concentration), and this is clearly less than required. Supplemental iron at the rate of 30 or 60 mg day^{-1} will maintain normal growth and haemoglobin values for a period of 40 weeks from birth (Matrone

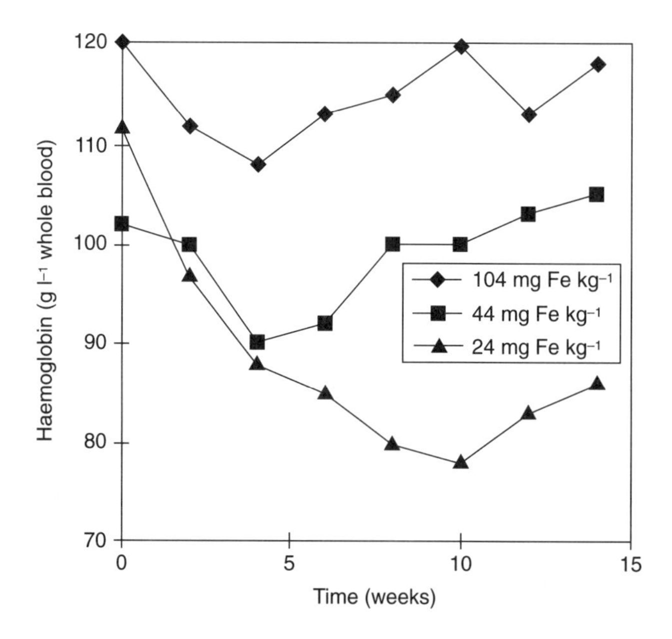

Fig. 13.3. Young animals given fixed, initially inadequate dietary Fe levels can show spontaneous recoveries from any anaemia (data for calves from Bremner *et al.*, 1976). This is because iron requirements for each unit of weight gain are constant while appetite increases as animals grow and the iron concentration needed in the diet therefore decreases with age.

et al., 1957). A later study with male calves indicated that a milk-substitute diet containing 40 mg Fe kg^{-1} DM is sufficient to prevent all but a mild anaemia, provided that the supplemental iron is in a soluble form (Bremner and Dalgarno, 1973). Experiments with weaned growing–finishing lambs show that 10 mg Fe kg^{-1} DM is inadequate and that their minimum requirements lie between 25 and 40 mg kg^{-1} DM (Lawlor *et al.*, 1965). Definitive experiments with older sheep and cattle or lactating cows have apparently not been reported, but their requirements are likely to be lower than those of younger animals.

Pigs

The iron requirement of young weaned pigs is influenced by availability, which increases if a given diet is fed in wet rather than dry form (Hitchcock *et al.*, 1974). Early estimates varied from 60 mg kg^{-1} DM (Matrone *et al.*, 1960) to 125 mg kg^{-1} DM (Ullrey *et al.*, 1960) for full growth and haemoglobin production in baby pigs. In a later study, 50 mg kg^{-1} DM of supplemental iron was reported as necessary to support maximum gain, and 100 mg kg^{-1} DM resulted in significantly higher haemoglobin and serum iron values in baby pigs fed on condensed cow's milk containing 2 mg Fe kg^{-1} DM (Hitchcock *et al.*, 1974). Reviewing all the literature available at the time, the Agricultural Research Council (ARC, 1981) concluded that pigs needed 60 mg Fe kg^{-1} DM up to 20 kg body weight, but less thereafter. Although there had been few further studies, the National Research Council (NRC, 1988) gave higher requirements of 40–100 mg Fe kg^{-1} DM for growing pigs (values decreased with age) and 80 mg Fe kg^{-1} DM for gilts and sows. Where copper sulphate is added to pig rations at the rate of 150–250 mg copper (Cu) kg^{-1} DM as a growth stimulant, iron (and zinc) requirements are increased (Bradley *et al.*, 1983), but 150 mg Fe kg^{-1} DM (with zinc) offsets the effect of 450 mg Cu kg^{-1} DM (Suttle and Mills, 1966). Although iron supplements can detoxify gossypol and prevent poisoning on diets rich in cotton-seed meal (Buitrago *et al.*, 1970), iron requirements cannot be said to increase, because gossypol does not impair iron metabolism.

Poultry

Early estimates of requirements for growth varied from 40 (Hill and Matrone, 1961) to 75–80 mg kg^{-1} DM (Davis *et al.*, 1962) for the first 4 weeks of life. Subsequent data also imply high growth requirements of 60–80 mg kg^{-1} DM for broiler chicks up to 21 days of age (McNaughton and Day, 1979). The uniformly high provision of 80 mg Fe g^{-1} DM for broilers up to 8 weeks of age recommended by NRC (1994) seems overgenerous, but their scaled reduction for turkey poults from 80 to 50 mg Fe kg^{-1} DM between 4 and 16 weeks of age seems sensible. An average hen's egg contains about 1 mg Fe and the daily demand for iron therefore rises with the onset of lay. However, the required concentrations for the most energy-rich diet given by NRC (1994) is only 56 mg Fe kg^{-1} DM and is 32% less for diets containing the least energy. All ordinary laying rations would be expected to cover these requirements, especially where ground limestone or oyster shell is used as a calcium supplement. These mineral supplements frequently contain 0.2–0.5% Fe. Despite their lack of practical or commercial significance, it seems curious that definitive studies of the dietary iron requirements of egg production have still not been made.

Toxicity of Iron

Free iron is cytotoxic because of its high redox potential and ability to generate free oxygen radicals via the Haber–Weiss reaction, and for this reason it is transported and stored tightly bound to protein. When tissue stores become excessive during chronic iron overload, sufficient reactive iron may be present to cause peroxidative damage at sites such as the liver (Kent and Bahu, 1979). The underlying pathogenic mechanism is peroxidative damage to lipid membranes (Gordeuk *et al.*, 1987), and the extent of injury will depend on the antioxidant status of the animal and particularly its vitamin E status (Ibrahim *et al.*, 1997). Vitamin E deficiency increases susceptibility to iron toxicity following the intramuscular injection of iron in piglets (Patterson and Allen, 1972) and, conversely, iron overload will reduce liver concentrations of vitamin E (Omara and Blakley, 1993). Iron-induced lipid peroxidation is also increased by the ingestion of readily peroxidized lipid sources, such as polyunsaturated fatty acids (PUFA) (Muntane *et al.*, 1995; Ibrahim *et al.*, 1997), and the ingestion of iron-rich soil with PUFA-rich grass in spring may present a hazard for grazing animals. High liver iron concentrations may also arise from exposure to plant toxins (e.g. lupinosis: Gardiner, 1961 and brassica haemolysins: Barry *et al.*, 1981), from the accelerated breakdown of erythrocytes after the haemolytic crisis in copper poisoning and presumably in conditions, such as vitamin E and selenium deficiency or brassica poisoning, in which erythrocyte lifespan is shortened. Copper deficiency also results in hepatic iron accumulation. Susceptibility to iron overload increases with age (Wu *et al.*, 1990).

There is nevertheless a high tolerence towards dietary iron in all species, protection being afforded by the powerful mucosal block to iron absorption. Calves reared on milk substitute tolerated 2 g Fe (as ferrous sulphate) kg^{-1} DM for 6 weeks, but spleen and liver iron concentrations were considerably raised (to 1.30 and 1.55 g kg^{-1} DM, respectively) and a further increase in dietary iron to 5 g Fe kg^{-1} DM depressed appetite and growth (Jenkins and Hidiroglou, 1987). Tolerance of iron has been exploited by giving pigs up to 3200 mg Fe kg^{-1} DM for 4 months to combat gossypol poisoning; they did not develop iron toxicity (Buitrago *et al.*, 1970). Similarly, massive daily doses (up to 0.1 g Fe kg^{-1} live weight (LW)) have been used in sheep to combat sporidesmin intoxication (Munday and Manns, 1989); the mechanism of protection is believed to involve inhibition of copper absorption, and herein lies another reason for avoiding unnecessarily high iron intakes. The grazing animal cannot avoid ingesting excess iron in the soil which contaminates pasture when it is in scarce supply in both dry and wet conditions. The diarrhoea which developed on pastures irrigated with iron-rich bore water in New Zealand was associated with copper depletion (Campbell *et al.*, 1974) (see Chapter 11). High iron intakes have not always caused reductions in liver copper in calves (McGuire *et al.*, 1985) or sheep and have even increased the toxicity of copper to sheep (Rosa *et al.*, 1986), possibly because the diets were low in sulphur (see Chapter 11). Although ingested soil iron is in a predominantly inert and insoluble form, sufficient can be retained to have a hepatoxic effect (N.F. Suttle, unpublished data) when concentrations in the liver reach 1000 mg Fe kg^{-1} DM. Blood indices, such as serum iron, TIBC and UIBC, did not reflect excessive iron intakes (1000 mg kg^{-1} DM) well in one study with calves (McGuire *et al.*, 1985).

As with all cumulative poisonings, total available dietary iron intake, rather than dietary iron concentration, will determine risk of chronic disorder. Both the clinical signs and the critical dietary level for chronic iron toxicity will differ from those associated with acute iron toxicity. The high concentrations quoted as tolerable for livestock (1000 mg kg^{-1} DM in cattle and poultry, 3000 mg kg^{-1} DM in pigs: NRC, 1980) fail to cause anorexia, vascular congestion and irritation in the gastrointestinal tract after brief exposure. However, lower concentrations may cause liver injury in the longer term, particularly if vitamin E status is low or there are endogenous sources of excess iron. If the iron is present in available forms, a marginal band of 750–1250 mg Fe kg^{-1} DM can be considered to separate the nutritionally acceptable from the potentially harmful iron concentration for ruminants and poultry; for pigs, the equivalent marginally unsafe band is 2500–3500 mg kg^{-1} DM. Where exogenous iron sources, such as oxides, hydroxides or ferrous carbonate, constitute a major proportion of the dietary iron, tolerable levels will be much higher.

References

Abbott, E.M., Parkins, J.J. and Holmes, P.J. (1988) Influence of dietary protein on the pathophysiology of haemonchosis in lambs given continuous infections. *Research in Veterinary Science* 45, 41–49.

Abou-Hussein, E.R.M., Raafat, M.A., Abou-Raya, A.K. and Shalaby, A.S. (1970) Manganese, iron and cobalt content in common Egyptian feedstuffs. *United Arab Republic Journal of Animal Production* 10, 245–254.

ARC (1980) *The Nutrient Requirements of Ruminant Livestock.* Commonwealth Agricultural Bureaux, Farnham Royal, UK, pp. 234–240.

ARC (1981) *The Nutrient Requirements of Pigs.* Commonwealth Agricultural Bureaux, Farnham Royal, UK, pp. 271–273.

Ashmead, D. (1979) The influence of chelated iron proteinate, fed to sows with no iron supplementation to their baby pigs. *Journal of Animal Science* 49 (suppl. 1), 235 (abstract).

Barry, T.N., Reid, T.C., Millar, K.R. and Sadler, W.A. (1981) Nutritional evaluation of kale (*Brassica oleracea*) diets 2. Copper deficiency, thyroid function and selenium status in young cattle and sheep fed kale for prolonged periods. *Journal of Agricultural Science, Cambridge* 96, 269–282.

Bassett, J.M., Burrett, R.A., Hanson, C., Parsons, R. and Wolfensohn, S.E. (1995) Anaemia in housed lambs. *Veterinary Record* 136, 137–140.

Baynes, R.D. (1997) The soluble form of the transferrin receptor and iron status. In: Fisher, P.W.F., L'Abbé, M.R., Cockell, K.A. and Gibson, R.S. (eds) *Trace Elements in Man and Animals – 9 – Proceedings of the Ninth International Symposium.* NRC Research Press, Ottawa, pp. 471–475.

Beard, J., Tobin, B. and Green, W. (1989) Evidence for thyroid hormone deficiency in iron-deficient, anaemic rats. *Journal of Nutrition* 119, 772–778.

Blaxter, K.L., Sharman, K.L. and MacDonald, A.M. (1957) Iron deficiency anaemia in calves. *British Journal of Nutrition* 11, 234–246.

Bradley, B.D., Graber, G., Condon, R.J. and Frobish, L.T. (1983) Effects of graded levels of dietary copper on copper and iron concentrations in swine tissues. *Journal of Animal Science* 56, 625–630.

Brady, P.S., Miller, E.R., Ku, P.K., Green, F.F. and Ullrey, D.E. (1976) Evaluation of an aminoacid–iron chelate hematinic. In: *Report of Swine Research.* Michigan State University Agricultural Experiment Station, p. 4.

Bremner, I. and Dalgarno, A.C. (1973) Iron metabolism in the veal calf: the availability of different iron compounds. *British Journal of Nutrition* 29, 229–243.

Bremner, I., Brockway, J.M., Donnelly, H.T. and Webster, A.J.F. (1976) Anaemia and veal calf production. *Veterinary Record* 99, 203–205.

Brock, J.H. and Mainou-Fowler, T. (1986) Iron and immunity. *Proceedings of the Nutrition Society* 45, 305–315.

Buitrago, J.A., Clawson, A.J. and Smith, F.H. (1970) Effects of dietary iron on gossypol accumulation in and elimination from porcine liver. *Journal of Animal Science* 31, 554–558.

Campbell, A.G., Coup, M.R., Bishop, W.H. and Wright, D.E. (1974) Effect of elevated iron intake on the copper status of grazing cattle. *New Zealand Journal of Agricultural Research* 17, 393–399.

Carleson, R.H., Swenson, M.J., Ward, G.M. and Booth, N.H. (1961) Effects of intramuscular iron dextran in newborn lambs and calves. *Journal of the American Veterinary Medical Association* 139, 457–461.

Chandra, R.K. (1976) Iron and immunocompetence. *Nutrition Reviews* 34, 129–132.

Confer, A.W., McGraw, R.D., Dierham, J.A., Morton, R.J. and Panciera, R.J. (1995) Serum antibody responses of cattle to iron-regulated outer membrane proteins of *Pasteurella haemolytica. Veterinary Immunology and Immunopathology* 47, 101–110.

Davis, P.N., Norris, L.C. and Kratzer, F.H. (1962) Iron deficiency studies in chicks using treated isolated soybean protein diets. *Journal of Nutrition* 78, 445–453.

Deming, J.G. and Czarnecki-Maulden, G.L. (1989) Iron bioavailability in calcium and phosphorus sources. *Journal of Animal Science* 67 (suppl. 1), 253 (abstract).

Fairweather-Tait, S.J. and Wright, A.J.A. (1984) The influence of previous iron intake on the estimation of bioavailability of Fe from a test meal given to rats. *British Journal of Nutrition* 51, 185–191.

Fox, T.E., Eagles, J. and Fairweather-Tait, S.J. (1997) Bioavailability of an iron glycine chelate for use as a food fortificant compared with ferrous sulphate. In: Fisher, P.W.F., L'Abbé, M.R., Cockell, K.A. and Gibson, R.S. (eds) *Trace Elements in Man and Animals – 9 – Proceedings of the Ninth International Symposium.* NRC Research Press, Ottawa, pp. 460–462.

Frieden, E. (1971) Ceruloplasmin: a link between copper and iron metabolism. *Advances in Chemistry Series* 100, 292–321.

Frieden, E. and Hsieh, H.S. (1976) Ceruloplasmin: the copper-transport protein with essential oxidase activity. In: Meister, A. (ed.) *Advances in Enzymology and Related Areas of Molecular Biology.* John Wiley & Sons, New York, 44, pp. 187–198.

Fritz, J.C., Pla, G.W., Roberts, T., Boehne, J.W. and Hove, E.L. (1970) Biological availability in animals of iron from common dietary sources. *Journal of Agricultural and Food Chemistry* 18, 647–652.

Furugouri, K. and Tohara, S. (1971) Studies on iron metabolism and anemia in piglets. 2. Blood volume and mean corpuscular constants. *Bulletin of National Institute of Animal Industry* 24, 75–82.

Gardiner, M.R. (1961) Lupinosis – an iron storage disease of sheep. *Australian Veterinary Journal* 37, 135–140.

Gipp, W.F., Pond, W.G., Kallfelz, F.A., Tasker, J.B., van Campen, D.R., Krook, L. and Visek, W.J. (1974) Effect of dietary copper, iron and ascorbic acid levels on hematology, blood and tissue copper, iron and zinc concentrations and Cu and Fe metabolism in young pigs. *Journal of Nutrition* 104, 532–541.

Gleed, P.T. and Sansom, B.F. (1982) Ingestion of iron in sow's faeces by piglets reared in farrowing crates with slatted floors. *British Journal of Nutrition* 47, 113–117.

Gordeuk, V.R., Bacon, B.R. and Brittenham, G.M. (1987) Iron overload: cause and consequences. *Annual Reviews of Nutrition* 7, 485–508.

Grassmann, E. and Kirchgessner, M. (1973) Katalase Aktivitat des Blutes von Saugferkeln und Mastkalbern bei mangelnder Eisenversorgung. *Zentralblatt für Veterinarmedizin* A 20, 481–486.

Green, L.E., Berriatua, E. and Morgan, K.L. (1993) Anaemia in housed lambs. *Research in Veterinary Science* 54, 306–311.

Green, L.E., Graham, M. and Morgan, K.L. (1997) Preliminary study of the effect of iron dextran on a non-regenerative anaemia of housed lambs. *Veterinary Record* 140, 219–222.

Gubler, C.J., Cartwright, G.E. and Wintrobe, M.M. (1957) Studies on copper metabolism. 20. Enzyme activities and iron metabolism in copper and iron deficiencies. *Journal of Biological Chemistry* 224, 533–546.

Hart, E.B., Elvehjem, C.A. and Steenbock, H. (1930) A study of the anemia of young pigs and its prevention. *Journal of Nutrition* 2, 277–294.

Henry, P.R. and Miller, E.R. (1995) Iron bioavailability. In: Ammerman, C.B., Baker, D.H. and Lewis, A.J. (eds) *Bioavailability of Nutrients for Animals.* Academic Press, New York, pp. 169–200.

Hibbs, J.R., Conrad, H.R., Vandersall, J.H. and Gale, C. (1963) Occurrence of iron deficiency anaemia in dairy calves at birth and its alleviation by iron dextran injection. *Journal of Dairy Science* 46, 1118–1124.

Hill, C.H. and Matrone, G. (1961) Studies on copper and iron deficiencies in growing chickens. *Journal of Nutrition* 73, 425–431.

Hitchcock, J.P., Ku, P.K. and Miller, E.R. (1974) Factors influencing iron utilization by the baby pig. In: Hoekstra, W.G., Suttie, J.W., Ganther, H.E. and Mertz, W. (eds) *Trace Element Metabolism in Animals – 2.* University Park Press, Baltimore, pp. 598–600.

Hostettler-Allen, R., Tappy, L. and Blum, J.W. (1993) Enhanced insulin-dependent glucose utilization in iron-deficient veal calves. *Journal of Nutrition* 123, 1656–1667.

Ibrahim, W., Lee, V.-S., Ye, C.-C., Szabo, J., Bruckner, G. and Chow, C.K. (1997) Oxidative stress and antioxidant status in mouse liver: effects of dietary lipid, vitamin E and iron. *Journal of Nutrition* 127, 1401–1406.

Jenkins, K. and Hidiroglou, M. (1987) Effect of excess iron in milk replacer on calf performance. *Journal of Dairy Science* 70, 2349–2354.

Kaneko, J.J. (1993) *Clinical Biochemistry of Domestic Animals*, 4th edn. Academic Press, New York.

Kent, G. and Bahu, R.M. (1979) Iron overload. In: MacSween, R.N.M., Anthony, P.P. and Schewr, P.J. (eds) *Pathology of the Liver.* Churchill Livingstone, Edinburgh, pp. 148–163.

Kolb, E. (1963) The metabolism of iron in farm animals under normal and pathologic conditions. A*dvances in Veterinary Science and Comparative Medicine* 8, 49–114.

Kreutzer, M. and Kirchgessner, M. (1991) Endogenous iron excretion: a quantitative means to control iron metabolism. *Biological Trace Element Research* 29, 77–92.

Kume, S.-E. and Tanabe, S. (1994) Effect of twinning and supplemental iron-saturated lactoferrin on iron status of newborn calves. *Journal of Dairy Science* 77, 3118–3123.

Lawlor, M.J., Smith, W.H. and Beeson, W.M. (1965) Iron requirement of the growing lamb. *Journal of Animal Science* 24, 742–747.

Leece, J.G. (1973) Effect of dietary regimen on cessation of uptake of macromolecules by piglet's intestinal epithelium and transport to the blood. *Journal of Nutrition* 103, 751.

Liebold, E.A. and Guo, B. (1992) Iron-dependent regulation of ferritin and transferrin receptor expression by the iron-responsive element binding protein. *Annual Review of Nutrition* 12, 345–368.

Lonnerdal, B. (1988) Trace elements in infancy: a supply/demand perspective. In: Hurley, L.S., Keen, C.L., Lonnerdal, B. and Rucker, R.B. (eds) *Proceedings of Sixth International Symposium on Trace Elements in Man and Animals.* Plenum Press, New York, pp. 189–195.

McGowan, J.P. and Crichton, A. (1924) Iron deficiency in pigs. *Biochemical Journal* 18, 265–272.

McGuire, S.O., Miller, W.J., Gentry, R.P., Neathery, M.W., Ho, S.Y. and Blackmon, D.M. (1985) Influence of high dietary iron as ferrous carbonate and ferrous sulphate on iron metabolism in young calves. *Journal of Dairy Science* 68, 2621–2628.

McNaughton, J.L. and Day, E.J. (1979) Effect of dietary Fe:Cu ratios on haematological and growth responses of broiler chickens. *Journal of Nutrition* 109, 559–564.

MAFF (1990) *UK Tables of the Nutritive Value and Chemical Composition of Foodstuffs.* Givens, D.I. (ed.) Rowett Research Services, Aberdeen, UK.

Matrone, G., Conley, C., Wise, G.H. and Waugh, R.K. (1957) A study of iron and copper requirements of dairy calves. *Journal of Dairy Science* 40, 1437–1447.

Matrone, G., Thomason, E.L., Jr and Bunn, C.R. (1960) Requirement and utilization of iron by the baby pig. *Journal of Nutrition* 72, 459–465.

Miller, E.R. (1961) Swine haematology from birth to maturity. *Journal of Animal Science* 20, 890–897.

Morris, E.R. and Ellis, R. (1976) Isolation of monoferric phytate from wheat bran and its biological value as an iron source to the rat. *Journal of Nutrition* 106, 753–760.

Munday, R. and Manns, E. (1989) Pretection by iron salts against sporidesmin intoxication in sheep. *New Zealand Veterinary Journal* 37, 65–68.

Muntane, J., Mitjavila, M.T., Rodriguez, M.C., Puig-Parellada, P., Fernandez, Y. and Mitjavila, S. (1995) Dietary lipid and iron status modulate lipid peroxidation in rats with induced adjuvant arthritis. *Journal of Nutrition* 125, 1930–1937.

NRC (1980) *Mineral Tolerance of Domestic Animals.* National Academy of Sciences, Washington, DC.

NRC (1988) *Nutrient Requirements of Pigs,* 9th revised edn. National Academy of Sciences, Washington, DC.

NRC (1994) *Nutrient Requirements of Poultry,* 9th revised edn. National Academy of Sciences, Washington, DC.

O'Dell, B.L. (1981) Roles for iron and copper in connective tissue biosynthesis. *Philosophical Transactions of the Royal Society, London* B294, 91–104.

Omara, F.O. and Blakley, B.R. (1993) Vitamin E is protective against iron toxicity and iron-induced hepatic vitamin E depletion in mice. *Journal of Nutrition* 123, 1649–1655.

Osborne, J.C. and Davis, J.W. (1968) Increased susceptibility to bacterial endotoxin of pigs with iron-deficiency anemia. *Journal of the American Veterinary Medical Association* 152, 1630–1632.

Patterson, D.S.P. and Allen, W.M. (1972) Biochemical aspects of some pig muscle disorders. *British Veterinary Journal* 128, 101–111.

Planas, J. and de Castro, S. (1960) Serum iron and total iron-binding capacity in certain mammals. *Nature London,* UK 187, 1126–1127.

Pollitt, E. (1993) Iron deficiency and cognitive function. *Annual Review of Nutrition* 13, 521–537.

Ramsay, W.N.M. and Campbell, E.A. (1954) Iron metabolism in the laying hen. *Biochemical Journal* 58, 313–317.

Robertson, A., Morrison, J.N., Wood, A.M. and Bremner, I. (1989) Effects of iron deficiency on metallothionein I concentrations in blood and tissues of rats. *Journal of Nutrition* 119, 439–445.

Rosa, I.V., Ammerman, C.B. and Henry, P.R. (1986) Interrelationships between dietary copper, zinc and iron on performance and tissue mineral concentration in sheep. *Nutrition Reports International* 34, 893–902.

Schrama, J.W., Schouten, J.M., Swinkels, J.W.G.M., Gentry, J.L., Reiling, G.de V. and Parmentier, M.K. (1997) Effect of haemoglobin status on humoral immune response of weanling pigs of different coping styles. *Journal of Animal Science* 75, 2588–2596.

Seamer, J. (1956) Piglet anaemia: a review of the literature. *Veterinary Reviews and Annotations* 2, 79–93.

Smith, I.M. and Hill, R. (1978) The effect on experimental mouse typhoid of chelated iron preparations in the diet. In: Kirchgessner, M. (ed.) *Proceedings of Third International Symposium on Trace Elements in Man and Animals.* Arbitskreis Tierernährungsforchung, Weihenstephan, pp. 383–386.

Stockman, R. (1893) The treatment of chlorosis by iron and some other drugs. *British Medical Journal* 1, 881–885, 942–944.

Strauss, R.G. (1978) Iron deficiency, infections and immune function: a reassessment. *American Journal of Clinical Nutrition* 31, 660–666.

Suttle, N.F. and Mills, C.F. (1966) Studies of the toxicity of copper to pigs. 1. Effects of oral supplements of zinc and iron salts on the development of copper toxicosis. *British Journal of Nutrition* 20, 135–148.

Suttle, N.F. and Peter, D.W. (1985) Rumen sulphide metabolism as a major determinant of copper availability in the diets of sheep. In: Mills, C.F., Bremner, I. and Chesters, J.K. (eds) *Proceedings of the Fifth International Symposium on Trace Elements in Man and Animals.* Commonwealth Agricultural Bureaux, Farnham Royal, UK, pp. 367–370.

Talbot, R.B. and Swenson, M.J. (1970) Blood volume of pigs from birth through 6 weeks of age. *American Journal of Physiology* 218, 1141–1144.

Teraguchi, S., Shin, K., Ozawa, K., Nakamura, S., Fukuwatori, Y., Tsyuki, S., Namahira, H. and Shimanura, S. (1995) Bacteriostatic effect of orally administered bovine lacto-ferrin on proliferation of *Clostridium* species in the gut of mice fed on bovine milk. *Applied and Environmental Microbiology* 61, 501–506.

Thomas, B., Thompson, A., Oyenuga, V.A. and Armstrong, R.H. (1952) The ash constituents of some herbage plants at different stages of maturity. *Empire Journal of Experimental Agriculture* 20, 10–22.

Ullrey, D.E., Miller, E.R., Thompson, O.A., Ackermann, I.M., Schmidt, D.A., Hoefer, J.A. and Luecke, R.W. (1960) The requirement of the baby pig for orally administered iron. *Journal of Nutrition* 70, 187–192.

Ullrey, D.E., Miller, E.R., Long, C.H. and Vincent, B.H. (1965) Sheep haematology from birth to maturity 1. Erythrocyte population, size and haemoglobin concentration. *Journal of Animal Science* 24, 141–145.

Underwood, E.J. (1977) *Trace Elements in Human and Animal Nutrition*, 4th edn. Academic Press, London, 545 pp.

Underwood, E.J. and Morgan, E.H. (1963) Iron in ruminant nutrition. 1. Liver storage iron, plasma iron and total iron-binding capacity levels in normal adult sheep and cattle. *Australian Journal of Experimental Biology and Medical Science* 41, 247–253.

Venn, J.A.J., McCance, R.A. and Widdowson, E.M. (1947) Iron metabolism in piglet anemia. *Journal of Comparative Pathology and Therapeutics* 57, 314–325.

Wein, E.M. and Van Campen, D.R. (1994) Enhanced Fe^{3+}-reducing capacity does not seem to play a major role in increasing iron absorption in iron-deficient rats. *Journal of Nutrition* 124, 2006–2015.

Weinberg, E.D. (1984) Iron witholding: a defence against infection and neoplasia. *Physiological Reviews* 64, 65–102.

Welch, R.M. and Van Campen, D.R. (1975) Iron availability to rats from soya beans. *Journal of Nutrition* 105, 253–256.

Wu, W.-H., Meydani, M., Meydani, S.N., Burklund, P.M., Blumbery, J.B. and Munro, H.N. (1990) Effect of dietary iron overload on lipid peroxidation, prostaglandin synthesis and lymphocyte proliferation in young and old rats. *Journal of Nutrition* 120, 280–289.

Zinoffsky, O. (1886) Ueber die Grosse des Hamoglobinmoleculs. *Hoppe-Seylers Zeitschrift für Physiologische Chemie* 10, 16.

Manganese 14

Introduction

It is almost 70 years since manganese was first shown to be essential for growth and fertility in mice (Kemmerer *et al.*, 1931) and rats (Orent and McCollum, 1931). Subsequently, two diseases of poultry, known as perosis or 'slipped tendon' and nutritional chondrodystrophy, were found to be prevented by manganese supplements (Lyons and Insko, 1937; Wilgus *et al.*, 1937). Some years elapsed before manganese deficiency in ruminants was demonstrated (Bentley and Phillips, 1951), but by this time it had become apparent that manganese was widely distributed in very low concentrations in the cells and tissues of the animal body and that it was necessary for the normal development of bone and proper functioning of reproductive processes in both males and females. The identification of specific biochemical roles for manganese proved elusive for many years, until the first manganese metalloprotein, pyruvate carboxylase, was discovered (Scrutton *et al.*, 1966 and 1972). A specific function for manganese (Mn) in the synthesis of the mucopolysaccharides of cartilage was then related to the activation of glycosyltransferases by the metal (Leach and Muenster, 1962; Leach, 1971), and a superoxide dismutase (MnSOD) was isolated from chicken liver mitochondria (Gregory and Fridovich, 1974) and found to contain 2 mg Mn mol^{-1}. More recently, severe manganese deprivation has been shown to impair immunity (Hurley and Keen, 1987) and central nervous system (CNS) function (Hurley, 1981).

Dietary Sources of Manganese

Forages

Pastures vary markedly in manganese concentration about a high mean value of 86 mg kg^{-1} dry matter (DM), and only 3% of the grass samples reported on had less than 20 mg Mn kg^{-1} DM (Minson, 1990). Mixed New Zealand pastures usually contained 140–200 mg kg^{-1} DM, but more than 400 mg Mn

© CAB *International* 1999. *Mineral Nutrition of Livestock*
(E.J. Underwood and N.F. Suttle)

kg^{-1} DM was found in some areas (Grace, 1973). An influence of species was suggested in a study of 17 grasses, grown on the same sandy loam soil (Beeson *et al.*, 1947), in which values ranged from 96 to 815 mg Mn kg^{-1} DM. However, differences due to species and state of maturity are generally small (Minson, 1990). Most high values probably arise from soil contamination because soil manganese concentrations, at around 1000 mg Mn kg^{-1} DM, are invariably much higher than those in the pasture they support. Contamination can also occur during sample processing if mills with steel blades are used. The acidity of the soil markedly influences manganese uptake by plants (Fig. 14.1). Mitchell (1957) noted that, when soil pH was increased marginally from 5.6 to 6.4 by liming, the manganese in clover decreased from 58 to 40 and in grass from 140 to 130 mg kg^{-1} DM. Low manganese levels were reported in Canadian maize silage (*c.* 22 mg kg^{-1} DM: Miltimore *et al.*, 1970; Buchanan-Smith *et al.*, 1974), and even lower mean values have been reported recently for maize silage harvested in the UK (14.6 ± 7.2 SD: MAFF, 1990) and USA (17.1 ± 20.3: Berger, 1995). Lucerne hays from Canada and the USA contain, on average, 43.7 and 22.7 mg Mn kg^{-1} DM, respectively, but values again vary widely (see Table 2.1).

Seeds and grains

Seeds and seed products vary widely in manganese content, mainly due to inherent species differences. Maize grain and, to a lesser extent, sorghum and barley are generally substantially lower in manganese than wheat or oats (Table 14.1). Barley commonly contains 15–28 mg Mn kg^{-1} DM (see Table 2.1). Manganese is highly concentrated in the outer layers of these grains so that the inclusion of the mill products, bran and pollard (middlings) (Table 14.1), in compound feeds markedly increases manganese intakes. Species

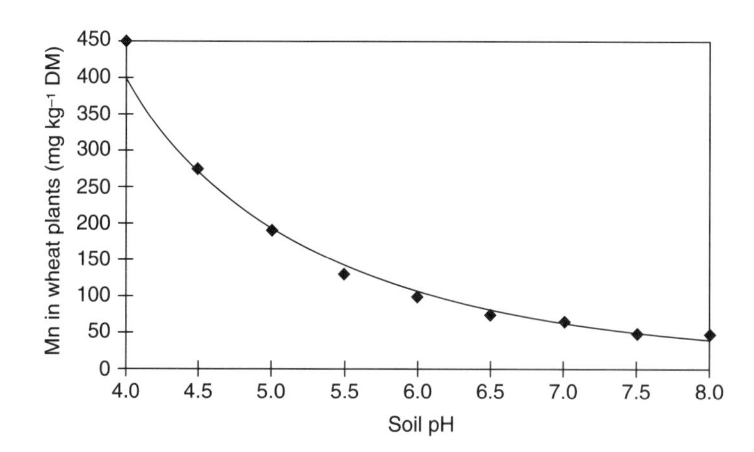

Fig. 14.1. Manganese concentrations in crops and pastures fall markedly when acid soils are limed (data for wheat plants in coordinated FAO trials: Silanpaa, 1982).

Table 14.1. Typical manganese contents in cereal grains and their mill products from three countries (mg kg^{-1} DM).

Source	Maize	Barley	Wheat	Oats	Sorghum	Wheat bran	Wheat pollard
North America	5	14	31	36	–	108	101
Australia	8	15	37	43	16	133	100
UK	5.9 (3.0)	18.5 (3.6)	35.6 (15.2)	45.2 (15.4)	8.3 (0.5)	88.7 (33.3)	–

SD – figures in parentheses.

differences in the manganese content of lupin seeds are even more striking. Seeds of the white lupin (*Lupinus albus*) contain 817 to 3397 mg Mn kg^{-1} DM, or some 10–15 times more than other common lupin species growing on the same sites (Gladstones and Drover, 1962; White *et al.*, 1981). Indeed, *Lupinus augustifolius* may sometimes provide inadequate manganese, levels as low as 6 mg kg^{-1} DM being found (White *et al.*, 1981). Field beans are also low in manganese (7.8 ± 1.5 mg kg^{-1} DM: MAFF, 1990).

Protein supplements

Protein concentrates of animal origin, such as blood-meal, meat-meal, feather-meal and fish-meal, contain less manganese (0.2–20 mg kg^{-1} DM) than protein supplements of plant origin, such as linseed, rapeseed and soybean meal (35–55 mg kg^{-1} DM: MAFF, 1990). Milk and milk products are low in this mineral, due to the generally low concentration of manganese in cow's milk (20–40 μg l^{-1}), which is insensitive to fluctuations in manganese intake.

Absorbability of Manganese

In the pioneer studies of Greenberg and Campbell (1940), only 3–4% of orally administered radiomanganese was absorbed by rats. When absorbability is generally low, as it is for manganese, variation in absorbability between dietary sources will have a profound effect on the incidence of deprivation and the dietary manganese intakes required to prevent it. When the animal has a marked influence on how much is absorbed and retained, as it does for manganese (Fig. 14.2), the assessment of absorbability (i.e. the dietary attribute, see p. 32 *et seq.*) is fraught with difficulties. The potential range in absorbability among feeds was illustrated by Johnson *et al.* (1991), working with human volunteers; comparing intrinsically and extrinsically labelled sources consumed with a basal meal low in manganese, absorptive efficiencies (expressed as percentages of dose) were 1.7, 2.2, 3.8, 5.2 and 8.9 for sunflower seeds, wheat, spinach, lettuce and manganese chloride ($MnCl_2$), respectively. Low absorbability for manganese in grains was associated with the formation of complexes with phytate and fibre. Furthermore, the method of labelling did not affect the differences between sources, indicating that

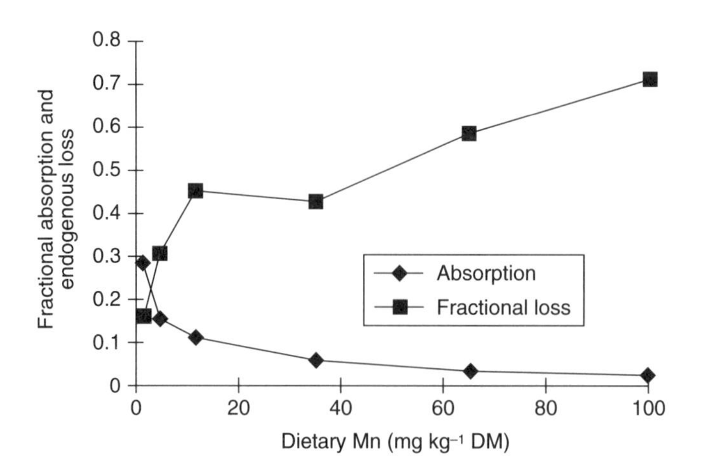

Fig. 14.2. Animals adjust to variations in dietary manganese supply by adjusting the proportions absorbed and excreted endogenously via the faeces (data for rats from Weigand *et al.,* 1986).

those factors which influenced the absorbability of inherent feed manganese could equally affect the absorbability of added inorganic manganese. There is further evidence for partial transference of the low manganese absorbability associated with phytate and fibre in chicks (Halpin and Baker, 1986), and the results of Fly *et al.* (1989) indicate that it may not be saturated by the addition of 1875 mg inorganic Mn kg^{-1} DM. However, the high absorbability of inorganic sources is not eliminated by the influences of phytate and fibre. Inorganic manganese (given as manganese sulphate ($MnSO_4$)) retained a high absorbability of 10.3% when added to a maize–soybean meal diet, in which only 2.8% of the inherent manganese was absorbed by chicks (Wedekind *et al.*, 1991). This range of values agrees well with that reported for humans (Johnson *et al.*, 1991) and may be applicable to all non-ruminant livestock. Requirements measured using fibre- and phytate-free diets with added inorganic manganese will grossly underestimate the needs of poultry on practical rations.

Effects of calcium and phosphorus

The addition of calcium and phosphorus to the diet had long been held to lower the absorbability of dietary manganese, but some groups now claim that those adverse effects, often obtained with supplements providing both calcium (Ca) and phosphorus (P), are attributable to phosphorus alone (Wedekind *et al.*, 1991). That the effects can be quantitatively significant is beyond doubt. Manganese absorption from inorganic sources can be reduced by about 50%, at dietary concentrations ranging from the suboptimal (< 12 mg Mn kg^{-1} DM: Baker and Odoho, 1994) to excessive (1000 mg Mn kg^{-1} DM:

Wedekind *et al.*, 1991), by the addition of 8 g *excess* inorganic P kg^{-1} DM to either semipurified (Baker and Oduho, 1994) or maize–soybean meal (Wedekind *et al.*, 1991) diets. Although excess calcium was added in both studies to avoid Ca:P imbalances, the same amount of excess calcium added without phosphorus had no effect on manganese absorption (Wedekind *et al.*, 1991). Provided such gross overfeeding of inorganic phosphorus is avoided, the major impairment of manganese utilization in practice is likely to come from phytate. Addition of phytase to phytate-rich diets improves the utilization of dietary manganese (Biehl *et al.*, 1995), and it would be surprising if the adverse effects of phytate were not sometimes enhanced by excess dietary calcium (see Chapter 4).

Absorbability from forages

There have been no reports on the absorbability of manganese from forages for ruminants, either as relative or absolute measurements. Since the main antagonist of manganese absorption – phytate – is broken down in the rumen, absorbability is probably higher than that achievable by non-ruminants on phytate-rich diets. The general level of absorption may reach the range 10–20% when manganese intakes are low. Bremner (1970) found that only about 10% of manganese in the rumen of sheep was in a soluble form, although over 50% of that in the diet was soluble in water. This may partly reflect the formation of insoluble manganous sulphide, which would dissociate in the acid abomasal digesta and become more absorbable at sites further down the gastrointestinal tract.

Metabolism

Absorption

Homeostasis is achieved initially by the regulation of absorption up to limits set by manganese source and dietary antagonists. The weanling rat can raise the efficiency of manganese absorption from 2% at 100 mg Mn kg^{-1} DM to 29% at a low dietary level (1.5 mg Mn kg^{-1} DM) in a semipurified diet (Fig. 14.2; Weigand *et al.*, 1986). Similar regulation probably occurs in farm animals. Manganese is well absorbed by the suckling or artificially reared animal. Calves reared on whole milk containing 0.75 mg Mn kg^{-1} DM retained 60% of an oral dose of ^{54}Mn, but a supplement of 15 mg Mn kg^{-1} DM reduced that figure to 16.3% (Carter *et al.*, 1974). Artificially reared piglets absorbed 35% of the manganese given as an extrinsic tracer in a milk formula feed (Atkinson *et al.*, 1993). Calves given a milk substitute low in manganese (0.75 mg kg^{-1} DM) retained 40% of their manganese intake (Kirchgessner and Neese, 1976), whereas those given a milk substitute containing 14.2 mg kg^{-1} DM retained only 4.9% of ingested manganese (Suttle, 1979). The description of manganese absorption as 'poor' in animals on solid diets (Underwood, 1981; McDowell, 1992) is partly a reflection of the substantial surplus of manganese provided by most practical rations. However, chicks (which receive solid diets from birth) retained < 5% of an oral dose of ^{54}Mn when the diet contained only 4 mg Mn kg^{-1} DM (Mathers and Hill, 1967); retention still fell when stable manganese was added to the dose and also fell from 4.5 to 2.8% as the chicks grew from 2 to 5 weeks of age. Since a purified diet of similar manganese concentration gave a low incidence of perosis (Hill, 1967), it would appear that poultry have a low 'ceiling' to their capacity for manganese aborption.

Transport and excretion

The small amounts of manganese absorbed are transported by transferrin (Davidsson *et al.*, 1989) to the liver, from where any surplus can be excreted partially via the bile, there being little scope for reabsorption. As the amounts of manganese absorbed increase, progressively larger proportions are excreted via the faeces (Fig. 14.2). Linear increases in biliary manganese concentrations have been reported in chicks when their deficent diet was supplemented with 7 or 14 mg Mn kg^{-1} DM (Halpin and Baker, 1986). Over 90% of the manganese absorbed from high-manganese diets (1000 mg Mn kg^{-1} DM) by chicks was excreted endogenously (Wedekind *et al.*, 1991). Biliary concentrations increased from 2.1 to 17.1 mg Mn l^{-1} in preruminant calves when the manganese level in their milk-replacer was increased from 40 to 1000 mg kg^{-1} DM (as $MnSO_4$) (Jenkins and Hidiroglou, 1991).

Tissue distribution

Manganese is one of the least abundant trace elements in all livestock tissues, concentrations in the whole body ranging from 0.5 to 3.9 mg kg^{-1} DM in the carcass of sheep and calves (Suttle, 1979; Grace, 1983). In adult sheep, manganese concentrations (mg kg^{-1} FW) were highest in the liver (4.2), followed by the pancreas (1.7) and kidney (1.2); elsewhere, values were < 0.3 (Grace, 1983). It is, therefore, hardly surprising that manganese is also one of the least abundant elements in milk, rarely exceeding 0.1 mg Mn l^{-1} in any species, except in colostrum. In quantitative terms, the gastrointestinal tract and skin contain most of the manganese found in the sheep carcass (Fig. 14.3).

Storage

There was no evidence of storage of manganese in liver or bone in lambs given diets ranging from 13 to 45 mg Mn kg^{-1} DM (Masters *et al.*, 1988) but increasing dietary levels from 123 to 473 mg Mn kg^{-1} DM (i.e. up to 30 times the requirement) in a grazing study increased tissue manganese by 25% in most organs and by 260% in the digestive tract (Grace and Lee, 1990). Data for chicks indicate that liver, kidney and skeletal manganese levels are responsive to dietary manganese if the test range is magnified tenfold (Fig. 14.4: Halpin and Baker, 1986). These increases in tissue manganese at very high intakes probably reflect a failure of homeostasis rather than successful storage. Because of the large contribution of the skeleton to body mass, a small rise in bone manganese concentration could constitute a significant

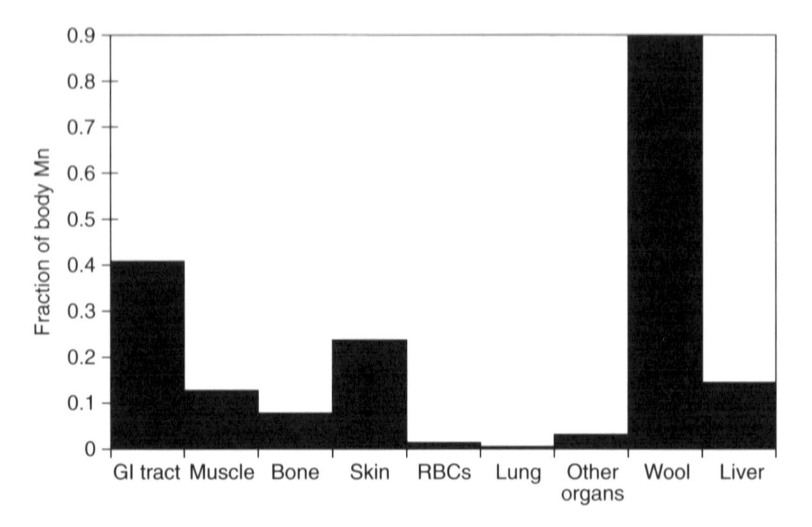

Fig. 14.3. Distribution of manganese in the shorn, empty body of mature sheep and the relatively large amount removed in the fleece (data from Grace, 1983), GI, gastrointestinal.

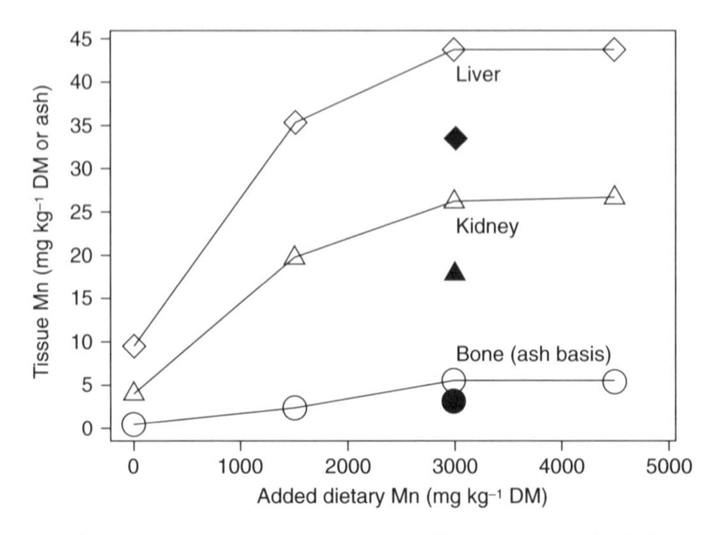

Fig. 14.4. Poultry can store excess manganese in their tissues but high dietary concentrations are required to cause appreciable increases particularly in bone (data from two experiments (open and closed symbols) of Halpin and Baker, 1986).

'passive' reserve (Leach and Harris, 1997). It would be surprising if any skeletal reserve in the pullet did not become available with the onset of lay and the accompanying bone resorption. There is little evidence that the fetus builds up a hepatic reserve of manganese (Graham *et al.*, 1994), but very high liver manganese concentrations (411–943 mg kg^{-1} DM) have been found in newborn calves following supplementation of the dam (Howes and Dyer, 1971).

Functions of Manganese

The functions of manganese can be linked to the metalloenzymes, which are activated by the element.

Cartilage development

Impaired glycosyl transferase activity reduces synthesis of mucopolysaccharides (vital structural components of cartilage) and causes skeletal defects in poultry deprived of manganese (see Leach, 1971, 1988). Manganese-deficient chicks have less proteoglycan in the cartilage of the tibial growth plate than manganese-replete chicks, and the carbohydrate composition of the monomers is changed (Liu *et al.*, 1994). In laying hens, subnormal egg production and poor shell formation may result from impaired mucopolysaccharide synthesis due to a deficient supply of manganese (Hill and Mathers, 1968), and a reduction in the hexosamine content of the shell matrix occurs (Longstaff and Hill, 1972).

Blood clotting

Manganese is also involved in the formation of prothrombin, a glycoprotein, through its activation of glycosyltransferases. The clotting response from vitamin K is reduced in manganese-deficient chicks (Doisey, 1973), but the precise interaction of manganese with vitamin K in the conversion of preprothrombin to prothrombin is not clear.

Lipid and carbohydrate metabolism

Pyruvate carboxylase probably sustains lipid as well as glucose metabolism, because the fat accumulation seen in manganese deficiency is also a feature of biotin deficiency in non-ruminants and biotin activates the same enzyme. Defects in lipid and carbohydrate metabolism have been reported in manganese-deficient rats and guinea-pigs (see Underwood, 1977), but little is known of these aspects of manganese metabolism in domestic livestock. An association between manganese and choline has long been recognized, and both these nutrients are necessary for complete protection against perosis in chicks (Jukes, 1940). A deficient diet reduces fat deposition and back-fat thickness in pigs (Plumlee *et al.*, 1956) and the body-fat content of goat kids born from deprived dams (Anke *et al.*, 1973b).

Resistance to oxidants

The superoxide dismutases (SOD) (there is also one containing copper (Cu) and zinc (Zn)) protect cells from damage by the free oxygen radical O_2^-. Manganese deficiency lowers MnSOD activity in the heart and increases the peroxidative damage caused by high dietary levels of polyunsaturated fatty acids (PUFA) (Malecki and Greger, 1996), but compensatory increases in CuZnSOD suggested overlapping roles for the two forms of SOD and possible interactions between dietary copper and manganese. Reductions in the activity of MnSOD have been related to structural changes in liver mitochondria and cell membranes (Bell and Hurley, 1974). Broiler chicks given a diet containing 17 mg Mn kg^{-1} DM developed manganese-responsive abnormalities in heart mitochondrial ultrastructure, accompanied by a reduction in MnSOD activity (Luo *et al.*, 1993). Mitochondria are responsible for 60% of cellular O_2 consumption and may be particularly vulnerable to free-radical damage (Leach and Harris, 1997). The importance of demand on trace-element-dependent pathways in determining requirement suggested in Chapter 3 can be further illustrated with respect to manganese. Tissue activities of MnSOD are low in lambs at birth (Paynter and Caple, 1984), and this probably reflects the low oxidative stress associated with the protected lifestyle of the fetus. The subsequent need to respire is associated with a particularly marked rise in lung MnSOD by 4 weeks of age. Most of the manganese in ovine heart is present as MnSOD (Paynter and Caple, 1984) and it is by far the most dominant dismutase in this tissue and also in muscle. Masters *et al.* (1988) noted significant reductions in MnSOD in heart and lung in their most depleted group. This led them to speculate that in circumstances of oxidant stress (e.g. from PUFA) or depletion of other dietary antioxidants (for example, Cu, selenium (Se) and vitamin E: see Fig. 15.2), manganese depletion may lead to dysfunction. The absence of such challenges in experiments with penned animals given semipurified diets may explain the apparently greater tolerance of ruminants towards experimental than naturally occurring manganese deficiency. The induction of MnSOD following the administration of tumour necrosis factor α (Wong and Goeddel, 1988) suggests a need for added protection against oxidative stress associated with the inflammatory responses to some infections.

Reproduction

Studies of manganese distribution among the tissues of the reproductive tract of normal and anoestrous ewes led Hidiroglou (1975) to suggest that this element has a possible role in the functioning of the corpus luteum. Lack of manganese may inhibit the synthesis of cholesterol and its precursors; this, in turn, may limit the synthesis of sex hormones and possibly other steroids, with consequent infertility (Doisey, 1973).

Biochemical Manifestations of Manganese Deprivation

The sequence of biochemical changes which precede the development of clinical signs of manganese deprivation in livestock is illustrated in Fig. 14.5. They differ from those for other elements in that depletion and deficiency phases are not easily distinguished by changes in manganese concentrations

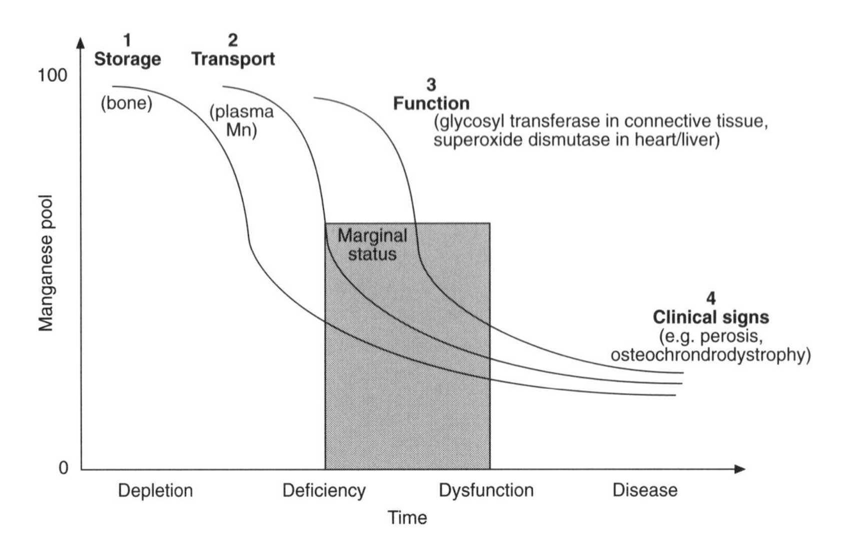

Fig. 14.5. Temporal stages of manganese deprivation leading to the appearance of clinical signs: see also Fig. 3.1 and related text.

at accessible sites (in the bloodstream or liver). Subnormal manganese concentrations in the bones and reduced activities of alkaline phosphatase have been found in the blood and soft tissues of deficient chicks and ducks (Van Reen and Pearson, 1955). However, bone alkaline phosphatase activity is not always subnormal, even when the characteristic bone changes of manganese deprivation are visible.

Occurrence and Clinical Manifestations of Manganese Deprivation

Manganese deprivation as a practical production problem is largely confined to avian species. Unequivocal evidence of recurring nutritional disabilities attributable to lack of dietary manganese in grazing sheep and cattle has not been reported, even in areas where plant-growth responses can be obtained from the application of manganese-containing fertilizers. Manganese deprivation is most likely to occur in ruminants reared on maize silage with maize or barley supplements. However, responses to manganese supplements in the growth of heifers and fertility of dairy cows have been observed in parts of Europe and may be due to the deficiencies conditioned by high contents of iron, calcium, phosphorus and potassium in the diets (Hartmans, 1974). Growth impairment results from both reduced feed consumption and decreased efficiency of feed use (Boyer *et al.*, 1942). The signs of deficiency vary with the degree and duration of the deficiency and with the age and productive function of the animal.

Skeletal abnormalities

If manganese deficiency occurs during fetal or embryonic development or early postnatal life, decreased endochondral bone growth can cause severe skeletal abnormalities and dwarfism (chondrodystrophy) (Liu *et al.*, 1994). In manganese-deprived pigs, the skeletal abnormalities are characterized by lameness and enlarged hock joints with crooked and shortened legs; in calves and sheep, by difficulty in standing and joint pains, with poor locomotion and balance; and in goats, by tarsal joint excrescences, leg deformities and ataxia. There are now several reports of congenital chondrodystrophy in calves associated with low manganese concentrations in the tissues of the newborn and the diet of the mother (e.g. Hidiroglou *et al.*, 1990; Valero *et al.*, 1990; Staley *et al.*, 1994), although unequivocal manganese responsiveness has yet to be demonstrated. Manganese deficiency in chicks (Gallup and Norris, 1939a), poults and ducklings is manifested as the disease perosis or 'slipped tendon' and in chick embryos by chondrodystrophy. Perosis is characterized by enlargement and malformation of the tibiometatarsal joint, twisting and bending of the tibia, thickening and shortening of the long bones and slipping of the gastrocnemius tendon from its condyles. With increasing severity of the condition, chicks are reluctant to move, squat on their hocks and soon die. Manganese deficiency in the laying hen is manifested by shortened, thickened wings and legs of the embryo and a 'parrot beak', resulting from a shortening of the lower mandible and globular contour of the head; mortality is high among manganese-deprived chicks.

Ataxia

Ataxia in the offspring of manganese-deficient animals was first observed in the chick (Caskey and Norris, 1940). This condition has since been intensively studied in rats and guinea-pigs and has been observed in kids born of deficient goats (Anke *et al.*, 1973a). Manganese deficiency in rats during late pregnancy produces an irreversible congenital defect in the young, characterized by ataxia and loss of equilibrium, arising from impaired vestibular function, due to a structural defect in the inner ear. This, in turn, arises from impaired mucopolysaccharide synthesis and hence bone development of the skull, particularly the otoliths of the embryos (Hurley *et al.*, 1960). The neonatal ataxia of manganese deficiency is thus secondary to the skeletal defects.

Reproductive disorders

Defective ovulation and testicular degeneration were observed in the earliest experiments with manganese-deficient rats and mice. Subsequently, numerous studies have indicated an impairment of reproductive function in manganese-deficient males and females of several farmed species (e.g. hens; Gallup and Norris, 1939b). Depressed or delayed oestrus and poor conception rates have been associated with manganese deprivation in cows, goats and ewes (Hignett, 1941; Wilson, 1966; Anke *et al.*, 1973a) but remain rare natural occurrences. Manganese-deficient ewes (Egan, 1972; Hidiroglou *et al.*,

1978) and cows (Rojas *et al.*, 1965) may require more services per conception and respond to manganese supplementation. However, identical twin cattle fed from the age of 1 or 2 months on rations containing only 16–21 mg Mn kg^{-1} DM for several years displayed no reduction in fertility compared with adequately supplemented controls (Hartmans, 1972). A study with ram lambs (Masters *et al.*, 1988) indicated that, when the dietary manganese concentration was reduced from 19 to 13 mg kg^{-1} DM for 11 weeks, there was a reduction in testicular growth relative to body growth. Manganese concentrations in the testes were low but unaffected by depletion, suggesting that the effect may have been caused by hormonal influences upon the testes. Lack of an effect of manganese on wool or body growth in that study confirms the particular sensitivity of reproductive processes to manganese deficiency.

Diagnosis of Manganese Deficiency

Manganese concentrations in the blood, bones, hair and liver decline only slightly in animals deprived of manganese. Reported blood manganese values are extremely variable, reflecting both individual variability and analytical inadequacies. Whole-blood concentrations substantially below 20 ng Mn ml^{-1} nevertheless suggest the possibility of a dietary deficiency in sheep and cattle (Hidiroglou, 1979). The study of Masters *et al.* (1988), in which ram lambs were given diets containing 13, 19, 30 or 45 mg kg^{-1} DM, provides new insights into the diagnostic value of manganese assays in different samples (Fig. 14.6). Liver manganese is frequently measured, because it is by far the richest of the tissues in manganese, but it showed no significant change in concentration about a mean of around 3 mg Mn kg^{-1} FW, whereas heart and lung both showed significant depletion from far lower maximal concentrations; plasma manganese fell significantly from 2.74 to 1.85 ng ml^{-1}. The diagnostic merits of plasma and heart manganese should be examined more closely, with levels of < 0.2 mg kg^{-1} FW and < 2 ng ml^{-1}, respectively, provisionally indicative of deficency in young lambs. Such low concentrations would, however, place great demands on analytical precision in the diagnostic laboratory. Attempts to use manganese in hair as an index of status have given conflicting results and widely varying individual values, but the manganese concentrations in wool and feathers may be more useful. The wool of lambs fed on a low-manganese diet for 22 weeks had on average only 6.1 mg Mn kg^{-1}, compared with 18.7 mg kg^{-1} DM in control lambs (Lassiter and Morton, 1968). Similarly, the skin and feathers of pullets fed on a low-manganese diet for several months averaged 1.2 mg Mn kg^{-1} DM, compared with 11.4 mg kg^{-1} DM in comparable birds on a manganese-rich diet (Mathers and Hill, 1968). The use of standard washing, solvent extraction and drying procedures should lessen variability and improve the diagnostic value of assaying manganese in appendages of the skin. Guidelines for the assessment of manganese status in farm livestock are summarized in Table 14.2. The manganese concentration in the whole diet is helpful in detecting

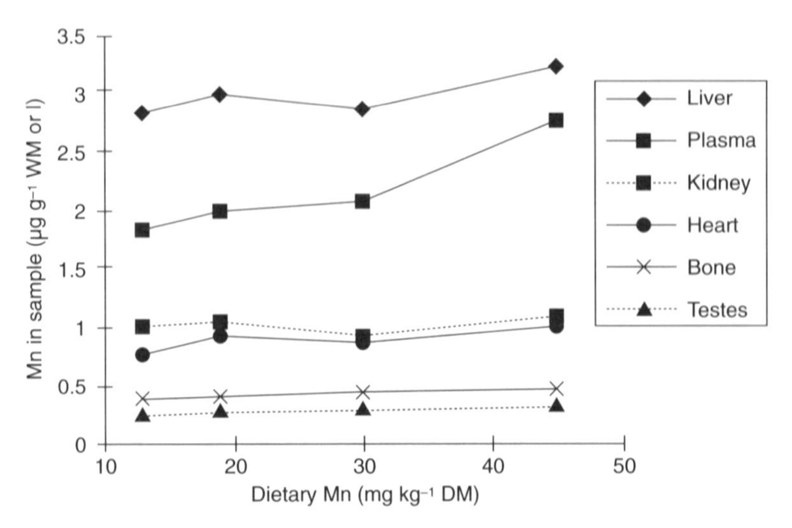

Fig. 14.6. Manganese concentrations in the heart are relatively more responsive to manganese deprivation than those in other soft tissues in lambs (data from Masters *et al.*, 1988).

Table 14.2. Marginal bands[a] for manganese concentrations in diet, blood and tissues for use in assessing risk of manganese disorders in farm livestock.

	Diet (mg kg^{-1} DM)[c]	Blood (µg l^{-1})[d]	Serum (µg l^{-1})[d]	Liver (mg kg^{-1} DM)[c]	Bone[b] (mg kg^{-1} DM)[c]	Heart (mg kg^{-1} DM)[c]	Pancreas (mg kg^{-1} DM)[c]
Cattle	10–20	20–70	5–6	5.0–7.5	1.0–1.4[R]	–	3.0–5.0
Sheep	8–20	12–20	1.8–2.0	8.0–9.0	0.3–0.42[Mf]	0.6–0.7	4–8
Goat	10–20	–	–	3.0–6.0	5.0–6.0[R]	1.5–2.0	–
Pigs	6–20	10–12	3–4	6.0–9.0	2.0–2.3[RcT]	–	2.0–3.0
Poultry	14–20	30–50	5–10	6–12	4–6[Tf]	–	10–15

[a] Values below bands indicate probability of positive response to manganese supplements: values within bands indicate possibility of such responses.
[b] R, rib; T, tibia; M, metatarsal; f, fat-free; c, cortical bone used for assay.
[c,d] Multiply by 0.0182 to obtain values in mmol and µmol, respectively.

possible deficiency in ruminants, but in non-ruminants the many factors with which manganese interacts must be considered.

Manganese Requirements

Estimates of manganese requirements vary with the species and strain of the animal, the chemical form in which the element is ingested and the chemical composition of the whole diet, particularly the calcium and phosphorus contents. Criteria of adequacy are also important, in particular whether the 'requirement' is for optimal growth, reproduction or the attainment of maximal manganese concentrations in a particular tissue.

Pigs

The manganese requirements of pigs for satisfactory reproduction are substantially higher than those needed for body growth. Pigs have been maintained on semipurified diets containing as little as 0.5 to 1.5 mg Mn kg^{-1} DM from normal weaning weights to varying market weights (Johnson, 1943; Plumlee *et al.*, 1956). However, tissue manganese depletion was marked, and continued deprivation throughout gestation and lactation causes skeletal and reproductive abnormalities; no abnormalities occurred with additional manganese at 40 mg kg^{-1} DM, but it is unlikely that supplementation at this rate was necessary. Grummer *et al.* (1950) found that a high-maize diet containing 12 mg Mn kg^{-1} DM was adequate for growth and skeletal development, with only a slight indication of improved reproductive performance from additional manganese. Similar results from other groups suggest that the maximum requirement for growing pigs is probably around 4 mg kg^{-1} DM and slightly higher (10 mg kg^{-1} DM) for sows on natural diets (NRC, 1988).

Poultry

The National Research Council (NRC, 1994) requirements for white-egg-laying strains are: starting chicks, 60; growing chickens (6–18 weeks), 30; laying hens, 17–25 (depending on food intake); and breeding hens, 33 mg Mn kg^{-1} DM. Requirements for brown-egg-laying strains were some 10% lower at all stages, because of their higher food intake, but scientific support for such a distinction is lacking. The requirements of turkeys were given by NRC (1977) as: 0–8 weeks, 55; 8 weeks to marketing or prebreeding, 25; and breeding hens, 35 mg Mn kg^{-1} DM; but these were increased, along with those of broilers, to a uniform 60 mg Mn kg^{-1} DM by NRC (1994). The increased recommendations were not based on new evidence but were the result of a renewed concern about the high needs demonstrated by old experiments in which large excesses of calcium and phosphorus were fed. In one study with turkey hens on a diet containing 3.22% Ca and 0.78% inorganic P, supplementation to provide between 54 and 108 mg Mn kg^{-1} DM was necessary for optimal reproductive performance (Atkinson *et al.*, 1967). The effect of excess dietary calcium and phosphorus in raising manganese requirements has been appreciated since the earliest studies with manganese in poultry. For example, 64% of the chicks fed on a ration containing 3.2% Ca, 1.6% P and 37 mg Mn kg^{-1} DM developed perosis, whereas no perosis was observed when the diet supplied normal levels of 1.2% Ca with 0.9% P (Schaible *et al.*, 1938). Where maize is the staple cereal in poultry diets, manganese supplementation is necessary to meet the bird's requirements for egg laying (when calcium intakes are necessarily high). However, the requirements of young broiler chicks for growth on maize–soybean meal diets are consistently met with 37.5 mg Mn kg^{-1} DM from natural sources (Wedekind *et al.*, 1991). The high NRC (1994) recommendations for broilers and turkeys are clearly a safe allowance to cope with poorly available sources and extreme variations in calcium and phosphorus intakes. Addition of phytase to poultry rations would probably obviate the need to supplement with manganese (Biehl *et al.*, 1995).

Cattle

The requirements of cattle are not precisely known, although it is clear that they are substantially lower for growth than for optimal reproductive performance and may be increased by high intakes of calcium and phosphorus. Early studies showed that 10 mg Mn kg^{-1} DM was adequate for growth but marginal for maximum fertility (Bentley and Phillips, 1951) and 20 mg kg^{-1} DM adequate for both growth and reproduction of heifers (Rojas *et al.*, 1965; Howes and Dyer, 1971). Hartmans (1974) used practical rations containing 16 and 21 mg Mn kg^{-1} DM for 2.5–3.5 years without any clinical evidence of deficiency or observable improvement from manganese supplementation of identical-twin heifers, but he recommended 25 mg Mn kg^{-1} DM for Dutch dairy cattle. The Agricultural Research Council (ARC, 1980) concluded that, while 10 mg Mn kg^{-1} DM was sufficient for growth in cattle, 20–25 mg kg^{-1} DM was sometimes needed for normal skeletal development and was probably sufficient for reproduction. Few pastures fail to meet manganese requirements.

Sheep and goats

A definitive study of the requirements of growing sheep has now been published (Masters *et al.*, 1988) and it showed that 13 mg Mn kg^{-1} DM was adequate for live-weight gain and wool growth but slightly more (16 mg kg^{-1} DM) was needed for testicular growth. These values are similar to those recommended for cattle, which is surprising, given that there is as much manganese in the fleece of a mature sheep as in the rest of the body (Fig. 14.3; Grace, 1983). Data on the manganese requirements of goats are meagre and ill-defined. Anke *et al.* (1973a) reported that female goats fed on diets containing 20 mg Mn kg^{-1} DM in the first year and 6 mg kg^{-1} DM in the second grew as well as those receiving an additional 100 mg Mn kg^{-1} DM, but reproductive performance was greatly impaired. The results of this experiment confirm the lower requirements for growth than for fertility but leave the minimum requirement for either purpose unresolved.

Control of Manganese Deprivation

Manganese deprivation in poultry can be prevented by the incorporation of manganese salts or oxides into the mineral supplements or whole ration. The supplementary manganese is readily transmitted by hens to the egg for use by the developing chick embryo and sufficient should be included in the mineral mixture to ensure that the requirements given above are met. All ordinary rations that are otherwise adequate for the growth, health and reproduction of pigs supply adequate manganese. Manganese supplementation is therefore not normally necessary for this species. In areas where 'simple' or 'conditioned' manganese deficiency is suspected in cattle, supplementation of the feed with manganese sulphate at the rate of 4 g for cows, 2 g for heifers or 1 g day^{-1} for calves is sufficient for either prevention or cure of deficiency. Treatment of pastures with 15 kg $MnSO_4$ ha^{-1} was effective in the Netherlands (Hartmans, 1974) but is rarely practised. Supplementation of rations based on maize and maize silage with 20 mg Mn kg^{-1} DM is recommended.

Inorganic sources

The importance of 'availability' of manganese in inorganic supplements was indicated by early work showing that the manganese in two ores, rhodochrosite (a carbonate ore) and rhodonite (a silicate ore), was relatively unavailable to poultry (Schaible *et al.* 1938; Gallup and Norris, 1939a, b). The commonly used manganese sources are $MnSO_4$, manganese oxide (MnO) and $MnCO_3$; of these, $MnSO_4$ has the highest 'availability' and values of the other two sources relative to $MnSO_4$ have been assessed at 30 and 55%, respectively, for poultry and 35 and 30%, respectively, for sheep (Henry *et al.*, 1995). Correspondingly higher levels of $MnCO_3$ and MnO should be used to correct dietary deficits.

Organic sources

In the face of powerful dietary constraints from phytate and fibre on manganese absorption, the potential for manganese sources of enhanced nutritive value, protected by complexation with peptides (the so-called 'chelates' or 'proteinates'), is great. Unfortunately, there is no indisputable evidence yet in the scientific literature that manganese can be effectively 'protected' in these ways. An early study with chicks claimed superior nutritive value for a methionine–manganese complex (MnMet) for chicks over MnO, the advantage widening from 30.1 to 74.4% when the diet contained phytate and fibre (Fly *et al.*, 1989). However, the use of a slopes:ratio method with only three points and vastly different (and both extreme) maxima (1285 mg Mn as MnMet vs. 1875 mg Mn as MnO kg^{-1} DM) invalidates the comparison. At the lower levels of manganese employed, the introduction of phytate/fibre lowered tibia manganese similarly with each source. Others have failed to find an advantage of MnMet over MnO for chicks (Scheideler, 1991). It should be noted that MnO is a relatively poor inorganic source, regularly surpassed by sources such as $MnSO_4$ and $MnCl_2$ (Ammerman *et al.*, 1995). Similar 'availabilities' were reported for a 'manganese–protein chelate' when compared with $MnSO_4$ in chicks, whether the diet was high or low in fibre and phytate (Baker and Halpin, 1987). Turning to ruminants, Henry *et al.* (1992) reported that MnMet was 1.2 times more 'available' than $MnSO_4$ for lambs. However, the advantage was only evident in one of the three tissues examined (bone;

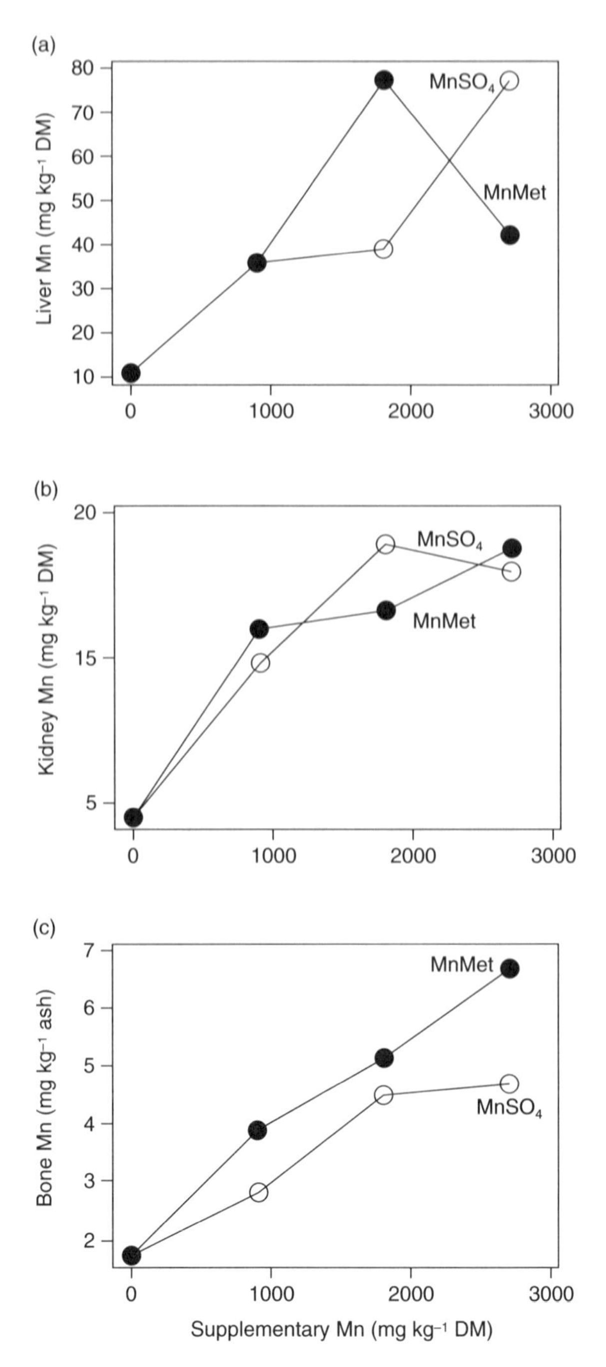

Fig. 14.7. Provision of lambs with manganese as Mn methionine (MnMet) rather than inorganic sulphate ($MnSO_4$) had little effect on tissue Mn except in bone at marginally toxic levels (data from Henry *et al.*, 1992): there is no convincing evidence of a difference in availability between sources.

Fig. 14.7c) and developed mostly over the final increment in dietary manganese (from 1800 to 2700 mg Mn kg^{-1} DM), which took values into the toxicological range. Provision of manganese in chelated or protected forms is unlikely to have any curative or preventive benefit over that provided by simpler, cheaper inorganic forms.

Toxicity

Ruminants

Animals dependent on pastures and forages are subject to wide differences in manganese intakes, which can vary from possibly deficient to potentially toxic. When sheep grazing New Zealand pastures containing 140–200 mg Mn kg^{-1} DM were given pellets providing 250 or 500 mg Mn day^{-1}, as $MnSO_4$, growth rate was significantly depressed and some reduction in heart and plasma iron (Fe) was evident, despite the fact that the pastures contained 1100–2200 mg Fe kg^{-1} DM (Grace, 1973). In a later study (Grace and Lee, 1990), a similar manganese supplement lowered liver iron and increased pancreatic zinc levels without retarding growth. A mutual metabolic antagonism between manganese and iron at the absorptive site has been known for many years (Matrone *et al.*, 1959). The anorexic effect of high manganese intakes is less when the element is given continuously by diet than by bolus administration (Black *et al.*, 1985), which would briefly raise ruminal manganese concentrations to very high levels. Furthermore, $MnSO_4$ would be more reactive than the form normally found when ruminants consume manganese-rich grass, where much of the manganese is likely to be present as inert oxides following soil contamination: sheep can tolerate 3000 mg Mn kg^{-1} DM for 21 days when given as MnO (Black *et al.*, 1995). The minimum level of manganese found necessary to depress appetite and growth rate in preruminant calves was 1000 mg Mn kg^{-1} DM (Jenkins and Hidiroglou, 1991); the corresponding values for weaned calves was 2600 mg kg^{-1} DM (Underwood, 1981) and for weaned lambs between 3000 and 4500 mg Mn kg^{-1} DM (Wong-Valle *et al.*, 1989).

Pigs and poultry

The toxic level of manganese for pigs has been variously estimated to be 500, 1250 and 4000 mg kg^{-1} DM and the range probably reflects the variable influence of fibre and phytate on manganese absorption. There is evidence of interference with iron absorption and a reduction in haemoglobin synthesis at high manganese intakes (ARC, 1981). The adverse interaction probably arises from competition with iron for binding sites on transferrin and other carriers, and the precise toxic threshold is therefore likely to depend on dietary iron concentration. The tolerable limit of 1000 mg Mn kg^{-1} DM set for pigs by ARC (1981) is a sensible compromise for diets of normal iron concentration and low in phytate. Poultry are tolerant towards manganese, no doubt reflecting their relatively poor efficiency of absorption; the safe limit has been placed at 2000 mg Mn kg^{-1} DM (NRC, 1980) but is bound to vary under the influence of dietary phytate.

References

Ammerman, C.B., Baker, D.H. and Lewis, A.J. (1995) In: *Bioavailability of Nutrients for Animals.* Academic Press, San Diego, pp. 383–398.

Anke, M., Groppel, B., Reisseg, W., Ludke, H., Grun, M. and Dittrich, G. (1973a) Manganmangel beim Wiederkauer. 3. Manganmangelbedingte Fortpflanzungs-, Skelett- und Nervenstorungen bei weiblichen Wiederkauern und ihren Nachkommen. *Archive für Tierernahrung* 23, 197–211.

Anke, M., Hennig, A., Groppel, B., Dittrich, G. and Grun, M. (1973b) Manganmangel beim Wiederkauer. 4. Der Einfluss des Manganmangels auf den Gehalt neugeborener Lammer an Fett, Protein, Mangan, Asche, Kalzium, Phosphor, Zink und Kupfer. *Archive für Tierernahrung* 23, 213–223.

ARC (1980) *The Nutrient Requirements of Ruminants.* Commonwealth Agricultural Bureaux, Farnham Royal.

ARC (1981) *The Nutrient Requirements of Pigs.* Commonwealth Agricultural Bureaux, Farnham Royal, Slough, UK, pp. 271–273.

Atkinson, R.L., Bradley, J.W., Couch, J.R. and Quisenberry, J.H. (1967) Effect of various levels of manganese on the reproductive performance of turkeys. *Poultry Science* 46, 472–475.

Atkinson, S.A., Shah, J.K., Webber, C.E., Gibson, I.L. and Gibson, R.S. (1993) A multi-element isotopic tracer assessment of true fractional absorption of minerals from formula with additives of calcium, phosphorus, zinc, copper and iron in young piglets. *Journal of Nutrition* 123, 1586–1593.

Baker, D.H. and Halpin, K.M. (1987) Research note: efficacy of a manganese–protein chelate compared with that of manganese sulphate for chicks. *Poultry Science* 66, 1561–1563.

Baker, D.H. and Odohu, G.W. (1994) Manganese utilisation in the chick: effects of excess phosphorus on chicks fed manganese deficient diets. *Poultry Science* 73, 1162–1165.

Beeson, K.C., Gray, L. and Adams, M.B. (1947) The absorption of mineral elements by forage plants. 1. The phosphorus, cobalt, manganese and copper content of some common grasses. *Journal of the American Society of Agronomy* 39, 356–362.

Bell, L.T. and Hurley, L.S. (1974) Histochemical enzyme changes in epidermis of manganese-deficient fetal mice. *Proceedings of the Society for Experimental Biology and Medicine* 145, 1321–1324.

Bentley, O.G. and Phillips, P.H. (1951) The effect of low manganese rations upon dairy cattle. *Journal of Dairy Science* 34, 396–403.

Berger, L.C. (1995) Why do we need a new NRC database? *Animal Feed Science and Technology* 53, 99–107.

Biehl, R.R., Baker, D.H. and DeLuca, H.F. (1995) 1 α-hydroxylated cholecalciferol compounds act additively with microbial phytase to improve phosphorus, zinc and manganese utilization in chicks fed soy-based diets. *Journal of Nutrition* 125, 2407–2416.

Black, J.R., Ammerman, C.B. and Henry, P.R. (1985) Effect of quantity and route of manganese monoxide on feed intake and serum manganese of ruminants. *Journal of Dairy Science* 68, 433–436.

Boyer, P.D., Shaw, J.H. and Phillips, P.H. (1942) Studies on manganese deficiency in the rat. *Journal of Biological Chemistry* 143, 417–425.

Bremner, I. (1970) Zinc, copper and manganese in the alimentary tract of sheep. *British Journal of Nutrition* 24, 769–783.

Buchanan-Smith, J.G., Evans, E. and Poluch, S.O. (1974) Mineral analyses of corn silage produced in Ontario. *Canadian Journal of Animal Science* 54, 253–256.

Carter, J.C., Miller, W.J., Neathery, M.W., Gentry, R.P., Stake, P.E. and Blackmon, D.M. (1974) Manganese metabolism with oral and intravenous ^{54}Mn in young calves as influenced by supplemental manganese. *Journal of Animal Science* 380, 1284–1290.

Caskey, C.D. and Norris, L.C. (1940) Micromelia in adult fowl caused by manganese deficiency during embryonic development. *Proceedings of the Society for Experimental Biology and Medicine* 44, 332–335.

Davidsson, L., Lonnerdahl, B., Sandstrom, B., Kunz, C. and Keen, C.L. (1989) Identification of transferrin as the major plasma carrier protein for manganese introduced orally or intravenously or after *in vitro* addition in the rat. *Journal of Nutrition* 119, 1461–1464.

Doisey, E.A., Jr (1973) Micronutrient controls on biosynthesis of clotting proteins and cholesterol. In: Hemphill, D.D. (ed.) *Trace Substances in Environmental Health – 6.* University of Missouri, Columbia, p. 193.

Egan, A.R. (1972) Reproductive responses to supplemental zinc and manganese in grazing Dorset Horn ewes. *Australian Journal of Experimental Agriculture and Animal Husbandry* 12, 131–135.

Fly, A.D., Izquierdo, O.A., Lowry, K.R. and Baker, D.H. (1989) Manganese bioavailability in a Mn–methionine chelate. *Nutrition Research* 9, 901–910.

Gallup, W.D. and Norris, L.C. (1939a) The amount of manganese required to prevent perosis in the chick. *Poultry Science* 18, 76–82.

Gallup, W.D. and Norris, L.C. (1939b) The effect of a deficiency of manganese in the diet of the hen. *Poultry Science* 18, 83–88.

Gladstones, J.S. and Drover, D.P. (1962) The mineral composition of lupins. 1. A survey of the copper, molybdenum and manganese contents of lupins in the South West of Western Australia. *Australian Journal of Experimental Agriculture and Animal Husbandry* 2, 46–53.

Grace, N.D. (1973) Effect of high dietary Mn levels on the growth rate and the level of mineral elements in the plasma and soft tissues of sheep. *New Zealand Journal of Agricultural Research* 16, 177–180.

Grace, N.D. (1983) Amounts and distribution of mineral elements associated with fleece-free empty body weight gains in grazing sheep. *New Zealand Journal of Agricultural Research* 26, 59–70.

Grace, N.D. and Lee, J. (1990) Effect of Co, Cu, Fe, Mn, Mo, Se and Zn supplementation on the elemental content of soft tissues and bone in sheep grazing ryegrass/white clover pasture. *New Zealand Journal of Agricultural Research* 33, 635–647.

Graham, T.W., Thurmond, M.C., Mohr, F.C., Holmberg, C.A., Anderson, M.L. and Keen, C.L. (1994) Relationship between maternal and foetal liver copper, iron, manganese and zinc concentrations and foetal development in California Holstein dairy cows. *Journal of Veterinary Diagnostic Investigation* 6, 77–87.

Greenberg, D.M. and Campbell, W.W. (1940) Studies in mineral metabolism with the aid of radioactive isotopes. 4. Manganese. *Proceedings of the National Academy of Sciences USA* 26, 448–452.

Gregory, E.M. and Fridovich, I. (1974) Superoxide dismutases: properties, distribution, and functions. In: Hoekstra, W.G., Suttie, J.W., Ganther, H.E. and Mertz, W. (eds) *Trace Element Metabolism in Animals – 2.* University Park Press, Baltimore, pp. 486–488.

Grummer, R.H., Bentley, O.G., Phillips, P.H. and Bohstedt, G. (1950) The role of manganese in growth, reproduction, and lactation in swine. *Journal of Animal Science* 9, 170–175.

Halpin, K.M. and Baker, D.H. (1986) Long-term effects of corn soyabean meal, wheat bran and fish meal on manganese utilisation in the chick. *Poultry Science* 65, 1371–1374.

Hartmans, J. (1972) Manganversuche mit monozygoten Rinderzwillingen. *Landwirtschaftliche Forschung Sonderheft* 27(ii), 1–11.

Hartmans, J. (1974) Tracing and treating mineral disorders in cattle under field conditions. In: Hoekstra, W.G., Suttie, J.W., Ganther, H.E. and Mertz, W. (eds) *Trace Element Metabolism in Animals – 2.* University Park Press, Baltimore, pp. 261–273.

Henry, P.R. (1995) Manganese bioavailability. In: Ammerman, C.B., Baker, D.H. and Lewis, A.J. (eds) *Bioavailability of Nutrients for Animals.* Chapter 11. Academic Press, New York, pp. 239–256.

Henry, P.R., Ammerman, C.B. and Littell, R.C. (1992) Relative bioavailability of manganese from a manganese methionine complex and inorganic sources for ruminants. *Journal of Dairy Science* 75, 3473–3478.

Hidiroglou, M. (1975) Mn uptake by the ovaries and reproductive tract of cycling and anestrous ewes. *Canadian Journal of Physiology and Pharmacology* 53, 969–972.

Hidiroglou, M. (1979) Manganese in ruminant nutrition. *Canadian Journal of Animal Science* 59, 217.

Hidiroglou, M., Ho, S.K. and Standish, J.F. (1978) Effects of dietary manganese levels on reproductive performance of ewes and on tissue mineral composition of ewes and day-old lambs. *Canadian Journal of Animal Science* 58, 35–41.

Hidiroglou, M., Ivan, M., Bryoa, M.K., Ribble, C.S., Janzen, E.D., Proulx, J.G. and Elliot, J.I. (1990) Assessment of the role of manganese in congenital joint laxity and dwarfism in calves. *Annales de Recherches Vétérinaires* 21, 281–284.

Hignett, S.L. (1941) Some aspects of bovine sterility. *Veterinary Record* 53, 21–25.

Hill, R. (1967) Vitamin D and manganese in the nutrition of the chick. *British Journal of Nutrition* 21, 507–512.

Hill, R. and Mathers, J.W. (1968) Manganese in the nutrition and metabolism of the pullet. 1. Shell thickness and manganese content of eggs from birds given a diet of low or high manganese content. *British Journal of Nutrition* 22, 635–643.

Howes, A.D. and Dyer, I.A. (1971) Diet and supplemental mineral effects on manganese metabolism in newborn calves. *Journal of Animal Science* 32, 141–145.

Hurley, L.S. (1981) Teratogenic effects of manganese, zinc and copper in nutrition. *Physiological Reviews* 61, 249–295.

Hurley, L.S. and Keen, C.L. (1987) Manganese. In: Mertz, W. (ed.) *Trace Elements in Human and Animal Nutrition,* Vol II. Academic Press, New York, p. 185.

Hurley, L.S., Wooten, E., Everson, G.J. and Asling, C.W. (1960) Anomalous development of ossification in the inner ear of offspring of manganese-deficient rats. *Journal of Nutrition* 71, 15–19.

Jenkins, K.J. and Hidiroglou, M. (1991) Tolerance of the pre-ruminant calf for excess manganese or zinc in milk replacer. *Journal of Dairy Science* 74, 1047–1053.

Johnson, P.E., Lykken, G.I. and Kortnta, E.D. (1991) Absorption and biological half-life in humans of intrinsic and extrinsic ^{54}Mn tracers from foods of plant origin. *Journal of Nutrition* 121, 711–717.

Johnson, S.R. (1943) Studies with swine on rations extremely low in manganese. *Journal of Animal Science* 2, 14–22.

Jukes, T.H. (1940) Effect of choline and other supplements on perosis. *Journal of Nutrition* 20. 445–458.

Kemmerer, A.R., Elvehjem, C.A. and Hart, E.B. (1931) Studies on the relation of manganese to the nutrition of the mouse. *Journal of Biological Chemistry* 92, 623–630.

Kirchgessner, M. and Neese, K.R. (1976) Copper, manganese and zinc content of the whole body and in carcase cuts of veal calves at different weights. *Zeitschrift für Lebensmittel. Unters-forschung* 161, 1–12.

Lassiter, J.W. and Morton, J.D. (1968) Effects of a low manganese diet on certain ovine characteristics. *Journal of Animal Science* 27, 776–779.

Leach, R.M., Jr (1971) Role of manganese in mucopolysaccharide metabolism. *Federation Proceedings* 30, 991–994.

Leach, R.M., Jr (1988) The role of trace elements in the development of cartilage matrix. In: Lonnerdal, B. and Rucker, R.B. (eds) *Trace Elements in Man and Animals – 6.* Plenum, New York, pp. 267–271.

Leach, R.M., Jr and Harris, E.D. (1997) Manganese. In: O'Dell, B.L. and Sunde, R.A. (eds) *Handbook of Nutritionally Essential Mineral Elements.* Marcel Dekker, New York, pp. 335–356.

Leach, R.M., Jr and Muenster, A.M. (1962) Studies on the role of manganese in bone formation. 1. Effect upon the mucopolysaccharide content of chick bone. *Journal of Nutrition* 78, 51–56.

Liu, A.C.-H., Heinrichs, B.S. and Leach, R.M., Jr (1994) Influence of manganese deficiency on the characteristics of proteoglycans of avian epiphyseal growth plate cartilage. *Poultry Science* 73, 663–669.

Longstaff, M. and Hill, R. (1972) The hexosamine and uronic acid contents of the matrix of shells of eggs from pullets fed on diets of different manganese content. *British Poultry Science* 13, 377–385.

Luo, X.G., Su, Q., Huang, J.C. and Liu, J.X. (1993) Effects on manganese (Mn) deficiency on tissue Mn-containing superoxide dismutase (MnSOD) activity and its mitochondrial ultrastructures in broiler chicks fed a practical diet. *Veterinaria et Zootechnica Sinica* 23, 97–101.

Lyons, M. and Insko, W.M., Jr (1937) *Chondrodystrophy in the Chick Embryo Produced by Manganese Deficiency in the Diet of the Hen.* Bulletin No. 371, Kentucky Agricultural Experiment Station.

McDowell, L.R. (1992) Manganese. In: *Minerals in Animal and Human Nutrition.* Academic Press, New York, pp. 246–264.

MAFF (1990) *UK Tables of Nutritive Values and Chemical Composition of Foodstuffs.* Edited by D.I. Givens. Rowett Research Services, Aberdeen, UK.

Malecki, E.A. and Greger, J.L. (1996) Manganese protects against heart mitochondrial lipid peroxidation in rats fed high levels of polyunsaturated fatty acids. *Journal of Nutrition* 126, 27–33.

Masters, D.G., Paynter, D.I., Briegel, J., Baker, S.K. and Purser, D.B. (1988) Influence of manganese intake on body, wool and testicular growth of young rams and on the concentration of manganese and the activity of manganese enzymes in tissues. *Australian Journal of Agricultural Research* 39, 517–524.

Mathers, J.W. and Hill, R. (1967) Factors affecting the retention of an oral dose of

radioactive manganese by the chick. *British Journal of Nutrition* 21, 513–517.
Mathers, J.W. and Hill, R. (1968) Manganese in the nutrition and metabolism of the pullet. 2. The manganese contents of the tissues of pullets given diets of high or low manganese content. *British Journal of Nutrition* 22, 635–643.
Matrone, G., Hartman, R.H. and Clawson, A.J. (1959) Studies of a manganese–iron antagonism in the nutrition of rabbits and baby pigs. *Journal of Nutrition* 67, 309–317.
Miltimore, J.E., Mason, J.L. and Ashby, D.L. (1970) Copper, zinc, manganese and iron variation in five feeds for ruminants. *Canadian Journal of Animal Science* 50, 293–300.
Minson, D.L. (1990) *Manganese Forage in Ruminant Nutrition.* Academic Press, New York, pp. 359–368.
Mitchell, R.L. (1957) The trace element content of plants. *Research, UK* 10, 357–362.
NRC (1977) *Nutrient Requirements of Poultry,* 7th edn. National Academy of Sciences, Washington, DC, 62 pp.
NRC (1980) *Mineral Tolerance of Domestic Animals.* USA National Academy of Sciences, Washington, DC.
NRC (1988) *Nutrient Requirements of Pigs,* 9th edn. National Academy of Sciences, Washington, DC, pp. 50–52.
NRC (1994) *Nutrient Requirements of Poultry,* 9th edn. National Academy of Sciences, Washington, DC.
Orent, E.R. and McCollum, E.V. (1931) Effects of deprivation of manganese in the rat. *Journal of Biological Chemistry* 92, 651–678.
Paynter, D. and Caple, I.W. (1984) Age-related changes in activities of the superoxide dismutase enzymes in tissues of the sheep and the effect of dietary copper and manganese on these changes. *Journal of Nutrition* 114, 1909–1916.
Plumlee, M.P., Thrasher, D.M., Beeson, W.M., Andrews, F.N. and Parker, H.E. (1956) The effects of a manganese deficiency upon the growth, development, and reproduction of swine. *Journal of Animal Science* 15, 352–367.
Rojas, M.A., Dyer, I.A. and Cassatt, W.A. (1965) Manganese deficiency in the bovine. *Journal of Animal Science* 24, 664–667.
Schaible, P.J., Bandemer, S.L. and Davidson, J.A. (1938) *The Manganese Content of Feedstuffs and its Relation to Poultry Nutrition.* Technical Bulletin No. 159, Michigan Agricultural Experiment Station.
Scheideler, S.E. (1991) Interaction of dietary calcium, manganese and manganese source (Mn oxide or Mn methionine complex) on chick performance and manganese utilisation. *Biological Trace Element Research* 29, 217–227.
Scrutton, M.C., Utter, M.F. and Mildvan, A.S. (1966) Pyruvate carboxylase. 6. The presence of tightly bound manganese. *Journal of Biological Chemistry* 241, 3480–3487.
Scrutton, M.C., Griminger, P. and Wallace, J.C. (1972) Pyruvate carboxylase: bound metal content of the vertebrate liver enzyme as a function of diet and species. *Journal of Biological Chemistry* 247, 3305–3313.
Silanpaa, M. (1982) *Micronutrients and the Nutrient Status of Soils: a Global Study.* FAO, Rome, p. 67.
Staley, G.P., Van der Lugt, J.J., Axsel, G. and Loock, A.H. (1994) Congenital skeletal malformations in Holstein calves associated with putative manganese deficiency. *Journal of the South African Veterinary Association* 65, 73–78.
Suttle, N.F. (1979) Copper, iron, manganese and zinc concentrations in the carcases of

lambs and calves and the relationship to trace element requirements for growth. *British Journal of Nutrition* 42, 89–96.

Underwood, E.J. (1977) *Trace Elements in Human and Animal Nutrition*, 4th edn. Academic Press, New York, 545 pp.

Underwood, E.J. (1981) Manganese. In: *Mineral Nutrition of Livestock*. Commonwealth Agricultural Bureaux, Farnham Royal, Slough, UK, pp. 170–195.

Valero, G., Alley, M.R., Badcoe, L.M., Manktellow, B.W., Merral, M. and Lowes, G.S. (1990) Chondrodystrophy in calves associated with manganese deficiency. *New Zealand Veterinary Journal* 38, 161–167.

Van Reen, R. and Pearson, P.B. (1955) Manganese deficiency in the duck. *Journal of Nutrition* 55, 225–234.

Wedekind, K.J., Titgemeyer, E.C., Twardock, R. and Baker, D.H. (1991) Phosphorus but not calcium affects manganese absorption and turnover in chicks. *Journal of Nutrition* 121, 1776–1786.

Weigand, E., Kirchgessner, M. and Helbig, V. (1986) True absorption and endogenous faecal excretion of manganese in relation to its dietary supply in growing rats. *Biological Trace Element Research* 10, 265–279.

White, C.L., Robson, A.D. and Fisher, H.M. (1981) Variation in the nitrogen, sulfur, selenium, cobalt, manganese, copper and zinc contents of grain from wheat and two lupin species grown in a Mediterranean climate. *Australian Journal of Agricultural Research* 32, 47–59.

Wilgus, H.S., Jr, Norris, L.C. and Heuser, G.F. (1937) The role of manganese and certain other trace elements in the prevention of perosis. *Journal of Nutrition* 14, 155–167.

Wilson, J.G. (1966) Bovine functional infertility in Devon and Cornwall: response to manganese therapy. *Veterinary Record* 9, 562–566.

Wong, G.H.W. and Goeddel, D.V. (1988) Induction of manganous superoxide dismutase by tumour necrosis factor: possible protective mechanism. *Science* 242, 941–944.

Wong-Valle, J., Henry, P.R., Ammerman, C.B. and Rao, P.V. (1989) Estimation of the relative bioavailability of manganese sources for sheep. *Journal of Animal Science* 67, 2409–2414.

15 Selenium

Introduction

For many years, biological interest in selenium was confined to its toxic effects on animals. Two naturally occurring diseases of livestock, 'blind staggers' and 'alkali disease', occurring in parts of the Great Plains of North America, were identified as manifestations of acute and chronic selenium poisoning, respectively (see Moxon, 1937). These discoveries gave a stimulus to investigation of selenium in soils, plants and animal tissues with a view to determining minimum toxic intakes and developing practical means of prevention and control. Not until 1957 was it discovered that selenium plays an essential physiological role in the higher animals, despite being present at lower concentrations in the tissues than most other essential elements. This discovery of essentiality arose from the work of Schwarz and Foltz (1957), who showed that the liver necrosis that develops in rats on certain diets could be prevented by supplements of selenium, and that of Patterson *et al.* (1957), who showed independently that this element also prevented exudative diathesis in chicks on similar diets. Within a few years, it was found that the muscular degeneration that occurs naturally in lambs and calves in parts of Oregon (Schubert *et al.*, 1961) and New Zealand was caused by a selenium deficiency and could be prevented by selenium therapy. Subsequently, selenium-deficient areas affecting the growth, health and fertility of livestock were discovered in many countries, and selenium metabolism and its relationship with vitamin E in animals became an active and rewarding field of research. Conclusive evidence that selenium is a dietary essential for growth in chicks, independent of, or additional to, its function as a substitute for vitamin E, was obtained by Thompson and Scott (1969). Four years later, a specific biochemical role for selenium emerged with the discovery that glutathione peroxidase (GPX) was a selenoprotein (Rotruck *et al.*, 1973) and the dependence of GPX activity of the tissues on selenium intakes became evident. In recent years, the essentiality of selenium has taken on a far more complex perspective, with the identification of over

© CAB *International* 1999. *Mineral Nutrition of Livestock*
(E.J. Underwood and N.F. Suttle)

30 distinctive selenoproteins, virtually all containing selenocysteine, and each with its own distribution, local function (see Table 15.1; Arthur and Beckett, 1994a; Sunde, 1994; Arthur, 1997) and selenium requirement (Lei *et al.*, 1998).

Dietary Sources of Selenium

The selenium concentrations occurring naturally in feeds and forages vary exceptionally widely, depending on the plant species, the part of the plant sampled, the season of sampling and the selenium status of the soils on which they have grown (Muth and Allaway, 1963; Miltimore *et al.*, 1975; Winter and Gupta, 1979; Grant and Sheppard, 1983).

Forages

In areas where selenium (Se)-responsive diseases in livestock occur, the pasture and forage concentrations are generally below 0.05 mg kg^{-1} dry matter (DM) and may be as low as 0.02 mg Se kg^{-1} DM, but pasture concentrations are not the best index of risk (Whelan *et al.*, 1994a). Selenium concentrations are reduced by regular applications of superphosphate fertilizer. Legumes tend to contain less selenium than grasses but the difference diminishes as soil selenium status declines (Minson, 1990). The importance of species is strikingly illustrated by the variously called 'accumulator', 'converter' or 'indicator' plants that occur in seleniferous areas; these are considered in the last section, on selenium poisoning.

Cereals and legume seeds

The cereal grains and other seeds also vary widely in selenium content between locations, with levels as low as 0.006 mg kg^{-1} DM in some deficient areas in Sweden and New Zealand and ranging from 0.05 to 3.06 mg Se kg^{-1} DM in Manitoba, Canada (Boila *et al.*, 1993). Species and varietal differences among the cereal grains tend to be obscured, but wheat grain can be higher in selenium than barley and oat grains (Miltimore *et al.*, 1975). Lupin-seed meals can be extremely low in selenium (0.02 mg kg^{-1} DM or less) (Moir and Masters, 1979).

Animal sources

The only protein concentrates in common use which can be relied on to improve substantially the selenium content of cereal-based pig and poultry rations are those of marine origin. In the Canadian study referred to earlier (Miltimore *et al.*, 1975), the richest sources of selenium were salmon and herring meals (1.9 mg Se kg^{-1} DM). Scott and Thompson (1971) found tuna fish-meal to contain as much as 5.1 and 6.2 mg Se kg^{-1} DM. Protein concentrates of animal but non-marine origin can also be rich sources of selenium but are more variable. Thus Moir and Masters (1979) found the selenium concentrations of 51 samples of meat-meal collected from eight abattoirs in Western Australia to range widely, from 0.11 to 1.14 mg kg^{-1} DM.

Availability

The dominant form of selenium in ordinary feeds and forages is protein-bound selenomethionine (SeM), together with much smaller amounts of selenocysteine and selenite. Tracer studies with radioisotopes indicate that the non-ruminant absorbs selenium from organic and inorganic sources with equally great efficiency (*c.* 90%; for review, see Henry and Ammerman, 1995), but this is misleading. When indicators of selenium function were used to assess availability to chicks, sources appeared to vary widely in potency, but ranking depended on the criterion used (Cantor *et al.*, 1975a, b; Fig. 15.1). Plant sources had a higher proportion of selenium 'available' for protection against one disorder (exudative diathesis (ED)) than the animal products tested (Fig. 15.1a; Cantor *et al.*, 1975a); other studies have yielded minimum values of 38% and 27% of that accorded to selenite for plant and animal sources, respectively (Henry and Ammerman, 1995), partially off-setting the lower content of selenium in plant sources. Tuna fish-meal compared favourably with selenite, a highly effective source against ED, in terms of availability to raise blood selenium (Fig. 15.1c) and was equal to selenite in terms of ED protection after acid digestion. Wheat was the most effective feed source against pancreatic fibrosis (PF) (Fig. 15.1b), and this was attributed to its SeM content, SeM being more effective than selenite or selenocysteine against PF (Fig. 15.1d). Differences in availability might therefore stem from differences in the release of feed selenium during digestion in the animal or from differences in the postabsorptive fate of the chemical forms in which selenium is absorbed; in either case, reliability of any extrinsic tracer is questionable.

Ruminants absorb tracer selenium far less efficiently and more variably than non-ruminants and a recent Canadian study begins to explain why this is so (Koenig *et al.*, 1997). Most of the ingested selenium leaves the sheep's rumen with insoluble particulate matter, particularly the bacterial fraction. The apparent absorption of selenium (AA_{Se}) was greater from a concentrate than from a lucerne hay diet (52.8 vs. 41.8%), due largely to an increase in absorption prior to the small intestine. Furthermore, these effects of diet were variably transferred to organic and inorganic selenium tracers (^{77}Se-enriched yeast and ^{82}Se-enriched selenite), leaving the organic source least absorbed from a forage diet (31.1%) at one extreme and the inorganic source most absorbed from a concentrate diet (48.9%) at the other. The contrasts were partly attributed to variable incorporation of feed and tracer selenium into the cell membranes of favoured species of rumen microbes. Alfaro *et al.* (1987) had previously noted a far greater discrepancy between AA_{Se} from an inorganic tracer (selenious acid, 60%) and a concentrate diet (29%). The influence of selenium intake on absorption is unclear. In one study with sheep, AA_{Se} from hay was higher (59.4 vs. 44.6%) when the selenium concentration was low (0.01 vs. 0.10 mg kg^{-1} DM: Krishnamurti *et al.*, 1997). In another with cattle, the AA_{Se} from hay + concentrates was 51% and independent of dietary level over the range 0.05–0.28 mg Se kg^{-1} DM (Harrison and Conrad, 1984). Similar calculations from data for the preruminant calf given

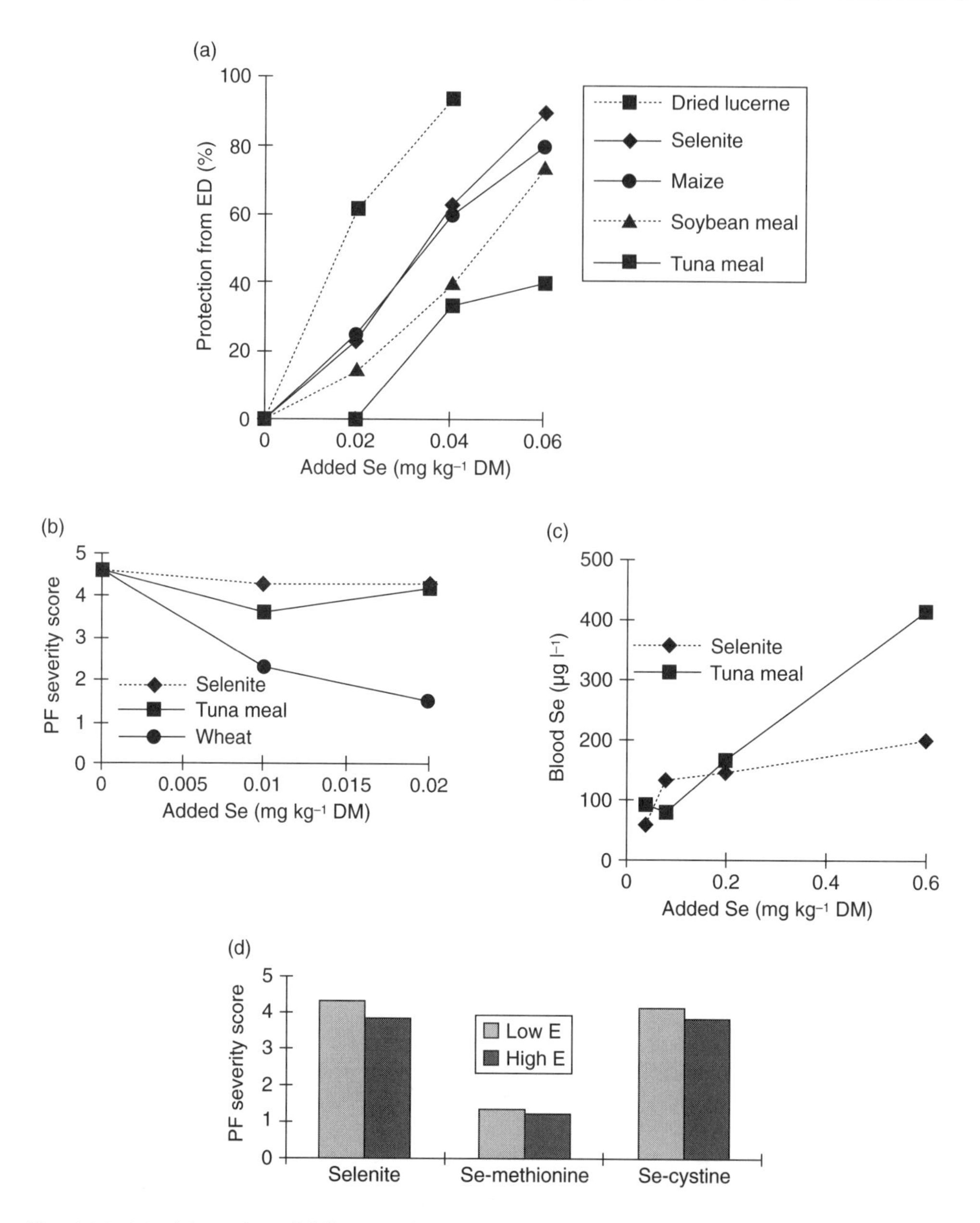

Fig. 15.1. Nutritive value of different selenium sources for chicks is determined by the test system as well as the source (data from Cantor *et al.*, 1975a, b).
(a) Protection against exudative diathesis (ED) is high from lucerne meal and low from a fish-meal with selenite intermediate.
(b) Protection against pancreatic fibrosis (PF) is equally poor with selenite and a fish-meal but good for a cereal source.
(c) Ability to raise blood selenium is higher for a fish-meal than for selenite.
(d) Protection against PF is good with selenomethionine, the major source of selenium in cereals, and may explain the good performance of wheat in (b).

selenite in a milk-replacer (Jenkins and Hidiroglou, 1986) indicate a higher AA_{Se} of 60%. The only measurements of true absorption (TA_{Se}) for cattle, given grass hays of marginal (0.035 mg kg^{-1} DM) or adequate (0.19) selenium concentration, yielded exceedingly low values of 10–16% (Koenig *et al.*, 1991a, b), but the preliminary enrichment used in the stable isotope study (4 mg ^{76}Se as selenite by stomach tube, for 5 days) may have perturbed metabolism. Another factor which may lower selenium absorption is the formation in the rumen of unavailable selenides. However, doubling the dietary copper concentration did not affect copper absorption in dairy cows (Koenig *et al.*, 1991b).

Metabolism

Little is known about the mechanisms whereby organic selenium is absorbed or whether the animal exerts any homeostatic control over the absorption of selenium. Selenate shares a common absorptive pathway with molybdate and sulphate (Cardin and Mason, 1975) and may be vulnerable to antagonisms from these anions in non-ruminants and ruminants (White and Somers, 1977; Pope *et al.*, 1979; Abdel-Rahim *et al.*, 1985) although little sulphate escapes from the rumen (Chapter 9).

Incorporation into selenoproteins

There are major differences in the postabsorptive metabolism of organic and inorganic sources of selenium. Selenium from parenterally administered selenite is rapidly incorporated into selenocysteine-rich proteins in plasma, particularly in selenium-depleted animals (Davidson and Kennedy, 1993), and may be made available for the synthesis of other selenoproteins through the activity of enzymes such as selenocysteine β lyase. Though well absorbed and retained, selenomethionine is slow to be converted to the selenocysteine needed for synthesis of functional proteins (Henry and Ammerman, 1995). Since most of the selenium in natural feeds is present as selenomethionine, parenterally administered isotopes of inorganic selenium may be misleading tracers for absorbed dietary selenium and give false estimates of fluxes between compartments and of faecal endogenous loss. The regulation of selenoproteins was reviewed by Burk and Hill (1993).

Cellular uptake

Selenium occurs in tissues at levels which vary with the species, the organ and the selenium status of the animal (Table 15.2); the richest tissue – kidney – can contain 15–20-fold higher concentrations than the poorest – muscle. There are corresponding differences in enzyme activities of the principal selenoproteins present, four GPXs. For example, in the rat and the chick, the liver and erythrocytes have the highest GPX1 activity, while in the lamb the erythrocytes are high but the liver is among the lowest in tissue enzyme activity. Selenium intake dramatically affects the GPX1 activity of body

Table 15.1. Selenoproteins which have been purified and/or cloned, their location and possible functions (after Arthur and Beckett, 1994a; Sunde, 1994).

Nomenclature	Selenoprotein	Principal location	Function
GPX1	Cystosolic GSH peroxidases (GPX)	Tissue cytosol, RBC	Storage, antioxidant
GPX2	Plasma GPX	Plasma, kidney, lung	Extracellular antioxidant
GPX3	Phospholipid hyperoxide GPX	Intracellular membranes, particularly testes	Intracellular antioxidant
GPX4	Gastrointestinal GPX	Intestinal mucosa	Mucosal antioxidant
ID1	Iodothyronine 5′-deiodinase Type I	Liver, kidney, muscle	Conversion of T_4 to T_3
ID2	Iodothyronine 5′-deiodinase Type II		Conversion of T_4 to T_3
ID3	Iodothyronine 5′-deiodinase Type III	Placenta	Conversion of T_4 to T_3
TRR	Thioredoxin reductase	Tissue cytosol	Redox/antioxidant
Sel P	Selenoprotein P	Plasma	Transport, antioxidant, storage, heavy metal detoxifier
Sel W	Selenoprotein W	Muscle	Antioxidant (?)
	Testes selenoprotein	Testes	Structural (?)

components and also the relative body distribution of the enzyme (Ganther *et al.*, 1976; Lei *et al.*, 1998). In the longer term, erythrocyte GPX1 activities increase in logarithmic fashion in relationship to selenium intake, eventually reaching a plateau. Incorporation of selenocysteine into erythrocyte GPX1 occurs at erythropoiesis (Wright, 1965) and, in the shorter term, there is a lag before the newly 'packaged' enzyme is released into the bloodstream and a further lag before it disappears, when the erythrocyte reaches the end of its normal lifespan (60–120 days, depending on the species). Other GPXs also show contrasting tissue responses to supplementation, earlier plateaux being reached for GPX4 in the thyroid and pituitary than in the liver or heart of young pigs (Lei *et al.*, 1998).

Excretion

Selenium can be lost from the body by exhalation, urinary excretion or faecal endogenous excretion. Biliary secretion of selenium can amount to 28% of intake (Langlands *et al.*, 1986); although most is reabsorbed, the remainder contributes significantly to faecal endogenous losses (FE_{Se}), which are primarily responsible for negative balances at low selenium intakes in sheep (Langlands *et al.*, 1986) and cattle (Koenig *et al.*, 1991). The selenium status of the animal influences FE_{Se} and the residual effects of an adequate dietary supply (0.16 mg Se kg^{-1} DM) can be manifested as increased faecal selenium excretion 84 days after transfer to a pasture of 'marginal' selenium concentration (0.06 mg kg^{-1} DM). Injected selenium accumulates principally in the liver and is extensively secreted via the bile (Langlands *et al.*, 1986; Archer and Judson, 1994). Assumptions that FE_{Se} will be independent of selenium intake (Koenig *et al.*, 1997) are clearly unsafe, and there is evidence that they increase with dry-matter intake (DMI) (Langlands *et al.*, 1986); both factors will contribute to the variability in AA_{Se} already referred to. While losses of selenium in urine are often regarded as small relative to those in faeces among ruminants, they can amount to 40–50% of selenium intake in sheep of low selenium status in negative balance (Langlands *et al.*, 1986) and are therefore largely unavoidable.

Maternal transfer

The lactating animal loses selenium via milk secretion. The concentrations of selenium in colostrum, like those of vitamin E, are four to five times higher than those in main milk, and both reflect the selenium status of the dam in sheep (Meneses *et al.*, 1994) and cattle (Conrad and Moxon, 1979). Transfer of selenium via the milk is more efficient than transfer via the placenta (Hidiroglou *et al.*, 1985; Zachara *et al.*, 1993), but supplementation of the mother during pregnancy can double the selenium status of her offspring at birth (Langlands *et al.*, 1990).

Functions of Selenium

Selenium is necessary for growth and fertility in animals and for the prevention of a variety of disease conditions, which show a variable response to vitamin E, for reasons which are becoming clearer as more is known about the functional forms of selenium and their localization. The selenoproteins about which most is known are given in Table 15.1. The four known peroxidases utilize glutathione as reducing substrate, and their multiplicity and ubiquity reflect the importance of controlling peroxidation, an essential biochemical reaction which – when unconstrained – can lead to chain reactions of free-radical generation and tissue damage. The task of terminating such reactions and protecting against peroxidation is shared by other tissue enzymes (e.g. the superoxide dismutases, copper–zinc (CuZn)- and manganese (Mn)SOD; catalase; glutathione-sulphur(S)-transferase) and by non-enzyme scavengers, such as vitamin E (MacPherson, 1994). The chances of selenium deficiency producing peroxidative damage may, therefore, depend on the degree of protection provided by other pathways and the rate of free-radical generation (Fig. 15.2). These interactions were clearly demonstrated in an experiment in which laying hens were reared on a diet inadequate in selenium and vitamin E and their newly hatched offspring given one of four treatments, neither, either or both antioxidants (Fig. 15.3). Peroxidation (as measured by thiobarbiturate-reactive substances (TBARS) in liver mitochondria) was maximal at hatching, possibly the time of maximal free-radical generation. Initially, only groups given vitamin E showed reduced

Table 15.2. Mean selenium concentrations (mg kg^{-1} DM) in the diet and tissues of normal pigs and pigs with nutritional muscular dystrophy (NMD) (from Lindberg, 1968).

	Diet	Kidney	Liver	Skeletal muscle	Heart	Pancreas
Healthy	0.126	11.47 ± 1.18	1.82 ± 0.16	0.52 ± 0.06	1.05 ± 0.10	1.42 ± 0.14
NMD	0.021	2.48 ± 0.29	0.20 ± 0.05	0.16 ± 0.08	0.19 ± 0.09	0.24 ± 0.13

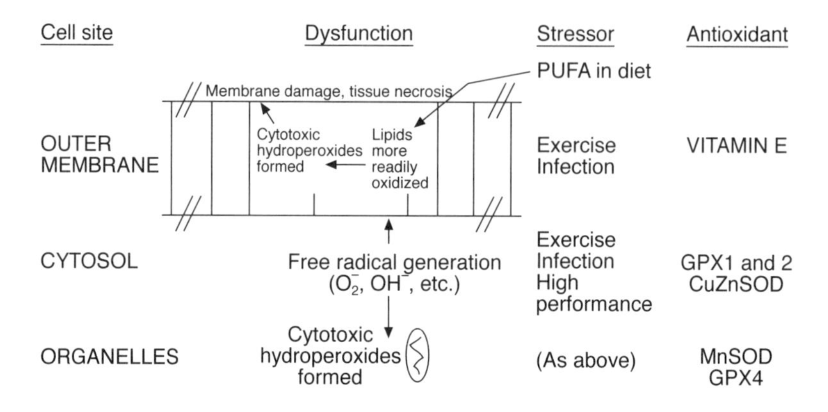

Fig. 15.2. The antioxidant function of selenium, as glutathione peroxidases (GPX), is complemented by other antioxidants, sometimes operating at other sites, and is 'stretched' by a variety of oxidant stressors: ultimate dysfunction in the form of damage to cell membranes therefore bears a complex relationship to selenium status (see Fig. 15.3).

peroxidation, but, as time passed and chicks became more deprived of selenium, both antioxidant supplements became necessary. Noguchi *et al.* (1973a) suggested that, in the associated disorder, ED, vitamin E functioned as a specific lipid-soluble antioxidant in the chick's cellular membranes, while selenium, as GPX, destroyed peroxides before they could reach the membranes. Recent work with rats has confirmed that vitamin E and selenium act synergistically (Awad *et al.*, 1994; Levander *et al.*, 1995). However, each nutrient cannot always fully compensate for a deficiency of the other; selenium may be particularly needed for male fertility (Marin-Guzman *et al.*, 1997) and particularly protective to the kidney (Liebovitz *et al.*, 1990).

Cytosolic peroxidase

The first peroxidase to be identified and studied in detail is now known as cytosolic or GPX1 to distinguish it from others. It is the predominant GPX and

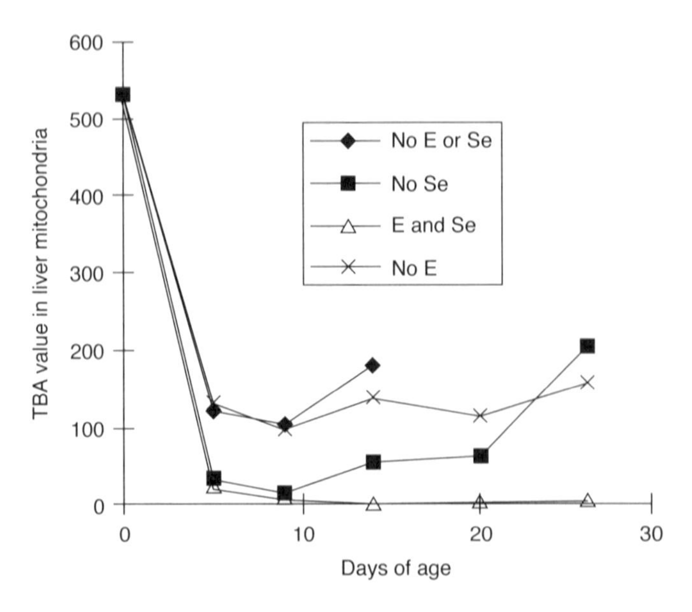

Fig. 15.3. Tissue peroxidation is influenced by interactions of selenium (Se) with other antioxidants, such as vitamin E, and with sources of oxidant stress: thus, thiobarbiturate (TBA)-like products of peroxidation are high in liver mitochondria of chicks deprived of vitamin E and Se at hatching (a stressor) and show only partial disappearance before disease (exudative diathesis) occurs after 13 days; combined supplementation with vitamin E and Se is needed to prevent peroxidation (data from Noguchi *et al.*, 1973b).

source of selenium in erythrocytes and liver, and all the selenium-responsive diseases are accompanied by decreases in blood and tissue GPX1 activities. Selenium is present in GPX1 in stoichiometric amounts, with 4 g atoms Se mol^{-1} (Flohe *et al.*, 1973), and the tetrameric enzyme catalyses the reduction of hydrogen peroxide (H_2O_2) and of hydroperoxides formed from fatty acids and other substances, according to the general reaction:

$$ROOH + 2GSH \rightarrow R\text{–}OH + HOH + GSSG$$

Whether GPX1 ever becomes a rate-limiting factor in protecting erythrocytes or tissues from peroxidative damage during selenium deficiency is questionable, because activities in erythrocytes and liver can fall to almost undetectable levels without obvious pathological or clinical changes (Arthur and Beckett, 1994a). It has therefore been proposed that GPX1 is essentially a storage selenoprotein (Sunde, 1994). It is hardly a suitable vehicle for transferring selenium to sites of synthesis of other selenoproteins and is more likely to constitute a safe depot for excess selenium, while providing a reserve of antioxidant capacity, which may be occasionally useful at times of maximal oxidant stress. Animals can survive without the gene for GPX1 (Cheng *et al.*, 1997), suggesting that the enzyme itself is non-essential.

Other peroxidases

The plasma or extracellular peroxidase, GPX3, is also tetrameric and synthesized principally in the lung and kidney, where its main functions may reside in protecting the renal proximal tubule from peroxidative damage. Substrate (glutathione (GSH)) concentrations in plasma are too low to support a major extracellular protective role for GPX3. The gastrointestinal peroxidase (GPX2) may also act locally to protect the intestinal mucosa from dietary hydroperoxides (Chu *et al.*, 1993). The family of immunogenically distinct peroxidases, performing similar functions at different sites, is completed by the phospholipid hydroperoxidase form (GPX4), a monomer associated with intracellular membranes (Fig. 15.2). This is spared during selenium deficiency, suggesting *a rate-limiting importance* (Weitzel *et al.*, 1990) *for GPX4 which may be responsible for the substitutive relationship between selenium and vitamin E* (Arthur and Beckett, 1994a). The functional independence of these newer peroxidases from GPX1 was demonstrated by the lack of effect of GPX1 gene deletion on their mRNA levels and activities (Cheng *et al.*, 1997).

Deiodinases

The first indications that selenium deficiency influenced iodine metabolism came from increases in the tetra- to triiodothyronine (T_4:T_3) ratio in selenium-depleted rats, a finding subsequently confirmed in cattle (Arthur *et al.*, 1988) and sheep (Donald *et al.*, 1994b). Eventually, a membrane-bound seleno-protein was identified as a type I iodothyronine deiodinase (ID1), capable of transforming T_4 to the physiologically active form, T_3 (for review, see Arthur and Beckett, 1994b). This deiodinase is located primarily in the liver and kidney and none is present in the thyroid in farm livestock; this means that T_3 is generated principally outside the thyroid. A second deiodinase (type II or ID2) can also form T_3 from T_4, but it is under feedback control from T_4 and therefore liable to be doubly inhibited in selenium deficiency. Species differ in the way in which they generate T_3, ID1 being the predominant deiodinase in ruminants and ID2 in non-ruminants (Nicol *et al.*, 1994). Type II is most abundant in brain and brown adipose tissue (BAT) and thermogenesis in the newborn is heavily dependent on BAT and T_3 generation. A third selenium-containing deiodinase (ID3) has been found in the placenta (Salvatore *et al.*, 1995). Skin and brain are probably responsible for maintaining local T_3 concentrations. Selenium deficiency may thus indirectly influence basic metabolic rate and a wide range of physiological processes, including parturition and survival during cold stress (see Chapter 12 for further details of thyroxine metabolism), with adverse effects on production.

Thioredoxin reductase

In many tissues, thioredoxin reductase (TRR) is as abundant as GPX1 and as a selenoprotein makes a major contribution to cell selenium concentrations while controlling cell redox state. The TRR system may allow selenium status to influence activity of glutathione transferase (GST), although GST does not contain selenium.

Other forms and functions

The principal constituent of plasma selenium is selenoprotein P, a novel molecule with up to ten selenium–cysteine (Se–Cys) residues incorporated into the peptide backbone. Selenoprotein P has the potential to complex heavy metals and may explain the protection which excess selenium provides against the toxicity of cadmium, mercury (Hill 1972) and lead (Rastocci *et al.*, 1970). The mutual metabolic antagonism between selenium and mercury is considered later in relation to their respective toxicities (p. 460 and Chapter 18). Another selenoprotein, first isolated from the heart and muscles (Whanger *et al.*, 1973), was subsequently called selenoprotein W (Vendeland *et al.*, 1993) and found to contain 1 g atom of selenium per mole as Se–Cys (Whanger *et al.*, 1977b), but its function is still unclear. Concentrations of W in muscle decline during selenium depletion but may be conserved in brain, while the provision of excess selenium (3 mg kg^{-1} DM) increases W in all lamb tissues except brain (Yeh *et al.*, 1997). Abnormal sperm development in the selenium-deficient rat was associated with shortage of a structural selenoprotein in mitochondrial capsules, and impaired viability of semen has been reported in selenium-deficient bulls (Slaweta *et al.*, 1988) and boars (Marin-Guzman *et al.*, 1997). Selenium appears to be involved in the metabolism of sulphydryl compounds (Sprinker *et al.*, 1971; Broderius *et al.*, 1973). Selenium deficiency also adversely affects the immune response, although much of the evidence comes from *in vitro* tests (Stabel and Spears, 1993).

Biochemical Manifestations of Selenium Deprivation

The sequence of biochemical changes in livestock deprived of selenium is illustrated generally in Fig. 15.4 and specifically for the chick in Figs 15.3 and 15.5 and differs in many respects from those described for other elements in previous chapters. Selenium status is not tightly regulated and evidence of dysfunction relies on the appearance of abnormal metabolites, which is influenced by the supply of other nutrients (Fig. 15.2).

Selenium and glutathione peroxidase in the blood, milk and tissues

Insufficient selenium intakes lead to slow reductions in blood selenium and GPX1 activity, after an initial lag, which reflect first the 'depletion' and then the 'deficiency' phases in the progress towards disorder (Chapter 3). Selenium is present in plasma at less than half the concentration in erythrocytes but responds more rapidly to change in selenium intake (Fig. 15.5) in both its GPX3 and selenoprotein P compartments. Early reductions in serum selenium may again indicate depletion and deficiency, but latterly the onset of *dysfunction* can be defined (see section on diagnosis). Selenium concentrations and GPX activities in tissues, such as muscle and liver, reflect dietary selenium supply and may correlate well with concentrations in blood (Ullrey, 1987), depending on the duration of deprivation. Since most of the selenium in liver

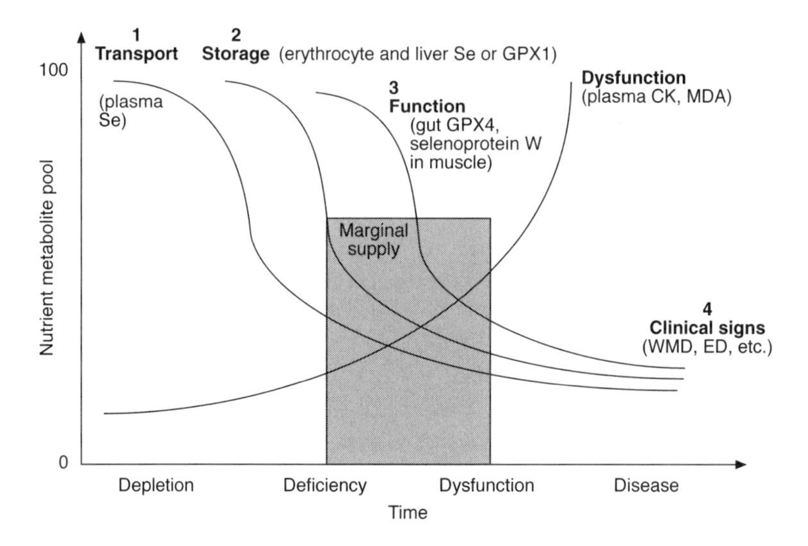

Fig. 15.4. General scheme for the sequence of biochemical changes in livestock deprived of selenium. Rapidity of transition from one phase to another is influenced by sources of oxidant stress, status of other antioxidants, notably vitamin E, and, in some species, by the source of selenium. GXP1 and GXP4 are two of several glutathione peroxidases; CK and MDA are abnormal constituents of plasma, creatine kinase and malonyldialdehyde, respectively; WMD, white muscle disease, and ED, exudative diathesis, are two of many possible clinical end-points in different species. See also Fig. 3.1 and related text.

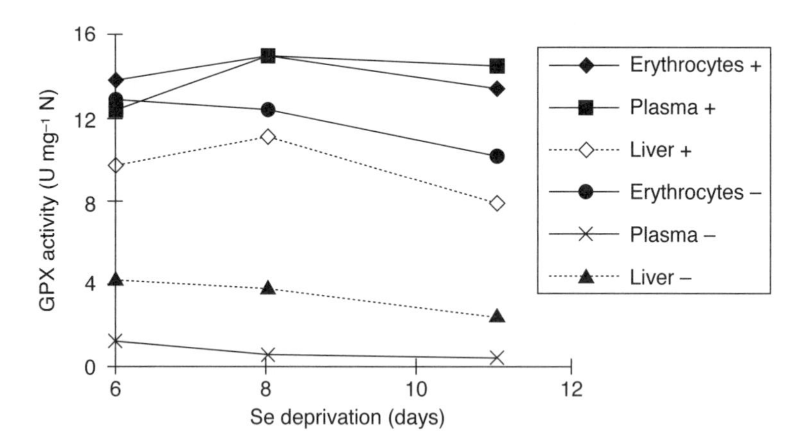

Fig. 15.5. Glutathione peroxidase activities decline at different rates at different sites when animals are deprived of selenium (–): note the faster declines in liver and plasma than in erythrocytes and that both precede the development of clinical disease (ED in Fig. 15.3: data from Noguchi *et al.*, 1973b).

is in the storage form, GPX1, reductions in concentrations reflect depletion more than deficiency. The selenium concentrations in milk and eggs are particularly sensitive to changes in selenium intakes by cows and hens. Hadjimarkos (1969) found the mean values for a 'Se-low' county and two 'Se-normal' counties in Oregon to be 0.005 and 0.05–0.07, respectively, for milk and 0.06 and 0.4–0.5 mg kg^{-1} DM, respectively, for eggs. The excessively low tissue selenium values typical of white muscle disease (WMD) in pigs, compared with those of normal animals, is shown in Table 15.2; similar changes occur in affected foals (Caple *et al.*, 1978). Subnormal tissue selenium concentrations in pigs affected with mulberry heart disease (MHD) have also been reported (Lindberg, 1968; Van Vleet *et al.*, 1970; Simesen *et al.*, 1979), although the values in liver and heart are higher than in those of pigs dying of hepatosis dietetica (HD) (Pedersen and Simesen, 1977; Table 15.3). This may indicate a greater role for oxidant stress and/or vitamin E deficit in MHD than in HD.

Indices of muscle damage

Abnormally high serum aspartate aminotransferase (AAT), lactic dehydrogenase (LDH) and creatine kinase (CK) activities occur in WMD, due to the leakage of these predominantly intracellular enzymes through damaged muscle membranes, the increase being roughly proportional to the amount of muscle damage. Increases in these 'muscle enzymes' provide unequivocal evidence of *membrane dysfunction* and values over ten times the normal have been observed in affected lambs and calves. However, individual variability is high (Table 15.4; Whanger *et al.*, 1977b) and the AAT value may be only moderately increased in animals with marked clinical symptoms (Oksanen, 1967). In subclinical cases, CK can decline markedly after the attainment of peak values (N.F. Suttle and D.W. Peter, unpublished data). 'Muscle enzyme' assays are only of use in diagnosing myopathies and cannot be used to assess the involvement of selenium if the primary deficiency is one of vitamin E, because selenium has minimal substitutive effect (Table 15.4) (Wise *et al.*, 1968; Oh *et al.*, 1976b; Whanger *et al.*, 1977a, b). Furthermore, low selenium status is not necessarily reflected by increases in AAT (Stevens *et al.*, 1985).

Table 15.3. Selenium content (mg kg^{-1} wet weight; mean ± SD) in livers and hearts of pigs with hepatosis diaetetica or mulberry heart disease and normal pigs (from Pedersen and Simesen, 1977).

	No.	Liver	Heart
Hepatosis diaetetica	12	0.068 ± 0.052	0.051 ± 0.041
Mulberry heart disease	11	0.141 ± 0.067	0.098 ± 0.044
Normal pigs	–	0.300 ± 0.100	0.164 ± 0.060

SD, standard deviation.

Table 15.4. Mean (± SE) blood selenium, erythrocyte glutathione peroxidase (GPX1) activity, plasma aspartate amino transferase (AAT), lactate dehydrogenase (LDH) and creatine kinase (CK) activities in normal and WMD lambs on low-Se hay + oat diets with and without selenium (from Whanger *et al.*, 1977b), showing only partial control of the disease by selenium administration.

Blood components or incidence	+ Se		No Se	
	Normal	WMD	Normal	WMD
WMD incidence (%)	56	44	44	56
CK (U ml^{-1})	97 ± 20[a]	3377 ± 1101[b]	97 ± 12[a]	4988 ± 930[b]
LDH (U ml^{-1})	609 ± 69[a]	3890 ± 1100[b]	562 ± 63[a]	4605 ± 1040[b]
AAT (U ml^{-1})	53 ± 6[a]	748 ± 260[b]	62 ± 5[a]	635 ± 92[b]
RBC GPX1 (U mg^{-1} Hb)	59 ± 7[a]	66 ± 5[a]	2.2 ± 0.5[b]	3.6 ± 1.0[b]
Blood Se (mg l^{-1})	0.14 ± 0.03	0.15 ± 0.04	< 0.02	< 0.02

Means within a row not sharing a common superscript letter are significantly different ($P < 0.01$).

Biochemical indices of peroxidation

Selenium deprivation becomes 'cell-threatening' when important intracellular selenoproteins, such as GXP2 and GXP4, are depleted to such an extent that free-radical generation increases and harmful lipid hydroperoxides are formed (Arthur, 1998) (Fig. 15.2). These can be distinguished by electron spin resonance (ESR) and detected directly by 'spin trapping' the free radical (Arthur *et al.*, 1988) or indirectly by measuring malonyldialdehyde (MDA), other TBARS (Walsh *et al.*, 1993) (Fig. 15.3; Noguchi *et al.*, 1973b; McMurray *et al.*, 1983) and F_2-isoprostanes (Lei *et al.*, 1998) – all end products of lipid peroxidation – in blood or tissues. Exhalation of ethane and pentane may also indicate peroxidation. These indicators of dysfunction give new insight into the stage of deprivation reached (Lei *et al.*, 1998). The techniques present difficulties and are non-specific, because peroxidation can arise from numerous causes, including vitamin E deprivation and exercise (Fig. 15.2).

Other indices of cellular dysfunction

Cells isolated from animals of low selenium status may behave abnormally during *in vitro* culture, particularly if stressors are applied. Thus erythrocytes may be more liable to *in vitro* haemolysis when exposed to peroxide (Siddons and Mills, 1981; Stevenson *et al.*, 1991). The decreased microbicidal activity and mitogen responses of leucocytes from selenium-deficient ruminants (for review, see MacPherson, 1994) may reflect increased intracellular generation of free radicals during *in vitro* challenge. However, such changes are again non-specific, are influenced by the nature of the challenge and may overestimate dysfunction *in vivo* (Chesters and Arthur, 1988).

Indices of secondary deprivation

Where animals respond to selenium because of a secondary impairment of iodine metabolism, low concentrations of T_3 may be present prior to treatment (< 2 nmol l^{-1}) and rise afterwards, but the T_3:T_4 ratio may not be helpful (Wichtel *et al.*, 1996) and the increased analytical costs are hard to justify.

Clinical Manifestations of Selenium Deprivation

All tissues are vulnerable to oxidant stress at certain stages of development and the clinical consequences of depriving livestock of the antioxidant selenium are extremely diverse.

Muscular degeneration in ruminants

Nutritional muscular dystrophy (NMD) or WMD is a degenerative rather than a dystrophic disease of striated muscles, which occurs, without neural involvement, in a wide range of animal species. Lesions are probably initiated by free-radical damage (Arthur, 1998). In some areas, the incidence is low, seasonal and sporadic and, in other areas, it is higher and more consistent, amounting to 10% or more of the flock or herd. Affected calves exhibit muscular stiffness, arrhythmia, tachycardia and abdominal breathing (Hidiroglou *et al.*, 1985). The similar disease in lambs is most common at 3–6 weeks but may occur at any age from birth to 12 months. Young cattle may develop WMD when they are turned out to graze spring pasture and older animals can develop an acute myopathy with myoglobinuria at this time (Allen *et al.*, 1975; Anderson *et al.*, 1976). The goat kid is believed to be more susceptible to WMD than either the lamb or calf (Rammell *et al.*, 1989). Affected lambs are disinclined to move about (hence the term 'stiff lamb disease') and can show respiratory distress; they lose condition, become prostrate and usually die. Very mildly affected animals may recover spontaneously. A bilaterally symmetrical distribution of the skeletal muscle lesions is characteristic, with the lesions most apparent in the thigh and shoulder muscles. The deep muscles overlying the cervical vertebrae may also be affected with typical white striations. Cardiac muscle can also be affected and abnormalities in the electrocardiogram develop early in affected lambs, becoming marked as death approaches (Godwin, 1968). In cattle, myocardial necrosis can cause sudden death (Bradley *et al.*, 1981). Calcification is a common feature of damaged muscle, and may be caused by mitochondrial calcium overload, due to impaired uptake of calcium by vesicles of the sarcoplasmic reticulum. This, in turn, is due to structural changes and diminished calcium-binding by proteins (Tripp *et al.*, 1993). Cardiac lesions repair by fibrosis, but skeletal muscles can repair completely.

Muscular degeneration in non-ruminants

White muscle disease has also been observed in foals (Higuchi *et al.*, 1989), adult horses (Owen *et al.*, 1977), in association with HD and MHD in pigs and with ED in chicks. The disease in all these species is both selenium- and vitamin E-responsive. A condition in turkeys, characterized by myopathies of the heart and gizzard and very high mortality at 5–6 weeks of age, similarly responds to selenium therapy (Scott *et al.*, 1967).

Exudative diathesis

The disease of poultry known as ED is characterized by a generalized oedema, which first appears on the breast, wing and neck and arises from abnormal permeability of the capillary walls. The greatest accumulation of fluid occurs under the ventral skin, giving it a greenish blue discoloration. Growth rate is subnormal and mortality of chicks is high. In commercial flocks consuming low-selenium grain, chicks are most commonly affected between 3 and 6 weeks of age, when they lose weight, reveal leg weakness and may become prostrate and die (Hartley and Grant, 1961). Exudative diathesis is completely prevented by either selenium or vitamin E (Noguchi *et al.*, 1973a) but by different mechanisms. A synthetic antioxidant, which will substitute for vitamin E and prevent encephalomalacia, will not prevent ED (Noguchi *et al.*, 1973b). The GPX3 activity of chick plasma is directly related to dietary selenium supply and the prevention of ED.

Pancreatic fibrosis

Severe selenium deficiency results in atrophy of the pancreas of chicks, accompanied by poor growth and feathering, even in the presence of high dietary vitamin E (Fig. 15.1d; Thompson and Scott, 1970). There is also a loss of appetite, which is regained within hours of selenium supplementation (Bunk and Combs, 1980). The pancreatic lesions have been characterized by Gries and Scott (1972) and Noguchi *et al.* (1973a); they become apparent at 6 days of age and return to normal within 2 weeks of selenium supplementation. Subnormal activities of pancreatic lipase and decreased hydrolysis of fat lead to impaired formation of the lipid bile micelles necessary for the absorption of lipid and vitamin E. There is thus a secondary vitamin E deficiency in the selenium-deficient chick, which can only be overcome by giving the vitamin with factors which promote its absorption (Thompson and Scott, 1970). Pancreatic fibrosis differs from ED in that the selenium in selenomethionine and wheat is much more protective than that in selenite or selenocystine (Fig. 15.1b, d), whereas with ED the opposite applies (Cantor *et al.*, 1975a, b). The explanation may involve a selenoprotein other than GPX which is essential to the integrity of the pancreatic acinar cell and more readily synthesized from selenomethionine than from other sources (Bunk and Combs, 1980), but selenomethionine is preferentially taken up by the pancreas (Cantor *et al.*, 1975a).

Hepatosis diaetetica in pigs

Spontaneous outbreaks of HD can occur wherever pigs are fed on grain rations naturally low in selenium, as in parts of Australia, New Zealand and Scandinavia. The disease is most common at 3–15 weeks of age and results in high mortality, associated with severe necrotic liver lesions, marked depletion of tissue selenium and an increase in liver-specific enzymes, such as ornithine carbamyltransferase (OCT), in the blood (Oksanen, 1967). Since the degenerative changes that may occur in the muscles do not affect OCT activity, blood OCT (or glutamate dehydrogenase) assays can be valuable in the diagnosis of HD. The disease can occur in the field singly or in any combination with WMD and MHD (Moir and Masters, 1979). A survey in Sweden revealed significant associations between MHD and HD incidence and between HD and WMD incidence (Grant, 1961). The mortality and lesions can be completely prevented by selenium supplements, but vitamin E supplements appear to be more effective in preventing the accompanying muscle degeneration and deposition of ceroid pigment in adipose tissue (Oksanen, 1967). A striking reduction in the incidence of HD occurred in Denmark after the legalization of the addition of selenium to pig feed (maximum 0.1 mg kg^{-1} DM) in that country in 1975 (Pedersen and Simesen, 1977).

'Mulberry heart disease'

This disease, first described by Lamont *et al.* (1950), is a dietetic microangiopathy of young, rapidly growing pigs, usually aged 1–4 months, which takes its name from the gross appearance of the heart. The haemorrhagic and necrotic lesions in the myocardium result in a vivid red mottling with transudation to the serous cavities and can cause death from acute cardiac failure. Post-mortem lesions include myocardial necrosis and microthromboses of the myocardial capillaries and have been attributed to eicosanoid imbalance and platelet aggregation. The value of vitamin E or selenium in the prevention of MHD (and HD) in pigs fed on diets high in polyunsaturated fatty acids (PUFA) was discovered some years later (Grant, 1961). Progress in understanding the pathogenesis of MHD has been hindered by the failure of experimental models to reproduce all lesions (Nolan *et al.*, 1995). Several groups of workers in Scandinavian and North American countries established that MHD and the associated disease conditions, HD and NMD, occur naturally in areas where cereal-based diets contain less than 0.05 mg Se kg^{-1} DM (Lindberg, 1968; Van Vleet *et al.*, 1970). Treatment of pregnant sows and their baby pigs with selenium and vitamin E preparations can prevent MHD and HD (Van Vleet *et al.*, 1973).

Blood disorders

Heinz-body anaemias have been reported in selenium-deficient steers (Morris *et al.*, 1984) and lambs (Suttle *et al.*, 1987) and attributed to peroxidative damage, due to low activities of GPX1 in erythrocytes. Postparturient haemoglobinuria can develop in dairy cows turned out to graze spring pasture; though not a specific consequence of selenium deficiency (see Chapter 5), there are circumstances in which low erythrocyte GPX1 activities are believed to be a contributory factor (Ellison *et al.*, 1986).

Reproductive disorders in poultry and pigs

In all animal species, selenium deficiency results in impaired reproductive performance in males and females. In laying hens, egg production and egg hatchability are both reduced (Cantor and Scott, 1974) and, in Japanese quail, the hatchability of fertile eggs and the viability of newly hatched chicks were both impaired (Jensen, 1968); hatchability is the most sensitive criterion of selenium deficiency (Latshaw *et al.*, 1977). In pigs, litter size (Mahan *et al.*, 1974), the conception rate of gilts to first service (Edwards *et al.*, 1977) and piglet mortality (Nielsen *et al.*, 1979) can all be improved by supplementary selenium. In boars, sperm development appears to be particularly sensitive to selenium supply, since a level which was adequate for growth (0.06 mg Se kg^{-1} DM) was associated with low sperm motility, a high proportion of 'tail abnormalities' and low fertilization rates (Marin-Guzman *et al.*, 1997).

Reproductive disorders in ruminants

In ewes, high embryonic mortality between 3 and 4 weeks after conception (i.e. around implantation) has been attributed to selenium inadequacy in parts of New Zealand (Hartley, 1963) and New South Wales (NSW) (Wilkins and Kilgour, 1982) in association with WMD and unthriftiness. In certain of these areas, 20–50% of ewes were infertile, losses of lambs were high and fertility was dramatically improved by the oral administration of selenium before mating. Neither vitamin E nor an antioxidant was effective for these purposes, as shown by the figures given in Table 15.5. Selenium-responsive unthriftiness can, however, occur with no drop in ewe fecundity (Langlands *et al.*, 1991b). In cows, a sodium selenite–vitamin E mixture injected 1 month before calving prevented losses from the birth of premature, weak or dead calves in parts of California (Mace *et al.*, 1963) and greatly reduced the incidence of retained placenta in a herd of cows in Scotland (Trinder *et al.*, 1969). Subsequent investigations revealed significantly lower blood selenium in affected than in unaffected herds, but selenium alone was less effective than a combined injection of selenite and vitamin E (Trinder *et al.*, 1973). Evidence that retained placenta can be an expression of selenium deficiency in the dairy cow was obtained by Ohio workers (Julien *et al.*, 1976a, b). Eger *et al.* (1985) showed that a single injection of 2.3 mg selenium prepartum

Table 15.5. Effects of selenium, vitamin E and antioxidant on ewe fertility (200 ewes per group) (from Hartley, 1963).

	Untreated	Selenium[a]	Vitamin E	Antioxidant
Barren ewes (%)	45	8	50	43
Lambs born per 100 ewes lambing	105	120	109	112
Lamb mortality (%)	26	15	16	14
Lambs marked per 100 ewes lambing	43	93	46	54

[a] Treatment confined to the mating period. Normal practice involves a further dose 1 month before lambing to reduce lamb mortality.

could reduce the incidence of retained placentae from 29.0 to 10.8%, a response equal to any achieved in four subsequent experiments from selenium given with vitamin E. However, herd incidence of retained placenta can vary tremendously from year to year with no clear indication of cause; differences in forage vitamin E may be involved. Harrison *et al.* (1984) reported concurrent reductions in the incidence of endometritis and cystic ovaries following selenium supplementation. Improvements in conception rates at first service following selenium supplementation have been reported in NSW (McClure *et al.*, 1986), but a satisfactory explanation of the mode of action of selenium in the bovine reproductive cycle has yet to appear. Calving intervals are improved by reductions in the incidence of retained placenta, but selenium must be able to increase conception rate directly, because it can do so in heifers which have not previously conceived (MacPherson *et al.*, 1987). Male fertility may also be adversely affected, and reduced viability of semen has been reported in selenium-deficient bulls (Slaweta *et al.*, 1988).

Lowered disease resistance

The most convincing evidence that selenium deficiency influences susceptibility to infection relates to mastitis in the dairy cow. Smith *et al.* (1984) found that injection of selenium reduced the duration but not the incidence of mastitis when dietary selenium was low. Inverse relationships between somatic cell counts in the milk and GPX1 in blood have been reported (Erskine *et al.*, 1989b) and experimental infection of the udder with *E. coli* gave rise to higher peak cell counts in selenium-depleted heifers than in their supplemented contemporaries (Erskine *et al.*, 1989a). The incidence of mastitis in cows of marginal selenium status can be reduced by massive supplements of vitamin E (2000–4000 IU day^{-1}: Weiss *et al.*, 1997). Similar associations in sows prompted a study of responses of polymorphonuclear cells to stimulation *in vitro*; selenium supplementation improved responses, but vitamin E had more widespread effects (Wuryastuti *et al.*, 1993). Responses to selenium and vitamin E given together, such as the decrease in diarrhoea and mortality in young calves reported by Spears *et al.* (1986), cannot safely be attributed to selenium alone. Prolonged selenium depletion did not impair resistance to viral infection in calves (Reffett *et al.*, 1988) or nematode infection in lambs (Jelinek *et al.*, 1988; McDonald *et al.*, 1989), but marginal depletion lowered the resistance of chicks to the protozoan parasite *Eimeria tenella* (Colnago *et al.*, 1994). Both vitamin E and selenium can improve antibody responses in non-ruminants and ruminants, but their additivity is variable (for reviews, see Stabel and Spears, 1993; MacPherson, 1994).

Subclinical Manifestations of Selenium Deficiency in Sheep

The term 'subclinical' is used to describe abnormalities which have no distinctive pathological features and which are generally mild in effect. In parts of Australia and New Zealand, a selenium-responsive 'ill-thrift' was reported in lambs and hoggets at pasture and in beef and dairy cattle of all ages (Hartley, 1967; McDonald, 1975). The condition varied from a subclinical growth deficit to visible unthriftiness with a rapid loss of weight and some mortality. No characteristic microscopic lesions were apparent, there was no increase in AAT and no consistent association with WMD and infertility. Striking increases in growth and wool yield from supplementary selenium were obtained in some cases. In economic terms, 'ill-thrift' in selenium-deficient sheep is probably the most important manifestation of dysfunction and needs to be broken down into its constituent parts.

Perinatal mortality

Selenium supplementation of the ewe increased the probability of lamb survival from 0.61 to 0.91 in NSW (Donald *et al.*, 1993), and the benefit accrued during the first 5 days of life. Administration of an iodine antagonist (thiocyanate) during pregnancy cut out the response to selenium, suggesting that it was a manifestation of impaired deiodination. Walker *et al.* (1979) had previously reported decreased susceptibility of naturally cold-stressed lambs to *Pasteurella multocida* infection following oral selenium supplementation, and Kott *et al.* (1983) improved lamb survival by injecting ewes with selenium during pregnancy in New Mexico. Selenium deprivation may restrict the thermogenic response of BAT to cold stress (see Chapter 12).

Growth retardation

Selenium supplementation significantly increased lamb weights at birth, mid-lactation and weaning by 4.0, 7.6 and 10.8%, respectively, in the NSW study (Langlands *et al.*, 1990). The importance of management factors was illustrated by the fact that responses were greater in merino than in cross-bred lambs and at the higher of two widely different stocking rates (6.3 or 12.5 ewes ha^{-1}). Furthermore, the selenium × breed and × stocking rate interactions were repeatable (Langlands *et al.*, 1991b). The importance of climate was indicated by a positive relationship between selenium responses and rainfall. Since vitamin E does not overcome such growth retardation, deiodinase impairment may be involved. Improvements in live weight are not confined to the young lamb. Whelan *et al.* (1994a) reported responses in three successive seasons in the same wethers at a site in Western Australia to various forms of selenium supplementation.

Wool production

Many workers have reported improvements in wool production following selenium supplementation, and Gabbedy (1971) reckoned these to be the most sensitive production index of selenium deficiency in Western Australia. The detailed observations of Langlands and his associates are again most illuminating. Fleece yields from ewes were increased by 3.8–7.5% each year in a 4-year study and significantly so in 3 of them (Langlands *et al.*, 1991a). Furthermore, their lambs produced 9.5% more wool without themselves being selenium-supplemented (Langlands *et al.*, 1990). The increase in ewe wool yield was greater in those which reared a lamb to weaning than in those which did not, reflecting the greater demand for selenium. Since thyroxine stimulates wool production, Donald *et al.* (1994a) tested the hypothesis that these effects may be related to induced iodine depletion. Fortnightly injections of T_4 or T_3 did not improve wool production in selenium-deficient sheep, but the irregular pattern of administration was unphysiological and T_3 concentrations in the blood were decreased by T_3 supplementation. Improvements of 22% in greasy wool yield have been reported at two successive shearings following selenium treatment (Whelan *et al.*, 1994b).

Cumulative effects

In New Zealand trials with lambs 5 months of age given selenium, mortality was reduced from 27 to 8% and highly significant weight gains were observed, giving a striking improvement in the total live weight produced (Fig. 15.6: Hartley, 1967). In eastern Australia (McDonald, 1975), treatment of merino lambs regularly from marking time reduced mortality from 17.5% to nil, increased weight gains by 1.9 kg/head at both weaning and 1 year of age and increased mean fleece weight by 14.4%. The increase in total annual fleece production of the treated groups (40 lambs) over the untreated groups (33 surviving lambs) was 47 kg or 39%. Smaller increases from selenium supplementation in weight gains, fleece weights or both have been reported in Western Australia (Gabbedy, 1971), Canada (Slen *et al.*, 1961), the western

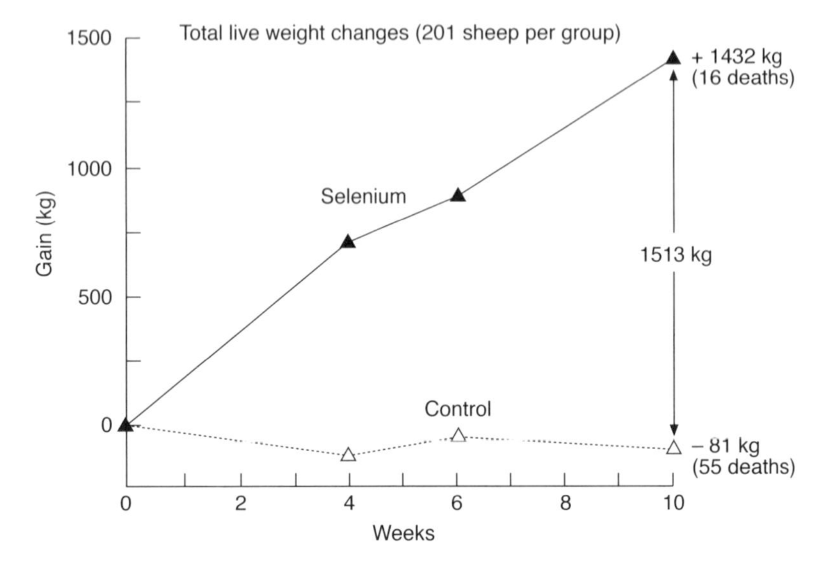

Fig. 15.6. The cumulative effects of treatment with selenium can be substantial where mortality is reduced: data for lambs by courtesy of Dr W.J. Hartley.

states of the USA (Oldfield *et al.*, 1963) and Scotland (Blaxter, 1963; Suttle *et al.*, 1984), but, in the latter case, mortality was not reduced.

Selenium-responsive 'ill-thrift' in cattle

Cattle can grow without restriction on pastures which provide insufficient selenium for sheep (Langlands *et al.*, 1989). Only three out of 21 herds on the NSW tablelands showed a mean improvement in weight gain when given selenium pellets, despite a very low blood selenium status in unsupplemented groups. Selenium-responsive herds were characterized by diarrhoea prior to treatment and it was suggested that vitamin E absorption might have been impaired (see also Rice and Kennedy, 1989). Growth retardation was not found in selenium-deprived cattle in Canada (Hidiroglou *et al.*, 1985). Two reports of growth responses to selenium in cattle have appeared, one from England (Gleed *et al.*, 1983) and the other from Australia (Koh and Judson, 1987); both were characterized by concomitant hypocupraemia, and maximal responses were only obtained if copper was given with selenium. Milk production can be impaired by selenium deficiency but only when blood and pasture selenium are exceedingly low (Tasker *et al.*, 1987). The effect is principally on milk-fat yield (Fraser *et al.*, 1987), raising the possibility of an effect on lipid metabolism via pancreatic dysfunction, but this has yet to be studied in selenium-deficient ruminants.

Occurrence of Selenium-responsive Disorders

The occurrence of disorders that respond to selenium supplementation does not bear a close correlation with soil or dietary selenium status, because selenium is but one of many factors influencing the underlying dysfunction and peroxidative damage to tissues. The major factors are as follows.

1. Selenium status.
2. Supply of other dietary antioxidants (notably vitamin E).
3. Supply of dietary oxidants (notably PUFA).
4. Endogenously generated oxidants (e.g. through exercise, infection, toxic chemicals).
5. Toxins (e.g. lupinosis).

Figure 15.7 illustrates the relative importance of each factor in contrasting sheep and cattle husbandry situations. If oxidant stress is minimal, impairment of selenium-dependent thyroid hormone funtion may become health-limiting through a reduction in IDI activity in BAT; this would obviously be most likely to occur when animals give birth outdoors in cold, wet and windy conditions (Fig. 15.7a). The higher incidence of WMD in lamb and calves given legumes rather than grass is unlikely to be explained in terms of selenium deprivation alone (Whanger *et al.*, 1972).

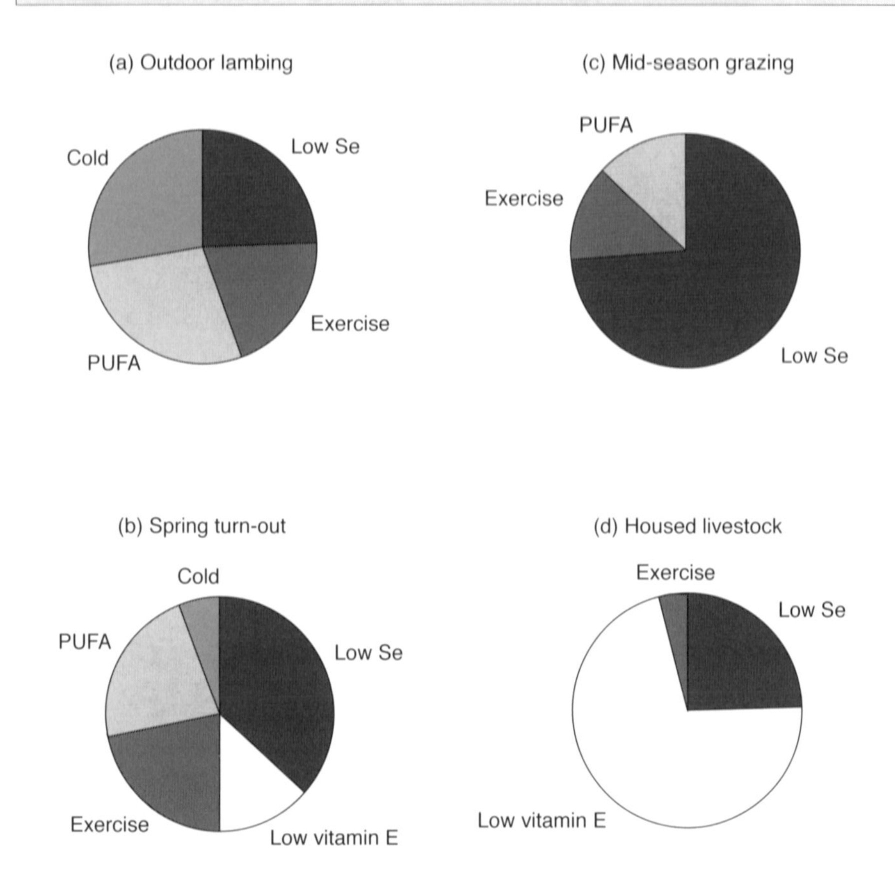

Fig. 15.7. Hypothetical illustration of the different relative contributions of selenium and vitamin E deprivation, oxidant stress (from cold exposure, muscular exercise or dietary polyunsaturated fatty acids (PUFA)) to the risk of Se-responsive disorders developing in sheep or cattle when housed or at pasture (a) lambs born outdoors, green sward, cold spring; (b) growing cattle turned out on to lush spring pasture; (c) continuously grazed sheep or cattle, mild temperatures; (d) housed sheep or cattle.

Selenium status

The possible shortcomings in dietary selenium supply can be indicated by surveys of soils, feeds and animals. Soils of granitic or volcanic origin are inherently low in selenium. The mountainous countries of northern Europe (Finland, Sweden and Scotland) are low in selenium, and geochemical mapping can confirm this (Selinus, 1988). The pumice soils of New Zealand in North Island and the tablelands of northern NSW are examples of volcanic soils low in selenium. The effects of soil origin are often confounded with those of climate and altitude. The selenium status of forages is lower at high than at low altitude (Jumba *et al.*, 1996), and this is probably due to the influence of rainfall. The selenium status of cattle (Langlands *et al.*, 1981) and sheep (Langlands *et al.*, 1991c) is negatively correlated with rainfall. Two factors may be responsible: first, the leaching of selenium from the soil and, secondly, the dilution of selenium taken up by more prolific plant growth when rainfall is high. A final factor that determines selenium levels in crops and forages is soil acidity, selenium uptake being higher on alkaline than on acid soils (Ullrey, 1974). The prevalence of low blood selenium levels in sheep (Anderson *et al.*, 1979) and cattle (Arthur *et al.*, 1979) in Scotland is attributable to a combination of granitic soils, high altitude, high rainfall and slightly acidic soils. Surveys of selenium concentrations of forages and crops in the USA revealed large areas of the Mid-West, such as the state of Michigan, where values were generally 'low' (< 0.05 mg Se kg^{-1} DM), and subsequent inclusion of selenium in the rations of pigs, dependent on locally grown grain, improved health (Ullrey, 1974). The selenium status of soils and crops is also low in many parts of Canada.

Vitamin E and other dietary antioxidants

The natural maize–soybean-meal diets commonly fed to pigs in the USA are uniformly low in vitamin E, unless supplemented. Selenium-responsive disorders, such as WMD, were commonplace, mortalities of 15–20% being recorded in Michigan, but the high concentration of selenium needed to prevent disorder (between 0.10 and 0.15 mg Se kg^{-1} DM: Ullrey, 1974) indicated that vitamin E deficiency had increased the risk of dysfunction. The natural forage diets for ruminants are rich in vitamin E when grazed but values fall during conservation, particularly if material is badly weathered (Fig. 15.7d; Miller *et al.*, 1995). Langlands *et al.* (1989) attributed the tolerance of grazing cattle in NSW of low blood selenium status to a generous supply of vitamin E (Fig. 15.7c). The perinatal period is one of maximum vulnerability to selenium deficiency for mother and offspring. Vitamin E is poorly transported via the placenta, and offspring rely on colostrum for their early supply. The mother, therefore, secretes large quantities of the vitamin in colostrum; in the case of the dairy cow, the amount is so large (*c.* 19 mg) that it causes a 47% drop in maternal vitamin E concentrations in plasma around parturition (Goff and Stabel, 1990). Increases in perinatal survival in the offspring of selenium-supplemented ewes have been reported (Donald *et al.*, 1993) (Fig. 15.7a). Storage of moist grain in silos with propionic acid as preservative can greatly reduce its vitamin E content, predisposing livestock fed on such grains to selenium-responsive disorders (Fig. 15.7d; Allen *et al.*, 1975). The treatment of grains and straws with alkali also destroys the vitamin and has been used in conjunction with low dietary selenium, to induce myopathy experimentally in calves (Walsh *et al.*, 1993a). Inclusion rates for vitamin E in mineral/vitamin supplements may often be too low to eliminate responses to selenium. Vitamin C is an antioxidant that can recycle vitamin E, and deficiencies are more likely to be found in diets for non-ruminants than for ruminants, increasing vulnerability to peroxidative damage.

Supply of dietary oxidants

High intakes of PUFA induce myopathy in both ruminants (ARC, 1980) and non-ruminants (ARC, 1981), but the precise role of selenium has been hard to elucidate because vitamin E was often given concurrently. Most cases of WMD in calves and acute paralytic myoglobinuria in older cattle occur shortly after turnout on to spring pasture, which is rich in PUFA (Fig. 15.7b; McMurray *et al.*, 1983). The myopathy occurs in the face of a rapid rise in plasma tocopherol and is selenium-responsive (McMurray and McEldowney, 1977; Arthur, 1988). Although hydrogenation of PUFA occurs in the rumen, increases in linolenic acid in plasma at turnout show that, in the short term, it is far from complete (McMurray *et al.*, 1983). The role of PUFA is one of free-radical generator, following formation of unstable lipid hydroperoxides in the tissues (Fig. 15.2). In a study of WMD in 39 goat kids, selenium deficiency, vitamin E deficiency and PUFA were thought to have contributed in different degrees to the development of myopathy (Rammell *et al.*, 1989). The higher risk of WMD on legume than on grass diets may be related to the higher rate of passage of legumes from the rumen, which allows more PUFA to escape hydrogenation. However, occurrence of selenium-responsive myopathies will be influenced more by PUFA intake in non-ruminant livestock than in ruminants.

Endogenous oxidants

Free-radical generation increases as metabolic rate increases and the need for GPXs to metabolize and for vitamin E to scavenge free radicals will therefore increase as metabolic activity increases. Arthur (1988) showed that the change to a PUFA-rich grass diet alone did not induce myopathy, but grazing did. Ullrey (1974) noted that the mingling of piglet litters precipitated WMD and he implicated the physical exertion of establishing a social order. He also noted that extremes of temperature and humidity increased WMD incidence. The importance of endogenous oxidants is indicated by the uneven distribution of myopathic lesions among fibre types, type I (slow-twitch, oxidative) being more vulnerable than type II (fast-twitch, glycolytic). Lack of previous exercise may be as important as sudden exercise to the incidence of myopathy at turnout, because there is less stimulus for the development of resistant type II muscle fibres in the confined animal (McMurray *et al.*, 1983). The localized generation of free radicals in neutrophils undergoing a respiratory burst after engulfing a pathogen may explain both the impaired phagocytic killing *in vitro* by cells from selenium-depleted calves (Boyne and Arthur, 1979) and the impaired immunocompetence *in vivo*, as discussed later.

Toxins

A selenium + vitamin E-responsive myopathy occurs in sheep grazing lupin stubbles contaminated with the fungus *Diaporthe toxica*; toxins from the fungus are believed to impair the metabolism of one or both antioxidants, because only certain forms (selenomethionine) or methods of administration (subcutaneous (s.c.) vitamin E) are effective (Smith and Allen, 1997).

The involvement of other dietary antioxidants and both exogenous and endogenous sources of oxidants in selenium-responsive, peroxidative dysfunction complicates both the diagnosis of disorder and the definition of requirements. If the biochemical dysfunction is one of poor deiodination rather than peroxidation, a new set of interactants, including goitrogens, is introduced and these may influence the occurrence of 'ill-thrift'.

Diagnosis of Selenium-responsive Disorders

The diagnosis of selenium deficiency presents considerable difficulties, because the clinical and pathological signs are not specific and selenium status is but one of many factors which determine the prevalence of disorders. The important thing is to recognize the limitations of deceptively easy biochemical measurements and the need for marginal bands of diagnostic uncertainty (Table 15.6).

Soil and diet

Soils containing < 0.5 mg Se kg^{-1} are likely to support crops and pastures with potentially inadequate selenium concentrations (< 0.05 mg kg^{-1} DM); the latter will be discussed more fully in the later section on requirements, but the failure to discriminate between a selenium-responsive herd and an

Table 15.6. Marginal bands[a] for indices of selenium deprivation in domestic livestock given diets adequate in vitamin E.[b]

	Blood	Serum	Liver	Muscle	Diet
	(nmol l^{-1})		(nmol kg^{-1} FW)		(mg kg^{-1} DM)
Sheep	500–900	250–500	250–450	300–400	0.03–0.05
Cattle	150–250	100–120	200–300[c]	250–300	0.02–0.04
Pigs[d]	1265–2530	760–1270	1500–3000	630–950	0.05–0.07
Poultry[d]	1075–1650	380–760	3200–6400	800–1000	0.06–0.10

[a] Mean values within bands for a population sample indicate a possibility of benefits in health and production from selenium supplementation if values are likely to be sustained.
[b] Ideally, diagnostic limits (like dietary requirements) should be set on the assumption that the supply of other nutrients is non-limiting; pragmatically, it could be argued that higher limits should be set for housed livestock, where dietary vitamin E may be limiting and more expensive than selenium to supplement.
[c] Values three times higher in fetus.
[d] Based on recommendations by Puls (1994).
Conversion factors: nmol × 0.0790 (for μg); mg × 12.665 (for μmol).

unresponsive herd with common low pasture selenium levels (0.02–0.03 mg Se kg^{-1} DM) should be noted (Wichtel *et al.*, 1996).

Diagnostic limits for element and enzyme in blood

Blood selenium concentrations have long been used to assess selenium status, because they can reflect a wide range in selenium intakes, but good correlations with GXP1 activity (e.g. Caple *et al.*, 1978; Thompson *et al.*, 1981), superficially the easier to measure, led to GPX1 becoming the more widely used. Participants in interlaboratory quality-control schemes will know that, like most enzyme assays, GPX1 activities were not easy to measure accurately or consistently until the advent of good commercial assay kits and standardized procedures (Ullrey, 1987). Growing cattle can tolerate exceedingly low blood selenium concentrations when they graze continuously. Initially, the prevalence of low blood selenium concentrations led Langlands *et al.* (1981) to speculate that 800,000 cattle in NSW were at risk from selenium deficiency. Subsequent dose:response trials showed that it was rare for growth to be constrained and the few herds which did respond (three out of 21) were not distinguishable on the basis of blood selenium, which was usually < 0.25 μmol (20 μg) l^{-1}, and it was suggested that a low vitamin E status was necessary for selenium to become limiting, even at grossly subnormal blood concentrations. Other workers found that cattle suffer no growth retardation, despite very low blood or plasma selenium status (< 20 μg l^{-1} plasma), but a low incidence of WMD was found (10–14% in 2 out of 3 years: Hidiroglou *et al.*, 1985). New Zealand work indicates that the lactating dairy cow is even more tolerant of selenium depletion, no depression of milk-fat yield occurring until blood selenium fell below a critical value of 0.15 μmol (12 μg) l^{-1} (Fraser *et al.*, 1987). A very different picture was painted with regard to fertility in the dairy cow, in which responsiveness to oral selenium has been claimed at blood selenium levels up to 87 μg l^{-1} (McClure *et al.*, 1986), but this was not confirmed by later studies (Wichtel *et al.*, 1994). In light of the above uncertainties and contrasts, marginal bands of blood (and plasma) selenium concentrations are given in Table 15.6 to separate the 'possibly responsive' from the 'normal' animal. Equivalent values for GPX1 depend on the assay conditions used in a particular laboratory; Table 15.7 gives values for growing lambs at a stated elemental selenium equivalence. It must be remembered that circulating levels of selenium and GPX are being relied upon as remote sensors of intracellular and membrane events of explosive potential.

Diagnostic limits for serum selenium

Serum (or plasma) selenium is the preferred measure of selenium status in pigs (Chavez, 1979; Lei *et al.*, 1998), sheep (Whelan *et al.*, 1994b) and cattle (Stevens *et al.*, 1985; Ullrey, 1987), particularly after a recent change in selenium nutrition (Thompson *et al.*, 1991), but it may correlate poorly with clinical disease. In Fig. 15.5, the reduction in plasma selenium was compete before disease (ED) was apparent in chicks. Serum selenium levels in

Table 15.7. Proposed marginal bands for blood glutathione peroxidase activity (GPX) for diagnosing and predicting selenium-responsive ill-thrift in lambs on improved hill pastures prior to weaning.

	Diagnosis	Prediction[a]
GPX (U g^{-1} Hb at 37°C)[b]	100–150	150–200
GPX (U g^{-1} cells at 30°C)	27–41	41–55
Blood selenium (ng ml^{-1})[b]	42–53	53–64

[a] Where selenium status regularly declines, higher marginal bands may be necessary to predict future growth retardation in lambs.
[b] The relationship between the two parameters was blood Se = 19.2 + 0.225 GPX ($r = 0.91$; 38 d.f.).

racehorses increased by 10% after a training jog (Gallagher and Stowe, 1980), due possibly to a concurrent decrease in plasma volume. Diagnostic limits determined by 'bent-stick' analysis of relationships with wool production – the most sensitive index of adequacy in sheep – indicate that, in merino ewes bearing lambs in eastern Australia, wool yield decreased when values fell below 40 and 70 μg Se l^{-1} (0.51 and 0.89 μmol l^{-1}) in plasma and blood, respectively (Langlands *et al.*, 1994). In another study in Western Australia with growing merino wether lambs, similar optimal levels were calculated (Whelan *et al.*, 1994a). For ewes not bearing lambs, the corresponding diagnostic values were much lower (< 0.25 μmol (20 μg) l^{-1} for plasma and 0.51 μmol (40 μg) l^{-1} for blood) and they maintained higher mean values than pregnant or lactating ewes on the same pasture (Langlands *et al.*, 1991a; Donald *et al.*, 1994a). If higher circulating concentrations of selenium are needed to sustain higher daily needs, then diagnostic limits may be higher for more productive breeds and pastures than for the merino grazing in Australia. In heifers, plasma selenium values as low as 0.12 μmol l^{-1} (10 μg l^{-1}) have been tolerated (Wichtel *et al.*, 1996), but in calves they have been associated with WMD (Hidiroglou *et al.*, 1985). In growing boars which developed selenium-responsive infertility, serum selenium had declined to 33 μg l^{-1} (Marin-Guzman *et al.*, 1997). The precise level of serum selenium attained at a given dietary concentration may be inversely related to growth rate in young pigs (Lei *et al.*, 1998). The multiplicity of factors influencing the relationship between serum selenium and performance yet again emphasizes the need for marginal bands when interpreting analytical results.

Diagnostic limits for tissue selenium

Diagnostic limits for tissue selenium have not been widely studied in the context of responsiveness to supplementation. The marginal bands given in Table 15.6 are therefore only tentative proposals.

Response to supplementation

For selenium, more than any other element, the *best diagnosis is usually afforded by a positive response to selenium supplementation* (see Chapter 19). Animals with selenium-induced iodine deficiency will not respond to iodine, and their plight may be exacerbated by exposure to thiouracil and thiocyanate goitrogens (see Chapter 12).

Prevention and Control of Selenium Deprivation

Selenium deprivation can be prevented in a number of ways, the method of choice depending on the conditions of husbandry (McPherson and Chalmers, 1984). Simultaneous use of two methods will give additive results without necessarily causing harm (Whelan *et al.*, 1994b), but such practices are rarely necessary and not generally recommended.

Discontinuous methods

Direct subcutaneous injections or oral drenching, usually with sodium selenite, in doses providing 0.1 mg kg^{-1} live weight (LW) at 1–3-monthly intervals, have been the most common means of preventing selenium-responsive diseases in grazing livestock (Meneses *et al.*, 1994). Selenite and selenate are of similar nutritive value when given in plethoric amounts with the diet to sheep (Henry *et al.*, 1988). The drenches or injections can be given when the animals are yarded for other management procedures (McDonald, 1975), such as anthelmintic dosing or vaccination, and can be combined in single products. For example, selenium is now commonly given in anthelmintic drenches to significant effect (Field *et al.*, 1988). The use of selenium injections or doses in the prescribed amounts does not result in excessive or dangerous concentrations in the edible tissues of treated animals (Cousins and Cairney, 1961; Doornenbal, 1975), even when given repeatedly with anthelmintic (Cooper *et al.*, 1989). It is, however, vital that selenium be given at responsive times (i.e. at mating, in late pregnancy and at weaning) and the optimum oral dose may be closer to 0.2 than 0.1 mg kg^{-1} LW (Langlands *et al.*, 1990). Pig diseases such as MHD and HD have been prevented by the injection of 0.06 mg Se kg^{-1} LW, together with vitamin E, into the sow and baby pig (Van Vleet *et al.*, 1973).

Continuous methods

It is possible to raise selenium intakes by importing grains and forages from high-selenium areas, but such a system requires controlled analyses of the feeds being blended (Allaway *et al.*, 1967). In New Zealand, selenium-containing mineral supplements have long been available commercially for incorporation into pig and poultry rations, to provide a total maximum of 0.15 mg Se kg^{-1} DM (Andrews *et al.*, 1968), and supplementation to 0.3 mg kg^{-1} DM is now permitted in the USA and highly effective in raising the selenium status of gilts and their offspring (Mahan and Kim, 1996). Trace mineralized salt, fortified with sodium selenite at the rate of 26–30 mg Se kg^{-1} and offered freely to ewes and lambs, decreased the incidence of WMD or plasma CK without increasing tissue concentrations above those found naturally in some unsupplemented lambs in different areas of the USA (Paulson *et al.*, 1968; Jenkins *et al.*, 1974; Ullrey *et al.*, 1978) but may not protect all individuals. A compressed, selenized salt block, containing 11.8 mg Se kg^{-1}, significantly raised the mean blood and plasma selenium concentrations of ewes and lambs but left 7–33% of ewes unprotected (Langlands *et al.*, 1990). The problem of individual variation in block consumption cannot be overcome by adding more selenium (Money *et al.*, 1986), although this may raise average block consumption by the flock (Langlands *et al.*, 1990). Given the uncertainty and small scale of many production responses to selenium, selenized mineral supplements given in loose or block form can be an attractive and cost-effective option.

Slow-release methods

Early studies in Australia showed that heavy ruminal pellets, consisting of 95% finely divided iron and 5% elemental selenium, released sufficient selenium to maintain adequate blood values for several months and prevented WMD in sheep (Kuchel and Buckley, 1969; Godwin *et al.*, 1970; Kuchel and Godwin, 1976). These pellets cannot be given to young lambs. Furthermore, they may not provide adequate protection beyond 12 months (Wilkins and Hamilton, 1980), and individual variation can be marked (Whelan *et al.*, 1994b). Pellets with an improved formulation are now available (Donald *et al.*, 1994c). In beef cattle, losses of 7–56% of administered pellets were observed in three herds out of 21 in one study (Langlands *et al.*, 1989). Regurgitation occurred within minutes of dosing and was often repeated upon redosing. Regurgitation was neither observed nor suspected in a New Zealand study with dairy heifers given two 30 g pellets containing 10% elemental selenium, and growth was improved in one herd (Wichtel *et al.*, 1994). Efficacy was demonstrated for 2–3 years in beef cows in a Canadian study (Hidiroglou *et al.*, 1985). Regurgitation of heavy pellets is not a problem in sheep (Langlands *et al.*, 1990). Controlled-release capsules can deliver selenium and anthelmintic simultaneously, protecting lambs for at least 180 days (Grace *et al.*, 1994). Sustained protection can also be provided by soluble-glass boluses to cattle (Koh and Judson, 1987), goats (Zervas, 1988) and sheep (Balla *et al.*, 1989). The parenteral administration of insoluble barium selenate at the high but safe dose of 1 mg Se kg^{-1} LW is effective, although the recommended site of injection and withdrawal period must be observed with care to avoid problems of high residues (Archer and Judson, 1994).

Indirect methods

The first indirect approach involved treating soil or foliage with selenite solutions (Watkinson and Davies, 1967a, b), but it was not widely practised because added selenium is poorly absorbed by most plants, especially from acid soils (Allaway *et al.* 1967), and residual effects are short-lived. Furthermore, high herbage selenium values immediately after application can pose a toxicity hazard. Annual applications of 10 g Se ha^{-1} to 1/4–1/3rd of the fields, accompanied by rotational grazing, is safer and just as effective (Watkinson, 1983; Millar and Meads, 1988). By far the most effective approach is to apply a less soluble form of selenium ($BaSeO_4$) at 10 g Se ha^{-1} in prill (granular) form, using slow-release technology. Such treatments are effective for 3 years, but the necessary improvements in selenium status of the grazing animal may take 6 weeks to materialize (Whelan *et al.* 1994a, b) and preliminary supplementation by injection to young lambs may sometimes be prudent. The selenium content of grains can be increased by soil treatment, but 100 g Se as selenite ha^{-1} were needed in one study to increase the selenium in barley grain from 0.01 to 0.05 mg kg^{-1} DM (Nielsen *et al.*, 1979) and the 'capture' rate was highly inefficient (< 0.2%).

Organic supplements

Superior availability of organic over inorganic sources of selenium has been claimed in pigs and ruminants. Selenium-enriched yeast raised concentrations of the element in the serum, milk and most tissues of gilts more than the same addition of selenium as selenite, and the selenium status of offspring was higher (Mahan and Kim, 1996). Conrad and Moxon (1979) found that 19% of the selenium provided to lactating cows as brewer's grains was transferred to the milk, whereas < 4.8% of that provided as selenite was transferred. Similar results have recently been obtained with a selenium yeast (Fisher *et al.*, 1995). These results are at variance with measurements of absorption using radioisotopes (Koenig *et al.*, 1997) but are probably the more reliable indicator of nutritive value. Selenium from selenomethionine restored appetite and liver GPX activities of selenium-deficient chicks faster then selenite (Bunk and Combs, 1980) and is particularly protective against PF in chicks (see Fig. 15.1d). However, there is no evidence yet that the provision of organic selenium offers any advantage in the practical nutrition of livestock, since inorganic selenium supplements quickly restore a normal status at lower cost.

Selenium Requirements

The minimum requirements for selenium for a given species vary with the form of selenium ingested, the criteria of adequacy employed and dietary composition, particularly its content of vitamin E. Selenium requirements for ruminants have been assessed by each of the classic methods – empirical experiment (Oh *et al.*, 1976a, b), field associations (ARC, 1980) and factorial modelling (Grace, 1994). The latter is useful in so far as it can cover wide ranges of productivity, identify important determinants of requirement and provide a basis for assessing the validity of isolated extreme values for TA_{Se} and FE_{Se}.

Factorial estimates for sheep and cattle

Results from a slightly modified model are presented in Table 15.8. They suggest that the requirement for sheep at all stages is around 0.03 mg Se kg^{-1} DM on a diet of low digestibility but higher (0.05) for a diet of high digestibility. This is consistent with field experiences, in which problems are associated with improvements in pasture quality (Wilkins *et al.*, 1982; Suttle *et al.*, 1994). The factorial model for cattle has to rely on some components obtained with sheep (Grace, 1994) and it generates requirements of a similar order in concentration terms (Table 15.8). If extremely low values of TA_{Se} for cattle (e.g. Koenig *et al.*, 1997) were representative, predicted requirements would become impossibly high.

Associations with field disorders

The Agricultural Research Council (ARC, 1980) review of field problems (mostly WMD) placed the requirement at 0.03 mg Se kg^{-1} DM for diets adequate in vitamin E, confirming Andrews *et al.*'s (1968) early assessment for grazing sheep in New Zealand and explaining the lack of response of flocks given 'low-selenium' hay and grain from Michigan (0.03–0.06 mg Se kg^{-1} DM) to supplementary selenium (Ullrey *et al.*, 1978). Dose:response trials in lambs grazing improved hill pastures in Scotland over 3 years showed few benefits from selenium supplementation, with herbage selenium commonly 0.02–0.05 mg kg^{-1} DM (Suttle *et al.*, 1999). The higher requirement indicated for optimal growth in lambs reared naturally indoors (0.04–0.06 mg Se kg^{-1} DM: Oh *et al.*, 1976b) may reflect a lower vitamin E status. Field experience with growing cattle shows them to be less susceptible than sheep to selenium deficiency (Langlands *et al.*, 1989), but Canadian

Table 15.8. Factorially derived† selenium requirements for cattle and sheep.

	Live weight (kg)	Milk yield or LWG (kg day^{-1})	DMI (kg day^{-1})		Requirement		
					Net	Gross (mg kg^{-1} DM)	
			q* 0.5	0.7	(µg day^{-1})	0.5	0.7
Lamb	20	0.1	0.7	0.4	11	0.031	0.055
		0.2	–	0.6	16	–	0.054
Beef	100	0.5	2.8	1.7	50	0.036	0.058
		1.0	–	2.4	75	–	0.062
	300	0.5	5.6	3.4	100	0.036	0.058
		1.0	8.3	4.7	125	0.030	0.054
Ewe	75	1.0	1.9	1.2	26	0.028	0.043
		3.0	3.7	2.4	38	0.020	0.032
Dairy		10	11.4	7.3	255	0.044	0.070
cow		30	–	15.1	515	–	0.068

q*, measure of diet quality or digestibility with 0.5 = poorly and 0.7 = highly digestible diets.
† Components for factorial model (see p. 11) were: M = 0.25 µg kg^{-1} LW, G = 50 µg kg^{-1} LWG, L = 6 and 13 µg kg^{-1} milk for sheep and cattle, respectively, and A = 0.50.

workers reported WMD in calves on pastures with 0.02–0.04 mg Se kg^{-1} DM (Hidiroglou *et al.*, 1985). Recent work in New Zealand confirms that herbage containing 0.02–0.03 mg Se kg^{-1} DM is marginal for heifers (Wichtel *et al.*, 1994, 1996). Sheep consume more food per unit body weight than cattle and this may place a higher burden on antioxidant defences and thus generate a greater need for selenium. The factorial model for selenium may be inadequate when there are sudden changes in local tissue requirements (e.g. ovulation, uterine contraction) at times of combined production demands. Wool production in sheep is more susceptible to selenium deficiency in growing ewes when first mated than in mature females (Wilkins and Kilgour, 1982; Wilkins *et al.*, 1982). The dairy cow conceives at peak lactation and may be required to mount an immune defence against udder infection (i.e. mastitis). The literature suggests that mastitis can respond to selenium supplementation at relatively high blood and dietary selenium concentrations, as does the incidence of retained placenta after the trauma of parturition. Increases in selenium requirements at such times are indicated (Table 15.8).

Pigs

Experiments with piglets from sows on diets deficient in selenium and low in vitamin E indicate that the addition of 0.05 mg Se kg^{-1} DM to a torula-yeast diet containing only 0.01–0.02 mg Se and 100 IU of vitamin E kg^{-1} DM maintained satisfactory growth rates, feed intakes and feed efficiency (Glienke and Ewan, 1977), but in an earlier study (Van Vleet *et al.*, 1973) 0.06 mg Se kg^{-1} DM was associated with high perinatal mortality. With purified diets for gilts, 0.03–0.05 mg Se kg^{-1} DM was adequate for first pregnancy, but 0.1 mg inorganic selenium and 22 IU vitamin E kg^{-1} DM was claimed to be necessary to maintain tissue selenium values (Piatkowski *et al.*, 1979). Raising levels from 0.06 to 0.07 mg Se kg^{-1} DM was sufficient to control myopathy in growing pigs in the study of Piper *et al.* (1975). Bearing in mind that the basal diets in such studies would contain mostly organic selenium and that diets were usually supplemented with the less available inorganic salts, ARC (1981) gave the requirement for growth in pigs as 0.16 mg Se kg^{-1} DM, a figure which recent research indicates to be close to optimal for antioxidant purposes (Lei *et al.*, 1998). The National Research Council (NRC, 1988) gives values which, when translated to concentrations, decrease from 0.28 to 0.10 mg Se kg^{-1} DM as pigs grow from 10 to 110 kg LW and to 0.05 mg Se kg^{-1} DM for lactating sows. A recent study with boars indicates that 0.06 mg Se kg^{-1} DM is insufficient for optimum male fertility (Marin-Guzman *et al.*, 1997).

Poultry

The influence of vitamin E on selenium requirements has been demonstrated most clearly in poultry. Thompson and Scott (1969) found that the selenium requirement (mg kg^{-1} DM) of chicks for growth was close to 0.05 when the purified diet contained no vitamin E, no more than 0.02 with 10 mg kg^{-1} DM added vitamin E and less than 0.01 with 100 mg kg^{-1} DM added. As much as

0.6 mg Se as selenite kg^{-1} DM is required to give complete protection from ED when the diet is low in vitamin E (see Fig. 15.1a). The dietary requirement of vitamin E-replete chicks to prevent hepatic microsomal peroxidation *in vitro* (Combs and Scott, 1974) and for complete protection from PF is approximately 0.06 mg kg^{-1} DM when the element is provided as sodium selenite (Thompson and Scott, 1970). The NRC (1994) gives selenium requirements of 0.1 mg kg^{-1} feed for growing chicks and laying hens. The requirements of turkey poults are higher than those of chicks; Scott *et al.* (1967) found 0.17 mg Se kg^{-1} DM to be necessary to prevent gizzard and heart myopathies when the diet was well supplied with vitamin E and as high as 0.28 mg kg^{-1} DM in diets marginal in the vitamin and in sulphur amino acids. Recent work with young female turkey poults indicates that 0.2 mg Se kg^{-1} DM may be required for maximal liver GPX1 activity for the 0–27-day period, twice the level required by the rat (Hadley and Sunde, 1997), but growth was not impaired with as little as 0.007 mg Se kg^{-1} DM in a diet adequate in vitamin E. Because of environmental concerns about excessive selenium in manures, the 0.20–0.28 mg Se kg^{-1} DM recommended by NRC (1994) should not be exceeded. With laying hens, 0.05 mg Se kg^{-1} DM prevented deficiency on a diet made up mostly of maize and torula yeast with no added vitamin E or antioxidants (Latshaw *et al.*, 1977).

Horses

The minimum selenium requirements of horses do not appear to have been studied. Cases of NMD have been observed in foals in areas where selenium-responsive diseases occur in grazing sheep and cattle (Andrews *et al.*, 1968; Caple *et al.*, 1978) and this condition has been associated with the consumption of feeds usually containing 0.04 mg Se kg^{-1} DM or less (Schougaard *et al.*, 1972; Wilson *et al.*, 1976; Higuchi *et al.*, 1989). It seems reasonable therefore to assume that the selenium allowance for growing horses approximates those for other non-ruminants, namely 0.10 mg kg^{-1} DM. A requirement of 2.4 μg Se kg^{-1} body weight was estimated by Stowe (1967), using a repletion–depletion technique with parenterally administered selenium, but it is difficult to convert this to an oral requirement.

Selenium Poisoning

Selenium is the most toxic of the essential trace elements. Problems can arise naturally and chronically or acutely from the careless administration of selenium supplements and the latter occur more commonly now, even in seleniferous areas (O'Toole and Raisbeck, 1995).

Natural occurrence

Seleniferous areas, in which chronic selenium poisoning occurs in livestock, have been identified in local areas of many countries, including the USA (Nebraska, Utah, Wyoming and South Dakota), the former Soviet Union,

Israel, Ireland, India and Australia. Toxicity can arise in several ways: from the consumption of accumulator or converter plants, such as *Astragalus* species of the vetch family; from the ingestion of normal forage species and cereals with high selenium levels due to the presence of high levels of available selenium in the soils; or from both together (Rosenfeld and Beath, 1964). Where accumulator plants occur, they play a dual role in the incidence of selenosis. They absorb selenium from forms relatively unavailable to other plant species and on their death return it to the soil in organic forms that are available to other species. The predominant forms of selenium in accumulator plants are the soluble organic compounds methylselenocysteine and selenocystathionine (Shrift, 1969), while selenomethionine predominates in other range species. Selenium accumulation is influenced by soil origin and pH. Seleniferous shales and a high pH in an arid climate are associated with high forage selenium levels, as is evident in parts of Ireland (Fleming and Walsh, 1957) and India (Arora *et al.*, 1975). Significant amounts of selenium are discharged into the atmosphere from the combustion of coal and municipal wastes and from ore-processing, but there is little evidence that contamination of soils and forages from industrial sources poses a toxic hazard to livestock (Lakin, 1972).

Clinical manifestations of chronic selenosis

Grazing livestock

Chronic selenium poisoning in grazing livestock was given the name 'alkali disease' and is characterized by dullness and lack of vitality, emaciation, roughness of coat, loss of hair, soreness and sloughing of the hooves, stiffness and lameness, due to erosion of the joints of the long bones. A similar disease was given the name 'blind staggers', although the affected stock were not always blind or staggering but were likely to collapse suddenly and die (Rosenfeld and Beath, 1964). 'Blind staggers' was associated more with the ingestion of 'selenium accumulators' and 'alkali disease' with consumption of grain and grass in seleniferous areas. Atrophy of the heart ('dish-rag' heart), cirrhosis of the liver and anaemia were reported in two early studies of 'blind staggers' and also associated with selenosis. No neuropathology was performed and the symptoms have never been reproduced experimentally, leading O'Toole and Raisbeck (1995) to suggest that other factors, such as polioencaphalomacia, alkaloid poisoning and starvation, were responsible. The lesions of experimental chronic selenosis are confined to the integument, and the lameness and pain from the condition of the hooves can be so severe that affected animals are unwilling to move, resulting in death from thirst and starvation. A recent histological study showed that the primary lesions were epidermal, tubules in the stratum medium of the hoof becoming replaced by islands of parakeratotic cellular debris and the germinal epithelium becoming disorganized. These features distinguished selenosis from laminitis, a common disorder of cattle and horses, in which the lesions are found in the dermis (O'Toole and Raisbeck, 1995). Tail-hair loss was associated with dyskeratosis and atrophic hair follicles. Selenium passes the placental and mammary barriers, so that calves and foals in seleniferous areas may be born with the typical deformed hooves or may develop them during the suckling period.

Pigs and poultry
Growing pigs exposed to high intakes of selenium can develop similar hoof lesions to those seen in ruminants, but, if the intake is sufficiently high to severely reduce appetite, lesions of the central nervous system (CNS) predominate, taking the form of bilateral malacia of grey matter in the spinal cord (Goehring *et al.*, 1984b). In all species, the incidence and severity of hoof lesions may be influenced by the quality and quantity of diet and the related rate of hoof growth. Impaired embryonic development is a feature of selenosis in adult female pigs. Young sows fed on a diet containing 10 mg Se kg^{-1} DM as selenite exhibited a subnormal conception rate and an increased proportion of piglets were dead, small or weak at birth (Wahlstrom and Olson, 1959). Growing chicks eat less food and grow slowly when given seleniferous diets and egg production and hatchability are reduced in laying hens, hatchability being particularly susceptible to increasing selenium intake (Ort and Latshaw, 1978). A proportion of the fertile eggs from hens on high-selenium diets produce grossly deformed embryos, characterized by missing eyes and beaks and distorted wings and feet.

Biochemical manifestations of chronic selenosis

The disturbances just described are accompanied by increased selenium concentrations in the tissues and fluids of the body. At intakes that ultimately give rise to adverse effects, there is a steady rise in tissue selenium concentrations over weeks or months, depending on the rate and continuity of intake, until saturation values are reached and excretion in the urine and faeces begins to keep pace with absorption (Cousins, 1960). Saturation values may reach 20–30 mg Se kg^{-1} DM in the liver, kidneys, hair and hooves of affected animals. High concentrations of the element provide indisputable evidence of an excessive intake, but levels in urine, blood and hair are highly variable. At selenium intakes sufficiently high to present risks of toxicity, the relationship between blood selenium and GPX1 activity breaks down, blood selenium continuing to increase, while GPX1 and hair selenium values plateau (Fig. 15.8; Goehring *et al.*, 1984a; Ullrey, 1987). The South Dakota workers have shown that the hair of cattle from normal areas generally contains 1–4 mg Se kg^{-1} DM, compared with 10–30 mg kg^{-1} DM for cattle on seleniferous range. They contend that, when hair of cattle and pigs consistently has less than 5 mg Se kg^{-1} DM, the diet is unlikely to contain sufficient selenium to induce chronic selenosis. Where exposure has been discontinuous, separating hair or hoof into proximal, medial and distal portions may give a better measure of exposure than the composite samples (J. Small and N.F. Suttle, unpublished data). The selenium concentrations in milk and eggs are particularly sensitive to high selenium intakes by cows and hens. Values ranging between 0.16 and

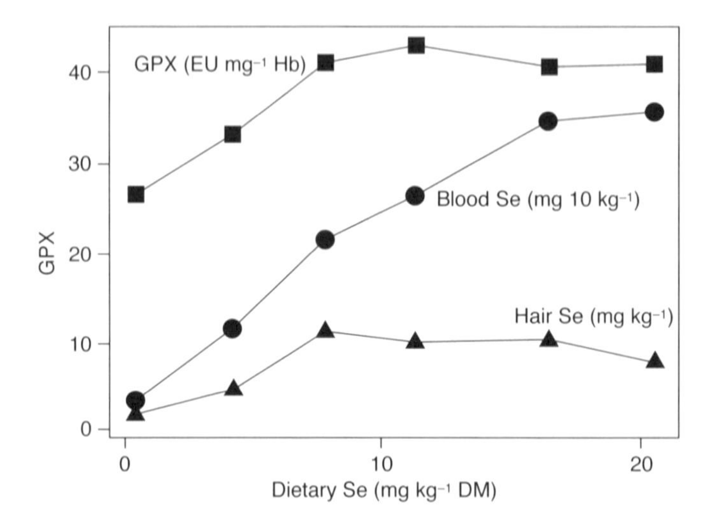

Fig. 15.8. Selenium concentrations in the hair and blood and glutathione peroxidase (GPX) activity in the blood of pigs exposed to excess selenium via the diet (data from Goehring *et al.*, 1984a).

1.27 mg l^{-1} have been reported for cow's milk from seleniferous rural areas in the USA (Rosenfeld and Beath, 1964). Increases of five- to ninefold (up to nearly 2 mg Se kg^{-1} DM) were obtained in egg selenium when 8 mg Se as selenite kg^{-1} DM was added to hens' diets for 42–62 weeks (Arnold *et al.*, 1973). Much earlier, Moxon and Poley (1938) had observed increases from 3.6 to 8.4 mg kg^{-1} DM in the yolk and from 11.3 to 41.3 mg kg^{-1} DM in the white of eggs from hens that had their selenium intakes raised from 2.5 to 10.0 mg kg^{-1} of ration. The biochemical changes in selenosis can be translated into the diagnostic guidelines given in Table 15.9.

Selenium tolerance

The tolerance of high selenium intakes shown by domestic livestock varies with the chemical form in which the element is ingested, the duration and continuity of intake, the criterion of tolerance that is applied, the nature of the whole diet and animal genotype, making tolerable levels hard to define for all species.

Source

Elemental selenium is well tolerated because of its insolubility and poor absorption. Selenides are less toxic than soluble selenites or selenates, and the selenium in natural seleniferous feeds can appear slightly more toxic than that in soluble salts. Dietary selenomethionine supplements produce higher selenium concentrations in animal tissues, especially muscle, than the same amount of selenium as dietary selenite (Scott and Thompson, 1971).

Table 15.9. Marginal bands[a] for biochemical indices of selenium status for use in assessing risk of chronic selenosis (mostly from Puls, 1994).

Species	Diet ($mg\ kg^{-1}$ DM)	Blood ($mg\ l^{-1}$)	Serum/plasma ($mg\ l^{-1}$)	Liver ($mg\ kg^{-1}$ FW)	Hair/fleece ($mg\ kg^{-1}$ DM)
Cattle	4–6	2.0–2.4	2.5–3.5	0.75–1.25	5–10[b]
Sheep	3–5	0.2–0.3	2.0–3.0	10–15	4–6
Pigs	4–8	2–3	–	1.5–3.0	4–6
Poultry	3–5	–	0.15–0.30	2.0–6.0	–

[a] Values above the band indicate definite risk of selenosis; values within bands indicate high previous selenium intakes, which, if continued, will probably lead to selenosis.
[b] 2–5 $mg\ kg^{-1}$ DM is the marginal range for bovine and equine hoof samples.

However, tissue selenium concentrations do not necessarily indicate predisposition to toxicity. In pigs exposed to similar levels of selenium as seleniferous grain or selenite, the former produced much higher concentrations in all tissues studied, except the kidney, and yet this was the better tolerated source (Goehring *et al.*, 1984a). O'Toole and Raisbeck (1995) gave steers supplements of selenium as selenite or selenomethionine, equivalent to dietary concentrations of 5, 10 and 25 mg Se kg^{-1} DM, for 120 days. Hoof lesions developed in four out of five steers given the higher levels of selenomethionine but only one out of four given the corresponding levels of selenite.

Pigs

The toxicity of selenium to the pig varies with the nature of the basal diet and rate of exposure. With a maize–soybean diet, 8 mg Se kg^{-1} DM as selenite was sufficient to impair appetite and growth during 5 weeks' exposure (Goehring *et al.*, 1984a) but was harmless when added to a ration consisting predominantly of wheat and oats (Goehring *et al.*, 1984b). In the former study, hoof lesions developed in some pigs given 12 mg Se kg^{-1} DM. It was concluded that the marginally toxic level for growing pigs was 4–8 mg Se kg^{-1} DM for maize–soybean diets but higher for other grain-based diets, whether the source of selenium was selenite or seleniferous wheat and oats (Goehring *et al.*, 1984b). Genetic differences in susceptibility to selenosis have been reported, with 'red'-haired pigs more susceptible than black- or white-haired breeds (Wahlstrom *et al.*, 1984).

Poultry

Adding selenite to the drinking-water of chicks impairs growth and appetite within 7 days with 4 mg Se or more l^{-1} (the level was equivalent to 7 mg Se kg^{-1} feed intake) (Cantor *et al.*, 1984). Mature hens can tolerate 10 mg Se kg^{-1} DM from seleniferous grain without signs of ill health, but the embryonic development of their eggs is adversely affected (Poley and

Moxon, 1938) and 5 mg selenite Se kg^{-1} DM is borderline for hatchability, the most sensitive criterion of toxicity (Ort and Latshaw, 1978).

Cattle

In young calves given added selenite in a milk-replacer for 45 days, growth and feed-conversion efficiency were impaired with 10 but not 5 mg added Se kg^{-1} DM (Jenkins and Hidiroglou, 1986). Ruminants should be more tolerant than preruminants, because of their lower efficiency of selenium absorption, but the precise tolerance of grazing cattle and horses on seleniferous range is difficult to establish, because of the wide range in selenium concentration of different herbage species, differences in grazing behaviour and the influence of other components of the diet. Where consumption is restricted to non-accumulator plants containing about 5 mg Se kg^{-1} DM, animals can continue for months before signs of chronic poisoning appear, although some effect on the performance of breeding stock would be expected. Echevarria *et al.* (1988) noted no toxic effects while feeding sheep on diets containing 9 mg Se as added selenite kg^{-1} DM for 30 days. High dietary protein contents reduce the toxicity of seleniferous diets, probably in part through the endogenous sulphate yielded by the metabolism of the sulphur-amino acids in the excess protein. Increasing sodium sulphate up to 0.87% of a sulphate-free diet progressively relieved the growth inhibition induced in young rats by 10 mg Se as selenate or selenite kg^{-1} DM (Halverson and Monty, 1960) but is probably less protective for ruminants. Other dietary factors that can ameliorate selenium toxicity are arsenic, mercury, silver, copper and cadmium.

Prevention and control of chronic selenosis

Selenium poisoning is incurable, but three possible methods exist for its prevention, each of which has practical limitations: (i) treatment of the soil to reduce selenium uptake by plants; (ii) treatment of the animal so that selenium absorption is reduced or excretion increased; and (iii) modification of the diet by the inclusion of substances that inhibit or antagonize the toxic effects of selenium within the body tissues. Various management possibilities also exist, including restricting the more seleniferous areas to grain production and 'exporting' the grain. Such seleniferous grain could help to raise selenium intakes by humans and animals in deficient areas.

Soil treatments

The addition of sulphur or gypsum (calcium sulphate ($CaSO_4$)) to soils in a seleniferous area in North America was unsuccessful in reducing selenium uptake by cereals, probably because these soils are mostly already high in sulphate. However, raising the soil inorganic S:Se ratio can depress selenium uptake by plants in some circumstances. Continued heavy dressings with superphosphate (containing $CaSO_4$) to increase pasture yields have been implicated as a cause of selenium deficiency in grazing sheep and cattle in parts of New Zealand (Andrews *et al.*, 1968) and Australia (McDonald, 1975). Applications of 1000 kg gypsum ha^{-1} greatly reduced the selenium concentrations in sugar cane in a recent Indian study (Dhillon and Dhillon, 1991).

Reducing retention

Urinary loss of selenium from the body can be enhanced by the administration of bromobenzene to cattle on seleniferous range (Moxon *et al.*, 1940), but this form of treatment has obvious practical limitations. The use of arsenic (As), first shown by Moxon (1938) to be a selenium antagonist in rats, also presents obvious difficulties. Provision of salt containing 25 mg As as sodium arsenite kg^{-1} to cattle on seleniferous range gave some protection, but observations by ranchers and further investigations indicated that this method of control was ineffective, probably because the arsenic intake was neither high enough nor regular enough (Dinkel *et al.*, 1957). A proprietary, high-sulphate mixture at the rate of 30 g day^{-1} protects buffaloes showing different stages of selenium toxicity (Degnala disease) (Arora *et al.*, 1975), but it is not clear how toxicity was alleviated.

Inhibitors

The discovery by Hill (1972, 1974) that dietary mercury affords protection against selenium toxicity in chicks opened up further possibilities for the prevention and control of selenosis in livestock, where strict dietary control of the animals can be achieved. A mutual protective effect of mercury and selenium was subsequently demonstrated in Japanese quail (El-Begearmi *et al.*, 1977) and the mechanisms confirmed in several investigations (see Underwood, 1977). Much remains to be learnt of the quantitative relationships between mercury and selenium and the mode of action or actions of the protective effects. However, when the mercury and selenium are both in inorganic forms, the most effective ratio in preventing selenium toxicity in chicks appears to be 1:1 (Hill, 1972).

Acute selenosis

Acute selenium poisoning is characterized by salivation, respiratory distress, oedema and pulmonary congestion, reflecting circulatory failure and degenerative changes in the heart, liver and kidney (Rosenfeld and Beath, 1964). Grace (1994) gives wide ranges for the acutely toxic selenium doses for ruminants (median lethal dose (LD_{50})) – 0.15–1.9 mg kg^{-1} LW for injected and 1.9–8.3 mg kg^{-1} LW for oral inorganic selenium. Toxicity has occurred at low injected doses in sheep where there was no history of selenium deficiency (0.45 mg Se kg^{-1} LW: Caravaggi *et al.*, 1970) and in cattle when other stressors (weaning and vaccination) were present (0.57 mg Se kg^{-1} LW: Shortridge *et al.*, 1971). Selenite, selenate and selenomethionine are more toxic than elemental selenium or selenides. Careful scaling of parenteral therapeutic doses to body weight is recommended. Since accumulator plants can contain 125–4800 mg Se kg^{-1} DM in the more toxic organic form, the possibility that 'blind staggers' is wholly or partly attributable to acute selenosis cannot be ruled out.

References

Abdel-Rahim, A.G., Arthur, J.R. and Mills, C.F. (1985) Selenium utilisation by sheep given diets differing in sulphur and molybdenum content. *Biological Trace Element Research* 8, 145–155.

Alfaro, E., Neathery, M.W., Miller, W.J., Gentry, R.P., Crowe, C.T., Fielding, A.S., Etheridge, R.E., Pugh, D.G. and Blackmon, D.M. (1987) Effects of ranging amounts of dietary calcium on selenium metabolism in dairy calves. *Journal of Dairy Science* 70, 831–836.

Allaway, W.H., Cary, E.E. and Ehlig, C.F. (1967) The cycling of low levels of selenium in soils, plants and animals. In: Muth, O.H. (ed.) *Selenium in Biomedicine*. AVI Publishing, Westport, Connecticut, pp. 273–296.

Allen, W.M., Bradley, R., Berrett, S., Parr, W.H., Swannack, K., Barton, C.R.Q. and MacPhee, A. (1975) Degenerative myopathy with myoglobinuria in yearling cattle. *British Veterinary Journal* 131, 292–306.

Anderson, P.H., Berrett, S. and Patterson, D.S.P. (1976) Some observations on 'paralytic myoglobinuria' of cattle in Britain. *Veterinary Record* 99, 316–318.

Anderson, P.H., Berett, S. and Patterson, D.S.P. (1979) The biological selenium status of livestock in Britain as indicated by sheep erythrocyte glutathione peroxidase activity. *Veterinary Record* 104, 235–238.

Andrews, E.D., Hartley, W.J. and Grant, A.B. (1968) Selenium-responsive diseases of animals in New Zealand. *New Zealand Veterinary Journal* 16, 3–17.

ARC (1980) *The Nutrient Requirements of Ruminants.* Commonwealth Agricultural Bureaux, Farnham Royal, Slough, UK, pp. 243–251.

ARC (1981) *The Nutrient Requirements of Pigs.* Commonwealth Agricultural Bureaux, Farnham Royal, Slough, UK, pp. 279–283.

Archer, J.A. and Judson, G.J. (1994) Selenium concentrations in tissues of sheep given a subcutaneous injection of barium selenate or sodium selenate. *Australian Journal of Experimental Agriculture* 34, 581–588.

Arnold, RL., Olson, O.E. and Carlson, C.W. (1973) Dietary selenium and arsenic additions and their effects on tissue and egg selenium. *Poultry Science* 52, 847–854.

Arora, S.P., Parvinder, K., Khirwar, S.S., Chopra, R.C. and Ludri, R.S. (1975) Selenium levels in fodders and its relationship with Degnala disease. *Indian Journal of Dairy Science* 28, 249.

Arthur, J.R. (1988) Effects of selenium and vitamin E status on plasma creatine kinase activity in calves. *Journal of Nutrition* 118, 747–755.

Arthur, J.R. (1997) Selenium biochemistry and function. In: Fischer, P.W., L'Abbé, M.R., Cockell, K.A. and Gibson, R.S. (eds) *Proceedings of the Ninth International Symposium on Trace Elements in Man and Animals (TEMA 9).* NRC Research Press, Ottawa, Canada, pp. 1–5.

Arthur, J.R. (1998) Free radicals and diseases of animal muscle. In: Reznick, A.Z. (ed.) *Oxidative Stress in Skeletal Muscle.* Birkhäuser Verlag, Basel, pp. 321–330.

Arthur, J.R. and Beckett, G.J. (1994a) New metabolic roles for selenium. *Proceedings of the Nutrition Society* 53, 615–624.

Arthur, J.R. and Beckett, G.J. (1994b) Roles of selenium in Type I iodothyronine 5′-deiodinase and in thyroid hormone and iodine metabolism. In: Burk, R.F. (ed.) *Selenium in Biology and Human Health*. Springer-Verlag New York Inc., New York, USA, pp. 93–115.

Arthur, J.R., Price, J. and Mills, C.F. (1979) Observations on the selenium status of cattle in the north-east Scotland. *Veterinary Record* 104, 340–341.

Arthur, J.R., Morrice, P.C. and Beckett, G.J. (1988) Thyroid hormone concentrations in selenium deficient and selenium-sufficient cattle. *Research in Veterinary Science* 45, 122–123.

Awad, J.A., Morrow, J.D., Hill, K.E., Roberts, L.J. and Burk, R.F. (1994) Detection and localisation of lipid peroxidation in selenium- and vitamin E-deficient rats using F2-isoprostanes. *Journal of Nutrition* 124, 810–816.

Balla, H.G., Herbert, E., Munro, C.S. and Suttle, N.F. (1989) Assessing the efficiency of a soluble glass bolus as a slow release source of copper, cobalt and selenium for sheep. *Proceedings of the Nutrition Society* 48, 86A.

Blaxter, K.L. (1963) The effect of selenium administration on the growth and health of sheep on Scottish farms. *British Journal of Nutrition* 17, 105–115.

Boila, R.J., Strothers, S.C. and Campbell, L.D. (1993) The concentration of selenium in the grain from wheat, barley and oats grown at selected locations throughout Manitoba. *Canadian Journal of Animal Science* 73, 217–221.

Boyne, R. and Arthur, J.R. (1979) Alterations of neutrophil function in selenium deficient cattle. *Journal of Comparative Pathology* 89, 151–158.

Bradley, R., Anderson, P.H. and Wilesmith, J.W. (1981) Changing patterns of nutritional myodegeneration (white muscle disease) in cattle and sheep in the period 1975–1985 in Great Britain. In: *Proceedings of the Sixth International Conference on Production Disease in Farm Animals.* Veterinary Research Laboratory, Stormont, Northern Ireland, pp. 248–251.

Broderius, M.A., Whanger, P.D. and Weswig, P.H. (1973) Tissue sulfhydryl groups in selenium-deficient rats and lambs. *Journal of Nutrition* 103, 336–341.

Bunk, M.J. and Combs, G.F. (1980) Effect of selenium on appetite in the selenium-deficient chick. *Journal of Nutrition* 110, 743–749.

Burk, R.F. and Hill, K.E. (1993) Regulation of selenoproteins. *Annual Review of Nutrition* 13, 65–81.

Cantor, A.H. and Scott, M.L. (1974) The effect of selenium in the hen's diet on egg production, hatchability, performance of progeny and selenium concentration in eggs. *Poultry Science* 53, 1870–1880.

Cantor, A.H., Scott, M.L. and Noguchi, T. (1975a) Biological availability of selenium in feedstuffs and selenium compounds for prevention of exudative diathesis in chicks. *Journal of Nutrition* 105, 96–105.

Cantor, A.H., Langevin, M.L., Noguchi, T. and Scott, M.L. (1975b) Efficacy of selenium in selenium compounds and feedstuffs for prevention of pancreatic fibrosis in chicks. *Journal of Nutrition* 105, 106–111.

Cantor, A.H., Nash, D. and Johnson, T.H. (1984) Toxicity of selenium in drinking water of poultry. *Nutrition Reports International* 29, 683–688.

Caple, I.W., Edwards, S.J.A., Forsyth, W.M., Whitely, P., Selth, R.H. and Fulton, L.J. (1978) Blood glutathione peroxidase activity in horses in relation to muscular dystrophy and selenium nutrition. *Australian Veterinary Journal* 54, 57–60.

Caravaggi, C., Clark, F.L. and Jackson, A.R.B. (1970) Acute selenium toxicity in lambs following intramuscular injection of sodium selenite. *Research in Veterinary Science* 11, 146.

Cardin, C.J. and Mason, J. (1975) Sulphate transport by rat ileum: effect of molybdate and other ions. *Biochemica et Biophysica Acta* 394, 46–54.

Chavez, E.R. (1979) Effect of dietary selenium on glutathione peroxidase activity in piglets. *Canadian Journal of Animal Science* 59, 67–75.

Cheng, W.-H., Ho, Y.-S., Ross, D.A., Valentine, B.A., Combs, G.F. and Lei, X.G. (1997) Cellular glutathione peroxidase knockout mice express normal levels of selenium-dependent plasma and phospholipid hydroperoxide glutathione peroxidases in various tissues. *Journal of Nutrition* 127, 1445–1450.

Chesters, J.K. and Arthur, J.R. (1988) Early biochemical defects caused by dietary trace element deficiencies. *Nutrition Research Reviews* 1, 39–56.

Chu, F.F., Doroshow, J.H. and Esworthy, R.S. (1993) Expression characterisation and tissue distribution of a new cellular selenium dependent glutathione peroxidase, GSHPx-G1. *Journal of Biological Chemistry* 268, 2571–2576.

Colnago, G.L., Jensen, L.S. and Long, P.L. (1994) Effect of selenium and vitamin E on the development of immunity to coccidiosis in chickens. *Poultry Science* 63, 1136–1143.

Combs, G.F. and Scott, M.L. (1974) Dietary requirements for vitamin E and selenium measured at the cellular level in the chick. *Journal of Nutrition* 104, 1292–1296.

Conrad, H.R. and Moxon, A.L. (1979) Transfer of dietary selenium to milk. *Journal of Dairy Science* 62, 404–411.

Cooper, B.S., West, D.M. and Pauli, J.V. (1989) Effects of repeated oral doses of selenium in sheep. *New Zealand Veterinary Journal* 37, 37.

Cousins, F.B. (1960) A fluorimetric microdetermination of selenium in biological material. *Australian Journal of Experimental Biology and Medical Science* 38, 11–15.

Cousins, F.B. and Cairney, I.M. (1961) Some aspects of selenium metabolism in sheep. *Australian Journal of Agricultural Research* 12, 927–943.

Davidson, W.B. and Kennedy, D.G. (1993) Synthesis of [^{75}Se] selenoproteins is greater in selenium-deficient sheep. *Journal of Nutrition* 123, 689–694.

Dhillon, K.S. and Dhillon, S.K. (1991) Accumulation of selenium in sugar cane (*Sacharum officinarum* Linn) in seleniferous areas of Punjab, India. *Environmental Geochemistry and Health* 13, 165–170.

Dinkel, C.A., Minyard, J.A., Whitehead, E.I. and Olsen, O.E. (1957) *Selenium and Cattle.* Circular No. 135, South Dakota Agricultural Experiment Station, p. 17.

Donald, G.E., Langlands, J.P., Bowles, J.E. and Smith, A.J. (1993) Subclinical selenium deficiency. 4. Effects of selenium, iodine and thiocyanate supplementation of grazing ewes on their selenium and iodine status and on the status and growth of their lambs. *Australian Journal of Experimental Agriculture* 33, 411–416.

Donald, G.E., Langlands, J.P., Bowles, J.E. and Smith, A.J. (1994a) Subclinical selenium deficiency. 5. Selenium status and the growth and wool production of sheep supplemented with thyroid hormones. *Australian Journal of Experimental Agriculture* 34, 13–18.

Donald, G.E., Langlands, J.P., Bowles, J.E. and Smith, A.J. (1994b) Subclinical selenium deficiency. 6. Thermoregulatory ability of perinatal lambs born to ewes supplemented with selenium and iodine. *Australian Journal of Experimental Agriculture* 34, 19–24.

Donald, G.E., Langlands, J.P., Bowles, J.E., Smith, A.J. and Burke, G.L. (1994c) Selenium supplements for grazing sheep. 3. Development of an intraruminal pellet with an extended life. *Animal Feed Science and Technology* 40, 295–308.

Doornenbal, H. (1975) Tissue selenium content of the growing pig. *Canadian Journal of Animal Science* 55, 325–330.

Echevarria, M.G., Henry, P.B., Ammerman, C.B. and Rao, P.V. (1988) Effects of time and dietary selenium concentration as sodium selenite on tissue selenium uptake by sheep. *Journal of Animal Science* 66, 2299–2305.

Edwards, M.J., Hartley, W.J. and Hansen, E.A. (1977) Selenium and lowered reproductive efficiency in pigs. *Australian Veterinary Journal* 53, 553–554.

Eger, S., Drori, D., Kadoori, I., Miller, K.N. and Schindler, H. (1985) Effects of selenium and vitamin E on incidence of retained placenta. *Journal of Dairy Science* 68, 2119–2122.

El-Begearmi, M.M., Sunde, M.L. and Ganther, H.E. (1977) A mutual protective effect of mercury and selenium in Japanese quail. *Poultry Science* 56, 313–322.

Ellison, R.S., Young, B.J. and Read, D.H. (1986) Bovine post-parturient haemoglobinuria: two distinct entities in New Zealand. *New Zealand Veterinary Journal* 34, 7–10.

Erskine, R.J., Eberhart, R.J., Grasso, P.J. and Scholz, R.W. (1989a) Induction of *Escherichia coli* mastitis in cows fed selenium-deficient or selenium-supplemented diets. *American Journal of Veterinary Research* 50, 2093–2100.

Erskine, R.J., Eberhart, R.J. and Hutchinson, L.J. (1989b) Blood selenium concentrations and glutathionine peroxidase activities in dairy herds with high and low somatic cell counts. *Journal of American Veterinary Medical Association* 190, 1417–1421.

Field, A.C., Suttle, N.F., Brebner, J. and Gunn, G. (1988) An assessment of the efficacy and safety of selenium and cobalt included in an anthelmintic for sheep. *Veterinary Record* 123, 97–100.

Fisher, D.D., Saxton, S.W., Elliot, R.D. and Beatty, J.M. (1995) Effects of selenium source on Se status of lactating cows. *Veterinary Clinical Nutrition* 2, 68–74.

Fleming, G.A. and Walsh, T.A. (1957) Selenium occurrence in certain Irish soils and its toxic effects on animals. *Proceedings of the Royal Irish Academy* 58B, 151.

Flohe, L., Gunzler, W.A. and Schock, H.H. (1973) Glutathione peroxidase: a selenoenzyme. *FEBS Letters* 32, 132–134.

Fraser, A.J., Ryan, T.J., Sproule, R., Clark, R.G., Anderson, D. and Pederson, E.O. (1987) The effect of selenium on milk production in dairy cattle. *Proceedings of the New Zealand Society of Animal Production* 47, 61–64.

Gabbedy, B.J. (1971) Effect of selenium on wool production, body weight and mortality of young sheep in Western Australia. *Australian Veterinary Journal* 47, 318–322.

Gallagher, K. and Stowe, H.D. (1980) Influence of exercise on serum selenium and peroxide reduction system of standard thoroughbreds. *American Journal of Veterinary Research* 41, 1333–1335.

Ganther, H.E., Hafeman, D.G., Lawrence, R.A., Serfass, R.E. and Hoekstra, W.G. (1976) Selenium and glutathione peroxidase in health and disease: a review. In: Prasad, A.S. (ed.) *Trace Elements in Human Health and Disease*, Vol. 2. Academic Press, New York, pp. 165–234.

Gleed, P.T., Allen, W.M., Mallenson, C.B., Rowlands, G.J., Sanson, B.F., Vagg, M.J. and Caswell, R.D. (1983) Effects of selenium and copper supplementation in the growth of beef steers. *Veterinary Record* 113, 388–392.

Glienke, L.R. and Ewan, R.C. (1977) Selenium deficiency in the young pig. *Journal of Animal Science* 45, 1334–1340.

Godwin, K.O. (1968) Abnormalities in the electrocardiograms of young sheep and lambs grazing natural pastures low in selenium. *Nature, UK* 217, 1275–1276.

Godwin, K.O., Kuchel, R.E. and Buckley, R.A. (1970) The effect of selenium on infertility in ewes grazing improved pastures. *Australian Journal of Experimental Agriculture and Animal Husbandry* 10, 672–678.

Goehring, T.B., Palmer, I.S., Olson, O.E., Libal, G.W. and Wahlstrom, R.C. (1984a) Effects of seleniferous grains and inorganic selenium on tissue and blood composition and growth performance of rats and swine. *Journal of Animal Science* 59, 725–732.

Goehring, T.B., Palmer, I.S., Olson, O.E., Libal, G.W. and Wahlstrom, R.C. (1984b) Toxic effects of selenium on growing swine fed corn–soya bean meal diets. *Journal of Animal Science* 59, 733–737.

Goff, J.P. and Stabel, J.R. (1990) Decreased plasma retinol, α-tocopherol and zinc concentrations during the periparturient period: effect of milk fever. *Journal of Dairy Science* 73, 3195–3199.

Grace, N.D. (1994) *Managing Trace Element Deficiencies.* New Zealand Pastoral Agriculture Research Institute, Palmerston North, pp. 9–24.

Grace, N.D., Venning, M. and Vincent, G. (1994) An evaluation of a controlled release system for selenium in lambs. *New Zealand Veterinary Journal* 42, 63–65.

Grant, A.B. and Sheppard, A.D. (1983) Selenium in New Zealand pastures. *New Zealand Veterinary Journal* 31, 131–136.

Grant, C.A. (1961) Morphological and aetiological studies of dietetic microangiopathy in pigs ('mulberry heart'). *Acta Veterinaria Scandinavica* 2 (suppl. 3), 107.

Gries, C.L. and Scott, M.L. (1972) Pathology of selenium deficiency in the chick. *Journal of Nutrition* 102, 1287–1296.

Hadjimarkos, D.M. (1969) Selenium: a caries enhancing trace element. *Caries Research* 3, 14.

Hadley, K.B. and Sunde, R.A. (1997) Determination of dietary selenium requirement in female turkey poults using glutathione peroxidase. In: Fisher, P.W., L'Abbé, M.R., Cockell, K.A. and Gibson, R.S. (eds) *Trace Elements in Man and Animals. 9.* Proceedings of the Ninth International Symposium. NRC Research Press, Ottawa, pp. 59–60.

Halverson, A.W. and Monty, K.J. (1960) An effect of dietary sulfate on selenium poisoning in the rat. *Journal of Nutrition* 70, 100–102.

Harrison, J.H. and Conrad, H.R. (1984) Effect of selenium intake on selenium utilisation by the non-lactating dairy cow. *Journal of Dairy Science* 67, 219–223.

Harrison, J.H., Hancock, D.D. and Conrad, H.R. (1984) Vitamin E and selenium for reproduction in the dairy cow. *Journal of Dairy Science* 67, 123–132.

Hartley, W.J. (1963) Selenium and ewe fertility. *Proceedings of the New Zealand Society of Animal Production* 23, 20–27.

Hartley, W.J. (1967) Levels of selenium in animal tissues and methods of selenium administration. In: Muth, O.H. (ed.) *Selenium in Biomedicine.* Avi Publishing, Westport, Connecticut, pp. 79–96.

Hartley, W.J. and Grant, A.B. (1961) A review of selenium responsive diseases of New Zealand livestock. *Federation Proceedings* 20, 679–688.

Henry, P.R. and Ammerman, C.B. (1995) Selenium bioavailability. In: Ammerman, C.B., Baker, D.H. and Lewis, A.J. (eds) *Bioavailability of Nutrients for Animals.* Academic Press, New York, pp. 303–331.

Henry, P.R., Echevarria, M.G., Ammerman, C.B. and Rao, P.V. (1988) Estimation of the relative bioavailability of inorganic selenium sources for ruminants using tissue uptake of selenium. *Journal of Animal Science* 66, 2306–2312.

Hidiroglou, M., Proulx, J. and Jolette, J. (1985) Intraruminal selenium for control of nutritional muscular dystrophy in the dairy cow. *Journal of Dairy Science* 68, 57–66.

Higuchi, T., Ichijo, S., Osame, S. and Ohishi, H. (1989) Studies on selenium and tocopherol in white muscle disease of foal. *Japanese Journal of Veterinary Science* 51, 52–59.

Hill, C.H. (1972) Interactions of mercury and selenium in chicks. *Federation Proceedings* 31, 692.

Hill, C.H. (1974) Reversal of selenium toxicity in chicks by mercury, copper, and cadmium. *Journal of Nutrition* 104, 593–598.

Jelinek, P.D., Ellis, T., Wroth, R.H., Sutherland, S.S., Masters, H.G. and Pettersen, D.S. (1988) The effect of selenium supplementation on immunity and the establishment of experimental *Haemonchus contortus* infection in weaner Merino sheep fed a low selenium diet. *Australian Veterinary Journal* 65, 214–217.

Jenkins, K.J. and Hidiroglou, M. (1986) Tolerance of the pre-ruminant calf for selenium in milk replacer. *Journal of Dairy Science* 69, 1865–1870.

Jenkins, K.J., Hidiroglou, M., Wauthy, J.M. and Proulx, J.E. (1974) Prevention of nutritional muscular dystrophy in calves and lambs by selenium and vitamin E additions to the maternal mineral supplement. *Canadian Journal of Animal Science* 54, 49–60.

Jensen, L.S. (1968) Selenium deficiency and impaired reproduction in Japanese quail. *Proceedings of the Society for Experimental Biology and Medicine* 128, 970–972.

Julien, W.E., Conrad, H.R., Jones, J.E. and Moxon, A.L. (1976a) Selenium and vitamin E and incidence of retained placenta in parturient dairy cows. *Journal of Dairy Science* 59, 1954–1959.

Julien, W.E., Conrad, H.R. and Moxon, A.L. (1976b) Selenium and vitamin E and incidence of retained placenta in parturient dairy cows. II. Prevention in commercial herds with prepartum treatment. *Journal of Dairy Science* 59, 1960–1962.

Jumba, I.O., Suttle, N.F., Hunter, E.A. and Wandiga, S.O. (1996) Effects of botanical composition, soil origin and composition on mineral concentrations in dry season pastures in Western Kenya. In: Appleton, J.D., Fuge, R. and McCall, G.J.H. (eds) *Environmental Geochemistry and Health.* Geological Society Special Publication No. 113, London, pp. 39–45.

Koenig, K.M., Buckley, W.T. and Shelford, J.A. (1991a) Measurement of endogenous faecal excretion and true absorption of selenium in dairy cows. *Canadian Journal of Animal Science* 71, 167–174.

Koenig, K.M., Buckley, W.T. and Selford, J.A. (1991b) True absorption of selenium in dairy cows: stable isotope methodology and effect of dietary copper. *Canadian Journal of Animal Science* 71, 175–183.

Koenig, K.M., Rode, L.M., Cohen, R.D.H. and Buckley, W.T. (1997) Effects of diet and chemical form of selenium on selenium metabolism in sheep. *Journal of Animal Science* 75, 817–827.

Koh, T.-S. and Judson, G.J. (1987) Copper and selenium deficiency in cattle: an evaluation of methods of oral therapy and an observation of a copper–selenium interaction. *Veterinary Research Communications* 11, 133–148.

Kott, R.W., Ruttie, J.L. and Southward, G.M. (1983) Effects of vitamin E and selenium injections on reproduction and pre-weaning lamb survival in ewes consuming diets marginally deficient in selenium. *Journal of Animal Science* 57, 553–558.

Krishnamurti, C.R., Ramberg, C.F., Shariff, M.A. and Boston, R.C. (1997) A compartmental model depicting short-term kinetic changes in selenium metabolism in ewes fed hay containing normal or inadequate levels of selenium. *Journal of Nutrition* 127, 95–102.

Kuchel, R.E. and Buckley, R.A. (1969) The provision of selenium to sheep by means of heavy pellets. *Australian Journal of Agricultural Research* 20, 1099–1107.

Kuchel, R.E. and Godwin, K.O. (1976) The prevention and cure of white muscle disease in lambs by means of selenium pellets. *Proceedings of the Australian Society of Animal Production* 11, 389–392.

Lakin, H.W. (1972) *Selenium Accumulation in Soils and its Absorption by Plants and Animals.* Bulletin No. 83, Geological Society of America.

Lamont, H.G., Luke, D. and Gordon, W.A.M. (1950) Some pig diseases. *Veterinary Record* 62, 737–747.

Langlands, J.P., Wilkins, J.F., Bowles, J.E., Smith, A.J. and Webb, R.F. (1981) Selenium concentration in the blood of ruminants grazing in northern New South Wales. 1. Analysis of samples collected in the National Brucellosis Eradication Scheme. *Australian Journal of Agricultural Research* 32, 511–521.

Langlands, J.P., Donald, G.E., Bowles, J.E. and Smith, A.J. (1986) Selenium excretion in sheep. *Australian Journal of Agricultural Research* 37, 201–209.

Langlands, J.P., Donald, G.E., Bowles, J.E. and Smith, A.J. (1989) Selenium concentrations in the blood of ruminants grazing in Northern New South Wales. 3. Relationship between blood concentration and the response in liveweight of grazing cattle given a selenium supplement. *Australian Journal of Agricultural Research* 40, 1075–1083.

Langlands, J.P., Donald, G.E., Bowles, J.E. and Smith, A.J. (1990) Selenium supplements for grazing sheep. 1. A comparison between soluble salts and other forms of supplement. *Animal Feed Science and Technology* 28, 1–13.

Langlands, J.P., Donald, G.E., Bowles, J.E. and Smith, A.J. (1991a) Subclinical selenium insufficiency. 1. Selenium status and the response in liveweight and wool production of grazing ewes supplemented with selenium. *Australian Journal of Experimental Agriculture* 31, 25–31.

Langlands, J.P., Donald, G.E., Bowles, J.E. and Smith, A.J. (1991b) Subclinical selenium insufficiency. 2. The response in reproductive performance of grazing ewes supplemented with selenium. *Australian Journal of Experimental Agriculture* 31, 33–35.

Langlands, J.P., Donald, G.E., Bowles, J.E. and Smith, A.J. (1991c) Subclinical selenium insufficiency. 3. The selenium status and productivity of lambs born to ewes supplemented with selenium. *Australian Journal of Experimental Agriculture* 31, 37–43.

Langlands, J.P., Donald, G.E., Bowles, J.E. and Smith, A.J. (1994) Selenium concentrations in the blood of ruminants grazing in Northern New South Wales. 4. Relationship with tissue concentrations and wool production of Merino sheep. *Australian Journal of Agricultural Research* 45, 1701–1714.

Latshaw, J.D., Ort, J.F. and Diesem, C.D. (1977) The selenium requirements of the hen and effects of a deficiency. *Poultry Science* 56, 1876–1881.

Lei, X.G., Dann, H.M., Ross, D.A., Cheng, W.-S., Combs, G.F. and Roneker, K.R. (1998) Dietary selenium supplementation is required to support full expression of three selenium-dependent glutathione peroxidases in various tissues of weanling pigs. *Journal of Nutrition* 128, 130–135.

Levander, O.A., Ager, A.L. and Beck, M.A. (1995) Vitamin E and selenium: contrasting and interacting nutritional determinants of host resistance to parasite and viral infections. *Proceedings of the Nutrition Society* 54, 475–487.

Liebovitz, B., Hu, M.-L. and Tappel, A.L. (1990) Dietary supplements of vitamin E, β-carotene, coenzyme Q10 and selenium protect tissues against lipid peroxidation in rat tissue slices. *Journal of Nutrition* 120, 97–104.

Lindberg, P. (1968) Selenium determination in plant and animal material, and in water: a methodological study. *Acta Veterinaria Scandinavica* Suppl. 23, 48 pp.

McClure, T.J., Eamens, G.J. and Healy, P.J. (1986) Improved fertility in dairy cows after treatment with selenium pellets. *Australian Veterinary Journal* 63, 144–146.

McDonald, J.W. (1975) Selenium-responsive unthriftiness of young Merino sheep in central Victoria. *Australian Veterinary Journal* 51, 433–435.

McDonald, J.W., Overend, D.J. and Paynter, D.I. (1989) Influence of selenium status in Merino weaners on resistance to trichostrongylid infection. *Research in Veterinary Science* 47, 319–322.

McMurray, C.H. and McEldowney, P.K. (1977) A possible prophylaxis and model for nutritional degenerative myopathy in young cattle. *British Veterinary Journal* 133, 535–542.

McMurray, C.H., Rice, D.A. and Kennedy, S. (1983) Nutritional myopathy in cattle: from a clinical problem to experimental models for studying selenium, vitamin E and polyunsaturated fatty acid interactions. In: Suttle, N.F., Gunn, R.G., Allen, W.M., Linklater, K.A. and Wiener, G. (eds) *Trace Elements in Animal Production and Veterinary Practice.* Occasional Publication No. 7, British Society of Animal Production, Edinburgh, pp. 61–76.

Mace, D.L., Tucker, J.A., Bills, C.B. and Ferreira, C.J. (1963) Reduction in incidence of birth of premature, weak and dead calves following sodium selenite and α-tocopherol therapy in pregnant cows. *Bulletin, California Department of Agriculture* 52(1), 21.

MacPherson, A. (1994) Selenium, vitamin E and biological oxidation. In: Garnsworthy, P.C. and Cole, D.J.A. (eds) *Recent Advances in Nutrition.* Nottingham University Press, Nottingham, pp. 3–30.

MacPherson, A. and Chalmers, J.S. (1984) Methods of selenium supplementation of ruminants. *Veterinary Record* 115, 544–546.

MacPherson, A., Kelly, E.F., Chalmers, J.S. and Roberts, D.J. (1987) The effect of selenium deficiency on fertility in heifers. In: Hemphill, D.D. (ed.) *Proceedings of the 21st Annual Conference on Trace Substances in Environmental Health.* University of Missouri, Columbia, pp. 551–555.

Mahan, D.C. and Kim, Y.Y. (1996) Effect of inorganic and organic selenium at two dietary levels on reproductive performance and tissue selenium concentrations in first parity gilts and their progeny. *Journal of Animal Science* 74, 2711–2718.

Mahan, D.C., Penhale, L.H., Cline, J.H., Moxon, A.L., Fetter, A.W. and Yarrington, J.T. (1974) Efficacy of supplemental selenium in reproductive diets on sow and progeny performance. *Journal of Animal Science* 39, 536–543.

Marin-Guzman, J., Mahan, D.C., Chung, Y.K., Pate, J.L. and Pope, W.F. (1997) Effects of dietary selenium and vitamin E on boar performance and tissue responses, semen quality and subsequent fertilization rates in mature gilts. *Journal of Animal Science* 75, 2994–3003.

Meneses, A., Batra, T.R. and Hidiroglou, M. (1994) Vitamin E and selenium in milk of ewes. *Canadian Journal of Animal Science* 71, 567–569.

Millar, K.R. and Meads, W.J. (1988) Selenium levels in the blood, liver, kidney and muscle of sheep after the administration of iron/selenium pellets or soluble-glass boluses. *New Zealand Veterinary Journal* 36, 8–10.

Miller, G.Y., Bartlett, P.C., Erskine, R.J. and Smith, K.L. (1995) Factors affecting serum selenium and vitamin E concentrations in dairy cows. *Journal of American Veterinary Medicine Association* 206, 1369–1373.

Miltimore, J.E., van Ryswyk, A.L., Pringle, W.L., Chapman, F.M. and Kalnin, C.M. (1975) Selenium concentrations in British Columbia forages, grains, and processed feeds. *Canadian Journal of Animal Science* 55, 101–111.

Minson, D.J. (1990) *Forage in Ruminant Nutrition.* Academic Press, New York, pp. 369–381.

Moir, D.C. and Masters, H.G. (1979) Hepatosis dietetica, nutritional myopathy, mulberry heart disease and associated hepatic selenium levels in pigs. *Australian Veterinary Journal* 55, 360–364.

Money, D.F.L., Meads, W.J. and Morrison, L. (1986) Selenised compressed salt blocks for selenium deficient sheep. *New Zealand Veterinary Journal* 34, 81–84.

Morris, J.G., Chapman, H.C., Walker, D.F., Armstrong, J.B., Alexander, J.D., Miranda, R., Sanchez, A., Sanchez, B., Blair-West, J.R. and Denton, D.A. (1984) Selenium deficiency in cattle associated with Heinz body anaemia. *Science* 223, 291–293.

Moxon. A.L. (1937) *Alkali Disease or Selenium Poisoning.* Bulletin No. 311, South Dakota Agricultural Experiment Station.

Moxon, A.L. (1938) The effect of arsenic on the toxicity of seleniferous grains. *Science, USA* 88, 81.

Moxon, A.L. and Poley, W.E. (1938) The relation of selenium content of grains in the ration to the selenium content of poultry carcass and eggs. *Poultry Science* 17, 77–80.

Moxon, A.L., Schaefer, A.E., Lardy, H.A., Dubois, K.P. and Olson, O.E. (1940) Increasing the rate of excretion of selenium from selenized animals by the administration of A-bromobenzene. *Journal of Biological Chemistry* 132, 785–786.

Muth, O.H. and Allaway, W.H. (1963) The relationship of white muscle disease to the distribution of naturally occurring selenium. *Journal of the American Veterinary Medical Association* 142, 1379–1384.

Nicol, F., Lefrane, H., Arthur, J.R. and Trayhurn, P. (1994) Characterisation and post-natal development of 5′-deiodinase activity in goat perinatal fat. *American Journal of Physiology* 267, R144–R149.

Nielsen, H.E., Danielsen, V., Simesen, M.G., Gissel-Nielsen, G., Hjarde, W., Leth, T. and Basse, A. (1979) Selenium and vitamin E deficiency in pigs. 1. Influence on growth and reproduction. *Acta Veterinaria Scandinavica* 20, 276–288.

Noguchi, T., Langevin, M.L., Combs, G.F., Jr and Scott, M.L. (1973a) Biochemical and histochemical studies of the selenium deficient pancreas in chicks. *Journal of Nutrition* 103, 444–453.

Noguchi, T., Cantor, A.H. and Scott, M.L. (1973b) Mode of action of selenium and vitamin E in prevention of exudative diathesis in chicks. *Journal of Nutrition* 103, 1502–1511.

Nolan, M.R., Kennedy, S., Blanchflower, W.J. and Kennedy, D.G. (1995) Lipid peroxidation, prostacyclin and thromboxane A2 in pigs depleted of vitamin E and selenium and supplemented with linseed oil. *British Journal of Nutrition* 74, 369–380.

NRC (1988) *Nutrient Requirements of Domestic Animals. Nutrient Requirements of Swine*, 9th edn. National Academy of Sciences, Washington, DC.

NRC (1994) *Nutrient Requirements of Domestic Animals. Nutrient Requirements of Poultry*, 9th edn. National Academy of Sciences, Washington, DC.

Oh, S.H., Sunde, R.A., Pope, A.L. and Hoekstra, W.G. (1976a) Glutathione peroxidase response to selenium intake in lambs fed a torula yeast-based, artificial milk. *Journal of Animal Science* 42, 977–983.

Oh, S.H., Pope, A.L. and Hoekstra, W.G. (1976b) Dietary selenium requirement of sheep fed a practical-type diet as assessed by tissue glutathione peroxidase and other criteria. *Journal of Animal Science* 42, 984–992.

Oksanen, H.E. (1967) Selenium deficiency: clinical aspects and physiological responses in farm animals. In: Muth, O.H. (ed.) *Selenium in Biomedicine.* Avi Publishing, Westport, Connecticut, pp. 215–229.

Oldfield, J.E., Schubert, J.R. and Muth, O.H. (1963) Implications of selenium in large animal nutrition. *Journal of Agricultural and Food Chemistry* 11, 388–390.

Ort, J.F. and Latshaw, J.D. (1978) The toxic level of sodium selenite in the diet of laying chickens. *Journal of Nutrition* 108, 1114–1120.

O'Toole, D. and Raisbeck, M.F. (1995) Pathology of experimentally induced chronic selenosis (alkali disease) in yearling cattle. *Journal of Veterinary Diagnostic Investigation* 7, 364–373.

Owen, R.apR., Moore, J.N., Hopkins, J.B. and Arthur, D. (1977) Dystrophic myodegeneration in adult horses. *Journal of American Veterinary Medicine Association* 171, 343–349.

Patterson, E.L., Milstrey, R. and Stokstad, E.L.R. (1957) Effect of selenium in preventing exudative diathesis in chicks. *Proceedings of the Society for Experimental Biology and Medicine* 95, 617–620.

Paulson, G.D., Broderick, G.A., Baumann, C.A. and Pope, A.L. (1968) Effect of feeding sheep selenium-fortified trace mineralized salt: effect of tocopherol. *Journal of Animal Sciences* 27, 195–202.

Pedersen, K.B. and Simesen, M.G. (1977) Om tilskud of selen og vitamin E-selen mangelsyndromet hos svin. *Nordisk Veterinaermedicin* 29, 161–165.

Piatkowski, T.L., Mahan, D.C., Cantor, A.H., Moxon, A.L., Cline, J.H. and Grifo, A.P., Jr (1979) Selenium and vitamin E in semi-purified diets for gravid and nongravid gilts. *Journal of Animal Science* 48, 1357–1365.

Piper, R.C., Froseth, J.A., McDowell, L.R., Kroening, G.H. and Dyer, I.A. (1975) Selenium–vitamin E deficiency in swine fed peas. *American Journal of Veterinary Research* 36, 273–281.

Poley, W.E. and Moxon, A.L. (1938) Tolerance levels of seleniferous grains in laying rations. *Poultry Science* 17, 72–76.

Pope, A.L., Moir, R.J., Somers, M., Underwood, E.J. and White, C.L. (1979) The effect of sulphur on Se absorption and retention in sheep. *Journal of Nutrition* 109, 1448–1455.

Puls, R. (1994) *Mineral Levels in Animal Health*, 2nd edn. Diagnostic data, Sherpa International, Clearbrook, BC, pp. 83–109.

Rammell, C.G., Thompson, K.G., Bentley, G.R. and Gibbons, M.W. (1989) Selenium, vitamin E and polyunsaturated fatty acid concentrations in goat kids with and without nutritional myodegeneration. *New Zealand Veterinary Journal* 37, 4–6.

Rastocci, S.C., Clausen, J. and Srivastava, K.C. (1970) Selenium and lead: mutual detoxifying effects. *Toxicology* 6, 377–378.

Reffett, J.K., Spears, J.W. and Brown, T.T. (1988) Effect of dietary selenium and vitamin E on the primary and secondary immune response in lambs challenged with parainfluenza virus. *Journal of Animal Science* 66, 1520–1528.

Rosenfeld, I. and Beath, O.A. (1964) *Selenium: Geobotany, Biochemistry, Toxicity and Nutrition.* Academic Press, New York, 411 pp.

Rotruck, J.T., Pope, A.L., Ganther, H.E., Swanson, A.B., Hafeman, D.G. and Hoekstra, W.G. (1973) Selenium: biochemical role as a component of glutathione peroxidase. *Science, USA* 179, 588–590.

Salvatore, D., Low, S.C., Berry, M., Maia, A.C., Harney, J.W., Croteau, W., St. Germain, D.L. and Larsen, P.J. (1995) Type 3 iodothyronine deiodinase cloning *in vitro* expression and functional analysis of the selenoprotein. *Journal of Clinical Investigation* 96, 2421–2430.

Schougaard, H., Basse, A., Gissel-Nielsen, G. and Simesen, M.G. (1972)

Ernaeringsmaessigt betinget muskeldvstrofi (NMD) hos fol. *Nordisk Veterinaermedicin* 24, 67–84.

Schubert, J.R., Muth, O.H., Oldfield, J.E. and Remmert, L.F. (1961) Experimental results with selenium in white muscle disease of lambs and calves. *Federation Proceedings* 20, 689–694.

Schwarz, K. and Foltz, C.M. (1957) Selenium as an integral part of factor 3 against dietary necrotic liver degeneration. *Journal of the American Chemical Society* 79, 3292–3293.

Scott, M.L. and Thompson, J.N. (1971) Selenium content of feedstuffs and effects of dietary selenium levels upon tissue selenium in chick and poults. *Poultry Science* 50, 1742–1748.

Scott, M.L., Olson, G., Krook, L. and Brown, W.R. (1967) Selenium-responsive myopathies of myocardium and of smooth muscle in the young poult. *Journal of Nutrition* 91, 573–583.

Selinus, O. (1988) Biogeochemical mapping of Sweden for goemedical and environmental research. In: Thornton, I. (ed.) *Proceedings of the Second International Symposium on Geochemistry and Health.* Science Reviews, Norwood, UK, pp. 13–20.

Shortridge, E.H., O'Hara, P.J. and Marshall, P.M. (1971) Acute selenium poisoning in cattle. *New Zealand Veterinary Journal* 19, 47–50.

Shrift, A. (1969) Aspects of selenium metabolism in higher plants. *Annual Review of Plant Physiology* 20, 475–494.

Siddons, R.C. and Mills, C.F. (1981) Glutathione peroxidase activity and erythrocyte stability in calves differing in selenium and vitamin E status. *British Journal of Nutrition* 46, 345–350.

Simesen, M.G., Nielsen, H.E., Danielsen, V., Gissel-Nielsen, G., Hjarde, W., Leth, T. and Basse, A. (1979) Selenium and vitamin E deficiency in pigs. 2. Influence on plasma selenium, vitamin E, ASAT and ALAT and on tissue selenium. *Acta Veterinaria Scandinavica* 20, 289–305.

Slaweta, R., Wasowiez, W. and Laskowska, T. (1988) Selenium content, glutathione peroxidase activity and lipid peroxide level in fresh bull semen and its relationship to motility of spermatozoa after freezing and thawing. *Journal of Veterinary Medicine, Animal Physiology, Pathology and Clinical Veterinary Medicine* 35, 455–460.

Slen, S.B., Demiruren, A.S. and Smith, A.D. (1961) Note on the effects of selenium on wool growth and body gains in sheep. *Canadian Journal of Animal Science* 41, 263–265.

Smith, G.M. and Allen, J.G. (1997) Effectiveness of α-tocopherol and selenium supplements in preventing lupinosis-associated myopathy in sheep. *Australian Veterinary Journal* 75, 341–348.

Smith, K.L., Harrison, J.H. and Hancock, D.D. (1984) Effect of vitamin E and selenium supplementation on incidence of clinical mastitis and duration of clinical symptoms. *Journal of Dairy Science* 67, 1293–1300.

Spears, J.W., Harvey, R.W. and Segerson, E.C. (1986) Effects of marginal selenium deficiency and winter protein supplementation on growth, reproduction and selenium status of beef cattle. *Journal of Animal Science* 63, 586–594.

Sprinker, L.H., Harr, J.R., Newberne, P.M., Whanger, P.D. and Weswig, P.H. (1971) Selenium deficiency lesions in rats fed vitamin E supplemented rations. *Nutrition Reports International* 4, 335–340.

Stabel, J.R. and Spears, J.W. (1993) Role of selenium in immune responsiveness and disease resistance. In: Kurfield, D.M. (ed.) *Human Nutrition – A Comprehensive Treatise,* Vol. 8. *Nutrition and Immunology.* Plenum Press, New York, pp. 333–355.

Stevens, J.B., Olsen, W.C., Kraemer, R. and Archaublau, J. (1985) Serum selenium concentrations and glutathione peroxidase activities in cattle grazing forages of various selenium concentrations. *American Journal of Veterinary Research* 46, 1556–1560.

Stevenson, L.M., Jones, D.G. and Suttle, N.F. (1991) Priority of antibody production for copper and selenium supplies in depleted cattle. In: Momcilovic, B. (ed.) *Proceedings of the Seventh International Symposium on Trace Elements in Man and Animals.* IMI, Zagreb, pp. 27-4–27-5.

Stowe, H.D. (1967) Serum selenium and related parameters of naturally and experimentally fed horses. *Journal of Nutrition* 93, 60–64.

Sunde, R.A. (1994) Intracellular glutathione peroxidases – structure, regulation and function. In: Burk, R.F. (ed.) *Selenium in Biology and Human Health.* Springer-Verlag, New York, pp. 45–77.

Suttle, N.F., Jones, D.G., Woolliams, J.A., Woolliams, C. and Weiner, G. (1984) Growth responses to copper and selenium in lambs of different breeds on improved hill pastures. *Proceedings of the Nutrition Society* 43, 103A.

Suttle, N.F., Jones, D.G., Woolliams, C. and Woolliams, J.A. (1987) Heinz body anaemia in copper and selenium deficient lambs grazing improved hill pastures. *British Journal of Nutrition* 58, 539–548.

Tasker, J.B., Bewick, T.D., Clark, R.G. and Fraser, A.J. (1987) Selenium response in dairy cattle. *New Zealand Veterinary Journal* 35, 139–140.

Thompson, J.N. and Scott, M.L. (1969) Role of selenium in the nutrition of the chick. *Journal of Nutrition* 97, 335–342.

Thompson, J.N. and Scott, M.L. (1970) Impaired lipid and vitamin E absorption related to atrophy of the pancreas in selenium deficient chicks. *Journal of Nutrition* 100, 797–809.

Thompson, K.G., Fraser, A.J., Harrop, B.M., Kirk, J.A., Bullians, J. and Cordes, D.O. (1981) Glutathione peroxidase activity and selenium concentration in bovine blood and liver as indicators of dietary selenium intake. *New Zealand Veterinary Journal* 29, 3–6.

Thompson, K.G., Ellison, R.S. and Clark, R.G. (1991) Monitoring selenium status – which test should we use? *New Zealand Veterinary Journal* 39, 152–154.

Trinder, N., Woodhouse, C.D. and Renton, C.P. (1969) The effect of vitamin E and selenium on the incidence of retained placentae in dairy cows. *Veterinary Record* 85, 550–553.

Trinder, N., Hall, R.J. and Renton, C.P. (1973) The relationship between the intake of selenium and vitamin E on the incidence of retained placentae in dairy cows. *Veterinary Record* 93, 641–644.

Tripp, M.J., Whanger, P.D. and Schmitz, J.A. (1993) Calcium uptake and ATPase activity of sarcoplasmic reticulum vesicles isolated from control and selenium deficient lambs. *Journal of Trace Elements and Electrolytes in Health and Disease* 7, 75–82.

Ullrey, D.J. (1974) The selenium deficiency pattern in animal agriculture. In: Hoekstra, W.G. and Ganther, H.E. (eds) *Proceedings of the Second International Symposium on Trace Elements in Man and Animals, Wisconsin.* University Park Press, Baltimore, pp. 275–294.

Ullrey, D.J. (1987) Biochemical and physiological indicators of selenium status in animals. *Journal of Animal Science* 65, 1712–1726.

Ullrey, D.J., Light, M.R., Brady, P.S., Whetter, P.A., Tilton, J.E., Henneman, H.A. and Mageo, W.T. (1978) Selenium supplements in salt for sheep. *Journal of Animal Science* 46, 1515–1521.

Underwood, E.J. (1977) *Trace Elements in Human and Animal Nutrition*, 4th edn. Academic Press, New York, 545 pp.

Van Vleet, J.F., Carlton, W. and Olander, H.J. (1970) Hepatosis dietetica and mulberry heart disease associated with selenium deficiency in Indiana swine. *Journal of the American Veterinary Medicine Association* 157, 1208–1219.

Van Vleet, J.F., Meyer, K.B. and Olander, H.J. (1973) Control of selenium–vitamin E deficiency in growing swine by parenteral administration of selenium–vitamin E preparations to baby pigs or to pregnant sows and their baby pigs. *Journal of the American Veterinary Medical Association* 163, 452–456.

Vendeland, S.C., Beilstein, M.A., Cheu, C.L., Jensen, O.L., Barofsky, E. and Whanger, P.D. (1993) Purification and properties of selenoprotein W from rat muscle. *Journal of Biological Chemistry* 268, 17103–17107.

Wahlstrom, R.C. and Olson, O.E. (1959) The effect of selenium on reproduction in swine. *Journal of Animal Science* 18, 141–145.

Wahlstrom, R.C., Goehring, T.B., Johnson, D.D., Libal, G.W., Olson, O.E., Palmer, I.S. and Thaler, R.C. (1984) The relationship of hair colour to selenium content of hair and selenosis in swine. *Nutrition Reports International* 29, 143–148.

Walsh, D.M., Kennedy, D.G., Goodall, E.A. and Kennedy, S. (1993a) Antioxidant enzyme activity in the muscles of calves depleted of vitamin E or selenium or both. *British Journal of Nutrition* 70, 621–630.

Walsh, D.M., Kennedy, S., Blanchflower, W.J., Goodall, E.A. and Kennedy, D.G. (1993b) Vitamin E and selenium deficiencies increase indices of lipid peroxidation in muscle tissue of ruminant calves. *International Journal of Vitamin Research* 63, 188–194.

Watkinson, J.H. (1983) Prevention of selenium deficiency in grazing animals by annual top-dressing of pasture with sodium selenate. *New Zealand Veterinary Journal* 31, 78–85.

Watkinson, J.H. and Davies, E.B. (1967a) Uptake of native and applied selenium by pasture species. 3. Uptake of selenium from various carriers. *New Zealand Journal of Agricultural Research* 10, 116–121.

Watkinson, J.H. and Davies, E.B. (1967b) Uptake of native and applied selenium by pasture species. 4. Relative uptake through foliage and roots by white clover and browntop. Distribution of selenium in white clover. *New Zealand Journal of Agricultural Research* 10, 122–133.

Weiss, W.P., Hogan, J.S., Todhunter, D.A. and Smith, K.L. (1997) Effect of vitamin E supplementation in diets with a low concentration of selenium on mammary gland health of dairy cows. *Journal of Dairy Science* 80, 1728–1737.

Weitzel, F., Ursini, F. and Wendel, A. (1990) Phospholipid hydroperoxide glutathione peroxidase in various mouse organs during selenium deficiency and repletion. *Biochemica Biophysica Acta* 1036, 88–94.

Whanger, P.D., Weswig, P.H., Oldfield, J.E., Cheeke, P.R. and Muth, O.H. (1972) Factors influencing selenium and white muscle disease: forage types, salts, amino acids and dimethyl sulfoxide. *Nutrition Reports International* 6, 21–37.

Whanger, P.D., Pedersen, N.D. and Weswig, P.H. (1973) Selenium proteins in ovine tissues. 2. Spectral properties of a 10,000 molecular weight selenium protein. *Biochemical and Biophysical Research Communications* 53, 1031–1035.

Whanger, P.D., Tripp, M.J. and Weswig, P.H. (1977a) Effects of selenium and vitamin E deficiencies in lambs on hepatic microsomal hemoproteins and mitochondrial respiration. *Journal of Nutrition* 107, 998–1005.

Whanger, P.D., Weswig, P.H., Schmitz, J.A. and Oldfield, J.E. (1977b) Effects of selenium and vitamin E on blood selenium levels, tissue glutathione peroxidase activities and white muscle disease in sheep fed purified or hay diets. *Journal of Nutrition* 107, 1298–1307.

Whelan, B.R., Barrow, N.J. and Peter, D.W. (1994a) Selenium fertilizers for pastures grazed by sheep. I. Selenium concentrations in whole blood and plasma. *Australian Journal of Agricultural Research* 45, 863−875.

Whelan, B.R., Barrow, N.J. and Peter, D.W. (1994b) Selenium fertilizers for pastures grazed by sheep. II. Wool and liveweight responses to selenium. *Australian Journal of Agricultural Research* 45, 875–886.

White, C.L. and Somers, M. (1977) Sulphur-selenium studies in sheep. 1. The effects of varying dietary sulphate and selenomethionine on sulphur, nitrogen and selenium metabolism in sheep. *Australian Journal of Biological Sciences* 30, 47–56.

Wichtel, J.J., Craigie, A.L., Varela-Alvarez, H. and Williamson, N.B. (1994) The effect of intra-ruminal selenium pellets on growth rate, lactation and reproductive efficiency in dairy cattle. *New Zealand Veterinary Journal* 42, 205–210.

Wichtel, J.J., Craigie, A.L., Freeman, D.A., Varela-Alvarez, H. and Williamson, N.B. (1996) Effect of selenium and iodine supplementation on growth rate and on thyroid and somatotropic function in dairy calves at pasture. *Journal of Dairy Science* 79, 1865–1872.

Wilkins, J.F. and Hamilton, B.A. (1980) Low release of selenium from recovered ruminal pellets. *Australian Veterinary Journal* 56, 87–89.

Wilkins, J.F. and Kilgour, R.J. (1982) Production responses to selenium in northern New South Wales. 1. Infertility in ewes and associated production. *Australian Journal of Experimental Agriculture and Animal Husbandry* 22, 18–23.

Wilkins, J.F., Kilgour, R.J., Gleeson, A.C., Cox, R.J., Geddes, S.J. and Simpson, I.H. (1982) Production responses to selenium in northern New South Wales. 2. Liveweight gain, wool production and reproductive performance in young Merino ewes given selenium and copper supplements. *Australian Journal of Experimental Agriculture and Animal Husbandry* 22, 24–28.

Wilson, T.M., Morrison, H.A., Palmer, N.C., Finley, G.G. and van Dreumel, A.A. (1976) Myodegeneration and suspected selenium/vitamin E deficiency in horses. *Journal of the American Veterinary Medical Association* 169, 213–217.

Winter, K.A. and Gupta, U.C. (1979) Selenium content of forages grown in Nova Scotia, New Brunswick and Newfoundland. *Canadian Journal of Animal Science* 59, 107–111.

Wise, W.R., Weswig, P.H., Muth, O.H. and Oldfield, J.E. (1968) Dietary interrelationship of cobalt and selenium in lambs. *Journal of Animal Science* 27, 1462–1465.

Wright, P.L. (1965) Life span of ovine erythrocytes as estimated from selenium-75 kinetics. *Journal of Animal Science* 11, 546–550.

Wuryastuti, H., Stowe, H.D., Bull, R.W. and Miller, E.R. (1993) Effects of vitamin E and selenium on immune responses of peripheral blood, colostrum and milk leukocytes of sows. *Journal of Animal Science* 71, 2464–2472.

Yeh, J.-Y., Pui-Ping, G., Beilstein, M.A., Forsberg, N.E. and Whanger, P.D. (1997) Selenium influences tissue levels of selenoprotein W in sheep. *Journal of Nutrition* 127, 394–402.

Zachara, B.A., Frafikowska, V., Lejman, H., Kimber, C. and Kaptur, M. (1993) Selenium and glutathione peroxidase in blood of lambs born to ewes injected with barium selanate. *Small Ruminant Research* 11, 135–141.

Zervas, G.P. (1988) Use of soluble glass boluses containing Cu, Co and Se in the prevention of trace-element deficiencies in goats. *Journal of Agricultural Science, Cambridge* 110, 155–158.

Zinc 16

Introduction

Zinc has been known to be essential for certain of the lower forms of life for over a century and for the higher plants since 1926. Todd *et al.* (1934) obtained the first unequivocal evidence that zinc is necessary for growth and health in rats and mice and shortly thereafter zinc deficiency was produced experimentally in pigs, poultry, lambs and calves. The deficiency was associated in all species with severe inappetence and growth depression, impaired reproductive performance and abnormalities of the skin and its appendages. The discovery that zinc prevented and cured parakeratosis (a thickening and hardening of the skin) in pigs and that this disease could occur on rations of commercial type in the presence of excess calcium (Tucker and Salmon, 1955) gave a great stimulus to studies of the nutritional physiology of zinc in farm animals; a further stimulus came from the demonstration of zinc deficiency in the chick by O'Dell and Savage (1957). In the 40 years after carbonic anhydrase was shown to be a zinc metalloenzyme (Keilin and Mann, 1940), a wide range of zinc-responsive pathological conditions were observed in animals and humans and prematurely linked to a wide array of other zinc metalloenzymes (Table 16.1), whose structure and function were reviewed by Riordan and Vallee (1976). However, the rate-limiting catalytic function(s) in animals affected by zinc deficiency has not been identified and structural and regulatory functions of zinc may have greater influence (Vallee and Falchuk, 1993; Cousins, 1996).

Natural Sources of Zinc

Forages

The mean zinc concentration in pastures is 36 mg kg^{-1} dry matter (DM); values vary widely (range 7 to 100 mg kg^{-1} DM), but a high proportion lie between 25 and 50 mg kg^{-1} DM (Minson, 1990). Improved mixed pastures in

© CAB *International* 1999. *Mineral Nutrition of Livestock*
(E.J. Underwood and N.F. Suttle)

Table 16.1. Some of the zinc metalloenzymes found in mammalian tissues and the functions they perform.

Enzyme	EC nomenclature	Function
Alcohol dehydrogenase	1.1.1.1	NAD^+-linked interconversion of alcohol and aldehyde
Alkaline phosphatase	3.1.3.1	Freeing PO_4 from bound forms e.g. monoesters
Carbonic anhydrase	4.2.1.1	Facilitation of CO_2 transport
Carboxypeptidases A and B	3.4.17.1 and 3.4.17.2	Hydrolysis of C-terminal amino acids from polypeptides e.g. in pancreatic digestion
Collagenase	3.4.24.3	Degradation of collagen fibrils
Leucine aminopeptidase	3.4.11.1	Liberation of amino acids from N-terminal end of proteins and polypeptides
Mannosidase		Hydrolysis of mannose
Superoxide dismutase	1.15.1.1	Destruction of the free radical O^{2-}

NAD^+, nicotinamide adenine dinucleotide; PO_4, phosphate.

New Zealand contained more zinc (Zn) in the North Island than in the South Island (38 vs. 22 mg kg^{-1} DM); tussock grassland contained 8–48 mg Zn kg^{-1} DM (Grace, 1972). Uncontaminated pastures in Scotland had from 25 to 35 mg Zn kg^{-1} DM (Mills and Dalgarno, 1972), but values five to 50 times higher were obtained for herbage exposed to industrial contamination. One study of North American forages revealed contrasting ranges of 20–60 mg Zn kg^{-1} DM for legumes and 10–30 mg kg^{-1} DM for the grasses (French *et al.*, 1957), but in another study levels were lower and the contrast much less (11–18 vs. 8–17 mg kg^{-1} DM: Price and Hardison, 1963). Similar low values occur in Western Australia, where herbage yield is increased by zinc fertilizers (Masters and Somers, 1980). Differences between species contribute little to reported variation in forage zinc (Minson, 1990), although Rhodes grass is often poor in zinc (Jumba *et al.*, 1995). State of maturity is more important, concentrations falling by almost 50%, irrespective of level of zinc fertilizer used, for successive cuts in one study (Gladstones and Loneragan, 1967). Hays, therefore, tend to be low in zinc (13–25 mg Zn kg^{-1} DM for lucerne) and silages slightly richer (12–45 mg kg^{-1} DM for maize; see Table 2.1).

Concentrates

The zinc content of cereal grains and other seeds varies little among plant species but can vary greatly in accordance with soil zinc status. The zinc concentrations of wheat, oats, barley and millet generally lie between 30 and 40 mg kg^{-1} DM (for barley, see Table 2.1), with slightly lower values common in maize grain and in all cereals grown on zinc-low soils, unless zinc-containing fertilizers are used (Underwood, 1962). Zinc resembles manganese in being highly concentrated in the outer layers of the grain. In a study of wheat from a low-zinc area, the following mean values were found: wheat, 16; bran,

49; pollard, 41; white flour 5 mg Zn kg^{-1} DM (Underwood, 1962). Cereal straws usually contain only one-third of the concentration found in the grain and frequently less than 12 mg Zn kg^{-1} DM (White, 1993). Protein sources, such as soybean, sesame-seed, cotton-seed and groundnut meals, are invariably higher in zinc than the cereal grains, and concentrations of 50–70 mg Zn kg^{-1} DM are common. Animal protein sources, such as meat-meal or tankage and fish-meal, are mostly richer in zinc than plant protein supplements, containing 80–120 mg Zn kg^{-1} DM. Even higher values (240–260 mg kg^{-1} DM) have been recorded for whale meal (Underwood, 1962). Meat-and-bone meal containing a high proportion of bone can be rich in zinc, due to the relatively high zinc concentration in bone, although such material could supply sufficient calcium to reduce zinc absorption. Normal cow's milk contains 3–5 mg Zn l^{-1}, most of which is associated with the casein, and dried skimmed milk is therefore a good source of zinc, with commonly between 30 and 40 mg kg^{-1} DM. Colostrum is even richer in zinc (14 mg l^{-1}) and the large amounts of zinc secreted after birth by cows causes a large temporary fall in plasma zinc (Goff and Stabel, 1990).

Absorbability

Zinc in cereals has a low and variable relative absorbability for pigs and poultry of around 60% when compared with inorganic sources (Baker and Ammerman, 1995). The zinc in vegetable protein concentrates has a higher relative value than cereal zinc (0.74–0.84 vs. 0.58 or less: Franz *et al.*, 1980). Unfortunately, absolute values of zinc absorbability (A_{Zn}) are needed to fully characterize zinc sources and few have been published, because they are so difficult to measure. Firstly, zinc is absorbed according to need, and tests that use excesses of zinc underestimate the potential of the source. Secondly, phytate, the major source of phosphorus in all grains (see Chapter 5), forms unabsorbable complexes with zinc in the gut. Thirdly, the extent of the zinc × phytate interaction depends on the chosen level of dietary calcium (see Fig. 16.6). Absorbability is therefore a characteristic of the complete diet rather than a specific component, even when that component provides most of the zinc. A small increase in dietary calcium (Ca) from 6.0 to 7.4 g Ca kg^{-1} DM reduced the availability of inorganic zinc for chicks by 3.8-fold (Wedekind *et al.*, 1994a) and would probably have the same effect on the zinc in feeds. Thus components that have a high intrinsic absorbability, such as phytate-free animal protein sources, will lose that attribute when mixed with phytate-rich feeds and calcium supplements. Few feeds or rations can be attributed with A_{Zn} coefficients and some of the rare values published may be misleading. A high reported A_{Zn} in a test cereal meal for rats (50%: House and Welch, 1989) may substantially overestimate A_{Zn} for cereals included in compound feeds for pigs and poultry, because of the inevitably low dietary calcium in a meal consisting solely of cereals. Others have accorded values as low as 17–20% for A_{Zn} in rats given cereals containing around 30 mg Zn kg^{-1} DM (Tidehag *et al.*, 1988) but rats utilize cereal zinc less than chicks (O'Dell *et al.*, 1972). There is a limit to the number of basic diets that can be tested.

The problem may eventually be eased by mathematical modelling; increasingly complex models for predicting the effects of calcium and phytate on zinc absorbability have been proposed (Davies *et al.*, 1985; Wing *et al.*, 1997), but none have allowed for the beneficial influence of dietary phytases. The harmful influence of the phytate antagonism could be reduced by minimizing the calcium and phosphorus levels in pig and poultry rations, decreasing concentrations of dietary phytate by plant breeding (Welch, 1997), increasing the exposure of phytate to plant phytases (e.g. by cereal choice and treatments such as soaking), the addition of microbial phytases (Lei *et al.*, 1993; Adeola *et al.*, 1995) and the use of hydroxylated vitamin D_3 to enhance the efficacy of phytase (Roberson and Edwards, 1994) (see Chapters 4 and 5). Rimbach and Pallauf (1992) found that the addition of 1000 U phytase kg^{-1} neutralized the effects of 5 g phytate kg^{-1} on the incidence of zinc deficiency in rats. Bran is rich in both phytate and phytase, and the value of such by-products as zinc sources and their influence on other sources is hard to predict. If plant protein sources are autoclaved or treated with chelating agents, such as ethylenediaminetetra-acetic acid (EDTA), much of the bound zinc becomes absorbable by pigs and poultry (Smith *et al.*, 1960; Vohra and Kratzer, 1964), but the amounts of chelating agent required are large and there may be enhanced urinary losses. Complexing zinc with ligands, such as picolinate, does not consistently improve zinc absorption (Cousins, 1996). There is no evidence yet of significant constraints upon zinc absorbability for ruminant livestock among forages or concentrates, due largely to the hydrolysis of phytate in the rumen; the general level of absorbability is probably uniformly high, at 60–70% of intake. However, if soybean protein is used in milk-replacers for the preruminant calf, zinc absorbability is reduced (Xu *et al.*, 1997).

Metabolism of Zinc

Absorption

Zinc is absorbed according to need in rats (Fig. 16.1; Wiegand and Kirchgessner, 1978) and ruminants (Suttle *et al.*, 1982) in an active, carrier-mediated (saturable) process (Davies, 1980), the details of which have yet to be resolved (Cousins, 1996). In the rat, the duodenum is the major site of zinc absorption (Davies, 1980). Mucosal induction of the metal-binding protein metallothionein limits zinc absorption at high intakes (Cousins, 1996). Enhancement of absorption during zinc depletion and inhibition during zinc overload both occur rapidly within a week of the change in zinc supply (Miller *et al.*, 1967; Stake *et al.*, 1975). Other dietary factors can set a ceiling on absorption by limiting the amount of zinc which is available for absorption (e.g. phytate) or interfering with the process of absorption. Absorption of zinc is impaired by non-starch polysaccharide in rats (Rubio *et al.*, 1994) and by elements such as copper and cadmium, which increase the mucosal binding of zinc by metallothionein (Bremner, 1993). The interaction between

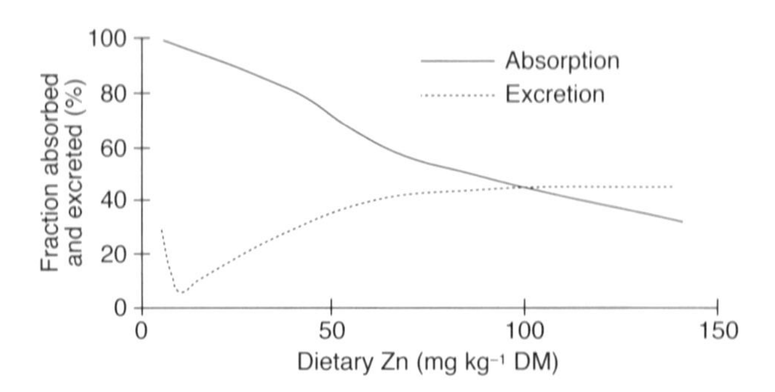

Fig. 16.1. Animals adjust to increasing dietary intakes of zinc by reducing the fractional adsorption of zinc and excreting a higher proportion of the absorbed zinc via the faeces (data from Wiegand and Kirchgessner, 1978).

copper and zinc assumes practical significance when copper is added to pig rations as a growth stimulant (see Chapter 11). Inhibition of zinc absorption from diets low in zinc by selenium may be important in seleniferous regions (House and Welch, 1989).

Transport and tissue distribution

Zinc is initially transported in the portal bloodstream loosely bound to plasma albumin, which accounts for about two-thirds of the plasma zinc. Zinc is also present in the blood plasma as an α_2-macroglobulin and as traces of metallothionein (MT). Induction of hepatic MT synthesis by zinc arriving at the liver plays a key role in removing zinc from the plasma and partitioning it between various pathways (Bremner, 1993). Thus glucocorticoids and cytokines (interleukins 1 and 6) reduce plasma zinc and increase hepatic zinc by inducing MT synthesis (Cousins, 1996). Mutant mice lacking the genes for MTI and -II have increased susceptibility to both deficiency and toxicity of zinc (Kelly *et al.*, 1996). Mild maternal zinc depletion lowers MTI in the plasma, erythrocytes and soft tissues of offspring (Morrison and Bremner, 1987). Although MT has a high affinity for zinc, other elements, such as cadmium and copper, also induce MT synthesis and have higher affinities for the metalloprotein; MT therefore lies at the heart of the many reported interactions between zinc, cadmium and copper (see Fig. 2.5 and Chapter 18). Mechanisms of zinc uptake by tissues beyond the liver are not well understood (Cousins, 1996), but the outcome in terms of distribution throughout the body is well known and is illustrated in Fig. 16.2 by Grace's (1983) data for sheep. The large amounts of zinc in wool are typical of the skin appendages in all species. Most of the zinc in the bloodstream (80%) is present in the erythrocytes, which contain about 1 mg Zn per 10^6 cells, of which over 85% is present as carbonic anhydrase and about 5% as copper–zinc superoxide dismutase (CuZnSOD); reticulocytes are particularly rich in zinc (*c.* 6 mg per 10^6 cells: Cousins, 1996). The blood of avian species is far richer in zinc than that of other species (see Table 16.3).

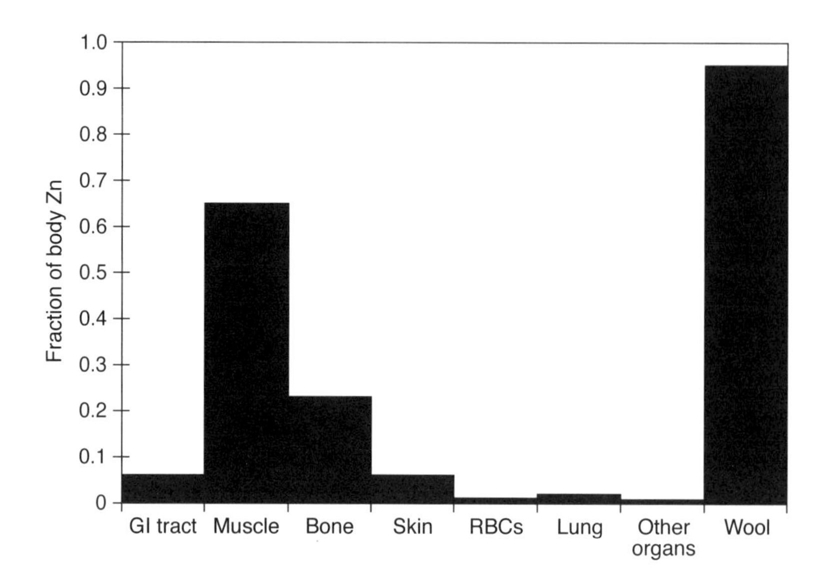

Fig. 16.2. Distribution of zinc in the shorn, empty body of mature sheep shown in relation to the large amounts removed in the fleece (data from Grace, 1983). GI, gastrointestinal.

Excretion

Excretion of zinc occurs predominantly via pancreatic secretions and the faeces, little zinc being voided in urine. However, the contribution of endogenous loss to homoeostasis is limited (Fig. 16.1) and substantial tissue accretion of zinc at high dietary concentrations (634 mg Zn kg^{-1} DM: Stake *et al.*, 1975) confirms failure of homoeostatis. Faecal endogenous losses fall dramatically in calves given a diet very low in zinc (2 mg kg^{-1} DM: Miller, W.J. *et al.*, 1968). The importance of excretory control at intermediate levels of zinc nutrition in sheep is questionable (Suttle *et al.*, 1982), and the reabsorption of secreted zinc is likely to be restricted by phytate in non-ruminants (Suttle, 1983).

Storage

Capacity to store zinc is poorly developed and short-term studies show little change in tissue zinc when dietary zinc supply is varied (for pigs, see Miller, E.R. *et al.*, 1968). Despite the absence of recognized stores, significant amounts of zinc may be redistributed from large pools in muscle and bone (Fig. 16.2) during deficiency, thus delaying the onset of clinical deficiency. Such factors may explain the tolerance of newborn lambs to maternal zinc depletion (Masters and Moir, 1983). There is no accumulation of a hepatic reserve of zinc in late pregnancy, at least in the sheep (Williams *et al.*, 1978), and supplementation of the mother will not increase the supply of zinc via the milk unless the diet is deficient in zinc (for data on the cow, see Miller, 1970). The largest tissue zinc responses to excessive supply are seen in bone in poultry (Wedekind *et al.*, 1992).

Functions of Zinc

The functions of zinc are numerous and go beyond the list of recognized zinc metalloenzymes given in Table 16.1. Links with vitamin A metabolism were provided by retinene reductase and alcohol dehydrogenase; these zinc metalloenzymes are necessary for the interconversion of vitamin A alcohol (retinol) to vitamin A aldehyde (retinene'), a process essential for normal vision. In recent years, further zinc-dependent enzymes have been discovered (Vallee and Falchuk, 1993), including an angiotensin-converting enzyme (ACE) (Reeves and O'Dell, 1986). Indirect effects of zinc deficiency on erythrocyte membrane composition and stability (Bettger, 1989) and essential fatty acid metabolism (Cunnane, 1988; Cunnane and Yang, 1997) have been reported. In the nutrition context, the most important functions are those which become limiting to health and production when livestock are deprived of zinc, and two functions stand out: gene expression and appetite.

Gene expression

Tetrahedral coordination of zinc to cysteine (Cys) and histidine (His) residues creates 'Zn-finger' domains in DNA-binding proteins (Berg, 1990), which influence transcription and hence cell replication (Chesters, 1992). Thus, the vital roles zinc plays in digestion, glycolysis, DNA synthesis, nucleic acid and protein metabolism may mostly derive from primary effects on gene expression, effects which are most marked when cells are rapidly dividing, growing or synthesizing (Chesters, 1992). Fetal growth is especially affected by lack of zinc (Hurley, 1980). There is increased expression of the gene for cholecystokinin (CCK) in the intestine of zinc-deficient animals, and responses to this appetite-regulating hormone may contribute to one of the earliest and most striking clinical features of zinc deprivation, namely anorexia (Table 16.2; Cousins, 1997).

Appetite

Zinc is involved in the regulation of appetite, but the mechanisms remain obscure and are probably multifactorial. O'Dell and Reeves (1989) suggested that appetite may be affected by changes in the concentrations of amino acid-derived neurotransmitters in the brain. The reduction in appetite in rats deprived of zinc is selective, carbohydrate being avoided and protein and fat preferred (Kennedy *et al.*, 1998). Again, gene expression may be involved, because expression of mRNA for pyruvate kinase (PK) is reduced. Since PK is highly regulated by insulin, sensitivity to this hormone may be reduced in zinc deprivation, making it difficult for animals to catabolize carbohydrate (Kennedy *et al.*, 1998).

Biochemical Manifestations of Zinc Deficiency

Blood and tissue zinc

The functional and structural abnormalities of zinc deprivation are associated with a wide variety of biochemical changes in the blood and tissues. An early decline in plasma or serum zinc has been observed in deficient animals of all species studied, including lambs (Mills *et al.*, 1967), calves (Miller and Miller, 1962) and baby pigs (Miller, E.R. *et al.*, 1968). As the deficient state develops, there is usually a small decline in the zinc concentration of most soft tissues, but the fall can be greater in pancreas and porcine liver (Table 16.3; Swinkels *et al.*, 1996). The most marked reductions occur in bone, hair, wool and feathers (Table 16.3), and the concentrations of zinc in these latter appendages are normally particularly high (100–200 mg kg^{-1} DM). The zinc concentrations in the male sex organs and secretions are also normally high and reflect the zinc status of the animal. Values of 105 ± 4.4 and 74 ± 5.0 mg Zn kg^{-1} DM were reported for the testes of normal and zinc-deficient rams, respectively (Underwood and Somers, 1969), and similar reductions were reported in goats, but not in young calves (Miller *et al.*, 1967).

Zinc enzymes

The first abnormality to be observed was subnormal carbonic anhydrase activity in the blood of zinc-deficient calves (Miller and Miller, 1962). Low alkaline phosphatase activity was subsequently reported in the serum of the zinc-deprived baby pig (Table 16.2; Miller, E.R. *et al.*, 1968) and in the blood, serum and bones of cows (Kirchgessner *et al.*, 1975), the serum of lambs (Saraswat and Arora, 1972) and ewes (Apgar, 1979) and the bones of turkey poults (Starcher and Kratzer, 1963). In the testes, intestine and liver of the zinc-deficient goat (Chhabra and Arora, 1993) CuZnSOD activities were

Table 16.2. Incorporation of pair-fed controls in studies of zinc deprivation shows that some abnormalities (e.g. poor growth) are due largely to poor appetite while others (parakeratosis and raised white cell count) are due solely to lack of zinc (data for the baby pig from Miller *et al.*, 1968a).

Dietary Zn (mg kg^{-1} DM) Food intake	12 *Ad libitum*		100 Restricted		100 *Ad libitum*
Live-weight gain (g day^{-1})	25	<	88	<	288
Food consumption (g day^{-1})	170	=	170	<	430
Gain/food	0.14	<	0.57	<	0.67
Parakeratosis (%)	100	>	0	=	0
Serum Zn (mg l^{-1})	0.22	<	0.57	<	0.67
Alkaline phosphatase (Sigma units)	0.4	<	5.5	<	7.0
Leucocyte count (10^3 mm^{-3})	17.0	>	12.9	=	11.8
Liver alcohol dehydrogenase (Δod min^{-1} mg^{-1} protein)	0.060	=	0.061	=	0.063

Table 16.3. Effects of zinc deprivation (D) on tissue zinc concentrations in the growing animal vary from organ to organ and from species to species when compared with Zn-replete controls (C).

		Plasma or serum (μmol l^{-1})[a]	Packed erythrocytes (μmol l^{-1})[a]	Muscle (mmol kg^{-1} DM)[a]	Liver	Pancreas	Appendages
Calf[1]	D	9.3	28.5	1.46	1.57		1.39
	C	13.0[b]	31.4	1.41	1.72		1.80[b]
Chick[2]	D	26.0	153.0	0.74[t]	0.85	1.07	2.17
	C	32.1	154.5	0.77[t]	0.80	1.39[b]	1.87
		(Whole blood)					
Kid[1]	D	23.3		2.35	2.13		1.27
	C	28.5[b]		2.31	2.35		1.53[b]
Pig[3]	D	3.4	–		1.64	1.56	–
	C	13.2	–		3.01[c]	3.00[c]	–
Lamb[4]		3.3	–	2.00	1.11	0.85	1.51
		18.3	–	2.60	1.62	1.35	1.83

[1] Miller *et al.* (1966); [2] Savage *et al.* (1964); [3] Miller *et al.* (Trial 3: 1968); [4] Ott *et al.* (1964).
[t], thigh muscle (in poultry, breast muscle had 60% less zinc).
[a] Multiply by 65.4 to obtain mg kg^{-1} DM or μg l^{-1}.
[b, c] Denotes significant effects ($P < 0.05$ or < 0.01) of zinc deprivation. Controls not pair fed.

reduced. Low plasma vitamin A values in the presence of adequate dietary vitamin A occur in zinc-deficient pigs (Stevenson and Earle, 1956), lambs (Saraswat and Arora, 1972) and young goats (Prasad and Arora, 1979; Chhabra and Arora, 1993), when compared with controls fed *ad libitum*. Liver alcohol dehydrogenase activity is sometimes reduced in zinc-deficient lambs and kids and could be related to the vitamin A depletion and night-blindness observed (Arora *et al.*, 1973; Chhabra and Arora, 1993), but the enzyme is not always affected (Table 16.2) and pair-feeding studies are needed to aid interpretation (Smith *et al.*, 1976). Some enzymes are present in excess of requirements and reduced activity would not necessarily cause dysfunction. The teratogenic effects of maternal zinc deficiency in the rat have been related to impaired activity of fetal thymidine kinase (Dreosti and Hurley, 1975) and to the vulnerability of DNA synthesis in the central nervous system (CNS) (Eckhert and Hurley, 1977). However, thymidine kinase is not a zinc-containing enzyme, other enzymes (e.g. DNA polymerases) are also reduced in activity and the fundamental dysfunction may be in gene expression (Chesters, 1992). Monitoring changes in the differential display of mRNA may provide better insight into the physiological role of zinc, but increases in mRNA for CCK (Cousins, 1997) may indicate attempts to limit the need for zinc by constraining appetite. Fundamental impairment in the G_1 phase of the cell cycle in zinc-deprived animals cannot be ruled out (Chesters and Boyne, 1997). A basic impairment of cell replication may explain enhanced urinary excretion of nitrogen and sulphur reported in zinc-deficient lambs by Somers and Underwood (1969) (Table 16.4).

Table 16.4. Effects of zinc deficiency on feed digestibility (%) and nitrogen and sulphur balances (g day^{-1}) in ram lambs (Somers and Underwood, 1969).

	Zn-deficient group (4)	Pair-fed controls (4)
Dry matter digestibility	64.5 ± 1.2	66.8 ± 0.6
Nitrogen in faeces	4.6 ± 0.2	4.5 ± 0.2
Sulphur in faeces	0.4 ± 0.04	0.4 ± 0.04
Nitrogen in urine	5.8 ± 0.30	3.7 ± 0.21
Sulphur in urine	0.63 ± 0.06	0.38 ± 0.02
Nitrogen balance	+2.0 ± 0.08	+4.2 ± 0.16
Sulphur balance	+0.25 ± 0.02	+0.49 ± 0.04

Uptake of heavy metals

Studies with rats marginally deprived of zinc have revealed a markedly increased uptake of mercury (Kul'kova *et al.*, 1993). Liver and kidney mercury (Hg) levels increased two- to threefold when dietary zinc was reduced from 40 to 6 mg kg^{-1} DM with 10 mg Hg kg^{-1} DM present. Marginal zinc deficiency may therefore be important in areas of heavy-metal pollution.

Clinical Manifestations of Zinc Deprivation

In all species, zinc deprivation is characterized clinically by inappetence, retardation or cessation of growth, lesions of the integument and its outgrowths – hair, wool or feathers – and decreased efficiency of feed utilization.

Anorexia

Loss of appetite is usually the earliest clinical sign of zinc deficiency, and studies in lambs have shown that the pattern of food intake changes from 'meal-eating' to 'nibbling' (Droke *et al.*, 1993a), indicating either an adaptive or metabolite-driven control of substrates for cell growth, development and function. This extreme sensitivity of appetite for solid foods to nutrient supply is unique to zinc, expressed in all species, and may reflect the pivotal role of zinc in cell replication. If impairment of appetite is bypassed by force-feeding a zinc-deficient diet, the demise of the animal is hastened (Flanagan, 1984). Pair-feeding studies show that many of the adverse effects of severe zinc deficiency are secondary to a loss of appetite (e.g. Miller *et al.*, 1967), including those on male fertility (Neathery *et al.*, 1973a; Martin and White, 1992), and pair-feeding is an essential feature of *in vivo* studies (Cunnane and Young, 1997).

Abnormalities of the skin and its appendages

Thickening, hardening and fissuring of the skin (parakeratosis) is a late sign of zinc deprivation in all species: in the chick there is a severe dermatitis, especially of the feet, and feathering is poor (Sunde, 1972, 1978); in young pigs, lesions are also most pronounced on the extremities; in calves, the muzzle, neck, ears, scrotum and back of the hind limbs are most affected; in the dairy cow, lesions appear mostly on hind limbs and teats (Schwarz and Kirchgessner, 1975); lambs exhibit open, parakeratotic lesions round the eyes, above the hoof and on the scrotum (Ott *et al.*, 1964). The lesions can also affect stratified epithelia lining the tongue and oesophagus and are similar to, but not attributable to, vitamin A deficiency (Smith *et al.*, 1976). The rate of healing of artificially inflicted skin wounds is retarded in the zinc-deficient animal but the effect was probably secondary to loss of appetite (Miller *et al.*, 1965). Wounds caused by ectoparasites or skin infections will obviously exacerbate the effects of parakeratosis. In horned lambs, the normal ring structure disappears from new horn growth and the horns are ultimately shed, leaving soft spongy outgrowths that continually haemorrhage (Mills *et al.*, 1967); changes in the structure of the hooves can also occur. Wool fibres lose their crimp and become thin and loose and the whole fleece may be shed. No further wool growth occurs until additional dietary zinc is supplied, when regrowth is immediate (Underwood and Somers, 1969). Posthitis and vulvitis can also occur in zinc-deficient lambs, associated with enlargement of the sebaceous glands (Demertzis, 1972). Excessive salivation is an early sign peculiar to ruminants (Mills *et al.*, 1967; Apgar *et al.*, 1993) and may reflect a combination of copious saliva production and reluctance to swallow.

Skeletal disorders

A reduction in the size and strength of the femur in deficient baby pigs has been observed, but comparisons with pair-fed controls indicated that the bone changes, like others (Table 16.2; Miller, E.R. *et al.*, 1968), were due to a reduced feed intake. Thickening and shortening of the long bones has been reported in chicks deprived of zinc (O'Dell and Savage, 1957; O'Dell *et al.*, 1958). Severe abnormalities of the head, limbs, vertebrae and body covering occur in chick embryos from the eggs of severely deprived hens (Kienholz *et al.*, 1961). Lack of zinc at this early growth stage evidently induces a gross disturbance of the skeleton as well as the integument. Bowing of the hind limbs, stiffness of the joints and swelling of the hocks occur in calves deprived of zinc (Miller and Miller, 1962).

Reproductive disorders

Suboptimal dietary zinc can reduce the size of litters, but abnormalities in fetal development or maternal behaviour, such as have been demonstrated in rats, were not observed in early studies with pigs (Hoekstra *et al.*, 1967). Decreased hatchability and a less marked disturbance of the embryo occur when hens are marginally deprived of zinc. Hypogonadism occurs in deprived bull calves (Pitts *et al.*, 1966), kids (Miller *et al.*, 1964) and ram lambs (Underwood and Somers, 1969). In the lambs, spermatogenesis practically ceased within 20 weeks on a semisynthetic diet containing 2.4 mg Zn kg^{-1} DM but recovered completely during a repletion period. Martin and

White (1992) concluded that anorexia reduced the secretion of gonadotrophin-releasing hormone from the hypothalamus in ram lambs deprived of zinc. Male goats which are sexually mature when exposed to severe zinc deprivation also show reductions in testicular size and loss of libido (Neathery *et al.*, 1973a). However, when single or combined deficiencies of zinc and vitamin A were compared in the goat, spermatogenesis was improved by supplements of vitamin A but not zinc, despite the prolonged feeding of a diet containing 15 mg Zn kg^{-1} DM, which induced skin lesions in some kids given either no supplementary zinc or vitamin A alone (Chhabra and Arora, 1993). Mild zinc deficiencies in pregnant ewes are not accompanied by any congenital malformation in the lambs, although the numbers born and their birth weights may be reduced (Egan, 1972; Masters and Fels, 1980; Mahmoud *et al.*, 1983). Feeding a diet very low in zinc (3 mg Zn kg^{-1} DM) during pregnancy reduced survival of the newborn lamb and pregnancy toxaemia occurred as a secondary consequence of anorexia in the ewe (Apgar *et al.*, 1993).

Immunological Consequences of Zinc Deprivation

Zinc deficiency is further characterized by impairment of the immune system. Atrophy of the thymus with subsequent reduction in the humoral immune capacity has been recorded in the young adult zinc-deficient A/g mouse (De Pasquale-Jardieu and Fraker, 1979). Splenic macrophages from severely zinc-deficient mice were less able to facilitate T-cell mitogenesis than those from pair-fed controls, but the effect was dependent on the duration of depletion (James *et al.*, 1987). There is also evidence of decreased cytokine production from T and B cells in mice infected with gut nematodes and given a diet almost devoid of zinc (Shi *et al.*, 1998). The important question is: how sensitive is the immune system to zinc deficiency compared with other zinc-dependent functions? Beach *et al.* (1981) concluded that in marginally depleted, mature mice, immunopathology was *secondary to loss of appetite.* Droke and Spears (1993) worked with severely or marginally zinc-deficient lambs and concluded that loss of appetite, poor growth and skin lesions occurred *before* susceptibility to infection increased. Furthermore, susceptibility to pneumonic *Pasteurella haemolytica* infection was not increased in zinc deficiency (Droke *et al.*, 1993b). However, a recent study with heifers indicated that responses to a subcutaneous injection of phytohaemagglutinin were impaired by zinc deprivation *before* there was any loss of appetite or fall in plasma zinc (Engle *et al.*, 1997).

Diagnosis of Zinc Disorders

Severe zinc disorders can readily be diagnosed from the combined evidence of clinical and pathological disorders and the biochemical defects just described, but the diagnosis of early stages or milder forms presents difficulties. Determination of zinc in the diet can be helpful, but variations in availability limit the value of this approach with pigs and poultry. The various biochemical abnormalities that follow zinc depletion vary in their time of onset and therefore in diagnostic value (Fig. 16.3).

Plasma or serum zinc

Zinc concentrations in blood serum or plasma are the most widely used indicator, but low values are usually an early change and a measure of *deficiency*, lacking certainty and sensitivity as a diagnostic criterion. However, a recent study reported a reduction in growth *before* any fall in plasma zinc in heifers (Engle *et al.*, 1997). In experiments with lambs, plasma zinc declined rapidly upon feeding diets low in zinc, but wool growth was maintained until values fell below 0.5 mg l^{-1} (7.7 μmol l^{-1}; White *et al.*, 1994) and liveweight gain until values were < 0.4 mg (6.2 μmol) l^{-1} (Mills *et al.*, 1967; White *et al.*, 1994); these values represent thresholds for *dysfunction* (White, 1993). Normal values mostly lie within the range 0.8–1.2 mg (12.3–18.5 μmol) Zn l^{-1}, but individual variability can be high and many factors other than dietary zinc affect concentrations. For example, values in pigs are higher in unfed than in fed animals, the difference widening as zinc intakes fall, until the unfed animal has a plasma zinc level twice that of the fed animal on the same low-zinc diet (Wedekind *et al.*, 1994b). Fasted values were the more closely related to growth but still more variable than bone zinc. Plasma zinc values are particularly susceptible to stress (Corrigall *et al.*, 1976). A significant decrease in plasma zinc was observed in Japanese quail as a consequence of 24-h fasting (Harland *et al.*, 1974), and

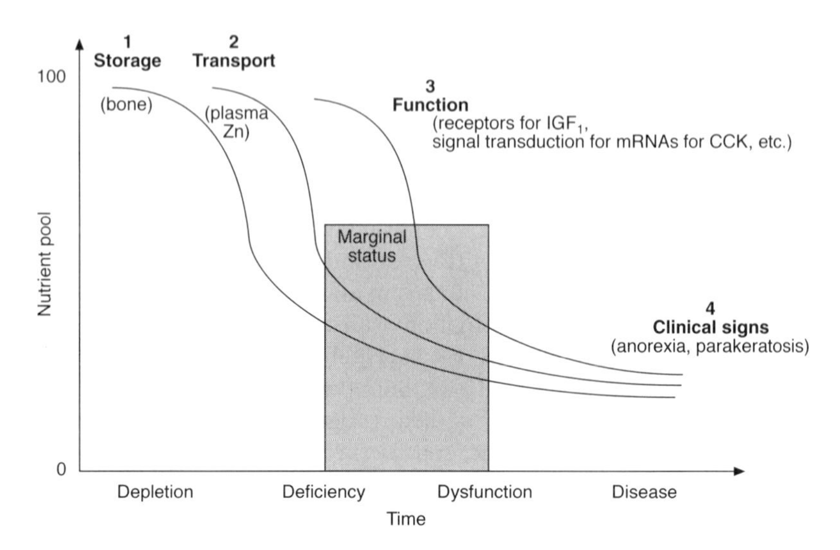

Fig. 16.3. Temporal stages of zinc deprivation leading to the appearance of clinical signs: see also Fig. 3.1 and related text, IGF_1, insulin-like growth factor 1.

there is usually a marked fall in cows during and immediately after parturition (Goff and Stabel, 1990), particularly in those with calving difficulties (Dufty *et al.*, 1977). A mean plasma value of 0.68 ± 0.23 mg l^{-1} was found 18–24 h after normal parturition, compared with 0.38 ± 0.14 mg Zn l^{-1} for cows with dystocia and stillborn calves. Hyperthermic stress has a depressing effect on serum zinc in cattle (Wegner *et al.*, 1973), as does microbial infection (Corrigall *et al.*, 1976; Orr *et al.*, 1990). Grace (1972) reported low plasma zinc values in healthy ewes grazing New Zealand pastures; mean values from 18 farms ranged from 0.53 to 0.89 mg l^{-1} and there was no correlation with pasture zinc concentrations, which ranged from 17 to 70 mg kg^{-1} DM. In contrast, Spais and Papasteriadis (1974) reported a good correlation between clinical symptoms and plasma zinc over the range 0.4–1.0 mg l^{-1} in cattle in Greece. Serum or plasma zinc values must obviously be used with caution in the diagnosis of zinc deficiency in farm animals. Since serum iron (Fe) declines and copper (Cu) rises under the influence of most stressors, abnormal Zn:Cu ratios but normal Zn:Fe ratios should distinguish such cases from those of true zinc deficiency.

Serum metalloproteins and enzymes

To overcome the difficulties presented by the interpretation of plasma zinc, the use of erythrocyte MT concentrations has been advocated; levels are reduced in zinc deficiency but unaffected by the stressors that affect plasma zinc (Bremner, 1991). Serum alkaline phosphatase activity falls during zinc deficiency (Apgar, 1979; Swinkels *et al.*, 1996) but follows a similar time course to serum zinc and is also affected by gut and bone disorders. The assay of enzymes such as ACE and mannosidase may bring a new dimension to the assessment of zinc status, since they show a later decrease than plasma zinc during deprivation (Fig. 16.4; Apgar *et al.*, 1993); increases in apoenzyme activity were claimed to have even greater potential, but the major change was a decrease in pair-fed controls given zinc (10 mg l^{-1}) via the drinking-water, deprived ewes showing little or no change with time.

Other indices

Salivary zinc concentrations have been studied in zinc deficiency but lack sensitivity for diagnostic purposes (Everett and Apgar, 1979). Hair and wool zinc concentrations reflect dietary intakes in all species studied, but individual variability is high and there is variation with age, breed, sampling site and seasonal conditions (Miller *et al.*, 1965a; Hidiroglou and Spurr, 1975; Grace and Sumner, 1986; White *et al.*, 1991). Subnormal hair and wool values can provide supporting evidence of a dietary deficiency, but they cannot be considered as sensitive diagnostic criteria. In view of the prominence and impact of anorexia as an early symptom of deficiency, there seemed much to commend Miller's (1970) suggestion that an immediate improvement in appetite following supplementation is good diagnostic evidence of a zinc-responsive disorder. The recent report that feed conversion efficiency rather than appetite can be the first abnormality to affect zinc-deprived cattle (Engle

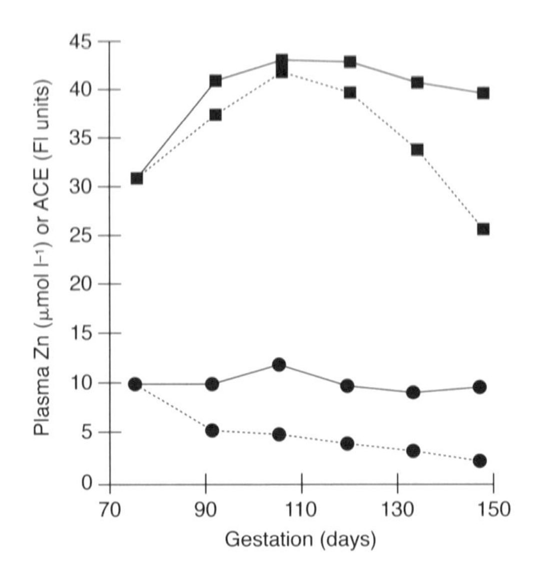

Fig. 16.4. Activity of angiotensin-converting enzyme (ACE) (■) declines at a later stage during zinc deprivation (broken line) than plasma Zn (●) concentration in pregnant ewes: Zn-supplemented ewes represented by solid line (data from Apgar *et al.*, 1993).

et al., 1997) requires further study. At post-mortem, assays of bone zinc can add to the diagnostic picture, which is summarized in Table 16.5, but liver zinc can increase greatly during infection, decreasing its reliability as a measure of zinc status.

Occurrence of Zinc Deprivation

Large areas of zinc-deficient soils exist in many countries (e.g. Western Australia), in which yields of pastures and crops are improved by applications of zinc-containing fertilizers. Zinc concentrations in both forages and grains are concurrently increased, but, even without these interventions, few reports of clinical signs of zinc deprivation in grazing animals appeared and it was generally assumed that the herbage supported by zinc-deficient soils carries enough zinc for the needs of animals. This assumption is only occasionally unsafe and the predominant risk of a disorder occurs among non-ruminants.

Non-ruminants

Commercial cereal-based diets containing plant protein sources, such as soybean, sesame, lupin and cotton-seed meals, cannot be relied on to provide sufficient zinc for chicks, because of the 'chelating' effects of phytate (Edwards *et al.*, 1959; Zeigler *et al.*, 1961). Feather fraying can occur in chicks on such diets, unless additional zinc is provided, especially during the first

Table 16.5. Marginal bands[a] for assessing risk of zinc deprivation in livestock from the most diagnostically useful biochemical indices of zinc status (for units in μmol, multiply values by 0.0153).

	Diet (mg kg^{-1} DM)	Serum (mg l^{-1})	'Coat' (mg kg^{-1} DM)	Pancreas (mg kg^{-1} WM)	Bone (mg kg^{-1} DM)
Cattle and goat	10–20	0.4–0.6	75–100	20–25	50–70 (rib)
Pig	25–35[b]	0.4–0.7	140–150	30–35	80–90
Poultry	50–70[b]	0.8–1.4	200–275	20–30	90–110
Sheep	10–20	0.4–0.6	80–100	< 18	

[a] Mean values for a population sample below the given ranges for more than one criterion indicate probable benefits from zinc supplementation in sufficient individuals to merit interventions. Values within bands indicate possibility of future benefits if zinc status does not improve.
[b] For cereal–vegetable protein diets; reduce range limits by 30% for animal protein sources or phytase-supplemented diets.

few days after hatching (Kienholz *et al.*, 1961; Sunde, 1978). Supplies of zinc from such diets are normally satisfactory for weaner pigs so long as dietary calcium is not in excess. Tucker and Salmon (1955) found that, when maize–soybean-meal diets were supplemented with 1.5% calcium carbonate ($CaCO_3$) or calcium phosphate ($Ca_3(PO_4)_2$), or with 2% bone-meal, growth was depressed and there was a high incidence of parakeratosis. Both disabilities were cured or prevented by additional dietary zinc. Mortality and lesions similar to those of parakeratosis and rectifiable by zinc supplements have been observed in growing pigs given 250 mg Cu kg^{-1} DM as copper sulphate as a growth stimulant (O'Hara *et al.*, 1960; Suttle and Mills, 1966). These findings illustrate the influence of dietary mineral balance on the incidence of zinc deprivation. It should be noted that the zinc content of seed and grains (and hence the prevalence of disorders) varies appreciably with their source. Risk of zinc deprivation in pigs and poultry decreases as the proportion of animal protein sources, such as meat-meal, fish-meal and milk by-products (casein), in the diet increases and is minimal when dietary calcium and copper are not abnormally high.

Ruminants

Nearly 40 years ago, typical abnormalities responsive to zinc therapy were observed in calves, yearlings and adult cattle under certain range conditions in what was then known as British Guiana (Legg and Sears, 1960). It is difficult to reconcile the findings with a simple zinc deficiency, because the pasture species contained 18–42 mg Zn kg^{-1} DM, which appears adequate. Similar findings were later reported on sparse grazings in Greece, which mostly contained 20–30 mg Zn kg^{-1} DM (Spais and Papasteriadis, 1974). More recently, Mahmoud *et al.* (1983) reported heavy ewe and lamb mortality in a flock maintained on hay from irrigated pastures of *Chloris guyana* (Rhodes grass), which was low in zinc (20 mg kg^{-1} DM) and protein (95 g

kg^{-1} DM); the problem was controlled by injecting zinc. Less severe conditions can occur in grazing sheep and cattle. Egan (1972) reported reproductive responses to supplemental zinc in grazing Dorset Horn ewes in some years in South Australia, where the pastures contained about 20 mg Zn kg^{-1} DM, and Mayland *et al.* (1980) obtained growth responses from supplemental zinc in cows and calves grazing forages in Idaho, most of which contained less than 17 mg Zn kg^{-1} DM. In parts of Western Australia, where the pastures frequently contain less than 20 mg Zn kg^{-1} DM in autumn and winter (Masters and Somers, 1980), administration of intraruminal zinc pellets (see Masters and Moir, 1980) before mating and during pregnancy increased the number of lambs born and reared and produced a small increase in lamb birth and weaning weights in one study (Masters and Fels, 1980). However, these responses were not repeatable (Masters and Fels, 1985) and restriction of zinc treatment to the second half of pregnancy had no beneficial effect. Early pregnancy is therefore the critical period in regard to the effects of zinc deprivation, as has been strikingly shown with rats given diets extremely low in zinc (Hurley and Swenerton, 1966; Hurley and Shrader, 1975).

Prevention and Control of Zinc Disorders

Housed livestock

With pigs and poultry and stall-fed milking cows, zinc deficiency can be prevented readily and cheaply by the incorporation of zinc salts into mineral supplements or whole mixed rations. Supplementation to supply 50 mg Zn kg^{-1} DM is normally sufficient, except with pigs receiving very large copper sulphate supplements as a growth stimulant, when an additional 150 mg Zn kg^{-1} DM may be necessary. An alternative approach for non-ruminants, discussed earlier, is to raise zinc availability in the diet by increasing phytase activity and thus removing the primary antagonist of zinc absorption, phytate.

Grazing livestock

Several methods are available for the prevention and control of zinc deficiency in grazing sheep and cattle. Treatment of the soils with zinc-containing fertilizers, where such fertilizers are required to increase crop and pasture yields, usually results in significantly increased forage and grain zinc concentrations. The amount of zinc required in Western Australia was 5–7 kg ha^{-1} of zinc sulphate, or its zinc equivalent as zinc ores, every 2–3 years, but amounts vary with the environment. Under more extensive range conditions, where fertilizer applications are uneconomic, the provision of salt-licks containing 1–2% Zn usually ensures sufficient intakes of zinc. These licks need to be consumed regularly if they are to be fully effective, because zinc is not stored. Oral drenching with zinc sulphate is effective (Legg and Sears, 1960), but the technique is costly in time and labour unless combined with drenching for other purposes. A heavy intraruminal zinc pellet has been designed which releases sufficient zinc to overcome seasonal deficiency in sheep (Masters and Moir, 1980). In young wethers consuming a very deficient diet (3.8 mg Zn kg^{-1}), the pellet maintained plasma and wool zinc concentrations and plasma alkaline phosphatase activities as effectively as zinc sulphate, providing 20.2 mg Zn kg^{-1} DM for at least 6 weeks. In a subsequent field experiment (Masters and Fels, 1980), the administration of zinc pellets before mating and during pregnancy improved ewe fertility.

Inorganic versus organic zinc supplements

The common forms of zinc used to supplement animal rations are the oxide (ZnO) and feed-grade sulphate ($ZnSO_4.H_2O$). Recent bioavailability comparisons in the chick suggest that feed-grade ZnO has only 44–78% of the availability possessed by $ZnSO_4$ when added to purified (Wedekind and Baker, 1990; Wedekind *et al.*, 1992) or practical (Sandoval *et al.*, 1997) diets, confirming an earlier finding with turkeys (Sullivan, 1961). Much speculation surrounds the merits of organic sources of zinc for livestock, notably the Zn:methionine (Met), Zn:lysine (Lys) and Zn:picolinate complexes, which are alleged to 'protect' zinc from dietary antagonists, such as phytate. However, ZnMet, given with or without picolinic acid, offered no advantage over $ZnSO_4$ when added to basal diets of marginal zinc status for young pigs (24–36 mg Zn kg^{-1} DM: Hill *et al.*, 1986). A recent comparison ranked availability of sources for pigs on maize–soybean rations as $ZnSO_4$ > ZnMet > ZnO > ZnLys (Fig. 16.5; Wedekind *et al.*, 1994b). Another recent study (Swinkels *et al.*, 1996) again showed no advantage of a zinc–mixed amino

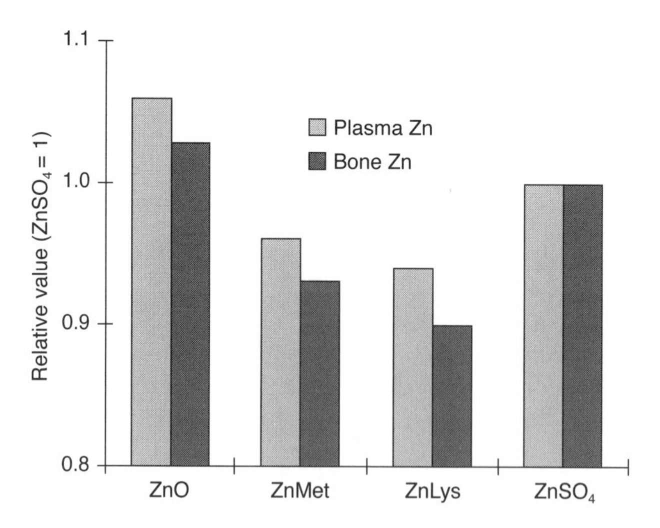

Fig. 16.5. Providing pigs with zinc complexed with lysine (ZnLys) or methionine (ZnMet) resulted in lower zinc concentrations in plasma and bone than when inorganic sources (the oxide, ZnO or sulphate, $ZnSO_4$) were added to a marginally inadequate maize-soybean meal diet (data from Wedekind *et al.*, 1994b): thus, claims for enhanced bioavailability of 'chelated' sources were not supported.

acid chelate over $ZnSO_4$ for any of six different indices of zinc status in repleting pigs; they were given a diet containing 8.4 g calcium and 200 g isolated soybean protein, which provided ample scope for 'protection', with both sources adding 45 mg Zn to a diet containing only 17 mg Zn kg^{-1} DM. A twofold advantage for ZnMet over $ZnSO_4$ has been briefly reported for chicks (Wedekind *et al.*, 1994a), but in general there is no evidence of consistent benefit from providing non-ruminants with zinc in such 'protected' forms rather than in simple organic salts (for a review, see Baker and Ammerman, 1995). Given the lack of evidence for dietary constraints on zinc availability in ruminants, it would be surprising if any benefits were attached to feeding organic as opposed to inorganic sources, and results to date show no consistent advantage (for reviews, see Spears, 1993; Baker and Ammerman, 1995). Higher availability was recently claimed for a ZnMet/ZnLys mixture over ZnO in young calves given vast excesses of zinc, on the basis of higher plasma and liver zinc concentrations, but diets supplemented with the chelates contained 18% more zinc (400 vs. 340 mg Zn kg^{-1} DM: Kincaid *et al.*, 1997).

Zinc Requirements

The minimum zinc requirements of farm animals vary with the species, breed, age and productive functions of the animal, with the composition of the diet, particularly the proportions of organic and inorganic constituents that affect zinc absorption, and with the chosen objective.

Pigs

Early experiments gave higher estimates of zinc requirements than recent studies. Weanling pigs fed on a soybean protein and maize diet, containing 16 mg Zn and 6.6 g Ca kg^{-1} DM, required supplementation to give a total of 41 mg Zn kg^{-1} DM to achieve freedom from parakeratosis and 46 mg Zn kg^{-1} DM to maximize growth rate (Smith *et al.*, 1960). Essentially similar results were obtained by Miller *et al.* (1970), with no difference between male and female requirements. However, Wedekind *et al.* (1994b) found that a maize–soybean finishing diet containing 27 mg Zn, 3.6 g phytate and 5.9 g Ca kg^{-1} DM was optimal for growth, despite a prior period of severe zinc depletion. The results of Hill *et al.* (1986) also pointed to a growth requirement of between 24 and 33 mg Zn kg^{-1} DM with a similar basal diet containing 6.5 g Ca kg^{-1} DM (phytate was not measured). These later estimates are much lower than the 50 mg Zn kg^{-1} DM recommended by the National Research Council (NRC, 1988). If the target was optimal plasma or bone zinc concentrations, Wedekind *et al.*'s (1994b) requirement rose to the NRC level. Calcium at about twice the normal requirement increases the incidence of parakeratosis on diets otherwise marginal in zinc and therefore increases the dietary zinc requirement (Tucker and Salmon, 1955; Hoefer *et al.*, 1960; Norrdin *et al.*, 1973). Maize/soybean rations containing 30–34 mg Zn and high in calcium (16 g kg^{-1} DM) must also be considered marginal in zinc for reproduction in sows (Hoekstra *et al.*, 1967) and the growth of the first generation (Pond and Jones, 1964). The influence of calcium and/or phytate on requirement can be taken into account by using molar ratios of (Ca × phytate):Zn or phytate:Zn, but the importance of objective remains, the optimum ratios being < 0.94 and < 6.5, respectively, for plasma and < 1.6 and < 11.3, respectively, for bone zinc (Fig. 16.6). Diets containing protein from animal sources, such as meat-meal or fish-meal, impose lower zinc requirements, except where dietary copper levels are very high.

Poultry

The dietary zinc requirement for growth in chicks is widely accepted as 35–40 mg kg^{-1} DM for maize–soybean protein diets and 20–35 mg kg^{-1} DM on diets in which the protein supplements come from animal sources (Moeller and Scott, 1958; O'Dell *et al.*, 1958; Dewar and Downie, 1984; Watkins and Southern, 1993). Excess dietary calcium does not increase these requirements to the extent that it does with pigs (Pensack *et al.*, 1958), possibly because the minimum requirement for calcium is sufficiently high to potentiate fully the zinc–phytate antagonism. The NRC (1977) gave the following estimates of zinc requirements (mg kg^{-1} DM): starting chickens (0–8 weeks), 40; growing chickens (8–18 weeks), 35; laying hens, 50; breeding hens, 65. Though recently confirmed (NRC, 1994), these amounts may be too low for the chick in its first few days after hatching if 'fraying' of the feathers is to be prevented (Englert *et al.*, 1966). In one experiment, the

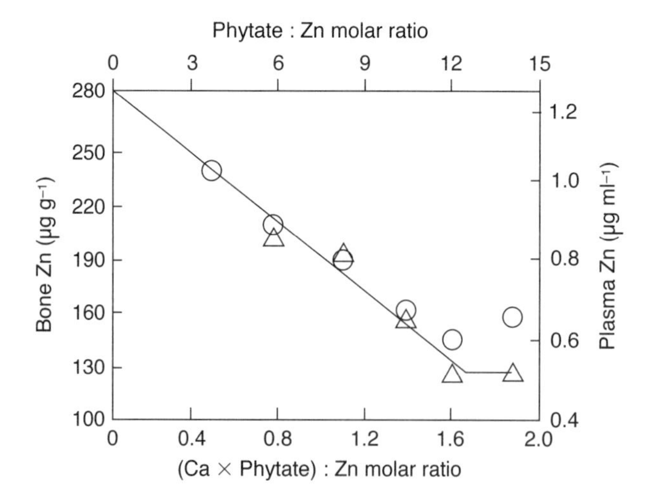

Fig. 16.6. The zinc requirements of pigs are better expressed in relationship to dietary phytate and calcium than as zinc concentrations but optimum levels depend on the chosen criterion of adequacy, plasma Zn (circle) or bone Zn (triangle) (Wedekind *et al.*, 1994b).

incidence of fraying was reduced from 40.6 to 5.6% by increasing dietary zinc in a maize–soybean diet from 44 to 104 mg Zn kg^{-1} DM for the first 2 days of life; there was no effect on weight gains or mortality (Sunde, 1972). Subsequently, Sunde (1978) obtained similar responses on a diet containing 38–55 mg Zn kg^{-1} DM by adding 23 mg Zn kg^{-1} DM for the first week. Adding zinc to the drinking-water at this early stage was also effective in reducing fraying, which varied in incidence between different strains of birds. The zinc requirements given for turkeys, pheasants and quail by the NRC (1977, 1994) are mostly higher than those given for chickens and hens, requirements for growing turkeys being placed as high as 70 mg kg^{-1} DM for the first 4 weeks of life, falling to 40 mg kg^{-1} DM by 11–14 weeks. The increase compared with the needs of young chicks is far higher than that reported by Dewar and Downie (1984) (75 vs. 20%) and probably unjustified; 50 mg Zn kg^{-1} DM seems a more realistic maximum requirement for young turkeys even for practical rations.

Sheep and cattle

The minimum zinc requirements of sheep and cattle deduced from feeding trials, based on semipurified diets, tend to be lower than is suggested from some field studies and they vary according to the chosen criterion of adequacy. Thus requirements of (mg kg^{-1} DM) 18–33 (Ott *et al.*, 1965), 15 (Mills *et al.*, 1967), 17 (Underwood and Somers, 1969), < 9 (Droke *et al.*, 1993a) and 7 (White, 1993) have been reported or indicated for growth in lambs and 10–14 for growth and normal plasma zinc values in calves (Mills *et al.*, 1967). White (1993) calculated that the minimum requirements for optimal wool growth and plasma zinc were approximately twice that for live-weight gain (Fig. 16.7). Reid *et al.* (1987) were unable to increase the growth of lambs given lucerne hay of 'marginal' zinc concentration (18 mg Zn kg^{-1} DM) by the use of fertilizers which increased levels by as much as 23 mg kg^{-1} DM. Requirements for male fertility were thought to be particularly high, since testicular growth and spermatogenesis in ram lambs were markedly subnormal at 17 mg Zn kg^{-1} DM and entirely normal at concentrations of 32 mg kg^{-1} DM (Underwood and Somers, 1969), in agreement with Ott *et al.* (1965). A more recent study has failed to place such a high absolute value on the requirement for optimal male fertility, although the relative need was still twice that for optimal growth (14 mg kg^{-1} DM: Martin and White, 1992). Comparable data for the requirements for dairy cattle are extremely limited. The growth of heifers was improved in a recent study by raising the zinc concentration of a predominantly hay diet from 17 to 40 mg kg^{-1} DM (Engle *et al.*, 1997). Ground maize was used as a supplement, raising the possibility of a rapid passage of undegraded phytate from the rumen. Lactating cows given a diet containing 16.6 mg Zn kg^{-1} DM did not become deficient (Neathery *et al.*, 1973b), and disorder induced by feeding a diet with 6 mg Zn kg^{-1} DM was prevented by raising the level to 28 mg kg^{-1} DM (Schwarz and Kirchgessner, 1975), indicating a requirement between 6 and 16.6 mg Zn kg^{-1} DM. Clinical zinc deficiency was produced in a quarter of the goats given a

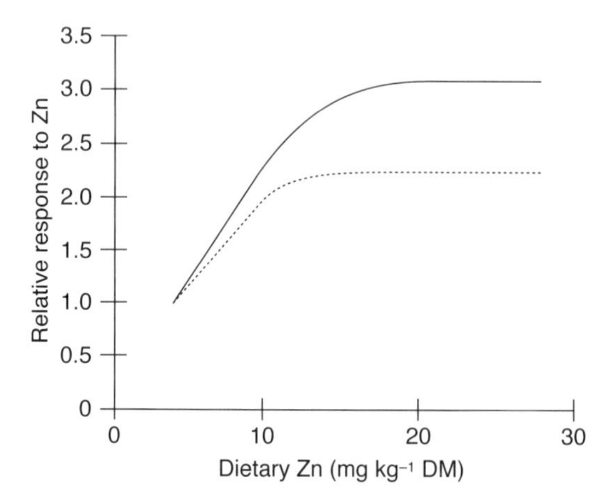

Fig. 16.7. Responses in wool growth (—) and live-weight gain (---) when zinc is added to a semi-purified diet low in zinc show that the optimum Zn level is much higher for wool growth (White, 1993).

diet consisting partly of wheat straw and containing 15 mg Zn kg^{-1} DM for 171–200 days (Chhabra and Arora, 1993).

An alternative approach is to examine the zinc concentrations associated with deficiency symptoms in the field, but results are just as variable. Demertzis and Mills (1973) observed lesions of infectious pododermatitis, responsive to supplementary zinc, in young bulls on rations containing 30–50 mg Zn kg^{-1} and signs of deficiency have been reported in cattle where the pastures or fodders contain 18–42 mg Zn kg^{-1} DM (Legg and Sears, 1960), 19–83 mg kg^{-1} DM (Dynna and Havre, 1963) and 20–30 mg kg^{-1} DM (Spais and Papasteriadis, 1974). The studies of beef cattle on range in Idaho (Mayland *et al.*, 1980) suggest that 12–25 mg Zn kg^{-1} DM were insufficient for optimum growth in calves. In parts of Western Australia, where the pastures frequently contain less than 20 mg Zn kg^{-1} DM, a mild deficiency in ewes is occasionally found (Masters and Fels, 1980). These field findings suggest that the zinc requirements for optimum growth and fertility of sheep and cattle can exceed 20 mg kg^{-1} DM. The weakness of this approach is that it focuses on a few exceptional cases when, year in, year out, sheep and cattle around the world graze pastures or consume feeds indoors with less than 20 mg Zn kg^{-1} DM and remain perfectly healthy (Grace, 1972; Price and Humphries, 1980; Pond, 1983; Masters and Fels, 1985; White *et al.*, 1991). Furthermore, when such incidents are reinvestigated, the initial diagnosis may be disputed (Hartmans, 1965). The relatively high zinc requirements in ruminants given natural as opposed to semipurified diets might be explained by the presence of factors that lower zinc availability in normal foodstuffs. However, suspicions that high dietary calcium impaired zinc utilization in ruminants have not been confirmed (Beeson *et al.*, 1977; Pond, 1983). The discrepancy may be explained by higher maintenance requirements on less digestible, roughage-based diets than on highly digestible, artificial diets or by localized increases in requirements of animals with chronic infections of gut, hoof or skin.

A third approach, involving the derivation of requirements by factorial modelling (see Chapter 1), was introduced by the Agricultural Research Council (ARC, 1980), since reasonable data were available for each production component of net requirement, i.e. weight gain, milk yield, wool growth and pregnancy. The resulting estimates (20–51 mg Zn kg^{-1} DM for sheep and 12–34 mg Zn kg^{-1} DM for cattle) relied heavily on the chosen values for the efficiency of absorption (0.55 for the milk-fed animal, 0.3 for weaned, growing animals and 0.2 for adults), which were based on the average of literature values. Subsequently, it became apparent that zinc was absorbed according to need and that absorptive efficiencies from experiments with zinc given in excess of need would underestimate the potential of the dietary zinc source, as with calcium (see Chapter 4). The first and only estimate of absorption from a low zinc forage (12 mg kg^{-1} DM) by mature sheep revealed a high absorptive efficiency of 0.75 (Suttle *et al.*, 1982). When all published estimates of the efficiency of zinc absorption in sheep and cattle are plotted against a measure of dietary excess (Zn intake ÷ Zn net requirement (I/R)) (Fig. 16.8), the outcome suggests that they can absorb zinc according to need from either inorganic or organic sources, up to a level of around 75% of that present. The ability of ruminal microorganisms to degrade

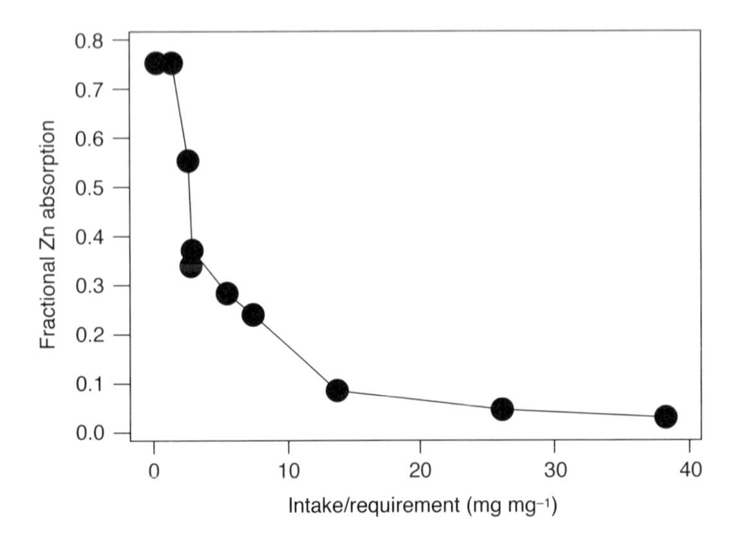

Fig. 16.8. Pooled data from experiments with sheep or cattle in which the true absorption of zinc was either calculated or is calculable confirm that values decline as Zn is provided in excess of net requirement (data from Miller *et al.*, 1967; Stake *et al.*, 1975; and Suttle *et al.*, 1982).

the principal antagonist of zinc, phytate, is largely responsible for the lack of constraint upon absorbability. The study of Suttle *et al.* (1982) also yielded a first direct estimate of faecal endogenous loss (FE_{Zn}) in the sheep of about 100 μg Zn kg^{-1} live weight (LW), which was much higher than the indirect estimate of 53 μg Zn kg^{-1} LW used by ARC (1980) and the solitary estimate of 27 μg kg^{-1} LW published for cattle (Hansard *et al.*, 1968). Values of 27–33 μg Fe_{Zn} kg^{-1} LW would be consistent with the total faecal zinc excretion observed in young calves (Miller, W.J. *et al.*, 1968) or dairy cows (Schwarz and Kirchgessner, 1975) given purified diets very low in zinc (2 mg kg^{-1} DM) if 0.25 of their dietary zinc intake had remained unabsorbed. However, the diets allowed clinical symptoms to develop and therefore failed to meet requirements.

In view of the comprehensive values that can be generated from a factorial model and in the hope of inspiring others to measure absorption and endogenous loss for zinc and other elements, revised factorial estimates of the minimum needs of sheep and cattle for zinc are given in Tables 16.6 and 16.7. The predicted requirements of around 10–15 mg Zn kg^{-1} DM for weaned ruminants are consistent with experimental findings and the absence of problems in the majority of livestock grazing pastures containing around 20 mg Zn kg^{-1} DM. For a given class of animal, minimum requirements increase when the required level of performance rises and highly digestible diets are fed. Even so, 18 mg Zn kg^{-1} DM is sufficient for all but the milk-fed

Table 16.6. Factorial estimates of zinc requirements for sheep at different dry-matter intakes (DMI).

	Live weight	Production rate (kg day^{-1})	Net[a] Requirement (mg day^{-1})	Gross[b] Requirement (mg day^{-1})	DMI (kg day^{-1})	Dietary requirement (mg Zn kg^{-1} DM)
Growth	5	0.15	4.6	5.4	0.2	27.0
	10	0.15	5.6	8.0	0.4	20.0
	20	0.15	6.6	9.4	0.64–0.57	14.7–16.5[c]
	40	0.075	6.8	9.7	1.0–0.6	9.7–16.2
		0.150	8.6	12.3	1.4–0.8	8.8–15.4
		0.300	12.2	17.4	1.1	15.8
Adult	50	–	6.0	8.6	0.8–0.5	10.8–17.2
Pregnancy (late)	75	2 fetuses	11.9	17.0	1.7–1.4	10.0–12.1
Lactation	75	1	15.7	22.4	1.9–1.5	11.8–14.9
		2	22.9	32.7	2.8–2.2	11.7–14.9
		3	30.1	43.0	3.7–2.4	11.6–17.9

[a] Net requirement consists of: maintenance, 0.1 mg Zn kg^{-1} LW; growth, 24 mg Zn kg^{-1} LW; pregnancy, 0.28 and 1.5 mg Zn day^{-1} for mid- and late stages; wool growth, 115 mg Zn kg^{-1} FW; milk, 7.2 mg Zn kg^{-1}.
[b] Absorption coefficients 0.70 except for youngest (milk-only diet) lambs.
[c] Higher values correspond to low DMI on highly digestible diets.

Table 16.7. Factorial estimates of zinc requirements for cattle.

	Live weight	Production rate (kg day^{-1})	Net[a] Requirement (mg day^{-1})	Gross[b] Requirement (mg day^{-1})	DMI (kg day^{-1})	Zn requirement (mg kg^{-1} DM)
Growth	40	0.5	16.0	18.8	1.0	18.7
	100	0.5	22.0	31.4	4.0–2.2	7.9–14.3[c]
		1.0	34.0	48.6	4.5–2.8	10.8–17.4
	200	0.5	32.0	45.7	6.0–3.5	7.6–13.1
		1.0	44.0	62.9	6.5–4.0	9.7–15.7
	300	0.5	43.0	61.4	8.0–4.5	7.7–13.6
		1.0	54.0	77.1	8.4–5.4	10.1–14.3
Adult	500	0	50.0	71.4	7.6–4.6	9.4–15.5
Pregnancy	500	90–100 days	51.1	73.0	9.0–6.6	8.1–11.1
		180–270 days	56.3	80.4	11.2–6.9	7.2–11.7
		kg milk day^{-1}				
Lactation	500	10	90.0	128.6	11.7–8.3	11.0–15.5
		20	129.5	185.0	17.8–12.7	10.4–14.6
		30	170.0	242.9	19.2–15.5	12.7–15.7

[a] Net requirement consists of: maintenance, 0.1 mg Zn kg^{-1} LW; growth, 24 mg Zn kg^{-1} LWG; pregnancy, 1.1 and 6.3 mg Zn day^{-1} for mid- and late stages; milk, 4 mg Zn kg^{-1}.
[b] Absorption coefficients 0.70 for all but the youngest (milk-fed) calf, for which the assumed value was 0.85.
[c] Higher values correspond to low DMI on highly digestible diets.

animal. Much higher requirements of around 40 mg Zn kg^{-1} DM have been calculated factorially for dairy cows (Weigand and Kirchgessner, 1982), but absorptive efficiency was assumed to decline from 40 to 25% as need increased. Suggestions that a level of 149 mg Zn kg^{-1} DM was inadequate for veal calves given soybean protein in a milk-replacer (Xu *et al.*, 1997) are misplaced, since growth and zinc concentrations in plasma and liver were all normal.

Zinc Toxicity

Livestock exhibit considerable tolerance to high intakes of zinc, the extent depending partly on the species but mainly on the nature of the diet, especially its contents of calcium, copper, iron and cadmium. Weanling pigs fed for several weeks on diets containing 1 g Zn kg^{-1} DM, as either the sulphate (Lewis *et al.*, 1957) or the carbonate (Brink *et al.*, 1959), suffered no ill effect, but at higher zinc concentrations growth and appetite were depressed and at 4 and 8 g kg^{-1} DM mortality was high (Brink *et al.*, 1959). Raising the dietary calcium from 7 to 11 g kg^{-1} DM ameliorated the toxic effects of 4 g Zn kg^{-1} DM on weanling pigs (Hsu *et al.*, 1975). At the slightly lower level of 3 g Zn (as ZnO) kg^{-1} DM, the growth of very young weanling pigs may actually be enhanced (Smith *et al.*, 1997). Broilers and layer hens also show a tolerance to zinc at 1–2 g Zn kg^{-1} DM and only slight growth and appetite depression at 4 g Zn kg^{-1} DM (Oh *et al.*, 1979). In the laying hen, appetite suppression at high zinc concentrations has been investigated as a means of curtailing egg production (Jackson *et al.*, 1986). The growth rate of preruminant calves was reduced when the zinc concentration in their milk-replacer reached 700 mg kg^{-1} DM (added as ZnO) (Jenkins and Hidiroglou, 1991). Weaned ruminants are slightly less tolerant of high zinc intakes than non-ruminants, due probably to the lack of protection from the phytate × zinc antagonism. Consumption by lambs of diets containing 1 g Zn kg^{-1} DM, as the oxide, reduced weight gains and decreased feed efficiency; 1.5 g kg^{-1} DM depressed feed consumption and 1.7 g kg^{-1} DM induced a depraved appetite (pica), characterized by excessive salt consumption and wood chewing (Ott *et al.*, 1966a). At the higher zinc concentrations, changes in rumen metabolism have been observed, probably through a toxic effect of the zinc on the rumen microorganisms (Ott *et al.*, 1966b), volatile fatty acid production and the acetic : propionic ratio being reduced.

High zinc intakes prevent facial eczema in sheep and cattle. Facial eczema is characterized by liver damage, photosensitization, weight loss and death. The disease occurs in sheep and cattle grazing pastures infected with the fungus *Pithomyces chartarum*, which produces the hepatotoxic mycotoxin, sporidesmin. The degree of protection afforded by oral zinc against the liver damage and photosensitivity increases with daily zinc intakes up to 2 g in sheep and 23 mg kg^{-1} body weight in cattle. Such intakes are some 50 times the normal zinc requirements in the absence of the mycotoxin and they induce signs of zinc toxicity (Towers *et al.*, 1975; Smith *et al.*, 1977), which initially limited their usefulness. However, an intraruminal bolus has been developed which affords safe and prolonged protection by releasing about 0.8 g Zn from ZnO per day for about 6 weeks (Munday *et al.*, 1997). The mechanism of protection is believed to involve the inhibition of free-radical generation by sporidesmin through a lowering of CuZnSOD activity. Prophylactic use of high zinc intakes to control swine dysentery has been studied, but the margin over the toxic dose is slender (Zhang *et al.*, 1995).

References

Adeola, O., Lawrence, B.V., Sutton, A.C. and Cline, T.R. (1995) Phytase-induced changes in mineral utilisation in zinc-supplemented diets for pigs. *Journal of Animal Science* 73, 3384–3391.

Apgar, J. (1979) Alkaline phosphatase activity and zinc level in plasma as indicators of zinc status in pregnant and lactating sheep. *Nutrition Reports International* 19, 371–376.

Apgar, J., Everett, G.A. and Fitzgerald, J.A. (1993) Dietary zinc deprivation effects parturition and outcome of pregnancy in the ewe. *Nutrition Research* 13, 319–330.

ARC (1980) *The Nutrient Requirements of Ruminant Livestock.* Commonwealth Agricultural Bureaux, Farnham Royal, UK, pp. 256–263.

Arora, S.P., Hatfield, E.E., Hinds, F.C. and Garrigus, U.S. (1973) Influence of dietary zinc on the activity of blood vitamin A alcohol dehydrogenase and carbonic anhydrase in lambs. *Indian Journal of Animal Sciences* 43, 140–144.

Baker, D.H. and Ammerman, C.B. (1995) Zinc bioavailability. In: Ammerman, C.B., Baker, D.H. and Lewis, A.J. (eds) *Bioavailability of Nutrients For Animals.* Academic Press, New York, pp. 367–398.

Beach, R.S., Gershwin, M.E. and Hurley, L.S. (1981) Nutritional factors and autoimmunity. I. Immunopathology of zinc deprivation in New Zealand mice. *Journal of Immunology* 126, 1999–2006.

Beeson, W.M., Perry, T.W. and Zurcker, T.D. (1977) Effect of supplemental zinc on growth and on hair and serum levels of beef cattle. *Journal of Animal Science* 45, 160–165.

Berg, J.M. (1990) Zinc fingers and other metal-binding domains: elements for interactions between molecules. *Journal of Biological Chemistry* 265, 6513–6518.

Bettger, W.J. (1989) The effect of dietary zinc deficiency on erythrocyte-free and membrane-bound amino acids. *Nutrition Research* 9, 911–919.

Bremner, I. (1991) A molecular approach to the study of copper and zinc metabolism. In: Momcilovic, B. (ed.) *Proceedings of the Seventh International Symposium on Trace Elements in Man and Animals.* IMI, Zagreb, pp. 1-1–1-3.

Bremner, I. (1993) Metallothionein in copper deficiency and copper toxicity. In: Anke, M., Meissner, D. and Mills, C.F. (eds) *Proceedings of the Eighth International Symposium on Trace Elements in Man and Animals.* Verlag Media Touristik, Gersdorf, pp. 507–515.

Brink, M.F., Becker, D.E., Terrill, S.W. and Jensen, A.H. (1959) Zinc toxicity in the weanling pig. *Journal of Animal Science* 18, 836–842.

Chesters, J.K. (1992) Trace element–gene interactions. *Nutrition Reviews* 50, 217–223.

Chesters, J.K. and Boyne, R. (1997) Interactions of mimosine and zinc deficiency on the transit of BHK cells through the cell cycle. In: Fischer, P.W.F., L'Abbe, M.R., Cockell, K.A. and Gibson, R.S. (eds) *Proceedings of the Ninth International Symposium on Trace Elements in Man and Animals (TEMA 9).* NRC Research Press, Ottawa, Canada, pp. 61–62.

Chhabra, A. and Arora, S.P. (1993) Effect of vitamin A and zinc supplement on alcohol dehydrogenase and superoxide dismutase activities of goat tissues. *Indian Journal of Animal Sciences* 63, 334–338.

Corrigall, W., Dalgarno, A.C., Ewen, L.A. and Williams, R.B. (1976) Modulation of plasma copper and zinc concentrations by disease status in ruminants. *Veterinary Record* 99, 396–397.

Cousins, R.B. (1996) Zinc. In: Filer, L.J. and Ziegler, E.E. (eds) *Present Knowledge in Nutrition,* 7th edn. International Life Science Institute–Nutrition Foundation, Washington, DC.

Cousins, R.B. (1997) Differential mRNA display, competitive polymerase chain reaction and transgenic approaches to investigate zinc-responsive genes in animals and man. In: Fischer, P.W.F., L'Abbe, M.R., Cockell, K.A. and Gibson, R.S. (eds) *Proceedings of the Ninth International Symposium on Trace Elements in Man and Animals (TEMA 9).* NRC Research Press, Ottawa, Canada, pp. 849–852.

Cunnane, S.C. (1988) Evidence that adverse effects of zinc deficiency on essential fatty acid composition in rats are independent of food intake. *British Journal of Nutrition* 59, 273–278.

Cunnane, S.C. and Yang, J. (1997) Disruption of the metabolism of polyunsaturated fatty acids (PUFA) during moderate zinc deficiency. In: Fischer, P.W.F., L'Abbe, M.R., Cockell, K.A. and Gibson, R.S. (eds) *Proceedings of the Ninth International Symposium on Trace Elements in Man and Animals.* NRC Research Press, Ottawa, pp. 604–608.

Davies, N.T. (1980) Studies on the absorption of zinc by rat intestine. *British Journal of Nutrition* 43, 189–203.

Davies, N.T., Carswell, A.J.P. and Mills, C.F. (1985) The effect of variation in dietary calcium intake on the phytate–zinc interaction in rats. In: Mills, C.F., Bremner, I. and Chesters, J.K. (eds) *Proceedings of the Fifth International Symposium on Trace Elements in Man and Animals.* Commonwealth Agricultural Bureaux, Farnham Royal, Slough, UK, pp. 456–457.

Demertzis, P.N. (1972) Posthitis and vulvitis in lambs with experimentally induced zinc deficiency. *Bulletin of the Hellenic Veterinary Medical Society* 3, 256–258.

Demertzis, P.N. and Mills, C.F. (1973) Oral zinc therapy in the control of infectious pododermatitis in young bulls. *Veterinary Record* 93, 219–222.

De Pasquale-Jardieu, P. and Fraker, P.J. (1979) The role of corticosterone in the loss in immune function in the zinc-deficient A/g mouse. *Journal of Nutrition* 109, 1847–1855.

Dewar, W.A. and Downie, J.N. (1984) The zinc requirements of broiler chicks and turkey poults fed on purified diets. *British Journal of Nutrition* 51, 467–477.

Dreosti, I.E. and Hurley, L.S. (1975) Depressed thymidine kinase activity in zinc-deficient rat embryos. *Proceedings of the Society for Experimental Biology and Medicine* 150, 161–165.

Droke, E.A. and Spears, J.W. (1993) *In vitro* and *in vivo* immunological measurements in growing lambs fed diets deficient, marginal or adequate in zinc. *Journal of Nutrition* 123, 71–90.

Droke, E.A., Spears, J.W., Armstrong, J.D., Kegley, E.B. and Simpson, R. (1993a) Dietary zinc affects serum concentrations of insulin and insulin-like growth factor I in lambs. *Journal of Nutrition* 123, 13–19.

Droke, E.A., Spears, J.W., Brown, T.T. and Quereshi, M.A. (1993b) Influence of dietary zinc and dexamethasone on immune responses and resistance to *Pasteurella haemolytica* challenge in growing lambs. *Nutrition Research* 13, 1213–1216.

Dufty, J.H., Bingley, J.B. and Cove, L.Y. (1977) The plasma zinc concentration of non-pregnant, pregnant and parturient Hereford cattle. *Australian Veterinary Journal* 53, 519–522.

Dynna, O. and Havre, G.N. (1963) Interrelationship of zinc and copper in the nutrition of cattle: a complex zinc–copper deficiency. *Acta Veterinaria Scandinavica* 4, 197–208.

Eckhert, C.D. and Hurley, L.S. (1977) Reduced DNA synthesis in zinc deficiency: regional differences in embryonic rats. *Journal of Nutrition* 107, 855–861.

Edwards, H.M., Jr, Dunahoo, W.S. and Fuller, H.L. (1959) Zinc requirement studies with practical rations. *Poultry Science* 38, 436–439.

Egan, A.R. (1972) Reproductive responses to supplemental zinc and manganese in grazing Dorset Horn ewes. *Australian Journal of Experimental Agriculture and Animal Husbandry* 12, 131–135.

Engle, T.E., Nockels, C.F., Kimberling, C.V., Weaber, D.L. and Johnson, A.B. (1997) Zinc repletion with organic and inorganic forms of zinc and protein turnover in marginally zinc-deficient calves. *Journal of Animal Science* 75, 3074–3081.

Englert, S.I., Jeffers, T.K., Sunde, M.L. and McGibbon, W.H. (1966) Differences among inbreds with respect to dietary zinc. *Poultry Science* 45, 1082–1083.

Everett, G.A. and Apgar, J. (1979) Effect of zinc status on salivary zinc concentrations in the rat. *Journal of Nutrition* 109, 406–411.

Flanagan, P.R. (1984) A model to produce pure zinc deficiency in rats and its use to demonstrate that dietary phytate increases the excretion of endogenous zinc. *Journal of Nutrition* 114, 493–502.

Franz, K.B., Kennedy, B.M. and Fellers, D.A. (1980) Relative bioavailability of zinc from selected cereals and legumes using rat growth. *Journal of Nutrition* 110, 2272–2283.

French, C.E., Smith, C.B., Fortmann, H.R., Pennington, R.P., Taylor, G.A., Hinish, W.W. and Swift, R.W. (1957) *Survey of Ten Nutrient Elements in Pennsylvania Forage Crops. 1. Red Clover.* Bulletin No. 624, Pennsylvania Agricultural Experiment Station.

Gladstones, J.S. and Loneragan, J.F. (1967) Mineral elements in temperate crop and pasture plants. 1. Zinc. *Australian Journal of Agricultural Research* 18, 427–466.

Goff, J.P. and Stabel, J.R. (1990) Decreased plasma retinol α-tocopherol and zinc concentration during the periparturient period: effect of milk fever. *Journal of Dairy Science* 73, 3195–3199.

Grace, N.D. (1972) Observations on plasma zinc levels in sheep grazing New Zealand pastures. *New Zealand Journal of Agricultural Research* 15, 284–288.

Grace, N.D. (1983) Amounts and distribution of mineral elements associated with fleece-free empty body weight gains in the grazing sheep. *New Zealand Journal of Agricultural Research* 26, 59–70.

Grace, N.D. and Sumner, R.M.W. (1986) Effect of pasture allowance, season and breed on the mineral content and rate of mineral uptake by wool. *New Zealand Journal of Agricultural Research* 29, 223–230.

Hansard, S.L., Mohammed, A.S. and Turner, J.W. (1968) Gestation age effects upon maternal–foetal zinc utilization in the bovine. *Journal of Animal Science* 27, 1097–1102.

Harland, B.F., Fox, M.R.S. and Fry, B.E., Jr (1974) Changes in plasma zinc related to fasting and dietary protein intake of Japanese quail. *Proceedings of the Society for Experimental Biology and Medicine* 145, 316–322.

Hartmans, J. (1965) The zinc supply of dairy cattle in the Netherlands. In: *Verslagen van Landbouwkundige Onderzoekingen*, No. 664. Pudoc, Wageningen, pp. 1–57.

Hidiroglou, M. and Spurr, D.T. (1975) Influence of cold exposure and diet change on the trace element composition of hair from Shorthorn cattle. *Canadian Journal of Animal Science* 55, 31–38.

Hill, D.A., Peo, E.R., Lewis, A.J. and Crenshaw, J.D. (1986) Zinc–amino acid complexes for swine. *Journal of Animal Science* 63, 121–130.

Hoefer, J.A., Miller, E.R., Ullrey, D.E., Ritchie, H.D. and Luecke, R.W. (1960) Interrelationships between calcium, zinc, iron and copper in swine feeding. *Journal of Animal Science* 19, 249–259.

Hoekstra, W.G., Faltin, E.C., Lin, C.W., Roberts, H.F. and Grummer, R.H. (1967) Zinc deficiency in reproducing gilts fed a diet high in calcium and its effect on tissue zinc and blood serum alkaline phosphatase. *Journal of Animal Science* 26, 1348–1357.

House, W.A. and Welch, R.M. (1989) Bioavailability of and interactions between zinc and selenium in rats fed wheat grain intrinsically labelled with ^{65}Zn and ^{75}Se. *Journal of Nutrition* 119, 916–921.

Hsu, F.S., Krook, L., Pond, W.G. and Duncan, J.R. (1975) Interactions of dietary calcium with toxic levels of lead and zinc in pigs. *Journal of Nutrition* 105, 112–118.

Hurley, L.S. (1981) Teratogenic effects of manganese, zinc and copper in nutrition. *Physiological Reviews* 61, 249–295.

Hurley, L.S. and Shrader, R.E. (1975) Abnormal development of preimplantation rat eggs after three days of maternal dietary zinc deficiency. *Nature, UK* 254, 427–429.

Hurley, L.S. and Swenerton, H. (1966) Congenital malformations resulting from zinc deficiency in rats. *Proceedings of the Society for Experimental Biology and Medicine* 123, 692–696.

Jackson, N., Gibson, S.W. and Stevenson, M.H. (1986) Effects of short and long-term feeding of zinc oxide supplemented diets on mature, female domestic fowl with special reference to tissue mineral content. *British Journal of Nutrition* 55, 333–349.

James, S.J., Swenseid, M. and Makinodan, T. (1987) Macrophage-mediated depression of T-cell proliferation in zinc-deficient mice. *Journal of Nutrition* 117, 1982–1988.

Jenkins, K.J. and Hidiroglou, M. (1991) Tolerance of the pre-ruminant calf to excess manganese and zinc in a milk replacer. *Journal of Dairy Science* 74, 1047–1053.

Jumba, I.O., Suttle, N.F., Hunter, E.A. and Wandiga, S.O. (1995) Effects of soil origin and mineral composition and herbage species on the mineral composition of forages in the Mount Elgon region of Kenya. 2. Trace elements. *Tropical Grasslands* 29, 47–52.

Keilin, D. and Mann, T. (1940) Carbonic anhydrase: purification and nature of the enzyme. *Biological Journal* 34, 1163–1176.

Kelly, E.J., Quaife, C.J., Froelick, G.J. and Palmiter, R.D. (1996) Metallothionein I and II protect against zinc deficiency and toxicity in mice. *Journal of Nutrition* 126, 1782–1790.

Kennedy, K.J., Rains, T.M. and Shay, N.F. (1998) Zinc deficiency changes preferred macronutrient intake in subpopulations of Sprague–Dewley outbred rats and reduces hepatic pyruvate kinase gene expression. *Journal of Nutrition* 128, 43–49.

Kienholz, E.W., Turk, D.E., Sunde, M.L. and Hoekstra, W.G. (1961) Effects of zinc deficiency in the diets of hens. *Journal of Nutrition* 75, 211–221.

Kincaid, R.L., Chew, B.P. and Cronrath, J.D. (1997) Zinc oxide and amino acids as sources of dietary zinc for calves: effects on uptake and immunity. *Journal of Dairy Science* 80, 1381–1388.

Kirchgessner, M., Schwarz, W.A. and Roth, H.P. (1975) Zur Aktivitat der alkalischen Phosphatase in Serum und Knochen von zinkdepletierten und -repletierten Kuhen. *Zeitschrift für Tierphysiologie Tierernahrung und Futtermittelkunde* 35, 191-200.

Kul'kova, J., Bremner, I., McGaw, B.A., Reid, M. and Beattie, J.H. (1993) Mercury–zinc interactions in marginal zinc deficiency. In: Anke, M., Meissner, D. and Mills, C.F. (eds) *Proceedings of the Eighth International Symposium on Trace Elements in Man and Animals.* Verslag Media Touristik, Gersdorf, Germany, pp. 635–637.

Legg, S.P. and Sears, L. (1960) Zinc sulphate treatment of parakeratosis in cattle. *Nature, UK* 186, 1061–1062.

Lei, X.G., Ku, P.K., Miller, E.R., Ullrey, D.E. and Yokoyama, M.T. (1993) Supplemental microbial phytase improved bioavailability of dietary zinc to weanling pigs. *Journal of Nutrition* 123, 1117–1123.

Lewis, P.K., Jr, Hoekstra, W.G. and Grummer, R.H. (1957) Restricted calcium feeding versus zinc supplementation for the control of parakeratosis in swine. *Journal of Animal Science* 16, 578–588.

Mahmoud, O.M., El Samani, F., Bakheit, R.O. and Hassan, M.A. (1983) Zinc deficiency in Sudanese desert sheep. *Journal of Comparative Pathology* 93, 591–595.

Martin, G.B. and White, C.L. (1992) Effects of dietary zinc deficiency on gonadotrophin secretion and testicular growth in young male sheep. *Journal of Reproduction and Fertility* 96, 497–507.

Masters, D.G. and Fels, H.E. (1980) Effect of zinc supplementation on the reproductive performance of grazing Merino ewes. *Biological Trace Element Research* 2, 281–290.

Masters, D.G. and Fels, H.E. (1985) Zinc supplements and reproduction in grazing ewes. *Biological Trace Element Research* 7, 89–93.

Masters, D.G. and Moir, R.J. (1980) Provision of zinc to sheep by means of an intra-ruminal pellet. *Australian Journal of Experimental Agriculture and Animal Husbandry* 20, 547–552.

Masters, D.G. and Moir, R.J. (1983) Effect of zinc deficiency on the pregnant ewe and developing foetus. *British Journal of Nutrition* 49, 365–372.

Masters, D.G. and Somers, M. (1980) Zinc status of grazing sheep: seasonal changes in zinc concentrations in plasma, wool and pasture. *Australian Journal of Experimental Agriculture and Animal Husbandry* 26, 20–24.

Mayland, H.F., Rosenau, R.C. and Florence, A.R. (1980) Grazing cow–calf responses to zinc supplementation. *Journal of Animal Science* 51, 966–974.

Miller, E.R., Luecke, R.W., Ullrey, D.E., Baltzer, B.V., Bradley, B.L. and Hoefer, J.A. (1968) Biochemical, skeletal and allometric changes due to zinc deficiency in the baby pig. *Journal of Nutrition* 95, 278–286.

Miller, E.R., Liptrap, D.O. and Ullrey, D.E. (1970) Sex influence on zinc requirement of swine. In: Mills, C.F. (ed.) *Trace Element Metabolism in Animals – 1.* Livingstone, Edinburgh, pp. 377–379.

Miller, J.K. and Miller, W.J. (1962) Experimental zinc deficiency and recovery of calves. *Journal of Nutrition* 76, 467–474.

Miller, W.J. (1970) Zinc nutrition of cattle: a review. *Journal of Dairy Science* 53, 1123–1135.

Miller, W.J., Pitts, W.J., Clifton, C.M. and Schmittle, S.C. (1964) Experimentally produced zinc deficiency in the goat. *Journal of Dairy Science* 47, 556–559.

Miller, W.J., Powell, G.W., Pitts, W.J. and Perkins, H.F. (1965a) Factors affecting zinc content of bovine hair. *Journal of Dairy Science* 48, 1091–1095.

Miller, W.J., Morton, J.D., Pitts, W.J. and Clifton, C.M. (1965b) The effect of zinc deficiency and restricted feeding on wound healing in the bovine. *Proceedings of the Society for Experimental Biology and Medicine* 118, 427–431.

Miller, W.J., Blackmon, D.M., Gentry, R.P., Powell, G.W. and Perkins, H.E. (1966) Influence of zinc deficiency on zinc and dry matter content of ruminant tissues and on excretion of zinc. *Journal of Dairy Science* 49, 1446–1453.

Miller, W.J., Blackmon, D.M., Gentry, R.P., Pitts, W.J. and Powell, G.W. (1967) Absorption, excretion and retention of orally administered zinc-65 in various tissues of zinc deficient and normal goats and calves. *Journal of Nutrition* 92, 71–78.

Miller, W.J., Martin, Y.G., Gentry, R.P. and Blackmon, D.M. (1968) ^{65}Zn and stable zinc absorption, excretion and tissue concentrations as affected by type of diet and level of zinc in normal calves. *Journal of Nutrition* 94, 391–401.

Mills, C.F. and Dalgarno, A.C. (1972) Copper and zinc status of ewes and lambs receiving increased dietary concentrations of cadmium. *Nature, UK* 239, 171–173.

Mills, C.F., Dalgarno, A.C., Williams, R.B. and Quarterman, J. (1967) Zinc deficiency and the zinc requirements of calves and lambs. *British Journal of Nutrition* 21, 751–768.

Minson, D.J. (1990) *Forages in Ruminant Nutrition.* Academic Press, New York, pp. 346–358.

Moeller, M.W. and Scott, H.M. (1958) Studies with purified diets. 3. Zinc requirement. *Poultry Science* 37, 1227–1228.

Morrison, J.N. and Bremner, I. (1987) Effect of maternal zinc supply on blood and tissue metallothionein I concentrations in suckling rats. *Journal of Nutrition* 117, 1588–1594.

Munday, R., Thompson, A.M., Fowke, E.A., Wesselink, C., Smith, B.L., Towers, N.R., O'Donnell, K., McDonald, R.M., Stirnemann, M. and Ford, A.J. (1997) A zinc-containing intraruminal device for faecal eczema control in lambs. *New Zealand Veterinary Journal* 45, 93–98.

Neathery, M.W., Miller, W.J., Blackmon, D.M., Pate, F.M. and Gentry, R.P. (1973a) Effects of long-term zinc deficiency on feed utilisation, reproductive characteristics and hair growth in the sexually mature male goat. *Journal of Dairy Science* 56, 98–105.

Neathery, M.W., Miller, W.J., Blackmon, D.M. and Gentry, R.P. (1973b) Performance and milk zinc from low zinc intake in dairy cows. *Journal of Dairy Science* 56, 212–217.

Norrdin, R.W., Krook, L., Bond, W.G. and Walker, E.F. (1973) Experimental zinc deficiency in weanling pigs on high and low calcium diets. *Cornell Veterinarian* 63, 264–290.

NRC (1977) *Nutrient Requirements of Domestic Animals. 1. Nutrient Requirements of Poultry*, 7th edn. National Academy of Sciences, Washington, DC, 62 pp.

NRC (1988) *Nutrient Requirements of Swine*, 9th edn. National Academy of Sciences, Washington, DC.

NRC (1994) *Nutrient Requirements of Domestic Animals. 1. Nutrient Requirements of Poultry*, 9th edn. National Academy of Sciences, Washington, DC, 62 pp.

O'Dell, B.L. and Reeves, P.G. (1989) Zinc status and food intake. In: *Zinc in Human Biology.* ILSI Press, Washington, DC, pp. 173–181.

O'Dell, B.L. and Savage, J.E. (1957) Symptoms of zinc deficiency in the chick. *Federation Proceedings* 16, 394.

O'Dell, B.L., Newberne, P.M. and Savage, J.E. (1958) Significance of dietary zinc for the growing chicken. *Journal of Nutrition* 65, 503–518.

O'Dell, B.L., Burpo, C.E. and Savage, J.E. (1972) Evaluation of zinc availability in foodstuffs of plant and animal origin. *Journal of Nutrition* 102, 653–660.

Oh, S.H., Nakane, H., Deagan, J.T., Whanger, P.D. and Arscott, G.H. (1979) Accumulation and depletion of zinc in chick tissue metallothioneins. *Journal of Nutrition* 109, 1720–1729.

O'Hara, P.J., Newman, A.P. and Jackson, R. (1960) Parakeratosis and copper poisoning in pigs fed a copper supplement. *Australian Veterinary Journal* 36, 225–229.

Orr, C.L., Hutcheson, D.P., Grainger, R.B., Cummins, J.M. and Mock, R.E. (1990) Serum copper, zinc, calcium and phosphorus concentrations of calves stressed by bovine respiratory disease and infectious bovine rhinotracheitis. *Journal of Animal Science* 68, 2893–2900.

Ott, E.A., Smith, W.H., Stob, M. and Beeson, W.M. (1964) Zinc deficiency syndrome in young lamb. *Journal of Nutrition* 82, 41–50.

Ott, E.A., Smith, W.H., Stob, M., Parker, H.E., Harrington, R.B. and Beeson, W.M. (1965) Zinc requirement of the growing lamb fed a purified diet. *Journal of Nutrition* 87, 459–463.

Ott, E.A., Smith, W.H., Harrington, R.B. and Beeson, W.M. (1966a) Zinc toxicity in ruminants. 1. Effect of high levels of dietary zinc on gains, feed consumption and feed efficiency of lambs. *Journal of Animal Science* 25, 414–418.

Ott, E.A., Smith, W.H., Harrington, R.B., Parker, H.E. and Beeson, W.M. (1966b) Zinc toxicity in ruminants. 4. Physiological changes in tissues of beef cattle. *Journal of Animal Science* 25, 432–438.

Pensack, J.M., Henson, J.N. and Bogdonoff, P.D. (1958) The effects of calcium and phosphorus on the zinc requirements of growing chickens. *Poultry Science* 37, 1232–1233.

Pitts, W.J., Miller, W.J., Fosgate, O.T., Morton, J.D. and Clifton, C.M. (1966) Effect of zinc deficiency and restricted feeding from two to five months of age on reproduction in Holstein bulls. *Journal of Dairy Science* 49, 995–1000.

Pond, W.G. (1983) The effect of dietary calcium and zinc levels on weight gain and blood and tissue mineral concentrations in growing Columbia- and Suffolk-sired lambs. *Journal of Animal Science* 56, 952–959.

Pond, W.G. and Jones, J.R. (1964) Effect of level of zinc in high calcium diets on pigs from weaning through one reproductive cycle and on subsequent growth of their offspring. *Journal of Animal Science* 23, 1057–1060.

Prasad, C.S. and Arora, S.P. (1979) Influence of dietary zinc on ß-carotene conversion and on the level of retinol-binding protein in the blood serum. *Indian Journal of Dairy Science* 32, 375.

Price, J. and Humphries, W.R. (1980) Investigation of the effect of supplementary zinc on growth rate of cattle in farms in N. Scotland. *Journal of Agricultural Science, Cambridge* 95, 135–139.

Price, N.O. and Hardison, W.A. (1963) *Minor Element Content of Forage Plants from the Central Piedmont Region of Virginia.* Bulletin No. 165, Virginia Agricultural Experiment Station.

Reeves, P.G. and O'Dell, B.L. (1986) The effects of dietary zinc deprivation on the activity of angiotensin-converting enzyme in serum of rats and guinea pigs. *Journal of Nutrition* 116, 128–134.

Reid, R.L., Jung, G.A., Stout, W.L. and Reaney, T.S. (1987) Effects of varying zinc concentrations on quality of alfalfa for lambs. *Journal of Animal Science* 64, 1725–1734.

Rimbach, G. and Pallauf, J. (1992) Effect of an addition of microbial phytase on zinc availability. *Zeitschrift für Ernuhrungwissenschaft* 31, 269–277.

Riordan, J.F. and Vallee, B.L. (1976) Structure and function of zinc metalloenzymes. In: Prasad, A.S. (ed.) *Trace Elements in Human Health and Disease*, Vol. 1. Academic Press, New York, pp. 227–256.

Roberson, K.D. and Edward, H.M. (1994) Effects of 1, 25-dihydroxycholecalciferol and phytase on zinc utilisation in broiler chicks. *Poultry Science* 73, 1312–1326.

Rubio, L.A., Grant, G., Dewey, P., Bremner, I. and Putzai, A. (1994) The intestinal true absorption of ^{65}Zn in rats is adversely affected by diets containing a Faba bean (*Vicia faba*) non-starch polysaccharide fraction. *Journal of Nutrition* 124, 2204–2211.

Sandoval, M., Henry, P.R., Ammerman, C.B., Miles, R.D. and Littell, R.C. (1997) Relative bioavailability of supplemental inorganic zinc sources for chicks. *Journal of Animal Science* 75, 3195–3205.

Saraswat, R.C. and Arora, S.P. (1972) Effect of dietary zinc on the vitamin A level and alkaline phosphatase activity in blood sera of lambs. *Indian Journal of Animal Sciences* 42, 358–362.

Schwarz, W.A. and Kirchgessner, M. (1975) Experimental zinc deficiency in lactating dairy cows. *Veterinary Medical Review* 1/2, 19–41.

Shi, H.N., Scott, M.E., Stevenson, M.M. and Koski, K.G. (1998) Energy restriction and zinc deficiency impair the function of murine T cells and antigen presenting cells during gastrointestinal nematode infection. *Journal of Nutrition* 128, 20–27.

Smith, B.L., Embling, P.P., Towers, N.R., Wright, D.E. and Payne, E. (1977) The protective effect of zinc sulphate in experimental sporidesmin poisoning of sheep. *New Zealand Veterinary Journal* 25, 121–127.

Smith, I.D., Grummer, R.H., Hoekstra, W.G. and Phillips, P.H. (1960) Effects of feeding an autoclaved diet on the development of parakeratosis in swine. *Journal of Animal Science* 19, 568–579.

Smith, J.C., Jr, McDaniel, E.G. and Chan, W. (1976) Alterations in vitamin A metabolism during zinc deficiency and food and growth restriction. *Journal of Nutrition* 106, 569–574.

Smith, J.W., Tokach, M.D., Goodband, R.D., Nelsson, J.L. and Richert, B.T. (1997) Effects of the interrelationship between zinc oxide and copper sulphate on growth performance of early weaned pigs. *Journal of Animal Science* 75, 1861–1866.

Somers, M. and Underwood, F.J. (1969) Studies of zinc nutrition in sheep. 2. The influence of zinc deficiency in ram lambs upon the digestibility of the dry matter and the utilization of the nitrogen and sulphur of the diet. *Australian Journal of Agricultural Research* 20, 899–903.

Spais, A.G. and Papasteriadis, A.A. (1974) Zinc deficiency in cattle under Greek conditions. In: Hoekstra, W.G., Suttie, J.W., Ganther, T.T.E. and Mertz, W. (eds) *Trace Element Metabolism in Animals – 2.* University Park Press, Baltimore, pp. 628–631.

Spears, J.W. (1993) Organic trace minerals in ruminant nutrition. *Animal Feed Science and Technology* 58, 151–163.

Stake, P.E., Miller, W.J., Gentry, R.P. and Neathery, N.W. (1975) Zinc metabolic adaptations in calves fed a high but non-toxic zinc level for varying time periods. *Journal of Animal Science* 40, 132–137.

Starcher, B. and Kratzer, F.H. (1963) Effect of zinc on bone alkaline phosphatase in turkey poults. *Journal of Nutrition* 79, 18–22.

Stevenson, J.W. and Earle, I.P. (1956) Studies on parakeratosis in swine. *Journal of Animal Science* 15, 1036–1045.

Sullivan, T.W. (1961) The availability of zinc in various compounds to broad-breasted bronze poults. *Poultry Science* 40, 340–346.

Sunde, M.L. (1972) Zinc requirement for normal feathering of commercial Leghorn-type pullets. *Poultry Science* 51, 1316–1322.

Sunde, M.L. (1978) Effectiveness of early zinc supplementation to chicks from five commercial egg strains. In: *Proceedings XVI World Poultry Congress, Rio de Janeiro, Brazil*, Vol. IV, p. 574.

Suttle, N.F. (1983) Assessment of the mineral and trace element status of feeds. In: Roberds, G.E. and Packham, R.G. (eds) *Proceeding of the Second Symposium of the International Network of Feed Information Centres.* Commonwealth Agricultural Bureaux, Farnham Royal, UK, pp. 211–237.

Suttle, N.F. and Mills, C.F. (1966) Studies on the toxicity of copper to pigs. 1. The effects of oral supplements of zinc and iron salts on the development of copper toxicosis. *British Journal of Nutrition* 20, 135–148.

Suttle, N.F., Lloyd-Davies, H. and Field, A.C. (1982) A model for zinc metabolism in sheep given a diet of hay. *British Journal of Nutrition* 47, 105–112.

Swinkels, J.W.G.M., Kornegay, E.T., Zhou, W., Lindemann, M.D., Webb, K.E. and Verstegen, M.W.A. (1996) Effectiveness of a zinc amino acid chelate and zinc sulphate in restoring serum and soft tissue zinc concentrations when fed to zinc-depleted pigs. *Journal of Animal Science* 74, 2420–2430.

Tidehag, P., Moberg, A., Sanzel, B., Hallmans, G., Sjostrom, R. and Wing, K. (1988) The availability of zinc, cadmium and iron from different grains measured as isotope absorption and mineral accumulation in rats. In: Hurley, L.S., Keen, C.L., Lonnerdal, B. and Rucker, R.B. (eds) *Trace Elements in Man and Animals 6*. Plenum Press, New York, p. 507.

Todd, W.R., Elvehjem, C.A. and Hart, E.B. (1934) Zinc in the nutrition of the rat. *American Journal of Physiology* 107, 146–156.

Towers, N.R., Smith, B.L., Wright, D.E. and Sinclair, D.P. (1975) Preventing facial eczema by using zinc. In: *Proceedings of the Ruakura Farmers' Conference*. Ruakura, New Zealand, pp. 57–61.

Tucker, H.F. and Salmon, W.D. (1955) Parakeratosis or zinc deficiency disease in the pig. *Proceedings of the Society for Experimental Biology and Medicine* 88, 613–616.

Underwood, E.J. (1962) A preliminary investigation of sources of zinc in Australian poultry diets. In: *Proceedings of the 12th World Poultry Congress, Sydney, Australia*, pp. 216–218.

Underwood, E.J. and Somers, M. (1969) Studies of zinc nutrition in sheep. 1. The relation of zinc to growth, testicular development and spermatogenesis in young rams. *Australian Journal of Agricultural Research* 20, 889–897.

Vallee, B.L. and Falchuk, K.H. (1993) The biochemical basis of zinc physiology. *Physiological Reviews* 73, 79–118.

Vohra, P. and Kratzer, F.H. (1964) Influence of various chelating agents on the availability of zinc. *Journal of Nutrition* 82, 249–255.

Watkins, K.L. and Southern, L.L. (1993) Effect of dietary sodium zeolite A on zinc utilisation by chicks. *Poultry Science* 72, 296–305.

Wedekind, K.J. and Baker, D.H. (1990) Zinc bioavailability of feed grade sources of zinc. *Journal of Animal Science* 68, 684–689.

Wedekind, K.J., Hortin, A.E. and Baker, D.H. (1992) Methodology for assessing zinc bioavailability: efficacy estimates for zinc-methionine, zinc sulphate and zinc oxide. *Journal of Animal Science* 70, 178–187.

Wedekind, K.J., Collings, G., Hancock, J. and Titgemeyer, E. (1994a) The bioavailability of zinc-methionine relative to zinc sulphate is affected by calcium level. *Poultry Science* 73 (suppl. 1), 114.

Wedekind, K.J., Lewis, A.J., Giesemann, M.A. and Miller, P.S. (1994b) Bioavailability of zinc from inorganic and organic sources for pigs fed corn–soybean meal diets. *Journal of Animal Science* 72, 2681–2689.

Wegner, T.N., Ray, D.E., Lox, C.D. and Stott, G.H. (1973) Effect of stress on serum zinc and plasma corticoids in dairy cattle. *Journal of Dairy Science* 56, 748–752.

Weigand, E. and Kirchgessner, M. (1978) Homeostatic adjustments in zinc digestion to widely varying zinc intake. *Nutrition and Metabolism* 22, 101–112.

Weigand, E. and Kirchgessner, M. (1982) Factorial estimation of the zinc requirement of lactating dairy cows. *Zeitschrift für Tierphysiologie, Tierernahrung und Futtermittelkunde* 49, 1–9.

Welch, R.M. (1997) Trace element interactions in food crops. In: Fischer, P.W.F., L'Abbe, M.R., Cockell, K.A. and Gibson, R.S. (eds) *Trace Elements in Man and*

Animals – 9. Proceedings of the Ninth International Symposium. NRC Research Press, Ottawa, Canada, pp. 6–9.

White, C.L. (1993) The zinc requirements of grazing ruminants. In: Robson, A.D. (ed.) *Zinc in Soils and Plants: Developments in Plant and Soil Sciences,* Vol. 55. Kluwer Academic Publishers, London, pp. 197–206.

White, C.L., Chandler, B.S. and Peter, D.W. (1991) Zinc supplementation of lactating ewes and weaned lambs grazing improved mediterranean pastures. *Australian Journal of Experimental Agriculture* 31, 183–189.

White, C.L., Martin, G.B., Hynd, P.T. and Chapman, R.E. (1994) The effect of zinc deficiency on wool growth and skin and wool histology of male Merino lambs. *British Journal of Nutrition* 71, 425–435.

Williams, R.B., McDonald, I. and Bremner, I. (1978) The accretion of copper and of zinc by the foetuses of prolific ewes. *British Journal of Nutrition* 40, 377–386.

Wing, K., Wing, A., Sjostrom, R. and Lonnerdal, B. (1997) Efficacy of a Michaelis–Menton model for the availabilities of zinc, iron and cadmium from an infant formula diet containing phytate. In: Fischer, P.W.F., L'Abbe, M.R., Cockell, K.A. and Gibson, R.S. (eds) *Proceedings of the Ninth International Symposium on Trace Elements in Man and Animals (TEMA 9).* NRC Research Press, Ottawa, Canada, pp. 31–32.

Xu, C., Wensing, T. and Beynen, A.C. (1997) The effect of dietary soybean versus skim milk protein on plasma and hepatic concentrations of zinc in veal calves. *Journal of Dairy Science* 80, 2156–2161.

Zhang, P., Duhamel, G.E., Mysore, J.V., Carlson, M.P. and Schneider, N.R. (1995) Prophylactic effect of dietary zinc in a laboratory mouse model of swine dysentry. *American Journal of Veterinary Research* 56, 334–339.

Zeigler, T.R., Leach, R.M., Jr, Norris, L.C. and Scott, M.L. (1961) Zinc requirement of the chick: factors affecting requirement. *Poultry Science* 40, 1584–1593.

Occasionally Beneficial Elements (Boron, Chromium, Lithium, Molybdenum, Nickel, Silicon, Tin, Vanadium) 17

The 14 elements discussed so far are undeniably essential to the health and well-being of farm livestock. It has not been difficult to demonstrate specific physiological functions for each element; severe disabilities arise when dietary concentrations are low and active transport mechanisms are then invoked to ameliorate the deficiency. It is argued that many other elements, including boron (B), chromium (Cr), lithium (Li), molybdenum (Mo), nickel (Ni), silicon (Si), tin (Sn) and vanadium (V) and others renowned for their toxicity (e.g. arsenic (As), cadmium (Cd) and lead (Pb): see Chapter 18) are also essential, albeit in 'ultratrace' concentrations (< 1 mg kg^{-1}: Nielsen, 1996). The customary evidence for 'essentiality' is threefold:

1. Animals are consistently less healthy when deprived of the element.
2. They show physiological responses to minute supplements of that element.
3. They have homeostatic mechanisms for modulating retention of the element.

It is further argued that, since each element is ubiquitous and reactive and often has specific functions in lower animals, it is only a matter of time before more will have their essentiality to animals confirmed (Nielson, 1996).

However, there are a number of important questions that must be answered before essentiality is unequivocally demonstrated:

1. Was the 'deficient' diet adequate with respect to all other essential elements?
2. Were the physiological responses life-enhancing?
3. Were the 'homeostatic' mechanisms any more than homeorhectic mechanisms providing defence against excess?

The first of these is particularly important. The clinical consequences of deprivation of the 'occasionally beneficial' and 'essentially toxic' elements are rarely specific (e.g. poor growth; low viability of newborn) and the same

© CAB *International* 1999. *Mineral Nutrition of Livestock*
(E.J. Underwood and N.F. Suttle)

abnormalities have progressively been attributed to each 'new' ultratrace element, which was usually neither measured nor added to the diet in earlier experimental demonstrations of 'essentiality'. Many of the biochemical consequences of deprivation are also shared (e.g. Li and Ni, impaired glycolysis). All the questions are important to farmers, who are sometimes tempted to purchase novel supplements because they contain rare elements from the far reaches of the periodic table, but do their stock need them on their 'table'? For the most part, the occasionally beneficial elements are so abundant in the farm environment that natural deficiencies are unlikely to arise.

BORON

Boron is an essential nutrient for plants but the evidence for essentiality to animals is equivocal.

Sources of Boron

Boron concentrations in soils in England and Wales showed relatively little variation about a high median value of 34 mg kg^{-1} dry matter (DM), 50% of the values falling between 20 and 40 (Archer and Hodgson, 1987), but little of that boron (*c.* 1 mg kg^{-1}) was extractable in hot water, the common test of availability to plants. Concentrations in plants are often lower than those in soils and are particularly low in cereal grains: higher levels are found during early cereal growth, concentrations falling from 7 to 2 mg B kg^{-1} DM in wheat plants between May and August (Silanpaa, 1982). Boron values tend to be higher in legumes than in grasses (Table 17.1) and Leach (1983) reported values from 35 to 66 mg B kg^{-1} DM in lucerne (*M. sativa*) grown under subtropical conditions and a variety of grazing treatments. However, consistently high levels of 38–40 mg kg^{-1} DM have been found in grass hays from Nevada (Green and Weeth, 1977; Weeth *et al.*, 1981). Root crops are relatively rich in boron, turnips, swedes and sugar beet contain roughly ten times more boron (13.5–20.8) than four cereal grains (1.4–1.7 mg kg^{-1} DM) grown at the same site in Finland (Fig. 17.1; Silanpaa, 1982). A diet of maize silage and soybean meal (90:10) for sheep contained 32 mg B kg^{-1} DM (Brown *et al.*, 1989), indicating high levels in at least one constituent, while two maize/ soybean rations for poultry contained 9.4 and 15.6 mg B kg^{-1} DM (Rossi *et al.*, 1993). Boron concentrations in bovine milk are reported to vary from 0.09–0.25 mg l^{-1} but increase greatly when boron intakes are increased (Kirchgessner *et al.*, 1967).

Metabolism of Boron

There is little evidence of homeostatic control of boron metabolism at high intakes as with all elements encountered in the anionic state (see comments

Table 17.1. Concentrations (mg kg^{-1} DM) of 'occasionally beneficial elements' commonly found in forages and cereals: ranges of means from surveys in different countries (from Kabata-Pendias and Pendias, 1992). Rb, rubidium.

Element	Grasses	Legumes	Cereals
B	4.9–7.4	14–78	0.7–7.3
Cr	0.1–0.35	0.2–4.2	0.01–0.55
Li	0.07–1.5	0.01–3.1	0.05
Mo	0.33–1.4	0.5–2.5	0.16–0.92
Ni	0.13–1.1	1.2–2.7	0.22–0.34
Rb	130	44–98	3–4
V	0.1–0.23	0.18–0.24	0.007–0.060

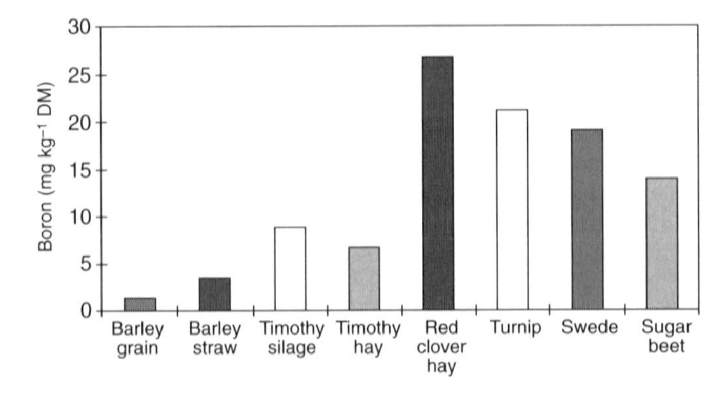

Fig. 17.1. Crops and forage species grown at the same location in Finland show considerable variation in boron concentration (data from Silanpaa, 1982).

on Cr, I, Mo, P and V). Plasma and urine boron concentrations are almost linearly related to boron intake over the range 0.1–4.5 g B 100 kg^{-1} live weight (LW) (Weeth *et al.*, 1981) and up to 69% is excreted in bovine urine (Green and Weeth, 1977). The plasma boron level associated with feeding only a boron-rich hay (40 mg B kg^{-1} DM) was 2.4 mg l^{-1}, far higher than that associated with poor conception (Small *et al.*, 1997). In a balance study with sheep, the faeces was the major route of boron excretion from the basal diet, but urinary excretion became the more important when borate (45 and 175 mg $head^{-1}$ day^{-1}) was provided (Brown *et al.*, 1989). In broilers, boron concentrations in muscle and liver increase proportionately as dietary boron concentrations increase (Fig. 17.2; Rossi *et al.*, 1993).

Essentiality of Boron

Beneficial responses from giving boron to animals were first noted in vitamin D-deficient chicks in 1981 and have recently been confirmed and extended (Hunt *et al.*, 1994). However, the basal diet was exceedingly low in boron

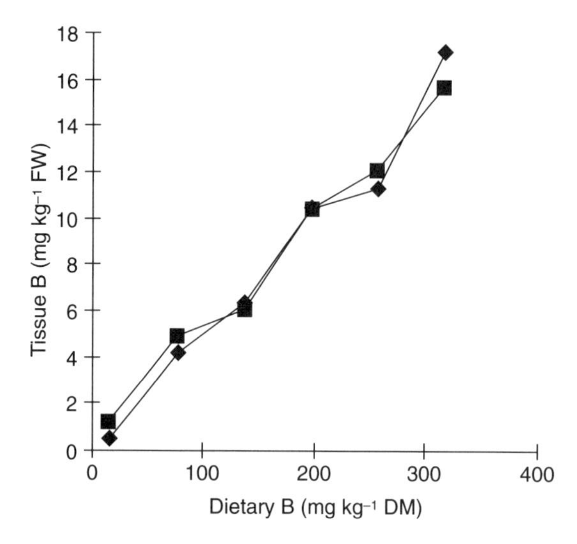

Fig. 17.2. Boron concentrations in the liver (diamond) and muscle (square) of broilers increase linearly with increases in dietary B level (data from Rossi *et al.*, 1993).

(< 0.18 mg B kg^{-1} DM) and the beneficial concentration so low (1.4 mg B kg^{-1} DM) that it would only be found in boron-deficient plants. Boron supplementation of the diet of men and women (particularly those receiving hormone supplements) has increased values for markers of copper, vitamin D and haematological status and has decreased bone turnover (Nielsen, 1994). However, boron deficiencies in farm livestock have yet to be reported. An association between poor conception and 'low' serum boron concentrations of 0.1–0.13 mg (9.2–13.6 μmol) l^{-1} has been reported recently within a beef cow herd (Small *et al.*, 1997), but association is not proof of essentiality. If deficiencies occur, they are most likely to do so in livestock fed grain-based diets. Thus, Rossi *et al.* (1993) found that the supplementation of practical broiler rations with 5 or 60 mg B kg^{-1} DM increased the growth of males and females by 3 to 10% in two experiments although the response was only significant in the males in one instance: that increase was accompanied by significant improvements in tibia weight and strength.

Toxicity of Boron

Boron is a well-tolerated element, more than 300 mg B kg^{-1} DM being required to produce the first signs of toxicity in poultry, a reduction in hatchability of hens' eggs (Puls, 1994). Loss of appetite and body weight occurred in heifers given 150–300 mg B l^{-1} in their drinking-water (Green and Weeth, 1977), but the treatments were equivalent to > 800 mg B kg^{-1} DM in the diet and drinking-water from natural sources is unlikely to contain such high levels. Plasma concentrations > 4 mg B l^{-1} indicate abnormally high boron intakes. Toxicity has occurred following the ingestion of borate fertilizer by cattle and is indicated by widespread increases in tissue levels from < 10 to > 100 mg B kg^{-1} DM (Puls, 1994).

CHROMIUM

Sources of Chromium

Chromium is far more abundant in soils than in crops. For soils, Puls (1994) gives ranges based on Canadian experience of 1–25 mg kg^{-1} DM, but data from England suggest a greater abundance (normal range 10–121 mg Cr kg^{-1} DM: Archer and Hodgson, 1987). In contrast, reported values for feedstuffs range from 0.01 to 4.2 mg Cr kg^{-1} DM, with cereals relatively poor and legumes relatively rich in chromium (Table 17.1). Significant amounts of soil-borne chromium will be ingested by grazing animals. The chromium concentrations reported in foods have decreased enormously over the last 30 years due to improvements in analytical techniques and evidence of contamination from stainless steel. Values of 40–60 μg Cr kg^{-1} DM have recently been reported for single batches of lupins, barley and hay fed to sheep in Australia (Gardner *et al.*, 1998), and 2.6 μg kg^{-1} DM in a maize/lucerne diet (Olsen *et al.*, 1996). However, typical rations for laying hens can contain > 500 μg Cr kg^{-1} DM (Lien *et al.*, 1996). The forms in which chromium is present in the diet may be more important than the amount presented, but they have yet to be fully and critically examined. Brewer's yeast contains an ethanol-extractable, low-molecular-mass complex (400 Da), while lucerne contains a much larger complex (2600 Da), which is insoluble in non-polar solvents (Starich and Blincoe, 1983). Wheat has also been shown to contain chromium in complex molecules.

Metabolism of Chromium

Little information is available on the metabolism of chromium in livestock other than the fact that organic sources can be absorbed some 20–30 times more efficiently than inorganic sources (Starich and Blincoe, 1983). Indeed, attention has mostly been focused on the inertness of ingested chromium. Chromium compounds, such as chromium sesquioxide (Cr_2O_3), have a long history of use as inert markers; when given orally in regular known doses, the chromium is almost totally excreted in the faeces and faecal analysis for chromium can provide information on food intake, digestibility of organic matter and the total faecal excretion of other minerals and nutrients (e.g. in sows: Saha and Galbraith, 1993). Chromium has an affinity for fibrous particles and this has been exploited by impregnating fibrous feeds with chromium (a process known as 'mordanting') when studying the digestion of fibre by ruminants (Uden *et al.*, 1980; Coleman *et al.*, 1984). Information on sites of absorption and rates of passage of digesta can be generated by using chromium in suitably cannulated or fistulated animals. Parker *et al.* (1990) reviewed the most recent development for use in grazing animals, the continuous provision of Cr_2O_3 from an intraruminal capsule. Radioisotopes of chromium in a soluble, chelated form (^{75}Cr ethylenediaminetetra-acetic acid ($^{75}CrEDTA$)) have also been used as inert markers (Downes and McDonald, 1964), showing that chelation of a mineral does not necessarily enhance its absorbability or availability. In the milk-fed ruminant, significant absorption and urinary excretion of CrEDTA may occur, indicating a temporary enhancement of gut permeability to the molecule (N.F. Suttle, unpublished data). One implication of the uses of chromium as an inert marker is that inorganic sources of chromium in the diet of ruminants will be very poorly absorbed, most solubilized chromium becoming irreversibly bound to indigestible material. Nevertheless, short-term ingestion of soil-borne chromium by sheep can raise concentrations in the kidney from 0.25 to 0.90 mg kg^{-1} DM (Suttle *et al.*, 1991). Since organic sources of chromium may be distinctively and usefully metabolized, the assessment of chromium status of both diet and livestock presents substantial difficulties. Increasing the dietary chromium concentration from 2.6 to 62.5 μg kg^{-1} DM with chromium picolinate (CrP) increased liver chromium by 50% but did not affect levels in the muscle of lambs (Olsen *et al.*, 1996). The adequacy of total chromium concentrations in blood or tissues as measures of sufficiency in terms of glucose tolerance and other potentially advantageous responses in livestock are only just beginning to receive attention (Subiyatno *et al.*, 1996).

Essentiality of Chromium

Essentiality for mammals was first indicated by Schwarz and Mertz (1959), who showed that supplements of trivalent chromium improved glucose tolerance in rats. Organic chromium in brewer's yeast has been termed 'glucose tolerance factor (GTF)', because it can enhance the sensitivity of cells to insulin (Mertz, 1974), and similar forms have been reported in wheat (Toepfer *et al.*, 1973). The chemical structure of GTF has yet to be defined, because purification causes loss of potency (Offenbacher *et al.*, 1997), but the molecule probably contains nicotinic acid. However, in a recent study (Kegley *et al.*, 1997a), chromium given to calves as chromium chloride ($CrCl_3$) or a chromium–nicotinic acid complex (CrNA) produced similar changes in insulin sensitivity in calves. Protein, nucleic acid and lipid metabolism may each be sensitive to chromium supply (Anderson, 1987), but this could reflect the manifold effects of insulin rather than further biochemical roles for chromium. The dietary requirements of livestock for chromium have not been defined but appear to be increased by stress. Strenuous exercise, transportation and infection may increase the dietary requirement by increasing the losses of chromium in urine (Anderson, 1987) and presumably by increasing glucose turnover. Exercise in sheep, sufficient to increase energy demand by 9% over a 10-week period, halved chromium concentrations in the liver (Gardner *et al.*, 1998). Two further factors make it hard to define the dietary circumstances in which chromium supplementation might benefit farm livestock. Firstly, concentrations of chromium in basal diets have rarely been reported, due to a preoccupation with responses to 'organic' chromium, usually given as the picolinate (CrP). Secondly, few direct comparisons have been made to establish the benefit of 'organic' or 'chelated' sources over inorganic supplements. Chromium is found in GTF-like forms in the livers of mammals given inorganic chromium (Yamamoto *et al.*, 1988), implying natural synthesis of insulin-potentiating forms of chromium. If this does occur, chromium may not need to be added in organic forms to all diets. Some of the metabolic responses of animals to chromium have been likened to those triggered by somatotrophin (Yang *et al.*, 1996), which raises the possibility that they are pharmacological rather than physiological.

Morbidity in ruminants

The most consistent benefits from chromium supplementation are found in recently transported feedlot calves. Appetite is often low and mortality and morbidity high in the first weeks after transportation. Infections with *Pasteurella haemolytica*, bovine viral diarrhoea, parainfluenza (PI_3), infectious rhinotracheitis (IBR) and respiratory syncytial viruses are common and calves usually receive multiple vaccines to protect them from these diseases. Addition of organic trivalent chromium to the diet has generally improved appetite, early weight gain and humoral immune function in recently transported, vaccinated calves (Chang and Mowat, 1992; Moonsie-Shageer and Mowat, 1993; Mowat *et al.*, 1993). In a 2 × 2 factorial study, in which an amino acid-chelated chromium source (adding 0.14 mg Cr to a diet containing 0.32 mg Cr kg^{-1} DM) and multiple vaccine were the treatments, significant improvements were mostly confined to the combined treatment with chromium and vaccine (Wright *et al.*, 1994). In another experiment with newly weaned calves, addition of 0.5 mg organic chromium kg^{-1} DM significantly improved antibody titres after vaccination against IBR (Burton *et al.*, 1994), but a similar study failed to confirm this finding (Kegley *et al.*, 1997b). It is interesting that multiple vaccines, particularly those directed against Gram-negative bacteria (e.g. *P. haemolytica*), are themselves 'stressful', in that they raise plasma cortisol and lower blood glucose in recently transported lambs (Suttle and Wadsworth, 1991). Thus, vaccination may add to the stress of transportation, enhancing the responses to the cortisol-limiting, insulin-facilitating chromium supplement. Similar factors may underlie the improved blastogenic (*in vitro*) responses of mononuclear cells from periparturient dairy cows given chromium (Burton *et al.*, 1993); humoral (*in vivo*) immune responses were not improved. The most remarkable response to be claimed for CrP to date (Villalobos *et al.*, 1997) is a reduction in the incidence of placental retention in a dairy herd in Mexico from an exceptional 56% to 16% ($P < 0.01$; $n = 25$). The herd history of this disorder had been linked to the stress of transportation of cows to delivery facilities, but the persistence of the problem in unsupplemented controls that were settled into facilities 9 weeks before expected parturition suggests other causes.

Insulin sensitivity in ruminants

Ruminants use acetate rather than glucose as a carbon source for lipogenesis and this may lead to unique insulin sensitivity (Gardner *et al.*, 1998). Supplementation of the diet of adult sheep with 1 mg Cr (as amino acid chelate) kg^{-1} DM increased by 30% the potential for glucose to be used for fat synthesis by increasing the activity of ATP-citrate lyase – a marker of the glucose–insulin axis – while decreasing fat depth over the twelfth rib by 20%. In another recent study of chromium responses during early lactation in dairy cows, addition of 5 mg Cr (as amino acid chelate) kg^{-1} DM to natural diets containing 0.8–1.6 mg kg^{-1} DM improved the milk yield of primiparous cows by 7 and 13% in two experiments (Yang *et al.*, 1996). Multiparous cows did not respond to chromium supplementation and it was suggested that nutritional, physiological and psychological stressors associated with the first lactation increased chromium requirements and induced a temporary state of deficiency. The increased milk yield was attributed to increased gluconeogenesis and a significant increase in blood glucose was found at one point (week 6 of lactation). Lactation curves indicated that the primary benefit occurred during the first 2 months after parturition. Further studies (Subiyatno *et al.*, 1996) suggested that gluconeogenesis or glycogenolysis was altered by supplementation with organic chromium, but there were complex effects of parity and physiological status (pregnant or lactating) on both the size and direction of hormonal responses. In 'unstressed' calves, a supplement of 0.37 mg Cr as CrP kg^{-1} DM lowered blood cholesterol and improved glucose tolerance but without improving growth, food consumption or feed conversion efficiency (Bunting *et al.*, 1994). Others have found that CrNA 'intensified the response to insulin' in young calves given a milk-replacer containing 0.31 mg Cr kg^{-1} DM, but a similar supplement of $CrCl_3$ (both sources provided a further 0.4 mg Cr kg^{-1} DM) was equally effective (Fig. 17.2), and neither source had improved growth after 63 days (Fig. 17.3; Kegley *et al.*, 1997b). In a study with CrNA alone, addition of 0.4 mg Cr kg^{-1} DM tended to improve growth in steers (10% improvement; $P < 0.1$) (Kegley *et al.*, 1997a). The circumstances in which cattle benefit from chromium supplementation and the size of that benefit may therefore be limited. Small supplements of CrP (providing 0.06 and 0.125 mg Cr kg^{-1} DM) to a maize/lucerne diet for lambs, low in Cr, lowered serum cholesterol but did not affect growth or carcass composition (Olsen *et al.*, 1996).

Pigs

Suggestions that 'conventional diets for pigs are moderately deficient in chromium and – that for maximum growth must be provided with a bioavailable source of chromium' (Page *et al.*, 1993) have not been confirmed (Ward

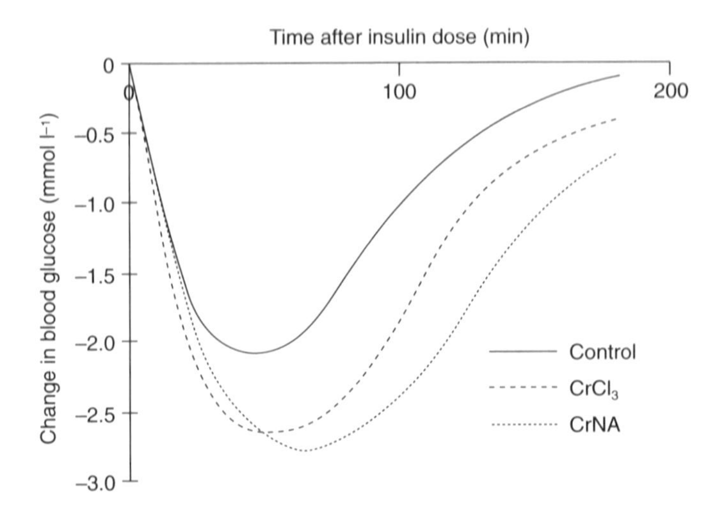

Fig. 17.3. Continuous dietary supplements of chromic chloride ($CrCl_3$) and chromium-nicotinic acid complex (CrNA) both accentuate the hypoglycaemic response of pre-ruminant calves to intravenous insulin (Kegley *et al.*, 1997b).

et al., 1997) and are hard to support even from the protagonists' data. Thus, in three experiments, 0.1 mg Cr as CrP kg^{-1} DM decreased growth in one trial and improved it in another, while in the third it contributed to a linear *decrease* in growth over the range 0.1–0.8 mg Cr kg^{-1}; 0.25 mg Cr kg^{-1} was marginally more consistent and beneficial, increasing growth in two experiments, but inorganic chromium was equally stimulatory (Page *et al.*, 1993). Cholesterol-lowering responses to CrP were consistent, as was the tendency for pigs to grow more latissimus dorsi muscle and deposit less rib fat when given organic chromium. Whether such repartitioning effects are seen on more complex and fibrous diets than a maize (73–80%) and soybean-meal (25–18%) mixture has still to be shown. Not all measures of muscling or fatness in pigs on maize–soybean diets are improved by CrP (Boleman *et al.*, 1995). The importance of level of chromium supplementation was confirmed in a recent study, in which addition of 0.2 mg Cr as CrP kg^{-1} DM to pig rations containing 0.2 mg Cr kg^{-1} DM marginally improved performance and carcass composition, but 0.4 mg CrP did not (Mooney and Cromwell, 1997). Comparisons with $CrCl_3$ were made, but at far higher levels (5 and 25 mg Cr kg^{-1} DM); had any differences between sources been found (in fact there were none), they would have been uninterpretable. There is a strange tendency for experiments with chelates in general to be poorly designed and to rely unduly on levels of probability which are *not* significant ($P > 0.05$). Chromium supplementation, whether as $CrCl_3$, CrP or CrNA, providing 0.2 mg Cr kg^{-1} DM, did not improve immune responses in pigs given a basal diet high in chromium (5.2 mg kg^{-1} DM) (Van Heugten and Spears, 1997).

Poultry

Live-weight gain of newly hatched turkey poults given a maize–soybean-meal diet low in crude protein (230 g kg^{-1}) was improved by 10% with the addition of 20 mg inorganic Cr (as $CrCl_3$) kg^{-1} DM and accompanied by a 60% increase in hepatic lipogenesis (Steele and Roseburgh, 1981). Fermentation by-products of Cr^{3+} (supplying 5 mg Cr kg^{-1} DM) have improved egg quality in hens and protected the egg interior from the harmful effects of vanadium (Jensen *et al.*, 1978). Addition of CrP to a maize/soybean-meal ration for laying hens to provide 0.2, 0.4 or 0.8 mg Cr kg^{-1} DM lowered cholesterol in the hens serum and egg yolk in a dose-dependent manner but the two higher levels reduced eggshell strength by 32% (Lien *et al.*, 1996).

Toxicity of Chromium

Bioreduction of Cr(VI) to the less toxic Cr(III) state is effected by many organisms (Starich and Blincoe, 1983) and chromium toxicity is rare, because even soluble sources, such as the chloride and chromate, are tolerated at concentrations of > 1000 mg kg^{-1} DM by farm livestock. Tissue concentrations are normally low (< 0.1 mg Cr kg^{-1} fresh weight (FW): Puls, 1994) and readily distinguished from the levels associated with exposure to excess chromium (> 10 mg Cr kg^{-1}).

LITHIUM

Lithium is now regarded as an essential dietary constituent for some species, but deficiencies are unlikely to occur naturally. Lithium ranks 27th in abundance among elements in the earth's crust and a mean level of 28 mg Li kg^{-1} DM has been reported for soils in eastern Europe, but values vary widely, according to soil type. Concentrations are much lower in crops, particularly cereal grains, while leguminous forages and field beans are richer sources (Table 17.1).

Metabolism of Lithium

The metabolism of lithium has not been widely studied, but, like the more abundant alkali metals, such as sodium and potassium, its high solubility and low molecular mass are likely to be associated with ready absorption, rapid turnover, extensive urinary excretion and minimal storage. Plasma concentrations are normally low (< 0.02 mg l^{-1}) but are linearly related to intake, a fact that has been exploited through the use of lithium salts as a marker for intake of supplementary foods at pasture (Vipond *et al.*, 1995); with dietary lithium concentrations increased to 162 mg kg^{-1}, plasma concentrations rose to 3–6 mg Li l^{-1}. Natural sources of lithium supported lower tissue lithium concentrations in rats than an inorganic source (Li_2CO_3) and were thought to be of lower availability (Patt *et al.*, 1978).

Essentiality of Lithium

Studies by Anke's group in Jena have repeatedly shown that goats deprived of lithium become unhealthy; they exhibit growth retardation, impaired fertility, low birth weights and reduced longevity on a diet containing 3.3 μg Li kg^{-1} DM, when compared with supplemented goats given 24 μg Li kg^{-1} DM (for summary, see Mertz, 1986). Reduced fertility has been confirmed in second- and third-generation lithium-deprived rats, and evidence of the conservation of relatively high tissue concentrations in endocrine organs in the face of depletion (57 μg Li kg^{-1} FW in adrenal, 140 μg Li kg^{-1} FW in pituitary) may indicate an endocrine function for lithium (Patt *et al.*, 1978). The low-lithium diet contained 5–15 μg Li kg^{-1} DM and consisted principally of ground yellow maize (67%). Any lithium-dependent pathway would have to be protected from competition from potassium and sodium, which are present in vastly greater concentrations in the tissues.

Toxicity of Lithium

Farm livestock are rarely exposed to excess lithium, but occasional reports of poisoning following exposure to industrial products (e.g. lithium greases) have been reported. The symptoms in mature beef cattle were severe depression, diarrhoea, ataxia and death and were reproduced by acute oral exposure to 500–700 mg Li (as lithium chloride (LiCl)) kg^{-1} LW (Johnson *et al.*, 1980). Serum concentrations rose to 40–60 mg Li l^{-1}, while soft-tissue concentrations rose from 5–20 to 50–120 mg kg^{-1} FW, with little difference between the sites examined (muscle, brain, heart, liver and kidney). Lower doses of lithium (250 mg Li kg^{-1} LW) caused a mild diarrhoea and are sufficiently unpleasant to elicit aversive behaviour in cattle (Ralphs and Chaney, 1993) and sheep (Thorhallsdottir *et al.*, 1990). Aversion to lithium salts has been exploited by training livestock to associate their aversion to lithium with concurrent intake of toxic plants (e.g. tall larkspur), which they subsequently avoid when given a choice under range conditions (Lane *et al.*, 1989). Livestock clearly recover well after brief exposure to sublethal doses of lithium, but their aversive behaviour is quickly overridden by social influences, such as the presence of untrained cohorts. In humans, lithium toxicity has arisen from the use of lithium as an antidepressant by subjects on salt-restricted diets (for citations, see Johnson *et al.*, 1980). It is therefore likely that both the toxic and the deficient dietary levels of lithium for livestock are influenced by sodium and potassium intake.

MOLYBDENUM

Early nutritional interest in molybdenum was centred upon its profound effects on the copper metabolism of ruminants (Underwood, 1977), but these are discussed fully in Chapter 11. The first indication of an essential role for molybdenum in animals came from the discovery that the flavoprotein enzyme, xanthine oxidase, contains molybdenum and that its activity depends on the presence of this metal (de Renzo *et al.*, 1953; Richert and Westerfield, 1953). Direct evidence that molybdenum is an essential mineral nutrient for laboratory animals soon followed.

Sources of Molybdenum

Soils vary widely in their molybdenum content and the extent to which they yield the element to growing crops and pastures. Soil molybdenum concentrations vary from 0.1 to 20 mg kg^{-1} DM, with sandy soils at the low extreme and those of marine origin at the high extreme. Approximately 10% of soil molybdenum is normally extractable, but the proportion rises with soil pH and, with it, the concentration in crops and pastures (see Fig. 2.5; COSAC, 1982). Cereals rarely contain > 1 mg Mo kg^{-1} DM (Table 17.1), but pasture levels vary widely, depending on soil conditions; a median value of 1.1 mg Mo kg^{-1} DM was reported for 20 improved hill pasture sites in Scotland, but the maximum value was 60 mg kg^{-1} DM (Suttle and Small, 1993). The literature regarding botanical species differences is inconsistent, but they do not appear to be large (Table 17.1). In British Columbia, mean values in forages ranged from 0.9 to 2.6 mg Mo kg^{-1} DM, with maize silage having the least, sedge the most and legumes and grasses similar intermediate values of 1.8 and 2.0 mg Mo kg^{-1} DM, repectively (Miltimore and Mason, 1971). Animal sources are mostly low in molybdenum, with the exception of marine products and milk from animals grazing molybdeniferous pastures.

Metabolism of Molybdenum

Molybdenum can be readily and rapidly absorbed by livestock. The hexavalent water-soluble forms, sodium and ammonium molybdate, and the molybdenum of high-Mo herbage, most of which is water-soluble, are highly absorbable by cattle (Ferguson *et al.*, 1943) and sheep (Grace and Suttle, 1979). However, at high copper (Cu):Mo ratios in fibrous, high-sulphur diets, ruminants may excrete the majority of ingested molybdenum in faeces (Fig. 17.4), due to the formation of insoluble thiomolybdates in the rumen (see Chapter 11). Metabolism of molybdate is quite different in the non-ruminant or milk-fed ruminant, being well absorbed from the stomach (van Campen and Mitchell, 1965) or abomasum (Miller *et al.*, 1972). Molybdate absorption across the intestinal mucosa is by an active, carrier-mediated process, which

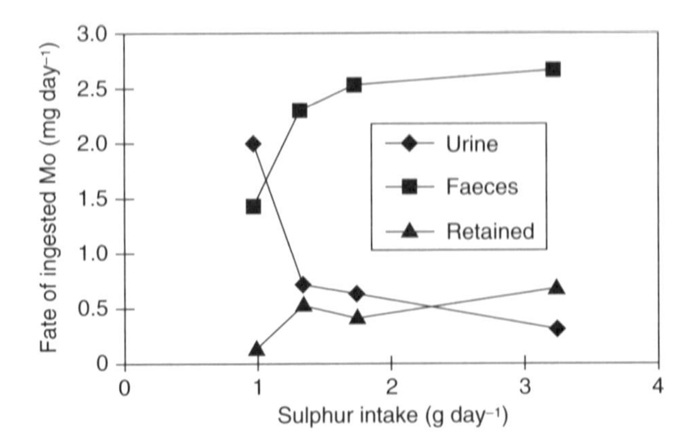

Fig. 17.4. Increases in daily sulphur intake can shift the route of excretion of a dietary molybdenum supplement (5 mg kg^{-1} DM) from urine to faeces while increasing Mo retention in sheep (Grace and Suttle, 1979).

is shared with and inhibited by sulphate (Mason and Cardin, 1977). Since little sulphur leaves the rumen as sulphate (see Chapter 9), the antagonism of molybdate absorption by sulphate will only occur as a natural physiological phenomenon in animals without a functional rumen. Absorbed molybdate is normally transported in plasma in the free ionic state but is stored in tissues as molybdopterin, bound to xanthine dehydrogenase (XDH) and aldehyde oxidase (AO) in the cytosol and to sulphite oxidase (SO) in mitochondrial membranes (Johnson, 1997). Excess molybdenum is predominantly excreted in urine. In the renal tubule, competitive inhibition of the reabsorption of molybdate by sulphate can increase molybdenum excretion (Bishara and Bray, 1978) and, at this stage, the ruminant and non-ruminant will express the same antagonism, because sulphur absorbed from the rumen as sulphide or from the intestine as microbial sulphur (S)-amino acids is metabolized to sulphate. It is the differences in sulphur metabolism in the gut between ruminants and non-ruminants that underlie their different susceptibilities to molybdenum toxicity and they may also affect tolerance of low molybdenum intakes.

Essentiality of Molybdenum

The molybdenum requirements of animals are extremely low. Chicks grow normally, reproduce and metabolize xanthine satisfactorily on diets containing only 0.2 mg Mo kg^{-1} DM (Higgins *et al.*, 1956). However, XDH, AO and SO all require molybdopterin cofactors (Johnson, 1997), and reduced activities, sufficient to impair function, have been observed in chicks when deficiency was exacerbated by high intakes of the molybdenum antagonist, tungsten (Higgins *et al.*, 1956; Johnson *et al.*, 1974). In tungstate-supplemented chicks, growth was retarded, anaemia developed, tissue molybdenum values fell and capacity to convert xanthine to uric acid was reduced. The development of anaemia probably reflects the importance of XDH as a source of ferroxidase activities, in both the intestinal mucosa (Topham *et al.*, 1982b) and liver (Topham *et al.*, 1982a); activities are inhibited by adding tungstate to rat diets, thereby impairing the release of iron bound to ferritin (see Chapter 13). High activities of XDH are found in liver, endothelial cells lining blood-vessels, macrophages, mast cells (Hellsten-Westing *et al.*, 1991) and the mammary gland. Inherited defects which reduce the activity of one or all of the above molybdenum-dependent enzymes in humans are life-threatening (Johnson, 1997). Large areas of molybdenum-deficient soils exist in which yield responses to applications of molybdenum occur in crops and pastures (Anderson, 1956). A significant growth response to added molybdenum (1 mg kg^{-1} DM) was reported in goat kids fed a semipurified diet containing 0.06 mg Mo kg^{-1} DM (Anke *et al.*, 1978) and attributed mainly to a 29% reduction in appetite. Continuation of the treatment through pregnancy and lactation caused reproductive failure and mortality or poor growth in subsequent generations; these defects may also have been secondary to loss of appetite. Loss of appetite may in turn reflect the needs of rumen microorganisms for molybdenum for cellulose digestion (Shariff *et al.*, 1990). However, many pastures grazed regularly by sheep and cattle are low in molybdenum and yet have no adverse effect on growth or the development of disabilities other than enhanced copper retention in the tissues (see Chapter 11). The biochemical activity of molybdenum hinges upon its ability to change between the quadri- and hexavalent states, giving redox potential which is linked to electron acceptors (cytochrome *c*, molecular oxygen (O_2), nicotinamide adenine dinucleotide (NAD^+)). Molybdeno-enzyme activity may thus be important as a cellular source of peroxide and free superoxide radicals, which can cause muscle damage during intense excerise in humans (Hellsten-Westing *et al.*, 1991) and trigger the inflammatory response to trauma (Friedl *et al.*, 1989), including invasion of the gut mucosa by parasitic nematodes, a perennial problem for most grazing livestock (Suttle *et al.*, 1992).

Toxicity of Molybdenum

Species differences

The tolerance of farm animals to high dietary molybdenum intakes varies with the species, the amount and chemical form of the ingested molybdenum, the copper status of the animal and the diet and the forms and concentration of sulphur in the diet. Cattle are the least tolerant species, followed closely by sheep, while pigs are the most tolerant of domestic livestock. The tolerance of horses is evident from their failure to show signs of toxicity on 'teart' pastures, which severely affect cattle, and is supported by experimental results (Strickland *et al.*, 1987). Pigs have tolerated dietary concentrations of 1000 mg Mo kg^{-1} DM for 3 months (Davis, 1950). This is 20–100 times the dietary concentrations that result in drastic scouring in cattle. Chick growth is inhibited at 200 mg Mo kg^{-1} DM (Gray and Daniel, 1954) and turkey poult growth is depressed at 300 mg kg^{-1} DM (Kratzer, 1952). The differences between ruminants and non-ruminants are probably explained by the ease with which powerful antagonists of copper metabolism, thiomolybdates, are generated in the rumen and then exert harmful effects on the intestinal mucosa (see Chapter 11). Differences between non-ruminant species are harder to explain, but they may also involve antagonisms of copper metabolism. Molybdenum is more toxic when given to a non-ruminant of low than of high copper status (Gray and Daniel, 1954), and the least tolerant monogastric species (avian) have low natural circulating levels of the copper-transport protein, ceruloplasmin. Furthermore, exposure of a non-ruminant (the rat) to molybdenum induces similar trichloroacetic acid (TCA)-insoluble copper complexes in plasma to those found in ruminants (N. Sangwan and N.F. Suttle, unpublished data) and is believed to restrict the tissue uptake of copper in certain circumstances.

Clinical signs

The clinical manifestations of molybdenum toxicity also vary among species. Growth retardation or weight loss and anorexia are invariable consequences of high intakes, but diarrhoea is a conspicuous feature only in cattle. Within a few days of being placed on some 'teart' pastures containing 20–100 mg Mo kg^{-1} DM, cattle begin to scour profusely and develop harsh, staring coats. Treatment with copper sulphate at the very high rate of 2 g day^{-1} for cows and 1 g day^{-1} for young stock or the intravenous injection of 200–300 mg Cu day^{-1} stops the scouring (Ferguson *et al.*, 1938), but much smaller supplements given via the drinking-water can also be effective (see Chapter 11). Scouring and weight losses are such dominant manifestations of molybdenosis in cattle that other disorders tend to be obscured. Cows conceive with difficulty and young bulls may exhibit a complete lack of libido, with testicular damage and little spermatogenesis (Thomas and Moss, 1952). These changes are comparable with those reported in molybdenotic rats. Sheep exposed to molybdenum can be affected by diarrhoea (Suttle and Field, 1968), and joint abnormalities, lameness, osteoporosis and spontaneous bone fractures have been observed in some areas (Pitt, 1976). Ruminants subjected to less severe exposure to molybdenum present symptoms that are generally indistinguishable from those caused by copper deprivation and are therefore attributable to induced copper deficiency (see Chapter 11).

Protection given by sulphur sources

The protection against molybdenum toxicity afforded by dietary sulphur and, to some extent, by protein is confined to non-ruminants. The marked reduction in molybdenum retention in sheep given supplementary sulphate by Dick (1956) was an artefact of the experiment, rather than evidence that sulphur can protect ruminants from molybdenum toxicity. Sheep were switched from a low- to a high-sulphur diet, having built up high blood and tissue molybdenum concentrations, with no evidence of toxicity while receiving a moderate molybdenum intake. When sulphur and molybdenum are added simultaneously to the diet, molybdenum retention increases, when compared with sheep given molybdate alone (Fig. 17.4; Grace and Suttle, 1979), and adverse effects mediated by impairment of copper metabolism are exacerbated. The situation is different in non-ruminants, where sulphur sources afford protection from the much higher molybdenum levels needed to cause toxicity. Sulphate limits molybdenum retention, both by reducing intestinal absorption and by increasing urinary excretion. The sulphate effect is not shared by other anions tested (Van Reen, 1959), but the protection afforded by high-protein diets, thiosulphate, cystine and methionine indicates that endogenous sulphate can be just as effective as dietary sulphate.

NICKEL

Sources of Nickel

Nickel is one of the less abundant elements in soils and crops. The median value for soils in England and Wales was 23.7, with a 'normal range' of 7.3–70.0 mg kg^{-1} DM, but only 5% was present in easily extractable forms (Archer and Hodgson, 1987). Concentrations of nickel in pasture grasses are lower than in soils, but legumes, such as lucerne, contain more (Table 17.1). In a study of wheat seed from 12 locations in North America, Welch and Cary (1975) reported nickel concentrations ranging widely from 0.08 to 0.35, with a low overall mean of 0.18 mg Ni kg^{-1} DM. Bovine milk normally contains about 0.02–0.05 mg Ni l^{-1} (Kirchgessner *et al.*, 1967), a value similar to that found in cereals, and the concentration is not influenced by dietary nickel intake. For example, O'Dell *et al.* (1970a) found most of the milk samples they examined to contain less than 0.1 mg Ni l^{-1}, and supplementation of the cow's ration with 365 or 1835 mg Ni day^{-1} as nickel carbonate did not increase nickel in the milk. However, there is a fourfold increase in nickel concentrations in the colostrum, compared with main milk (Kirchgessner *et al.*, 1967).

Metabolism of Nickel

Lower organisms undoubtedly require nickel and, since these include the anaerobic rumen microbes, nickel should influence rumen metabolism. Supplementation of diets containing 0.26–0.85 mg Ni kg^{-1} DM with 5 mg Ni (as $NiCl_3$) has increased ruminal urease, growth rate and feed conversion efficiency of lambs and steers given diets high in energy and low in protein (Spears, 1984), but this may not indicate a causal relationship. Supplementation of a similar basal diet has increased urease activity, without affecting growth (Spears *et al.*, 1986), the utilization of urea or nitrogen retention in lambs (Milne *et al.*, 1990). Other potentially limiting, nickel-dependent activities in the rumen microorganisms may have been involved, including methanogenesis and sulphate reduction (Nielsen, 1996). Nickel supplementation has also frequently altered patterns of VFA production in the rumen, but the changes have been inconsistent (Milne *et al.*, 1990). Nickel is poorly absorbed (1–5%) and transported in bound form, chiefly to albumin. Nickel concentrations in plasma are low (< 0.017 μmol l^{-1}), but tissue concentrations vary from organ to organ, being much higher in the kidney (2.78 μmol kg^{-1}) than in the liver (0.77 μmol kg^{-1}) of rat pups born of dams given adequate nickel; depletion reduced these concentrations by 39 and 44%, respectively (Stangl and Kirchgessner, 1996).

Essentiality of Nickel

Nickel was first shown to be essential for chicks fed on a highly purified diet under strict environmental control (Nielsen and Sauberlich, 1970). Depigmentation of the shank skin, thickened legs, swollen hocks, growth retardation, ultrastructural liver changes and anaemia were reported in chicks given a basal diet providing only 2–15 μg Ni kg^{-1}, and an additional 50 μg Ni kg^{-1} alleviated all abnormalities (Nielsen *et al.*, 1975b). Nickel was also found to improve growth and fertility in rats (Nielsen and Ollerich, 1974; Nielsen *et al.*, 1975a) and pigs (Anke *et al.*, 1974). The symptoms of deficiency have recently been confirmed in first-generation rat offspring as growth retardation, impaired reproduction and anaemia, accompanied by lowered liver activities of many enzymes involved in the citric acid cycle and release into the bloodstream of the liver aminotransferases (Stangl and Kirchgessner, 1996). Similar clinical and biochemical abnormalities were seen in the nickel-deficient goat reared on diets with < 100 μg Ni kg^{-1} DM (Anke *et al.*, 1991). Although various enzyme changes have been observed in deficient animals, a specific functional role for nickel has not been demonstrated. The basal diets used in these studies have all been exceedingly low in nickel (< 0.1 mg kg^{-1} DM) and a mere 1 mg Ni kg^{-1} DM was usually sufficient to restore normality. Naturally occurring nickel deficiency in grazing livestock has never been reported and appears unlikely to occur, in view of the low animal requirements and the relatively high concentrations commonly present in pasture plants. The low levels of nickel in milk (0.2–0.5 mg kg^{-1} DM) presumably equal or exceed the needs of suckling offspring for that element. Deficiency is more likely to occur on cereal-based diets than on forage-based diets.

Toxicity of Nickel

There is a wide margin between the beneficial and toxic doses of nickel for all species, particularly non-ruminants. The susceptibility of the ruminant may be a reflection of the susceptibility of the rumen microflora to relatively low levels of nickel when administered in soluble forms (e.g. > 50 mg Ni as $NiCl_3$ kg^{-1} DM). With less soluble forms, such as $NiCO_3$, up to 250 mg Ni kg^{-1} is tolerated; above this, disturbance of rumen function leads to inappetence, but no specific toxicity signs or pathology, other than nephritis (O'Dell *et al.*, 1970b). Poultry tolerate up to 400 mg Ni as the sulphate or acetate kg^{-1} DM (Weber and Reid, 1968). Small increases in dietary nickel (5 mg kg^{-1} DM) can cause sixfold increases in nickel levels in the bovine kidney to 0.3 mg Ni kg^{-1} DM, suggesting a gross dietary excess (Spears *et al.*, 1986). In toxicity, kidney concentrations reach 38 mg Ni kg^{-1} DM, but the predominant route of excretion is via the faeces (O'Dell *et al.*, 1971).

SILICON

Sources of Silicon

As the second most abundant element in the earth's crust, silicon is likely to present problems of nutritional excess rather than deficiency. Grasses commonly contain silicon in macroelement proportions (20–40 g Si kg^{-1} DM) and the coarser fodder or range species up to twice that range (Fig. 17.5; Smith and Urquhart, 1975). The leaf is richer in silicon than the stem of the rice plant. Cereal grains are equally rich in silicon, particularly the more fibrous types, such as oats (Carlisle, 1986). Leguminous forages contain much less silicon (Fig. 17.5), sometimes as little as 1.8 g kg^{-1} DM, but still macro- rather than trace-element proportions. Most animal products are low in silicon (e.g. eggs, with 3 mg Si kg^{-1} DM) and the most marginal food is milk, with only 1 mg Si l^{-1}. Silicon occurs in foods in inorganic form as silica (SiO_2) and monosilicic acid and also in organic forms, such as pectin and mucopolysaccharides. Silicon is also a ubiquitous contaminant of the environment.

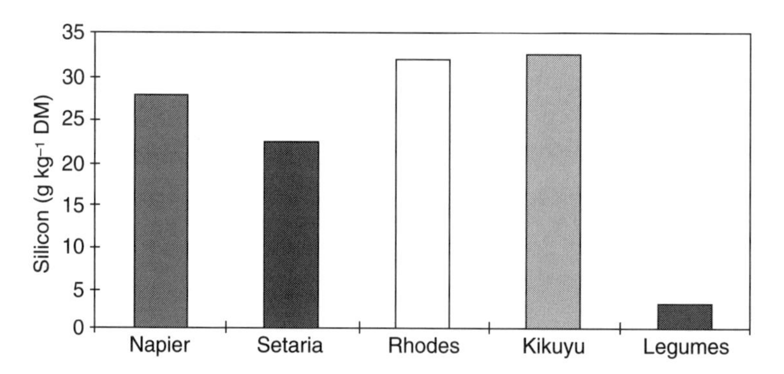

Fig. 17.5. Silicon concentrations are high in tropical grasses but much lower in legumes (Jumba, 1989).

Metabolism of Silicon

The analysis of silicon in feeds is complicated by the difficulty of 'digesting' the element, even with strong acids, so it is hardly surprising that most of the silicon that is ingested can resist digestion in the gut and be excreted in faeces. Jones and Handreck (1965) found that sheep given dry graminaceous or leguminous diets excreted 99% of the ingested silicon in their faeces; the absorbed silicon was mostly recovered in urine. However, in grazing situations, the urine can become the predominant route of excretion, and silicon in green feeds may be far more absorbable than that in dry feeds (Nottle and Armstrong, 1966). Silicon is believed to be absorbed and transported in the bloodstream as monosilicic acid and to readily enter tissues and cross membranes in this free form (Carlisle, 1986). Silicon is uniformly distributed within the cells of soft tissues, such as liver. Some silicon is present in plants as fine grains of amorphous silica (opal) or phytoliths; the grains can be fine enough to be absorbed and may be trapped in lymph nodes or excreted in urine unchanged (Carlisle, 1986). Absorption, therefore, does not guarantee a supply of utilizable silicon. The silicon that is absorbed is rapidly excreted and concentrations in blood do not reflect absorbed load.

Essentiality of Silicon

Chicks reared in a plastic isolator with a filtered air supply on a diet containing only 1 mg Si kg^{-1} DM became decidedly unhealthy, compared with those given 250 mg added Si kg^{-1} DM. Growth rate was reduced by 30–50%, wattles did not develop, combs were stunted and development of skull and long bones was impaired (Carlisle, 1986). Silicon is involved in glycoaminoglycan formation in cartilage and connective tissue; in bone, this leads to reduced calcification, while, in cartilage, defects occur at articular surfaces. Connective tissues, such as tendon, and organs rich in connective tissue (e.g. aorta and trachea) have high silicon concentrations, compared with other tissues and organs (11–17 vs. 2–4 mg kg^{-1} FW). It must be stressed, however, that silicon deficiency will not develop on crop or forage-based diets. Risks of deficiency in the suckled or milk-fed animal are probably avoided by environmental contamination.

Anti-nutritional Effects of Excess Silicon

Silicon toxicity is not a recognized condition, but high dietary intakes of silicon have a variety of adverse effects on health and nutrition. In grazing livestock, the abrasive effects of substantial intakes of silicate-rich soil particles has been associated with excessive tooth wear (Healy and Ludwig, 1965). Complexation of elements such as magnesium, iron and manganese by silicon has been described (Carlisle, 1986) and may explain the reduction in digestibility reported when silicates are added to the diet of ruminants (Smith and Nelson, 1975; Smith and Urquart, 1975). Such effects would be additional to the low digestibility, which is the inevitable consequence of consuming coarse forages, with high dietary concentrations (6–8%) of largely indigestible and unabsorbable silicon. The high urinary concentrations of silicon (some present as phytoliths) which accompany high intakes of silicon by grazing animals may trigger urolithiasis. The risk, however, is governed by the volume of urine excreted and physicochemical properties, such as pH, which influence the initiation and growth of the calculus, rather than silicon intake *per se*. Thus, urolithiasis in sheep in Western Australia was associated with inland rather than coastal regions and with the dry season, when low water intakes gave rise to maximal silica concentrations in the urine (Nottle and Armstrong, 1966). Siliceous calculi consist of polysilicic acid–protein complexes (Keeler, 1963), and these are less likely to form when the urine is high in phosphate and of low pH (Emerick, 1987). Reciprocal antagonism between silicon and molybdenum has been reported in chicks whereby supplements of one element reduce blood and tissue concentrations of the other (Carlisle and Curran, 1979).

TIN

Sources of Tin

There have been few studies of tin in farming contexts. Surface soils generally contain little tin (0.9–1.7 mg kg^{-1} dry weight (DW): Kabata-Pendias and Pendias, 1992) and levels in plants are consequently low. An early (1948) study showed that pastures in Scotland contained 0.3–0.4 mg Sn kg^{-1} DM, while later (1970) studies of cereals found much higher values of 5.6–7.9 mg kg^{-1} DM (Underwood, 1977). The volatility of tin means that losses may occur during sample reduction (e.g. by ashing), leading to underestimation of the amounts initially present.

Metabolism and Toxicity of Tin

Tin is poorly absorbed and concentrations in most tissues are < 1 mg kg^{-1} DM. Tin is of low toxicity to mammals, > 1000 mg kg^{-1} DM being required to restrict growth and induce anaemia in rats (Underwood, 1977) and reduce egg production in laying hens when given as the oxide (Puls, 1994).

Essentiality of Tin

Using the familiar combination of purified diet and isolator environment, Schwarz *et al.* (1970) showed that rat growth was poor (1.1 g live-weight gain

(LWG) day^{-1}) unless 1–2 mg Sn kg^{-1} DM was given as stannic sulphate, 0.5 mg being insufficient to sustain optimal growth. The concentration of tin in the basal diet was not stated, presumably because it was below detection limits for the assay. No specific biochemical roles have been reported for tin and deficiency does not give rise to characteristic pathology, but it was suggested that the physicochemical properties of Sn(IV) raised possibilities for contributions of tin to the tertiary structure of proteins and oxidation–reduction reactions at physiological pH. However, the growth rate of the isolated rats given tin remained 45% below that of conventionally reared rats, raising the possibility of multiple nutritional inadequacies.

VANADIUM

Vanadium has been studied mostly for its toxicological and pharmacological properties.

Sources of Vanadium

Vanadium is abundant in soils, particularly in the finer clay fractions (65–200 mg V kg^{-1} DM), but only a small fraction (< 10%) is extractable (Berrow *et al.*, 1978) and levels in pasture and crops are usually low (< 0.1 mg kg^{-1} DM) (Table 17.1). Since grazing animals cannot avoid consuming soil when they graze and contamination of the herbage varies from *c.* 2% in summer to 10% or more of dry-matter intake (DMI) in winter, soil ingestion will be a major source of vanadium intake for pastured animals. Levels in the environment may be increasing due to the combustion of vanadium-rich oils (Davison *et al.*, 1997) and use of vanadium-rich dietary phosphate supplements (Sell *et al.*, 1982). Prolonged use of phosphatic fertilizers did not increase the vanadium (or arsenic, chromium or lead) content of soils or maize crops in Minnesota (Goodroad and Campbell, 1979), but contamination of pasture with ash from petroleum products has been linked to toxicity in grazing cattle (Lillie, 1970, cited by Hansard *et al.*, 1978).

Metabolism of Vanadium

The reactivity of vanadium stems from the wide range of oxidation states which the element can enter, movements between the tetravalent and pentavalent state being the major contributor to its redox potential in mammalian tissues (Nielsen, 1996; Davison *et al.*, 1997). Vanadium appears to be readily transported across membranes by non-specific anion channels and binds to proteins, such as transferrin and lactoferrin. Vanadate can substitute for phosphate, mimicking the effects of cyclic AMP. As peroxovanadate, it influences free-radical generation and mimics insulin (Nielsen, 1996). It is, therefore, hardly surprising that animals absorb very little vanadium. Sheep absorbed only 1.6% of their dietary supply of vanadium, whether the diet contained 2.6 or 202.6 mg V kg^{-1} DM (Patterson *et al.*, 1986). Intravenously administered dioxovanadate was metabolized differently from the orally dosed element, raising questions as to the physiological significance of *in vitro* studies with the element. Tissue concentrations of vanadium are normally low (< 10 μg V kg^{-1} FW) but increase substantially in liver and bone when large dietary supplements of vanadate are given (Fig. 17.6; Hansard *et al.*, 1978). Of more practical significance is the fact that simulated soil ingestion studies with sheep have shown that the ingestion of as little as 4.9 or 8.3 mg soil V kg^{-1} DM for 41 days can raise kidney vanadium concentrations from undetectable levels (< 0.2 mg kg^{-1} DM) to means of 0.65 or 1.85 mg kg^{-1} DM (Suttle *et al.*, 1991); ingested soil vanadium is therefore partially absorbed.

Essentiality of Vanadium

Female goats given a diet containing < 10 μg V kg^{-1} DM showed a higher abortion rate when compared with control goats given 2 mg V kg^{-1} DM (Anke *et al.*, 1989). Vanadium-depleted goats produced less milk and there was a higher mortality among their offspring than in those of control goats. Vanadium deprivation in the rat has tended to decrease growth, while increasing thyroid weight; dietary iodine concentration affects tissue responses to dietary vanadium (Nielsen, 1996).

Toxicity of Vanadium

Vanadium toxicity causes diuresis, naturesis, diarrhoea and hypertension, with irreversible damage to both liver and kidney at high doses (Nechay, 1984). Species vary widely in their susceptibility to vanadium toxicity, with poultry the most and sheep the least susceptible of farmed species. As little as 5 mg V, added as ammonium vanadate (NH_4VO_3) kg^{-1} DM, retarded the growth of broiler chicks given a maize–soybean-meal diet, but previous work had indicated a tolerance of 13–20 mg V kg^{-1} DM from various sources, including superphosphates (Cervantes and Jensen, 1986). Hen body weight and egg production were both reduced during 140 days' exposure to a diet containing 50 mg added V as calcium orthovanadate, while 100 mg added V caused up to 56% mortality, without specific pathology (Kubena and Phillips, 1983). Growth suppression, mortality, enteritis and cystitis were recorded in weanling pigs given 200 mg V as vanadate kg^{-1} DM for 10 weeks (Van Vleet *et al.*, 1981). Lambs tolerated 200 mg added V, as ammonium metavanadate, for at least 84 days; higher levels (400 and 800) caused immediate diarrhoea and anorexia (Hansard *et al.*, 1978). Similar adverse effects have been reported in poultry with as little as 28.5 mg V, added as NH_4VO_3 (Sell *et al.*, 1982). Albumen quality was particularly sensitive to vanadium supplementation, as little as 6 and 7 mg V added as dicalcium phosphate lowering egg albumin content. Dietary factors, such as high iron, S-amino acid and chromium concentrations, reduce the toxicity of vanadium (Nielsen, 1996). Tissue concentrations of vanadium are normally relatively low (< 0.25 mg kg^{-1} FW: Puls, 1994) but are increased in lambs given diets with 202 mg V kg^{-1} DM to 3.3, 2.8 and 11.1 mg kg^{-1} in bone, liver and kidney, respectively, without signs of toxicity (Fig. 17.6; Hansard *et al.*, 1978); the increase in muscle vanadium was exceedingly small in comparison (0.04 up to 0.41 mg kg^{-1} DM).

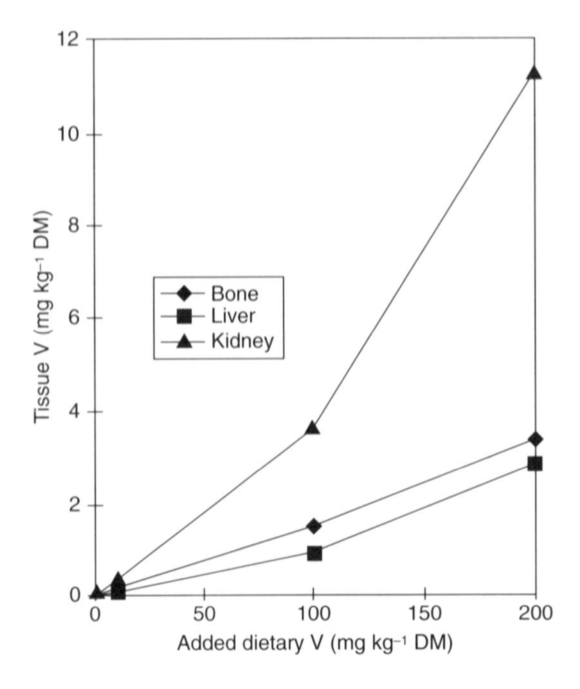

Fig. 17.6. Exposure of lambs to vanadate via the diet cause proportional increases in vanadium in liver and bone: large increases in kidney V suggest that the urinary route of excretion is important for dealing with dietary excesses (data from Hansard *et al.*, 1978).

References

Anderson, R.A. (1956) Molybdenum as a fertilizer. *Advances in Agronomy* 8, 163–202.

Anderson, R.A. (1987) Chromium. In: Mertz, W. (ed.) *Trace Elements in Human and Animal Nutrition*. Academic Press, New York, pp. 225–244.

Anke, M., Grun, M., Dittrich, G., Groppel, B. and Hennig, A. (1974) Low nickel rations for growth and reproduction in pigs. In: Hoekstra, W.G., Suttie, J.W., Ganther, H.E. and Mertz, W. (eds) *Trace Element Metabolism in Man and Animals – 2.* University Park Press, Baltimore, pp. 715–718.

Anke, M., Green, M., Partschefeld, M. and Groppel, B. (1978) Molybdenum deficiency in ruminants. In: Kirchgessner, M. (ed.) *Proceedings of the Third International Symposium on Trace Element Metabolism in Man and Animals.* Arbeitskreis für Tierernahrungsforschung, Weihenstephan, pp. 230–233.

Anke, M., Groppel, B., Grun, K., Langer, M. and Arnhold, W. (1989) The essentiality of vanadium for animals. In: *Proceedings of the Sixth International Trace Element Symposium,* Vol. 1. Friedrich Schiller University, Jena, pp. 17–27.

Anke, M., Groppel, B. and Krause, U. (1991) The essentiality of the toxic elements cadmium, arsenic and nickel. In: Momcilovic, B. (ed.) *Proceedings of the Seventh International Symposium on Trace Elements in Man and Animals, Dubrovnik.* IMI, Zagreb, pp. 11-6–11-8.

Archer, F.C. and Hodgson, J.H. (1987) Total and extractable trace element content of soils in England and Wales. *Journal of Soil Science* 38, 421–431.

Berrow, M.L., Wilson, M.J. and Reaves, G.A. (1978) Origin of extractable titanium and vanadium in the horizons of Scottish podzols. *Geoderma* 21, 89–103.

Bishara, H.N. and Bray, A.C. (1978) Competition between molybdate and sulphate for renal tubular reabsorption in sheep. *Proceedings of the Australian Society of Animal Production* 12, 123.

Boleman, S.L., Boleman, S.J., Bidner, T.D., Southern, L.L., Ward, T.L., Pontif, J.E. and Pike, M.M. (1995) Effects of chromium picolinate on growth, body condition and tissue accretion in pigs. *Journal of Animal Science* 73, 2033–2042.

Brown, T.F., McCormick, M.E., Morris, D.R. and Zerinque, L.K. (1989) Effects of boron on mineral balance in sheep. *Nutrition Research* 9, 503–512.

Bunting, L.D., Fernandez, J.M., Thompson, D.L. and Southern, L.L. (1994) Influence of chromium picolinate on glucose usage and metabolic criteria in growing Holstein calves. *Journal of Animal Science* 72, 1591–1599.

Burton, J.L., Mallard, B.A. and Mowat, D.N. (1993) Effects of supplemental chromium on immune responses of peri-parturient and early lactation dairy cows. *Journal of Animal Science* 71, 1532–1539.

Burton, J.L., Mallard, B.A. and Mowat, D.N. (1994) Effects of supplemental chromium on antibody responses of newly weaned feedlot calves to immunisation with infectious bovine rhinotracheitis and parainfluenza 3 virus. *Canadian Journal of Veterinary Research* 58, 148–151.

Carlisle, E.M. (1986) Silicon. In: Mertz, W. (ed.) *Trace Elements in Human and Animal Nutrition,* Vol. 2, 5th edn. Academic Press, New York, pp. 373–390.

Carlisle, E.M. and Curran, M.J. (1979) A silicon–molybdenum interaction *in vivo. Federation Proceedings* 38, 553.

Cervantes, H.M. and Jensen, L.S. (1986) Interaction of monensin with dietary vanadium, potassium and protein and its effect on hepatic rubidium and potassium in chicks. *Poultry Science* 65, 1591–1597.

Chang, X. and Mowat, D.N. (1992) Supplemental chromium for stressed growing feeder calves. *Journal of Animal Science* 70, 559–565.

Coleman, S.W., Evans, B.C. and Horn, G.W. (1984) Some factors influencing estimates of digesta turnover rates using markers. *Journal of Animal Science* 58, 979–986.

COSAC (1982) *Trace Element Deficiency in Ruminants.* Report of a Study Group from the Scottish Agricultural Colleges (SAC) and Scottish Agricultural Research Institutes, SAC, Edinburgh, p. 49.

Davis, G.K. (1950) Influence of copper on metabolism of phosphorus and molybdenum. In: *Copper Metabolism: a Symposium on Animal, Plant and Soil Relationships.* Johns Hopkins University Press, Baltimore, Maryland, pp. 216–229.

Davison, A., Kowalski, L., Xuefeng, Y. and Siu-Sing, T. (1997) Vanadium as a modulator of cellular regulation the role of redox reactivity. pp. 229–233. In: Fischer, P.W.F., L'Abbé, M.R. and Cockell, K.A. (eds) *Proceedings of the Ninth International Symposium on Trace Elements in Man and Animals, Banff.* NRC Research Press, Ottawa.

de Renzo, E.C., Kaleita, E., Heytler, P., Oleson, J.J., Hutchings, B.L. and Williams, J.H. (1953) The nature of the xanthine oxidase factor. *Journal of the American Chemical Society* 75, 753.

Dick, H.T. (1956) Molybdenum in animal nutrition. *Soil Science* 81, 229–258.

Downes, A.M. and McDonald, I.W. (1964) The chromium–Si complex of ethylene diaminotetracetic acid as soluble rumen marker. *British Journal of Nutrition* 18, 153–163.

Emerick, R.J. (1987) Phosphate inhibition of protein-polysilicic acid complex formation *in vitro*, a factor in preventing silica urolithiasis. *Journal of Nutrition* 117, 1924–1928.

Ferguson, W.S., Lewis, A.H. and Watson, S.J. (1938) Action of molybdenum in milking cattle. *Nature, UK* 141, 553.

Ferguson, W.S., Lewis, A.H. and Watson, S.J. (1943) The teart pasture of Somerset: the cause and cure of teartness. *Journal of Agricultural Science, Cambridge* 33, 44–51.

Friedl, H.P., Till, G.O., Trentz, O. and Ward, P.A. (1989) Roles of histamine, complement and xanthine oxidase in thermal injury of skin. *American Journal of Pathology* 135, 203–217.

Gardner, G.E., Pethick, D.W. and Smith, C. (1998) Effect of chromium chelavite supplementation on the metabolism of glycogen and lipid in adult Merino sheep. *Australian Journal of Agricultural Research* 49, 137–145.

Goodroad, L.L. and Campbell, A.C. (1979) Effects of phosphorus fertilizer and lime on the As, Cr, Pb and V content of soils and plants. *Journal of Environmental Quality* 8, 493–496.

Grace, N.D. and Suttle, N.F. (1979) Some effects of sulphur intake on molybdenum metabolism in sheep. *British Journal of Nutrition* 41, 125–136.

Gray, L.F. and Daniel, L.J. (1954) Some effects of excess molybdenum on the nutrition of the rat. *Journal of Nutrition* 53, 43–51.

Green, G.H. and Weeth, H.J. (1977) Responses of heifers to boron in water. *Journal of Animal Science* 46, 812–818.

Hansard, S.L.I., Ammerman, C.B., Fick, K.R. and Miller, S.M. (1978) Performance and vanadium content of tissues in sheep as influenced by dietary vanadium. *Journal of Animal Science* 46, 1091–1095.

Healy, W.B. and Ludwig, T.G. (1965) Wear of sheeps teeth. 1. The role of ingested soil. *New Zealand Journal of Agricultural Research* 8, 737–752.

Hellsten-Westing, Y., Sollevi, A. and Sjodin, B. (1991) Plasma accumulation of hypoxanthine, uric acid and creatine kinase following exhaustive runs of differing duration in man. *European Journal of Applied Physiology* 62, 380–384.

Higgins, E.S., Richert, D.A. and Westerfeld, W.W. (1956) Molybdenum deficiency and tungstate inhibition studies. *Journal of Nutrition* 59, 539–559.

Hunt, C.D., Herbel, J.L. and Idso, J.P. (1994) Dietary boron modifies the effects of vitamin D on indices of energy substrate utilisation and mineral metabolism in the chick. *Journal of Bone and Mineral Research* 9, 171–181.

Jensen, L.S., Maurice, D.V. and Murray, M.W. (1978) Evidence for a new biological role of chromium. *Federation Proceedings* 37, 404.

Johnson, J.H., Crookshank, H.R. and Smalley, H.E. (1980) Lithium toxicity in cattle. *Veterinary and Human Toxicology* 22, 248–251.

Johnson, J.I., Rajagopalan, K.V. and Cohen, H.J. (1974) Effect of tungsten on xanthine oxidase and sulfite oxidase in the rat. *Journal of Biological Chemistry* 249, 859–866.

Johnson, J.L. (1997) Molybdenum. In: O'Dell, B.L. and Sunde, R.A. (eds) *Handbook of Nutritionally Essential Mineral Elements.* Marcel Dekker, New York, pp. 413–438.

Jones, L.H.P. and Handreck, K.A. (1965) The relation between the silica content of the diet and the excretion of silica by sheep. *Journal of Agricultural Science, Cambridge* 65, 129–134.

Jumba, I.O. (1989) Tropical soil–plant interactions in relation to mineral imbalances in grazing livestock. PhD thesis, University of Nairobi, Chapter 5.

Kabata-Pendias, A. and Pendias, H. (1992) *Trace Elements in Soils and Plants*, 2nd edn. CRC Press, Boca Raton, Florida.

Keeler, R.F. (1963) Silicon metabolism and silicon–protein matrix interrelationship in bovine urolithiasis. *Annals of New York Academy of Science* 104, 592–611.

Kegley, E.B., Spears, J.W. and Eisemann, J.H. (1997a) Performance and glucose metabolism in calves fed a chromium–nicotinic acid complex or chromium chloride. *Journal of Dairy Science* 80, 1744–1750.

Kegley, E.B., Spears, J.W. and Brown, T.T. (1997b) Effect of shipping and chromium supplementation on performance, immune response and disease resistance of steers. *Journal of Animal Science* 75, 1956–1964.

Kirchgessner, M., Friesecke, H. and Koch, G. (1967) *Nutrition and the Composition of Milk.* Crosby Lockwood, London, p. 129.

Kratzer, F.H. (1952) Effect of dietary molybdenum upon chicks and poults. *Proceedings of the Society for Experimental Biology and Medicine* 80, 483–486.

Kubena, L.F. and Phillips, T.D. (1983) Toxicity of vanadium in female leghorn chickens. *Poultry Science* 62, 47–50.

Lane, M.A., Ralphs, M.H., Olsen, J.D., Provenuza, F.D. and Pfister, J.A. (1989) Conditioned taste aversion: potential for reducing cattle loss to larkspur. *Journal of Range Management* 43, 127–131.

Leach, G.J. (1983) Influence of rest interval, grazing duration and mowing on the growth, mineral content and utilisation of a lucerne pasture in a sub-tropical environment. *Journal of Agricultural Science, Cambridge* 101, 169–183.

Lien, T.F., Chen, S.-Y., Shiaw, S.-P., Froman, P. and Hu, C.Y. (1996) Chromium picolinate reduces laying hen serum and egg yolk cholesterol. *Professional Animal Scientist* 12, 77–80.

Mason, J. and Cardin, C.J. (1977) The competition of molybdate and sulphate ions for a transport system in the ovine small intestine. *Research in Veterinary Science* 22, 313–315.

Mertz, W. (1974) Chromium as a dietary essential for man. In: Hoekstra, W.G., Suttie, J.W., Ganther, H.E. and Mertz, W. (eds) *Proceedings of the Second International Symposium of Trace Elements Metabolism in Animals.* University Park Press, Baltimore, Maryland, pp. 185–198.

Mertz, W. (1986) Lithium. In: Mertz, W. (ed.) *Trace Elements in Human and Animal Nutrition*, Vol. 2, 5th edn. Academic Press, New York, pp. 391–398.

Miller, J.K., Moss, B.R., Bell, M.C. and Sneed, N. (1972) Comparison of ^{99}Mo metabolism in young cattle and swine. *Journal of Animal Science* 34, 846–850.

Milne, J.S., Whitelaw, F.G., Price, J. and Shand, W.J. (1990) The effect of supplementary nickel on urea metabolism in sheep given a low protein diet. *Animal Production* 50, 507–512.

Miltimore, J.E. and Mason, J.L. (1971) Copper to molybdenum ratio and molybdenum and copper concentrations in ruminant feeds. *Canadian Journal of Animal Science* 51, 193–200.

Mooney, K.W. and Cromwell, G.L. (1997) Efficacy of chromium picolinate and chromium chloride as potential carcase modifiers in swine. *Journal of Animal Science* 75, 2661–2671.

Moonsie-Shageer, S. and Mowat, D.N. (1993) Effects of level of supplemental chromium on performance, serum constituents and immune status of stressed feeder calves. *Journal of Animal Science* 71, 232–238.

Mowat, D.N., Chang, X. and Yang, W.Z. (1993) Chelated chromium for stressed feeder calves. *Canadian Journal of Animal Science* 73, 49–55.

Nechay, B.R. (1984) Mechanisms of action of vanadium. *Annual Reviews of Pharmacology and Toxicology* 24, 501–524.

Nielsen, F.H. (1994) Biological and physiological consequences of boron deprivation in humans. *Environmental Health Perspectives* 102 (suppl.), 59–63.

Nielsen, F.H. (1996) Other trace elements. In: Filer, L.J. and Ziegler, E.E. (eds) *Present Knowledge in Nutrition*, 7th edn. International Life Sciences Institute, Nutrition Foundation, Washington DC.

Nielsen, F.H. and Ollerich, D.A. (1974) Nickel: a new essential trace element. *Federation Proceedings* 33, 1767–1772.

Nielsen, F.H. and Sauberlich, H.E. (1970) Evidence of a possible requirement for nickel by the chick. *Proceedings of the Society for Experimental Biology and Medicine* 134, 845–849.

Nielsen, F.H., Myron, D.R., Givand, S.H., Zimmerman, T.J. and Ollerich, D.A. (1975a) Nickel deficiency in rats. *Journal of Nutrition* 105, 357–362.

Nielsen, F.H., Myron, D.R., Givand, S.H. and Ollerich, D.A. (1975b) Nickel deficiency and nickel–rhodium interaction in chicks. *Journal of Nutrition* 105, 1607–1619.

Nottle, M.C. and Armstrong, J.M. (1966) Urinary excretion of silica by grazing sheep. *Australian Journal of Agricultural Research* 17, 165–173.

O'Dell, G.D., Miller, W.J., King, W.A., Ellers, J.C. and Jurecek, H. (1970a) Effect of nickel supplementation on production and composition of milk. *Journal of Dairy Science* 53, 1545–1548.

O'Dell, G.D., Miller, W.J., King, W.A., Moore, S.L. and Blackmon, D.M. (1970b) Nickel toxicity in the young bovine. *Journal of Nutrition* 100, 1447–1454.

O'Dell, G.D., Miller, W.J., Moore, S.L., King, W.A., Ellers, J.C. and Jurecek, H. (1971) Effect of dietary nickel level on excretion and nickel content of tissue in male calves. *Journal of Animal Science* 32, 769–773.

Offenbacher, E.G., Pi-Sunyer, F.X. and Stoecker, B.J. (1997) Chromium. In: O'Dell, B.L. and Sunde, R.A. (eds) *Handbook of Nutritionally Essential Mineral Elements*. Marcel Dekker, New York, pp. 389–411.

Olsen, Q.R., Rule, D.C., Field, R.A., Snowder, G.D. and Hu, C.Y. (1996) Dietary chromium picolinate does not influence growth or carcass composition in feedlot lambs. *Sheep and Goat Research Journal* 12, 22–24.

Page, T.G., Southern, L.L., Ward, T.L. and Thompson, D.L. (1993) Effect of chromium picolinate on growth and serum and carcass traits of growing–finishing pigs. *Journal of Animal Science* 71, 656–662.

Parker, W.J., Morris, S.T., Garrick, D.J., Vincent, G.L. and McCutcheon, S.N. (1990) Intraruminal chromium controlled-release capsules for measuring intake in ruminants: a review. *Proceedings of the New Zealand Society of Animal Production* 50, 437–442.

Patt, E.L., Pickett, E.E. and O'Dell, B.L. (1978) The effect of dietary lithium levels on tissue lithium concentrations, growth rate and reproduction in the rat. *Bioinorganic Chemistry* 9, 299–310.

Patterson, B.W., Hansard, S.L., Ammerman, C.B., Henry, P.R., Zech, L.A. and Fisher, W.R. (1986) Kinetic model for whole-body vanadium metabolism: studies in sheep. *American Journal of Physiology* 251, R325–R332.

Pitt, M.A. (1976) Molybdenum toxicity: interactions between copper, molybdenum and sulphate. In: *Agents and Actions* 6(6). Birkauser Verlag, Basle, pp. 758–769.

Puls, R. (1994) *Mineral Levels in Animal Health*, 2nd edn. Sherpa International, Clearbrook, British Columbia.

Ralphs, M.H. and Chaney, C.D. (1993) Influence of cattle age, lithium chloride dose level and food type in the retention of food aversions. *Journal of Animal Science* 71, 373–379.

Richert, D.A. and Westerfield, W.W. (1953) Isolation and identification of the xanthine oxidase factor as molybdenum. *Journal of Biological Chemistry* 203, 915–923.

Rossi, A.F., Miles, R.D., Damron, B.L. and Flunker, L.K. (1993) Effects of dietary boron supplementation on broilers. *Poultry Science* 72, 2124–2130.

Saha, D.C. and Galbraith, R.L. (1993) A modified chromic oxide indicator ratio technique for accurate determination of nutrient digestibility. *Canadian Journal of Animal Science* 73, 1001–1004.

Schwarz, K. and Mertz, W. (1959) Chromium (III) and the glucose tolerance factor. *Archives of Biochemistry and Biophysics* 85, 292–295.

Schwarz, K., Milne, D.B. and Vinyard, E. (1970) Growth effect of tin compounds in rats maintained in a trace-element controlled environment. *Biochemical and Biophysical Research Communications* 40, 22–28.

Sell, J.L., Arthur, J.A. and Williams, I.L. (1982) Adverse effects of dietary vanadium, contributed by dicalcium phosphate, on albumen quality. *Poultry Science* 61, 2112–2116.

Shariff, M.A., Boila, R.J. and Wittenberg, K.M. (1990) Effects of dietary molybdenum on rumen dry matter disappearance in cattle. *Canadian Journal of Animal Science* 70, 319–323.

Silanpaa, M. (1982) *Micronutrient and the Nutrient Status of Soils: a Global Study.* FAO, Rome, p. 91.

Small, J.A., Charmley, E., Rodd, A.V. and Freeden, A.H. (1997) Serum mineral concentrations in relation to estrus and conception in beef heifers and cows fed conserved forage. *Canadian Journal of Animal Science* 77, 55–62.

Smith, G.S. and Nelson, A.B. (1975) Effects of sodium silicate added to rumen cultures on forage digestion, with interactions of glucose, urea and minerals. *Journal of Animal Science* 41, 891–899.

Smith, G.S. and Urquhart, N.S. (1975) Effect of sodium silicate added to rumen cultures on digestion of silicous forages. *Journal of Animal Science* 41, 882–890.

Spears, J.W. (1984) Nickel as a 'newer trace element' in the nutrition of domestic animals. *Journal of Nutrition* 59, 823–835.

Spears, J.W., Harvey, R.W. and Samsell, L.J. (1986) Effects of dietary nickel and protein on growth, nitrogen metabolism and tissue concentrations of nickel, iron, zinc, manganese and copper in calves. *Journal of Nutrition* 116, 1873–1882.

Stangl, G.I. and Kirchgessner, M. (1996) Effect of nickel deficiency on various metabolic parameters of rats. *Animal Physiology and Animal Nutrition* 75, 164–174.

Starich, G.H. and Blincoe, C. (1983) Dietary chromium – forms and availabilities. *The Science of the Total Environment* 28, 443–454.

Steele, N.C. and Roseburgh, W. (1981) Effect of trivalent chromium on hepatic lipogenesis by the turkey poult. *Poultry Science* 60, 617–622.

Strickland, K., Smith, F., Woods, M. and Mason, J. (1987) Dietary molybdenum as a putative copper antagonist in the horse. *Equine Veterinary Journal* 19, 50–54.

Subiyatno, A., Mowat, D.N. and Yang, W.Z. (1996) Metabolite and hormonal responses to glucose or propionate infusions in periparturient dairy cows supplemented with chromium. *Journal of Dairy Science* 79, 1436–1445.

Suttle, N.F. and Field, A.C. (1968) The effect of intake of copper, molybdenum and sulphate on copper metabolism in sheep. I. Clinical condition and distribution of copper in the blood of the pregnant ewe. *Journal of Comparative Pathology* 78, 351–362.

Suttle, N.F. and Small, J. (1993) Evidence for delayed availability of copper in supplementation trials with lambs on molybdenum-rich pastures. In: Anke, M., Meissner, D. and Mills, C.F. (eds) *Proceedings of the Eighth Symposium on Trace Elements in Man and Animals.* Verlag Media Touristik, Gersdorf, pp. 651–655.

Suttle, N.F. and Wadsworth, I. (1991) Physiological responses to vaccination in sheep. *Proceedings of the Sheep Veterinary Society* 11, 113–116.

Suttle, N.F., Brebner, J. and Hall, J. (1991) Faecal excretion and retention of heavy metals in sheep ingesting topsoil from fields treated with metal-rich sewage sludge. In: Momcilovic, B. (ed.) *Proceedings of the Seventh International Symposium on Trace Elements in Man and Animals.* IMI, Zagreb, pp. 32-7–32-8.

Suttle, N.F., Knox, D.P., Angus, K.W., Jackson, F. and Coop, R.L. (1992) Effects of dietary molybdenum on nematode and host during *Haemonchus contortus* infection in lambs. *Research in Veterinary Science* 52, 230–235.

Thomas, J.W. and Moss, S. (1952) The effect of orally administered molybdenum on growth, spermatogenesis and testis histology of young dairy bulls. *Journal of Dairy Science* 34, 929–934.

Thorhallsdottir, A.G., Proveuza, F.D. and Ralph, D.F. (1990) Social influences on conditioned food aversions in sheep. *Applied Animal Behaviour Science* 25, 45–50.

Toepfer, E.W., Mertz, W., Royinski, E.E. and Polansky, M.M. (1973) Chromium in foods in relation to biological activity. *Journal of Agricultural and Food Chemistry* 21, 69–73.

Topham, R.W., Walker, M.C. and Calisch, M.P. (1982a) Liver xanthine dehydrogenase and iron mobilisation. *Biochemical and Biophysical Research Communications* 109, 1240–1246.

Topham, R.W., Walker, M.C., Calisch, M.P. and Williams, R.W. (1982b) Evidence for the participation of intestinal xanthine oxidase in the mucosal processing of iron. *Biochemistry* 21, 4529–4535.

Uden, P., Colucci, P.E. and van Soest, P.J. (1980) Investigation of chromium, cerium and cobalt as markers in digesta. Rate of passage studies. *Journal of Science in Food and Agriculture* 31, 625–632.

Underwood, E.J. (1977) *Trace Elements in Human and Animal Nutrition*, 4th edn. Academic Press, New York, pp. 449–451.

van Campen, D.R. and Mitchell, E.A. (1965) Absorption of Cu^{64}, Zn^{65}, Mo^{99} and Fe^{65} from ligated segments of the rat gastrointestinal tract. *Journal of Nutrition* 86, 120–126.

Van Heugten, E. and Spears, J.W. (1997) Immune response and growth of stressed weanling pigs fed diets supplemented with organic and inorganic forms of chromium. *Journal of Animal Science* 75, 409–416.

Van Reen, R. (1959) The specificity of the molybdate–sulfate interrelationship in rats. *Journal of Nutrition* 68, 243–250.

Van Vleet, J.F., Boon, G.D. and Ferrans, V.J. (1981) Induction of lessions of selenium-vitamin E deficiency in weanling swine fed silver, cobalt, tellurium, zinc, cadmium and vanadium. *American Journal of Veterinary Research* 42, 789–799.

Villalobos, F., Romero, R.C., Farrago, C.M.R. and Rosado, A.C. (1997) Supplementation with chromium picolinate reduces the incidence of placental retention in dairy cows. *Canadian Journal of Animal Science* 77, 329–330.

Vipond, J.E., Horgan, G. and Anderson, D. (1995) Estimation of food intake in sheep by blood assay for lithium content following ingestion of lithium-labelled food. *Animal Science* 60, 513.

Ward, T.L., Southern, T.L. and Bidner, T.D. (1997) Interactive effects of dietary chromium tripicolinate and crude protein level in growing–finishing pigs provided with inadequate pen space. *Journal of Animal Science* 75, 1001–1008.

Weber, C.W. and Reid, B.L. (1968) Nickel toxicity in growing chicks. *Journal of Nutrition* 95, 612–618.

Weeth, H.J., Speth, C.F. and Hanks, D.R. (1981) Boron content of plasma and urine as indicators of boron intake in cattle. *American Journal of Veterinary Research* 42, 474–477.

Welch, R.M. and Cary, E.E. (1975) Concentration of chromium, nickel and vanadium in plant materials. *Journal of Agricultural and Food Chemistry* 23, 479–482.

Wright, A.J., Mowat, D.N. and Mallard, B.A. (1994) Supplemental chromium and bovine respiratory disease vaccines for stressed feeder calves. *Canadian Journal of Animal Science* 74, 287–295.

Yamamoto, A., Wada, O. and Suzuki, H. (1988) Purification and properties of biologically active chromium complex from bovine colostrum. *Journal of Nutrition* 118, 39–45.

Yang, W.Z., Mowat, D.N., Subiyatno, A. and Liptrap, R.M. (1996) Effects of chromium supplementation on early lactation performance of Holstein cows. *Canadian Journal of Animal Science* 76, 221–230.

Essentially Toxic Elements (Aluminium, Arsenic, Cadmium, Fluorine, Lead, Mercury)

18

This chapter will cover those elements which are known principally for their toxic properties, while occasionally fulfilling the criteria of an essential element for livestock. The criteria of essentiality are discussed more fully at the beginning of the previous chapter and will be reconsidered at the end of this chapter. The soil acts as a major source and reservoir of each element and is therefore a major determinant of risk of specific imbalances occurring in livestock. These risks have been enhanced by contamination of the environment, and the insidious nature of some adverse effects has been a matter of growing concern for some time (Chowdury and Chandra, 1987; Fox, 1987; Scheuhammer, 1987).

ALUMINIUM

Aluminium is the most abundant mineral in most soils, and it is hardly surprising that it presents problems of excess but not deficiency to livestock under farming conditions.

Sources of Aluminium

Aluminium constitutes 3–6% of most soils, but concentrations in the soil solution and groundwater remain low because the element is present largely in insoluble siliceous complexes. Concentrations of aluminium (Al) in uncontaminated crops and forages are usually much lower (50–100 mg kg^{-1} dry matter (DM)), but trees, ferns and tropical plants may contain 3–4 g Al kg^{-1} DM (Underwood, 1977). The principal source of aluminium exposure for grazing livestock is from soil-contaminated pasture, and aluminium may constitute up to 0.5% of dry-matter intake (DMI) when pasture is sparse. The

© CAB *International* 1999. *Mineral Nutrition of Livestock*
(E.J. Underwood and N.F. Suttle)

act of grazing itself can increase the degree of contamination of pasture with soil and aluminium in autumn (Robinson *et al.*, 1984). High concentrations in the finest soil particles (Brebner *et al.*, 1985) ensure that aluminium is a ubiquitous contaminant of the farm environment. Aluminium may also enter the diet through the use of contaminated mineral supplements (e.g. soft phosphate, 70 g Al kg^{-1} DM), feed-pelleting agents, such as bentonite (110 g Al kg^{-1} DM), or 'aids to digestion', such as zeolite (60 g Al kg^{-1} DM).

Metabolism of Aluminium

Studies of aluminium metabolism have been made with simple, relatively soluble salts, such as aluminium chloride ($AlCl_3$), and yet the element remains poorly absorbed and retained, urinary excretion only becoming important when intakes are excessive (Alfrey, 1986). Thus, aluminium concentrations in the major soft tissues normally occupy the low range of 2–4 mg kg^{-1} DM and exposure of lambs to 2000 mg Al kg^{-1} DM for 2 months raised values by no more than 2 mg kg^{-1} DM, with liver the most – and brain the least – affected tissue (Valdivia *et al.*, 1982). Much higher concentrations (10–80 mg Al kg^{-1} DM) have been reported in silicon-rich tissues, such as the lung, thymus, aorta and trachea, raising suggestions of related functions for the two elements (Carlisle and Curran, 1993). They may, however, reflect common forms (aluminium silicates) and routes (inhalation, ingestion) of entry for the two elements.

Toxicity of Aluminium

Concerns over the pathogenicity of aluminium in the aged or renal-dialysis patient obscure the fact that aluminium, even when ingested in reactive forms, such as $AlCl_3$, is of low toxicity. At dietary concentrations above 2 g Al kg^{-1} DM, appetite is depressed in lambs (Valdivia *et al.*, 1982), but at that level no clinical abnormalities developed during 2 months' exposure. Addition of 1 g Al (as aluminium sulphate ($Al_2(SO_4)_3$)) kg^{-1} DM to the diet of laying hens for 4 months did not reduce egg production (Hahn and Guenter, 1986). Furthermore, aluminium at these high doses has been used as an antidote to fluoride toxicity (see section on fluoride). If aluminium does present a hazard to farm livestock, it is likely to arise in free-range animals, both ruminant and non-ruminant, through an antagonism of phosphorus metabolism following the ingestion of aluminium-rich soil (Krueger *et al.*, 1985). Such an antagonism has been implicated in phosphorus deficiency in cattle grazing tropical pastures (McDowell, 1992) and phosphorus absorption decreased when aluminium (as $AlCl_3$) was added to the diet of lambs (Valdivia *et al.*, 1982); furthermore, appetite was depressed, particularly if the diet was low in phosphorus. However, no evidence of hypophosphataemia or poor bone mineralization has been found during long-term, soil-ingestion studies with diets adequate in phosphorus (N.F. Suttle, unpublished data) or in short-term studies with cattle (Allen *et al.*, 1986). Furthermore, the addition of 1.3 g Al kg^{-1} DM, in the form of soil with high sorptive capacity for phosphorus, has *increased* the efficiency of phosphorus absorption by lambs (Garcia-Bojalil *et al.*, 1988). High intakes of aluminium have been implicated in the aetiology of hypomagnesaemic tetany in cattle, and intraruminal doses of $Al_2(SO_4)_3$, equivalent to 4 g Al kg^{-1} DM, cause rapid reductions in serum magnesium (Allen *et al.*, 1984). The hypomagnesaemia preceded appetite suppression and was not merely a secondary phenomenon. Furthermore, additions of 2.9 mg Al as the citrate kg^{-1} DM reduced plasma magnesium, while increasing urinary calcium excretion and causing only a small reduction in appetite (Allen *et al.*, 1986). However, similar concentrations of aluminium given as soil had no effect on plasma magnesium. Grazing livestock can clearly withstand considerable and sustained intakes of soil-borne aluminium (Robinson *et al.*, 1984; Allen *et al.*, 1986; Sherlock, 1989). However, the absorbability and antagonicity of aluminium from tropical shrubs used as forages merit investigation.

Essentiality of Aluminium

Evidence for the essentiality of aluminium as a dietary constituent has only been obtained through the use of rigorously purified diets and environments. Female goats given a diet containing 162 μg Al kg^{-1} DM for five pregnancies had a raised abortion rate (14%), and their offspring showed poor growth, incoordination and weakness in hind legs: life expectancy was decreased in comparison with a group given 25 mg Al kg^{-1} DM (Angelow *et al.*, 1993). Addition of 30 mg Al kg^{-1} DM to the low-aluminium diet of day-old chicks increased their growth rate by 22% over the first 18 days of life, but clinical abnormalities in unsupplemented chicks were not reported (Carlisle and Curran, 1993). Aluminium influences cultured cells *in vitro* (Nielsen, 1996), but the physiological significance of such effects is unclear.

ARSENIC

Arsenic was once surpassed only by lead as a toxicological hazard to farm livestock (Selby *et al.*, 1977). It is a powerful inhibitor of sulphydryl enzyme systems and keto acid oxidation, and most incidents of toxicity arose from the exploitation of the toxicity of arsenic towards pests (rodenticides) and the common pathogens of livestock (including coccidia and cestodes) and crops (insecticides and fungicides). Arsenic is also toxic to humans and limits have therefore been placed upon the acceptable arsenic concentrations in drinking-water and edible meat by many countries. In view of its long and colourful history as a poison, claims that arsenic might be an essential dietary constituent (Nielsen and Uthus, 1984) were hard to swallow!

Sources of Arsenic

The natural abundance of arsenic (As) in agricultural soils is low, a median value of 10.4 mg As kg^{-1} DM being reported for England and Wales, with a scatter of high values (maximum 140) associated with mining areas (Archer and Hodgson, 1987). However, in a study confined to south-west England, values as high as 727 mg As kg^{-1} were recorded and an area of 722 km^2 was reckoned to be contaminated with arsenic (Thornton, 1996). Plants take up little arsenic and usually contain < 0.5 mg As kg^{-1} DM, but values may be increased by soil contamination or the application of arsenical pesticides. Marine sources are much richer, fish-meals containing 2–20 mg As kg^{-1} DM (Ammerman *et al.*, 1973). The maximum acceptable concentration (MAC) of arsenic in drinking-water in the European Community is 50 μg l^{-1}.

Metabolism of Arsenic

Like all elements that exist primarily in the anionic state, inorganic arsenic is well absorbed and excreted principally and extensively in the urine. Tissue accretion of arsenic is therefore slow and occurs mainly in liver, kidney, skin and its appendages (Selby *et al.*, 1977). In their review of the literature, Doyle and Spaulding (1978) reported arsenic concentrations of 0.02–0.19 μg g^{-1} wet weight for liver from cattle, pigs and sheep, 0.01–0.15 μg g^{-1} for kidney and 0.01–0.11 μg g^{-1} for muscle (multiply by 13.35 for μmol kg^{-1}). Much higher values were reported for poultry (0.70, 0.28 and 0.09 μg g^{-1} in liver, kidney and muscle, respectively) and were attributed to the widespread use of arsenicals in poultry feeds. Implementation of brief withdrawal periods prior to slaughter prevents such increases in tissue arsenic. The high arsenic concentrations produced in the liver and thigh muscle of broiler chickens (1.23 and 0.41 μg g^{-1} wet weight, respectively) by feeding arsanilic acid at 'growth-stimulant' levels (90 mg kg^{-1} DM) for 42 days were reduced to background concentrations (0.14 and 0.23 μg g^{-1}, respectively) 7 days after removing the source of arsenic from the diet (Proudfoot *et al.*, 1991). Significant differences in arsenic concentrations between muscles were reported, with levels 60% or more lower in breast than in thigh muscle. More importantly, no growth stimulation was evident by the end of the study, and the practice of adding arsenicals to poultry rations is questionable.

Toxicity of Arsenic

The incidence of arsenic toxicity has fallen markedly following the withdrawal of most arsenic-containing products from the market-place. There are a few areas, such as south-west England (Thornton, 1996), where the disturbance of arsenic-rich deposits during mining (of tin, for example) lead to pollution of the soil, atmosphere or groundwater. In the latter case, the disturbance of natural drainage systems by pipe-laying operations can further change the distribution of arsenic in the affected area. Current toxicity problems relate to the continued use of organoarsenicals at high doses for the control of infections and at low doses as growth stimulants. For example, arsanilic acid is used at 45–100 mg As kg^{-1} DM as a growth stimulant for pigs and at 200–250 mg As kg^{-1} DM for the control of swine dysentery, while 400–2000 mg As kg^{-1} DM can cause chronic toxicity. Excessive intakes of arsenic can result from errors in feed formulation, extended duration of treatment or treatment of dehydrated and debilitated animals. The symptoms of arsenic toxicity vary with the rate and duration of exposure and the source of arsenic involved (Bahri and Romdane, 1991). Water-soluble inorganic salts, such as arsenites and arsenates, are the most toxic, particularly the trivalent forms, while the insoluble elemental form is non-toxic. Accidental exposure to large oral doses of inorganic arsenic cause acute toxicity, in which the primary symptom is gastroenteritis. A rapid drop in blood pressure causes collapse, accompanied by muscular twitching, convulsions and gastrointestinal haemorrhage, in sheep (Sharman and Angus, 1991). Other metals can have similar effects, and confirmation of acute arsenic poisoning is provided by the presence of excess arsenic in the tissues and body fluids and in the gut contents, if exposure has been recent. Where exposure has primarily involved the lungs or skin, the focus of pathogenicity shifts to the liver and kidney, which both show severe necrosis.

Chronic toxicity is most commonly caused by overexposure to organic arsenic sources that are of medium (e.g. phenylarsonic compounds) to low (arsanilic acid) toxicity. Early signs are those of a neurological disorder, such as incoordination, swaying and ataxia ('drunken hog syndrome'), but affected animals remain alert and continue to eat and drink. Later, extreme agitation may be shown by pigs, which scream with their noses pressed to the ground when they are disturbed. Demyelination of some peripheral nerves may be evident histologically. The biochemical confirmation of arsenic poisoning depends on the species and the characteristics of arsenic exposure, notably pattern and source (Puls, 1994).

Essentiality of Arsenic

Arsenic deprivation has decreased growth, impaired fertility and increased mortality in goats and miniature pigs (Anke *et al.*, 1991). Adverse effects of arsenic deficiency on chicks and laboratory animals have also been reported, although they may only occur when the diet is imbalanced with respect to nutrients other than arsenic (Nielsen, 1996). In view of the toxicity of arsenic to lower forms of animal life, the benefits of adding arsenic to the diet may arise from pharmacological effects on the gut microflora, but the levels of arsenic used to demonstrate essentiality are far lower than those used for prophylaxis. Estimates of arsenic requirement, at 25–50 μg kg^{-1} DM, are well below those likely to be found in any commercial ration or natural forage.

CADMIUM

Cadmium is a highly reactive and toxic element, which is sparsely distributed in most agricultural ecosystems. Once absorbed by animals or humans, however, cadmium is poorly excreted, and increasing efforts are being made to limit the entry of cadmium (Cd) into the human food-chain.

Sources of Cadmium

Soils and crops

Archer and Hodgson (1987) gave a median value of only 0.5 mg Cd kg^{-1} DM for the soils of England and Wales, which is similar to that reported subsequently for New Zealand soils (Bramley, 1990). Uptake by plants is generally poor, particularly from clay soils, and forages and crops grown on normal soils usually contain < 1 mg Cd kg^{-1} DM. Occasionally, soils become enriched with cadmium from cadmium-rich fertilizers or the dispersal of wastes from the mining and smelting of metals such as zinc and lead; values > 2.4 mg Cd kg^{-1} DM are regarded as anomalously high, and, in England and Wales, the maximum value recorded was 10.5 mg Cd kg^{-1} DM (Archer and Hodgson, 1987). Cadmium is strongly retained in the topsoil and, if contamination continues, soil cadmium concentrations will slowly increase (van Bruwaene *et al.*, 1984). Cadmium concentrations in plant tissue will also rise, but the increase depends on soil pH, plant species and the part of the plant sampled (van Bruwaene *et al.*, 1984). Morcombe *et al.* (1994a) found higher uptake of cadmium by sheep grazing pastures on acidic, sandy soils than on alkaline, clay soils.

Superphosphates

The most important exogenous sources of cadmium are superphosphate fertilizers, which vary in cadmium concentration from < 5 to 134 mg Cd kg^{-1}, depending on country or region of origin (Bramley, 1990). Regular use of cadmium-rich, Pacific sources of superphosphate in Australia and New Zealand has led to increases in the cadmium concentrations in soils and pastures sufficient to raise kidney cadmium concentrations in grazing lambs above MAC (> 1 mg kg^{-1} fresh weight (FW): Bramley, 1990; Morcombe *et al.*, 1994b). Pasture concentrations can be contained within acceptable limits by avoiding cadmium-rich sources of superphosphate and probably by liming, which reduces cadmium uptake by cereals (Oliver *et al.*, 1996). Ingestion of fertilizer and topsoil during grazing adds to the body burden of cadmium in the grazing animal.

Municipal sewage sludges

Sewage sludges contain variable and occasionally excessive cadmium concentrations (up to 20 mg kg^{-1} DM). In Europe, European Community directives and national guidelines aim to restrict the accumulation of cadmium in sludge-amended soils to < 3 mg kg^{-1} DM (sampling depth, 20 cm). In the grazing situation, ingestion of topsoil enriched with sludge cadmium is again likely to contribute to the tissue accumulation of cadmium. Blood, liver and kidney cadmium levels were increased during studies with sheep which simulated the soil-ingestion pathway (Brebner *et al.*, 1993), but not sufficiently to present a hazard via the human food-chain. If cadmium from sludges is retained in soils as avidly as that from superphosphate and is as available to plants, the total cadmium intake by grazing animals is again likely to raise liver and kidney cadmium to unacceptable levels in some areas. However, concurrent contamination with zinc may limit the uptake of cadmium, both from sewage sludges and from mining and smelting sources. Irrigation water that is contaminated by municipal sewage can also be an important source of cadmium. Under the flooded-rice culture practised in the Po valley, soil cadmium concentrations have risen by up to 6% per annum (van Bruwaene *et al.*, 1984).

Metabolism of Cadmium

There is a vast literature on biochemical responses of animals to cadmium, but most of it has little or no nutritional significance, because of the high dietary concentrations used or the total bypassing of the protective gut mucosa (i.e. parenteral administration) during experiments.

Absorption

Cadmium is poorly absorbed from solid diets by all species, particularly ruminants, which usually take up less than 1% of the cadmium ingested as inorganic salts (Neathery *et al.*, 1974; van Bruwaene *et al.*, 1984). Inorganic cadmium sources added to milk are well absorbed (Bremner, 1978), but cadmium is not transferred naturally via the dam's milk (Suttle *et al.*, 1997). Any cadmium that is taken up is largely retained. Cows dosed orally with ^{109}Cd had only retained 0.75% of the dose after 14 days, but 0.13% was still present in the tissues after 131 days (van Bruwaene *et al.*, 1982). Similar results have been reported for sheep (Scheuhammer, 1987; Houpert *et al.*, 1995); they absorbed 0.15–0.5% of an oral cadmium dose, but the half-life for elimination from the body was 100–150 days. In a short-term (4-day) study with Japanese quail, the proportion of an oral cadmium dose retained increased with size of dose (Scheuhammer, 1987), but this may reflect the fact that protective mechanisms had yet to become fully operative. Deficiencies of calcium, iron and zinc each increase the absorption and retention of cadmium (Kollmer and Berg, 1989).

Retention

The principal sites of cadmium retention are the gastrointestinal tract, liver and kidney (Neathery *et al.*, 1974; Houpert *et al.*, 1995) and, in the latter organ, the half-life is measured in years. It was once believed that mammals lacked homeostatic mechanisms for controlling cadmium retention and limiting toxicity, but this is no longer the case. When ewes were exposed to diets containing cadmium-enriched soils for 3 years (dietary cadmium up to 0.4 mg kg^{-1} DM), liver cadmium concentrations reached a plateau after about 6 months (Fig. 18.1; Brebner *et al.*, 1993). Similarly, liver cadmium concentrations in heifers showed no further increase when exposure to 5 mg Cd (added as $CdCl_2$) kg^{-1} DM continued from 394 days (parturition) throughout lactation for a further 160 days (Smith *et al.*, 1991b). Lee *et al.* (1996) have recently reported curvilinear increases in liver cadmium over time (28 months) in lambs grazing pastures of slightly raised cadmium content (0.5 mg Cd kg^{-1} DM). In a 191-day study with lambs, liver cadmium concentrations were not linearly related to dietary cadmium concentration, which ranged up to 60 mg kg^{-1} DM (Doyle *et al.*, 1974). Control of cadmium retention is achieved largely by metallothionein, a cysteine-rich protein that is the major cadmium-binding protein in the body and believed to transport cadmium from liver to kidney. Exposure to cadmium induces the synthesis of metallothionein at many sites, including the gut mucosa and liver, in what is believed to be a protective role, but Cd–metallothionein is nephrotoxic (Bremner, 1978).

Interactions

Cadmium shares physiological properties, such as metallothionein induction, with other elements, notably copper and zinc, having similar electron configurations in the outer valency shell of the atom (Hill and Matrone, 1969). These elements are thus often mutually antagonistic, exposure to cadmium

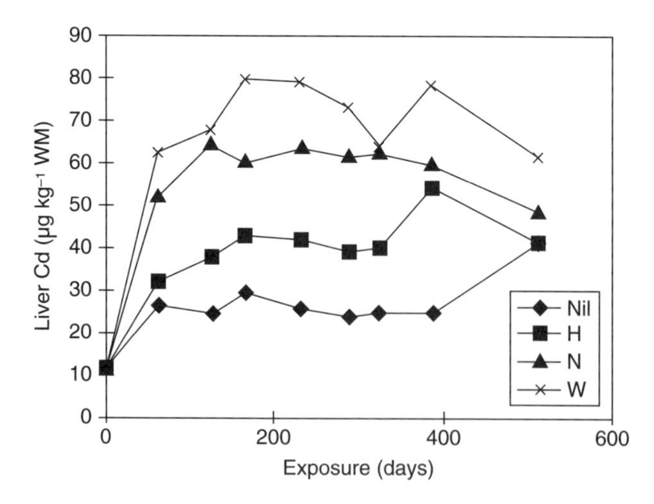

Fig. 18.1. Early high rates of cadmium accumulation in the liver of lambs were not sustained during long-term exposure via three ingested soils (H, N and W) enriched with Cd from sewage sludge, suggesting homeostatic adjustment to loading with the element (Stark *et al.*, 1998).

lowering copper status and exposure to zinc lowering cadmium status (Bremner, 1978). In other respects, however, the metabolism of cadmium is quite distinctive, and the element is not readily transferred across either ovine (Mills and Dalgarno, 1972; Suttle *et al.*, 1997) or bovine (Neathery *et al.*, 1974; Smith *et al.*, 1991a) placenta and mammary gland or across avian oviduct (Sell, 1975) in the way that copper and zinc are. The administration of selenium lowers the toxicity of cadmium by shifting the distribution of tissue cadmium from metallothionein towards high-molecular-mass proteins (Bremner, 1978). Cadmium is also involved in interactions with macronutrients and macrominerals. Addition of phytate (5 g kg^{-1} DM) to a diet supplemented with 5 mg Cd kg^{-1} DM caused a threefold increase in liver cadmium in rats (Rimbach *et al.*, 1996). The effect was believed to be a secondary response to a lowering of calcium absorption and induction of calcium-binding protein (CaBP) in the gut mucosa; CaBP has a strong affinity for cadmium. Supplementing a diet of moderate sulphur (S) concentration (1.9 g S kg^{-1} DM) with a further 4 g S reduced cadmium accumulation in the liver and kidney of sheep by 60% (Smith and White, 1997); molybdenum was less effective and the two antagonists did not act additively, providing a contrast with their synergistic antagonism of copper (see Chapter 11).

Toxicity of Cadmium to Livestock

Puls (1994) gives normal, high and toxic levels of cadmium for different species, and those of agricultural interest are summarized in Table 18.1. Blood cadmium concentrations are only marginally increased above normal (> 0.01 μg ml^{-1}), which is analytically problematic and diagnostically unhelpful in assessing cadmium toxicity. Excretion of metallothionein in urine is indicative of cadmium exposure. Diagnostic problems arise because the symptoms of cadmium toxicity and the level of exposure required for their manifestation are highly variable, due partly to the involvement of interactions with copper and zinc; furthermore, susceptibility varies from species to species (see Chapter 2: Interactions).

Ruminants

In early studies employing high dietary cadmium concentrations (> 40 mg Cd kg^{-1} DM), many of the symptoms of cadmium toxicity in ruminants were similar to those of zinc deficiency and were prevented by zinc supplementation (Powell *et al.*, 1964); they included loss of appetite, poor growth, retarded testicular development and parakeratosis in sheep. Paradoxically, exposure to diets containing 5–60 mg added Cd kg^{-1} DM increased zinc concentrations in the liver and kidney of sheep (Doyle and Pfander, 1975). Later studies, using a less severe cadmium challenge and diets barely adequate in copper, revealed other abnormalities, such as anaemia, impaired bone mineralization, loss of wool crimp, abortion and stillbirths, which are associated with copper deficiency. Exposure of pregnant ewes and

Table 18.1. Guidelines for assessing cadmium exposure and likelihood of cadmium toxicity in farm livestock from cadmium concentrations in diet, liver or kidney: normal (N), high (H) and toxic (T) ranges are given (chiefly from Puls, 1994).

		Diet	Liver	Kidney
		(mg kg^{-1} DM)[a]	(mg kg^{-1} FW)[a]	
Cattle and Sheep	N	0.1–0.2	0.02–0.05	0.03–0.10
	H[b]	0.5–5.0	0.1–1.5	1.0–5.0
	T	> 50	50–160	100–250
Pigs	N	0.1–0.8	0.1–0.5	0.1–1.0
	H[b]	1.0–5.0	1.0–5.0	2.0–5.0
	T	> 80	> 13	< 270
Poultry	N	0.1–0.8	0.05–0.5	0.1–1.5
	H[b]	1.0–5.0	1.0–50	2.0–10.0
	T	> 40	> 15	> 70

[a] Multiply by 8.9 to obtain units in mmol.
[b] High dietary cadmium levels may lead to unacceptably high concentrations in offal; maximum acceptable concentrations (MAC) vary from country to country but are often 1 mg Cd kg^{-1} FW and likely to fall. Cadmium may also induce copper or zinc deficiencies if the respective dietary concentrations are marginal.

subsequently their lambs to diets containing as little as 3.0–3.4 mg Cd kg^{-1} DM lowered the copper status of the lambs substantially (Mills and Dalgarno, 1972; Dalgarno, 1980) – sufficiently to impair wool crimp, a sign of clinical copper deficiency. However, the effect of cadmium on copper status was not discernible in the presence of excess zinc (Zn) (750 mg Zn kg^{-1} DM) (Campbell and Mills, 1979). In another study, 3 mg Cd kg^{-1} DM retarded lamb growth (Bremner and Campbell, 1980); raising the dietary zinc level from 30 to 150 mg kg^{-1} DM counteracted the adverse effect of cadmium more effectively than an increase in dietary copper from 4.5 to 15 mg kg^{-1} DM, suggesting that induced zinc deficiency, rather than copper deficiency, was the factor limiting growth. In cattle, a supplement of 5 mg Cd kg^{-1} DM fed throughout pregnancy reduced liver copper concentrations in the newborn calf by 29%, while 1 mg Cd kg^{-1} DM was sufficient to reduce liver copper in the dam by 40% (Smith *et al.*, 1991a). The high susceptibility of ruminants to copper deficiency makes them more vulnerable than other species to the antagonistic effects of cadmium towards copper. However, Doyle *et al.* (1974) found that lambs tolerated 15 mg Cd kg^{-1} DM (30–60 mg retarded growth) when given a diet which maintained high liver copper concentrations (mean 688 mg Cu kg^{-1} DM).

Non-ruminants

Similar interactions can occur in non-ruminants. When rats on a marginally copper (Cu)-deficient diet (2.6 mg Cu kg^{-1} DM) were given one of four cadmium levels at each of three zinc levels (Fig. 18.2; Campbell and Mills, 1974), each antagonist was most inhibitory towards copper when the other

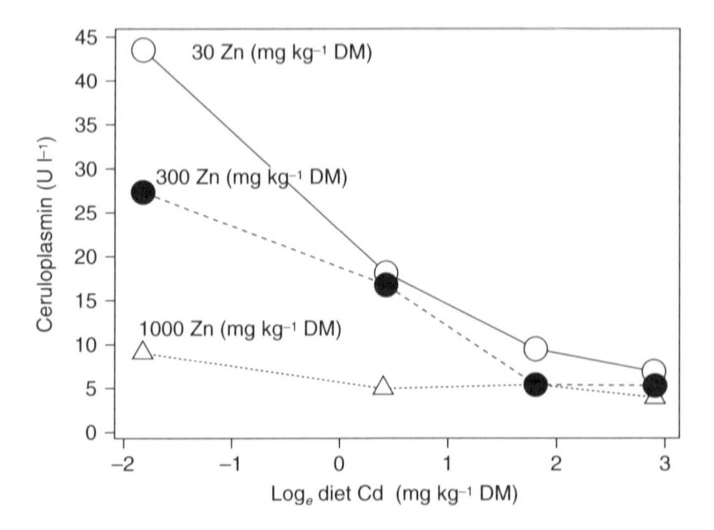

Fig. 18.2. When rats are concurrently exposed to two copper antagonists (cadmium and zinc, Cd and Zn) there is a negative interaction between the antagonists, each being less antagonistic towards copper in the presence than the absence of the other (data from Campbell and Mills, 1974).

was not added; cadmium was by far the more potent antagonist, with a supplement of 6 mg Cd kg^{-1} DM having the same inhibitory effect as 1000 mg Zn kg^{-1} DM. When rats were given the diet without added zinc, a mere 1.5 mg Cd kg^{-1} DM was sufficient to induce hypocupraemia and reduce bone mineralization. Supplee (1961, cited by Ammerman *et al.*, 1973) demonstrated a cadmium × zinc antagonism in turkey poults, 2 mg Cd kg^{-1} DM inducing hock and feather abnormalities on a low-zinc (10 mg Zn kg^{-1} DM) but not a high-zinc (60 mg Zn kg^{-1} DM) diet; the higher zinc level did not fully protect against 20 mg added Cd kg^{-1} DM. Most studies have been conducted with copper and zinc in adequate supply from the diet or body stores, and relatively high cadmium concentrations were required to cause toxicity, the major cause of ill health being nephrotoxicity caused by cadmium *per se*. In such circumstances, the toxicity of cadmium is best diagnosed by the presence of elevated concentrations of cadmium in the kidney and the presence of histological kidney damage (Scheuhammer, 1987). In avian species, dietary concentrations of 75 mg Cd kg^{-1} DM commonly cause kidney damage and also delayed maturation of the testes in males (Schcuhammer, 1987). In the laying hen, exposure to diets containing 48 (Leach *et al.*, 1979) and 60 mg Cd kg^{-1} DM (Sell, 1975) has reduced egg production.

Duration of exposure

Because cadmium is a cumulative poison, chronic toxicity is more likely to occur in aged breeding stock than in young stock slaughtered for meat consumption. Five- to eightfold increases in kidney and liver cadmium levels were reported in cattle exposed to pastures treated with sewage sludge by irrigation and soil incorporation for 8 years (Fitzgerald *et al.*, 1985). Cadmium concentrations showed larger proportionate increases than any other element, but liver copper and zinc concentrations increased, no clinical or histopathological abnormalities were recorded and the maximal kidney cadmium level recorded (32 mg kg^{-1} FW) remained well below the level usually associated with kidney damage in humans (200 mg kg^{-1} FW).

Livestock as Sources of Cadmium Exposure for Humans

Duration of exposure to cadmium has an economically important influence on the acceptability of carcasses for human consumption. Kidney cadmium concentrations increase with exposure time in sheep, following essentially linear relationships (Fig. 18.3; Petersen *et al.*, 1991; Lee *et al.*, 1996; Stark *et al.*, 1998) and may exceed MAC (Langlands *et al.*, 1988; Petersen *et al.*, 1991). Cadmium accumulates to a lesser extent in liver (Fig. 18.3; Petersen *et al.*, 1991) and hardly at all in muscle; the chances of anyone consuming enough liver or kidney of consistently high cadmium content to harm the kidney seem remote. However, cadmium ingested in kidney is largely present as metallothionein, passes straight to a kidney of the consumer and is

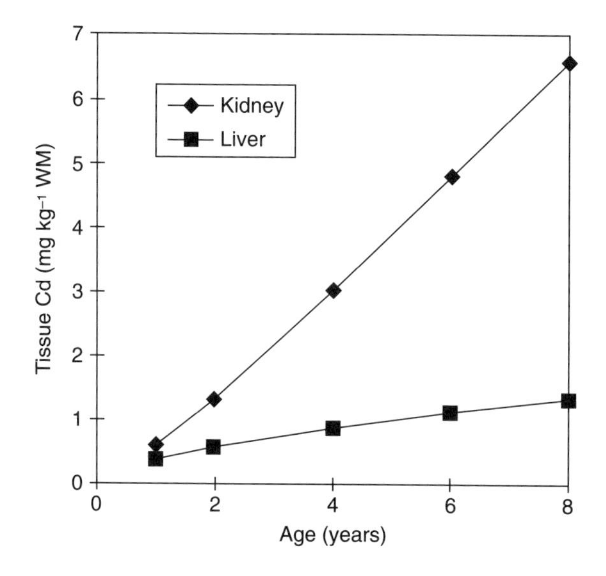

Fig. 18.3. Continuous exposure of sheep to pastures enriched with cadmium from superphosphates in Western Australia causes linear increases in Cd concentrations in kidney and liver over time; values in kidney can exceed maximum acceptable concentrations for human consumption (Peterson *et al.*, 1991).

sufficiently heat-stable to survive cooking. Current restrictions (MAC) are therefore likely to remain and even decrease. All sources of cadmium accumulation in the farm ecosystem must therefore be carefully monitored and practical steps taken to minimize them.

Essentiality of Cadmium

The only evidence that dietary cadmium may be essential for the normal development and health of livestock comes from studies with female goats given a diet containing < 15 μg Cd kg^{-1} DM when compared with a control group given 250 μg Cd kg^{-1} DM (Anke *et al.*, 1991). Reproduction was impaired in cadmium-depleted females and those offspring born alive were less likely to survive than those from cadmium-supplemented females. Subsequently, cadmium deprivation was reported to limit the growth of rats (Nielsen, 1996). Rations made from natural foodstuffs are always likely to meet any dietary requirement for cadmium.

FLUORINE

Biological interest in fluorine was at first confined to its toxic effects. Chronic endemic fluorosis in humans and farm animals had been identified in several countries by 1930, although severe fluorosis in cattle during and after volcanic eruptions in Iceland has a much longer history. The early literature on fluoride toxicity in its more acute forms was presented in the classical monograph by Roholm (1937). Industrial contamination of herbage and the growing use of fluorine-containing phosphates as mineral supplements introduced further fluorine hazards to livestock (Jacob and Reynolds, 1928). This stimulated interest in the distribution of fluorine in soils, plants and animal tissues, the means of prevention and control of chronic fluorosis and the factors affecting the tolerance of this element by different animal species. As the principal sources of fluorine in the farm environment were recognized and controlled, the emphasis of research shifted to verification of the essentiality of fluorine in the diet, not only for dental health but for growth.

Sources of Fluorine

Pastures and crops

Most plant species have a limited capacity to absorb fluorine from the soil, in which most of the fluorine is present as calcium fluoride (CaF_2). The fluorine (F) concentrations of pastures and forages are therefore characteristically low, unless they have been contaminated by deposition of fumes and dusts of industrial origin or by irrigation with fluoride-rich (often geothermal) waters (Shupe, 1980). Since the soil usually contains far higher fluoride concentrations

(30–100 mg F kg^{-1} DM) than the plant, ingestion of fluoride-rich soil on overgrazed pastures can significantly contribute to fluoride intakes. Analysis of uncontaminated pastures in England revealed a range of 2–16 about a mean of 5.3 mg F kg^{-1} DM (Allcroft *et al.*, 1965). The concentration of fluorine in 107 'clean' samples of lucerne hay from areas throughout the USA ranged widely, from 0.8 to 36.5, about a median of 2.0 mg F kg^{-1} DM (Suttie, 1969). Cereals and their by-products usually contain only 1–3 mg F kg^{-1} DM (McClure, 1949). Uncontaminated feed can rarely be incriminated in chronic fluorosis in farm livestock.

Rock phosphate supplements

The principal sources of fluoride for livestock are commercial feeds that contain fluoride-rich phosphate supplements. Thus, 10% of 168 samples of dairy feed from seven different states of the USA contained more than 30 mg F kg^{-1} DM, and some samples had over 200 mg F kg^{-1} DM (Suttie, 1969). Different fluoroapatite (rock phosphate) sources vary widely in their fluorine content, depending on their origin. Continental sources of rock phosphate contain 3–4% F, whereas the Pacific and Indian Ocean island deposits usually contain only about half that concentration. The island deposits can apparently be used safely in the rations of cows, pigs and poultry (Snook, 1962). The high-fluoride rock phosphates can be injurious to livestock when used over long periods in the amounts ordinarily required as calcium and phosphorus supplements (Phillips *et al.*, 1934) and are now routinely thermally 'defluorinated', without reducing the availability of phosphorus (see Chapter 5). A variety of chemically processed phosphorus (P) sources are available to the feed industry and they vary widely in the P:F ratio (Table 18.2). In the USA, a minimum acceptable P:F ratio of 100:1 has been set, to minimize the risk of fluorosis; sources with high P:F ratios are generally the more expensive (Thompson, 1980).

Animal by-products

The bones of mature animals, even in the absence of abnormal exposure to fluoride, are very much higher in fluoride than the soft tissues, and bone-meals can therefore constitute a significant source of fluorine for farm animals, typically containing 700 mg F kg^{-1} DM. Low P:F ratios, coupled with high cost, make bone-meal an unattractive source of phosphorus (Thompson, 1980). Meat meal, or tankage, is a significant source of fluorine only when it contains a high proportion of bone; some commercial meat meals contain as much as 200 mg F kg^{-1} DM. Other components of animal origin are invariably low in fluorine, because the soft tissues and fluids of the body rarely contain more than 2–4 mg F kg^{-1} DM. Milk and milk products contain even less fluoride. Cow's milk contains 0.1–0.3 mg F l^{-1} or 1–2 mg kg^{-1} DM and levels are little affected by fluoride intake.

Drinking-water

Chronic fluorosis is enzootic in sheep, cattle, goats and horses in parts of India, Australia, Argentina, Africa and the USA, as a consequence of the

Table 18.2. Comparison of phosphorus (P) sources in terms of their fluoride (F) concentration, P:F ratio and relative cost as P sources (after Thompson, 1980).

	F ($g\ kg^{-1}$ DM)	P:F ratio	F provided in adding 2.5 g P kg^{-1} DM ($mg\ kg^{-1}$ DM)	Relative cost[a]
Monocalcium phosphate (DCP)	1.4–1.6	132	19	100
Precipitated DCP	0.5	360	7	94
Defluorinated phosphate	1.6	112	22	94
Phosphoric acid (wet process)	2.0	118	21	72
Phosphoric acid (furnace process)	0.1	> 2000	1	156
Sodium phosphate	0.1	> 2000	1	200
Ammonium polyphosphate	1.2	120	21	107
Monammonium phosphate (MAP)				
Feed grade	1.8	133	19	107
'Low F'	0.9	266	9	128
Fertilizer grade	22.0	10	250	77
Diammonium phosphate (DAP)				
Feed grade	1.6	125	20	113
Fertilizer grade	20.0	10	250	88
Bone-meal	0.2–3.5	35–6000	15	250
Curaçao rock phosphate	5.4	26	96	86
Colloidal (soft) phosphate	12.0–14.0	6–7	361	39
Triple superphosphate	20.0	10.5	240	67
Fluoride rock phosphate	37.0	3.5	711	13

[a] Costs per unit of P at 1979 prices; DCP = 100.

consumption of waters abnormally high in fluorine, usually from deep wells or bores. Surface waters from such areas commonly contain less than 1 mg F l^{-1}, whereas the bore waters often contain 5–15 mg F l^{-1} and as much as 40 mg l^{-1} when evaporation has occurred in troughs or bore drains before consumption by stock (Harvey, 1952). Elsewhere, drinking-water is not normally a significant source of fluorine, containing 0.1–0.6 mg F l^{-1}.

Industrial contamination

In parts of North Africa, a chronic fluorosis, known as 'darmous', occurs from the contamination of the herbage and water-supplies with high-fluoride phosphatic dusts blown from rock-phosphate deposits and quarries. The further industrial processing of such phosphates, aluminium reduction, brick and tile production and steel manufacturing also release significant amounts of fluoride into the atmosphere, and forages downwind from production sites become contaminated by fumes and dusts. The introduction of stringent controls on industrial emissions can greatly reduce fluorosis problems of this nature (e.g. Lloyd, 1983).

Metabolism of Fluorine

The hazard presented by a particular fluorine source is determined not only by the quantity ingested but by the way it is metabolized.

Absorption

Unfortunately, there is little information on the availability of fluorine in natural feeds, but it is probably high, in keeping with other halides. The fluoride in hay is as available as that in soluble sources, such as sodium fluoride (NaF) (Shupe *et al.*, 1962; Table 18.3), which are almost completely absorbed from the gastrointestinal tract. The fluorine of less soluble compounds, such as bone-meal, CaF_2, cryolite (Na_3AlF_6), rock phosphate (RP) and defluorinated rock phosphate (DFP), is less well absorbed (see Table 18.3). Fluoride retention in the bones of growing sheep is much lower from DFP than from dicalcium phosphate (DCP) (Hemingway, 1977; Suttie, 1980). Suttie (1980) found that the fluoride in RP, DFP and DCP was 70, 50 and 20% as available as that in NaF, as judged by skeletal fluoride accretion in lambs given diets containing 60 mg F kg^{-1} DM as each source for 10 weeks. If representative, these results will mean that animals will tolerate five times more fluorine in the form of DCP than as NaF, and yet the latter provides the standard whereby safe limits for livestock are set! Where DFP is used at normal rates, there should be no difficulty in achieving safe fluorine levels in animal rations (Suttie, 1978). The fact that increases in bone fluoride concentrations are 75% less when cattle were given fluorine as polluted soil rather than NaF (both sources provided 40 mg F kg^{-1} DM: Oelschlager *et al.*, 1970) suggests that the soil fluorine was poorly absorbed.

Postabsorption

Absorbed fluoride is partially excreted via the urine and the kidney of exposed animals contains higher fluoride concentrations than any other soft tissue, though still less than 2.5 mg kg^{-1} DM in one study (Shupe, 1980). Retention is largely confined to the skeleton but there is a negative exponential relationship between fluoride intake and bone fluoride concen-

Table 18.3. Mean alkaline phosphatase activity and fluoride content of the bones of groups of four heifers given different amounts and forms of fluoride for 588 days (Shupe *et al.*, 1962).

	Fluoride content (mg kg^{-1} DM)			Phosphatase activity[a]	
Treatment	Diet	Metatarsal	Ribs[b]	Metatarsal	Ribs[b]
Low-F hay	10	344	423	73	85
Low-F hay + CaF_2	69	931	1287	104	116
Low-F hay + NaF	68	1880	2753	146	248
High-F hay	66	2130	2861	163	191

[a] mg P hydrolysed in 15 min g^{-1} bone.
[b] 11th, 12th and 13th ribs.

tration after a fixed exposure time, at least in the rib of sheep (Masters *et al.*, 1992). The relationship between duration of exposure at a fixed intake and bone fluoride concentration in cattle also follows the law of diminishing returns (Fig. 18.4; Shupe, 1980). Exposure of the pregnant and lactating animal to fluoride raises fluoride concentrations in the blood of the neonate and the milk of the dam, but the increases are far less than those seen in the dam's bloodstream. Addition of 30 mg F l^{-1} to the drinking-water of ewes raised their plasma fluoride from < 0.03 to 0.15–0.64 mg (7.9–33.7 μmol) F l^{-1}; levels in the newborn lamb rose from 9.0 to 300 μg (0.5–15.8 μmol) F l^{-1} and those in the ewe's milk from 0.13 to 0.40 mg (6.8–21.1 μmol) F l^{-1} (Wheeler *et al.*, 1985).

Toxicity of Fluorine

Fluorine is largely a cumulative poison and clinical signs of toxicity may not appear for many weeks or months in animals ingesting moderate amounts of fluoride. Animals are protected by two physiological mechanisms – fluorine excretion in the urine and fluorine deposition in the bones or eggs. During this latent period in cattle, neither milk production (Suttie and Kolstad, 1977) nor the digestibility and utilization of energy and protein of the ration are significantly depressed (Shupe *et al.*, 1962). There may, however, be transient subclinical changes in sensitive organs, such as the testes, and development of spermatozoa may be impaired in poultry (Mehdi *et al.*, 1983) and rabbits (Kumar and Susheela, 1995). Fluorine deposition in the skeleton proceeds

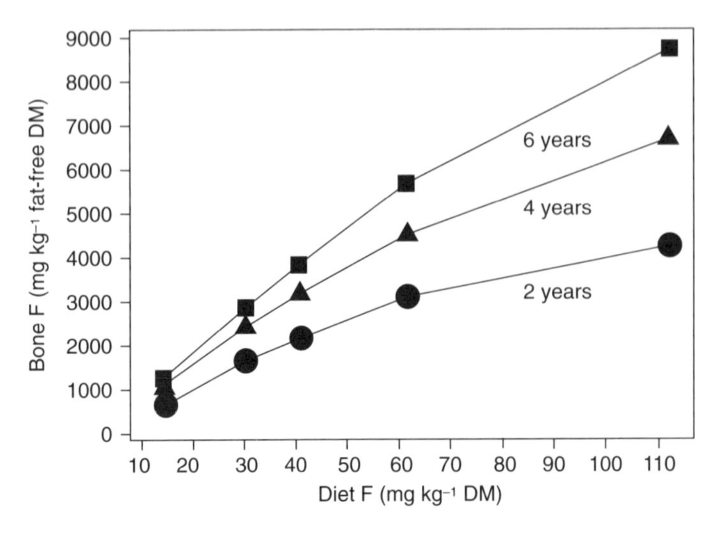

Fig. 18.4. Fluorine concentrations in the skeleton of dairy cows increase with dietary F concentration and duration of exposure but both responses are curvilinear (data from Shupe, 1980).

rapidly at first and then more slowly, until eventually a saturation stage is reached. Beyond this saturation point, which is marked by values of the order of 15–20 g F kg^{-1} dry bone in cattle (30–40 times greater than normal) and 2.3 g F kg^{-1} (five times greater than normal) in hens (Table 18.4), 'flooding' of the susceptible soft tissues with fluorine occurs, metabolic breakdown takes place and death ensues. Chronic skeletal abnormalities are usually detected before this terminal stage is reached. Tolerance to dietary fluoride depends on the species and age of the animal, the chemical form of the fluorine, the duration and continuity of intake and the nature and amount of the diet being consumed. Direct signs of fluorosis are rare in newborn or suckling animals, because placental and mammary transfer of fluorine is limited (Forsyth *et al.*, 1972; Wheeler *et al.*, 1985). General undernutrition tends to accentuate the toxic effects of fluorine (Suttie and Faltin, 1973).

Mink

Mink are the most resilient species of farmed animal studied and they can tolerate some 600 mg F as NaF kg^{-1} DM of a fish/liver/chicken diet for over a year, but 1000 mg F caused unthriftiness, hyperexcitability and mortality in adult females and their offspring, as well as dental and skeletal lesions (Aulerlich *et al.*, 1987).

Poultry

Poultry are also a tolerant species, and maximum safe dietary levels of 300–400 and 500–700 mg F kg^{-1} DM as RP have been reported for growing chicks and laying hens, respectively (Gerry *et al.*, 1949). Tolerance levels similar to those given for chicks have been reported for growing female turkeys, but in the young male 200 mg F kg^{-1} DM as NaF decreased weight gains (Anderson *et al.*, 1955). Increased proportions of dietary fat enhance the growth-retarding effect of excess fluorine in chicks (Bixler and Muhler, 1960), possibly by delaying gastric emptying (McGown and Suttie, 1974). The distribution of fluoride in hens exposed to sodium fluoride at two rates is shown in Table 18.4. The lower level (100 mg F kg^{-1} DM) caused large increases in tibial fluoride concentrations but little accumulation elsewhere

Table 18.4. Effect of level of fluoride exposure as NaF over 122 days on the concentrations of fluoride in body fluids, tissues and products of hens (Hahn and Guenter, 1986).

Dietary F	Plasma[a]	Tibia	Liver	Kidney	Muscle	Egg	
(mg kg^{-1} DM)	(mg l^{-1})	(mg kg^{-1} fat-free DM)[b]				Yolk	Shell
16	0.7	538	5.2	3.4	4.0	4.3	3.9
100	2.8	2247	5.5	3.8	4.0	3.3	12.7
1300	10.1	2600	19.2	31.8	6.7	18.4	307.2
1300 + 1040 Al	6.0	2228	10.5	11.9	5.5	7.5	55.5

[a] Ionic F.
[b] Multiply by 52.63 for μmol kg^{-1} fat-free DM.

and no fall in egg production. The higher level (1300 mg kg^{-1} DM) caused fluoride to accumulate in the eggshell, kidney, liver, plasma and muscle (in decreasing order) (Hahn and Guenter, 1986). An early loss of feathers occurred, as though moulting had been induced, slight diarrhoea was induced and appetite and egg production decreased by 50% during the 112-day trial, but mortality did not increase. The survival of the laying hen may be attributable to the deposition of fluoride in each egg that is laid, but hatchability may be impaired.

Pigs

Pigs are also tolerant of fluoride. No abnormalities were reported by Forsyth *et al.* (1972) when they fed up to 450 mg F kg^{-1} DM as NaF to young female pigs through two pregnancies. Accumulation of fluoride was noted in their newborn and 4-week-old offspring, particularly if the calcium level in the maternal diet was reduced (from 12 to 5 g kg^{-1} DM), but there was no 'generation build-up'.

Sheep

Downward revision of the safe fluoride allowances for breeding ewes was suggested by Wheeler *et al.* (1985), who found evidence of marginal toxicity on treatments equivalent to 25 mg added F kg^{-1} diet. The widely used figures set by the National Research Council (NRC, 1980) were 60 mg F kg^{-1} DM for ewes and 150 mg kg^{-1} DM for fattening lambs. Subsequent work supported the higher tolerance for lambs (Milhaud *et al.*, 1984).

Cattle

The limit of tolerance for dairy cows is approximately 40 mg F kg^{-1} DM when ingested as NaF. Signs of fluorosis appeared within 3–5 years at 50 mg F kg^{-1} DM from this source (Suttie *et al.*, 1957). However, in a study beginning with young calves and lasting for 7 years, the tolerance for soluble fluoride was no more than 30 mg F kg^{-1} DM (Shupe *et al.*, 1963b). Tolerance of dairy cows to the fluoride in RP and CaF_2 (and presumably to soil fluoride) was approximately twice that to the fluoride in NaF, each source providing 65 mg F kg^{-1} DM, but there was little difference in toxicity between NaF and fluoride-contaminated hay (Shupe *et al.*, 1962). Tolerance of fluoride in DFP may be five times that in NaF, the source used experimentally to define tolerances. Thus, while many processed phosphorus sources appear to provide discomfortingly high additions of fluoride to cattle diets, they are probably safe and can only be improved upon at considerable cost (Table 18.2).

Diagnosis of Chronic Fluorosis in Ruminants

The need to combine clinicopathological information and biochemical data in a flexible framework when diagnosing mineral imbalances is nowhere more apparent than in the diagnosis of fluorosis in dairy cattle. Table 18.5 summarizes the vast experience of Shupe (1980) on the subject, and there are important general points to make before considering the individual contribution which each criterion can make and the problems it presents. A particular criterion can reflect:

- the severity of past daily exposure (incisor score);
- the current severity of daily exposure (fluorine in milk or blood);
- the duration as well as severity of exposure (fluorine in bone and urine, molar and hyperostosis score).

Each criterion is open to wide individual variation in expression, with clinicopathological features (the most important in welfare and economic terms) the most variable.

Dental lesions

In young animals exposed to excess fluorine before eruption of permanent teeth, the teeth become modified in size, shape, colour, orientation and

Table 18.5. A guide to the diagnosis of fluorosis in dairy cattle based on chronic experimental exposure studies and field cases (after Shupe, 1980).

		Fluorosis state				
		Normal	Marginal	Mild	Moderate	Severe
F in diet (mg kg^{-1} DM)[a]		< 15	15–30	30–40	40–60	60–109
Incisor score		0–1	0–2	2–3	3–4	4–5
F in milk (mg l^{-1})[a]		< 0.12	< 0.12	0.12–0.15	0.15–0.25	> 0.25
F in blood (mg l^{-1})[a]		< 0.30	< 0.30	0.3–0.4	0.4–0.5	0.5–0.6
	Age					
F in rib bone	2	0.4–0.7	0.71–1.60	1.6–2.1	2.1–3.0	3.0–4.2
(g kg^{-1} fat-free	4	0.7–1.1	1.14–2.40	2.4–3.1	3.1–4.5	4.5–6.6
DM)[b]	6	0.65–1.22	1.22–2.80	2.8–3.8	3.8–5.6	5.6–8.7
F in urine	2	2.3–3.8	3.8–8.0	8.0–10.5	10.5–14.7	14.7–19.9
(mg l^{-1})[a]	4	3.5–5.3	5.3–10.3	10.3–13.3	13.3–18.5	18.5–25.6
	6	3.5–6.0	6.0–11.3	11.3–14.8	14.8–21.0	21.0–30.1
Molar score	2	0–1	0–1	0–1	0–1	0–3
	4	0–1	0–1	0–1	1–2	1–4
	6	0–1	0–1	0–1	1–3	1–5
Periosteal	2	0	0–1	0–1	0–2	0–3
hyperostosis	4	0	0–1	0–1	0–3	0–4
score	6	0	0–1	0–2	0–4	0–5

[a] Multiply by 52.6 to obtain values in μmol l^{-1} or kg^{-1}.
[b] To obtain approximate values kg^{-1} rib ash, multiply by 1.67 (assuming 60% ash in sample); since rib and tail vertebra values are similar on an ash basis, the same factor is required for generating guidelines for vertebral ash; metacarpal and metatarsal bones contain 25% less F than rib on a fat-free DM basis; multiply further by 52.6 for mmol kg^{-1}.

structure, because fluoride impairs the function of ameloblasts – the enamel-forming organs – and the abnormal enamel matrix fails to calcify (Suttie, 1980). The incisors become pitted and blackened, with erosion of the hypoplastic enamel, the molars become abraded and there may be exposure of the pulp cavities, due to fracture or wear. The uneven molar 'table' impairs the mastication of food and is particularly debilitating (Suttie, 1980). The detailed histological and physical characteristics of the incisors in bovine dental fluorosis have been described by Shearer *et al.* (1978a, b). Such dental defects do not arise if first exposure to fluorine occurs after the eruption of the permanent teeth, which, in sheep, varies from 12 months of age for the first incisors to 30 months of age for the last molars.

Skeletal abnormalities

The development of skeletal abnormalities during fluorosis has been reviewed by Shupe (1980) and Suttie (1980). Exposure of immature stock can decrease endochondral bone growth, but the lesion in adult stock is predominantly one of excessive periosteal bone formation (hyperostosis) and the affected bones are often visibly enlarged and have porous, chalky white surfaces instead of an ivory sheen. In cross-section, the shaft is often irregularly overgrown and lacking its normal compact appearance. These abnormalities can occur at any age and are not necessarily accompanied by dental lesions. Exostoses of the jaw and long bones become visible in severely exposed animals and the joints may become thickened and ankylosed, due to calcification of tendons at their points of attachment to the bone. Stiffness and lameness then become apparent, movement difficult and painful and foraging therefore restricted. Growth may become subnormal and weight losses may occur, together with a reduction in milk production and fertility, secondary to the reduced feed consumption (Phillips and Suttie, 1960). These manifestations of fluorosis are less prominent in young animals and in pen-fed animals. The poor lamb and calf crops characteristic of fluorosis areas arise primarily from mortality of the newborn, due to the impoverished condition of the mothers, rather than from a failure of the reproductive process itself (Harvey, 1952). Thyroid enlargement and anaemia have been associated with fluorine exposure in the field (Hillman *et al.*, 1979), but they have not been seen in experimental studies with cattle (Suttie, 1980), pregnant ewes (Wheeler *et al.*, 1985) or mink (Aulerlich *et al.*, 1987).

Bone and tooth composition

In unexposed adult animals, the whole bones commonly contain 0.2–0.5 g F kg^{-1} and values rarely exceed 1.2 g kg^{-1} fat-free DM (FFDM). The teeth contain less than half these concentrations, with higher levels in the dentin than in the less metabolically active enamel, and there is marked variation in concentration within the same fluorotic tooth (Shearer *et al.*, 1978b). Fluorine is incorporated more rapidly into active areas of bone growth than static regions (epiphyses > shaft), into cancellous bones (ribs, vertebrae and sternum) faster than compact cortical bones (metacarpals) (see Table 18.3) and into the surface areas (periosteal and endosteal) of the shaft in preference to the inner regions. The differences in fluorine concentrations within and among the bones and teeth can therefore influence the diagnosis of fluorosis. Fluoride determinations on tail bones obtained by biopsy provide a measure of fluoride accumulation in cattle (Burns and Allcroft, 1962). Suttie (1967) found that values in the metacarpus (FFDM) were approximately 50% of those in tail-bone ash, although levels in the two types of bone were highly correlated. Bovine tail vertebrae and rib contain similar fluoride concentrations on an ash basis, but levels in metacarpals are some 25% lower; values of 4–6 g F kg^{-1} ash at the former sites were associated with a low incidence of lameness in UK cases of fluorosis (Burns and Allcroft, 1966). In dairy cattle, experimental fluorine toxicosis was associated with values in excess of 5.5 g kg^{-1} FFDM in compact bone and 7.0 g kg^{-1} DM in cancellous bone, with concentrations between 4.5 and 5.5 g F kg^{-1} DM indicating a marginal status (Suttie *et al.*, 1958). The toxic thresholds for fluorine in the cancellous bones of sheep have been placed at 7 g kg^{-1} FFDM or 11 g F kg^{-1} ash (Puls, 1994). The calcium (Ca):P ratio of fluorotic bone is normal, the magnesium content increased and there may be a precipitation as CaF_2.

Urine

In sheep and cattle not exposed to excess fluorine, the urinary concentration rarely exceeds 10 mg l^{-1}. In long-term fluorosis experiments with dairy cows, the urinary fluoride concentrations were as follows: normal animals, less than 5 mg F l^{-1}; those with borderline toxicity, 20–30 mg F l^{-1}, and those with systemic toxicity, over 35 mg F l^{-1} (Suttie *et al.*, 1961; Shupe *et al.*, 1963a). Urinary fluorine can come from the release of the element from the skeleton after animals are removed from a high-fluoride supply, particularly when the diet cannot sustain mineralization of the skeleton (e.g. during lactation). High urinary fluoride levels therefore reflect either current ingestion or previous exposure to high intakes and can be easily measured, using an ion-specific electrode. Samples are best taken in the morning on more than one occasion, and expression on a specific-gravity or creatinine basis improves the diagnostic value of results obtained (Shupe, 1980).

Blood

Most of the fluoride found in whole blood (*c.* 75%) is present in the plasma, and concentrations are therefore higher in plasma than in whole blood. Plasma fluoride values are related to the current rate of fluoride ingestion (Milhaud *et al.*, 1984). During their formative period, the teeth are sensitive to small changes in plasma fluoride concentration: when concentrations approach 0.5 mg l^{-1} or more, severe dental lesions appear in young cattle; at values between 0.2 and 0.5 mg F l^{-1}, less severe damage occurs; and below 0.2 mg l^{-1}, few adverse effects are apparent (Suttie *et al.*, 1972). In a long-term study of dairy cows continuously or periodically exposed to high fluoride intakes, the plasma values increased from 0.1 to 1.0 mg l^{-1} (Suttie *et al.*, 1972). On the basis of these findings, 0.2 mg F (10.5 μmol) l^{-1} can be regarded as a critical plasma concentration in cattle. However, plasma values are difficult to interpret, because there is a marked diurnal variation and the values respond so rapidly to changes in fluoride intake that samples must be taken very soon after the actual ingestion of the fluoride if they are to reflect recent exposure. Sheep appear to tolerate higher circulating levels of fluoride than cattle. In lambs given up to 2.5 mg F kg^{-1} live weight (LW) in a milk diet, plasma values reached 0.6 mg (31.5 μmol) F l^{-1} with no obvious signs of ill health (Milhaud *et al.*, 1984), while in pregnant ewes a similar level was only associated with marginal evidence of toxicity (Wheeler *et al.*, 1985). However, the sheep studies were both of relatively short duration (4–8 months).

Other indices

Bone alkaline phosphatase activity is increased in fluorotic chicks (Motzok and Branion, 1958) and cows (Shupe *et al.*, 1962). The influence of different amounts and forms of dietary fluorine on the fluoride content and alkaline phosphatase activity of the bones of heifers was presented in Table 18.3.

Continuity of intake

The importance of continuity of fluorine intake and the age of the animal is illustrated by experience with sheep consuming water-borne fluoride. Artesian bore water containing 5 mg F l^{-1} induced severe dental abnormalities and other signs of fluorosis in sheep in the hot climatic conditions of Queensland (Harvey, 1952), whereas in the cooler conditions of South Australia, where very little water was drunk during the wet winter months, no ill effects were observed in mature sheep given 20 mg F l^{-1} as NaF in the drinking-water (Peirce, 1954). During periods of low intake, the exchangeable skeletal stores are depleted and the fluoride excreted in the urine. When compared on the basis of total yearly intake, skeletal fluoride storage is similar for continuous and intermittent exposure (Suttie, 1980). However, alternating periods of high and low intakes can be more damaging to animals than continuous intakes, because of rapid increases in tissue concentration during periods of high intake (Suttie *et al.*, 1972). With short-term exposure to high intakes (*c.* 90 mg F kg^{-1} DM), systemic reactions, such as weight loss and unthriftiness, due to decreased appetite, can arise in dairy cattle (Suttie *et al.*, 1972).

Treatment and Prevention of Fluorosis

Both the dental and skeletal lesions of chronic fluorosis are essentially irreversible, the latter being beyond the scope of remodelling. If livestock cannot be removed from the source of exposure, steps can be taken to delay the progression of the disease. Aluminium salts protect against high intakes of fluoride by sheep (Becker *et al.*, 1950) and cattle (1% as aluminium sulphate

($Al_2(SO_4)_3$) (Allcroft *et al.*, 1965), apparently through reducing fluoride absorption from the intestinal tract. Similar additions of aluminium sulphate greatly reduced the fluorosis in hens given 1300 mg F kg^{-1} DM and reduced fluoride concentrations at most sites, particularly eggshell, liver and kidney (Table 18.4). Calcium salts function similarly in rats (Weddle and Muhler, 1954), while borate and silicate limit the effects of excess fluorine on the skeleton of pigs (Seffner and Tuebener, 1983).

Essentiality of Fluorine

The discovery made nearly 70 years ago that the incidence of dental caries in human populations was significantly higher where the water-supplies were virtually free from fluoride than in areas where the water contained 1.0–1.5 mg F l^{-1} focused attention on the beneficial as well as the toxic effects of fluorine. Dental caries does not present a health problem in farm animals, but fluorine has long been recognized as a constituent of not only bones and teeth but also the soft tissues and fluids of the body. There is as yet no evidence that fluorine performs any specific essential function, but provision of fluorine stimulated the growth of rats fed on a purified diet in plastic isolators with minimum atmospheric contamination (Schwarz and Milne, 1972; Milne and Schwarz, 1974); in the latter study, potassium fluoride (KF) was added at 2.5 mg F kg^{-1} DM to a diet containing less than 0.04 mg F kg^{-1} DM. Others failed to confirm the need for dietary fluorine for optimum growth (Maurer and Day, 1957; Doberenz *et al.*, 1964). Controversy continued with the report that female mice on a diet containing 0.1–0.3 mg F kg^{-1} DM developed a progressive infertility in two successive generations, accompanied by a severe anaemia in mother and offspring, unless they were given 50–200 mg F l^{-1} as NaF in the drinking-water (Messer *et al.*, 1974). Other workers reported no such effects (Weber and Reid, 1974). Tao and Suttie (1976) suggested that the apparent essentiality could have been due to a pharmacological effect of fluoride in improving iron utilization in a marginally iron-sufficient basal diet, since fluoride could enhance the intestinal absorption of iron in rats (Ruliffson *et al.*, 1963). The only evidence of essentialilty of fluoride in farm livestock comes from a recent study by Anke *et al.* (1997b), who reported skeletal abnormalities in female goats and poor growth in their offspring after ten generations on a diet containing < 0.3 mg F kg^{-1} DM. All natural diets are likely to provide sufficient fluorine to meet any dietary requirement.

LEAD

Lead poisoning is one of the most frequently reported causes of poisoning in farm livestock, with cattle the most commonly affected (Blakley, 1984). Young calves are particularly vulnerable; in Northern Ireland, lead poisoning once accounted for 4.5% of all deaths in calves, while, in western Canada, 50.9% of the lead poisoning cases in bovines were in animals < 6 months of age (Blakley, 1984). Cases were more common in dairy than in beef breeds in Canada. These effects of species, breed and age are attributable to the access cattle are inadvertently given to point sources of lead (e.g. old batteries, tins of paint) in the vicinity of farm buildings, the trapping of heavy objects in the adult reticulum and the lack of a functional and protective rumen microflora in the youngest calves. Cases of lead poisoning are mostly acute in form and the 20,000 cases which once occurred in cattle every year, worldwide, were seen as inevitable consequences of the vast amounts of lead which were redistributed from the earth's crust each year by industrial processes (1.5×10^6 tons: R.P. Botts, cited by Bratton, 1984). The effects of chronic lead exposure on tissue lead concentrations in farm livestock are receiving increasing attention, due to concern over the subclinical effects of raised blood lead concentrations in young children. All sources of lead (Pb) intake are being scrutinized and MAC of 1–2 mg Pb kg^{-1} FW have been defined for edible meat and offal.

Sources of Lead

Soil lead concentrations vary widely; in the UK, the median value was 36.8 mg Pb kg^{-1} DM, but the range extended from 4.5 to 2900 mg Pb kg^{-1} DM (Archer and Hodgson, 1987). Mining activity (Moffat, 1993) and dispersal of sewage sludge (Suttle *et al.*, 1991) can raise soil lead concentrations to 500 mg kg^{-1} DM or more. Although soils can contain appreciable concentrations of lead in ethylenediaminetetra-acetic acid (EDTA)-extractable forms (Archer and Hodgson, 1987), lead is poorly taken up by plants and concentrations in pasture and crops rarely exceed 5 mg kg^{-1} DM. The principal threat to livestock therefore comes from the soil and the vulnerable animals are those consuming soil while grazing or foraging on contaminated land. Automotive exhaust fumes cause local accumulations beside major roads and probably contribute to atmospheric deposition, which affects vast areas, including the upland peat soils of Scotland, in which levels of > 50 mg Pb kg^{-1} DM have been frequently recorded (Bacon *et al.*, 1992). When lead-contaminated pasture is ensiled, downward migration of lead in the silo can lead to fourfold increases in lead concentration in the lowermost layers, sufficient, at 119 mg Pb kg^{-1} DM, to cause clinical lead toxicity in a dairy herd (Coppock *et al.*, 1988).

Metabolism of Lead

Lead is commonly regarded as a cumulative poison, with duration of exposure as important as level of exposure, but there is evidence that livestock exert some control over the amounts which they retain.

Absorption

Lead is poorly absorbed by non-ruminant and ruminant species alike; sheep and rabbits were reported to absorb only 1% of the lead they ingested (Blaxter, 1950) and rats only 0.14% (Quarterman and Morrison, 1975). However, there can be considerable variation about these low mean values, depending on dietary composition. In lambs, raising the dietary sulphur concentration from low (0.7) to moderate (2.3) or high (3.8 g S kg^{-1} DM) levels increased the mean survival time from 6 to 15 or 30 weeks, respectively, when 200 mg Pb kg^{-1} DM was added as the acetate to their diet (Quarterman *et al.*, 1977). Precipitation of unabsorbable lead sulphide in the sulphide-rich rumen at the higher sulphur intakes probably reduced the toxicity of lead. Lead absorption increases in iron deficiency (Morrison and Quarterman, 1987) and can be decreased by raising dietary calcium and phosphorus concentrations in avian species (Scheuhammer, 1987) and lambs (Morrison *et al.*, 1977). Lead has an affinity for the calcium-binding protein which is induced in calcium deficiency and by vitamin D (Fullmer *et al.*, 1985). The apparent absorption of lead decreases with increasing duration of exposure or dietary lead concentration (Fick *et al.*, 1976).

Tissue accretion

There is further evidence that lead is not a simple cumulative poison, since tissue lead levels are often non-linear functions of daily and cumulative lead intakes, although patterns of accretion vary widely from tissue to tissue. In blood and liver, lead concentrations show early rapid increases in lambs at a constant daily rate of exposure before reaching a plateau (Brebner *et al.*, 1993). During medium-term exposure (7 weeks), most lamb tissues retain diminishing proportions of the ingested lead as dietary concentrations increase, liver showing a more marked curvilinearity than brain or bone, for example (Fig. 18.5; Fick *et al.*, 1976). Thus, animals adapt to lead exposure and blood and liver lead concentrations are poor indices of long-term exposure. Kidney lead concentrations show distinctive changes during extreme lead exposure, increasing markedly with dietary levels of 1000 mg Pb kg^{-1} DM (Fig. 18.5). Stronger linear increases in kidney lead with duration of exposure (Stark *et al.*, 1998) confirm the limits of adaptation. Lead accumulates in bone more than in soft tissues at lower levels of exposure (Fig. 18.5), particularly at growth points in the immature animal and at times of rapid bone turnover in the mature animal. Four- to fivefold greater increases in bone lead have been reported in laying than in non-laying birds exposed to lead (Finley and Dieter, 1978). Calcium and phosphorus supplements delay the release of lead from the skeleton when exposure to lead has ceased (Quarterman *et al.*, 1978).

Maternal transfer

When cows (Pinault and Milhaud, 1985) or ewes (Carson *et al.*, 1974; Suttle *et al.*, 1997) are exposed to lead during gestation, raised lead concentrations in the liver or blood of the newborn are evidence of placental transfer of lead.

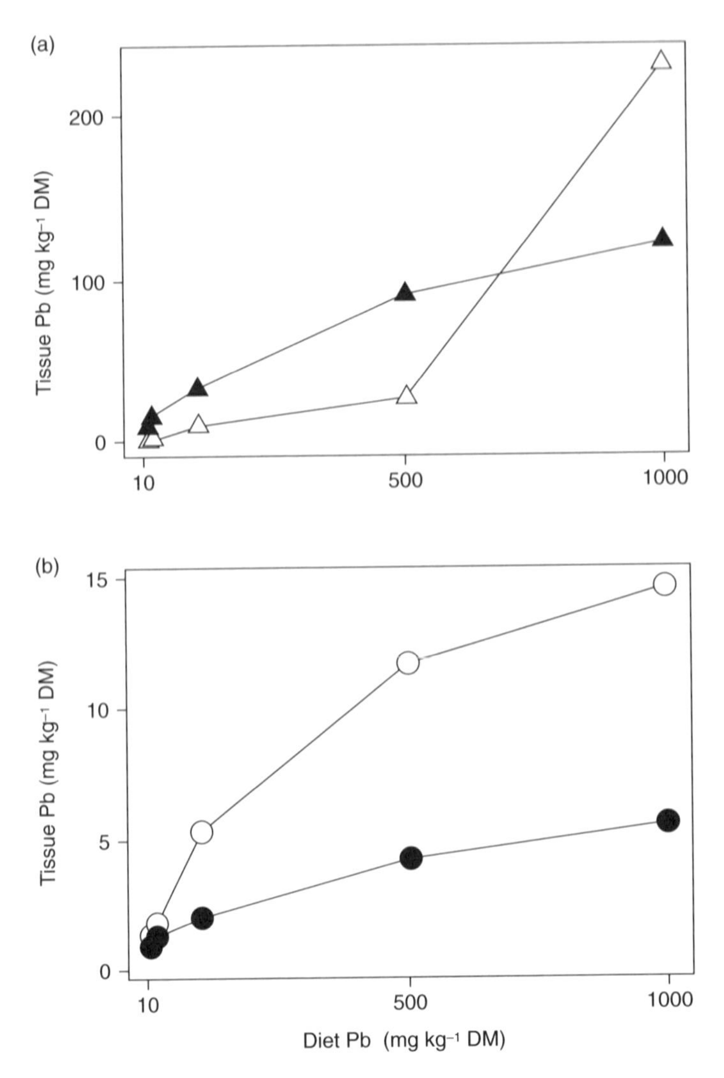

Fig. 18.5. Tissue lead concentrations increase with dietary Pb concentration and duration of exposure but both responses are curvilinear (see text for the latter effect): limits to liver accumulation are particularly marked and partition may shift towards the kidney at the very high exposure levels (data from Fick *et al.*, 1976). Δ, kidney; ▲, bone; ○, liver; ●, brain.

Raised lead levels in milk indicate that maternal transfer will continue during lactation. Although the increases in concentration are small, lead is well absorbed from milk (Morrison and Quarterman, 1987) and risk to the suckling or other consumer may arise.

Toxicity of Lead

Symptoms

Acute and chronic lead toxicity are clinically distinguishable conditions. Acute toxicity is characterized by gastrointestinal haemorrhage and anaemia, together with liver necrosis and kidney damage. In chronic lead exposure, lesions of the alimentary tract and anaemia are not seen, but osteoporosis may accompany hydronephrosis, particularly in young lambs (Butler *et al.*, 1957; Quarterman *et al.*, 1977). The first visible sign of malaise in chronic lead toxicity was a loss of appetite. 'Lead lines' (probably porphyrin pigments arising from erythrocyte breakdown) may eventually be seen in the gingiva adjoining the teeth. Suckling lambs in a former lead-mining area were particularly vulnerable to lead-induced osteodystrophy between 6 and 10 weeks of age (Butler *et al.*, 1957) and recent work in the same area has shown that blood lead concentrations peak when the condition occurs (Fig. 18.6; Moffat, 1993). A similar peak in blood lead has been reported in the offspring of housed ewes on constant lead intakes for over a year and was associated with a peak in milk lead concentrations in mid-lactation (Suttle *et al.*, 1997). Mobilization of lead from a demineralizing maternal skeleton (see Chapter 4) may be responsible. Risks of lead toxicity to mother and lamb may both be heightened at that time. Conversely, factors that increase bone mineralization, such as dietary supplements of calcium and phosphorus, may reduce lead toxicity, partly by increasing the safe capture and 'dilution' of lead in an increased volume of newly formed bone.

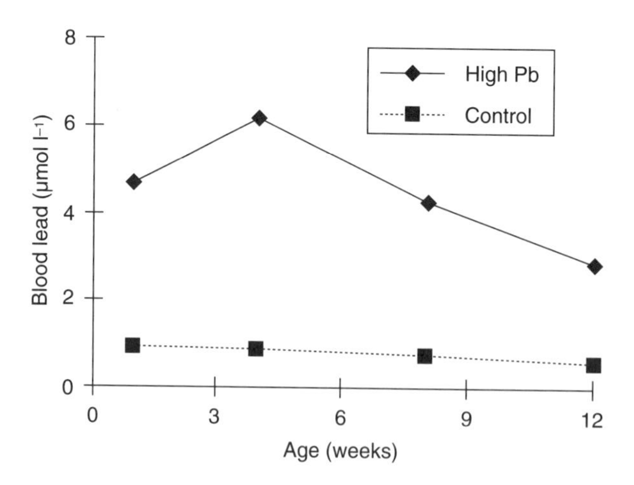

Fig. 18.6. Outbreaks of osteodystrophy in suckling lambs in former lead-mining areas are preceded by high blood Pb concentrations: efficient absorption of Pb from milk and ingested Pb-rich soil may be responsible (Moffat, 1993).

Tolerance

In both acute and chronic toxicity, maximum tolerable doses cannot easily be defined. A dose of 5 mg Pb kg^{-1} LW day^{-1} killed milk-fed calves within 7 days but was apparently harmless to similar calves given a diet of hay and grain (Zmudzki *et al.*, 1984), confirming the protective influence of a functional rumen. Calves given $PbCO_3$ while on a milk-replacer diet tolerated 2.6 mg kg^{-1} LW day^{-1} for 7 weeks (Lynch *et al.*, 1976a) but succumbed to 3.9 mg kg^{-1} LW day^{-1} after 12 weeks (Lynch *et al.*, 1976b). High tolerance may have been due to the insoluble source used. Weaned calves showed only a slight reduction in appetite and growth while receiving 1000 mg Pb as $PbSO_4$ kg^{-1} DM for 4 weeks (Neathery *et al.*, 1987). Fick *et al.* (1976) reported no adverse effects of feeding 1000 mg Pb kg^{-1} DM to weaned lambs for 12 weeks, whereas Quarterman *et al.* (1977) reported rapid weight loss in some individuals after 6 weeks with 200 mg Pb kg^{-1} DM and a moderate sulphur intake. Avian species are more susceptible than ruminants, chicken and quail showing reductions in egg production with 200 and 10 mg Pb kg^{-1} DM, respectively, present in the diet (Scheuhammer, 1987). Toxicity is likely to be reduced when diets rich in calcium are fed.

Subclinical effects

Publicity given to the alleged effects of small increases in blood lead concentrations on the learning ability and behaviour of children prompted a study of the effect of chronic exposure of ewes for 26 weeks, commencing 1 month prior to mating and continuing during pregnancy and into lactation (120 or 230 mg elemental Pb orally day^{-1}) on the behaviour and reproduction of the ewe, but no defects were found, despite increases in mean blood lead concentrations from 64 to 302 μg l^{-1} (van Gelder *et al.*, 1973). Similar levels of exposure throughout gestation raised blood concentrations in lambs to 250 μg l^{-1} and slowed their learning behaviour (Carson *et al.*, 1974). In the study of Lynch *et al.* (1976b) referred to above, the normal age-related increase in aminolaevulinic acid dehydratase (ALAD) activity in erythrocytes was totally inhibited by as little as 0.65 mg Pb kg^{-1} LW day^{-1} and was accompanied by a rise in blood lead to 911 μg l^{-1}, although calf growth was unimpaired. Early reductions in ALAD were reported in cows given mixed sources of lead equivalent to 100 mg Pb kg^{-1} DM with no harmful effect on live weight or milk yield (Pinault and Milhaud, 1985). Blood levels peaked at 350 μg l^{-1} and did so sooner when exposure began during lactation than when it began during gestation (13 vs. 9 weeks). Increases in zinc protoporphyrin (ZPP) concentrations were also found, indicating that lead exposure had been sufficient to inhibit ferrochelatase activity and to affect haem synthesis.

Diagnosis of Lead Toxicity

Diagnosis of lead toxicity is confirmed by the presence of high lead concentrations in kidney (Table 18.6) and histological evidence of kidney

Table 18.6. Guidelines for assessing lead exposure and likelihood of lead toxicity in livestock from Pb concentrations in diet, blood and tissues; normal (N), high (H) and toxic (T) levels are given.

Group	Status	Diet	Bone	Liver	Kidney	Blood
		(mg kg^{-1} DM)[a]		(mg kg^{-1} FM)[a]		(mmol l^{-1})[b]
Cattle	N	1–6	1–7	0.1–0.5	0.1–0.5	0.05–0.25
and	H[c]	20–1000	30–75	0.8–2.0	0.7–4.0	0.5–1.5
sheep	T[d]	> 2000	> 75	> 8.0	> 20.0	> 2.0
Pigs	N	2–8	1?	0.2–0.5	0.2–1.2	0.2–1.5
	H[c]	20–750	100–500	5.0–35.0	5.0–25.0	1.5–4.8
	T[d]	> 750	> 500	> 35.0	> 25.0	> 4.8
Poultry	N	1–10	< 50	0.1–5.0	0.1–5.0	0.2–1.0
	H[c]	20–200	150–400	5.0–18.0	5.0–20.0	1.0–2.0
	T[d]	> 200	> 400	> 18.0	> 20.0	> 2.0

[a] Multiply by 4.826 to obtain values in μmol kg^{-1} DM or fresh matter (FM).
[b] Multiply by 207.2 to obtain values in μg l^{-1}.
[c] Values within the 'high' category indicate unusually high levels, but not high enough to harm animal health.
[d] Values in 'toxic' category will cause harm but not necessarily death.

damage at post-mortem. In acute cases, high lead concentrations may be found in digesta and faeces. Surviving cohorts are likely to have raised blood lead concentrations (Table 18.6) and early dysfunction is indicated by a fall in erythrocyte ALAD (Bratton and Zmudski, 1984; Bratton *et al.*, 1986) and a rise in ZPP concentrations. The ALAD activity is slightly reduced by heavy exposure to other metals, such as cadmium and selenium.

Livestock as Sources of Lead Exposure for Humans

Lead does not accumulate in muscle during lead exposure (Stark *et al.*, 1998) and there is no risk to humans consuming carcass meat from livestock exposed to lead. There is, however, a small possible risk from the consumption of offal and a greater risk that carcasses may be rejected because lead levels in offal exceed national MAC (2 and 1 mg kg^{-1} FW in liver and kidney, respectively, in Europe). Chronic exposure to lead may therefore raise kidney lead concentrations to levels deemed unacceptably high for human consumption. Simulation of high levels of ingestion (100 g kg^{-1} DM) of soils containing 592 and 841 mg Pb kg^{-1} DM by lambs caused mean kidney lead values to rise to 9.0 and 9.7 mg Pb kg^{-1} DM (Brebner *et al.*, 1993), similar to those attained after feeding 100 mg Pb as acetate kg^{-1} DM (Fick *et al.*, 1976), albeit for a longer period (160 vs. 84 days). However, no increase in kidney lead occurred in cattle grazing on sludge-amended soils, containing 79–135 mg Pb kg^{-1} DM, for 8 years (Fitzgerald *et al.*, 1985).

Essentiality of Lead

Lead can display certain characteristics of an essential nutrient. By maintaining young rats under rigorously clean conditions and feeding highly purified diets containing < 200 μg Pb kg^{-1} DM, growth was depressed in comparison with a lead-supplemented group (Schwarz, 1974). Kirchgessner and his associates confirmed and extended these findings in the offspring of depleted dams in a series of studies using diets with only 18–45 μg Pb kg^{-1} DM (Nielsen, 1996). Disturbances of iron metabolism and anaemia were prominent features in the lead-depleted rat. More recently, artificially reared piglets given a diet containing 31 μg Pb kg^{-1} DM gained 15% less weight than those supplemented with 800 μg Pb kg^{-1} DM; impaired lipid metabolism was suspected (Kirchgessner *et al*., 1991). However, a unique biochemical role for lead has yet to be identified.

MERCURY

The 1970s saw an explosion of interest in the toxicity of mercury, following events in Minamata Bay, Japan. Industrial discharges of inorganic mercury into the bay were converted to organic (methylated) mercury by lower marine organisms and entered the human food-chain in toxic amounts for a coastal community heavily dependent on fish in their diet. Stringent restrictions were placed on acceptable mercury levels in food (0.5 mg kg^{-1} DM) by the Food and Drugs Administration (FDA) and renewed interest was taken in the residual effects of the long-standing exploitation of fungicidal and other toxic properties of the element in agriculture. The methylation of mercury is not, however, a modern phenomenon but goes back to the dawn of creation and is a fundamental property of anaerobic microorganisms, expressed in any environment naturally or unnaturally endowed with mercury. Puls (1994) has published an extensive bibliography on the subject and here only the essential features pertaining to the toxicity of mercury to livestock and their possible importance as a source of mercury in the human diet will be considered.

Sources of Mercury

The sources of mercury that pose threats to livestock or enter the food-chain via livestock do not originate in the farm environment. Soils contain little mercury (Hg), Archer and Hodgson (1987) reporting a maximum value of 2.12 mg Hg kg^{-1} DM for 305 samples from England and a median value of 0.09 mg kg^{-1} DM. Pastures and crops also contain little mercury (< 0.1 mg kg^{-1} DM). The only mercury-rich component in normal diets is likely to be fish-meal, but the mercury may be present in the more available form of methylmercury. However, the ingestion of mercury in fish has led to less accretion of mercury in tissues than was anticipated. Chang *et al.* (1977) incorporated sufficient raw fish in the diet of pigs to raise dietary mercury to levels of 3–8 mg kg^{-1} DM. No signs of toxicity were visible clinically or at post-mortem and there were no teratogenic effects. A similar level of mercury given as synthetic methylmercury caused fatty infiltration of the liver. Fish is rich in selenium and it has been suggested that this attenuates the threat from mercury in fish products (Ganther *et al.*, 1972). The use of mercury-based fungicides for dressing cereal seed raises mercury concentrations to potentially hazardous levels (*c.* 20 mg Hg kg^{-1} DM) and such materials will cause toxicity if fed to cattle (Boyd, 1985) or other species. Accidental consumption of mercurous chloride, intended for the treatment of club root in brassica crops, has recently caused the deaths of dairy heifers (Simpson *et al.*, 1997). The topical administration of mercury-containing drugs for the treatment of skin infections has also resulted in deaths from mercury poisoning.

Metabolism of Mercury

Absorption

The metabolism of mercury is dominated by the contrasting behaviour of inorganic and organic forms. Inorganic forms are poorly absorbed, the range being variously quoted as 1–3% (Kostial *et al.*, 1978) and 5–15% of intake (Clarkson, 1987). Organic forms are fat-soluble and much more absorbable. Neathery *et al.* (1974) recorded an apparent absorption of 59% for radio-labelled methylmercury chloride in lactating dairy cows, and values as high as 90% have been reported for phenyl derivatives (Clarkson, 1987). Very young mammals absorb mercury efficiently, even from inorganic sources (30–40% of intake: Kostial *et al.*, 1978).

Retention

Mercury absorbed in organic forms is avidly retained, little appearing in urine (1.1%) or milk (0.17%) and most (72%) being found in muscle (Neathery *et al.*, 1974). Mercury has an affinity for sulphydryl groups, which may explain its tissue distribution, toxicity and ability to interact with other micronutrients (Scheuhammer, 1987). Absorbed mercury is transported primarily in association with the glutathione-rich erythrocytes, which contain 20 times more mercury than the plasma in exposed animals. Mercury also has an affinity for cysteine-rich molecules, such as metallothionein, from which it can displace bound zinc and cadmium. The addition of selenium to the diet affords substantial protection from inorganic and organic mercury toxicity in poultry and there is in fact a mutual antagonism between these elements (El-Begearmi *et al.*, 1977). The mercury–selenium interaction may be mediated by the induction of the cysteine-rich selenoprotein P (see Chapter 15; Chang and Suber, 1982). However, vitamin E supplements also afford protection against mercury toxicity, suggesting that exposure to mercury creates an oxidative stress at the membrane level.

Toxicity of Mercury

Organic forms of mercury are far more toxic than inorganic forms and responsible for most incidents of mercury poisoning. For example, when weanling pigs were given 0.5, 5.0 or 50 mg Hg kg^{-1} DM in either form, the lowest level in the organic form (CH_3HgCl) caused fatty liver degeneration and the highest was lethal after 3–4 weeks, whereas the highest inorganic level caused only subclinical (fatty liver) disorder after rearing to 82 kg live weight (Chang *et al.*, 1977). The symptoms of toxicity were similar to those found in other farmed species exposed to organic mercury: anorexia and inability to drink; blindness, ataxia and hypersensitivity; recumbency and death. Post-mortem examination revealed generalized congestion and a haemorrhagic gastroenteritis. In chickens, there are loss of appetite and signs of wing and muscle weakness (Scheuhammer, 1987); demyelination has also been reported. Chang *et al.* (1977) found high levels of mercury in hair (7.1), kidney (1.0) and liver (0.6 mg kg^{-1} DM) in affected pigs but no elevation of levels in the central nervous system (CNS). Higher mercury concentrations were found in clinically normal pigs exposed to inorganic mercury than in those poisoned by organic mercury. Tissue mercury concentrations must therefore be interpreted cautiously and in conjunction with pathological evidence. Exceedingly high concentrations of mercury were found in the kidneys of heifers poisoned with mercuric chloride ($HgCl_2$) (300–350 mg kg^{-1} DM: Simpson *et al.*, 1997); the associated clinical symptoms were a bloody diarrhoea, excessive thirst, salivation, extreme depression and an unsteady gait. The hazard presented by inorganic mercury should not be minimized; as little as 2 mg Hg kg^{-1} DM has impaired testicular development in young quail and 8 mg Hg kg^{-1} DM has depressed the fertility of eggs produced by young hens (Scheuhammer, 1987). Diagnostic guidelines have been presented by Puls (1994).

Essentiality of Mercury

There have been no claims to date that mercury is an essential constituent of the diet for any species. Chang *et al.* (1977) reported a tendency for the growth of pigs to be increased by the addition of 50 mg Hg as $HgCl_2$ kg^{-1} DM to the diet. They attributed the growth response to bactericidal activity, akin to that underlying the growth response to antibiotics added to similar diets.

RUBIDIUM

Essentiality of Rubidium

A brief note on rubidium (Rb) is given to mark the latest addition to the list of elements which some regard as 'essential'. Feeding adult female goats on a diet containing < 280 μg Rb as opposed to 10 mg Rb kg^{-1} DM resulted in abortion, lower birth weight and raised mortality among kids and low weaning weights (Anke *et al.*, 1997a).

ESSENTIALITY RECONSIDERED

In view of the similarities in the clinical consequences of depriving livestock of most of the essentially toxic (ETE) and occasionally beneficial elements (OBE) (see also Chapter 17) and the single species (goats) and source (Jena workers) of most of the positive responses to supplementation in livestock, it is worthwhile putting forward some explanations of apparent essentiality to stimulate investigations by other groups. Five possibilities spring to mind and most are of a pharmacological nature.

1. *Stimulation of the immune system.* In keeping with their predominantly pathogenic qualities, both lead and cadmium stimulate cells of the immune system at low concentrations *in vitro* (0.2–1.0 mg l^{-1} and 50–500 nmol l^{-1}, respectively: Borella and Giardino, 1991). If such mitogenic properties were induced by small dietary supplements of potentially toxic elements (PTE), the adjuvant effect might improve the resistance of livestock to miscellaneous infections, improving longevity and fertility in long-term studies.
2. *Alterations in the gut microflora.* In view of the sensitivity of rapidly dividing cells to PTE, their addition to the diet of ruminants may cause subtle changes in gut microflora, particularly those first exposed to the element in the rumen. It is the bactericidal effects of fluorine which are responsible for its beneficial effects on dental caries in childhood when added at low concentrations to the drinking-water. Rumen microbes may have higher requirements for some OBE and ETE than the host: when these needs are not met, cellulose digestion and appetite may be impaired, leading to the common losses which accompany inanition.
3. *Antagonism towards antinutritional factors in the diet.* Given the prolonged exposure often necessary to produce clinical symptoms of 'deficiency' for the rarely beneficial elements, it is possible that the benefits from supplementation arise from an anomaly in the basal diet, which is counteracted by a number of PTE. For example, EDTA is commonly used to remove minerals from protein sources, such as soybean proteins, for the specially prepared, deficient diets. If prolonged exposure to residual traces of EDTA is harmful, addition of heavy metals that chelate with EDTA may be beneficial to the PTE-deprived animal.

4. *Deranged protective mechanisms.* Livestock historically exposed to PTE may constitutively express proteins or invoke mechanisms which ensure that their tissues, particularly 'front-line' epithelial cells lining the gastrointestinal tract, are protected from those PTE. The normal situation is for those protective mechanisms to be activated. When an 'exotic' situation is created by removing the inductive PTE from the diet, however, the protective capacity may be freed to bind essential and useful elements, thus impairing performance.
5. *Impurities in mineral supplement.* The ultratrace quantities of the 'newer essential elements' which animals appear to require places high demands on the purity of the specific mineral sources used in each investigation of essentiality; these have rarely been published and, where 10–25 mg kg^{-1} DM of the ETE or OBE are added, a small impurity (of F in $AlCl_3$, for example) might lead to 'ghost' responses to the impurity.

Of these five possible explanations only the third involves an artefact of the experiment; the remainder would be consistent with essentiality if the response could not be duplicated by another PTE.

References

Alfrey, A.C. (1986) Aluminium. In: Mertz, W. (ed.) *Trace Elements in Human and Animal Nutrition.* Academic Press, New York, pp. 399–413.

Allcroft, R., Burns, K.N. and Herbert, C.N. (1965) *Fluorosis in Cattle. 2. Development and Alleviation: Experimental Studies.* Animal Disease Surveys Report No. 2, Part 2, HMSO, London, 58 pp.

Allen, V.G., Robinson, D.L. and Hembry, F. (1984) Effects of ingested aluminium sulphate on serum magnesium and the possible relationship to hypomagnesaemic tetany. *Nutrition Reports International* 29, 107–115.

Allen, V.G., Horn, F.P. and Fontenot, J.P. (1986) Influence of ingestion of aluminium, citric acid and soil on mineral metabolism of lactating beef cows. *Journal of Animal Science* 62, 1396–1403.

Ammerman, C.B., Fick, K.R., Hansard, S.L. and Miller, S.M. (1973) *Toxicity of Certain Minerals to Domestic Animals: a Review.* Animal Science Research Report AL73–6, University of Florida, Gainsville.

Anderson, J.O., Hurst, J.S., Strong, D.C., Nielsen, H., Greenwood, D.A., Robinson, W., Shupe, J.L., Binns, W., Bagley, R. and Draper, C. (1955) Effect of feeding various levels of sodium fluoride to growing turkeys. *Poultry Science* 34, 1147–1153.

Angelow, L., Anke, M., Groppel, B., Glei, M. and Muller, M. (1993) Aluminium: an essential element for goats. In: Anke, M., Meissner, D. and Mills, C.F. (eds) *Proceedings of the Eighth International Symposium on Trace Elements in Animals and Man, Dresden.* Verlag Media Touristik, Gersdorf, pp. 699–704.

Anke, M., Groppel, B. and Krause, V. (1991) The essentiality of the toxic elements cadmium, arsenic and nickel. In: Momcilovic, B. (ed.) *Proceedings of the Seventh International Symposium on Trace Elements in Man and Animals (TEMA 7), Dubrovnik.* IMI, Zagreb, pp. 11-6–11-8.

Anke, M., Gurtler, H., Angelow, L., Gottschalk, J., Drobner, C., Anke, S., Illing-Gunther, H., Muller, M., Arnhold, W. and Shafter, V. (1997a) Rubidium – an essential element for animals and humans. In: Fischer, P.W.F., L'Abbé, M.R., Cockell, K.A. and Gilson, R.S. (eds) *Proceedings of the Ninth Symposium on Trace Elements in Man and Animals.* NRC Research Press, Ottawa, pp. 189–191.

Anke, M., Gurtler, H., Neubert, H., Glei, M., Anke, S., Jaritz, M., Freytag, H. and Shafter, V. (1997b) Effects of fluorine-poor diets in generations of goats. In: Fischer, P.W.F., L'Abbé, M.R., Cockell, K.A. and Gilson, R.S. (eds) *Proceedings of the Ninth Symposium on Trace Elements in Man and Animals.* NRC Research Press, Ottawa, pp. 192–194.

Archer, F.C. and Hodgson, J.H. (1987) Total and extractable trace element contents of soils in England and Wales. *Journal of Soil Science* 38, 421–431.

Aulerlich, R.J., Napolitano, A.C., Bursian, S.J., Olson, B.A. and Hochstein, J.R. (1987) Chronic toxicity of dietary fluorine to mink. *Journal of Animal Science* 65, 1759–1767.

Bacon, J.R., Berrow, M.L. and Shand, C.A. (1992) Isotopic composition as an indicator of origin of lead accumulation in surface soils. *International Journal of Environmental Analytical Chemistry* 46, 71–76.

Bahri, L.E. and Romdane, S.B. (1991) Arsenic poisoning in livestock. *Veterinary and Human Toxicology* 33, 259–264.

Becker, D.E., Griffith, J.M., Hobbs, C.S. and MacIntire, W.H. (1950) The alleviation of fluorine toxicosis by means of certain aluminum compounds. *Journal of Animal Science* 9, 647.

Bixler, D. and Muhler, J.C. (1960) Retention of fluoride in soft tissues of chickens receiving different fat diets. *Journal of Nutrition* 70, 26–30.

Blakley, B.R. (1984) A retrospective study of lead poisoning in cattle. *Veterinary and Human Toxicology* 26, 505–507.

Blaxter, K.L. (1950) Lead as a nutrional hazard to farm livestock. II. Absorption and excretion of lead by sheep and rabbits. *Journal of Comparative Pathology* 60, 140–159.

Borella, P. and Giardino, A. (1991) Lead and cadmium at very low doses affect *in vitro* immune response of human lymphocytes. *Environmental Research* 55, 165–177.

Boyd, J.H. (1985) Organomercuric poisoning in fat cattle. *Veterinary Record* 116, 443–444.

Bramley, R.G.V. (1990) Cadmium in New Zealand agriculture. *New Zealand Journal of Agricultural Research* 33, 505–519.

Bratton, G.R. and Zmudski, J. (1984) Laboratory diagnosis of Pb poisoning in cattle: a reassessment and review. *Veterinary and Human Toxicology* 26, 387–392.

Bratton, G.R., Childress, M., Zmudski, J., Womac, C., Rowe, L.D. and Tiffany-Castiglioni, E. (1986) Delta aminolaevulonic acid dehydratase (EC 4.3.1.24) activity in erythrocytes from cattle administered low concentrations of lead acetate. *American Journal of Veterinary Research* 47, 2068–2074.

Brebner, J., Thornton, I., McDonald, P. and Suttle, N.F. (1985) The release of trace elements from soils under conditions of simulated rumenal and abomasal digestion. In: Mills, C.F., Bremner, I. and Chesters, J.K. (eds) *Proceedings of the Fifth International Symposium on Trace Elements in Animals and Man.* Commonwealth Agricultural Bureaux, Farnham Royal, UK, pp. 850–852.

Brebner, J., Hall, J., Smith, S., Stark, B., Suttle, N.F. and Sweet, N. (1993) Soil ingestion is an important pathway for the entry of potentially toxic elements (PTE) from sewage sludge-treated pasture into ruminants and the food chain. In: Allan, R.J. and Nriagu, J.O. (eds) *Proceedings of the Ninth International Conference on Heavy Metals in the Environment,* Vol. 1. CEP, Edinburgh, pp. 446–449.

Bremner, I. (1978) Cadmium toxicity. *World Review of Nutrition and Dietetics* 32, 165–197.

Bremner, I. and Campbell, J.K. (1980) The influence of dietary copper intake on the toxicity of cadmium. *Annals of the New York Academy of Sciences* 355, 319–332.

Burns, K.N. and Allcroft, R. (1962) The use of tail bone biopsy for studying skeletal deposition of fluorine in cattle. *Research in Veterinary Science* 3, 215–218.

Burns, K.N. and Allcroft, R. (1966) *Fluorosis in Cattle: Occurrence, Diagnosis and Alleviation.* Report of the Fourth International Meeting, World Association for Buiatrics, pp. 94–115.

Butler, E.J., Nisbet, D.I. and Robertson, J.M. (1957) Osteoporosis in lambs in a lead mining area. 1. A study of the naturally occurring disease. *Journal of Comparative Pathology* 67, 378–398.

Campbell, J.K. and Mills, C.F. (1974) Effects of dietary cadmium and zinc on rats maintained on diets low in copper. *Proceedings of the Nutrition Society* 33, 15A–17A.

Campbell, J.K. and Mills, C.F. (1979) The toxicity of zinc to pregnant sheep. *Environmental Research* 20, 1–13.

Carlisle, E.M. and Curran, M.J. (1993) Aluminium: an essential element for the chick. In: Anke, M., Meissner, D. and Mills, C.F. (eds) *Proceedings of the Eighth International Symposium on Trace Elements in Man and Animals.* Verlag Media Touristik, Gersdorf, pp. 695–698.

Carson, T.L., Van Gelder, G.A., Karas, G.C. and Buck, W.B. (1974) Slowed learning in lambs prenatally exposed to lead. *Archives of Environmental Health* 29, 154–156.

Chang, C.W.J. and Suber, R. (1982) Protective effect of selenium on methyl mercury toxicity: a possible mechanism. *Bulletin of Environmental Contamination and Toxicology* 29, 285–289.

Chang, C.W.J., Nakamura, R.M. and Brooks, C.C. (1977) Effect of varied dietary levels and forms of mercury on swine. *Journal of Animal Science* 45, 279–285.

Chowdury, B.-A. and Chandra, R.K. (1987) Biological and health implications of toxic heavy metal and essential trace element interactions. In: *Progress in Food and Nutrition Science*, Vol. II. Pergamon Press, Oxford, pp. 55–113.

Clarkson, T.W. (1987) Mercury. In: Mertz, W. (ed.) *Trace Elements in Human and Animal Nutrition*, Vol. 7. Academic Press, New York, p. 417.

Coppock, R.W., Wagner, W.C. and Reynolds, R.D. (1988) Migration of lead in a glass-lined, bottom-loading silo. *Veterinary and Human Toxicology* 30, 458–459.

Dalgarno, A.C. (1980) The effect of low level exposure to dietary cadmium on cadmium, zinc, copper and iron content of selected tissues of growing lambs. *Journal of Science in Food and Agriculture* 31, 1043–1049.

Doberenz, A.R., Kurnick, A.A., Kurtz, E.B., Kemmerer, A.R. and Reid, B.L. (1964) Effect of a minimal fluoride diet on rats. *Proceedings of the Society for Experimental Biology and Medicine* 117, 689–693.

Doyle, J.J. and Pfander, W.H. (1975) Interactions of cadmium with copper, iron, zinc and manganese in ovine tissues. *Journal of Nutrition* 105, 599–606.

Doyle, J.J. and Spaulding, J.E. (1978) Toxic and essential elements in meat – a review. *Journal of Animal Science* 47, 398–419.

Doyle, J.J., Pfander, W.H., Grebing, S.E. and Pierce, J.O. (1974) Effect of dietary cadmium on growth, cadmium absorption and cadmium tissue levels in growing lambs. *Journal of Nutrition* 104, 160–166.

El-Begearmi, M.M., Sunde, M.L. and Ganther, H.E. (1977) A mutual protective effect of mercury and selenium in Japanese quail. *Poultry Science* 56, 313–322.

Fick, K.R., Ammerman, C.B., Miller, S.M., Simpson, C.F. and Loggins, P.E. (1976) Effect of dietary lead on performance, tissue mineral composition and lead absorption in sheep. *Journal of Animal Science* 42, 515–523.

Finley, M.T. and Dieter, M.P. (1978) Influence of laying on lead accumulation in Mallard ducks. *Journal of Toxicology and Environmental Health* 4, 123–129.

Fitzgerald, P.R., Peterson, J. and Lue-Hing, C. (1985) Heavy metals in tissues of cattle exposed to sludge-treated pasture for eight years. *American Journal of Veterinary Research* 46, 703–707.

Forsyth, D.M., Pond, W.G. and Krook, L. (1972) Dietary calcium and fluoride interactions in swine: *in utero* and neonatal effects. *Journal of Nutrition* 102, 1639–1646.

Fox, M.R.S. (1987) Assessment of cadmium, lead and vanadium status of large animals as related to the human food chain. *Journal of Animal Science* 65, 1744–1752.

Fullmer, C.S., Edelstein, S. and Wasserman, R.H. (1985) Lead-binding properties of intestinal calcium-binding protein. *Journal of Biological Chemistry* 260, 6816–6819.

Ganther, H.E., Goudie, C., Sunde, M.I., Kopecky, M.J., Wagner, P., Ott, S.-H. and Hoekstra, W.G. (1972) Selenium: relation to decreased toxicity of methylmercury added to diets containing tuna. *Science* 175, 1122–1124.

Garcia-Bojalil, C.M., Ammerman, C.B., Henry, P.R., Littell, R.C. and Blue, W.G. (1988) Effects of dietary phosphorus, soil ingestion and dietary intake level on performance, phosphorus utilisation and serum and alimentary tract mineral concentrations in lambs. *Journal of Animal Science* 66, 1508–1519.

Gerry, R.W., Carrick, C.W., Roberts, R.E. and Hauge, S.M. (1949) Raw rock phosphate in laying rations. *Poultry Science* 28, 19–23.

Hahn, P.H.B. and Guenter, W. (1986) Effect of dietary fluoride and aluminium on laying hen performance and fluoride concentration in blood, soft tissue, bone and egg. *Poultry Science* 65, 1343–1349.

Harvey, J.M. (1952) Chronic endemic fluorosis of Merino sheep in Queensland. *Queensland Journal of Agricultural Science* 9, 47–141.

Hemingway, R.G. (1977) Fluorine retention in growing sheep: a comparison between two phosphorus supplements as sources of fluorine. *Proceedings of the Nutrition Society* 36, 82A.

Hill, C.H. and Matrone, G.D. (1969) Usefulness of chemical parameters in the study of *in vivo* and *in vitro* interactions of transition elements. *Federation Proceedings* 29, 1474–1479.

Hillman, D., Bolenbaugh, D.L. and Convey, E.M. (1979) Hypothyroidism and anaemia related to fluoride in dairy cattle. *Journal of Dairy Science* 62, 416–423.

Houpert, P., Mehennaoni, S., Joseph-Enriquez, B., Federspiel, B. and Milhaud, G. (1995) Pharmokinetics of cadmium following intravenous and oral administration to non-lactating ewes. *Veterinary Research* 26, 145–154.

Jacob, K.D. and Reynolds, D.S. (1928) The fluorine content of phosphate rock. *Journal of Official Agricultural Chemists* 11, 237–258.

Kirchgessner, M., Plass, D.L. and Reichlmayr-Lais, A.M. (1991) Lead deficiency in swine. In: Momcilovic, B. (ed.) *Proceedings of the Seventh International Symposium on Trace Elements in Man and Animals (TEMA 7), Dubrovnik.* IMI, Zagreb, pp. 11-20–11-21.

Kollmer, W.E. and Berg, D. (1989) The influence of a zinc-, calcium- or iron-deficient diet on the resorption and kinetics of cadmium in the rat. In: Southgate, D.A.T., Johnson, I.T. and Fenwick, G.R. (eds) *Nutrient Availability: Chemical and*

Biological Aspects. Royal Society of Chemistry Special Publication No. 72, Cambridge, pp. 287–289.

Kostial, K., Jugo, S., Rabar, I. and Maljkovic, T. (1978) Influence of age on metal metabolism and toxicity. *Environmental Health Perspectives* 25, 81–86.

Krueger, G.L., Morris, T.K., Suskind, R.R. and Widner, E.M. (1985) The health effects of aluminium compounds in mammals. *CRC Critical Reviews in Toxicology* 13, 1–24.

Kumar, A. and Susheela, A.K. (1995) Effects of chronic fluoride toxicity on the morphology of ductus epididymus and the maturation of spermatozoa in the rabbit. *International Journal of Experimental Pathology* 76, 1–11.

Langlands, J.P., Donald, G.E. and Bowles, J.E. (1988) Cadmium concentrations in liver, kidney and muscle in Australian sheep and cattle. *Australian Journal of Experimental Agriculture* 28, 291–297.

Leach, R.M., Jr, Wang, K.W. and Baker, D.E. (1979) Cadmium and the food chain: the effect of dietary cadmium on tissue composition in chicks and laying hens. *Journal of Nutrition* 109, 437–443.

Lee, J., Rounce, J.R., MacKay, A.D. and Grace, N.D. (1996) Accumulation of cadmium with time in Romney sheep grazing ryegrass–white clover pasture: effect of cadmium from pasture and soil intake. *Australian Journal of Agricultural Research* 47, 877–894.

Lloyd, M.K. (1983) Environmental toxicity. In: Suttle, N.F., Gunn, R.G., Allen, W.M., Linklater, K.A. and Wiener, G. (eds) *Trace Elements in Animal Production and Veterinary Practice.* British Society of Animal Production Occasional Publication No. 7, Edinburgh, pp. 119–124.

Lynch, G.P., Jackson, E.D., Kiddy, C.A. and Smith, D.F. (1976a) Responses of young calves to oral doses of lead. *Journal of Dairy Science* 59, 1490–1494.

Lynch, G.P., Smith, D.F., Fisher, M., Pike, T.L. and Weinland, B.T. (1976b) Physiological responses of calves to cadmium and lead. *Journal of Animal Science* 42, 410–421.

McClure, F.J. (1949) Fluorine in foods: survey of recent data. *Public Health Reports* 64, 1061–1074.

McDowell, L.R. (1992) Aluminium, arsenic, cadmium, lead and mercury. In: *Minerals in Animal and Human Nutrition.* Academic Press, New York, pp. 355–356.

McGown, E.L. and Suttie, J.W. (1974) Influence of fat and fluoride on gastric emptying time. *Journal of Nutrition* 104, 909–915.

Masters, D.G., White, C.L., Peter, D.W., Purser, D.B., Roe, S.P. and Barnes, M.J. (1992) A multi-element supplement for grazing sheep. II. Accumulation of trace elements in sheep fed different levels of supplement. *Australian Journal of Agricultural Research* 43, 809–817.

Maurer, R.L. and Day, H.G. (1957) The non-essentiality of fluorine in nutrition. *Journal of Nutrition* 62, 561–573.

Mehdi, A.W.R., Al-Soudi, K.A., Al-Jiboori, N.A.J. and Al-Hiti, M.K. (1983) Effect of high fluoride intake on chicken performance, ovulation, spermatogenesis and bone fluoride content. *Fluoride* 16, 37–43.

Messer, H.H., Armstrong, W.D. and Singer, L. (1974) Essentiality and function of fluoride. In: Hoekstra, W.G., Suttie, J.W., Ganther, H.E. and Mertz, W. (eds) *Trace Element Metabolism in Animals – 2.* University Park Press, Baltimore, pp. 425–437.

Milhaud, G., Cazieux, A. and Enriquez, B. (1984) Experimental studies on fluorosis in the suckling lamb. *Fluoride* 17, 107–114.

Mills, C.F. and Dalgarno, A.C. (1972) Copper and zinc status of ewes and lambs receiving increased dietary concentrations of cadmium. *Nature, London* 239, 171–173.

Milne, D.B. and Schwarz, K. (1974) Effect of different fluorine compounds on growth and bone fluoride levels in rats. In: Hoekstra, W.G., Suttie, J.W., Ganther, H.E. and Mertz, W. (eds) *Trace Element Metabolism in Animals −2.* University Park Press, Baltimore, pp. 710–714.

Moffat, W.H. (1993) Long term residual effects of lead mining on man and grazing livestock within a rural community in southern Scotland. PhD thesis, University of Edinburgh, pp. 152–214.

Morcombe, P.W., Petterson, D.S., Ross, P.J. and Edwards, J.R. (1994a) Soil and agronomic factors associated with cadmium accumulations in kidneys of grazing sheep. *Australian Veterinary Journal* 71, 404–406.

Morcombe, P.W., Petterson, D.S., Masters, H.G., Ross, P.J. and Edwards, J.R. (1994b) Cadmium concentrations in kidneys of sheep and cattle in Western Australia. 1. Regional distribution. *Australian Journal of Agricultural Research* 45, 851–862.

Morrison, J.N. and Quarterman, J. (1987) The relationship between iron status and lead absorption in rats. *Biological Trace Element Research* 14, 115–126.

Morrison, J.N., Quarterman, J. and Humphries, W.R. (1977) The effect of dietary calcium and phosphate on lead poisoning in lambs. *Journal of Comparative Pathology* 87, 417–429.

Motzok, I. and Branion, H.D. (1958) Influence of fluorine on phosphatase activities of plasma and tissues of chicks. *Poultry Science* 37, 1469–1471.

Neathery, M.W., Miller, W.J., Gentry, R.P., Stake, P.E. and Blackmon, D.M. (1974) Cadmium^{109} and methyl mercury-203 metabolism, tissue distribution and secretion into milk in dairy cows. *Journal of Dairy Science* 57, 1177–1184.

Neathery, M.W., Miller, W.J., Gentry, R.P., Crowe, C.T., Alfaro, E., Fielding, A.S., Pugh, D.G. and Blackmore, D.M. (1987) Influence of high dietary deed on selenium metabolism in dairy calves. *Journal of Dairy Science* 70, 645–652.

Nielsen, F.H. (1996) Other trace elements. In: Filer, L.J. and Ziegler, E.E. (eds) *Present Knowledge in Nutrition*, 7th edn. International Life Sciences Institute, Nutrition Foundation, Washington, DC.

Nielsen, F.H. and Uthus, E.O. (1984) Arsenic. In: Frieden, E. (ed.) *Biochemistry of the Ultratrace Elements.* Plenum Press, New York, pp. 319–340.

NRC (1980) *Mineral Tolerance of Domestic Animals.* National Academy of Sciences, Washington, DC.

Oelschlager, W.K., Loeffler, K. and Opletova, L. (1970) Retention of fluorine in bones and bone sections of oxen in respect of equal increases in fluorine in the form of soil, flue dust from an aluminium reduction plant and sodium fluoride. *Landwirsch Forsch* 23, 214–224.

Oliver, D.P., Tiller, K.G., Conyers, M.K., Slattery, W.J., Alston, A.M. and Merry, R.H. (1996) Effectiveness of liming to minimise uptake of cadmium by wheat and barley grain grown in the field. *Australian Journal of Agricultural Research* 47, 1181–1193.

Peirce, A.W. (1954) Studies on fluorosis of sheep. 2. The toxicity of water-borne fluoride for mature grazing sheep. *Australian Journal of Agricultural Research* 5, 545–554.

Petersen, D.S., Masters, H.G., Spiejers, E.J., Williams, D.E. and Edwards, J.R. (1991) Accumulation of cadmium in sheep. In: Momcilovic, B. (ed.) *Proceedings of the Seventh International Symposium on Trace Elements in Man and Animals (TEMA 7), Dubrovnik.* IMI, Zagreb, pp. 26-13–26-14.

Phillips, P.H. and Suttie, J.W. (1960) The significance of time in intoxication of domestic animals by fluoride. *Archives of Industrial Health* 21, 343–345.

Phillips, P.H., Hart, E.B. and Bohstedt, G. (1934) *Chronic Toxicosis in Dairy Cows due to the Ingestion of Fluorine*. Research Bulletin Wisconsin Agricultural Experiment Station No. 123.

Pinault, L. and Milhaud, G. (1985) Evaluation of subclinical lead poisoning in dairy cattle. *Veterinary Pharmacology and Toxicology*, Chapter 61, 715–724.

Powell, G.W., Miller, W.I., Morton, J.D. and Clifton, C.M. (1964) Influence of dietary cadmium level and supplemental zinc on cadmium toxicity in the bovine. *Journal of Nutrition* 84, 205–211.

Proudfoot, F.G., Jackson, E.D., Hulan, H.W. and Salisbury, C.D.C. (1991) Arsanilic acid as a growth promoter for chicken broilers when administered via either the feed or drinking water. *Canadian Journal of Animal Science* 71, 221–226.

Puls, R. (1994) *Mineral Levels in Animal Health: Diagnostic Data*, 2nd edn. Sherpa International, Clearbrook, British Columbia.

Quarterman, J. and Morrison, J.N. (1975) The effects of dietary calcium and phosphorus on the retention and excretion of lead in rats. *British Journal of Nutrition* 34, 351–362.

Quarterman, J., Morrison, J.N., Humphries, W.R. and Mills, C.F. (1977) The effect of dietary sulphur and of castration on lead poisoning in lambs. *Journal of Comparative Pathology* 87, 405–416.

Quarterman, J., Morrison, J.N. and Humphries, W.R. (1978) The influence of high dietary calcium and phosphate on lead uptake and release. *Environmental Research* 17, 60–67.

Rimbach, G., Pallaul, J., Brandt, K. and Most, E. (1996) Effect of phytic acid and microbial phytase on Cd accumulation, Zn status and the apparent absorption of Ca, P, Mg, Fe, Zn, Cu and Mn in growing rats. *Annals of Nutrition and Metabolism* 39, 361–370.

Robinson, D.L., Hemkes, O.J. and Kemp, A. (1984) Relationships among forage aluminium levels, soil contamination on forages and availability of elements to dairy cows. *Netherland Journal of Agricultural Science* 32, 73–80.

Roholm, K. (1937) *Fluorine Intoxication*. H. K. Lewis, London.

Ruliffson, W., Burns, L.V. and Hughes, J.S. (1963) The effect of fluoride ion on ^{59}Fe iron levels in the blood of rats. *Transactions of Kansas Academy of Science* 66, 52.

Scheuhammer, A.M. (1987) The chronic toxicity of aluminium, cadmium, mercury and lead in birds: a review. *Environmental Pollution* 46, 263–295.

Schwarz, K. (1974) New essential trace elements (Sn, V, F, Si): progress report and outlook. In: Hoekstra, W.G., Suttie, J.W., Ganther, H.E. and Mertz, W. (eds) *Proceedings of the Second International Symposium on Trace Element Metabolism in Animals*. University Press, Baltimore, pp. 355–380.

Schwarz, K. and Milne, D.B. (1972) Fluorine requirements for growth in the rat. *Bioinorganic Chemistry* 1, 331–336.

Seffner, W. and Tuebener, W. (1983) Antidotes in experimental fluorosis in pigs: morphological studies. *Fluoride* 16, 33–37.

Selby, L.A., Case, A.A., Osweiler, G.D. and Mayes, H.M. (1977) Epidemiology and toxicology of arsenic poisoning in domestic animals. *Environmental Health Perspectives* 19, 183–189.

Sell, J.L. (1975) Cadmium in the laying hen: apparent absorption, tissue distribution and virtual absence of transfer into eggs. *Poultry Science* 54, 1674–1678.

Sharman, G.A.M. and Angus, K.W. (1991) Inorganic and organic poisons. In: Martin, W.B. and Aitken, I.D. (eds) *Sheep Diseases*. Blackwell Scientific, London, pp. 317–319.

Shearer, T.R., Kolstad, D.L. and Suttie, J.W. (1978a) Bovine dental fluorosis: histologic and physical characteristics. *American Journal of Physiology* 212, 1165–1168.

Shearer, T.R., Kolstad, D.L. and Suttie, J.W. (1978b) Electron probe microanalysis of fluorotic bovine teeth. *American Journal of Veterinary Research* 39, 1393–1398.

Sherlock, J.C. (1989) Aluminium in foods and the diet. In: Massey, R.C. and Taylor, D. (eds) *Aluminium in Food and the Environment.* Royal Society of Chemistry Special Publication, London, pp. 68–76.

Shupe, J.L. (1980) Clinicopathologic features of fluoride toxicosis in cattle. *Journal of Animal Science* 51, 746–757.

Shupe, J.L., Miner, M.L., Harris, L.E. and Greenwood, D.A. (1962) Relative effects of feeding hay atmospherically contaminated by fluoride residue, normal hay plus calcium fluoride, and normal hay plus sodium fluoride to dairy heifers. *American Journal of Veterinary Research* 23, 777–787.

Shupe, J.L., Harris, L.E., Greenwood, D.A., Butcher, J.E. and Nielsen, H.M. (1963a) The effect of fluorine on dairy cattle. 5. Fluorine in the urine as an estimator of fluorine intake. *American Journal of Veterinary Research* 24, 300–306.

Shupe, J.L., Miner, M.L., Greenwood, D.A., Harris, L.E. and Stoddard, G.E. (1963b) The effect of fluorine on dairy cattle. 2. Clinical and pathologic effects. *American Journal of Veterinary Research* 24, 964–979.

Simpson, V.R., Stuart, N.C., Munro, R., Hunt, A. and Livesey, C.T. (1997) Poisoning of dairy heifers by mercurous chloride. *Veterinary Record* 140, 549–552.

Smith, G.M. and White, C.L. (1997) A molybdenum–sulphur–cadmium interaction in sheep. *Australian Journal of Agricultural Research* 48, 147–154.

Smith, R.M., Griel, L.C., Muller, L.D., Leach, R.M. and Baker, D.H. (1991a) Effects of dietary cadmium chloride throughout gestation on blood and tissue metabolites of primigravid and neonatal dairy cattle. *Journal of Animal Science* 69, 4078–4087.

Smith, R.M., Griel, L.C., Muller, L.D., Leach, R.M. and Baker, D.H. (1991b) Effects of long term dietary cadmium chloride on tissue, milk and urine mineral concentrations of lactating dairy cows. *Journal of Animal Science* 69, 4088–4096.

Snook, L.C. (1962) Rock phosphate in stock feeds: the fluorine hazard. *Australian Veterinary Journal* 38, 42–47.

Stark, B.A., Livesey, C.T., Smith, S.R., Suttle, N.F., Wilkinson, J.M. and Cripps, P.J. (1998) *Implications of Research on the Uptake of PTEs from Sewage Sludge by Grazing Animals.* Report to the Department of the Environment, Transport and the Regions (DETR) and the Ministry of Agriculture, Fisheries and Food (MAFF). WRc, Marlow, UK.

Suttie, J.W. (1967) Vertebral biopsies in the diagnosis of bovine fluoride toxicosis. *American Journal of Veterinary Research* 28, 709–712.

Suttie, J.W. (1969) Fluoride content of commercial dairy concentrates and alfalfa forage. *Journal of Agricultural and Food Chemistry* 17, 1350–1352.

Suttie, J.W. (1978) Effects of fluorides on animals. In: *First International Minerals Conference, St Petersburg, Florida.* International Minerals and Chemical Corporation, Illinois, p. 87.

Suttie, J.W. (1980) Nutritional aspects of fluoride toxicosis. *Journal of Animal Science* 51, 759–766.

Suttie, J.W. and Faltin, E.C. (1973) Effects of sodium fluoride on dairy cattle: influence of nutritional state. *American Journal of Veterinary Research* 34, 479–483.

Suttie, J.W. and Kolstad, D.L. (1977) Effects of dietary fluoride ingestion on ration intake and milk production. *Journal of Dairy Science* 60, 1568–1573.

Suttie, J.W., Miller, R.F. and Phillips, P.H. (1957) Studies of the effects of dietary NaF on dairy cows. 1. The physiological effects and the developmental symptoms of fluorosis. *Journal of Nutrition* 63, 211–224.

Suttie, J.W., Phillips, P.H. and Miller, R.F. (1958) Studies of the effects of dietary sodium fluoride on dairy cows. 3. Skeletal and soft tissue fluorine deposition and fluorine toxicosis. *Journal of Nutrition* 65, 293–304.

Suttie, J.W., Gesteland, R. and Phillips, P.H. (1961) Effects of dietary sodium fluoride on dairy cows. 6. In young heifers. *Journal of Dairy Science* 44, 2250–2258.

Suttie, J.W., Carlson, J.R. and Faltin, E.C. (1972) Effects of alternating periods of high- and low-fluoride ingestion on dairy cattle. *Journal of Dairy Science* 55, 790–804.

Suttle, N.F., Brebner, J. and Hall, J. (1991) Faecal excretion and retention of heavy metals in sheep ingesting topsoil from fields treated with metal rich sludge. In: Momcilovic, B. (ed.) *Proceedings of the Seventh International Symposium on Trace Elements in Man and Animals (TEMA 7), Dubrovnik.* IMI, Zagreb, pp. 32-7–32-8.

Suttle, N.F., Brebner, J., Stark, B., Sweet, N. and Hall, J.W. (1997) Placental and mammary transfer of lead and cadmium by ewes exposed to lead and cadmium enriched, sewage-sludge treated soils. In: Fischer, P.W.F., L'Abbé, M.R., Cockell, K.A. and Gilson, R.S. (eds) *Proceedings of the Ninth International Symposium on Trace Elements in Man and Animals (TEMA 9), Banff.* NRC Research Press, Ottawa, pp. 168–170.

Tao, S. and Suttie, J.W. (1976) Evidence for a lack of an effect of dietary fluoride level on reproduction in mice. *Journal of Nutrition* 106, 1115–1122.

Thompson, D.J. (1980) Industrial considerations related to fluoride toxicity. *Journal of Animal Science* 51, 767–772.

Thornton, I. (1996) Sources and pathways of arsenic in the geochemical environment. In: Appleton, J.D., Fuge, R. and McCall, G.J.H. (eds) *Environmental Geochemistry and Health.* Geological Society Special Publication No. 113, London, pp. 153–162.

Underwood, E.J. (1977) *Trace Elements in Human and Animal Nutrition* 4. Academic Press, New York, pp. 430–432.

Valdivia, R., Ammerman, C.B., Henry, P.R., Feaster, J.P. and Wilcox, C.J. (1982) Effect of dietary aluminium and phosphorus on performance, phosphorus utilisation and tissue mineral composition in sheep. *Journal of Animal Science* 55, 402–410.

van Bruwaene, R., Gerber, G.B., Kirchmann, R. and Colard, J. (1982) Transfer and distribution of radioactive cadmium in dairy cows. *International Journal of Environmental Studies* 9, 47–51.

van Bruwaene, R., Kirchmann, R. and Impens, R. (1984) Cadmium contamination in agriculture and zootechnology. *Experientia* 40, 43–50.

van Gelder, G.A., Carson, T., Smith, R.M. and Buck, W.B. (1973) Behavioural toxicologic assessment of the neurologic effect of lead in sheep. *Clinical Toxicology* 6, 405–418.

Weber, C.W. and Reid, B.L. (1974) Effect of low-fluoride diets fed to mice for six generations In: Hoekstra, W.G., Suttie, J.W., Ganther, H.E. and Mertz, W. (eds) *Trace Element Metabolism in Animals –2.* University Park Press, Baltimore, pp. 707–709.

Weddle, D.A. and Muhler, J.C. (1954) The effects of inorganic salts on fluorine storage in the rat. *Journal of Nutrition* 54, 437–444.

Wheeler, S.M., Brock, T.B. and Teasdale, D. (1985) Effects of adding 30 mg fluoride/l drinking water given to pregnant ewes and their lambs upon physiology and wool growth. *Journal of Agricultural Science, Cambridge* 105, 715–726.

Zmudzki, J., Bratton, G.R., Womac, C. and Rowe, L.D. (1984) The influence of milk diet, grain diet and method of dosing on lead toxicity in young calves. *Toxicology and Applied Pharmacology* 76, 490–497.

19 Design of Supplementation Trials for Assessing Mineral Deprivation

The purpose of this final chapter is to provide a basis for bridging the gaps in knowledge which complicate assessments of the extent to which shortages of minerals in the diet limit livestock production. The sources of uncertainty should by now be familiar to the reader, because they are common to nearly all the minerals – major and trace – likely to become limiting under practical farming conditions; they are fivefold.

1. Uncertainty regarding the effective mineral supply, i.e. amounts ingested, absorbed and utilized.
2. Uncertainty regarding the minimal demands of the livestock species, given the universal ability to conserve losses and redistribute what is retained.
3. Uncertainty surrounding the abrupt changes which physiological and environmental factors can cause in demand.
4. Uncertainty regarding the rate-limiting metabolic pathway when supply fails to meet demand.
5. Uncertainty arising from the poor correlations between conventional clinical biochemistry criteria and the activity of the rate-limiting pathway.

There is widespread agreement that the soundest diagnosis of a mineral deficiency is provided by a clinical or production response to supplements of the mineral or minerals thought to be lacking (Suttle, 1987; Clark *et al.*, 1989; Towers *et al.*, 1993). It is only recently, however, that scientists have begun to lay down guidelines for the design and interpretation of supplementation trials (Clark *et al.*, 1985; Grace, 1994) and to make fuller use of their data. Countless trials have been destined to confirm the null hypothesis before a record was taken, because of inadequate numbers and the failure to consult an experienced but pragmatic statistician before starting. The essential requirements for a good supplementation trial are given below.

© CAB *International* 1999. *Mineral Nutrition of Livestock*
(E.J. Underwood and N.F. Suttle)

Controls

Unless a sufficient number of animals are left unsupplemented as controls, the benefits from supplementation cannot be gauged. Only naïve or unscrupulous investigators conduct trials without untreated animals. 'Before and after treatment' comparisons are notoriously unreliable, because any spontaneous recoveries (e.g. acquired immunity to gut parasites) or new constraints (e.g. drought) will give false positives or negatives.

Adequate Group Size

Minimum statistical requirements

The group sizes needed to demonstrate statistically significant differences in performance can be calculated quite simply from previous records showing the variability in the chosen criterion of performance. Whether the variables are continuous (e.g. live weight) or discrete (e.g. mortality), the appropriate analysis (e.g. ANOVA or chi-square) can be run on simulated data sets. White (1996) has illustrated the effect of group size on the probability of discounting as 'non-significant' a 5% improvement in body weight with different degrees of 'background' variation in body weight within groups (Fig. 19.1). Although such a goal is small, it may well be economically significant and representative of the maximum benefit likely to be attained when deprivation is marginal (Kumagai and White, 1995). Minimum group sizes of 20 have been recommended for ascertaining the responsiveness of lambs to cobalt supplementation (Clark *et al.*, 1989). Table 19.1 illustrates the even greater importance of group size regarding the attainment of statistical significance for a discrete variable, such as calf mortality or ewe infertility. Even when the incidence of infertility is high (40%) and halved by the hypothetical treatment, groups of over 40 animals are needed for statistical significance by the chi-square test (>3.84 for $P<0.05$ with 1 degree of freedom (d.f.)). When performance is good (10% infertility) but still improvable (incidence halved to 5%), over 200 animals are needed in each group to prove the point. The contrast between statistical and economic significance is highlighted with discrete, 'all-or-nothing' variables like these. A non-significant ($P>0.05$) increase of 5% in the calf or lamb crop every other year would be welcomed by most farmers and repay the outlay on treatments faster than a consistent significant 5% improvement in growth.

Unavoidable and avoidable constraints

Group sizes are subject to some unavoidable constraints, such as flock or herd size, the available grazing area or housing, the handling facilities and the need to return stock to their normal environment. However, it may be neither necessary nor advisable to return stock the same day that they are gathered. Gathering animals the day prior to recording and/or treatment and fasting overnight might reduce errors in weighing and blood analysis due to

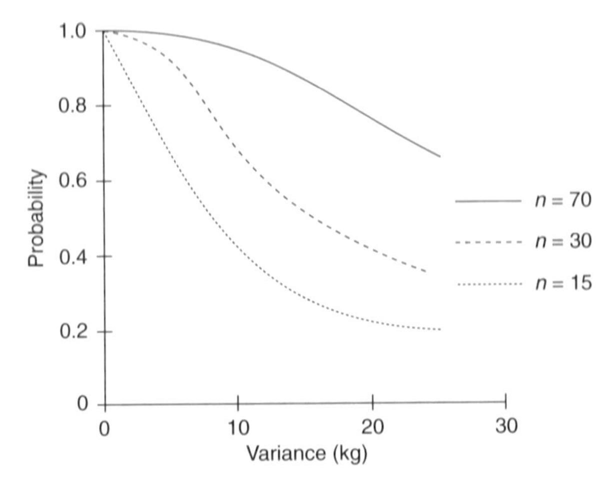

Fig. 19.1. Minimum acceptable group sizes for dose–response trials can be calculated from data on the variance of the response trait and the anticipated benefit from treatment: here, the assumed benefit is one of 2 kg in live-weight gain in lambs; unless variance is low, large groups are required to correctly reject the null hypothesis (from White, 1996).

Table 19.1. Hypothetical illustration of the group size needed to obtain a significant improvement in incidence of a disorder (say, mortality or infertility) with a particular treatment (for $P < 0.05$, chi-square must be > 3.84).

	Hypothetical group		Group size					
	Control	Treated	10	20	40	80	160	320
	Incidence (%)		Chi-square					
Expt A	40	20	0.95	1.91	3.81	7.62	15.24	30.48
Expt B	10	5	–	0.36	0.72	1.44	2.88	5.76

the recent ingestion of food, while increasing the number of animals that can be sampled in a working day. Prolonged yarding may, however, adversely affect the assessment of cobalt status via serum B_{12} assays (Clark *et al.*, 1989). Other constraints may be entirely avoidable. For example, group size is sometimes restricted by the excessive number of observations made on or samples taken from each individual and the time and cost of subsequently processing any samples. Priority should be given to providing sufficient animals to confirm or deny responsiveness in the performance criterion. The number of blood samples needed to attain statistical significance for a given biochemical test is likely to be smaller than that required for a performance criterion and the two parameters may be poorly correlated. In some countries, there are increasing pressures to reduce the number of animals used in experiments on welfare grounds. 'Restricted' or 'licensed' procedures (e.g. blood sampling) need be applied to only a small proportion of those subjected to unrestricted farm procedures (e.g. weighing).

Accuracy

Statistical significance is determined by the ratio of accountable (e.g. treatment + litter effects) to unaccountable (i.e. random) variation. Random variation can be reduced by rigorous attention to the detail and consistency of all procedures and observations. The principles of good experimental and good laboratory practice (GEP and GLP) are becoming increasingly familiar. While some requirements may seem pedantic, it takes little time to regularly check the accuracy of both field (e.g. weighings) and laboratory (e.g. analyses) results against known standards and adjusting either equipment or results accordingly. Electronic recording and data transference are more reliable than the human ear, hand and memory. All statistical inputs should be double-checked and attention given to the outliers routinely 'flagged' by modern statistical packages. The discarding of biologically impossible weight gains, for example, is duty, not deceit; so too is the reanalysis of duplicate samples with statistically verified discordance.

Repetition

Repetition of observations and samplings can provide a safeguard against diurnal, day-to-day, seasonal and year-to-year variation.

Diurnal variation

Awareness of diurnal fluctuations is a prerequisite for deciding and adhering to an optimum time of observation (before or after feeding, morning or afternoon milking). If the sequence of a prolonged set of samplings is repeated (e.g. in individually penned animals), the sequence might usefully be recorded and entered as a variable in the statistical model.

Day-to-day variation

Repeated observations on successive days are increasingly used to improve the reliability of the beginning and end-points of an observation period, e.g. body weights. In heavily fleeced sheep, live weight can be increased by several kilograms by heavy rainfall prior to a particular weighing.

Seasonal variation

Seasonality affects the growth of crops and pastures, the physiological processes of animals and the life cycle of their pathogens; it is therefore a feature of all supplementation trials. Seasonal influences can be sought by dividing an experiment into periods, although period effects are not necessarily synonymous with true season effects; they may indicate the waning efficacy of a treatment or differences in composition between batches of diet, for example. Knowledge of the seasonality of responses to mineral supplementation can influence the decision whether or not to routinely supplement, the timing of any treatment and the method of supplementation. Different treatments can rarely be applied sequentially to the same group without becoming confounded with seasonal influences.

Year-to-year variation

From the livestock-producers' viewpoint, the repeatability of responses to supplementation from year to year is of paramount importance. From the supplement-producers' viewpoint, any variation is commercially exploitable, because tales of 'the good years' are more easily told and recalled than those of 'the bad years'. The classic example of year-to-year variation in response to a mineral supplement was provided by Lee (1951) in his studies of 'coast disease' in South Australia. He and his colleagues showed that positive responses to cobalt supplementation were obtained in only 8 of 13 years at the same site. Furthermore, the effects of deficiency varied in severity from marginal growth depression to 100% mortality. In a sequential study of growth responses in lambs to copper, cobalt and selenium over 3 years, the significance of copper responsiveness may have been underestimated by year-to-year fluctuations, copper status being highest in the year when copper was under test (Suttle *et al.*, 1999).

Making Use of Individuality

Regardless of the steps taken by experimenter and farmer to minimize the variation between individuals in a group, herd or flock, some will persist in both supplemented and unsupplemented individuals. Instead of being regarded as a nuisance by all concerned, much can be learned from individual variation in performance and it can be used to increase the likelihood of obtaining significant responses, to predict future responses and to form the basis of cost–benefit analysis. A further benefit might be reduction in the group size necessary to demonstrate significant benefits from supplementation. More can be done than simply pairing or stratifying the population according to a production parameter, such as live weight or milk yield.

Pretreatment variation

The typical picture presented by a suspected marginal mineral disorder is one of irregular performance, rather than consistent failure. Variation in performance may be associated with variation in mineral status. Before the trial even begins, a correlation between past performance and initial mineral status is worth looking for, though not necessarily indicative of cause and effect. Individuals may have been performing badly for a variety of reasons (Table 19.2).

Table 19.2. Seven possible reasons for poor growth in a group of young growing animals on first inspection, listed in order of probability.

Order	Reason for poor growth
1	Too little digestible, utilizable non-mineral nutrients provided by the diet
2	Maternal undernutrition (in the case of newly weaned or of suckled offspring)
3	Parasitic, microbial or viral infection
4	Environmental stressors (wind, temperature, water-supply)
5	Low genetic potential
6	Too little of one mineral in the diet
7	Too little of two or more minerals in the diet

Covariance with early mineral-dependent performance

Good correlations between past growth rate and initial mineral status would only be expected either if the last two reasons for poor performance given in Table 19.2 applied or if maternal performance (second reason) had been restricted by deprivation of the same mineral that their offspring lacked. In the first case, one would expect random variation in subsequent overall performance to be reduced by using the initial ranking or prior growth rate of the individual as a covariate and for the relationship to be *negative*, i.e. the poorest initial performers showing the greatest response to supplementation. Similar approaches can be taken with all parameters of production. A simple example was provided by Joyce and Brunswick (1975) in their studies of responses of lactating cows to sodium supplementation on a lucerne diet, low in sodium; individuals with the poorest yields prior to treatment showed the largest responses to sodium, while previously good yielders failed to respond. It is surprising how little use has been made of this obvious tactic in field trials. It would be surprising if the use of conception-rate records from previous matings as covariates did not greatly increase the chances of detecting and quantifying the effects of mineral deprivations on fertility. In Table 19.3, influences of copper supplementation on the growth of lambs only became significant when initial live weight was used as a covariate. However, significance was not obtained later in the season (6–12 weeks), when growth was slower and the supplement was no longer delivering copper (Suttle, 1994).

Covariance with early mineral-independent performance

If individual performance prior to treatment was restricted only by factors other than supply of the supplemented mineral (Table 19.2), the relationship with subsequent performance is likely to be positive when the influence continues (e.g. poor milk supply, poor genotype, chronic infection) but negative when it does not continue (e.g. weaning, acute infection) and compensatory growth is expressed. Whatever mechanism underlies a significant correlation, covariance analysis will improve the chances of obtaining a significant response to treatment. In all cases where the final

Table 19.3. Treatment with copper oxide needles (CuO) failed to improve average weight gain of a group of 25 lambs compared with their untreated twins on molybdeniferous pasture (50–60 mg Mo kg^{-1} DM), but within-group regressions of live weight (kg) after 6 weeks on initial live weight were different.

CuO	Live-weight gain (g day^{-1})	Regression (kg units)	
		Intercept (SD)	Coefficient (SD)
0	173^{a}	8.1^{b} (± 2.19)	0.95^{d} (± 0.162)
+	191^{a}	1.1^{c} (± 2.27)	1.51^{e} (± 2.19)
	21.6 (SED)		

Means with different superscripts within columns differ significantly.
SD, standard deviation.

measure of performance is proportional to performance prior to treatment, the old practice of calculating improvement by difference between the initial and final value (e.g. live-weight gain) will be *less* effective than covariance analysis.

What covariance has to offer

Covariance analysis has more to offer than simply improving the chances of getting significant treatment effects. It is not something to be forgotten once P becomes < 0.05. Significant positive relationships between trial performance and mineral status can be found *within* treated or untreated groups which do not differ significantly in their mean trial performances. This indicates that the population distribution contained a 'tail' of responsive animals that alone benefited from supplementation. Such populations are marginally responsive and cost–benefit analysis can be applied to ascertain whether all, none or part of the population merits treatment. Such decisions will be influenced by the cost and complexity of treatment and the cost of reliably ascertaining the mineral status of the entire population.

Making Use of Group Responses

Clarke *et al.* (1985) proposed the use of the relationship of differences *between* treated and untreated groups to a measure of mineral status in the animal in previous experiments to predict responses on 'unknown' farms from the same index of status. Plasma vitamin B_{12} concentrations were plotted against the average live-weight benefit after treating lambs with cobalt or vitamin B_{12}. The relationship for a more complete data set from New Zealand (Clark *et al.*, 1989) is shown in Fig. 19.2 and can be used to improve the interpretation of plasma vitamin B_{12} results (see Chapter 10), as well as predicting benefit in terms of performance. The exponential nature of the relationship is probably explained by the fact that, as vitamin B_{12} status declines, more and more individuals show larger and larger benefits. Mean

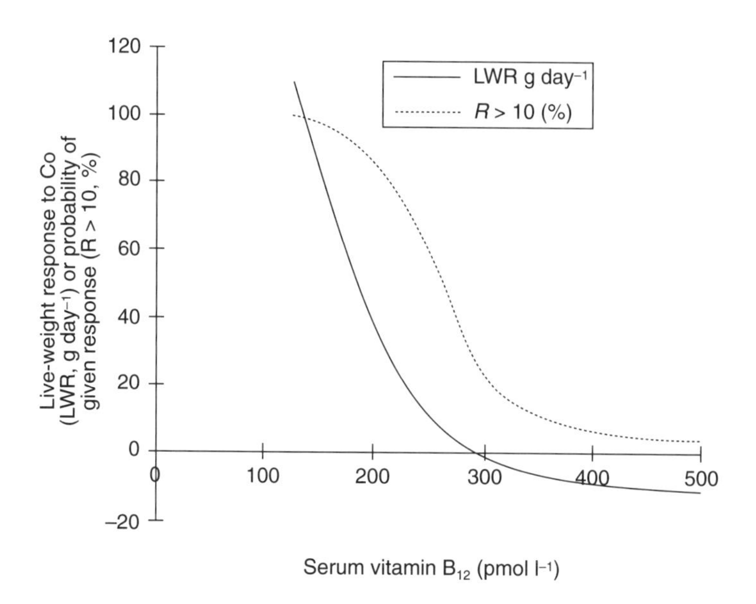

Fig. 19.2. Collation of data from cobalt supplementation trials with lambs in New Zealand allowed the interpretation of serum and liver vitamin B_{12} concentrations to be given an economic base (Clark *et al.*, 1985, 1989) but that base varies within and between countries.

values for groups are not necessarily the best basis for deriving such relationships, since they may be unduly influenced by extreme individuals when group size is small: the merit of median or geometric mean values should always be held in view. The New Zealand approach has been used in Australia with selenium (Fig. 19.3; Langlands *et al.*, 1989) and again emphasizes the continuous nature of the distribution of responses and the need at the very least for use of marginal bands in classifying responsiveness (see Chapter 3). As blood selenium declined linearly below 316 nmol (25 μg) l^{-1}, so the probability of obtaining a growth response in calves increased, but large responses (90 g day^{-1}) remained improbable. The same group used a 'bent-stick' model to predict responses in wool growth from blood selenium (see Chapter 15). These simple strategies allow the animal to be the arbiter of 'deprivation' and begin to address *all* the uncertainties set out at the start of the chapter.

Choosing the Best Treatment(s)

The would-be experimenter is presented with a bewildering array of possible treatments (for a review of trace element treatments, see Judson, 1996). The choice of treatment(s) for diagnostic trials is (are) governed by different criteria from those which a farmer might use routinely. The additional supply

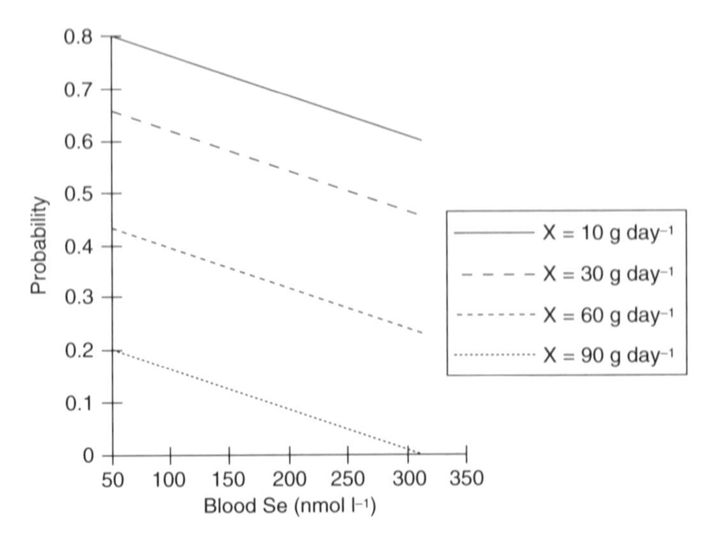

Fig. 19.3. Collation of data from large numbers of dose–response trials can allow prediction of the likely benefit from supplementation in other areas from a common measure of trace element status: here, probability of different growth responses in cattle is predicted from blood selenium (Langlands *et al.*, 1989).

of the mineral suspected to be limiting performance must be sufficient to alleviate any constraint throughout the period of risk and yet present no risk of excess when first administered. The product used experimentally may be more expensive than that used subsequently, in order to guarantee efficacy. There are, however, various options to be considered.

The case for and against multiple treatments

The first decision that must be made when using the supplementation trial to diagnose mineral-responsive disorders is how selective to be. To some extent, this may depend on the weight of evidence from previous analyses indicating the mineral status of soil, diet and animal blood or tissues, but, given the uncertainty of the respective relationships with animal performance, there is much to commend the 'blunderbuss' approach of providing a mixture of all the commonly needed minerals (necessarily by diet for the major elements) and measuring the cumulative impact on performance (White *et al.*, 1992). The major advantage of the multi-element approach is that it usually provides a quick answer to the question of whether or not any one mineral or combination of minerals is likely to be limiting performance, coupled with an indication of the order of the benefit which is likely to arise once those constraints are lifted (Kumagai and White, 1995). However, four difficulties arise with multiple treatments.

1. No evidence will be provided as to which mineral component(s) has been responsible for any improvement in health and performance.

2. The flock or herd at pasture must be divided up to provide matched replicates, with minerals either given or withheld, to provide proof of efficacy.
3. Uptake of a free-access mineral mixture (the common method of multiple supplementation) may vary substantially within treated groups (see Chapter 3).
4. Some minerals not needed in the mixture may be provided to such an excess that performance is impaired, masking responses to any component that is lacking.

The case for and against single treatments

The case for single treatments is that they should provide an unequivocal answer to a simple question: 'To what extent is a deficit of the supplemented mineral currently limiting performance?' In the case of trace elements, the mineral supplement can be given individually by mouth or by injection, allowing the effects of a known dose to be assessed within a grazing population. Failure can arise if the chosen method of treatment is inadequate for some members of the treated group, but safeguards can be incorporated into the design (e.g. scaling dose carefully to live weight; redosing if criteria of mineral status fall into or below the 'marginal' category; discarding data for exceptional individuals which show no early evidence of improved mineral status, on the assumption of maladministration). The disadvantages of single treatments are:

- the cost of failure to select the single most limiting element at an early stage;
- incomplete estimates of responsiveness when other elements are also limiting performance.

Examples of the single-treatment approach are given in the relevant sections of the chapters dealing with specific minerals.

The case for clustered treatments

Compromises can be reached by increasing the number of treatment groups to three: a fully supplemented group (A), a group (B) deprived of only one mineral and an unsupplemented group (C): then A − C gives the total response to minerals and B − C the minimal response to the selectively withdrawn mineral. If the most limiting macromineral is likely to be sodium (Na), this would be clearly demonstrated by the contrast between Na + calcium (Ca) + phosphorus (P) + magnesium (Mg) vs. Ca + P + Mg vs. 0, with the minerals given as continuous dietary supplements. Similarly, the most limiting trace element might be ascertained by selectively withdrawing copper (Cu), cobalt (Co) or selenium (Se) from the set Cu + Co + Se given as boluses by mouth or injection (Suttle, 1987). Some workers have provided salt for control groups and looked for increments in performance following the addition of further minerals by mouth (P, Co) or injection (Cu, iodine (I), Se) in all possible combinations (McDowell *et al.*, 1982). However, the responses to individual elements may be so small that only combined supplements give significant responses (McDowell *et al.*, 1982; Masters and Peter, 1990).

Re-examining unresponsive populations

The failure of either multiple, clustered or single mineral treatments to improve performance in field trials does not mean that the potential for responses should be dismissed for all time. The next step is to investigate other causes for the poor performance that prompted the supplementation trials in the first instance; lack of digestible organic matter and chronic parasitic infections are the most likely alternative causes of ill-thrift in grazing livestock. If such constraints are identified and removed, mineral supplies may become truly limiting at the new, increased level of performance. However, prior evidence of marginal mineral status should be obtained before re-embarking on costly and laborious supplementation trials.

Specific methods of supplementation

Where possible, individually dosed treatments are to be preferred, because of the certainty and uniformity with which they can be administered, but efficacy is not guaranteed. For example, in the trials of MacPherson *et al.* (1999), injections of vitamin B_{12} were preferred to cobalt bullets for raising vitamin B_{12} status, but the chosen dose (1 mg) was inadequate when repeated in lambs that had grown from 20 to 30 kg between doses during the first 6 weeks of the experiment. The cobalt status of untreated lambs declined steadily and responses were more likely to have been obtained after the second dose, had it not been inadequate. The scaling of repeated doses to body weight in growing animals is therefore recommended. In the same study, failure to scale selenium injections to live weight may have led to the smallest individuals being overdosed (Suttle *et al.*, 1999), and failure to repeat an oral cupric oxide needles treatment may have underestimated the degree to which copper deficiency limited late lamb growth on improved Scottish hill pastures (Table 19.3; Suttle *et al.*, 1999). Mineral-responsive disorders can only be diagnosed by deploying treatments of a type and frequency that achieve sustained alleviation of deficiencies in all treated animals.

References

Clark, R.G., Wright, D.F. and Millar, K.R. (1985) A proposed new approach and protocol to defining mineral deficiencies using reference curves: cobalt deficiency in young sheep is used as a model. *New Zealand Veterinary Journal* 32, 1–5.

Clark, R.G., Wright, D.F., Millar, K.R. and Rowland, J.D. (1989) Reference curves to diagnose cobalt deficiency in sheep using liver and serum vitamin B_{12} levels. *New Zealand Veterinary Journal* 37, 7–11.

Grace, N.D. (1994) *Managing Trace Element Deficiencies.* New Zealand Pastoral Agricultural Research Institute, Palmerston North, New Zealand, pp. 63–68.

Joyce, J.P. and Brunswick, I.C.F. (1975) Sodium supplementation of sheep and cattle fed lucerne in New Zealand. *New Zealand Journal of Experimental Agriculture* 3, 299–304.

Judson, G.J. (1996) Trace element supplements for sheep at pasture. In: Masters, D.G. and White, C.L. (eds) *Detection and Treatment of Mineral Problems in Grazing Sheep.* ACIAR Monograph No. 37, Canberra, pp. 57–80.

Kumagai, H. and White, C.L. (1995) The effect of supplementary minerals, retinol and α-tocopherol on the vitamin status and productivity of pregnant Merino ewes. *Australian Journal of Agricultural Research* 46, 1159–1174.

Langlands, J.P., Donald, G.E., Bowles, J.E. and Smith, A.J. (1989) Selenium concentration in the blood of ruminants grazing in northern New South Wales. 3. Relationship between blood concentration and the response in liveweight of grazing cattle given a selenium supplement. *Australian Journal of Agricultural Research* 40, 1075–1083.

Lee, H.J. (1951) Cobalt and copper deficiencies affecting sheep in South Australia. Part 1. Symptoms and distribution. *Journal of Agriculture, South Australia* 54, 475–490.

McDowell, L.R., Bauer, B., Galdo, E., Koger, M., Loosli, J.K. and Conrad, J.H. (1982) Mineral supplementation of beef cattle in the Bolivian tropics. *Journal of Animal Science* 55, 964–970.

MacPherson, A., Suttle, N.F., Linklater, K.A. and Rice, D.A. (1999) The influence of trace element status on the pre-weaning growth of lambs on improved hill pastures in Scotland. 2. Cobalt. *Journal of Agricultural Science* (in press).

Masters, D.G. and Peter, D.W. (1990) Marginal deficiencies of cobalt and selenium in weaner sheep: response to supplementation. *Australian Journal of Experimental Agriculture* 30, 337–341.

Suttle, N.F. (1987) The absorption, retention and function of minor nutrients. In: Hacker, J.B. and Ternouth, J.H. (eds) *Nutrition in Herbivores.* Academic Press, Sydney, pp. 333–362.

Suttle, N.F. (1994) Meeting the copper requirements of ruminants. In: Garnsworthy, P.C. and Cole, D.J.A. (eds) *Recent Advances in Nutrition – 1994.* Nottingham University Press, Nottingham, pp. 173–188.

Suttle, N.F., MacPherson, A., Linklater, C.A. and Phillips, P. (1999) The influence of trace element status on the pre-weaning growth of lambs on improved hill pastures in Scotland. I. Copper, molybdenum, sulphur and iron. *Journal of Agricultural Science* (in press).

Towers, N.R., Gravett, I., Smith, J.F., Smeaton, P.C. and Knight, T.W. (1993) In: Grace, N.D. (ed.) *The Mineral Requirements of Grazing Ruminants.* Occasional Publication No. 9, New Zealand Society of Animal Production, pp. 142–149.

White, C.L. (1996) Understanding the mineral requirements of sheep. In: Masters, D.G. and White, C.L. (eds) *Detection and Treatment of Mineral Problems in Grazing Sheep.* ACIAR Monograph No. 37, Canberra, pp. 15–30.

White, C.L., Masters, D.G., Peter, D.W., Purser, D.B., Roc, S.P. and Barnes, M. (1992) A multi-element supplement for grazing sheep. I. Intake, mineral status and production response. *Australian Journal of Agricultural Research* 43, 795–808.

Appendices

Appendix Table 1. Chemical symbols, atomic numbers and atomic weights of elements of nutritional significance and their grouping in the periodic table.

Name	Symbol	Atomic number	Atomic weight	Group	Name	Symbol	Atomic number	Atomic weight	Group
Aluminium	Al	13	26.98	IIIb	Mercury	Hg	80	200.59	IIb
Arsenic	As	33	74.92	Vb	Molybdenum	Mo	42	95.94	VIa
Barium	Ba	56	137.34	IIa	Nickel	Ni	28	58.71	VIII
Boron	B	5	10.81	IIIb	Nitrogen	N	7	14.007	Vb
Cadmium	Cd	48	112.40	IId	Oxygen	O	8	15.9994	VIb
Calcium	Ca	20	40.08	IIa	Phosphorus	P	15	30.974	Vb
Carbon	C	6	12.011	IVb	Potassium	K	19	39.102	Ia
Cerium	Ce	58	140.12	IIIa	Rubidium	Rb	37	85.47	Ia
Caesium	Cs	55	132.91	Ia	Ruthenium	Ru	44	101.1	VIII
Chloride	Cl	17	35.453	VIIb	Selenium	Se	34	78.96	VIb
Chromium	Cr	24	52.00	VIa	Silicon	Si	14	28.09	IVb
Cobalt	Co	27	58.93	VIII	Silver	Ag	47	107.870	Ib
Copper	Cu	29	63.54	Ib	Sodium	Na	11	22.9898	Ia
Fluorine	F	9	19.00	VIIb	Strontium	Sr	38	87.62	IIb
Hydrogen	H	1	1.0080	VIIb	Sulphur	S	16	32.064	VIb
Iodine	I	53	126.90	VIIb	Technetium	Tc	43	(99)	VIIa
Iron	Fe	26	55.85	VIII	Tin	Sn	50	118.69	IVb
Lanthanum	La	57	138.91	IIIa	Titanium	Ti	22	47.90	IVa
Lead	Pb	82	207.19	IVb	Tungsten	W	74	183.85	VIa
Lithium	Li	3	6.939	Ia	Vanadium	V	23	50.94	Va
Magnesium	Mg	12	24.312	IIa	Zinc	Zn	30	65.37	IIb
Manganese	Mn	25	54.94	VIIa					

Appendix Table 2. Common ingregients of mineral supplements in Europe and their mineral content prior to mixing. Source: International Association of the European (EU) Manufacturers of Major, Trace and Specific Feed Mineral Materials (EMFEMA), Brussels.

			Content (%)		
Mineral	Source	Formulae	Theoretical	Actual	Other elements
Ca	Calcium carbonate	$CaCO_3$	40	37–39.5	
	Oyster shells	$CaCO_3$	40	*c.* 37	
Mg	Magnesium oxide	MgO		50.5–52.0	
	Magnesium sulphate	$MgSO_4.7H_2O$		10.0	13 S
	Magnesium chloride	$MgCL_2.6H_2O$		12.0	35 Cl
	Magnesium hydroxide	$Mg(OH)_2$		36–38	
	Magnesium phosphate	$MgHPO_4.nH_2O$		24–28	13–15 P
Na	Marine salt	NaCl		39.2	
	Vacuum-dried salt	NaCl		38.9–39.2	
	Rock salt	NaCl		36.4	
	Bicarbonate	$NaHCO_3$		27.0	
	Sodium sulphate (anhydrous)	Na_2SO_4		32.0	22 S
P	'di' calcium phosphate (anhydrous)	$CaHPO_4$		18–20.5	25–27 Ca
	'di' calcium phosphate (dihydrate)	$CaHPO_4.2H_2O$		17–18.6	23–25 Ca
	'mono-di' calcium phosphate	$CaHPO_4.Ca(H_2PO_4)_2.H_2O$		20–22	19.5–23.0 Ca
	'mono' calcium phosphate	$Ca(H_2PO_4)_2.H_2O$		22–22.7	15–17.5 Ca
	Phosphate-containing CaCO3	$Ca(H_2PO_4)_2.xH_2O + CaCO_3$		18	21 Ca
	Calcium-magnesium phosphate	$(Ca,Mg)PO_4.nH_2O$		18	15 Ca; 9 Mg
	Na–Ca–Mg phosphate	$(Na,Ca,Mg)PO_4.nH_2O$		17.5	7.5–8.5 Ca 4.5–5.5 Mg 12.0–14.5 Na
	Mono ammonium phosphate	$NH_4H_2PO_4$		26	11 N
	Mono sodium phosphate	$NH_2PO_4.nH_2O$		24	20 Na
Co	Cobaltous sulphate heptahydrate	$CoSO_4.7H_2O$		21	
Cu	Copper sulphate	$CuSO_4.5H_2O$		25	
Zn	Zinc sulphate	$ZnSO_4.H_2O$		35	
	Zinc oxide	ZnO		75	
Fe	Ferrous sulphate monohydrate	$FeSO_4.H_2O$		29.5–30.0	
	Ferrous sulphate heptahydrate	$FeSO_4.7H_2O$		19.5	
Mn	Manganous oxide	MnO		52–63	
	Manganous sulphate monohydrate	$MnSO_4.H_2O$		31–32	
Mo	Sodium molybdate dihydrate	$Na_2MoO_4.2H_2O$		39.5–40	
I	Calcium iodate monohydrate	$Ca(IO_2)_2.H_2O$		61.6	
	Sodium iodate	NaI		84.7	
	Potassium iodide (may be stabilized with stearates)	KI		76.5	
Se	Sodium selenite	Na_2SeO_3		45	

Index